# UNIVERSE

## THE DEFINITIVE VISUAL GUIDE

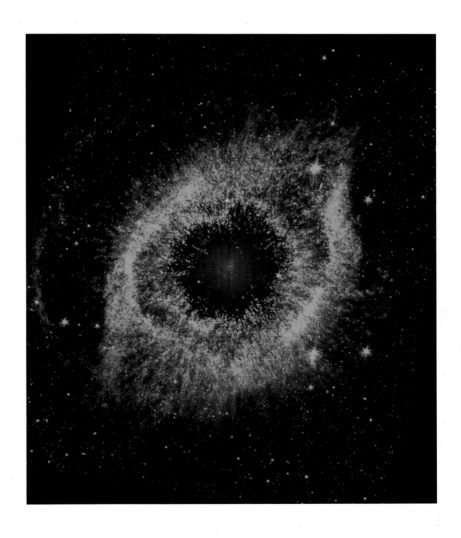

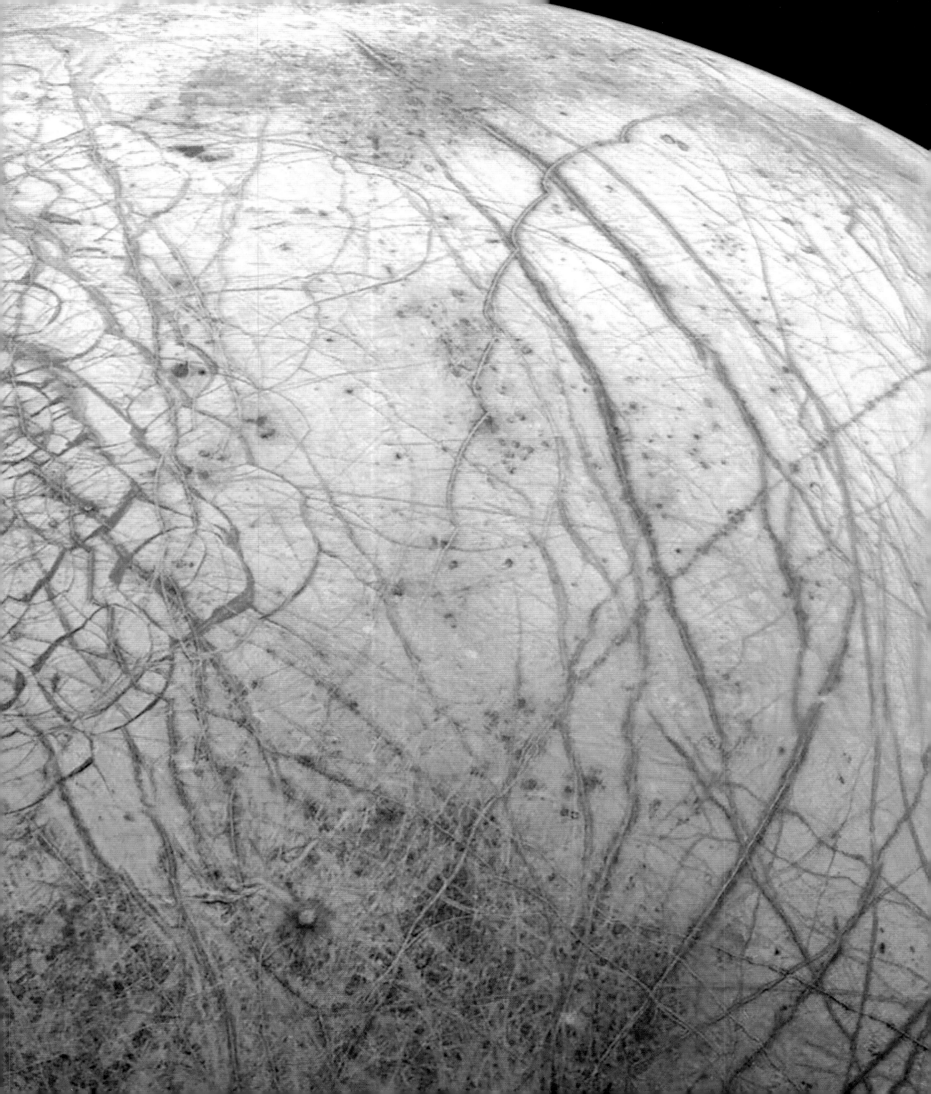

# DK SMITHSONIAN UNIVERSE

## THE DEFINITIVE VISUAL GUIDE

General Editor
**Martin Rees**

## DK | Penguin Random House

**THIS EDITION**

**DK LONDON**

**PROJECT EDITOR** Miezan van Zyl
**US EDITOR** Kayla Dugger
**PROJECT ART EDITOR** Steve Woosnam-Savage
**MANAGING EDITOR** Angeles Gavira Guerrero
**MANAGING ART EDITOR** Michael Duffy
**JACKET DESIGN DEVELOPMENT MANAGER** Sophia MTT
**PRODUCTION EDITOR** Gillian Reid
**SENIOR PRODUCTION CONTROLLER** Meskerem Berhane
**ASSOCIATE PUBLISHING DIRECTOR** Liz Wheeler
**PUBLISHING DIRECTOR** Jonathan Metcalf
**ART DIRECTOR** Karen Self

**DK DELHI**

**SENIOR EDITOR** Suefa Lee
**EDITOR** Ishita Jha
**PROJECT ART EDITOR** Rupanki Arora Kaushik
**ART EDITOR** Nobina Chakravorty
**DTP DESIGNERS** Pawan Kumar, Ashok Kumar
**ASSISTANT PICTURE RESEARCHER** Nimesh Agrawal
**SENIOR MANAGING EDITOR** Rohan Sinha
**MANAGING ART EDITOR** Sudakshina Basu
**PICTURE RESEARCH MANAGER** Taiyaba Khatoon
**PRE-PRODUCTION MANAGER** Balwant Singh
**PRODUCTION MANAGER** Pankaj Sharma

**PREVIOUS EDITIONS**

**SENIOR EDITOR** Peter Frances
**PROJECT EDITORS** Georgina Garner, Rob Houston, Gill Pitts, Martyn Page, David Summers, Miezan van Zyl
**EDITORS** Joanna Chisholm, Ben Hoare, Giles Sparrow
**PROOF READERS** Steve Setford, Jane Simmonds, Nikky Twyman
**INDEXERS** Hilary Bird, Jane Parker
**SENIOR ART EDITORS** Mabel Chan, Spencer Holbrook, Peter Laws
**PROJECT ART EDITORS** Dave Ball, Sunita Gahir, Alison Gardner, Mark Lloyd, Duncan Turner
**DESIGNERS** Kenny Grant, Jerry Udall
**DESIGN ASSISTANT** Marilou Prokopiou
**PICTURE RESEARCHER** Louise Thomas
**ILLUSTRATORS** Anbits, Combustion Design and Advertising, Fanatic Design, JP Map Graphics, Moonrunner Design, Pikaia Imaging, Planetary Visions, Precision Illustration
**PRODUCTION CONTROLLERS** Heather Hughes, Mary Slater
**PRODUCTION EDITORS** John Goldsmid, Adam Stoneham
**MANAGING EDITOR** Camilla Hallinan
**MANAGING ART EDITOR** Michelle Baxter
**PUBLISHER** Sarah Larter
**ART DIRECTORS** Philip Ormerod, Bryn Walls
**ASSOCIATE PUBLISHING DIRECTOR** Liz Wheeler
**PUBLISHING DIRECTOR** Jonathan Metcalf
**CONSULTANT FOR REVISED EDITION** Andrew K. Johnston, Center for Earth and Planetary Studies, National Air and Space Museum, Smithsonian Institution, USA.

**WARNING**

Looking at the Sun with the naked eye, binoculars, or a telescope can cause eye damage. Advice on safe viewing of the Sun is provided on page 85 of this book (see Solar Telescopes). The authors and publishers cannot accept any liability to readers failing to follow this advice.

This American Edition, 2020
First American Edition, 2005
Published in the United States by DK Publishing
1450 Broadway, Suite 801, New York, NY 10018

Copyright © 2005, 2012, 2020 Dorling Kindersley Limited
DK, a Division of Penguin Random House LLC
22 23 24 10 9 8 7 6 5
009–316672–Sep/2020

A catalog record for this book
is available from the Library of Congress.
ISBN 978-1-4654-9995-0

DK books are available at special discounts when purchased in bulk for sales promotions, premiums, fund-raising, or educational use. For details, contact: DK Publishing Special Markets, 1450 Broadway, Suite 801, New York, NY 10018
SpecialSales@dk.com

Printed in China

For the curious

**www.dk.com**

*Jacket* Mars; *Endpapers* the Orion Nebula; *Half-title page* the Helix Nebula; *Title page* Jupiter's moon Europa; *Contents page* the Carina Nebula

## Smithsonian

Established in 1846, the Smithsonian is the world's largest museum and research complex, dedicated to public education, national service, and scholarship in the arts, sciences, and history. It includes 19 museums and galleries and the National Zoological Park. The total number of artifacts, works of art, and specimens in the Smithsonian's collection is estimated at 155 million.

# CONTENTS

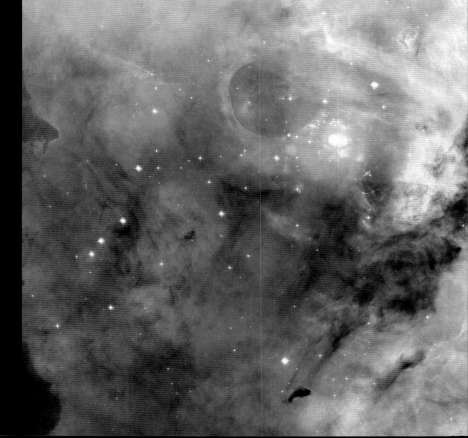

## GUIDE TO THE UNIVERSE

## THE NIGHT SKY

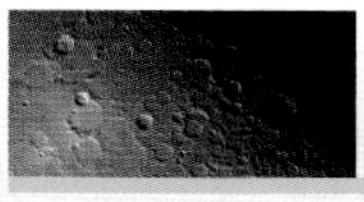

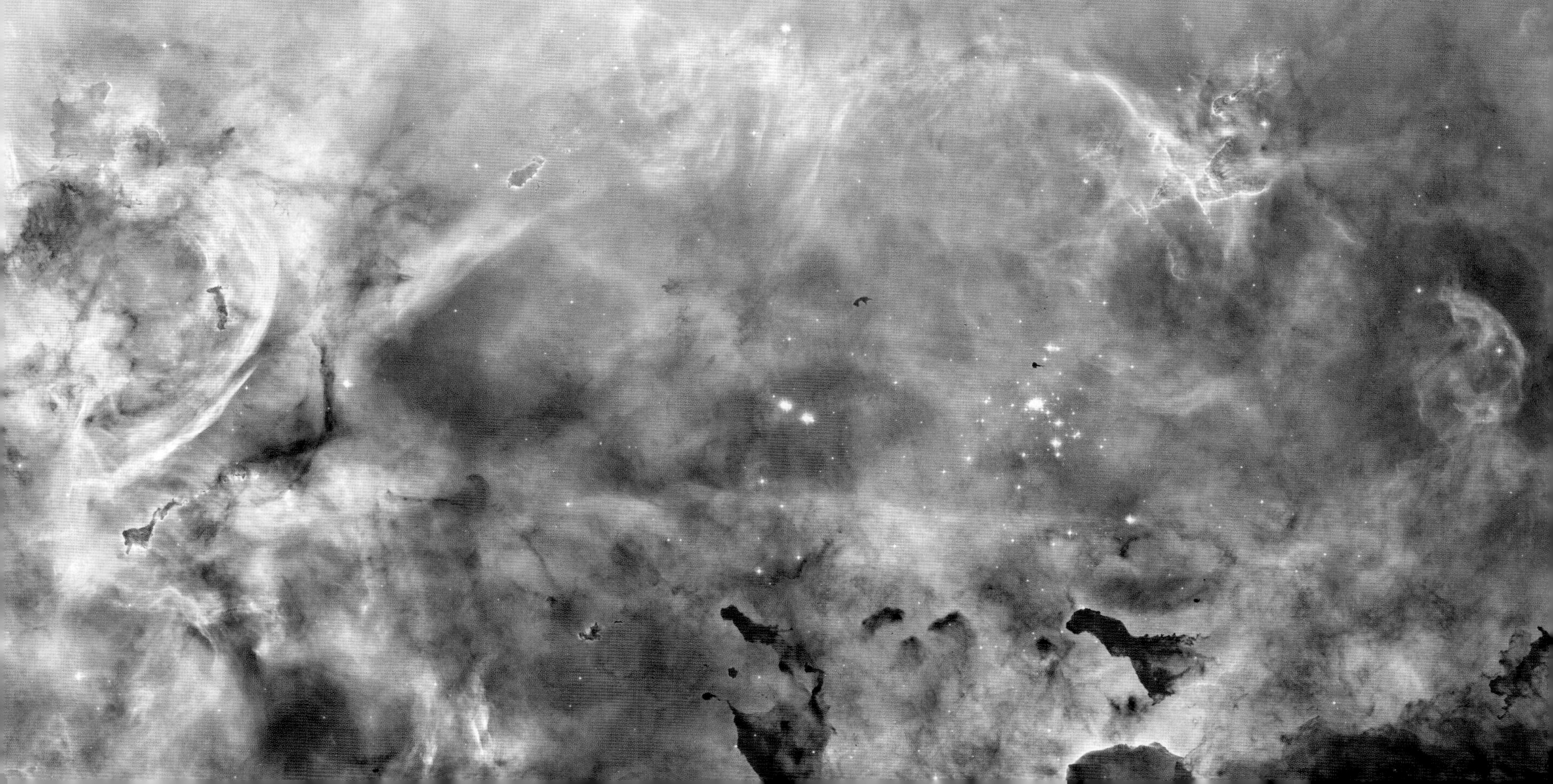

# ABOUT THIS BOOK

*Universe* is divided into three main sections. The INTRODUCTION is an overview of the basic concepts of astronomy. GUIDE TO THE UNIVERSE looks, in turn, at the solar system, the Milky Way (our home galaxy), and the regions of space that lie beyond. Finally, THE NIGHT SKY is a guide to the sky for the amateur skywatcher.

## INTRODUCTION

This section is about the universe and astronomy as a whole. It is subdivided into three parts. WHAT IS THE UNIVERSE? looks at different kinds of objects in the universe and the forces governing how they behave and interact. THE BEGINNING AND END OF THE UNIVERSE covers the origin and history of the universe, while THE VIEW FROM EARTH explains what we see when we look at the sky.

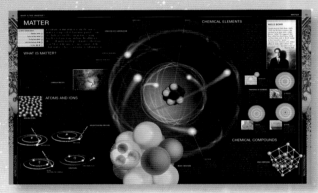

**△ WHAT IS THE UNIVERSE?**
This section begins by looking at some basic questions about the size and shape of the universe. It goes on to explain concepts such as matter and radiation, the motion of objects in space, and the relationship between time and space.

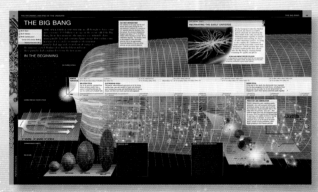

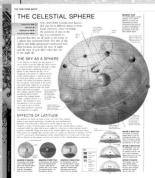

**△ THE BEGINNING AND END OF THE UNIVERSE**
The universe is thought to have originated in an event known as the Big Bang. This section describes the Big Bang in detail and looks at how the universe came to be the way it is now, as well as how it might end.

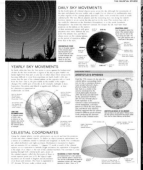

**△ THE VIEW FROM EARTH**
This section presents a simple model for making sense of the changing appearance of the sky. It also contains practical advice on looking at the sky with the naked eye, telescopes, and binoculars.

## GUIDE TO THE UNIVERSE

This part of the book focuses on specific regions of space, starting from the Sun and then moving outward to progressively more distant reaches of the universe. It is divided into three sections, covering the solar system, the Milky Way, and features beyond the Milky Way. In each section, introductory pages describe features in a general way and explain the processes behind their formation. These pages are often followed by detailed profiles of actual features (such as individual stars), usually arranged in order of their distance from Earth.

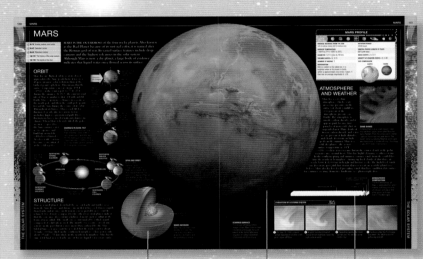

**△ THE SOLAR SYSTEM**
This section is about the Sun and the many bodies in orbit around it. It covers the eight planets one by one and then looks at asteroids, comets, and meteors, as well as the remote regions on the margins of the solar system. For most planets, profiles of individual surface features or moons are also included.

artwork of planet's interior structure

main image shows planet as it appears from space

illustrations show atmospheric composition for each planet

color-coded panel contains references to other relevant sections

**THE MILKY WAY ▷**
The subject of this section is the Milky Way and the stars, nebulae, and planets that it contains. Pages such as those shown here describe how particular types of features are formed.

## THE NIGHT SKY

This section is an atlas of the night sky. It is divided into two parts. The first (THE CONSTELLATIONS) is a guide to the 88 regions into which astronomers divide the sky. It contains illustrated profiles of all the constellations arranged according to their position in the sky, with the most northerly ones first and the southernmost last. The second part (the MONTHLY SKY GUIDE) is a month-by-month guide containing a summary of the highlights for each month, detailed star charts, and charts showing the positions of the planets.

detailed chart

text describes features of interest

**THE CONSTELLATIONS ▷**
Each constellation profile is illustrated with a chart, two locator maps, and one or more photographs. A more detailed guide to the section can be found on pp.348–349.

## THEMED PANELS

Three types of color-coded panels are used to present a more detailed focus on selected subjects. These panels appear both on explanatory pages and in feature profiles.

### MYTHS AND STORIES ▷

As well as being studied scientifically, objects in the night sky have featured in myths, superstitions, and folklore, which form the subject of this type of panel.

**EXPLORING SPACE**
### ARISTOTLE'S SPHERES

Until the 17th century CE, the idea of a celestial sphere surrounding Earth was not just a convenient fiction—many people believed it had a physical reality. Such beliefs date back to a model of the universe developed by the

*sphere of "fixed" stars*

**MYTHS AND STORIES**
### ASTROLOGY AND THE ECLIPTIC

Astrology is the study of the positions and movements of the Sun, Moon, and planets in the sky in the belief that these influence human affairs. At one time, when astronomy was applied mainly to devising calendars, astronomy and astrology were intertwined, but their aims and methods have now diverged. Astrologers pay little attention to constellations, but measure the positions of the Sun and planets in sections of the ecliptic that they call "Aries" and "Taurus," for example. However, these sections no longer match the constellations of Aries, Taurus, and so on.

STARGAZER

JAN

### JOHANNES KEPLER

The German astronomer Johannes Kepler (1571–1630) discovered the laws of planetary motion. His first law states that planets orbit the Sun in elliptical paths. The next states that the closer a planet comes to the Sun, the faster it moves, while his third law describes the link between a planet's distance from the Sun and its orbital period. Newton used Kepler's

### ◁ EXPLORING SPACE

This type of feature is used to describe the study of space either from Earth's surface or from spacecraft. Individual panels describe particular discoveries or investigations.

### ◁ BIOGRAPHY

Profiles of notable astronomers and pioneers of spaceflight, as well as a brief summary of their achievements, appear in this type of panel.

## CONTRIBUTORS

**Martin Rees**  General editor

**Robert Dinwiddie**  What Is the Universe?, The Beginning and End of the Universe, The View From Earth, The Solar System, The Milky Way

**Philip Eales**  The Milky Way

**David Hughes**  The Solar System

**Iain Nicolson**  Glossary

**Ian Ridpath**  The View From Earth, The Night Sky

**Robin Scagell**  The View from Earth

**Giles Sparrow**  The Solar System, Beyond the Milky Way

**Pam Spence**  The Milky Way

**Carole Stott**  The Solar System

**Kevin Tildsley**  The Milky Way

**David Rothery**  The Solar System

*name or astronomical catalog number of feature (features without a popular name are identified by number)*

**EMISSION NEBULA**
# Carina Nebula

| | |
|---|---|
| **CATALOG NUMBER** | |
| NGC 3372 | |
| **DISTANCE FROM SUN** | |
| 8,000 light-years | |
| **MAGNITUDE 1** | |

CARINA

*locator map shows constellation in which feature can be found and its position within the constellation*

*table of summary information (varies between sections)*

*selected features are described in double-page feature profiles*

### △ FEATURE PROFILES

Throughout the Guide to the Universe, introductory pages are often followed by profiles of a selection of specific objects. For example, the introduction to star formation (left) is followed by profiles of actual star-forming regions in the Milky Way (above).

*themed panel (see above)*

### BEYOND THE MILKY WAY ▷

This section looks at features found beyond our own galaxy, including other galaxies and galaxy clusters and superclusters, the largest known structures in the universe.

*locator shows constellation in context*

*artwork of figure depicted by constellation*

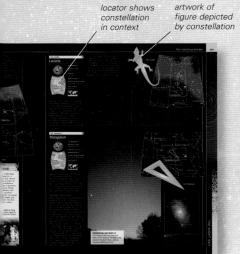

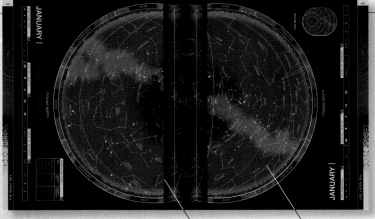

### △ MONTHLY SKY GUIDE

This section includes two charts for each month of the year for observers in northern and southern latitudes. The section is described in more detail on pp.428–429.

*chart on this page shows view looking north, with view to south on facing page*

*lines on chart show reference points for observers at different latitudes*

## THE GREEK ALPHABET

Astronomers use a convention for naming some stars in which Greek letters are assigned to stars according to their brightness. These letters appear on some of the charts in this book.

| | | | |
|---|---|---|---|
| α | alpha | ν | nu |
| β | beta | ξ | xi |
| γ | gamma | ο | omicron |
| δ | delta | π | pi |
| ε | epsilon | ρ | rho |
| ζ | zeta | σ | sigma |
| η | eta | τ | tau |
| θ | theta | υ | upsilon |
| ι | iota | φ | phi |
| κ | kappa | χ | chi |
| λ | lambda | ψ | psi |
| μ | mu | ϖ | omega |

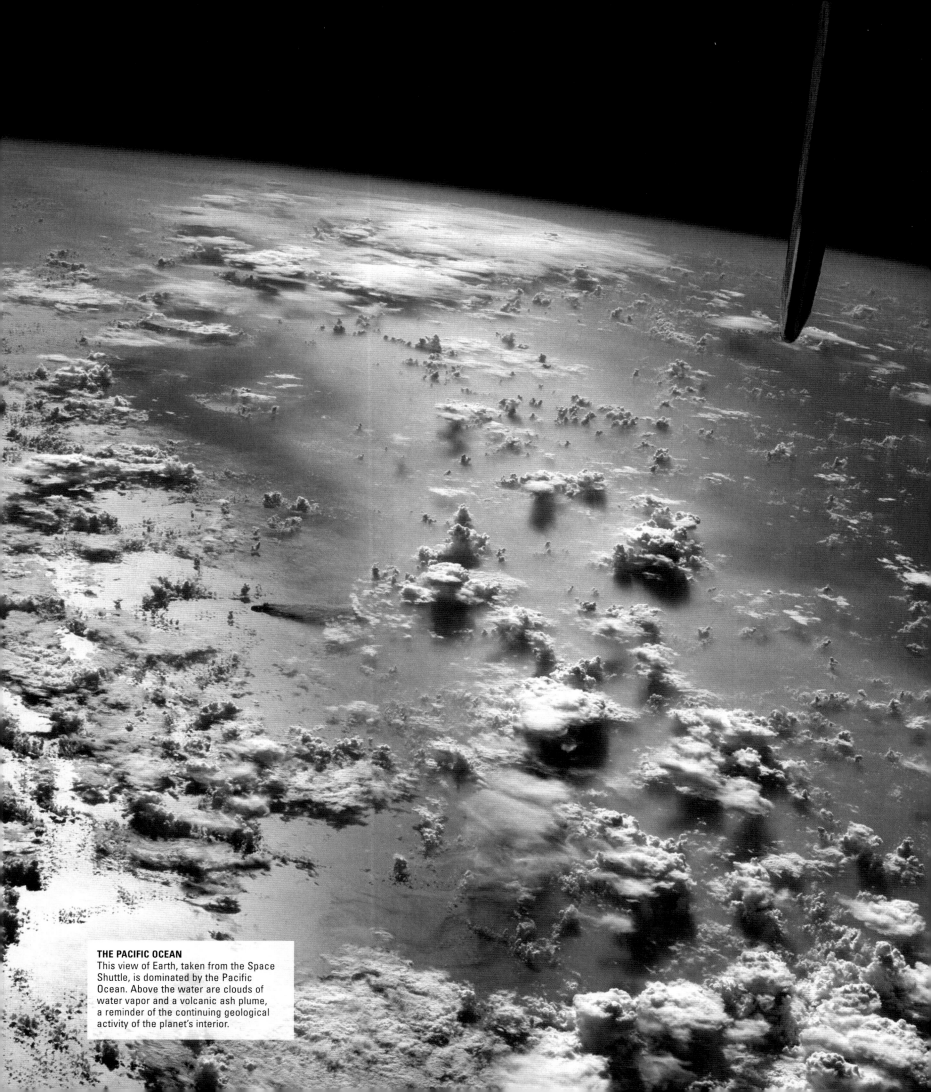

**THE PACIFIC OCEAN**
This view of Earth, taken from the Space
Shuttle, is dominated by the Pacific
Ocean. Above the water are clouds of
water vapor and a volcanic ash plume,
a reminder of the continuing geological
activity of the planet's interior.

**KENNEDY SPACE CENTER**
Many of humankind's first ventures into space were set in motion on the launchpads of Kennedy Space Center. This remains the busiest launch and landing site of the US space program, and it was also the main base for the Space Shuttle.

**THE FLORIDA COAST**
The islands and reefs of the Florida Keys are seen here from Earth orbit. The reefs are partly made of living organisms in the form of corals. To date, life has not been found anywhere other than on Earth, but the search for alien life will be perhaps the most fascinating quest of the 21st century.

THE NIGHT SKY has always evoked mystery and wonder. Since antiquity, astronomers have tried to understand the patterns of the "fixed stars" and the motions of the Moon and planets. The motive was partly a practical one, but there has always been a more "poetic" motivation, too—a quest to understand our place in nature. Modern science reveals a cosmos far vaster and more varied than our ancestors could have envisioned.

No continents on Earth remain to be discovered. The exploratory challenge has now broadened to the cosmos. Humans have walked on the Moon; uncrewed spacecraft have beamed back views of all the planets; and some people now living may one day walk on Mars.

The stars, fixed in the "vault of heaven," were a mystery to the ancients. They are still unattainably remote, but we know that many of them are shining even more brightly than the Sun. Within the last decade, we have learned something remarkable that was long suspected: many stars are, like our Sun, encircled by orbiting planets. The number of known planetary systems already runs into hundreds—there could, all together, be a billion in our galaxy. Could some of these planets resemble the Earth and harbor life? Even intelligent life?

All the stars visible to the unaided eye are part of our home galaxy—a structure so vast that light takes a hundred thousand years to cross it. But this galaxy, the Milky Way, is just one of billions visible through large telescopes. These galaxies are hurtling away from each other, as though they had all originated in a "big bang" 13 or 14 billion years ago. But we don't know what banged, nor why it banged.

The beauty of the night sky is a common experience of people from all cultures; indeed, it is something that we share with all generations since prehistoric times. Our modern perception of the "cosmic environment" is even grander. Astronomers are now setting Earth in a cosmic context. They seek to understand how the cosmos developed its intricate complexity—how the first galaxies, stars, and planets formed and how, on at least one planet, atoms assembled into creatures able to ponder their origins. This book sets humanity's concept of the cosmos in its historic context and presents the latest discoveries and theories. It is a beautiful "field guide" to our cosmic habitat: it should enlighten and delight anyone who has looked up at the stars with wonder and wished to understand them better.

MARTIN REES

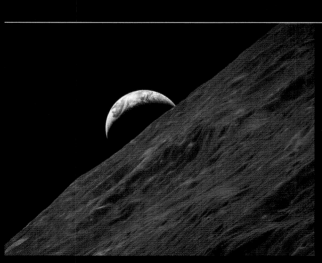

**THE MOON**
**1.3 light-seconds from Earth**

The Earth is seen here rising above the horizon of its satellite, the Moon. Our home planet's delicate biosphere contrasts with the sterile moonscape on which the Apollo astronauts left their footprints.

# The Sun

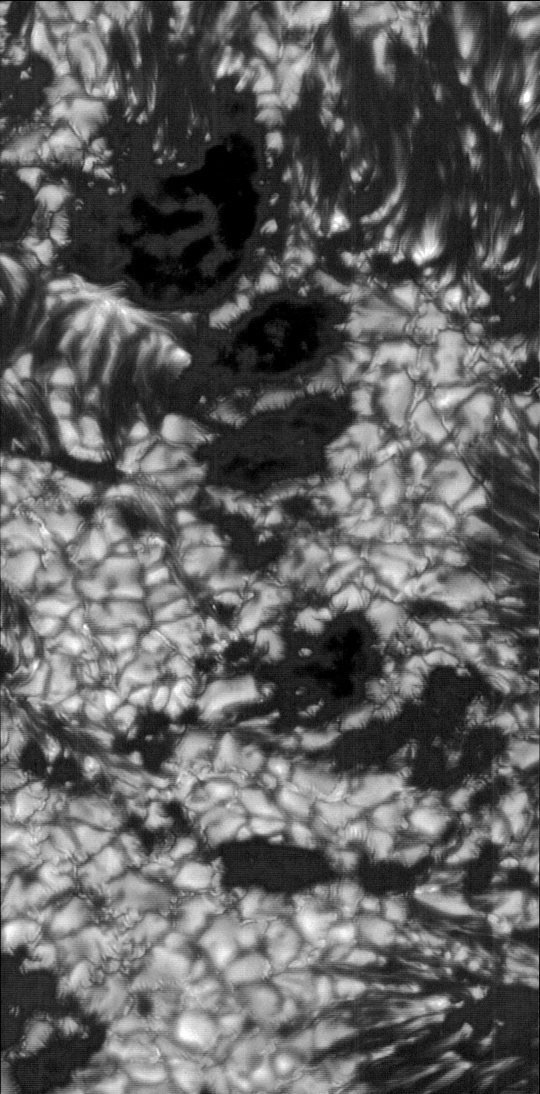

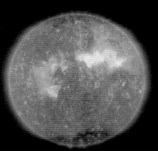

### OUR LOCAL STAR
The Sun dominates the solar system. Our chief source of heat and light, it also holds Earth and the rest of the planets in their orbits. This ultraviolet image reveals the dynamic activity in the ultra-hot corona above the Sun's visible surface.

### A SOLAR FLARE
The Sun usually appears to the unaided eye as a bright but featureless disk. However, during a total solar eclipse, when light from the disk is blocked out by the Moon, violent flares in the outer layers of the atmosphere can be seen more clearly.

### ULTRA-HOT CORONA
The gas in the Sun's corona is heated to several million degrees, causing it to emit X-rays, which show up in this image taken by the Japanese YOHKOH satellite. The dark areas are regions of low-density gas that emit a stream of particles, known as the solar wind, into space.

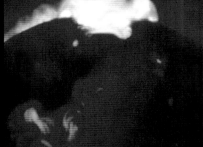

### PROMINENCES
In the corona, electrified gas called plasma forms into huge clouds known as prominences, flowing through the Sun's magnetic field. As the prominence in this image erupts, it hurls plasma out of the Sun's atmosphere and into space.

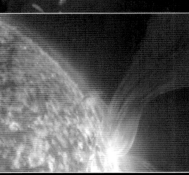

### SUNSPOTS ON THE SOLAR SURFACE
#### 8 light-minutes from Earth

These regions, cooler and darker than the rest of the Sun's surface, are sustained by strong magnetic fields. Some sunspots are large enough to engulf the Earth. Sunspot numbers vary in cycles that take about 11 years to complete, and peaks in the cycle coincide with disturbances, such as aurorae, in our own atmosphere.

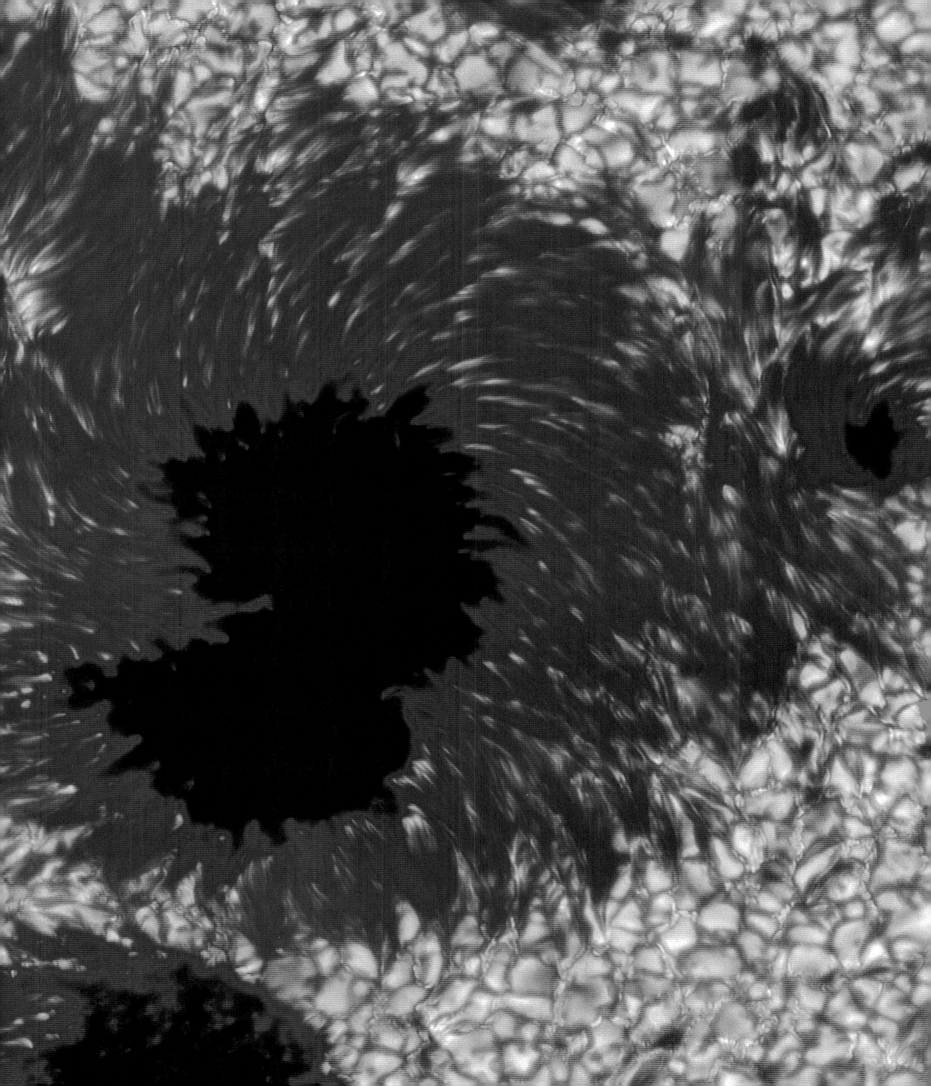

### CANYONS ON MARS
**4 light-minutes from Earth**

Mars is one of the solar system's four inner rocky planets. This image (with exaggerated vertical scale) shows part of the Valles Marineris, a vast canyon system.

### SATURN AND ITS RINGS
**71 light-minutes from Earth**

There are rings of dust and ice particles in nearly circular orbits around all four of the giant gas planets, but those around Saturn are especially beautiful. This close-up was taken by the Cassini spacecraft.

### IO
**34 light-minutes from Earth**

Jupiter has 79 known moons—and there are almost certainly others yet to be discovered. Io, Jupiter's innermost moon, is seen here moving in front of the turbulent face of the planet.

### 433 EROS
**3.8 light-minutes from Earth**

A vast number of asteroids are in independent orbit around the Sun. Eros is marked by the impact of much smaller bodies. This image was taken by the NEAR Shoemaker craft from only 60 miles (100 km) above the surface.

# The planets

 **JUPITER'S GREAT RED SPOT**
**34 light-minutes from Earth**

The gas giant Jupiter is more massive than all of the other planets in the solar system combined. Its mysterious swirling vortex, the Great Red Spot, has been known since the 17th century, but our knowledge of Jupiter improved greatly when the planet was visited by uncrewed spacecraft in the 1970s. This image of the Great Red Spot was taken in 1979 by Voyager 1 using filters that exaggerate its colors.

# Stars and galaxies

### THE CENTER OF OUR GALAXY
**25,000 light-years from Earth**

The center of our galaxy, the Milky Way, is thought to harbor a black hole as heavy as 3 million Suns. This image reveals flare-ups in X-ray activity close to the event horizon, the point of no return for any objects or light that approach the black hole.

### CENTAURUS A
**15 million light-years from Earth**

Not all galaxies exist in isolation; occasionally, they interact. Centaurus A is far more "active" than our own galaxy. It has an even bigger black hole than the Milky Way's, and its gravity may have captured and "cannibalized" a smaller neighbor.

### THE WHIRLPOOL GALAXY
**31 million light-years from Earth**

The Whirlpool is involved in another case of galaxy interaction. A spinning, disk-like galaxy, viewed face-on, its spiral structure may have been induced by the gravitational pull of a smaller satellite galaxy (at the top of this picture).

### THE ORION NEBULA
**1,500 light-years from Earth**

The Orion Nebula is a vast cloud of glowing dusty gas within the Milky Way, inside which new stars are forming. The nebula contains bright blue stars much younger than the Sun and some protostars whose nuclear power sources have not yet ignited.

# The limits of time and space

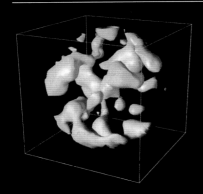

**GALAXY SUPERCLUSTERS**
This image, generated by plotting the positions of 15,000 galaxies, depicts the main "topographic" features of our cosmic environment out to 700 million light-years from Earth. The yellow blobs are superclusters of galaxies, which are interspersed with black voids.

**LARGE-SCALE STRUCTURES**
This view of the sky, in infrared light, shows how galaxies outside the Milky Way are distributed in clusters and filamentary structures. The galaxies are color-coded according to brightness, with bright ones in blue and faint ones in red.

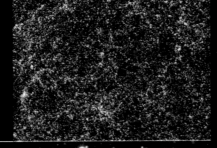

**DISTANT CLUSTER OF GALAXIES**
This massive cluster of galaxies is one of the most distant known to astronomers, some 8.5 billion light-years from Earth. Superimposed on the optical picture is an X-ray image revealing hot gas (shown in purple) that pervades the cluster.

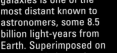 **DWARF GALAXIES BURSTING INTO LIFE**
**9 billion light-years from Earth**

Tiny young galaxies brimming with stars in the process of formation, some 9 billion light-years away, are seen in this image taken at near-infrared wavelengths by the Hubble Space Telescope. They stand out in the image because energy from the new stars has caused oxygen in the gas around them to light up like a neon sign. This phase of rapid star birth is thought to represent an important stage in the formation of dwarf galaxies, the most numerous type of galaxy in the universe.

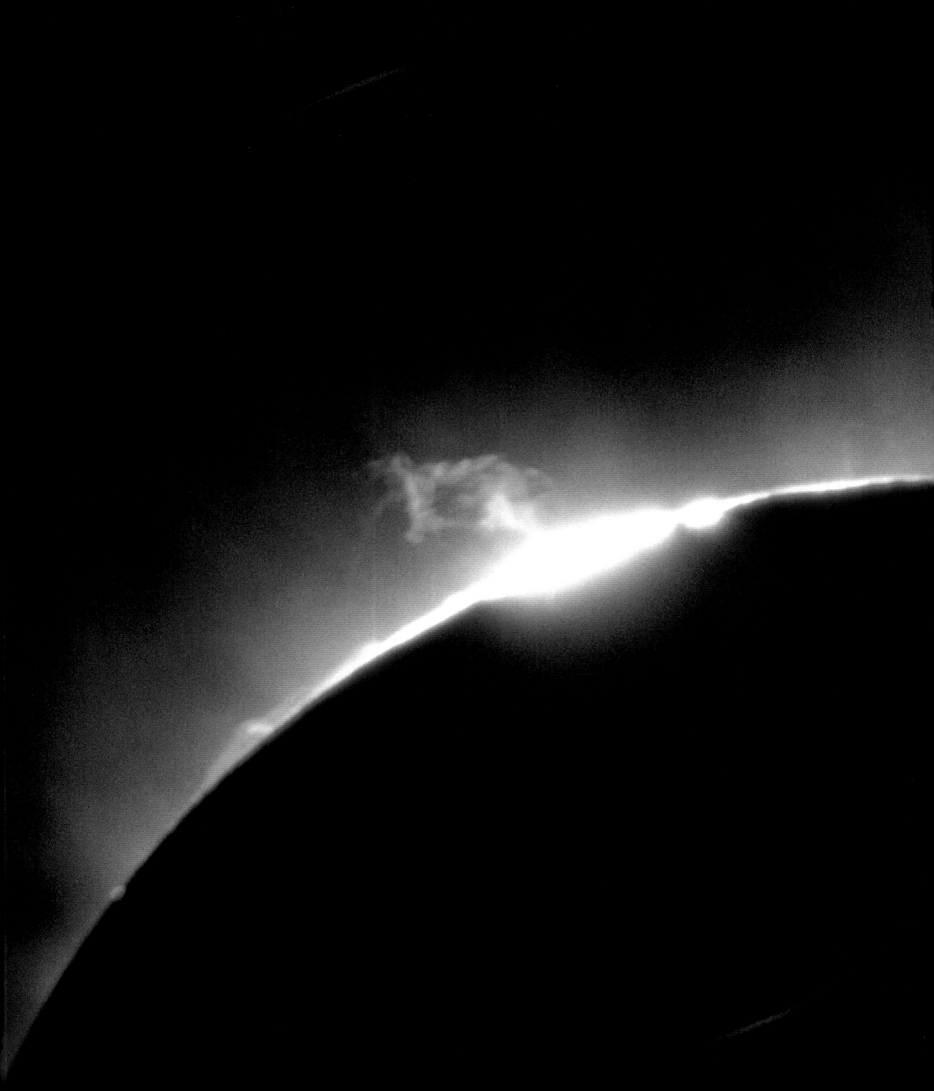

# INTRODUCTION

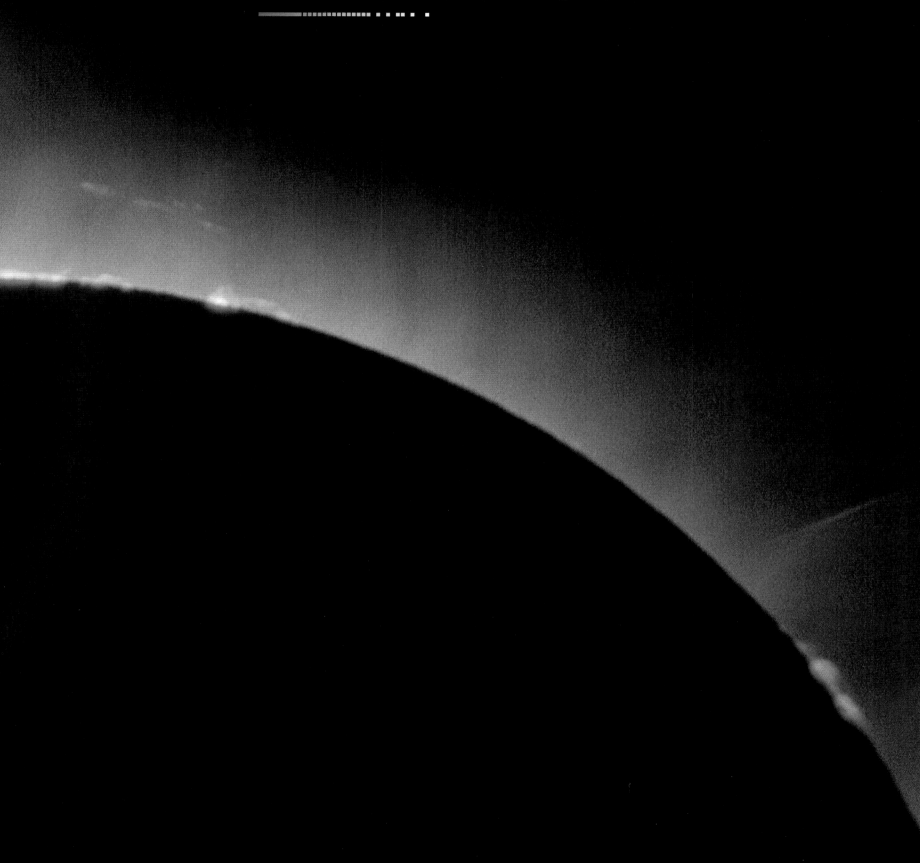

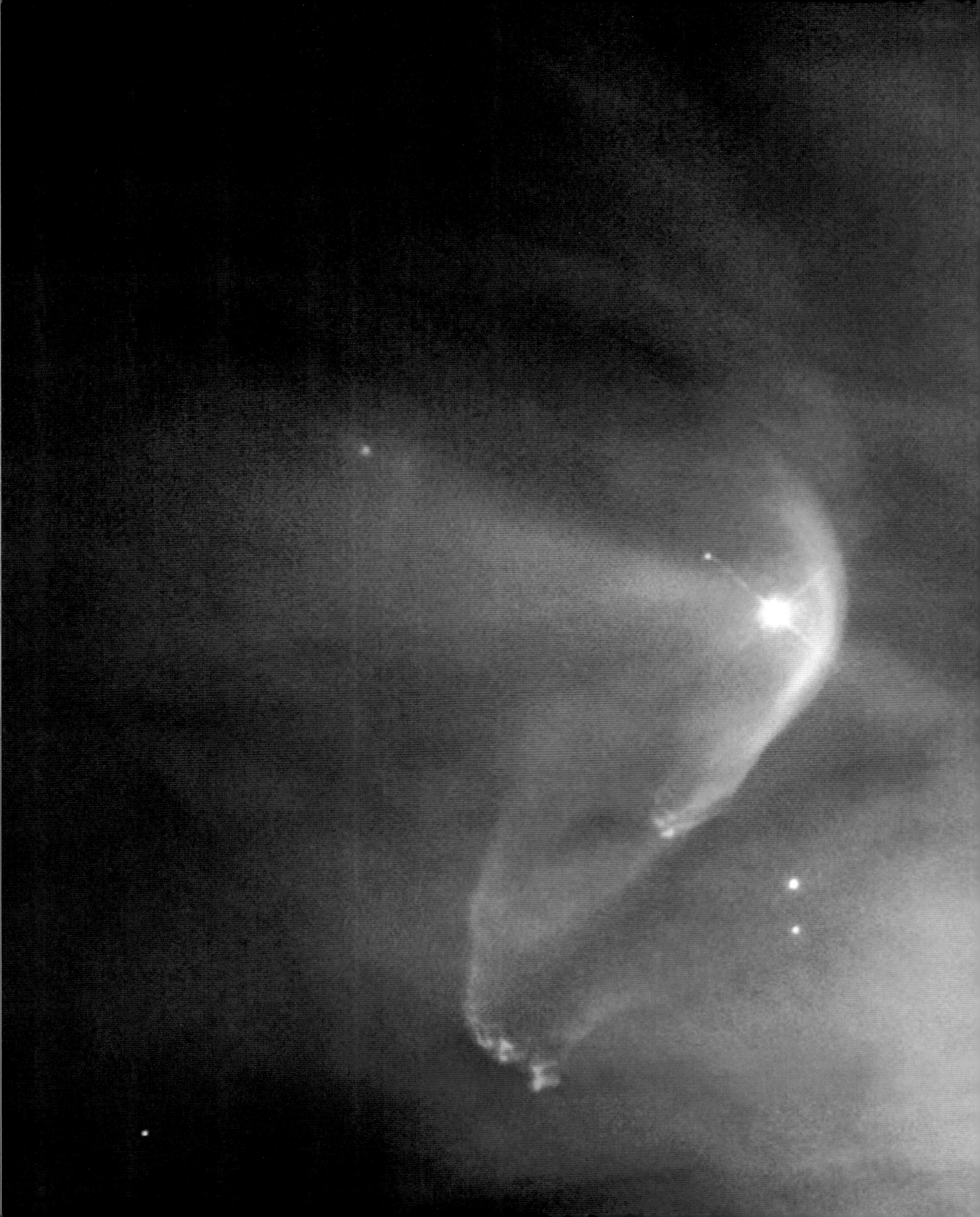

# WHAT IS THE UNIVERSE?

*"There are grounds for cautious optimism that we may now be near the end of the search for the ultimate laws of nature."*

Stephen Hawking

THE UNIVERSE IS ALL OF EXISTENCE—all of space and time and all the matter and energy within it. The universe is unknowably vast, and ever since it formed, it has been expanding, carrying distant regions apart at speeds up to—and in some cases exceeding—the speed of light. The universe encompasses everything from tiny subatomic particles to vast galaxy superclusters, yet it seems that all are governed by the same basic laws. All visible matter (which is only a small percentage of the total matter) is built from the same subatomic blocks, and the same fundamental forces govern all interactions between these elements. Knowledge of these cosmic operating principles—from general relativity to quantum physics—informs cosmology, the study of the universe as an entity. Cosmologists hope to answer questions such as "How big is the universe?", "How old is it?", and "How does it work, on the grandest scale?"

**BOW SHOCK AROUND A STAR**
This mysterious image from the Orion Nebula shows how matter and radiation interact on a stellar scale. A star surrounded by gas and dust has met a fierce wind of particles blowing from a bright young star (out of frame). Around the star, a crescent-shaped gaseous bow shock has built up, like water flowing past the prow of a boat.

# THE SCALE OF THE UNIVERSE

**EVERYTHING IN THE UNIVERSE** is part of something larger. The scale of the Earth and its moon may be relatively easy for the human mind to grasp, but the nearest star is unimaginably remote, and the farthest galaxies are billions of times more distant yet. Cosmologists, who study the size and structure of the universe, use mathematical models to build a picture of the universe's vast scale.

## THE SIZE OF THE UNIVERSE

Cosmologists may never determine exactly how big the universe is. It could be infinite. Alternatively, it might have a finite volume, but even a finite universe would have no center or boundaries and would curve in on itself. So, paradoxically, an object traveling off in one direction would eventually reappear from the opposite direction. What is certain is that the universe is expanding and has been doing so since its origins in the Big Bang, 13.8 billion years ago (see p.48). By studying the patterns of radiation left from the Big Bang, cosmologists can estimate the minimum size of the universe should it turn out to be finite. Some parts must be separated by at least tens of billions of light-years. Because a light-year is the distance that light travels in a year, (5.879 trillion miles, or 9.461 trillion km), the universe is bewilderingly big.

galaxy NGC 147

galaxy NGC 185

the Andromeda Galaxy, 2.65 million light-years from the Milky Way

Andromeda I

Andromeda II

Andromeda III

Triangulum Galaxy

the stellar neighborhood lies in the Orion Arm of the Milky Way, some 26,000 light-years from its center

galactic nucleus

Alpha Centauri

Sun

Sirius

orbit of Pluto

asteroid belt

Sun

Earth

orbit of Neptune

Earth

the Moon moves around the Earth in a slightly elliptical orbit

5,000 light-years

5 light-years

1 light-hour

0.5 light-seconds

### THE MILKY WAY

The solar system and its stellar neighbors are a tiny part of the Milky Way Galaxy, a disk of 100 to 400 billion stars and enormous clouds of gas and dust. The Milky Way is over 150,000 light-years across and has a supermassive black hole in its central nucleus.

### THE STELLAR NEIGHBORHOOD

The closest star system to the Sun, Alpha Centauri, lies 4.37 light-years, or 26 trillion miles (41 trillion km) away. Within 16 light-years of the Sun are 59 stars. These include 13 star systems, each containing two or three stars. An example is Sirius, the brightest star in the night sky, which is actually a system of two stars of markedly different size.

### THE SOLAR SYSTEM

The Earth–Moon system is part of the solar system, comprising our local star, the Sun, and all the objects that orbit it, including comets 0.7 light-years or more. Neptune, the outermost planet, is on average 2.8 billion miles (4.5 billion km) from the Sun.

### THE EARTH AND MOON

Earth has an average diameter of 7,920 miles (12,740 km), while the diameter of the Moon's orbit around Earth is about 480,000 miles (770,000 km). A space probe sent to the Moon typically takes just a few days to get there.

### REMOTE OBJECT

This embryonic galaxy, called SPT0615-JD (right), is one of the most distant known. The light from it, detected by the Hubble and Spitzer Space Telescopes, began its journey toward us 13.3 billion years ago

### VIEW FROM EARTH

The Milky Way Galaxy is a complex 3-D structure, but from our position within it, it appears as a 2-D band across the sky (above).

## THE LOCAL GROUP OF GALAXIES

The Milky Way is one of a cluster of galaxies, called the Local Group, that occupies a region 10 million light-years across. It contains more than 50 known galaxies, only one of which—the Andromeda Galaxy—of similar size to the Milky Way. Most others are small (dwarf) galaxies.

## THE LOCAL SUPERCLUSTER

The Local Group of galaxies, together with some nearby galaxy clusters, such as the giant Virgo Cluster, is contained within a vast structure called the Virgo Supercluster. It is 110 million light-years across and (if dwarf galaxies are included) contains tens of thousands of galaxies.

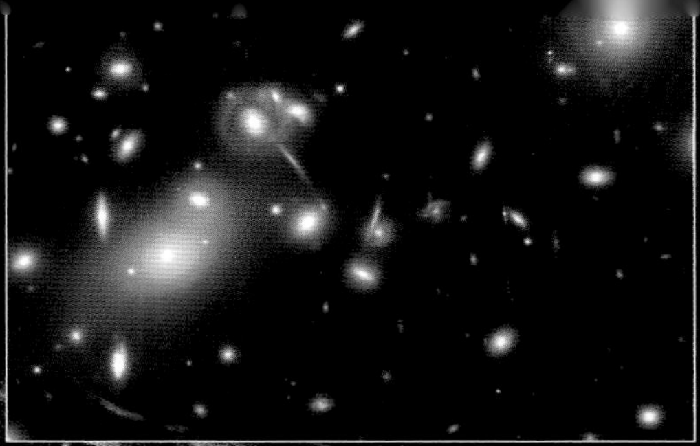

## DISTANT GALAXY CLUSTER

The vast galaxy cluster Abell 2218 (left) is visible from Earth even though it is more than 2 billion light-years away.

## LARGE-SCALE STRUCTURE

Galaxy superclusters clump into knots or extend as filaments that can be billions of light-years long, with large voids separating them. However, at the largest scale, the density of galaxies, and thus all visible matter, in the universe is very uniform.

Ursa Minor
dwarf galaxy

Milky
Way

250,000
light-years

Leo A

10 million
light-years

# THE OBSERVABLE UNIVERSE

Although the universe has no edges and may be infinite, the part of it that scientists have knowledge of is bounded and finite. Called the observable universe, it is the spherical region around Earth from which light has had time to reach us since the universe began. The boundary that separates this region from the rest of the universe is called the cosmic light horizon. Light reaching Earth from an object very close to this horizon must have been traveling for most of the age of the universe—that is, approximately 13.8 billion years. This light must have traveled a distance of around 13.8 billion light-years to reach Earth. Such a distance can be defined as a "lookback" or "light-travel-time" distance between Earth and the distant object. However, the true distance is much greater, because since the light arriving at Earth left the object, the object has been carried farther away by the universe's expansion (see p.45).

100 million
light-years

region
observable from
both planets

Planet X

Earth

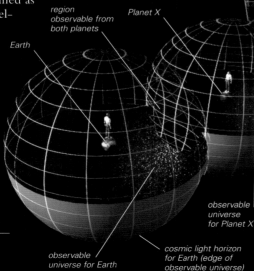

## OVERLAPPING OBSERVABLE UNIVERSES

Earth and Planet X—an imaginary planet with intelligent life, located tens of billions of light-years away—would have different observable universes. These may overlap, as shown here, or they may not.

observable
universe for Earth

cosmic light horizon
for Earth (edge of
observable universe)

observable
universe
for Planet X

## FROM HOME PLANET TO SUPERCLUSTERS

The universe has a hierarchy of structures. Earth is part of the solar system, nested in the Milky Way, which in turn is part of the Local Group. The Local Group is just part of one of millions of galaxy superclusters that extend in sheets and filaments throughout the observable universe.

# CELESTIAL OBJECTS

THE UNIVERSE CONSISTS of energy, space, and matter. Some of the matter drifts through space as single atoms or simple gas molecules. Other matter clumps into islands of material, from dust motes to giant suns, or implodes to form black holes. Gravity binds all of these objects into the great clouds and disks of material known as galaxies. Galaxies in turn fall into clusters and finally form the biggest celestial objects of all—superclusters.

## GAS, DUST, AND PARTICLES

Much of the ordinary matter of the universe exists as a thin and tenuous gas within and around galaxies and as an even thinner gas between galaxies. The gas is made mainly of hydrogen and helium atoms, but some clouds inside galaxies contain atoms of heavier chemical elements and simple molecules. Mixed in with the galactic gas clouds is dust—tiny solid particles of carbon or substances such as silicates (compounds of silicon and oxygen). Within galaxies, the gas and dust make up what is called the interstellar medium. Visible clumps of this medium, many of them the sites of star formation, are called nebulae. Some, called emission nebulae, produce a brilliant glow as their constituent atoms absorb radiant energy from stars and reradiate it as light. In contrast, dark nebulae are visible only as smudges that block out starlight. Particles of matter also exist in space in the form of cosmic rays—highly energetic subatomic particles traveling at high speed through the cosmos.

**DARK NEBULA**
Also known as the Boogeyman Nebula, LDN 1622 is an extensive cloud of dust and gas. It lies about 500 light-years away.

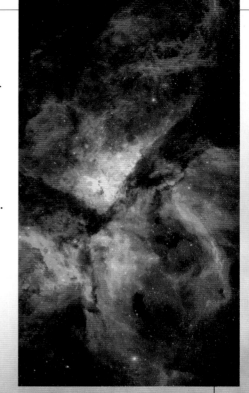

**STAR-FORMING NEBULA**
The Carina Nebula, a giant cloud of gas, is a prominent feature of the sky in the Southern Hemisphere and is visible to the naked eye. Different colors in this image represent temperature variations in the gas.

**GLOWING GAS**
This ocean of glowing gas is an active region of star formation in the Omega Nebula, an emission nebula. Clouds of gas and dust may give birth to stars and planets, but they are also cast off by dying stars, eventually to be recycled into the next stellar generation.

# STARS AND BROWN DWARFS

The universe's light comes mainly from stars—hot balls of gas that generate energy through nuclear fusion in their cores. Stars form from the condensation of clumps of gas and dust in nebulae and sometimes occur in pairs or clusters. Depending on their initial mass, stars vary in color, surface temperature, brightness, and life span. The most massive stars, known as giants and supergiants, are the hottest and brightest, but last for only a few million years. Low-mass stars (the most numerous) are small, dim, red, and may live for billions of years; they are called red dwarfs. Smaller still are brown dwarfs. These are failed stars, not massive or hot enough to sustain the type of fusion that occurs in stars, and emit only a dim glow. Brown dwarfs may account for much of the ordinary matter in the universe.

**BROWN DWARF**
The dot to the right of center in this picture is a brown dwarf called Gliese 229b. The bigger, brighter object is the red dwarf star Gliese 229, around which it orbits.

**SUPERGIANT STAR**
The supergiant Betelgeuse is so vast, it can be seen as a blob in large modern telescopes, though it is some 650 light-years away.

**DOUBLE STAR**
Albireo is a binary, or double star, consisting of a bright yellow-orange main star and a dimmer, bluish companion.

**GLOBULAR CLUSTER**
Star clusters such as M3, above, are ancient objects that orbit galaxies. M3 has about half a million stars.

# STAR REMNANTS

Stars do not last forever. Even the smallest, longest-lived red dwarfs eventually fade away. Stars of medium mass, such as the Sun, expand into large, low-density stars called red giants before they blow off most of their outer layers. They then collapse to form white dwarf stars that gradually cool and fade. The expanding shells of blown-off matter surrounding such stars are called planetary nebulae (although they have nothing to do with planets). More massive stars have even more spectacular ends, disintegrating in explosions called supernovae. The expanding shell of ejected matter may be seen for thousands of years and is called a supernova remnant. Not all of the star's material is blown off, however. Part of the core collapses to a compact, extremely dense object called a neutron star. The most massive stars of all collapse to black holes (see p.26).

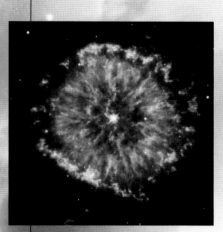

**PLANETARY NEBULA**
This glowing cloud of gas, called NGC 6751, was ejected several thousand years ago from the hot white dwarf star visible at its center.

**SUPERNOVA REMNANT**
The Veil Nebula is the shock wave from a star that exploded 5,000–15,000 years ago. It is 2,600 light-years away, and its material may one day form new stars.

# PLANETS AND SMALLER BODIES

The solar system (our own star, the Sun, and everything that orbits it) is thought to have formed from dust and gas that condensed into a spinning disk called a protoplanetary disk. The central material became the Sun, while the outer matter formed planets and other small, cold objects. A planet is a sphere orbiting a star and, unlike a brown dwarf, produces no nuclear fusion. As thousands of planets have now been detected orbiting stars elsewhere in our galaxy (see pp.296–299), it now seems almost certain that planets are common throughout the universe. In the solar system, the planets are either gas giants, such as Jupiter, or smaller, rocky bodies, such as Earth and Mars. Still smaller objects fall into five categories. Moons are objects that orbit planets or asteroids. Asteroids are rocky bodies of about 150 ft (50 m) to 600 miles (1,000 km) across. Comets are chunks of ice and rock a few miles in diameter that mostly orbit in the far reaches of the solar system. Ice dwarfs are similar but are up to a few hundred miles across. Meteoroids are the remains of shattered asteroids or dust from comets.

**IO**

**EUROPA**

**GANYMEDE**

**CALLISTO**

**GALILEAN MOONS**
Other than Earth's own Moon, these four large moons orbiting the planet Jupiter were the first ever discovered, by Galileo Galilei in 1610.

**PLANET EARTH**
Our home planet seems unusual in having surface water and supporting life. We do not know how rare this is in the universe.

coma

gas tail

dust tail

**COMET IKEYA–ZHANG**
A few comets travel in orbits that bring them close to the Sun. Frozen chemicals in the comet then vaporize to produce a glowing coma (head) and long tails of dust and gas. This bright comet was visible in 2002.

# GALAXIES

The solar system occupies just a tiny part of an enormous, disk-shaped structure of stars, gas, and dust called the Milky Way Galaxy. Until around a hundred years ago, our galaxy was thought to comprise the whole universe; few people imagined that anything might exist outside of the Milky Way. Today, we know that just the observable part of our universe contains more than 1 trillion separate galaxies. They vary in size from dwarf galaxies, a few hundred light-years across and containing about 100 million stars, to giants spanning several hundred thousand light-years and containing as many as 100 trillion stars. As well as stars, galaxies contain clouds of gas, dust, and dark matter (see opposite), all held together by gravity. They come in five shapes: spiral, barred spiral, elliptical (spherical to football-shaped), lenticular (lens-shaped), and irregular. Astronomers identify galaxies by their number in one of several databases of celestial objects. For example, NGC 1530 indicates galaxy 1530 in a database called the New General Catalog (NGC).

**QUASAR**
Some, if not most, galaxies are thought to have been quasars earlier in their life. Quasars are extremely luminous galaxies powered by matter falling into a massive, central black hole.

**SPIRAL GALAXY**
This image taken by the Spitzer Space Telescope shows a nearby spiral galaxy called M81. The sensor captured infrared radiation rather than visible light, and the image highlights dust in the galaxy's nucleus and spiral arms.

galactic nucleus, or core

spiral arm

**BARRED SPIRAL**
In a barred spiral galaxy, such as NGC 1530 (above), the spiral arms radiate from the ends of the central barlike structure rather than from the nucleus.

# BLACK HOLES

A black hole is a region of space containing, at its center, some matter squeezed into a point of infinite density, called a singularity. Within a spherical region around the singularity, the gravitational pull is so great that nothing, not even light, can escape. Black holes can therefore be detected only from the behavior of material around them; those discovered so far typically have a disk of gas and dust spinning around the hole, throwing off hot, high-speed jets of material or emitting radiation (such as X-rays) as matter falls into the hole. There are two main types of black holes: supermassive and stellar. Supermassive black holes, which can have a mass equivalent to billions of suns, exist in the centers of most galaxies, including our own. Their exact origin is not yet understood, but they may be a by-product of the process of galaxy formation. Stellar black holes form from the collapsed remains of exploded supergiant stars (see p.267) and may be very common in all galaxies.

black hole shadow—
dark region cast by and enclosing the sphere of trapped light and matter

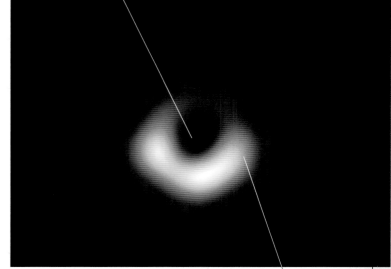

**STELLAR BLACK HOLE**
The black hole SS 433 is situated in the center of this false-color X-ray image. It is detectable because it is sucking in matter from a nearby star and blasting out material and X-ray radiation, visible here as two bright yellow lobes.

**GALACTIC BLACK HOLE**
This image, captured in 2019 by the Event Horizon Telescope, was the first to show an overt black hole—a supermassive black hole at the center of galaxy M87.

spinning disk of dust and gas

# GALAXY CLUSTERS

Galaxies are bound by gravity to form clusters of about 100 to several thousand. Clusters vary from 3 to 35 million light-years across. Some have a concentrated central core and a well-defined spherical structure; others are irregular in shape and structure. The cluster of galaxies that contains our own galaxy is called the Local Group. The neighboring Virgo Cluster is a large, irregular cluster of about 1,300 galaxies lying 50 million light-years away. Chains of a dozen or so galaxy clusters are linked loosely by gravity and make up superclusters, which can be 500 million light-years or more in extent. The Laniakea Supercluster, of which the Local Group is a part, is 520 million light-years long. Superclusters in turn are arranged in broad sheets and filaments, separated by voids of up to several hundred million light-years across.

**HICKSON COMPACT GROUP 87**
This cluster includes a face-on spiral galaxy in the center of the image, two closer oblique spirals, and an elliptical galaxy at lower right.

**RICH CLUSTER**
One of the most massive galaxy clusters known, Abell 1689 is thought to contain hundreds of galaxies (colored gold here).

# DARK MATTER AND DARK ENERGY

There is far more matter in the universe than is contained in stars and other visible objects. The invisible mass is called "dark matter." Its composition is unknown. Some might take the form of MACHOs (massive compact halo objects), which include brown dwarfs and some types of black holes. However, it seems probable that the majority consists of WIMPs (weakly interacting massive particles) or some other type of as-yet undiscovered subatomic particle. Evidence that dark matter exists includes the fact that many galaxies would fly apart rather than rotate unless they contain large amounts of unseen matter. Even if all the dark matter deduced from observations is included, the density of the universe is not sufficient to satisfy theories of its evolution. To find a solution, cosmologists have proposed the existence of "dark energy," a force that counteracts gravity and causes the universe to expand faster (see p.58). The exact nature of dark energy is still speculative.

**DARK MATTER DISTRIBUTION**
This image from a computer simulation shows the way in which dark matter (red clumps and filaments) must be distributed within the galaxy superclusters in our local universe.

**DISTORTED GALAXY**
Nicknamed "the Tadpole," this galaxy lies 420 million light-years away. Like any galaxy, it is a vast, spinning wheel of matter bound together by gravity. In clusters, gravity can also rip galaxies apart. The streamer of stars emerging from this galaxy is thought to have been torn out by the gravity of a smaller galaxy passing close by.

EXPLORING SPACE

## THE SEARCH FOR DARK MATTER

To find dark matter, scientists are investigating some of the several forms it could take. Underground detectors search for evasive particles, such as WIMPs and neutrinos. Neutrinos are so tiny, they were once thought to be massless, but they do have a minute mass. There are so many in the cosmos that their combined mass could account for about 1 percent of the universe's dark matter. WIMPs, if detected, could account for far more.

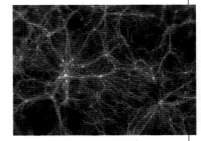

**WIMP DETECTOR**
These photomultiplier tubes are part of the XENON1T detector at Gran Sasso, Italy. They function to detect light flashes as WIMPs smash into atoms in a large tank of liquid xenon.

# MATTER

EXAMINED AT THE TINIEST SCALE, the universe's matter is composed of elementary particles, some of which, governed by various forces, group together to form atoms and ions. In addition to these well-understood types of matter, other forms exist. Most of the universe's mass consists of this "dark matter," whose exact nature is still unknown.

**EMPTY SPACE**
*Most of an atom is empty—
the protons, neutrons,
and electrons are all shown
here much larger than
their real size relative
to the whole atom*

**STRUCTURE OF A CARBON ATOM**
At the center of the atom is the nucleus, which contains protons and neutrons. Electrons move around within two regions, called shells, surrounding the nucleus. The shells appear fuzzy because electrons do not move in defined paths.

**OUTER ELECTRON SHELL**
*Region in which four
electrons orbit*

**INNER ELECTRON SHELL**
*Region within which
two electrons orbit*

## WHAT IS MATTER?

Matter is anything that possesses mass—that is, anything affected by gravity. Most matter on Earth is made of atoms and ions. Elsewhere in the universe, however, matter exists under a vast range of conditions and takes a variety of forms, from thin interstellar medium (see p.228) to the matter in neutron stars and black holes (see p.267). Not all of this matter is made of atoms, but all matter is made of some kind of particle. Certain types of particles are elementary—that is, they are not made of smaller subunits. The elementary particles within ordinary matter are quarks and electrons, which make up atoms and ions. Most of the universe's matter, however, is not ordinary matter, but dark matter (see p.27), perhaps composed partly of neutrinos, theoretical WIMPs (weakly interacting massive particles), or both.

**LUMINOUS MATTER**
These illuminated gas clouds in interstellar space are made of ordinary matter, composed of atoms and ions.

## ATOMS AND IONS

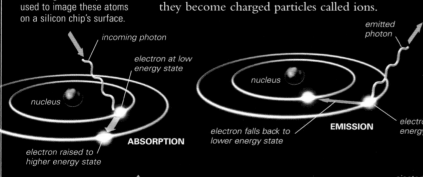

**IMAGING ATOMS**
A device called a scanning-tunneling microscope was used to image these atoms on a silicon chip's surface.

The quarks and electrons that make up atoms occur within specific regions of each atom. The quarks are bound in groups of three by gluons, which are massless particles of force. The quark groups form particles called protons and neutrons. These are clustered in a compact region at the center of the atom called the nucleus. Most of the rest of an atom is empty space, but moving around within this space are electrons. These carry a negative electrical charge and have a very low mass—nearly all the mass in an atom is in the protons and neutrons. Atoms always contain equal numbers of protons (positively charged) and electrons (negatively charged) and so are electrically neutral. If they lose or gain electrons, they become charged particles called ions.

**ABSORPTION AND EMISSION**
The electrons in atoms can exist in different energy states. By moving between energy states, they can either absorb or emit packets, or quanta, of energy. These energy packets are called photons.

*incoming photon*

*electron at low energy state*

*nucleus*

*nucleus*

*emitted photon*

*electron falls back to lower energy state*

**EMISSION**

*electron at high energy state*

*nucleus*

**ABSORPTION**

*electron raised to higher energy state*

*red quark*

*gluon*

*green quark*

*inner-shell electron*

*incoming high-energy photon*

*ejected electron (charge -1 )*

*nucleus*

*nucleus*

*empty shell*

*blue quark*

*electron in outer shell*

**IONIZATION**
One way an atom may become a positive ion is by the electron's absorbing energy from a high-energy photon and, as a result, being ejected along with its charge from the atom.

**ION (CHARGE +1)**

*neutron*

**ATOM
(NEUTRAL, NO CHARGE)**

*neutron*

*proton*

*proton*

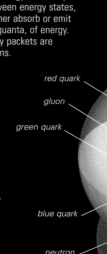

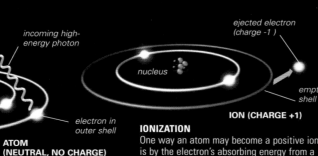

INTRODUCTION

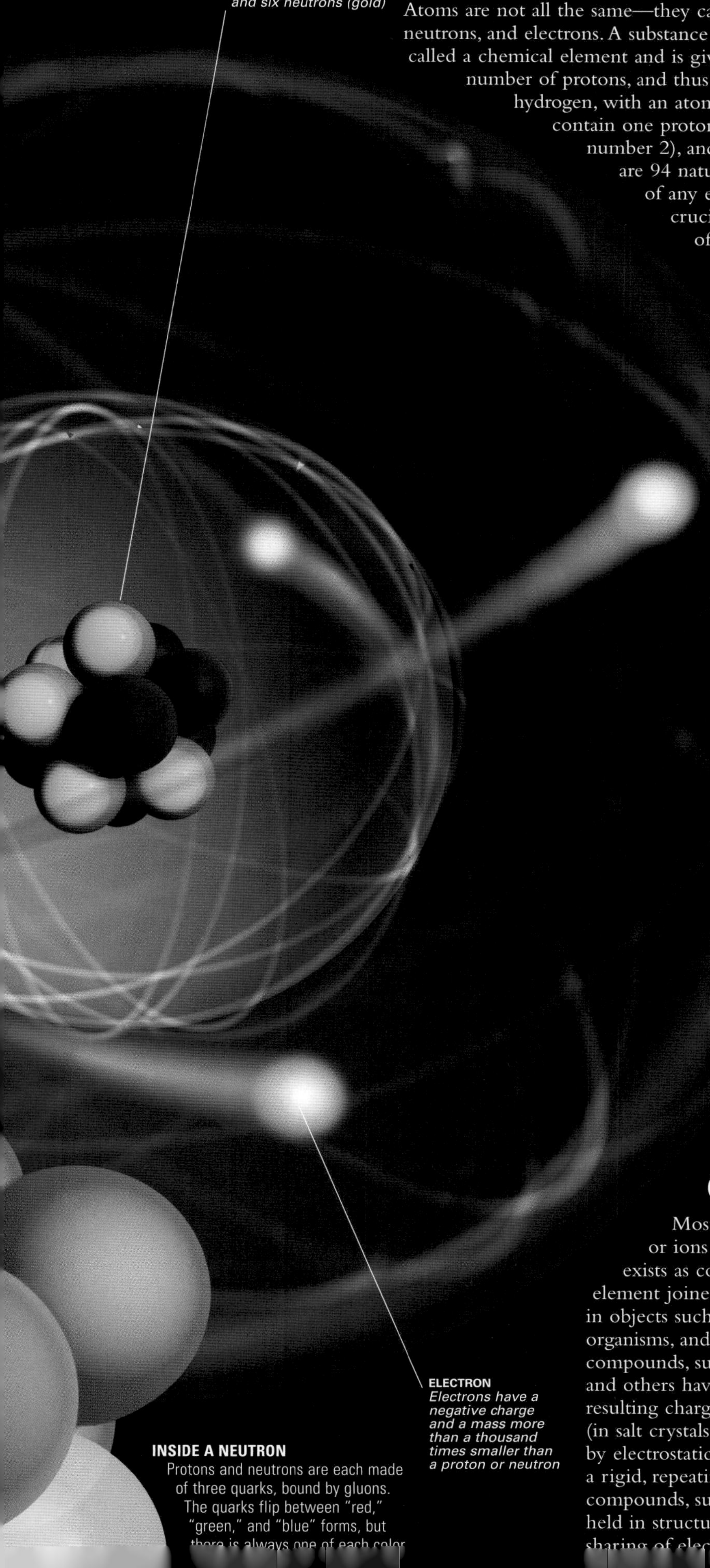

**NUCLEUS**
*A tightly bound ball of six protons (purple) and six neutrons (gold)*

# CHEMICAL ELEMENTS

Atoms are not all the same—they can hold different numbers of protons, neutrons, and electrons. A substance made of atoms of just one type is called a chemical element and is given an atomic number equal to the number of protons, and thus electrons, in its atoms. Examples are hydrogen, with an atomic number of 1 (all hydrogen atoms contain one proton and one electron), helium (atomic number 2), and carbon (number 6). Altogether, there are 94 naturally occurring elements. The atoms of any element are all the same size and, crucially, contain the same configuration of electrons, which is unique to that element and gives it specific chemical properties. The universe once consisted almost entirely of the lightest elements, hydrogen and helium. Most of the others, including such common ones as oxygen, carbon, and iron, have largely been created in stars and star explosions.

## NIELS BOHR

Danish physicist Niels Bohr (1885–1962) was the first to propose that electrons in an atom move within discrete "orbits." He suggested that these orbits have fixed energy levels and that atoms emit or absorb energy in fixed amounts ("quanta") as electrons move between orbits. Bohr's orbits are today called orbitals; they are substructures of electron shells.

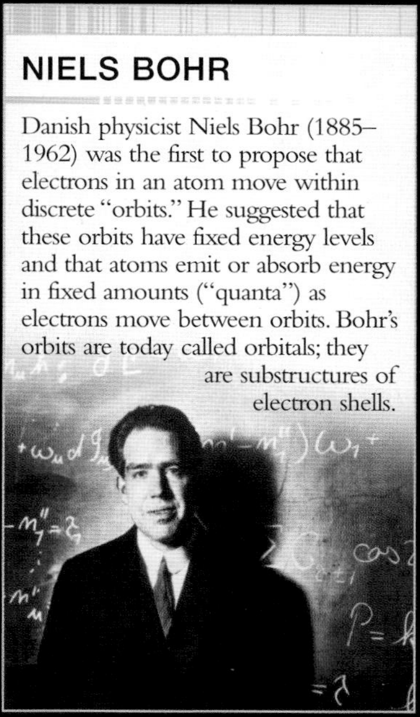

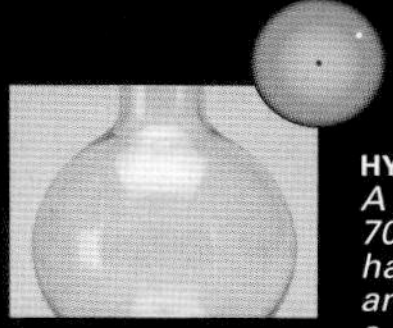

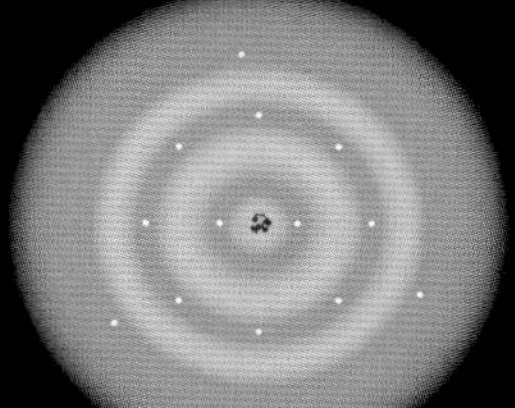

**HYDROGEN**
*A colorless gas at 70°F (21°C). Its atoms have just 1 proton and 1 electron in a single shell.*

**PROPERTIES OF ELEMENTS**
Elements vary markedly in their properties, as shown by the four examples here. These properties are determined by the elements' different atomic structures.

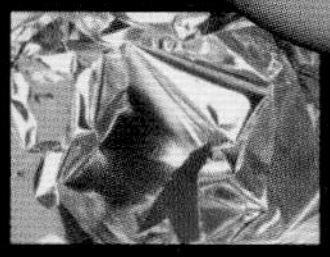

**ALUMINUM**
*A solid metal at 70°F (21°C). Its atoms have 13 protons, 14 neutrons, and 13 electrons in 3 shells.*

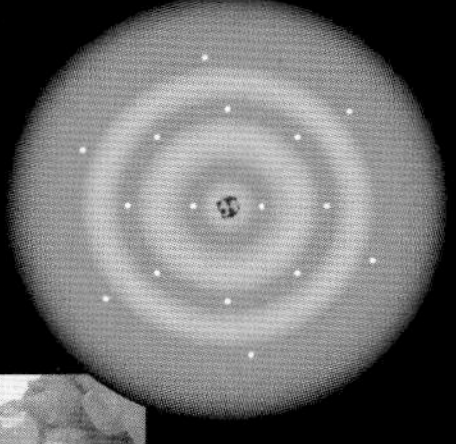

**SULFUR**
*A yellow, brittle solid at 70°F (21°C). Its atoms have 16 protons, 16–18 neutrons, and 16 electrons in 3 shells.*

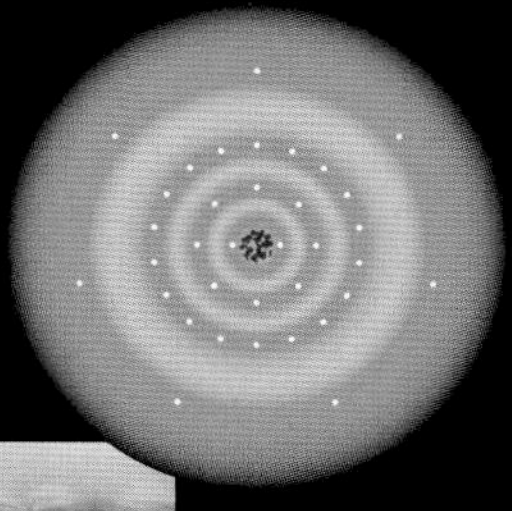

**BROMINE**
*A fuming brown liquid at 70°F (21°C). Its atoms have 35 protons, 44 or 46 neutrons, and 35 electrons in 4 shells.*

# CHEMICAL COMPOUNDS

Most matter in the universe consists of unbound atoms or ions of a few chemical elements, but a significant amount exists as compounds, containing atoms of more than one element joined by chemical bonds. Compounds occur in objects such as planets and asteroids, in living organisms, and in the interstellar medium. In ionic compounds, such as salts, some atoms have gained and others have lost electrons. The resulting charged ions are typically (in salt crystals, for example) bonded by electrostatic forces and arranged in a rigid, repeating structure. In covalent compounds, such as water, the atoms are held in structures called molecules by the sharing of electrons between them.

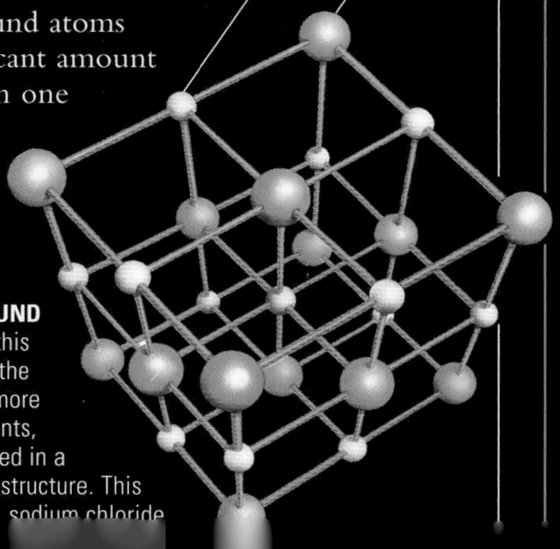

*sodium ion*

*chloride ion*

**IONIC COMPOUND**
Compounds of this type consist of the ions of two or more chemical elements, typically arranged in a repeating solid structure. This example is salt, sodium chloride.

**ELECTRON**
*Electrons have a negative charge and a mass more than a thousand times smaller than a proton or neutron*

**INSIDE A NEUTRON**
Protons and neutrons are each made of three quarks, bound by gluons. The quarks flip between "red," "green," and "blue" forms, but there is always one of each color

# STATES OF MATTER

Ordinary matter exists in four states, called solid, liquid, gas, and plasma. These differ in the energy of the matter's particles (molecules, atoms, or ions) and in the particles' freedom to move relative to one another. Substances can transfer between states—by losing or gaining heat energy, for instance. The constituents of a solid are locked by strong bonds and can hardly move, whereas in a liquid they are bound only by weak bonds and can move freely. In a gas, the particles are bound very weakly and move with greater freedom, occasionally colliding. A gas becomes a plasma when it is so hot that collisions start to knock electrons out of its atoms. A plasma therefore consists of ions and electrons moving extremely energetically. Because stars are made of plasma, it is the most common state of ordinary matter in the universe; the gaseous state is the second most common.

**SOLID, LIQUID, AND GAS**
On Earth, water can sometimes be found as a liquid, in solid form (ice or snow), and as a gas (water vapor), all in close proximity.

# FORCES INSIDE MATTER

The bonds that hold solids, liquids, gases, and plasma together are based on the electromagnetic (EM) force. This force attracts particles of unlike electrical charge. It is one of three forces that control the small-scale structure of matter. The others are the strong nuclear force, composed of fundamental and residual parts; and the weak nuclear force or interaction. Together with a fourth force, gravity, these are the fundamental forces of nature. The EM, weak, and strong forces are mediated by force-carrier particles, which belong to a group of particles called the bosons. (A force-carrier particle for gravity, called the graviton, is also hypothesized.) The EM force, as well as binding atoms in solids and liquids, also holds electrons within atoms. The strong force holds together protons, neutrons, and atomic nuclei. The weak force brings about radioactive decay and other nuclear interactions.

**PLASMA**
Plasma exists naturally in stars but can also be artificially created. In a plasma ball, electricity is induced to flow from a charged metal ball through a gas to the surface of a glass sphere, creating plasma streamers.

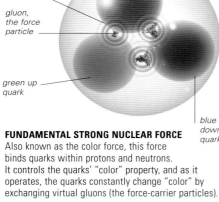

neutron
red down quark
fundamental strong nuclear force
gluon, the force particle
green up quark
blue down quark

**FUNDAMENTAL STRONG NUCLEAR FORCE**
Also known as the color force, this force binds quarks within protons and neutrons. It controls the quarks' "color" property, and as it operates, the quarks constantly change "color" by exchanging virtual gluons (the force-carrier particles).

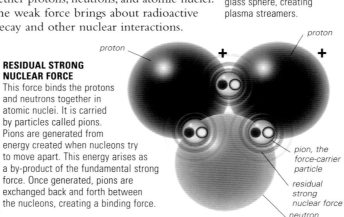

proton

**RESIDUAL STRONG NUCLEAR FORCE**
This force binds the protons and neutrons together in atomic nuclei. It is carried by particles called pions. Pions are generated from energy created when nucleons try to move apart. This energy arises as a by-product of the fundamental strong force. Once generated, pions are exchanged back and forth between the nucleons, creating a binding force.

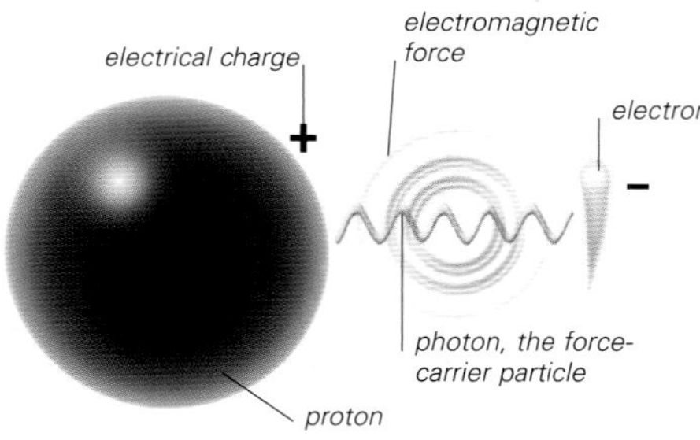

proton

pion, the force-carrier particle
residual strong nuclear force
neutron

**STEVEN WEINBERG**

American physicist Steven Weinberg (b. 1933) is best known for his theory that two of the fundamental forces—the weak interaction and the electromagnetic force—are unified, or work in an identical way, at extremely high energy levels, such as those existing just after the Big Bang (see p.48). Weinberg's so-called electroweak theory was confirmed by particle accelerator experiments in 1973. He and his colleagues received the 1979 Nobel Prize for physics.

electrical charge
electromagnetic force
electron
photon, the force-carrier particle
proton

**ELECTROMAGNETIC FORCE**
Within an atom, the electromagnetic (EM) force holds the electrons within the shells surrounding the nucleus. It attracts the negatively electrically charged electrons toward the positively charged nucleus and keeps electrons apart. The force carrier for the EM force is the photon.

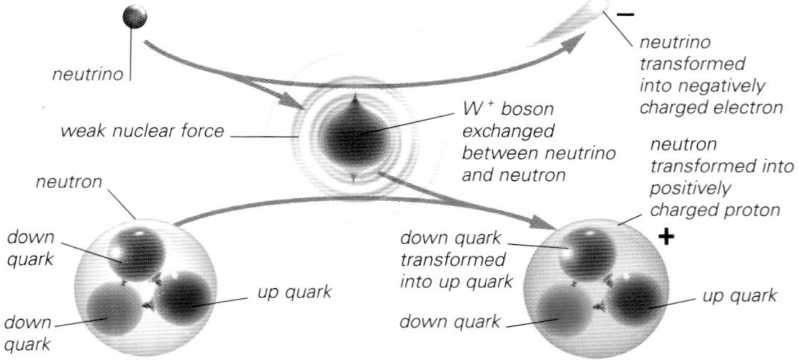

neutrino
weak nuclear force
neutron
down quark
down quark
up quark

$W^+$ boson exchanged between neutrino and neutron

neutrino transformed into negatively charged electron
neutron transformed into positively charged proton
down quark transformed into up quark
down quark
up quark

**WEAK INTERACTION, OR WEAK NUCLEAR FORCE**
This force governs radioactive decay, among other interactions. Its force-carrier particles are $W^+$, $W^-$, and $Z^0$ bosons. The interaction illustrated above is called neutrino-neutron scattering. Here, a $W^+$ boson controls the changing of a neutrino into an electron and the transformation of a down quark into an up quark, converting a neutron into a proton.

# PARTICLE PHYSICS

For decades, physicists have sought a better understanding of matter and the fundamental forces. Research has centered on smashing particles together in devices called particle accelerators. These experiments have identified hundreds of different particles (most of them highly unstable), which vary in properties such as mass, electric charge, and "spin." Known particles and their interactions are currently explained by a theory called the standard model of particle physics, which includes a particle classification (see table, right). One recently discovered addition to the model is the Higgs boson. This is the force carrier for an energy field that permeates space and endows other particles with mass through its interactions with them. However, the hypothetical particle called the graviton—proposed to be a force carrier for gravity—does not fit comfortably into the model. This is because the best theory of gravity (general relativity, see pp.42–43) is incompatible with aspects of the standard model. New theories such as string theory (see panel, below) attempt to unite gravity with particle physics.

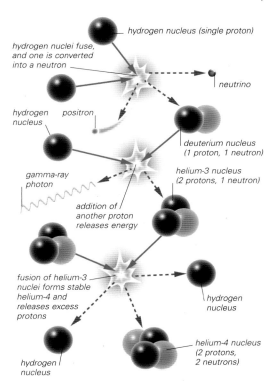

**COLLISION EVENT**
The results of a collision between xenon ions in the Large Hadron Collider are shown here as a computer graphic. The blue tracks represent baryon interactions.

## CLASSIFICATION OF PARTICLES

Physicists distinguish composite particles, which have an internal structure, from elementary particles, which do not. They also divide particles into fermions and bosons. Fermions (leptons, quarks, and baryons) are the building blocks of matter. Bosons (gauge bosons and mesons) are primarily force-carrier particles.

**ELEMENTARY PARTICLES**

Leptons and quarks form matter, while gauge bosons carry forces. Quarks feel the strong nuclear force, but leptons do not.

**LEPTONS**

— electron, charge −1
— neutrino, charge 0

Six different leptons exist, but the 2 above are the only stable ones and are those that occur in ordinary matter.

**QUARKS**

— up, charge +2/3
down, charge −1/3

There are 6 "flavors" of quark, but only 2 occur in ordinary matter: "up" and "down." Each can exist in any of 3 "colors."

**GAUGE BOSONS**

These are force-carrier particles. Some shown are hypothetical.

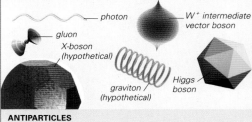
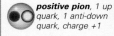

— photon
gluon
X-boson (hypothetical)
W+ intermediate vector boson
graviton (hypothetical)
Higgs boson

**ANTIPARTICLES**

Most particles have an antimatter equivalent that has the same mass but whose charge and other properties are opposite.

positron (antielectron), charge +1
anti-up quarks, charge −2/3
antineutrino
**antiproton**, 1 anti-down and 2 anti-up quarks, charge −1
**antineutron**, 1 anti-up and 2 anti-down quarks, charge 0

**COMPOSITE PARTICLES**

Also known as hadrons, these are composed of quarks, antiquarks, or both, bound by gluons.

**BARYONS**
Relatively large-mass particles containing 3 quarks.

**proton**, 1 down and 2 up quarks, charge +1

**neutron**, 1 up and 2 down quarks, charge 0

**MESONS**
Particles containing a quark and an antiquark.

**positive pion**, 1 up quark, 1 anti-down quark, charge +1

Hundreds of other baryons and mesons exist.

**EXOTIC PARTICLES**
Further particles have been hypothesized that do not have a place in this particle classification. They include magnetic monopoles and WIMPs (weakly interacting massive particles).

# NUCLEAR FISSION AND FUSION

Twentieth-century physicists learned that atomic nuclei are not immutable but can break up or join together. In nature, unstable atomic nuclei can spontaneously disassemble, giving off particles and energy, measured as radioactivity. Similarly, in the artificial process of nuclear fission, large nuclei are intentionally split into smaller parts, with huge energy release. On a cosmic scale, a more important phenomenon is nuclear fusion. In this process, atomic nuclei join, forming a larger nucleus and releasing energy. Fusion powers stars and has created most or all atoms of the majority of naturally occurring chemical elements. The most common fusion reaction in stars joins hydrogen nuclei (protons) into helium nuclei. In this and other fusion reactions, the products of the reaction have a slightly lower mass than the combined mass of the reactants. The lost mass converts into huge amounts of energy, in accordance with Einstein's famous equation $E = mc^2$ that links energy (E), mass (m), and the speed of light (c) (see p.41).

hydrogen nucleus (single proton)
hydrogen nuclei fuse, and one is converted into a neutron
neutrino
hydrogen nucleus
positron
gamma-ray photon
deuterium nucleus (1 proton, 1 neutron)
helium-3 nucleus (2 protons, 1 neutron)
addition of another proton releases energy
fusion of helium-3 nuclei forms stable helium-4 and releases excess protons
hydrogen nucleus
hydrogen nucleus
helium-4 nucleus (2 protons, 2 neutrons)
hydrogen nucleus

**FUSION REACTION IN THE SUN**
In stars the size of the Sun or smaller, the dominant energy-producing fusion process is called the proton–proton chain. This chain of high-energy collisions fuses hydrogen nuclei (free protons), via several intermediate stages, into helium-4 nuclei. Energy is released in the form of gamma-ray photons and in the movement energy of the helium nuclei. Positrons and neutrinos are also produced.

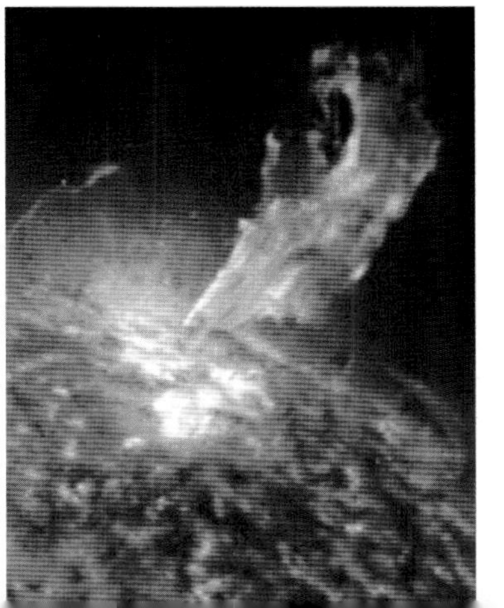

**THE HEAT OF FUSION**
All solar energy comes from nuclear fusion in the Sun's core. The energy gradually migrates to the Sun's surface and into space through heat transfer by convection, conduction, and radiation.

EXPLORING SPACE

## STRING THEORY

For decades, physicists have sought a "theory of everything" that will unify the four fundamental forces of nature and provide an underlying scheme for how particles are constructed. A leading contender is string theory, which proposes that each elementary particle consists of a tiny vibrating filament called a string. The vibrational modes, or frequencies, of these strings lend particles their varied properties. Although it sounds bizarre, many leading physicists are enthusiastic adherents of string theory.

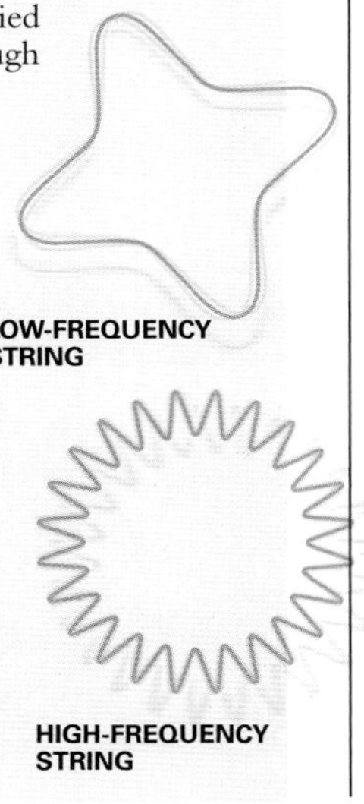

**LOW-FREQUENCY STRING**

**VIBRATING STRINGS**
A string is closed, like a loop, or open, like a hair. The two closed strings shown here are vibrating at different resonant frequencies, just as the strings on a guitar have rates at which they prefer to vibrate.

**HIGH-FREQUENCY STRING**

INTRODUCTION

**NEUTRINO OBSERVATORY**
High-energy processes in the universe produce neutrinos—fast-moving particles that rarely interact with matter. To detect them, scientists created the IceCube Neutrino Observatory in Antarctica. Eighty-six holes drilled in the ice contain over 5,000 optical sensors. In 2017, IceCube detected a super-high-energy neutrino that was traced to a galaxy 5.7 billion light-years away.

# RADIATION

RADIATION IS ENERGY IN THE FORM of waves or particles that are emitted from a source and can travel through space and some types of matter. Electromagnetic (EM) radiation includes light, X-rays, and infrared radiation. Particulate radiation consists of fast-moving charged particles such as cosmic rays and particles emitted in radioactive decay. EM radiation is vastly more significant in astronomy.

electric field strength
amplitude
magnetic field strength
wavelength

**HOW WAVES TRAVEL**
An EM wave consists of oscillating electrical and magnetic fields arranged perpendicular to each other and carrying energy forward.

## ELECTROMAGNETIC RADIATION

Energy in the form of EM radiation is one of the two main components of the universe, the other being matter (see p.28). This type of radiation is produced by the motion of electrically charged particles, such as electrons. A moving charge gives rise to a magnetic field. If the motion is constant, then the magnetic field varies and in turn produces an electric field. By interacting with each other, the two fields sustain one another and move through space, transferring energy. As well as visible light, EM radiation includes radio waves, microwaves, infrared (heat) radiation, ultraviolet radiation, X-rays, and gamma rays. All these phenomena travel through space at the same speed—called the velocity of light. This speed is very nearly 186,000 miles (300,000 km) per second or 670 million mph (1 billion kph).

## WAVELIKE BEHAVIOR

In most situations, EM radiation acts as a wave—a disturbance moving energy from one place to another. It has properties such as wavelength (the distance between two successive peaks of the wave) and frequency (the number of waves passing a given point each second). This wavelike nature is shown by experiments such as the double-slit test (see below), in which light waves diffract (spread out) after passing through a slit and also interfere with each other as their peaks and troughs overlap. The forms of EM radiation differ only in wavelength, but this affects other properties, such as penetrating power and ability to ionize atoms (see p.28).

**INTERFERING WAVES**
The slit experiment is analogous to disturbing two points on the surface of a liquid. The ripples interfere to corrugate the liquid's surface.

## PARTICLELIKE BEHAVIOR

EM radiation behaves mainly like a wave, but it can also be considered to consist of tiny packages or "quanta" of energy called photons. Photons have no mass but carry a fixed amount of energy. The energy in a photon depends on its wavelength— the shorter the wavelength, the more energetic the photon. For example, photons of blue (short-wavelength) light are more energetic than photons of red (long-wavelength) light. A classic demonstration of light's particlelike properties is provided by a phenomenon called the photoelectric effect (see below). If a blue light shines on a metal surface, it causes electrons to be ejected from the metal, whereas even a very bright red light has no such effect.

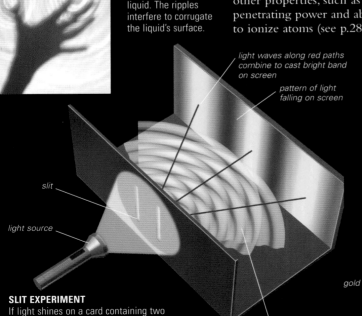

light waves along red paths combine to cast bright band on screen

pattern of light falling on screen

slit

light source

**SLIT EXPERIMENT**
If light shines on a card containing two slits, diffraction spreads the light waves out like ripples emanating in arcs from each slit. The two wave trains then interfere to produce a light and dark pattern on the screen.

light waves forming interference pattern

low-energy photon of red light

low-energy electron

gold foil

**RED LIGHT**
When red light shines on a metal surface, no electrons are ejected, even if the light is intensely bright.

high-energy photon of blue light

**BLUE LIGHT**
Blue light shining on the same surface causes electrons to be ejected, because the blue photons are more energetic.

ultra-high-energy ultraviolet photon

electron ejected at higher energy

**ULTRAVIOLET LIGHT**
Shining ultraviolet light on the metal surface causes electrons to be ejected at very high energy.

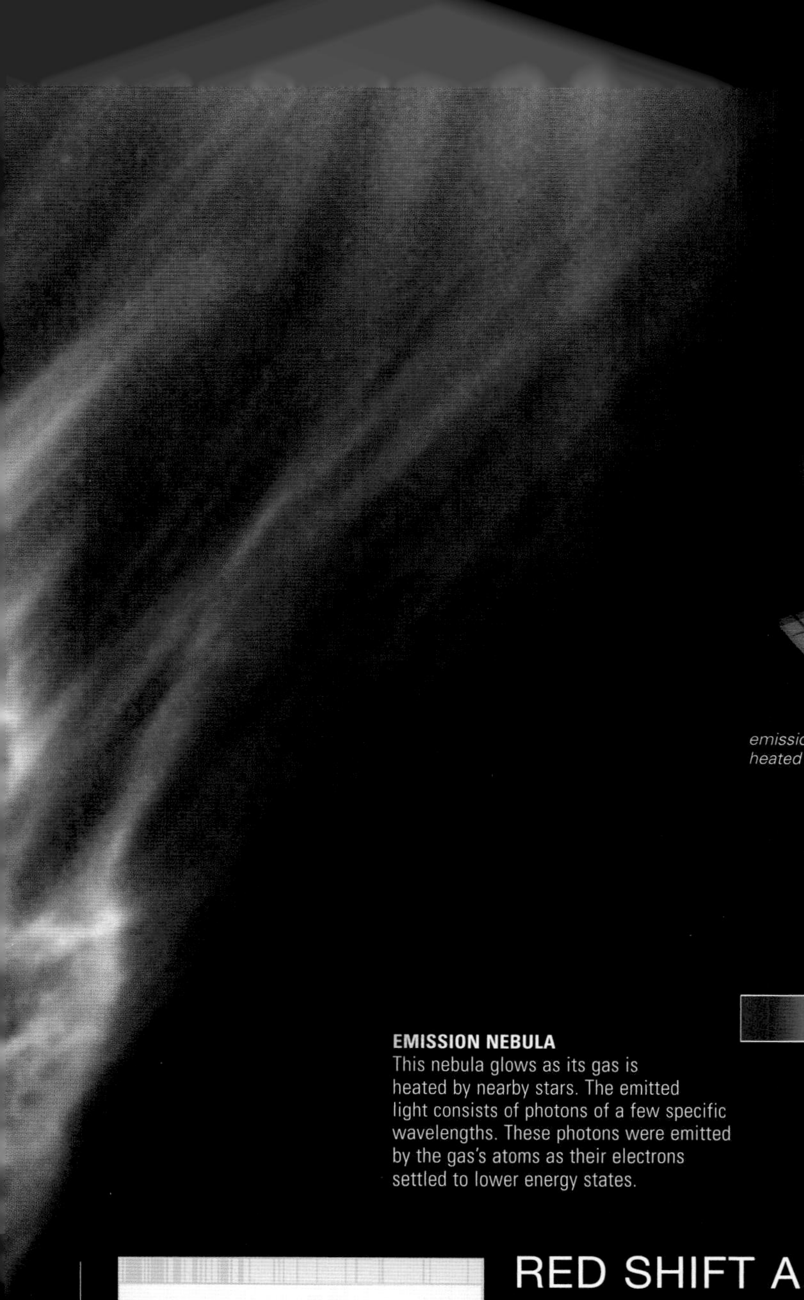

# ANALYZING LIGHT

The radiation output of celestial objects is a mixture of wavelengths. When passed through a prism, the light is split into its component wavelengths, giving a record called a spectrum. A star's spectrum usually contains dark lines called absorption lines, caused by photons being absorbed at certain wavelengths by atoms in the star's atmosphere. They can be used to establish what chemical elements are present. The spectrum of a nebula can also reveal its composition. When heated by radiation from a nearby star, its atoms emit their own light. The resulting spectrum, called an emission spectrum, consists of a series of bright lines characteristic of different elements.

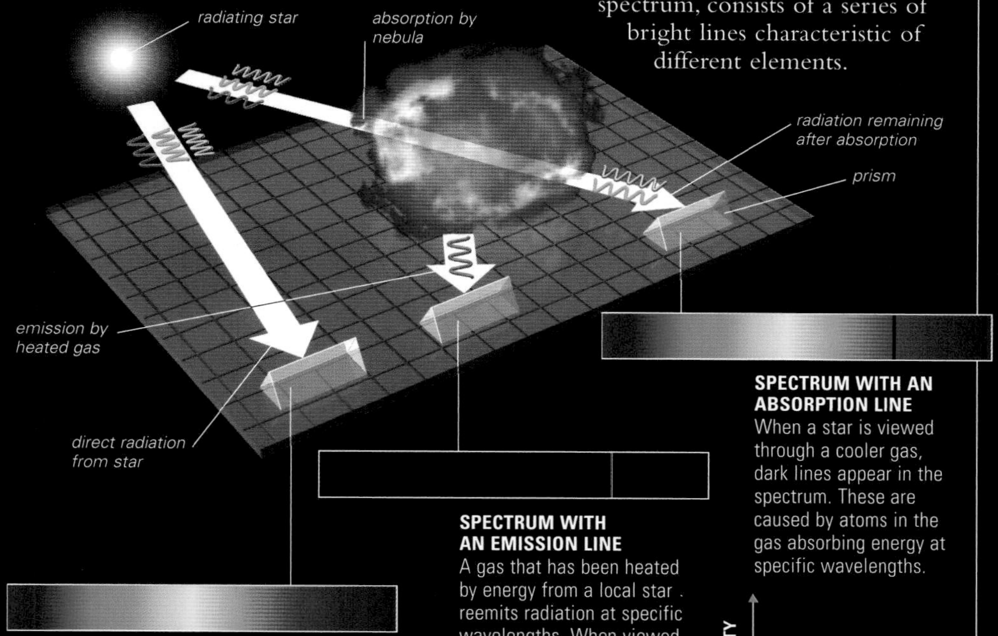

radiating star

absorption by nebula

radiation remaining after absorption

prism

emission by heated gas

direct radiation from star

### SPECTRUM WITH AN ABSORPTION LINE
When a star is viewed through a cooler gas, dark lines appear in the spectrum. These are caused by atoms in the gas absorbing energy at specific wavelengths.

### EMISSION NEBULA
This nebula glows as its gas is heated by nearby stars. The emitted light consists of photons of a few specific wavelengths. These photons were emitted by the gas's atoms as their electrons settled to lower energy states.

### CONTINUOUS SPECTRUM
A hot, dense gas such as a star produces a continuous light spectrum from its surface, with all different light wavelengths (colors) represented.

### SPECTRUM WITH AN EMISSION LINE
A gas that has been heated by energy from a local star . reemits radiation at specific wavelengths. When viewed obliquely, this produces an emission-line spectrum.

### RADIATION FROM HOT OBJECTS
Not only is the total radiation greater from hotter objects, but the wavelength of peak intensity is also shorter (toward the blue end of the light spectrum). Astronomers can calculate the temperature of stars by measuring the peak of the star's spectrum.

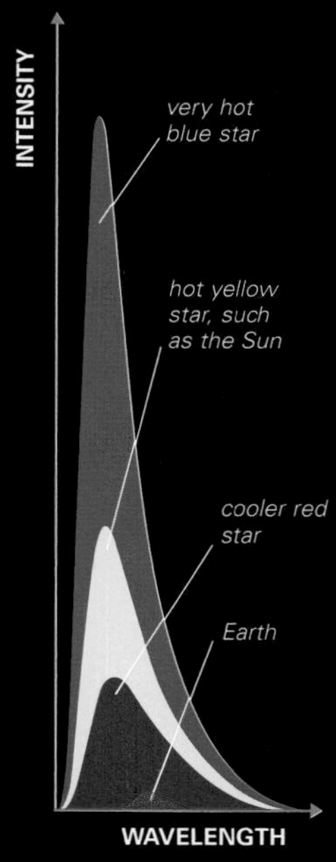

INTENSITY

very hot blue star

hot yellow star, such as the Sun

cooler red star

Earth

WAVELENGTH

## LOUIS DE BROGLIE

The French physicist Louis de Broglie (1892–1987) received the Nobel Prize in 1929. He found that particles of matter, such as electrons, have wavelike properties. The dual nature of matter and light (each has both particlelike and wavelike properties) is called wave- particle duality.

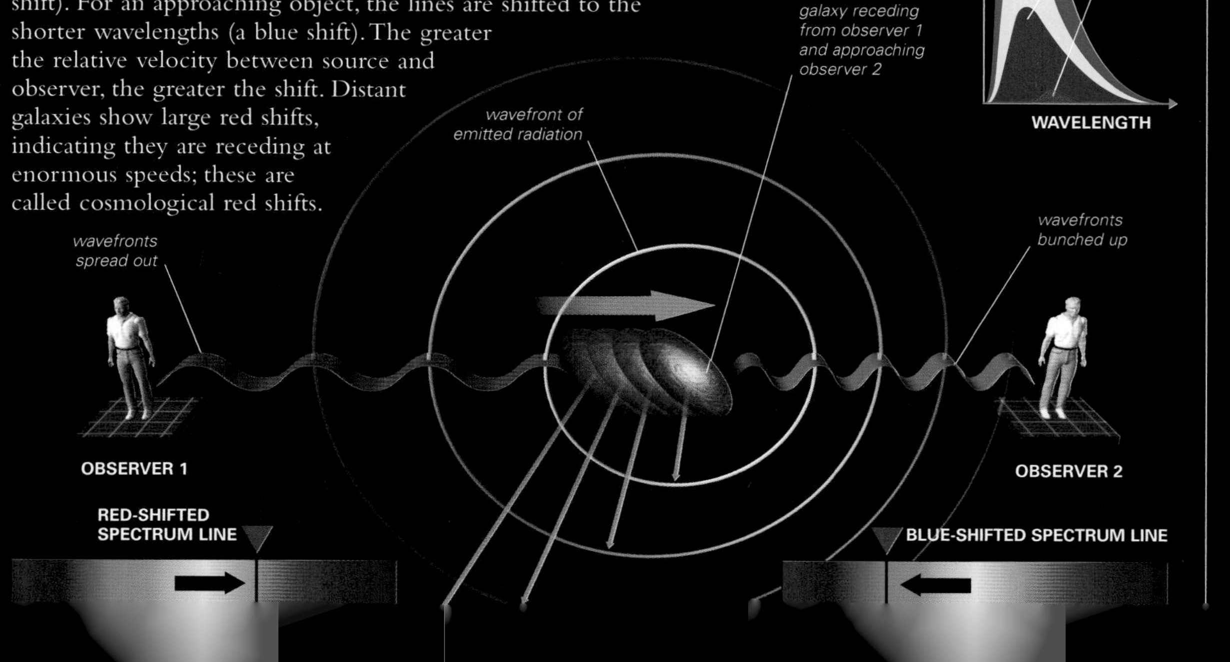

### SHIFTING WAVELENGTHS
Shifts occur because of a phenomenon called the Doppler effect. The wavefronts of light from a receding object are stretched out, increasing their wavelength, while those of an approaching object are squashed up

# RED SHIFT AND BLUE SHIFT

The spectrum of radiation received by an observer shifts if the source of the radiation is moving relative to the observer—and celestial objects are always moving. Astronomers can detect the shifts by measuring the position of spectral lines, which occur at characteristic places. Observers watching an object moving away see its spectral lines shifted toward longer wavelengths (a red shift). For an approaching object, the lines are shifted to the shorter wavelengths (a blue shift). The greater the relative velocity between source and observer, the greater the shift. Distant galaxies show large red shifts, indicating they are receding at enormous speeds; these are called cosmological red shifts.

galaxy receding from observer 1 and approaching observer 2

wavefront of emitted radiation

wavefronts spread out

wavefronts bunched up

OBSERVER 1

OBSERVER 2

RED-SHIFTED SPECTRUM LINE

BLUE-SHIFTED SPECTRUM LINE

# ACROSS THE SPECTRUM

Celestial objects can emit radiation across the EM spectrum, from radio waves through visible light to gamma rays. Some complex objects, such as galaxies and supernova remnants, shine at nearly all these wavelengths. Cool objects tend to radiate photons with less energy and are therefore only visible at longer wavelengths. Toward the gamma-ray end of the spectrum, photons are increasingly powerful. High-energy X-rays and gamma rays originate only from extremely hot sources, such as the gas of galaxy clusters (see p.327) or violent events, such as the swallowing of matter by black holes (see p.267). To detect all this radiation and form images, astronomers need a range of instruments—each type of radiation has different properties and must be collected and focused in a particular way. Radiation at many wavelengths does not penetrate to Earth's surface and is detectable only by orbiting observatories above the atmosphere.

*light shield*

*primary mirror*

*solar array*

## RADIO WAVES TELESCOPE ARRAY

Radio waves can be many feet (meters) long. To create sharp images from such long waves, astronomers collect and focus them using telescopes with huge dish antennae. They might use a single dish or an entire array. The Atacama Large Millimeter Array, or ALMA (right), in northern Chile is the world's largest array. It consists of 66 high-precision antennas that can be moved across a desert plateau into various configurations and their data combined to form a single, fine-detailed image.

*parabolic dish reflects all incoming radio waves to the subreflector*

*receiver is in the recess at center of dish*

*subreflector focuses the radio waves onto receiver*

## MICROWAVES SPACE OBSERVATORY

Most microwaves are absorbed by Earth's atmosphere, so microwave observatories must be placed in space. Between 2009 and 2013, the European Space Agency's Planck Observatory (above) mapped the cosmic microwave background radiation (see p.54) across the whole sky. This is the oldest EM radiation in the universe, released soon after the Big Bang. The observatory carried out its survey from a stable orbit around the Sun 900,000 miles (1.5 million km) from Earth.

*red denotes a fractionally higher temperature*

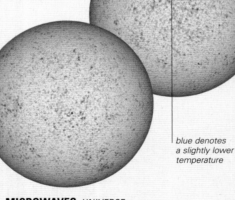

*blue denotes a slightly lower temperature*

## INFRARED MOUNTAINTOP TELESCOPE

Little infrared radiation from space reaches sea level on Earth, but some penetrates down to the height of mountaintops. Some infrared telescopes, such as NASA's Spitzer Space Telescope (see p.247), have been launched into space, but most infrared astronomy is conducted from mountain observatories. This one, named VISTA (Visible and Infrared Survey Telescope for Astronomy), is located at the Paranal Observatory in northern Chile at an altitude of 8,645 ft (2,635 m). It is the world's largest telescope dedicated to surveying in the near infrared (the band of infrared wavelengths closest to visible light). With a 13.5-ft (4.1-m) main mirror, VISTA achieves great resolution. It can be used for detecting, imaging, or peering into a variety of celestial objects, from dim galaxies and galaxy clusters to brown dwarfs and dusty nebulae.

## RADIO WAVES PROTOPLANETARY DISK

This image produced by ALMA (see above) shows a protoplanetary disk (disk of dust and gas) rotating around a young, distant star named HL Tauri. The dark rings may signify that planets are forming in the disk. To create such an image, a radio dish, or an array, scans a tiny area of sky, gradually building an image by recording variations in radio intensity. Other objects that produce radio emissions include hydrogen clouds in galaxies or (from the synchrotron radiation they emit) active galaxies (see p.320) and pulsars (see p.267).

## MICROWAVES UNIVERSE

The lack of microwave sources in the nearby universe is fortunate, as it makes it easier to observe the cosmic background radiation, which reaches Earth at microwave wavelengths. The pattern of microwaves from the whole sky, as measured by the Planck Observatory, is here projected onto two hemispheres.

## INFRARED NEBULA

This infrared image of the Orion Nebula—a bright, star-forming emission nebula—was taken with the VISTA telescope. VISTA's infrared vision means that it can peer deep into the normally hidden dusty regions of the nebula and reveal the many active young stars buried there.

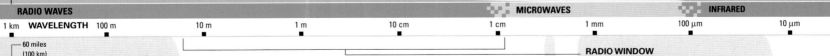

| RADIO WAVES | | | | | | MICROWAVES | | INFRARED | |
|---|---|---|---|---|---|---|---|---|---|
| 1 km | WAVELENGTH 100 m | 10 m | 1 m | 10 cm | 1 cm | 1 mm | 100 µm | 10 µm |

**HEIGHT IN EARTH'S ATMOSPHERE**

- 60 miles (100 km)

- 30 miles (50 km)

- 0

*opaque atmosphere at long radio wavelengths*

**ATMOSPHERIC ABSORPTION**
Only certain types of EM radiation—visible light and some radio waves—can pass through Earth's atmosphere. Others are absorbed to varying extents and can only be detected from space or at high altitudes. Gray areas indicate the altitude at which different wavelengths are absorbed.

*transparent atmosphere at shorter radio wavelengths*

**RADIO WINDOW**
*Radiation with wavelengths between 0.4 in–36 ft (1 cm–11 m) passes readily through the atmosphere. This part of the spectrum, which includes some radio waves and some microwaves, is called the "radio window."*

*opaque atmosphere*

## EXPLORING SPACE

# COMBINING THE VISIBLE AND INVISIBLE

Astronomers have developed telescopes that can gather information from EM radiation other than visible light, but they still face a problem of how to visualize the invisible. The most popular technique uses computers to create "false-color" images—pictures that show the object in particular wavelengths of radiation, sometimes color-coded, but often just using varying intensities of a single color. These images of a supernova remnant

**1** Visible light image from Hubble Telescope and Very Large Telescope in Chile.

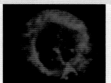
**2** False-color X-ray image of E0102 taken by the Chandra X-ray Observatory.

known as E0102 show radiation in visible light and in X-rays, revealing the temperature of different regions. Combining the images shows the overall structure.

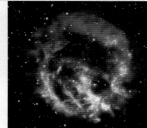
**COMPOSITE IMAGE**
Here, the visible light data has been combined with the false-color X-ray data.

**VISIBLE LIGHT** OPTICAL TELESCOPE
Optical telescopes with the largest mirrors achieve the brightest, sharpest images and the greatest power (see p.82). They range from those of amateur astronomers, such as this example with an 8.5-in (21.5-cm) mirror, to those of large observatories, with mirrors up to 34 ft (10.4 m) wide. The Extremely Large Telescope (ELT) in Chile, with a 130-ft (39.3-m) wide mirror, will be the world's largest optical telescope once it is completed (expected 2024).

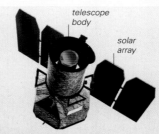
**ULTRAVIOLET**
ORBITING OBSERVATORY
Between 2003 and 2012, NASA's Galaxy Evolution Explorer (GALEX) (above) surveyed sources of ultraviolet (UV) radiation. UV radiation originates from hot sources such as young stars, the Sun's corona, neutron stars, supernova remnants, and some galaxies.

**X-RAYS**
ORBITING OBSERVATORY
X-rays are highly energetic and so powerful that they pass through conventional mirrors. To focus X-rays, telescopes such as the Chandra X-ray Observatory (above) use a nest of curved "grazing incidence" mirrors of polished metal. X-rays glance off these mirrors, like ricocheting bullets, toward the focal point.

**GAMMA RAYS** ORBITING TELESCOPE
Gamma rays are the most energetic EM waves, emitted by the most violent cosmic events. The Fermi Gamma-Ray Space Telescope (above) was launched in 2008 to study gamma rays from phenomena such as supernovae, black holes, pulsars, and gamma-ray bursts.

**VISIBLE LIGHT** GALAXY
The spiral galaxy M90, which lies 30 million light-years away, is shown here as it appears to human eyes through a large telescope. This galaxy is similar in size to the Milky Way. The image was taken at Kitt Peak National Observatory in Arizona.

**ULTRAVIOLET** SUPERNOVA REMNANT
Wispy tendrils of gas and hot dust glow brightly in this UV image of the Cygnus Loop supernova remnant, taken by GALEX. It is all that remains from a massive stellar explosion that occurred between 5,000 and 8,000 years ago.

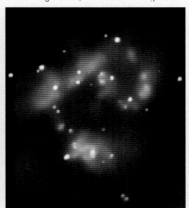
**X-RAY** GALAXY
The orange-pink regions in this Chandra Observatory image of two colliding galaxies (called the Antennae, see p.317) are X-ray-emitting "superbubbles" of hot gas. The point X-ray sources (bright spots) are black holes and neutron stars.

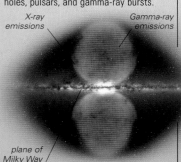
**GAMMA RAY** GALAXY STRUCTURES
In 2010, using data from the Fermi Telescope, scientists discovered two gamma-ray-emitting lobes (shown here colored purple) extending for 25,000 light-years above and below the plane of our galaxy. These structures may be linked to a massive, energetic burst of star formation near the galaxy's center.

VISIBLE | ULTRAVIOLET | X-RAYS | GAMMA RAYS

1 μm | 100 nm | 10 nm | 1 nm | 0.1 nm | 0.01 nm | 0.001 nm | 0.0001 nm | 0.00001 nm

transparent atmosphere

**OPTICAL WINDOW**
Wavelengths of radiation between 300 and 1,100 nm (nanometers) pass easily through the atmosphere (the visible light spectrum extends from 400 to 700 nanometers).

opaque atmosphere

# GRAVITY, MOTION, AND ORBITS

GRAVITY IS THE ATTRACTIVE FORCE that exists between every object in the universe, the force that both holds stars and galaxies together and causes a pin to drop. Gravity is weaker than nature's other fundamental forces, but because it acts over great distances and between all bodies possessing mass, it has played a major part in shaping the universe. Gravity is also crucial in determining orbits and creating phenomena such as planetary rings and black holes.

## DISKS AND RINGS
The disk- and ringlike structures common in celestial objects are maintained by gravity. Examples include Saturn's rings (pictured), spiral galaxies, and the disks around black holes. Every particle in Saturn's rings is held in orbit through gravitational interactions with billions of other particles and with Saturn itself.

## NEWTONIAN GRAVITY

The scientific study of gravity began with Galileo Galilei's demonstration, in about 1590, that objects of different weight fall to the ground at exactly the same, accelerating rate. In 1665 or 1666, Isaac Newton realized that whatever force causes objects to fall might extend into space and be responsible for holding the Moon in its orbit. By analyzing the motions of several heavenly objects, Newton formulated his law of universal gravitation. It stated that every body in the universe exerts an attractive force (gravity) on every other body and described how this force varies with the masses of the bodies and their separation. To this day, Newton's law remains applicable for understanding and predicting the movements of most astronomical objects.

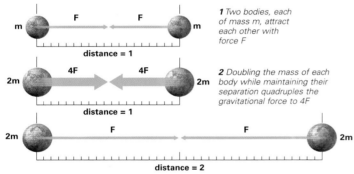

*1 Two bodies, each of mass m, attract each other with force F*

*2 Doubling the mass of each body while maintaining their separation quadruples the gravitational force to 4F*

*3 Doubling the separation between the bodies reduces the force by a factor of 4, back to F*

## ISAAC NEWTON

The English mathematician and physicist Isaac Newton (1642–1727) was one of the greatest-ever scientific intellects. As well as his discoveries in the fields of gravity and motion, he co-discovered the mathematical technique of calculus. In 1705, Newton became the first scientist to be knighted for his work.

### MASS AND DISTANCE
Any two bodies are attracted by a force of gravity proportional to the mass of one multiplied by the mass of the other. The force is also inversely proportional to the square of their separation.

## NEWTON'S LAWS OF MOTION

From his studies of gravity and the motions of heavenly bodies, and again extending concepts first developed by Galileo, Newton formulated his three laws of motion. Before Galileo and Newton, people thought that an object in motion could continue moving only if a force acted on it. In his first law of motion, Newton contradicted this idea: he stated that an object remains in uniform motion or at rest unless a net force acts on it. (A net force is the sum of all forces acting on an object.) Newton's second law states that a net force acting on an object causes it to accelerate (change its velocity) at a rate that is directly proportional to the magnitude of the force. It also states that the smaller the mass of an object, the higher the acceleration it experiences for a given force. Newton's third law states that for every action, there is an equal and opposite reaction—for example, Earth's gravitational pull on the Moon is matched by the pull of the Moon on Earth.

### FIRST LAW OF MOTION
The first law states that an object remains in a state of rest or moves at a constant speed in a straight line unless acted on by a net force.

*constant, uniform motion*   *altered motion*

*force*

### SECOND LAW OF MOTION
When an object of low mass and one of greater mass are subjected to a force of the same magnitude, the low-mass object accelerates at a higher rate.

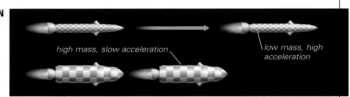

*high mass, slow acceleration*

*low mass, high acceleration*

### THIRD LAW OF MOTION
To every action, there is an equal and opposite reaction. The forward thrust of a rocket is a reaction to the backward blast of combusted fuel.

*action: backward blast of fuel*   *reaction*

## WEIGHT AND FREE FALL

The size of the gravitational force acting on an object is called its weight. An object's rest mass (measured in pounds or kilograms) is constant, while its weight (measured in newtons) varies according to the local strength of gravity. A mass of 2.2 lb (1 kg) weighs 9.8 newtons on Earth but only 1.65 newtons on the Moon. Weight can be measured, and the feeling of weight experienced, only when the gravity producing it is resisted by a second, opposing force. A person standing on Earth feels weight not so much from the pull of gravity as from the opposing push of the ground on his or her feet. In contrast, a person orbiting Earth is actually falling toward Earth under gravity. Such a person is in "free fall" and experiences apparent weightlessness. This is due not to lack of gravity but to the absence of a force opposing gravity.

### WEIGHTLESSNESS
Astronauts in training must frequently experience apparent weightlessness. Here, a plane is plunging sharply from high altitude, putting the trainee astronauts inside into free fall.

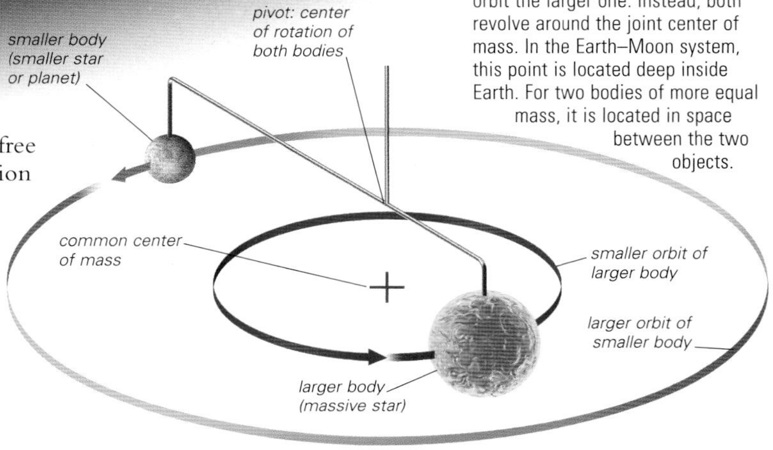

# SHAPES OF ORBITS

When one object is in orbit around another object of higher mass, it is in free fall toward the larger body. It experiences a constant gravitational acceleration toward the larger object that deflects what would otherwise be its straight-line motion into a curved trajectory. The direction of its motion and the direction of acceleration both constantly change, producing its curved path. All closed orbits in nature have the shape of an ellipse (a stretched circle). The degree to which an ellipse varies from a perfect circle is called its eccentricity. Many solar system orbits (such as the Moon's around Earth) are not very eccentric—that is, they are almost circular. Others, such as Pluto's orbit around the Sun, are much more eccentric and highly elongated. Some celestial bodies follow open, nonreturning orbits along curves with shapes called parabolas and hyperbolas.

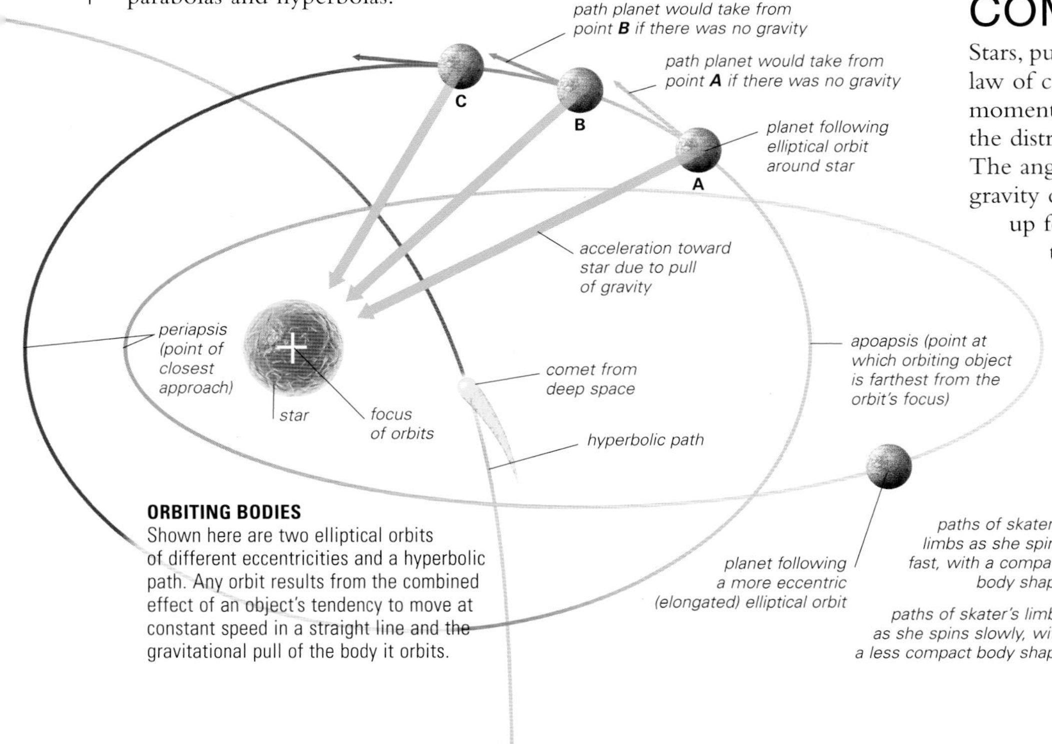

**COMMON CENTER OF MASS**
In an orbital system of two bodies, the smaller body does not simply orbit the larger one. Instead, both revolve around the joint center of mass. In the Earth–Moon system, this point is located deep inside Earth. For two bodies of more equal mass, it is located in space between the two objects.

smaller body (smaller star or planet)

pivot: center of rotation of both bodies

common center of mass

smaller orbit of larger body

larger orbit of smaller body

larger body (massive star)

path planet would take from point **B** if there was no gravity

path planet would take from point **A** if there was no gravity

planet following elliptical orbit around star

C

B

A

acceleration toward star due to pull of gravity

periapsis (point of closest approach)

star

focus of orbits

comet from deep space

hyperbolic path

apoapsis (point at which orbiting object is farthest from the orbit's focus)

planet following a more eccentric (elongated) elliptical orbit

**ORBITING BODIES**
Shown here are two elliptical orbits of different eccentricities and a hyperbolic path. Any orbit results from the combined effect of an object's tendency to move at constant speed in a straight line and the gravitational pull of the body it orbits.

# COMPACT, ROTATING BODIES

Stars, pulsars, galaxies, and planets all rotate, governed by the law of conservation of angular momentum. An object's angular momentum is related to its rotational energy, which depends on the distribution of mass in the object and on how fast it spins. The angular momentum of any rotating object stays constant, so if gravity causes the object to contract, its spin rate increases to make up for the redistribution of mass. Compact, fast-rotating objects therefore tend to form from diffuse, slowly rotating ones.

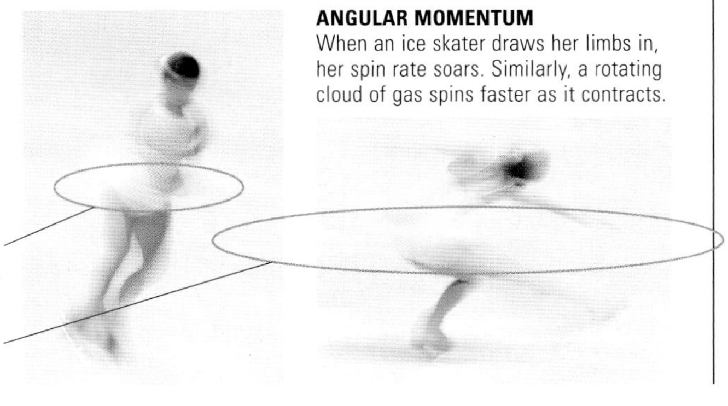

**ANGULAR MOMENTUM**
When an ice skater draws her limbs in, her spin rate soars. Similarly, a rotating cloud of gas spins faster as it contracts.

paths of skater's limbs as she spins fast, with a compact body shape

paths of skater's limbs as she spins slowly, with a less compact body shape

# SPACE AND TIME

MOST PEOPLE SHARE SOME COMMON-SENSE NOTIONS about the world. One is that time passes at the same rate for everyone. Another is that the length of a rigid object does not change. In fact, such ideas, which once formed a bedrock for the laws of physics, are an illusion: they apply only to the restricted range of situations with which people are most familiar. In fact, time and space are not absolute, but stretch and warp depending on relative viewpoint. What is more, the presence of matter distorts both space and time to produce the force of gravity.

## PROBLEMS IN NEWTON'S UNIVERSE

Problems with the Newtonian view of space and time (see p.38) first surfaced toward the end of the 19th century. Up to that time, scientists assumed that the positions and motions of objects in space should all be measurable relative to some nonmoving, absolute "frame of reference," which they thought was filled with an invisible medium called "the ether." However, in 1887, an experiment to measure Earth's motion through this ether, by detecting variations in the velocity of light sent through it in different directions, produced some unexpected results.

First, it failed to confirm the existence of the ether. Second, it indicated that light always travels at the same speed relative to an observer, whatever that observer's own movements. This finding suggested that light does not follow the same rules of relative motion that govern everyday objects such as cars and bullets. If a person were to chase a bullet at half of the bullet's speed, the rate at which the bullet moved away from him or her would halve. However, if a person were to chase a light wave at half the speed of light, the wave would continue moving away from him or her at exactly the same velocity.

**CONSTANT SPEED OF LIGHT**
Light leaves both the ceiling lights and the headlights of the moving cars at the same speed relative to its source. Paradoxically, light from both sources reaches an observer standing in the tunnel at, again, exactly the same speed.

### ALBERT EINSTEIN

The work of German-born mathematician and physicist Albert Einstein (1879–1955) made him the most famous scientist of the 20th century. Although he won the Nobel Prize for his work on the particlelike properties of light (see p.34), he is more famous for his theories of special relativity (1905) and general relativity (1915). These theories introduced a revolutionary new way of thinking about space, time, mass, energy, and gravity.

## SPECIAL RELATIVITY

In 1905, Albert Einstein rejected the idea that there is any absolute or "preferred" frame of reference in the universe. In other words, everything is relative. He also rejected the idea that time is absolute, suggesting that it need not pass at the same rate everywhere. To replace the old ideas, he formed the special theory of relativity, called "special" because it is restricted to frames of reference in constant, unchanging motion (because they are not being accelerated by a force, see p.38). He based the entire theory on two principles. The first principle, called the principle of relativity, states that the same laws of physics apply equally in all constantly moving frames of reference. The second principle states that the speed of light is constant and independent of the motion of the observer or source of light. Einstein recognized that this second principle conflicts with accepted notions about how velocities add together, and further, that combining it with the first principle seems to lead to perplexing, nonintuitive results. He perceived, however, that human intuition about time and space could be wrong.

direction of Observer 2's motion

path of Observer 1's ball, as seen by Observer 2

Observer 1

path of Observer 1's ball, as seen by Observer 1

### VIEWPOINT ONE

From Observer 1's point of view, the green ball within his or her own frame of reference travels up and down. If Observer 1 looks across at the red ball in a frame of reference in relative motion, the ball seems to follow an arc.

**MOVING FRAMES OF REFERENCE**
Here, we see two travelers—effectively two moving reference frames—passing each other. Each tosses a ball up. By the principle of relativity, the laws of physics apply in each reference frame, so each traveler observes the behavior of the two balls as predicted by those laws. Although the two travelers see different motions for each ball, neither traveler's point of view is superior to the other's—both are equally valid, and there is no preferred frame of reference.

direction of Observer 1's motion

# EFFECTS OF SPECIAL RELATIVITY

The results that flow from the principles of special relativity are remarkable. Using thought experiments, Einstein showed that for the speed of light to be the same in all reference frames, measures of space and time in one frame must be transformed to those in another. These transformations show that when an object moves at high speed relative to an observer, the observer sees less of its length—an effect called Lorentz contraction. Also, time for such an object appears to run more slowly—an effect called time dilation. So measurements of time and space vary between moving reference frames. Einstein also showed that an object gains mass when its energy increases and loses it when its energy decreases. This led him to realize that mass and energy have an equivalence, which he expressed in the famous equation E (energy) = m (mass) × c² (the speed of light squared).

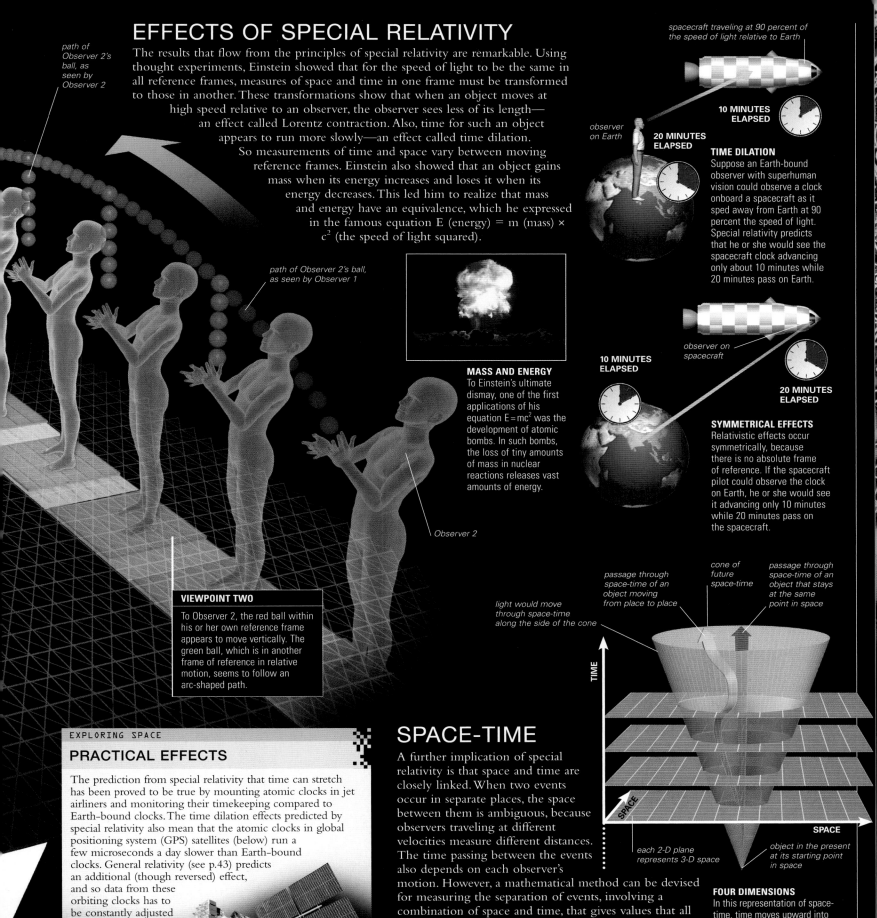

*path of Observer 2's ball, as seen by Observer 2*

*path of Observer 2's ball, as seen by Observer 1*

*Observer 2*

*spacecraft traveling at 90 percent of the speed of light relative to Earth*

*observer on Earth*

**20 MINUTES ELAPSED**

**10 MINUTES ELAPSED**

### TIME DILATION
Suppose an Earth-bound observer with superhuman vision could observe a clock onboard a spacecraft as it sped away from Earth at 90 percent the speed of light. Special relativity predicts that he or she would see the spacecraft clock advancing only about 10 minutes while 20 minutes pass on Earth.

*observer on spacecraft*

**10 MINUTES ELAPSED**

**20 MINUTES ELAPSED**

### SYMMETRICAL EFFECTS
Relativistic effects occur symmetrically, because there is no absolute frame of reference. If the spacecraft pilot could observe the clock on Earth, he or she would see it advancing only 10 minutes while 20 minutes pass on the spacecraft.

### MASS AND ENERGY
To Einstein's ultimate dismay, one of the first applications of his equation E=mc² was the development of atomic bombs. In such bombs, the loss of tiny amounts of mass in nuclear reactions releases vast amounts of energy.

### VIEWPOINT TWO
To Observer 2, the red ball within his or her own reference frame appears to move vertically. The green ball, which is in another frame of reference in relative motion, seems to follow an arc-shaped path.

EXPLORING SPACE

## PRACTICAL EFFECTS

The prediction from special relativity that time can stretch has been proved to be true by mounting atomic clocks in jet airliners and monitoring their timekeeping compared to Earth-bound clocks. The time dilation effects predicted by special relativity also mean that the atomic clocks in global positioning system (GPS) satellites (below) run a few microseconds a day slower than Earth-bound clocks. General relativity (see p.43) predicts an additional (though reversed) effect, and so data from these orbiting clocks has to be constantly adjusted to maintain accuracy.

## SPACE-TIME

A further implication of special relativity is that space and time are closely linked. When two events occur in separate places, the space between them is ambiguous, because observers traveling at different velocities measure different distances. The time passing between the events also depends on each observer's motion. However, a mathematical method can be devised for measuring the separation of events, involving a combination of space and time, that gives values that all observers can agree on. This led to the idea that events in the universe should no longer be described in three spatial dimensions, but rather in a four-dimensional world incorporating time, called space-time. In this system, the separation between any two events is described by a value called a space-time interval.

*passage through space-time of an object moving from place to place*

*cone of future space-time*

*passage through space-time of an object that stays at the same point in space*

*light would move through space-time along the side of the cone*

**TIME**

**SPACE**

**SPACE**

*each 2-D plane represents 3-D space*

*object in the present at its starting point in space*

### FOUR DIMENSIONS
In this representation of space-time, time moves upward into the future, while the three spatial dimensions are reduced to two-dimensional planes. The cone represents the effective limits of space-time for any object—its boundary is defined by the speed of light.

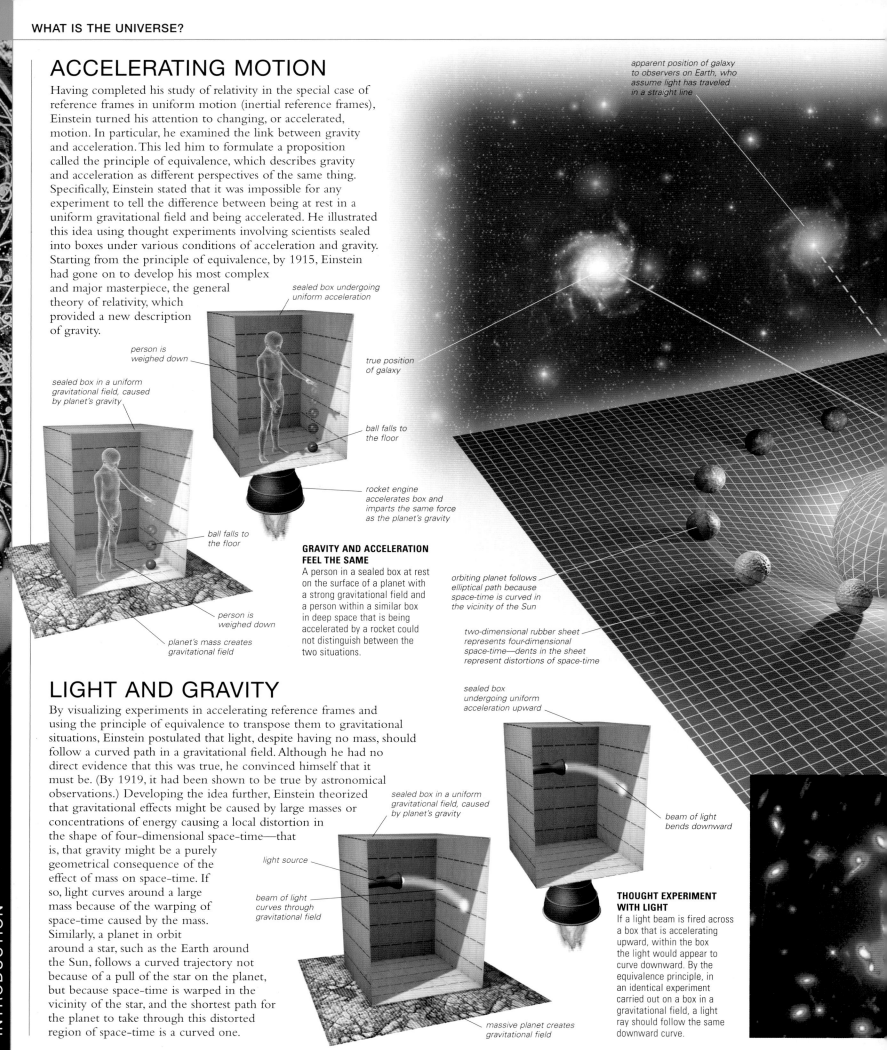

# ACCELERATING MOTION

Having completed his study of relativity in the special case of reference frames in uniform motion (inertial reference frames), Einstein turned his attention to changing, or accelerated, motion. In particular, he examined the link between gravity and acceleration. This led him to formulate a proposition called the principle of equivalence, which describes gravity and acceleration as different perspectives of the same thing. Specifically, Einstein stated that it was impossible for any experiment to tell the difference between being at rest in a uniform gravitational field and being accelerated. He illustrated this idea using thought experiments involving scientists sealed into boxes under various conditions of acceleration and gravity. Starting from the principle of equivalence, by 1915, Einstein had gone on to develop his most complex and major masterpiece, the general theory of relativity, which provided a new description of gravity.

apparent position of galaxy to observers on Earth, who assume light has traveled in a straight line

sealed box undergoing uniform acceleration

person is weighed down

sealed box in a uniform gravitational field, caused by planet's gravity

true position of galaxy

ball falls to the floor

rocket engine accelerates box and imparts the same force as the planet's gravity

ball falls to the floor

**GRAVITY AND ACCELERATION FEEL THE SAME**
A person in a sealed box at rest on the surface of a planet with a strong gravitational field and a person within a similar box in deep space that is being accelerated by a rocket could not distinguish between the two situations.

orbiting planet follows elliptical path because space-time is curved in the vicinity of the Sun

person is weighed down

planet's mass creates gravitational field

two-dimensional rubber sheet represents four-dimensional space-time—dents in the sheet represent distortions of space-time

# LIGHT AND GRAVITY

By visualizing experiments in accelerating reference frames and using the principle of equivalence to transpose them to gravitational situations, Einstein postulated that light, despite having no mass, should follow a curved path in a gravitational field. Although he had no direct evidence that this was true, he convinced himself that it must be. (By 1919, it had been shown to be true by astronomical observations.) Developing the idea further, Einstein theorized that gravitational effects might be caused by large masses or concentrations of energy causing a local distortion in the shape of four-dimensional space-time—that is, that gravity might be a purely geometrical consequence of the effect of mass on space-time. If so, light curves around a large mass because of the warping of space-time caused by the mass. Similarly, a planet in orbit around a star, such as the Earth around the Sun, follows a curved trajectory not because of a pull of the star on the planet, but because space-time is warped in the vicinity of the star, and the shortest path for the planet to take through this distorted region of space-time is a curved one.

sealed box undergoing uniform acceleration upward

sealed box in a uniform gravitational field, caused by planet's gravity

light source

beam of light curves through gravitational field

beam of light bends downward

**THOUGHT EXPERIMENT WITH LIGHT**
If a light beam is fired across a box that is accelerating upward, within the box the light would appear to curve downward. By the equivalence principle, in an identical experiment carried out on a box in a gravitational field, a light ray should follow the same downward curve.

massive planet creates gravitational field

## DENTED SPACE-TIME
Space-time can be visualized as a rubber sheet in which massive objects make dents. In this view, planets orbit the Sun because they roll around the dent it produces. Similarly, light passing by a massive object has its straight-line path deflected by following the local curvature of space-time. Remember, however, that it is four-dimensional space-time, not just space, that is warped.

# GENERAL RELATIVITY AT WORK

Einstein encapsulated his theory of how mass distorts space-time in his set of "field equations." Physicists have used these equations to find that it is in the strongest gravitational fields—where massive, dense objects distort space-time most strongly—that reality departs furthest from that predicted by Newton (see p.38). For instance, Mercury is so close to the Sun that it always moves in a strong gravitational field (or in strongly curved space-time). Its orbit is distorted in a way that Newton could not account for, but which general relativity explains perfectly (see p.110). General relativity also provides a framework for models of the universe's structure, development, and eventual fate. It predicts that the universe must be either expanding or contracting. Before the introduction of general relativity, space and time were thought of only as an arena in which events took place. After general relativity, physicists realized that space and time are dynamic entities that can be affected by mass, forces, and energy. However, while general relativity accurately describes the universe on a large scale, it has little to say about how it works at the tiniest, subatomic scale.

object with large mass

warped space-time

## PINCHED SPACE
Instead of a two-dimensional sheet, four-dimensional space-time can also be visualized as a three-dimensional volume that is narrowed or "pinched in" around large masses.

white dwarf star

relatively weak gravity

moderately deep gravitational well

### WHITE DWARF STAR
A white dwarf is a very dense, planet-sized star that can be thought of as producing a smaller but deeper dent in space-time than a star like the Sun does.

intense gravity close to star

relatively weak gravity

deep, steep gravitational well

massive, dense neutron star

### NEUTRON STAR
A neutron star is an exceedingly dense stellar remnant that makes a very deep dent in space-time. A neutron star significantly deflects light passing by, but cannot capture it.

intense gravity

relatively weak gravity

event horizon, beyond which nothing, not even light, can break free of the gravitational field

extremely intense gravity

gravitational well of infinite depth, with steepness (gravity) increasing to infinity

singularity at the center of the black hole

### BLACK HOLE
In a black hole, all the mass is concentrated into an infinitely dense point at the center, called a singularity. A singularity produces an infinite distortion in space-time—a bottomless gravitational well. Any light that passes a boundary called the "event horizon" near the entrance to this well cannot return.

distortion of space-time caused by the Sun's mass deflects light from distant galaxy

space-time around the Sun is warped by the Sun's mass, creating a so-called a "gravitational well"

telescope on Earth

Sun

# GRAVITATIONAL WAVES

One of Einstein's major predictions based on general relativity was the existence of gravitational waves. These waves can be regarded as disturbances in the curvature of space-time that propagate outward from their source—accelerated masses—at the speed of light. For over 100 years, gravitational waves remained hypothetical and unobserved, but in early 2016, it was announced that the first direct observation of such waves had been made, originating from a pair of merging black holes some 1.3 billion light-years from Earth. The waves were detected by a collaboration between scientists working on an experiment at a US observatory called LIGO (Laser Interferometer Gravitational-Wave Observatory) and others operating the Virgo interferometer, an instrument in northern Italy that is also designed to identify gravitational waves. Since 2016, several further instances of gravitational waves have been observed emanating from merging black holes or, in one case, from merging neutron stars (see p.267).

### MERGING STARS
In this artist's rendition, two neutron stars orbit each other at high speed as a prelude to merging. Such an event is predicted to produce both gravitational waves and a burst of gamma rays—both detected in August 2017 by the LIGO-Virgo project.

### GRAVITY BENDING LIGHT
The effect of gravity on the path of light is not obvious unless an observer looks deep into space at the universe's most massive objects—clusters of galaxies. This image shows galaxies as white blobs. Their combined gravity bends light so much that the images of more distant galaxies appear as blue streaks, stretched and squashed by the galaxy cluster's gravity.

# EXPANDING SPACE

A CRUCIAL PROPERTY of the universe is that it is expanding. It must be growing, because distant galaxies are quickly receding from Earth and more distant ones are receding even faster. Assuming that the universe has always been expanding, it must once have been smaller and denser—a fact that strongly supports the Big Bang theory of its origin.

## MEASURING EXPANSION

For the relatively nearby universe, the expansion can be calculated by comparing the distances to remote galaxies and the speeds at which they are receding. The galaxies' velocities are measured by examining the red shifts in their light spectra (see p.35). Their distances are calculated by detecting a class of stars called Cepheid variables in the galaxies and measuring the stars' cycles of magnitude variation (see pp.282, 313). For more remote parts of the universe, the expansion rate is measured by methods such as analysis of tiny fluctuations in the cosmic microwave background radiation (see pp.51, 54). The result is a number known as the Hubble constant—an expression of the universe's expansion rate. The value of the constant is about 50,000 mph (81,000 kph) per million light-years for the relatively nearby universe and a little less for the most remote observable objects. This means, for example, that two relatively nearby galaxies situated 1 billion light-years apart will be receding from each other at 50 million mph (81 million kph). On a familiar time scale, this is actually a very gradual expansion—an increase in the galaxies' distance of 1 percent takes tens of millions of years.

*recessional velocity (measured by red shift)*

*distance from Earth (measured by variable stars)*

### HUBBLE CONSTANT
The recession velocity of remote galaxies rises with distance, and this relationship forms a straight line on a graph. Estimates of the line's slope yield values of the Hubble constant.

## THE NATURE OF EXPANSION

Several notable features have been established about the universe's expansion. First, although all distant galaxies are moving away, neither Earth nor any other point in space is at the center of the universe. Rather, everything is receding from everything else, and there is no center. Second, at a local scale, gravity dominates over cosmological expansion and holds matter together. The scale at which this happens is surprisingly large—even entire clusters of galaxies resist expansion and hold together. Third, it is incorrect to think of galaxies and galaxy clusters moving away from each other "through" space. A more accurate picture is that of space itself expanding and carrying objects with it. Finally, the expansion rate almost certainly varies. Cosmologists are greatly interested in establishing how the expansion rate may change in the future. The future rate of expansion will decide the eventual fate of the universe (see pp.58–59).

### LOCAL GRAVITY
The galaxies above are not moving apart. They will continue to collide despite cosmological expansion. Galaxy clusters are also held together by gravity.

the universe 6 billion years ago was much smaller

galaxies close together

free gas and dust not yet absorbed into galaxies

6 BILLION YEARS AGO

3 BILLION YEARS AGO

some galaxies evolve into spiral shapes

PRESENT DAY

galaxies becoming less crowded

galaxy cluster, bound by gravity, does not expand

3 BILLION YEARS IN THE FUTURE

# TIME AND EXPANDING SPACE

**PEERING INTO DEEP SPACE**
This Hubble "deep-field" photograph shows a jumble of galaxies viewed at different distances. Each appears as it existed billions of years ago.

young, blue galaxy 4 billion light-years away, pictured as it was 4 billion years ago

diffuse, young galaxy not yet condensed into a tight spiral

elliptical galaxy, 6 billion light-years away

spiral galaxy, 3 billion light-years distant

The continued expansion of space, combined with the constant speed of light, turns the universe into a giant time machine. The light from a remote galaxy has taken billions of years to reach Earth, so astronomers see the galaxy as it was billions of years ago. In effect, the deeper astronomers look into space, the farther they peer into the universe's history. In the remotest regions, they see only incompletely formed galaxies as they looked soon after the Big Bang. The most dim and distant galaxies are now receding from Earth at a rate faster than the speed of light. We can still see light that was emitted from them in the distant past, but eventually they will fade from view. At greater distances yet, beyond the observable universe (see p.23), there may exist other objects that have moved away so fast that light from them has never reached Earth.

## EDWIN HUBBLE

American astronomer Edwin Hubble (1889–1953) is famous for being the first to prove that the universe is expanding, though several other astronomers—such as the Belgian Georges Lemaître—made important contributions to this discovery. Hubble is also noted for his earlier proof that galaxies are external to the Milky Way and for his system of galaxy classification. The Hubble Space Telescope and the Hubble constant are both named after him.

## LOOKBACK DISTANCE

The expansion of space complicates the expression of distances to very remote objects, particularly those that we now observe as they existed more than 5 billion years ago. When astronomers describe the distance to such faraway objects, by convention they use the "lookback" or "light-travel-time" distance. This is the distance that light from the object has traveled through space to reach us today, and it tells us how long ago the light left the object. But because space has expanded in the interim, the distance of the galaxy when the light began its journey toward Earth is less than the lookback distance. Conversely, the true distance to the remote object (called the "comoving" distance) is greater than the lookback distance. These distinctions need to be remembered when, for example, a galaxy is stated as being 10 billion light-years away.

**DIVERGING WORLDS**
An object described as being 11 billion light-years away (lookback distance) has a greater true distance (comoving distance) due to the effects of the universe's expansion.

voids between galaxy clusters progressively enlarge and become almost empty of dust and gas

**1.** Eleven billion years ago, a photon of light departs distant galaxy X traveling toward the Milky Way. The two galaxies are separated by 4 billion light-years of space.

**2.** Six billion years later, the photon has not yet reached its destination, because space has expanded, carrying the galaxies much farther apart.

**3.** The photon reaches the Milky Way, where an observer sees X as it was 11 billion years ago, 11 billion light-years away (lookback distance). Meanwhile, X's true (comoving) distance has increased to 18 billion light-years.

photon leaves galaxy X

**11 BILLION YEARS AGO**    Milky Way    distant galaxy X receding

**5 BILLION YEARS AGO**    photon travels toward Milky Way    galaxy X still receding

**PRESENT DAY**    photon arrives    lookback distance    true, comoving distance

**ACCELERATING EXPANSION**
This is a conceptual interpretation of how a region of space may have changed over a 9-billion-year period. As space has expanded, the galaxies within it have been carried apart, evolving as they go. This interpretation shows expansion speeding up—a scenario now accepted by most cosmologists.

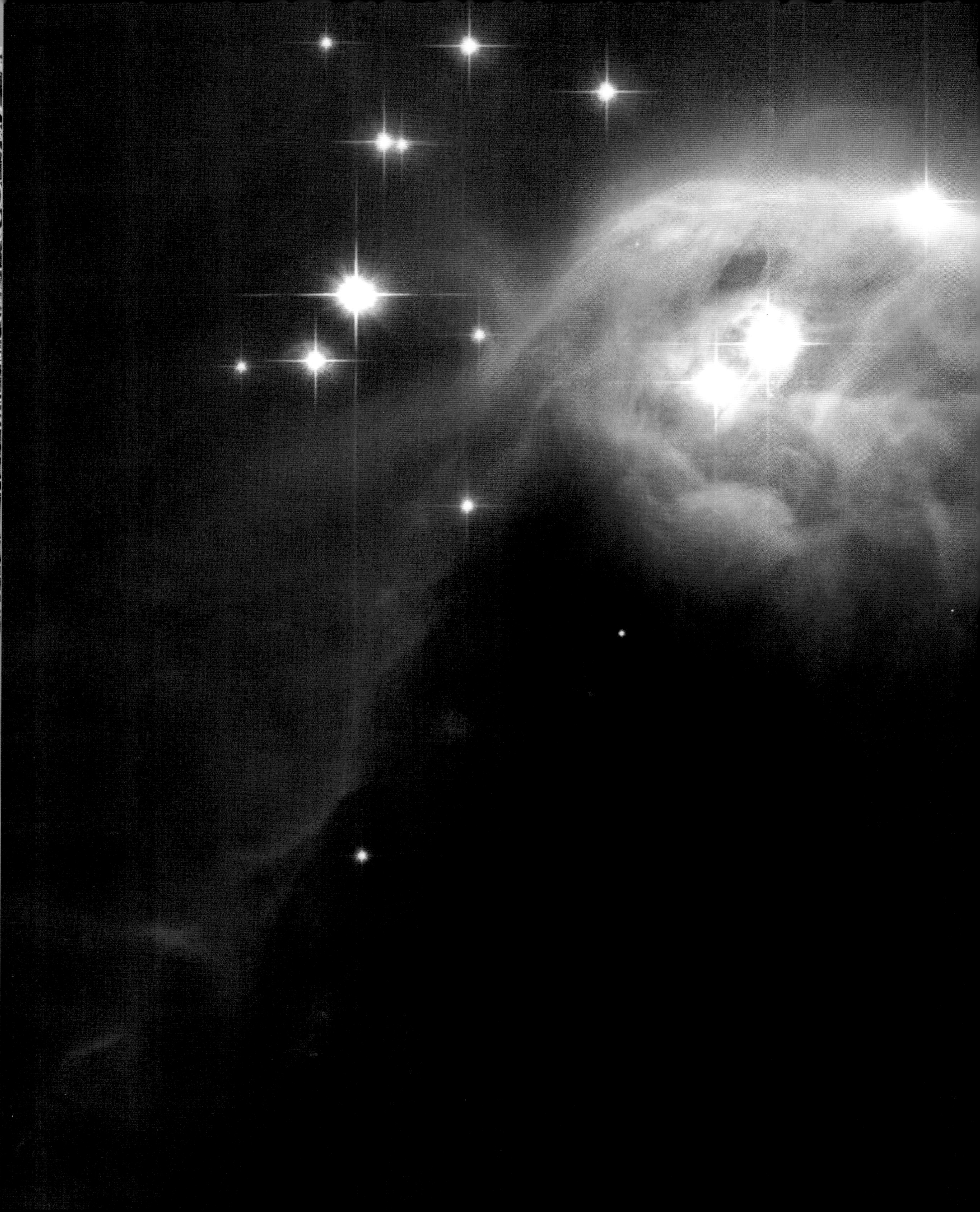

*"Some say the world will end in fire,*
*Some say in ice."*

Robert Frost

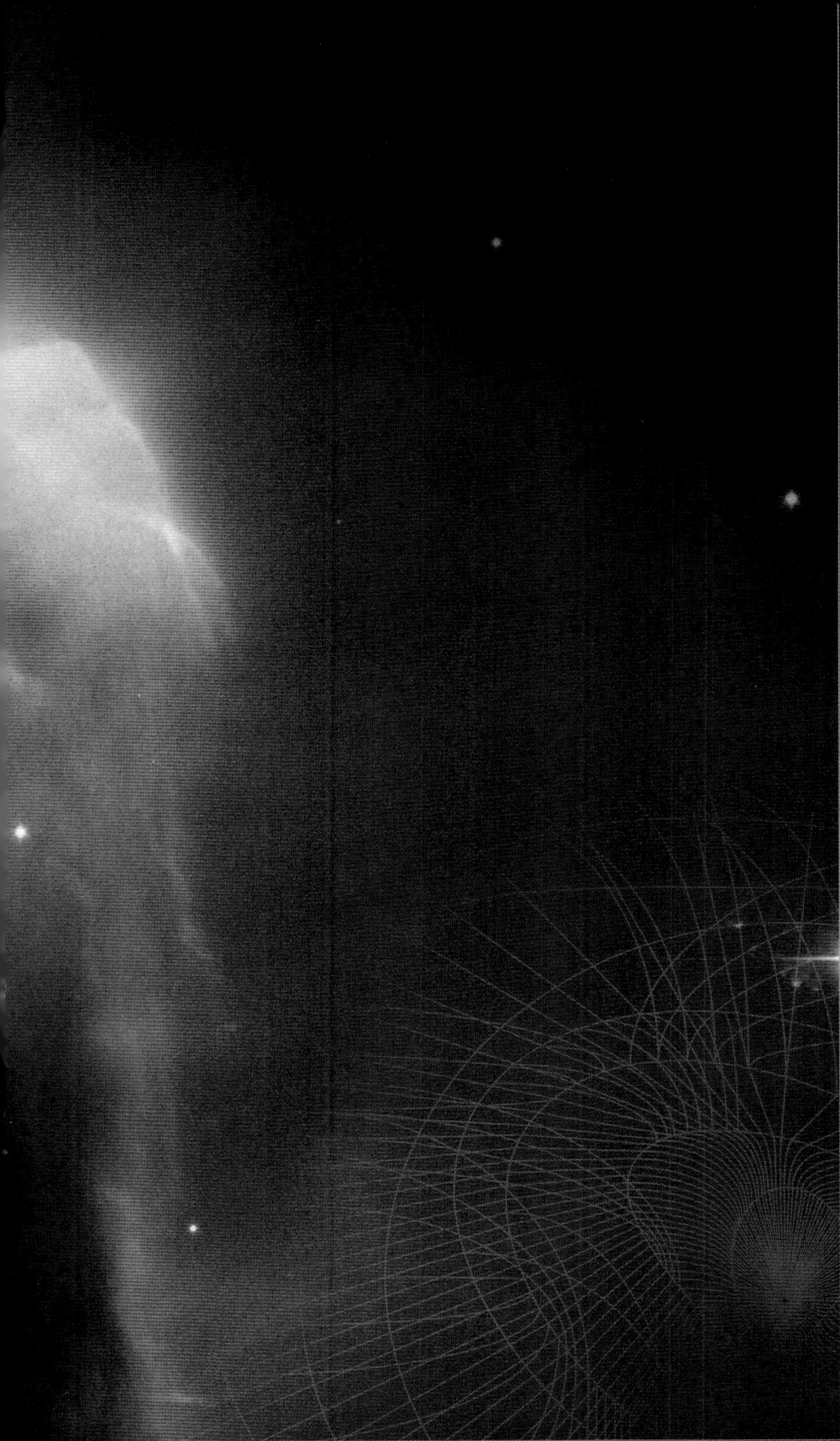

THE STORY OF THE UNIVERSE can be traced back to its very first instants, according to the Big Bang theory of its origins. In the Big Bang model, the universe was once infinitely small, dense, and hot. The Big Bang began a process of expansion and cooling that continues today. It was not an explosion of matter into space, but an expansion of space itself, and in the beginning, it brought time and space into existence. The Big Bang model does not explain all features of the universe, however, and it continues to be refined. Nonetheless, scientists use it as a framework for mapping the continuing evolution of the universe, through events such as the decoupling of matter and radiation (when the first atoms were formed and the universe became transparent) and the condensation of the first galaxies and the first stars. Study of the Big Bang and the balance between the universe's gravity and a force called dark energy can even help predict how the universe will end.

**CRADLE OF STAR BIRTH**
This pillar of gas and dust is the Cone Nebula, one of the most active cradles of star formation in the Milky Way. The clouds of material giving birth to these stars were once parts of stars themselves. The recycling of material in the life cycles of stars has been key to the universe's enrichment and evolution.

# THE BEGINNING AND END OF THE UNIVERSE

# THE BIG BANG

TIME, SPACE, ENERGY, AND MATTER are all thought to have come into existence 13.8 billion years ago, in the event called the Big Bang. In its first moments, the universe was infinitely dense, unimaginably hot, and contained pure energy. But within a tiny fraction of a second, vast numbers of elementary particles had appeared, created out of energy as the universe cooled. Within a few hundred thousand years, these particles had combined to form the first atoms.

## IN THE BEGINNING

The Big Bang was not an explosion in space, but an expansion of space, which happened everywhere. Physicists do not know what happened in the first instant after the Big Bang, known as the Planck epoch, but at the end of this period, they believe that gravity split from the other forces of nature, followed by the strong nuclear force (see p.30). Many believe this event triggered "inflation"—a short but rapid expansion. If inflation did occur, it helps to explain why the universe seems so smooth and flat. During inflation, a fantastic amount of mass–energy came into existence in tandem with an equal but negative amount of gravitational energy. By the end of inflation, matter had begun to appear.

*singularity at the start of time*

### THE FIRST MICROSECOND
The timeline on this page and the next shows some events during the first microsecond (1 millionth of a second or $10^{-6}$ seconds) after the Big Bang. Over this period, the universe's temperature dropped from about $10^{34}$°C (10 billion trillion trillion degrees) to a mere $10^{11}$°C (100 billion degrees). The timeline refers to the diameter of the observable universe: this is the approximate historical diameter of the part of the universe we can currently observe.

### THE PLANCK EPOCH
No current theory of physics can describe what happened in the universe during this time.

### GRAND UNIFIED THEORY EPOCH
During this epoch, matter and energy were completely interchangeable. Three of the fundamental forces of nature were still unified.

| DIAMETER | $3 \times 10^{-26}$ ft/$10^{-26}$ m | 33 ft/10 m | $10^5$ m (62 miles/100 km) |
|---|---|---|---|
| TEMPERATURE | $10^{28}$K (18,000 trillion trillion °F/10,000 trillion trillion °C) | | $10^{22}$K (18 billion trillion °F/10 |

### INFLATION EPOCH
Part of the universe expanded from billions of times smaller than a proton to something between the size of a marble and a football field.

### ELECTROWEAK EPOCH
This period, which followed separation out of the strong nuclear force, saw vast numbers of quark and antiquark pairs forming from energy and annihilating back to energy. Gluons and other more exotic particles also appeared.

| TIME | *A thousand-billionth of a yoctosecond $10^{-36}$ seconds* | *A hundred-millionth of a yoctosecond $10^{-32}$ seconds* | *1 yoctosecond $10^{-24}$ seconds* |
|---|---|---|---|

*a ten-trillionth of a yoctosecond $10^{-43}$ seconds*

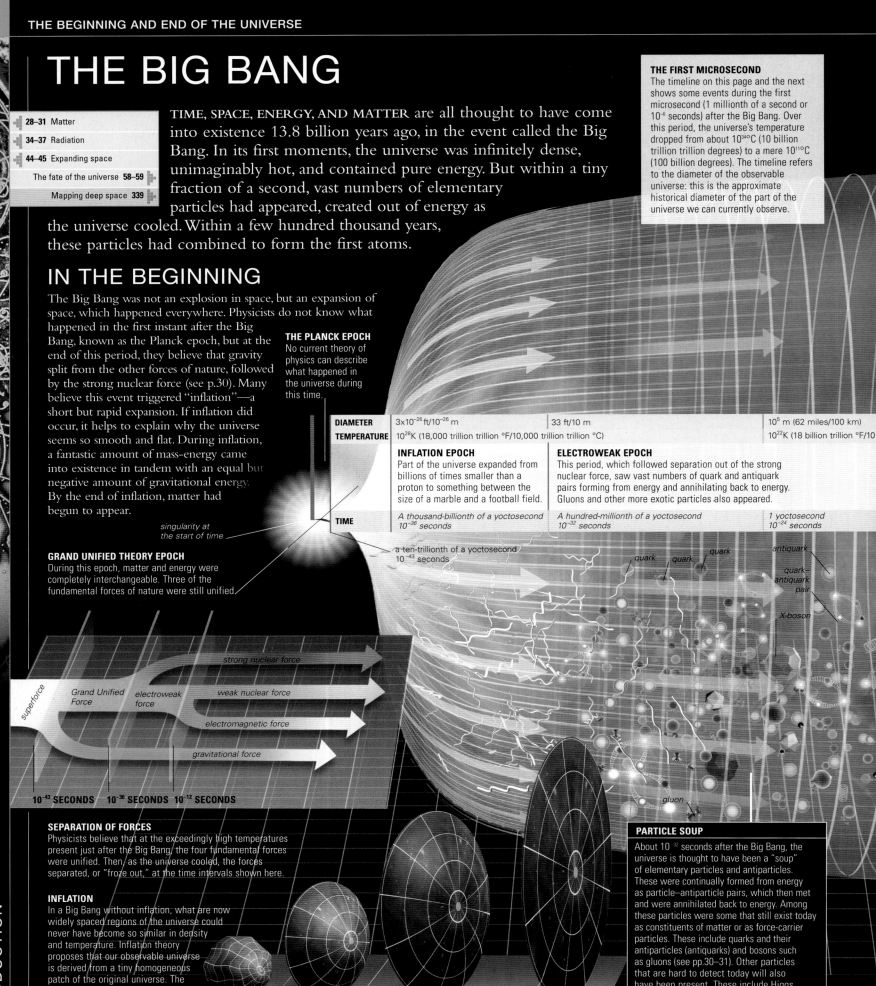

*quark*   *quark*   *quark*   *antiquark*

*quark– antiquark pair*

*X-boson*

*gluon*

superforce

Grand Unified Force   electroweak force   strong nuclear force   weak nuclear force   electromagnetic force   gravitational force

$10^{-43}$ SECONDS    $10^{-36}$ SECONDS    $10^{-12}$ SECONDS

### SEPARATION OF FORCES
Physicists believe that at the exceedingly high temperatures present just after the Big Bang, the four fundamental forces were unified. Then, as the universe cooled, the forces separated, or "froze out," at the time intervals shown here.

### INFLATION
In a Big Bang without inflation, what are now widely spaced regions of the universe could never have become so similar in density and temperature. Inflation theory proposes that our observable universe is derived from a tiny homogeneous patch of the original universe. The effect of inflation is like expanding a wrinkled sphere—after the expansion, its surface appears smooth and flat.

WRINKLED    SMOOTHER    VERY SMOOTH    EXTREMELY SMOOTH AND FLAT

### PARTICLE SOUP
About $10^{-32}$ seconds after the Big Bang, the universe is thought to have been a "soup" of elementary particles and antiparticles. These were continually formed from energy as particle–antiparticle pairs, which then met and were annihilated back to energy. Among these particles were some that still exist today as constituents of matter or as force-carrier particles. These include quarks and their antiparticles (antiquarks) and bosons such as gluons (see pp.30–31). Other particles that are hard to detect today will also have been present. These include Higgs bosons, which impart mass to other particles, and possibly gravitons (hypothetical gravity-carrying particles).

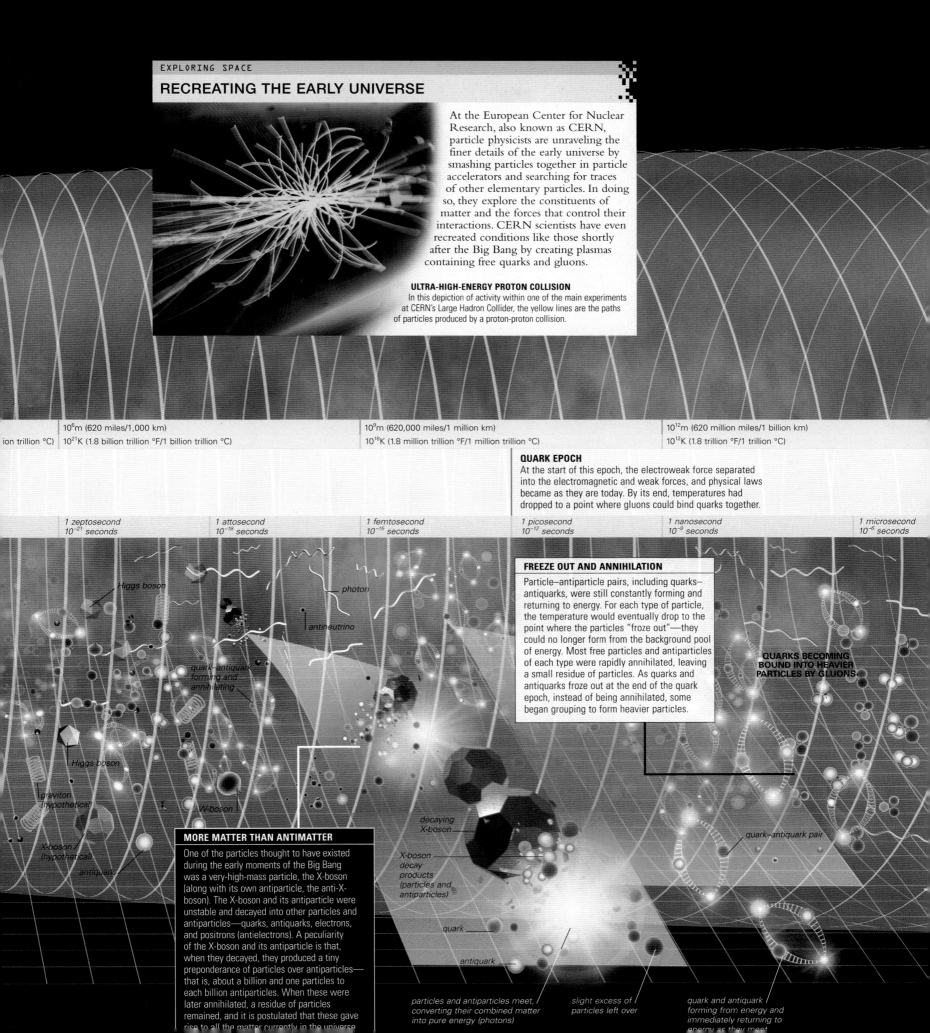

# RECREATING THE EARLY UNIVERSE

At the European Center for Nuclear Research, also known as CERN, particle physicists are unraveling the finer details of the early universe by smashing particles together in particle accelerators and searching for traces of other elementary particles. In doing so, they explore the constituents of matter and the forces that control their interactions. CERN scientists have even recreated conditions like those shortly after the Big Bang by creating plasmas containing free quarks and gluons.

**ULTRA-HIGH-ENERGY PROTON COLLISION**
In this depiction of activity within one of the main experiments at CERN's Large Hadron Collider, the yellow lines are the paths of particles produced by a proton-proton collision.

ion trillion °C) | $10^6$m (620 miles/1,000 km) | $10^9$m (620,000 miles/1 million km) | $10^{12}$m (620 million miles/1 billion km)
$10^{21}$K (1.8 billion trillion °F/1 billion trillion °C) | $10^{18}$K (1.8 million trillion °F/1 million trillion °C) | $10^{12}$K (1.8 trillion °F/1 trillion °C)

**QUARK EPOCH**
At the start of this epoch, the electroweak force separated into the electromagnetic and weak forces, and physical laws became as they are today. By its end, temperatures had dropped to a point where gluons could bind quarks together.

1 zeptosecond $10^{-21}$ seconds | 1 attosecond $10^{-18}$ seconds | 1 femtosecond $10^{-15}$ seconds | 1 picosecond $10^{-12}$ seconds | 1 nanosecond $10^{-9}$ seconds | 1 microsecond $10^{-6}$ seconds

Higgs boson

photon

antineutrino

quark–antiquark forming and annihilating

**FREEZE OUT AND ANNIHILATION**
Particle–antiparticle pairs, including quarks–antiquarks, were still constantly forming and returning to energy. For each type of particle, the temperature would eventually drop to the point where the particles "froze out"—they could no longer form from the background pool of energy. Most free particles and antiparticles of each type were rapidly annihilated, leaving a small residue of particles. As quarks and antiquarks froze out at the end of the quark epoch, instead of being annihilated, some began grouping to form heavier particles.

**QUARKS BECOMING BOUND INTO HEAVIER PARTICLES BY GLUONS**

Higgs boson

graviton (hypothetical)

W-boson

decaying X-boson

X-boson decay products (particles and antiparticles)

quark–antiquark pair

X-boson (hypothetical)

antiquark

**MORE MATTER THAN ANTIMATTER**
One of the particles thought to have existed during the early moments of the Big Bang was a very-high-mass particle, the X-boson (along with its own antiparticle, the anti-X-boson). The X-boson and its antiparticle were unstable and decayed into other particles and antiparticles—quarks, antiquarks, electrons, and positrons (antielectrons). A peculiarity of the X-boson and its antiparticle is that, when they decayed, they produced a tiny preponderance of particles over antiparticles—that is, about a billion and one particles to each billion antiparticles. When these were later annihilated, a residue of particles remained, and it is postulated that these gave rise to all the matter currently in the universe

quark

antiquark

particles and antiparticles meet, converting their combined matter into pure energy (photons)

slight excess of particles left over

quark and antiquark forming from energy and immediately returning to energy as they meet

# THE EMERGENCE OF MATTER

About 1 microsecond ($10^{-6}$ or one millionth of a second) after the Big Bang, the young universe contained, in addition to vast quantities of radiant energy, or photons, a seething "soup" of quarks, antiquarks, and gluons. Also present were the class of elementary particles called leptons (mainly electrons, neutrinos, and their antiparticles) forming from energy and then being annihilated back to energy. The stage was set for the next processes of matter formation that led to our current universe. First, quarks and gluons met to make heavier particles—particularly protons and a smaller number of neutrons. Next, the neutrons combined with some of the protons to form atomic nuclei, mainly those of helium. The remaining protons, destined to form the nuclei of hydrogen atoms, stayed uncombined. Finally, after about 400,000 years, the universe cooled sufficiently for electrons to combine with the free protons and helium nuclei—so forming the first atoms.

## THE NEXT HALF-MILLION YEARS
The timeline on these two pages shows events from 1 microsecond to about 500,000 years after the Big Bang. The temperature dropped from $10^{11}$°K (180 billion °F/100 billion °C) to 4,500°F (2,500°C). Today's observable universe expanded from about 50 light-hours (100 billion km) to many millions of light-years wide.

| DIAMETER | 60 billion miles/100 billion km | 10 light-years (1 light-year = 5.88 million miles/9.46 trillion km) | 100 light-years |
|---|---|---|---|
| TEMPERATURE | $10^{11}$K (180 billion °F/100 billion °C) | $10^{10}$K (18 billion °F/10 billion °C) | $10^9$K (1.8 billion °F/1 billion °C) |

**HADRON EPOCH**
Around the beginning of this epoch, quarks and antiquarks began combining to form particles called hadrons. These included baryons (protons and neutrons), antibaryons, and mesons.

**LEPTON EPOCH**
During this epoch, leptons (electrons, neutrinos, and their antiparticles) were very numerous. By its end, the electrons annihilated with positrons (antielectrons).

**BIG BANG NUCLEOSYNTHESIS**
Neutrons gradually converted into protons as the universe cooled, but when there was about one neutron for every seven protons, most remaining neutrons combined with protons to make helium nuclei, each with two protons and two neutrons.

| TIME | 1 microsecond $10^{-6}$ seconds—1 millionth of a second | 1 second | 10 seconds |
|---|---|---|---|

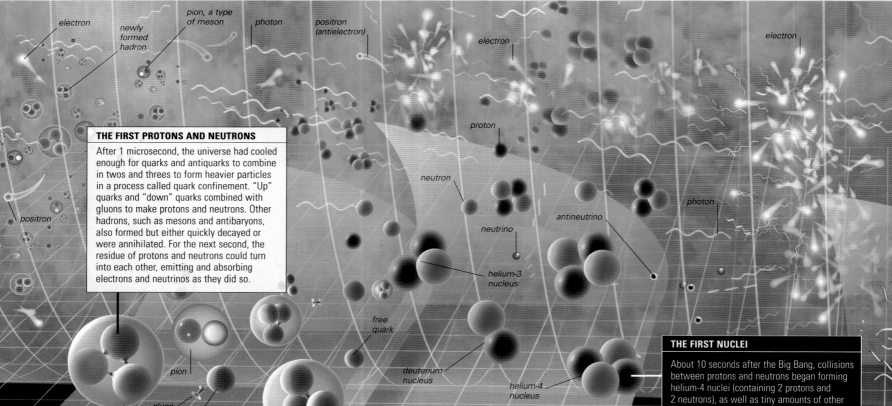

electron

newly formed hadron

pion, a type of meson

photon

positron (antielectron)

electron

electron

proton

neutron

neutrino

antineutrino

photon

helium-3 nucleus

positron

pion

gluon

free quark

free quark

deuterium nucleus

helium-4 nucleus

## THE FIRST PROTONS AND NEUTRONS

After 1 microsecond, the universe had cooled enough for quarks and antiquarks to combine in twos and threes to form heavier particles in a process called quark confinement. "Up" quarks and "down" quarks combined with gluons to make protons and neutrons. Other hadrons, such as mesons and antibaryons, also formed but either quickly decayed or were annihilated. For the next second, the residue of protons and neutrons could turn into each other, emitting and absorbing electrons and neutrinos as they did so.

## THE FIRST NUCLEI

About 10 seconds after the Big Bang, collisions between protons and neutrons began forming helium-4 nuclei (containing 2 protons and 2 neutrons), as well as tiny amounts of other atomic nuclei, such as helium-3 (2 protons and 1 neutron), lithium (3 protons and 4 neutrons), and deuterium (1 proton and

# EVIDENCE FOR THE BIG BANG

The strongest evidence for the Big Bang is the radiation it left, called the cosmic microwave background radiation (CMBR). George Gamow (see panel, opposite) predicted the radiation's existence in 1948. Its detection in the 1960s was confirmation, for most cosmologists, of the Big Bang theory. Other observations help support the theory.

**BACKGROUND RADIATION** The spectrum of the CMBR, discovered by Arno Penzias and Robert Wilson (below), indicates a uniformly hot early universe.

**EXPANSION** If the universe is expanding and cooling, it must once have been much smaller and hotter.

**BALANCE OF ELEMENTS** Big Bang theory exactly predicts the proportion of light elements (hydrogen, helium, and lithium) seen in the universe today.

**GENERAL RELATIVITY** Einstein's theory predicts that the universe must either be expanding or contracting—it cannot stay the same size.

**DARK NIGHT SKY** If the universe were both infinitely large and old, Earth would receive light from every part of the night sky and it would look bright—much brighter even than the densest star field (above). The fact that it is not is called Olbers' paradox. The Big Bang resolves the paradox by proposing that the universe has not always existed.

---

10,000 light-years

$10^8$K (180 million °F/100 million °C)

## PHOTON EPOCH

During this relatively lengthy epoch, the ocean of matter particles (comprising mainly electrons, protons, and helium nuclei) were in a continual state of interaction with photons (radiant energy), making the universe "foggy."

1,000 seconds

100 million light-years
4,000K (6,740°F/3,727°C)

## RECOMBINATION EPOCH

During this epoch, atoms formed as atomic nuclei combined with electrons. About nine hydrogen atoms were made for each helium atom. A few atoms of lithium and deuterium (heavy hydrogen) also formed. Photons were now free to travel through the universe.

380,000 years

---

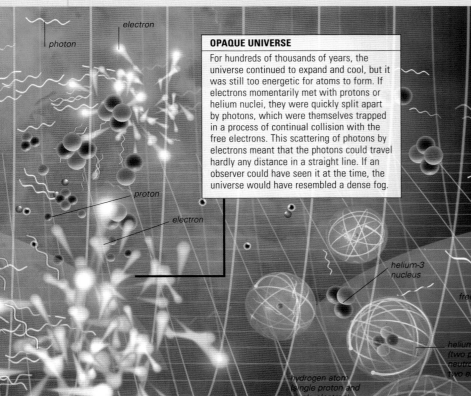

electron

photon

### OPAQUE UNIVERSE

For hundreds of thousands of years, the universe continued to expand and cool, but it was still too energetic for atoms to form. If electrons momentarily met with protons or helium nuclei, they were quickly split apart by photons, which were themselves trapped in a process of continual collision with the free electrons. This scattering of photons by electrons meant that the photons could travel hardly any distance in a straight line. If an observer could have seen it at the time, the universe would have resembled a dense fog.

proton

electron

helium-3 nucleus

free photon

helium atom (two protons, two neutrons, and two electrons)

hydrogen atom (single proton and single electron)

helium-4 nucleus

hydrogen atom—nine times more numerous than any other atoms

### THE FIRST ATOMS

Some 380,000 years after the Big Bang, when the temperature had dropped to about 6,737°F (3,725°C), protons and atomic nuclei began to capture electrons, forming the first atoms. Electrons were now bound up in atoms, so they no longer scattered photons. Matter and radiation therefore became "decoupled," and the photons were released to travel through the universe as radiation—the universe became transparent. These first free photons are still detectable as the cosmic microwave background radiation (CMBR).

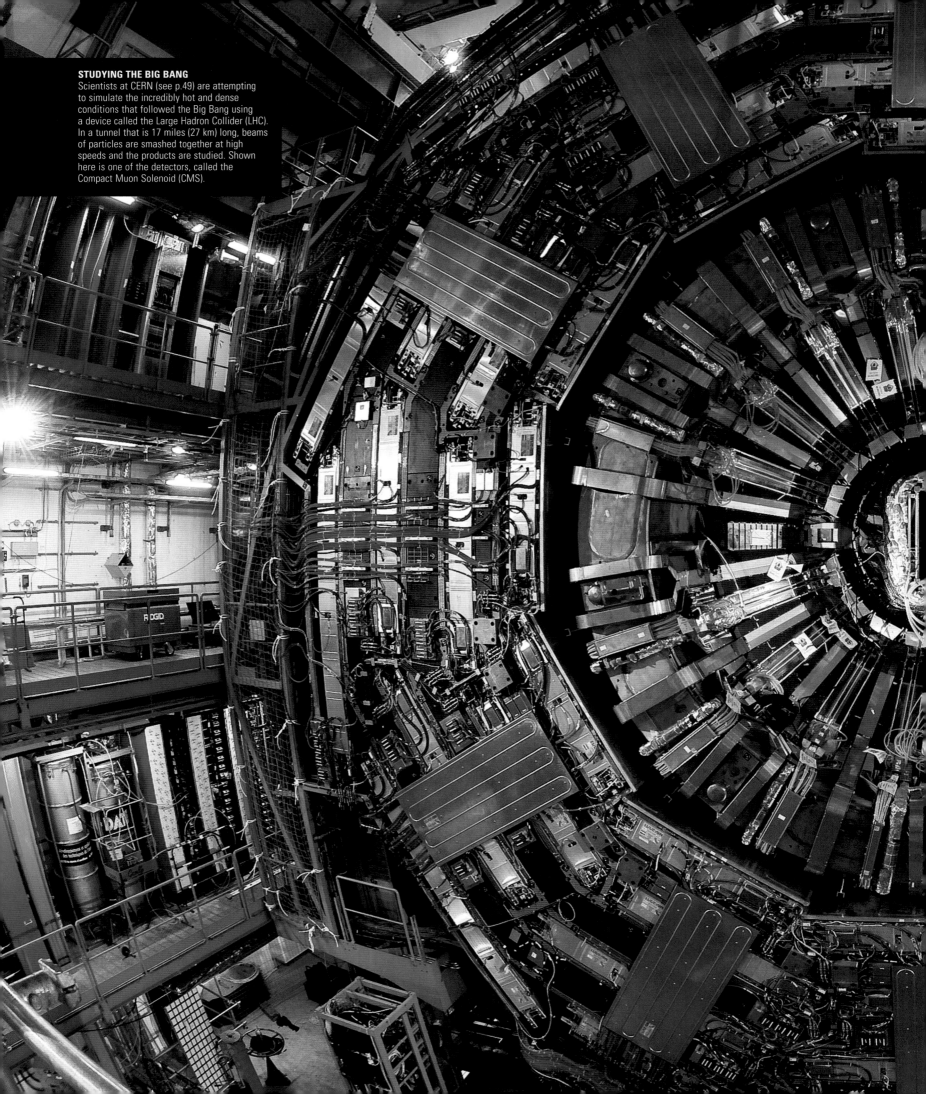

**STUDYING THE BIG BANG**
Scientists at CERN (see p.49) are attempting to simulate the incredibly hot and dense conditions that followed the Big Bang using a device called the Large Hadron Collider (LHC). In a tunnel that is 17 miles (27 km) long, beams of particles are smashed together at high speeds and the products are studied. Shown here is one of the detectors, called the Compact Muon Solenoid (CMS).

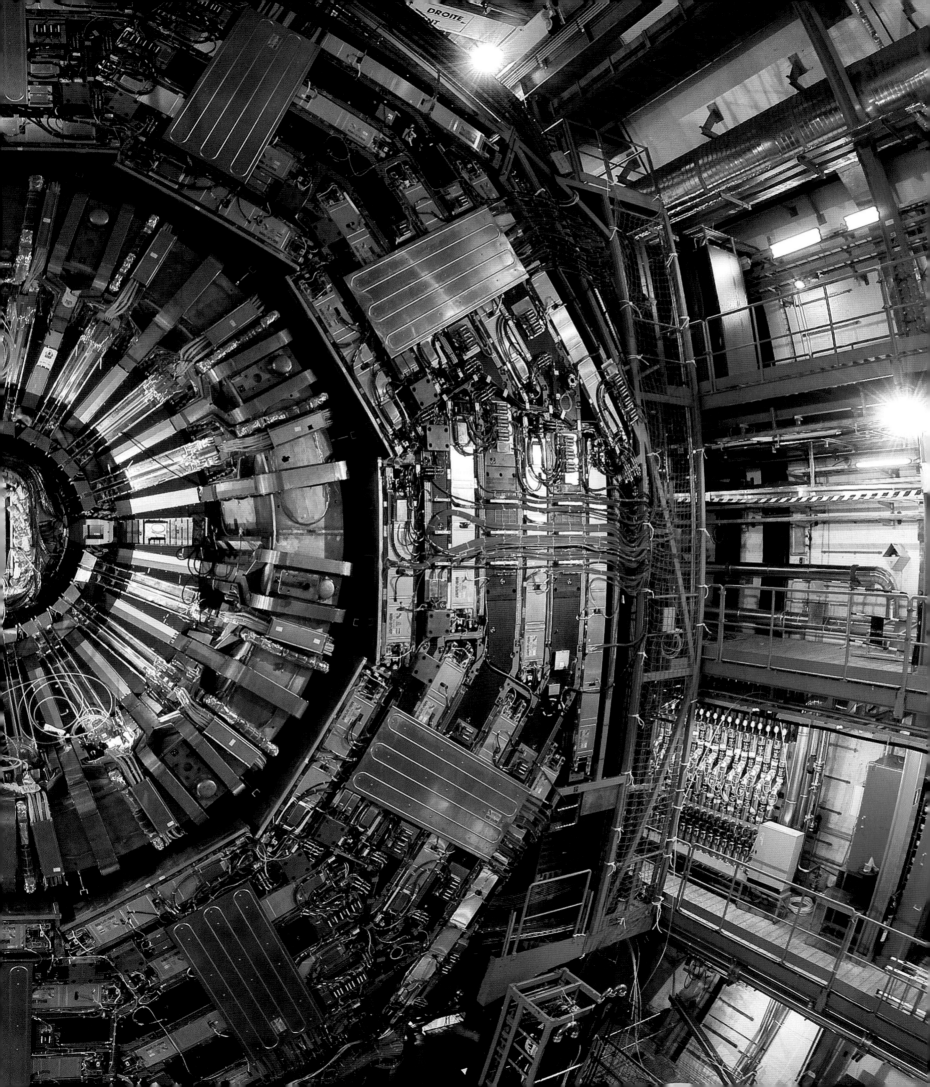

# OUT OF THE DARKNESS

THE PERIOD FROM THE BIRTH of atoms 380,000 years after the Big Bang to the ignition of the first stars hundreds of millions of years later is known as the "dark ages" of the universe. What happened in this era, and the subsequent "cosmic renaissance" as starlight filled the universe, is an intricate puzzle. Astronomers are solving it by analyzing the relic radiation of the Big Bang and using the world's most powerful telescopes to peer to the edges of the universe.

## THE AFTERMATH OF THE BIG BANG

At an age of 400,000 years, the universe was full of photons of radiation streaming in all directions and of atoms of hydrogen and helium, neutrinos, and other dark matter. Although it was still hot, at 5,400°F (3,000°C), and full of radiation, astronomers see no light if they try to peer back to that moment. The reason is that as the universe has expanded, it has stretched the wavelengths of radiation by a factor of a thousand. The photons reach Earth not as visible light, but as low-energy photons of cosmic microwave background radiation (CMBR). Their wavelength, once characteristic of the fireball of the universe, is now that of a cold object with a temperature of -454°F (-270°C)—only 5°F (3°C) above absolute zero.

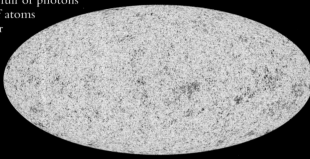

**INFANT UNIVERSE**
This Planck Observatory image is an all-sky picture of tiny fluctuations in the temperature of the CMBR, which relate to early irregularities in matter density. In effect, it is a snapshot of the infant universe.

## THE DARK AGES

Earth will never receive visible light from the period before the first stars ignited, a few hundred million years after the Big Bang, but cosmologists can reconstruct what happened during that time using other data, such as those of the CMBR. The CMBR reveals tiny fluctuations in the density of matter at the time the first atoms formed. Cosmologists think that gravity working on these ripples caused the matter to begin forming into clumps and strands. These irregularities in the initial cloud of matter probably laid the framework of present-day large-scale objects, such as galaxy superclusters (see pp.336–337). The development of such structures over billions of years has been simulated with computers. These simulations rely on assumptions about the density and properties of matter, including dark matter, in the infant universe, as well as the influence of dark energy (a force opposing gravity, see p.58). Some simulations closely resemble the distribution of matter seen in the universe today.

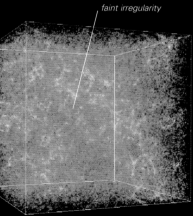

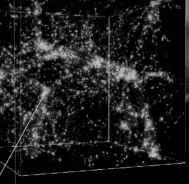

*faint irregularity*    *matter filament*    *denser filament of matter containing galaxy clusters*

*knot of matter has become a galaxy supercluster*

**UNIVERSE AT 500,000 YEARS OLD**
This computer simulation of the development of structure in the universe starts with matter almost uniformly dispersed in a cube that is 140 million light years high, wide, and deep.

**1.3 BILLION YEARS OLD**
A billion years later, considerable clumping and filament formation has occurred. To compensate for the cosmic expansion since the previous stage, the cube has been scaled to size.

**5 BILLION YEARS OLD**
A further 4 billion years later, and (again, after rescaling) the matter has condensed into some intricate filamentous structures interspersed with sizable bubbles or voids of empty space.

**13.8 BILLION YEARS OLD**
The matter distribution in the simulation now resembles the kind of galaxy-supercluster structure seen in the local universe (within a few billion light years).

## EARLY GALAXIES

Astronomers are still trying to pinpoint when the very first stars ignited and in what types of early galactic structures this may have occurred. Recent infrared studies, with instruments such as the Spitzer Space Telescope and Very Large Telescope, have revealed what seem to be very faint galaxies with extremely high red shifts existing as little as 400 million years after the Big Bang. Their existence indicates that well-developed precursor knots and clumps of condensing matter may have existed as little as 100 to 300 million years after the Big Bang. It is within these structures that the first stars probably formed.

**MOST DISTANT GALAXY**
This image from the Hubble telescope shows GN-Z11, the most distant galaxy known, appearing as it existed just 400 million years after the Big Bang.

## THE FIRST STARS

The first stars, which may have formed only 180 million years after the Big Bang, were made almost entirely of hydrogen and helium—virtually no other elements were present. Physicists think that star-forming nebulae that lacked heavy elements condensed into larger gas clumps than those of today. Stars forming from these clumps would have been very large and hot, with perhaps 100 to 1,000 times the mass of the Sun. Many would have lasted only a few million years before dying as supernovae. Ultraviolet light from these stars may have triggered a key moment in the universe's evolution—the reionization of its hydrogen, turning it from a neutral gas back into the ionized (electrically charged) form seen today. Alternatively, radiation from quasars (see p.320) may have reionized the universe.

**DEATH OF MEGASTARS**
The first, massive stars may have exploded as "hypernovae"—events associated today with black hole formation and violent bursts of gamma rays. These artist's impressions depict one model of hypernova development.

*200-solar-mass "megastar"*

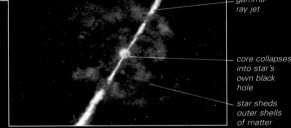

gamma-ray jet

core collapses into star's own black hole

star sheds outer shells of matter

**IONIZING POWER OF STARS**
These young, high-mass stars in the Orion Nebula ionize the gas around them, causing it to glow. Ionized hydrogen between galaxy clusters today may have been created by the far fiercer radiation of the first generation of stars and hypernovae.

## COSMIC CHEMICAL ENRICHMENT

During the course of their lives and deaths, the first massive stars created and dispersed new chemical elements into space and into other collapsing protogalactic clumps. A zoo of new elements, such as carbon, oxygen, silicon, and iron, was formed from nuclear fusion in the hot cores of these stars. Elements heavier than iron, such as barium and lead, were formed during their violent deaths. Second- and third-generation stars, smaller than the primordial megastars, later formed from the enriching interstellar medium. These stars created more of the heavier elements and returned them to the interstellar medium via stellar winds and supernova explosions. Galactic mergers and the stripping of gas from galaxies (see p.327) led to further intergalactic mixing and dispersion. These processes of recycling and enrichment of the cosmos continue today. In the Milky Way Galaxy, the new heavier elements have been essential to the formation of objects from rocky planets to living organisms.

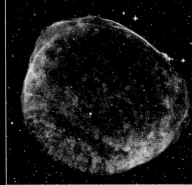

**STARDUST**
Supernova remnant SN 1006 is a sphere of enriched material expanding into space. Elements heavier than iron have mostly been

**ABUNDANCE OF ELEMENTS**
The ordinary (atom-based) matter of the universe initially consisted of hydrogen and helium, with a trace of lithium. Today, it still consists mainly of hydrogen and helium, but stellar processes have boosted the

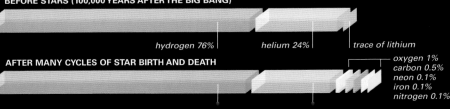

BEFORE STARS (100,000 YEARS AFTER THE BIG BANG)

hydrogen 76%   helium 24%   trace of lithium

AFTER MANY CYCLES OF STAR BIRTH AND DEATH

oxygen 1%
carbon 0.5%
neon 0.1%
iron 0.1%
nitrogen 0.1%

# LIFE IN THE UNIVERSE

THE ONLY KNOWN LIFE IN THE COSMOS is that on Earth. Life on Earth is so ubiquitous, however, and the universe so enormous, that many scientists think there is a very good chance that life also exists elsewhere. Much depends on whether the development of life on Earth was a colossal fluke— the product of an extremely improbable series of events—or, as many believe, not so unexpected given what is suspected about primordial conditions on the planet.

## LIVING ORGANISMS

What exactly constitutes a living organism? Human ideas on this are heavily reliant on the study of life on Earth, since scientists have no experience of the potential breadth of life beyond. Nonetheless, biologists have agreed on a few basic features that distinguish life from nonlife anywhere in the cosmos—as a bare minimum, a living entity must be able to replicate itself and, over time, to evolve. Beyond that, the definition of life is not universally agreed on. As an illustration, there is uncertainty about whether viruses are living. Although they self-replicate, viruses lack some characteristics that most biologists consider essential to life—notably, they do not exist as cells or possess their own biochemical machinery. It is also uncertain that other characteristics common to life on Earth, such as carbon chemistry or the use of liquid water, must inevitably be a feature of extraterrestrial life. Disagreements over such matters add complexity to discussions of the likelihood of life beyond Earth.

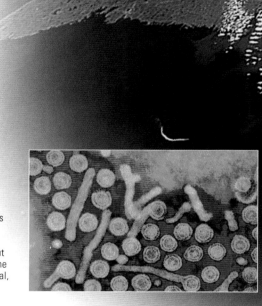

**VIRUS PARTICLES**
Viruses, such as this hepatitis virus, are on the border between living and nonliving matter. They self-replicate but can do so only by hijacking the metabolic machinery of animal, plant, or bacterial cells.

## ORIGINS OF LIFE

Most scientists agree that the beginnings of life on Earth were linked to the accumulation of simple organic (carbon-containing) molecules in a "primordial soup" in Earth's oceans not long after their formation. The molecules originated from reactions of chemicals in Earth's atmosphere, stimulated by energy, perhaps from lightning. Within the soup, over millions of years the organic compounds reacted to form larger and more complex molecules, until a molecule appeared with the capacity to replicate itself. By its nature, this molecule—a rudimentary gene—became more common. Through mutations and the mechanism of natural selection, variants of this gene developed more sophisticated survival adaptations, eventually evolving into a bacterialike cell—the precursor of all other life on Earth.

Many evolutionary biologists would say that the decisive event was the appearance of the self-replicator, after which living organisms would inevitably follow.

**SUBZERO LIFE FORM**
This so-far-unclassified life form was found living deep in the Antarctic ice sheet. Life can exist in a wider range of conditions than once thought.

**STROMATOLITES**
Some of the earliest remains of life are fossil stromatolites—mineral mounds built billions of years ago in shallow seas by cyanobacteria (blue-green algae). Stromatolites still grow on the Australian coast (left).

### RECREATING PRIMORDIAL EARTH

In 1953, American chemist Stanley Miller (1930–2007) recreated what he thought was Earth's primordial atmosphere in a flask. He sent sparks, simulating lightning, into the gas mixture, which lacked oxygen. The result was many different amino acids—some of the basic building blocks of life.

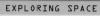

**STANLEY MILLER**
Here, Stanley Miller recreates the experiment he first conducted as a graduate student. It showed that amino acids could have formed in Earth's oxygen-free early atmosphere.

# HOW RARE IS LIFE?

Until about 30 years ago, the ranges of conditions thought essential to life, such as those of temperature and humidity, were thought to be narrow. Since then, scientists have found extremophiles (organisms that thrive in extreme conditions) living in adverse environments on Earth. Organisms may live deep in ice sheets or in boiling-hot water around vents in the ocean floor. Some exist in communities divorced from sunlight and live on energy from chemical sources. Bacteria are even found living 2 miles (3 km) deep in the Earth's crust, living on hydrogen, which they convert to water. Extremophiles have encouraged the idea that life can exist in a wide range of conditions. Some scientists are still hopeful that extraterrestrial life will be found in the solar system, although exploration of the most likely location, Mars, has proved negative so far. Beyond the solar system, many scientists think that life must be widespread. At these remote distances, scientists are most interested in whether intelligent, contactable life exists. In the 1960s, American radio astronomer Frank Drake (b. 1930) developed an equation for predicting the number of civilizations in the galaxy capable of interstellar communication. Because few of the factors in the equation can be estimated accurately, applying it (see panel, right) can have any outcome from less than one to millions, depending on the estimated values. Nevertheless, it is not unreasonable to suggest that at least a few such civilizations may exist in the Milky Way.

**LIFE ON EUROPA?**
Jupiter's moon Europa is covered with ice. There is probably an ocean of liquid water under this icy crust, which could conceivably harbor life.

## ALIEN CIVILIZATIONS?

Applying the Drake Equation involves estimating factors, such as the fraction of stars that develop planets, then multiplying all the factors. The example below uses only moderately optimistic estimates (some are just guesses).

**RATE OF STAR BIRTH** A fair estimate would be two new stars per year in the Milky Way.

*99% of new stars develop planets*

**STARS WITH PLANETS** Perhaps 99 percent of these stars develop planetary systems.

*0.4 planets will be habitable*

**HABITABLE PLANETS** On average, maybe only 0.4 planets per system are habitable.

*90% of habitable planets develop life*

**PLANETS WITH LIFE** Life may well develop on 90% of habitable planets.

*90% of life-bearing planets bear only simple life*      *10%*

**INTELLIGENT LIFE** Possibly about 10% of new instances of life develop intelligence.

*90% of intelligent life never talks to the stars*      *10%*

**COMMUNICATING LIFE** Possibly only 10% of such life develops interstellar communications.

*some civilizations die before contact*

**LIFE SPAN OF CIVILIZATION** These civilizations might, on average, last 10,000 years.

*70 civilizations alive today*

### CONCLUSION
Using the estimates above, one might expect there to be about $2 \times 0.99 \times 0.4 \times 0.9 \times 0.1 \times 0.1 \times 10,000 =$ approximately 70 alien civilizations in our galaxy that, in theory, we should be able to communicate with. However, some of the estimates may be wildly wrong.

**RECOGNIZING LIFE**
If humans ever encounter extraterrestrial life, it is by no means certain that we would immediately recognize it. Not everyone would see life, rather than just discoloration, in this algal bloom growing in the North Atlantic (above right).

# LOOKING FOR LIFE

Attempts to identify extraterrestrial life forms follow a number of approaches. Within the solar system, scientists analyze images of planets and moons for signs of life and send probes to feasible locations, such as Mars and Saturn's moon Titan.

Outside the solar system, the main focus of the search is SETI (the search for extraterrestrial intelligence)—a set of programs that involve scanning the sky for radio signals that look like they were sent by aliens. A search has also begun for Earth-like planets around nearby stars (see pp.296–299). Finally, CETI (communication with extraterrestrial intelligence) involves broadcasting the presence of humans by sending signals toward target stars. In 1974, a CETI message in binary code was sent toward the M13 star cluster, 21,000 light-years away. In 1999, the more elaborate "Encounter 2001" message was sent from a Ukrainian radio telescope toward some nearby Sun-like stars. Even if aliens pick up this message, we can expect no reply for at least a century.

**ARECIBO DISH**
The Arecibo Telescope in Puerto Rico is the world's second-largest single-dish radio telescope. It has been used for SETI and in one CETI attempt.

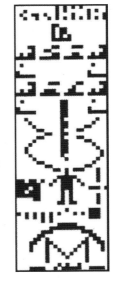

**MESSAGE TO ALIENS**
The Arecibo Telescope message contains symbols of a human body, DNA, the solar system, and the Arecibo dish itself.

# THE FATE OF THE UNIVERSE

ALTHOUGH IT IS POSSIBLE THAT THE UNIVERSE will last forever, the types of structures that exist in it today, such as planets, stars, and galaxies, almost certainly will not. At some distant point in the future, our galaxy and others will either be ripped apart; suffer a protracted, cold death; or, in the least likely scenario, be crushed out of existence in a reverse of the Big Bang. Which of these fates befalls the universe depends to a considerable extent on the nature of dark energy—a mysterious, gravity-opposing force recently found to be playing a major part in the universe's large-scale behavior.

## BIG CRUNCH AND BIG CHILL

Until recently, cosmologists assumed that the universe's expansion rate (see pp.44–45) must be slowing due to the "braking" effects of gravity. They also believed that a single factor—the universe's mass-energy density—would decide which of two basic fates awaited it. Cosmologists measure the density of both mass and energy together since Einstein demonstrated that mass and energy are equivalent and interchangeable (see p.41). They calculated that if this density was above a critical value, gravity would eventually cause the universe to stop expanding and collapse in a fiery, all-annihilating implosion (a "Big Crunch"). If, however, the universe's density was below or exactly on the critical value, the universe would expand forever, albeit with its expansion rate gradually slowed by gravity. In this case, the universe would end in a lengthy, cold death (a "Big Chill"). Research aimed at resolving this issue found that the universe has properties suggesting that it is extremely close to being "flat" (opposite), with a density of exactly the critical value. Even though some of the mass-energy in the universe needed to render it flat seemed hard to locate, its density must be near the critical value, and so its most likely fate was eternal expansion. However, in the late 1990s, models of the fate of the universe were thrown into confusion by new findings indicating that the universe's expansion is not slowing down at all.

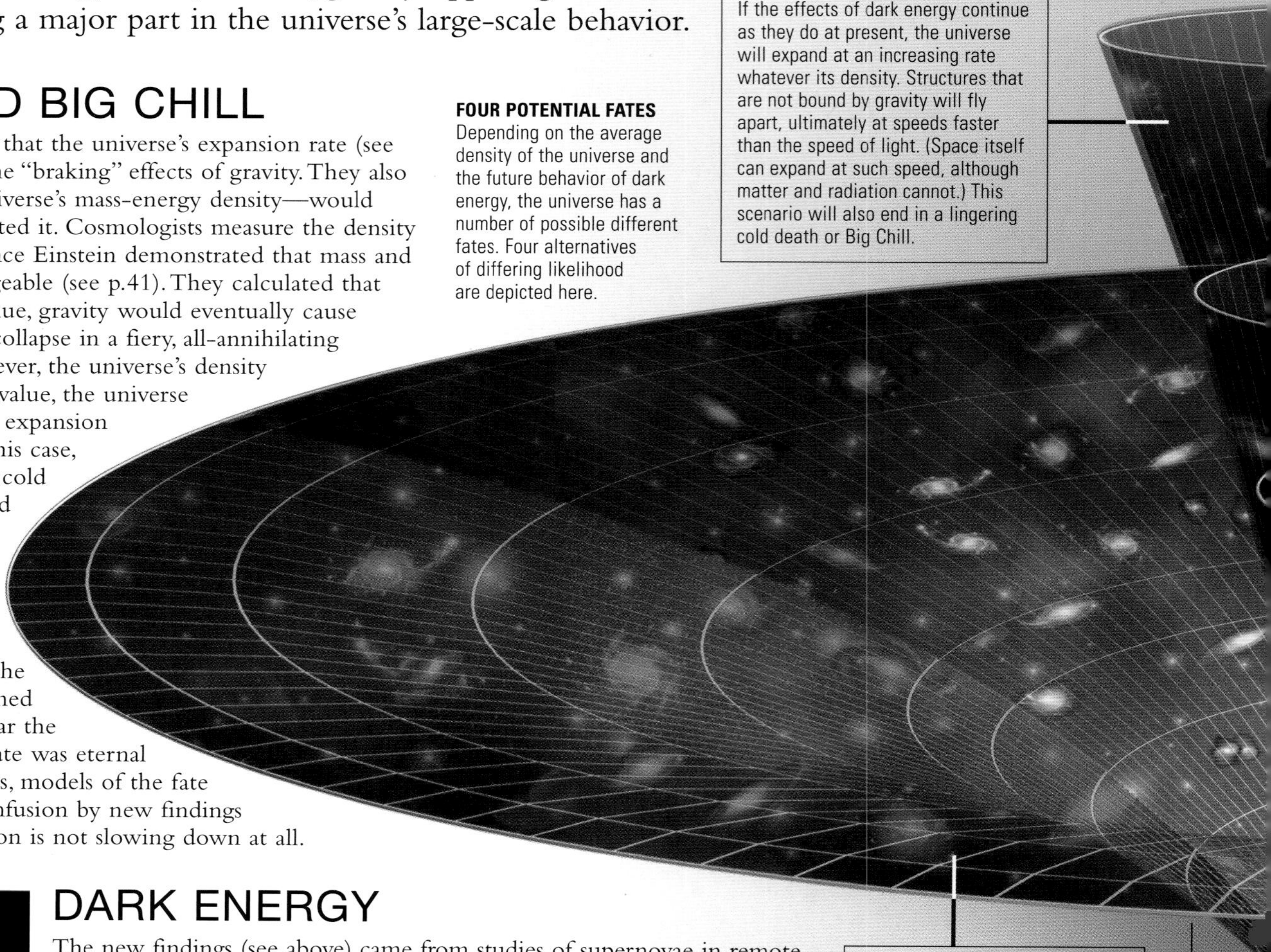

**FOUR POTENTIAL FATES**
Depending on the average density of the universe and the future behavior of dark energy, the universe has a number of possible different fates. Four alternatives of differing likelihood are depicted here.

### BIG CHILL
If the universe has a mass-energy density close to or just less than the critical value, and should the effects of dark energy tail off, the universe might continue to expand at a rate that slowly decreases but never comes to a complete halt. Over unimaginably long periods of time, it suffers a lingering cold death or "Big Chill."

### MODIFIED BIG CHILL
If the effects of dark energy continue as they do at present, the universe will expand at an increasing rate whatever its density. Structures that are not bound by gravity will fly apart, ultimately at speeds faster than the speed of light. (Space itself can expand at such speed, although matter and radiation cannot.) This scenario will also end in a lingering cold death or Big Chill.

## DARK ENERGY

The new findings (see above) came from studies of supernovae in remote galaxies. The apparent brightness of these exploding stars can be used to calculate their distance, and by comparing their distances with the red shifts of their home galaxies, scientists can calculate how fast the universe was expanding at different times in its history. The calculations showed that the expansion of the universe is accelerating and that some repulsive force is opposing gravity, causing matter to fly apart. This force has been called dark energy, and its exact nature is uncertain, though it appears similar to a gravity-opposing force, the "cosmological constant," proposed by Albert Einstein as part of his theory of general relativity (see pp.42–43). The existence of dark energy also accounts for the missing mass-energy in the universe required to make it flat (above) and modifies the number of possible fates for the universe.

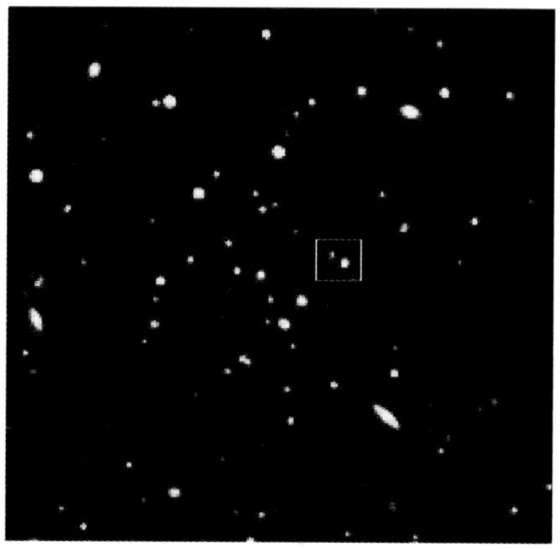

**SUPERNOVAE CLUES**
Type Ia supernovae, like that depicted here, all have the same intrinsic brightness. Thus, their apparent brightness reveals their distance.

**SUPERNOVA DISCOVERY**

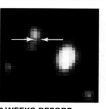
**3 WEEKS BEFORE**

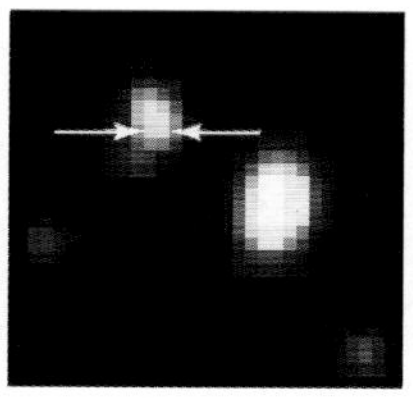

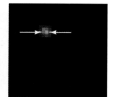
**AFTER SUPERNOVA**

**DIFFERENCE**

### BIG RIP
If the strength of dark energy increased, it could overcome all the fundamental forces and totally disintegrate the universe in a "Big Rip." This could happen 20–30 billion years from now. First, galaxies would be torn apart, then solar systems. A few months later, stars and planets would explode, followed shortly by atoms. Time would then stop.

### DARK ENERGY DOMINANCE
Dark energy provides 68 percent of the mass-energy density of the universe. Atom-based matter (in stars and the interstellar medium) and neutrinos contribute just 4.9 percent.

**MASS-ENERGY CONTRIBUTIONS**    neutrinos 0.1%, heavy elements 0.1%, photons 0.0005%

dark energy about 68%    dark matter about 27%    hydrogen and helium 4.7%

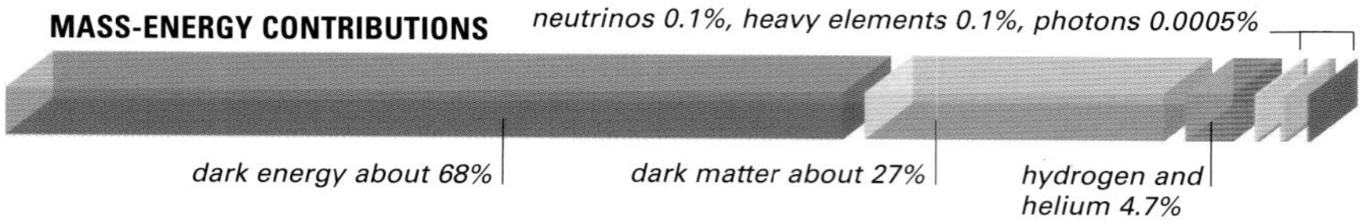

## THE GEOMETRY OF SPACE

Cosmologists base their ideas on the fate of the universe partly on mathematical models. These indicate that, depending on its mass-energy density, the universe has three possible geometries, each with a different space-time curvature that can be represented by a 2-D shape. Before the discovery of dark energy, there was a correspondence between these geometries and the fate of the universe. A positively curved or "closed" universe was envisaged to end in a Big Crunch and a negatively curved or "open" universe in a Big Chill. A "flat" universe would also end in a Big Chill, but one in which the universe's expansion eventually slows to a virtual standstill. With the discovery of dark energy, the correspondence no longer holds. If dark energy remains constant in intensity, any type of universe may expand forever. If dark energy is capable of reversing, any type of universe could end in a Big Crunch. Currently, the most favored view is that the universe is flat and will undergo an accelerating expansion. A cataclysmic "Big Rip" scenario, in which increasing dark energy tears the universe apart, is less likely.

*Big Crunch*

**FLAT UNIVERSE**
If the density of the universe is exactly on a critical value, it is "flat." In a flat universe, parallel lines never meet. The 2-D analogy is a plane. The universe is thought to be flat or nearly flat.

**CLOSED UNIVERSE**
If the universe is denser than a critical value, it is positively curved or "closed" and is finite in mass and extent. In such a universe, parallel lines converge. The 2-D analogy is a spherical surface.

**OPEN UNIVERSE**
If the universe is less dense than a critical value, it is negatively curved or "open" and infinite. The 2-D analogy of such a universe is a saddle-shaped surface on which parallel lines diverge.

TIME

*present day*

BIG BANG

**BIG CRUNCH**

In this version of doomsday, all matter and energy collapse to an infinitely hot, dense singularity in a reverse of the Big Bang. This scenario currently looks the least likely unless the effect of dark energy reverses in future. Even if it did happen, the earliest it could do so would be tens of billions of years from now.

## A COLD DEATH

If the universe peters out in a Big Chill, its death will take a long time. Over the next $10^{12}$ (1 trillion) years, galaxies will exhaust their gas in forming new stars. About $10^{25}$ (10 trillion trillion) years in the future, most of the universe's matter will be locked up in star corpses such as black holes and burned-out white dwarfs circling and falling into the supermassive black holes at the centers of galaxies. At $10^{32}$ (1 followed by 32 zeros) years from now, protons will start decaying to radiation (photons), electrons, positrons, and neutrinos. All matter not in black holes will fall apart. After another $10^{67}$ years, black holes will start evaporating by emitting particles and radiation, and in about $10^{100}$ years, even supermassive black holes will evaporate. The utterly cold, dark universe then will be nothing but a diffuse sea of photons and elemental particles.

**FINAL SURVIVORS**
At a very distant stage of the Big Chill, all the universe's matter, even that in black holes, will have decayed or evaporated to radiation. Apart from some very long-wavelength photons, the only constituents of the universe will be neutrinos, electrons, and positrons.

*photon*

*neutrino*

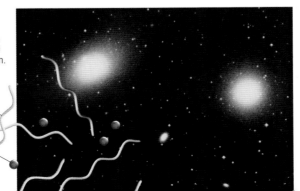

**FATE OF GALAXIES**
A trillion years from now, the universe will contain just old, fading galaxies. All their gas and dust will be used up, and most of the stars will be dying.

OBJECTS IN THE UNIVERSE—galaxies, stars, planets, nebulae—are scattered across three dimensions of space and one of time. Viewed from widely separated locations in the universe, their relative positions look completely different. To find objects in space, study their movements, and make celestial maps, astronomers need an agreed reference frame, and for most purposes the frame used is Earth itself. The prime element of this Earth-based view is the celestial sphere—an imaginary shell around Earth to which astronomers pretend the stars are attached. Apparent movements of celestial objects on this sphere can be related to the actual movements of Earth, the planets (as they orbit the Sun), the Moon (as it orbits Earth), and the stars as they move within the Milky Way. Understanding the celestial sphere, and conventions for naming and finding objects on it, are essential first steps in astronomy.

**MOVEMENT ON THE SKY**
This photograph, obtained over a four-hour period from the Las Campanas Observatory in Chile, looks toward the south celestial pole. The circular, clockwise star trails across the sky are a feature of the Earth-based view of the cosmos, since they result solely from the Earth's rotation.

# THE VIEW FROM EARTH

# THE CELESTIAL SPHERE

FOR CENTURIES, humans have known that stars lie at different distances from Earth. However, when recording the positions of stars in the sky, it is convenient to pretend that they are all stuck to the inside of a sphere that surrounds Earth. The idea of this sphere also helps astronomers understand how their location on Earth, the time of night, and the time of year affect what they see in the night sky.

## THE SKY AS A SPHERE

To an observer on Earth, the stars appear to move slowly across the night sky. Their motion is caused by Earth's rotation, although it might seem that the sky is spinning around our planet. To the observer, the sky can be imagined as the inside of a sphere, known as the celestial sphere, to which the stars are fixed and relative to which the Earth rotates. This sphere has features related to the real sphere of the Earth. It has north and south poles, which lie on its surface directly above Earth's North and South Poles, and it has an equator (the celestial equator), which sits directly above Earth's equator. The celestial sphere is like a celestial version of a globe—the positions of stars and galaxies can be recorded on it, just as cities on Earth have their positions of latitude and longitude on a globe.

the Sun and planets are not fixed on the celestial sphere, but move around on or close to a circular path called the ecliptic

celestial equator—a circle on the celestial sphere concentric with Earth's equator

### IMAGINARY GLOBE
The celestial sphere is purely imaginary, with a specific shape but no precise size. Astronomers use exactly defined points and curves on its surface as references for describing or determining the positions of stars and other celestial objects.

north celestial pole lies directly above Earth's North Pole

celestial sphere

Earth's axis of spin

stars are fixed to the sphere's surface and appear to move in the opposite direction to Earth's spin

vernal or spring equinox (first point of Aries)

Earth's spin

Earth's North Pole

Earth

Sun's motion

autumnal equinox (first point of Libra), one of two points of intersection between celestial equator and ecliptic

south celestial pole lies below Earth's South Pole

## EFFECTS OF LATITUDE

An observer on Earth can view, at best, only half of the celestial sphere at any instant (assuming a cloudless sky and unobstructed horizon). The other half is obscured by Earth's bulk. In fact, for an observer at either of Earth's poles, a specific half of the celestial sphere is always overhead, while the other half is never visible. For observers at other latitudes, Earth's rotation continually brings new parts of the celestial sphere into view and hides others. This means, for example, that over the course of a night, an observer at a latitude of 60°N or 60°S can see up to three-quarters of the celestial sphere for at least some of the time, and an observer at the equator can see every point on the celestial sphere at some time.

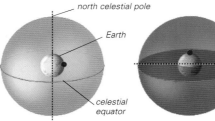

north celestial pole

Earth

celestial equator

### OBSERVER AT EQUATOR
For a person on the equator, Earth's rotation brings all parts of the celestial sphere into view for some time each day. The celestial poles are on the horizon.

### OBSERVER AT NORTH POLE
For this observer, the northern half of the celestial sphere is always visible and the southern half is never visible. The celestial equator is on the observer's horizon.

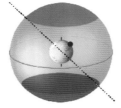

### OBSERVER AT MIDLATITUDE
For this observer, a part of the celestial sphere is always visible, a part is never visible, and Earth's rotation brings other parts into view for some of the time each day.

**KEY**

stars always visible

stars never visible

stars sometimes visible

● position of observer

········· observer's horizon

north celestial pole

circumpolar area

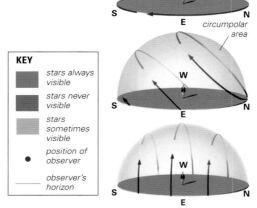

### MOTION AT NORTH POLE
At the poles, all celestial objects seem to circle the celestial pole directly over-head. The motion is counter-clockwise at the North Pole and clockwise at the south.

### MOTION AT MIDLATITUDE
At midlatitudes, most stars rise in the east, cross the sky obliquely, and set in the west. Some (circumpolar) objects never rise or set, but circle the celestial pole.

### MOTION AT EQUATOR
At the equator, stars and other celestial objects appear to rise vertically in the east, move overhead, and then fall vertically and set in the west.

# DAILY SKY MOVEMENTS

As the Earth spins, all celestial objects move across the sky, although the movements of the stars and planets become visible only at night. For an observer in midlatitudes, stars in polar regions of the celestial sphere describe a daily circle around the north or south celestial pole. The Sun, Moon, planets, and the remaining stars rise along the eastern horizon, sweep in an arc across the sky, and set in the west. This motion has a tilt to the south (for observers in the Northern Hemisphere) or to the north (Southern Hemisphere)—the lower the observer's latitude, the steeper the tilt. Stars have fixed positions on the sphere, so the pattern of their movement repeats with great precision once every sidereal day (see p.66). The planets, Sun, and Moon always move on the celestial sphere, so the period of repetition differs from that of the stars.

**CIRCUMPOLAR STARS**
Stars in the polar regions of the celestial sphere describe perfect part-circles around the north or south celestial pole during one night, as shown by this long-exposure photograph.

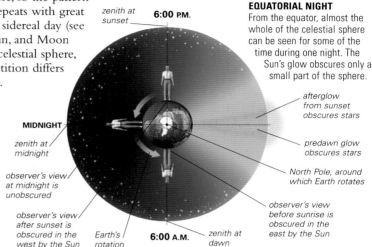

**EQUATORIAL NIGHT**
From the equator, almost the whole of the celestial sphere can be seen for some of the time during one night. The Sun's glow obscures only a small part of the sphere.

zenith at sunset    6:00 P.M.

afterglow from sunset obscures stars

MIDNIGHT

predawn glow obscures stars

zenith at midnight

observer's view at midnight is unobscured

North Pole, around which Earth rotates

observer's view after sunset is obscured in the west by the Sun

Earth's rotation    6:00 A.M.    zenith at dawn

observer's view before sunrise is obscured in the east by the Sun

# YEARLY SKY MOVEMENTS

As Earth orbits the Sun, the Sun seems to move against the background of stars. As the Sun moves into a region of the sky, its glare washes out fainter light from that part, so any star or other object there temporarily becomes difficult to view from anywhere on Earth. Earth's orbit also means that the part of the celestial sphere on the opposite side to Earth from the Sun—that is, the part visible in the middle of the night—changes. The visible part of the sky at, for example, midnight in June, September, December, and March is significantly different—at least for observers at equatorial or midlatitudes on Earth.

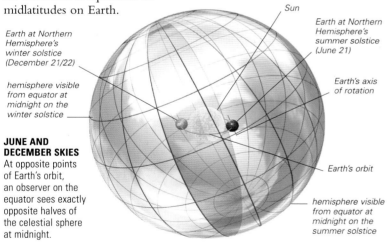

Earth at Northern Hemisphere's winter solstice (December 21/22)

Sun

Earth at Northern Hemisphere's summer solstice (June 21)

hemisphere visible from equator at midnight on the winter solstice

Earth's axis of rotation

**JUNE AND DECEMBER SKIES**
At opposite points of Earth's orbit, an observer on the equator sees exactly opposite halves of the celestial sphere at midnight.

Earth's orbit

hemisphere visible from equator at midnight on the summer solstice

## ARISTOTLE'S SPHERES

Until the 17th century CE, the idea of a celestial sphere surrounding Earth was not just a convenient fiction—many people believed it had a physical reality. Such beliefs date back to a model of the universe developed by the Greek philosopher Aristotle (384–322 BCE) and elaborated by the astronomer Ptolemy (85–165 CE). Aristotle placed Earth stationary at the universe's center, surrounded by several transparent, concentric spheres to which the stars, planets, Sun, and Moon were attached. Ptolemy supposed that the spheres rotated at different speeds around Earth, producing the observed motions of the celestial bodies.

sphere of "fixed" stars

**ARISTOTELIAN MODEL OF THE UNIVERSE**
Stars are fixed to the outer sphere. Working inward, the other spheres around Earth carry Saturn, Jupiter, Mars, the Sun, Venus, Mercury, and the Moon.

# CELESTIAL COORDINATES

Using the celestial sphere concept, astronomers can record and find the positions of stars and other celestial objects. To define an object's position, astronomers use a system of coordinates, similar to latitude and longitude on Earth. The coordinates are called declination and right ascension. Declination is measured in degrees and arc-minutes (60 arc-minutes = 1 degree/1°) north or south of the celestial equator, so it is equivalent to latitude. Right ascension, the equivalent of longitude, is the angle of an object to the east of the celestial meridian. The meridian is a line passing through both celestial poles and a point on the celestial equator called the first point of Aries or vernal equinox point (see p.65). An object's right ascension can be stated in degrees and arc-minutes or in hours and minutes. One hour is equivalent to 15°, because 24 hours make a whole circle.

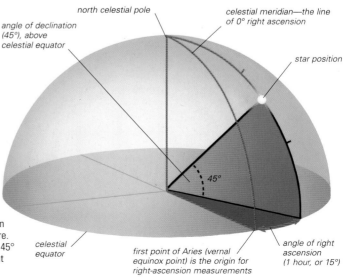

north celestial pole

celestial meridian—the line of 0° right ascension

angle of declination (45°), above celestial equator

star position

45°

**RECORDING A STAR'S POSITION**
The measurement of a star's position on the celestial sphere is shown here. This star has a declination of about 45° (sometimes written +45°) and a right ascension of about 1 hour, or 15°.

celestial equator

first point of Aries (vernal equinox point) is the origin for right-ascension measurements

angle of right ascension (1 hour, or 15°)

# CELESTIAL CYCLES

TO AN OBSERVER ON EARTH, celestial events occur within the context of cycles determined by the motions of the Earth, Sun, and Moon. These cycles provide us with some of our basic units for measuring time, such as days and years. They include the apparent daily motions of all celestial objects across the sky, the annual apparent movement of the Sun against the celestial sphere, the seasonal cycle, and the monthly cycle of lunar phases. Other related cycles produce the dramatic but predictable events known as lunar and solar eclipses.

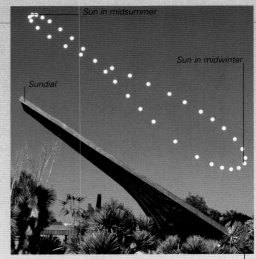

**THE SUN'S ANALEMMA**
To produce this image, the Sun was photographed, above a sundial, at the same time of day on 37 occasions throughout 1 year. The vertical change in its position is due to Earth's tilt. The horizontal drift is due to Earth changing its speed on its elliptical orbit around the Sun. The resulting figure-eight pattern is called an analemma.

## ASTROLOGY AND THE ECLIPTIC

Astrology is the study of the positions and movements of the Sun, Moon, and planets in the sky in the belief that these influence human affairs. At one time, when astronomy was applied mainly to devising calendars, astronomy and astrology were intertwined, but their aims and methods have now diverged. Astrologers pay little attention to constellations, but measure the positions of the Sun and planets in sections of the ecliptic that they call "Aries" and "Taurus," for example. However, these sections no longer match the constellations of Aries, Taurus, and so on.

**STARGAZER**
This 17th-century illustration, taken from a treatise written in India on the zodiac, depicts a stargazer using an early form of mounted telescope.

## THE SUN'S CELESTIAL PATH

As the Earth travels around the Sun, to an observer on Earth, the Sun seems to trace a path across the celestial sphere known as the ecliptic. Because of the Sun's glare, this movement is not obvious, but the Sun moves a small distance each day against the background of stars. The band of sky extending for 9 degrees (see p.63) on either side of the Sun's path is called the zodiac (see opposite) and incorporates parts or all of 24 constellations (see p.72). Of these, the Sun passes through 13 constellations, of which 12 form the "signs of the zodiac," well-known to followers of astrology (see panel, left). The Sun spends a variable number of days in each of these 13 constellations. However, the Sun currently passes through each constellation on dates very different from traditional astrological dates. For example, someone born between March 21 and April 19 is said to have the sign Aries, although the Sun currently passes through Aries between April 19 and May 14. This disparity is partly caused by a phenomenon called precession.

## PRECESSION

The Earth's axis of rotation is tilted from the ecliptic plane by 23.5°. The tilt is crucial in causing seasons (see opposite). At present, the axis points at a position on the northern celestial sphere (the north celestial pole) close to the star Polaris, but this will not always be so. Like a spinning top, Earth is executing a slow "wobble," which alters the direction of its axis over a 25,800-year cycle. The wobble, called precession, is caused by the gravity of the Sun and Moon. It also causes the south celestial pole, the celestial equator, and two other reference points on the celestial sphere, called the equinox points, to change their locations gradually. The coordinates of stars and other "fixed" objects, such as galaxies (see p.63), therefore change, so astronomers must quote them according to a standard "epoch" of around 50 years. The current standard was exactly correct on January 1, 2000.

**ISLAMIC ZODIAC**
This Islamic depiction of part of the celestial sphere includes several constellations that are also well-known zodiacal "star signs," such as Scorpius and Leo. The illustration decorates a 19th-century manuscript from India that brought together Islamic, Hindu, and European knowledge of astronomy.

**MIDNIGHT SUN**
This multiple-exposure photograph (below) shows the path of the Sun around midnight near the summer solstice in Iceland. Since the photograph was taken in polar latitudes, Earth's angle of tilt ensures the Sun does not set.

path of north celestial pole across the sky every 25,800 years

Deneb

Alderamin, Pole Star in 8000 CE

Vega, Pole Star in 13,000 CE

Polaris (current north Pole Star)

25,800-year wobble of Earth's axis

Earth's axis of rotation

angle of tilt remains the same throughout precession

equator

rotation of Earth around its axis

**EARTH'S WOBBLE**
Precession causes Earth's spin axis to trace out the shape of a cone. As it does so, both the north and south celestial poles trace out circular paths on the celestial sphere, in a 25,800-year cycle.

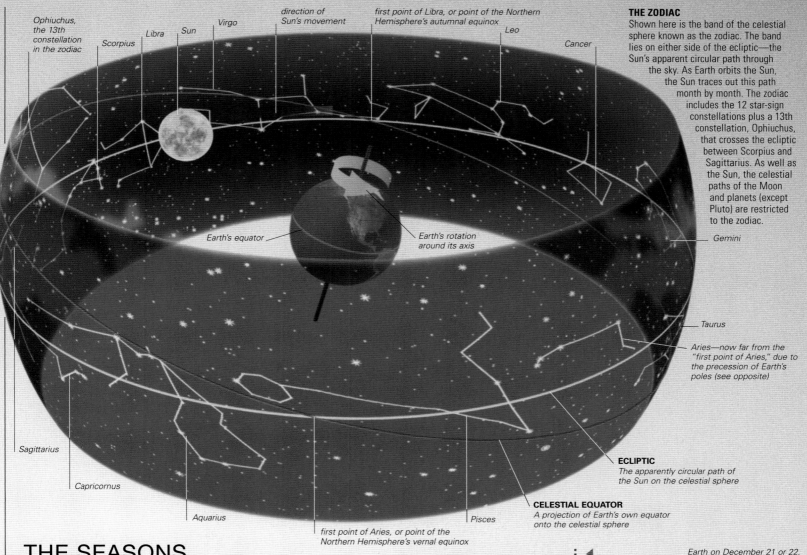

Ophiuchus, the 13th constellation in the zodiac

Scorpius

Libra

Sun

Virgo

direction of Sun's movement

first point of Libra, or point of the Northern Hemisphere's autumnal equinox

Leo

Cancer

Earth's equator

Earth's rotation around its axis

Sagittarius

Capricornus

Aquarius

first point of Aries, or point of the Northern Hemisphere's vernal equinox

Pisces

**THE ZODIAC**
Shown here is the band of the celestial sphere known as the zodiac. The band lies on either side of the ecliptic—the Sun's apparent circular path through the sky. As Earth orbits the Sun, the Sun traces out this path month by month. The zodiac includes the 12 star-sign constellations plus a 13th constellation, Ophiuchus, that crosses the ecliptic between Scorpius and Sagittarius. As well as the Sun, the celestial paths of the Moon and planets (except Pluto) are restricted to the zodiac.

Gemini

Taurus

Aries—now far from the "first point of Aries," due to the precession of Earth's poles (see opposite)

**ECLIPTIC**
The apparently circular path of the Sun on the celestial sphere

**CELESTIAL EQUATOR**
A projection of Earth's own equator onto the celestial sphere

# THE SEASONS

Earth's orbit around the Sun takes 365.25 days and provides a key unit of time, the year. Earth's seasons result from the tilt of its axis relative to its orbit. Due to Earth's tilt, one or other of its hemispheres is normally pointed toward the Sun. The hemisphere that tilts toward the Sun receives more sunlight and is therefore warmer. Each year, the Northern Hemisphere reaches its maximum tilt toward the Sun around June 21—summer solstice in the Northern Hemisphere and winter solstice in the Southern Hemisphere. For some time around this date, the north polar region is sunlit all day, while the south polar region is in darkness. Conversely, around December 21, the situation is reversed. Between the solstices are the equinoxes, when Earth's axis is broadside to the Sun and the periods of daylight and darkness are equal for all points on Earth. Earth's tilt also defines the tropics. The Sun is overhead at midday on the Tropic of Cancer (23.5°N) around June 21, above the Tropic of Capricorn (23.5°S) around December 21, and directly above the equator at midday during the equinoxes.

**SOLSTICES AND EQUINOXES**
At the solstices, in June and December, one hemisphere has its longest day, the other its shortest. At the equinoxes, in March and September, the lengths of day and night are equal everywhere on Earth.

Earth on March 20 or 21, the Northern Hemisphere's vernal or spring equinox

midday sun overhead at Tropic of Cancer

Sun

Earth on June 21 or 22, the Northern Hemisphere's summer solstice

Earth on December 21 or 22, the Northern Hemisphere's winter solstice

midday sun overhead at Tropic of Capricorn

Earth's orbit

Earth on September 22 or 23, the Northern Hemisphere's autumnal equinox

23.5° angle of tilt

axis of spin

Tropic of Cancer, 23.5°N

solar radiation

Tropic of Capricorn, 23.5°S

direction of Earth's spin

**SUNLIGHT INTENSITY**
The intensity of solar radiation is greatest within the tropics. Toward the poles, the Sun's rays impinge at an oblique angle. They must pass through a greater thickness of atmosphere, and they are spread over a wider area of ground.

# MEASURING DAYS

**SOLAR TIME**
Solar time is the way of gauging time from the Sun's apparent motion across the sky, as measured by a sundial. One solar day is subdivided into 24 hours.

Every day, Earth rotates once, and most locations on its surface pass from sunlight to shadow and back, producing the day–night cycle. However, there are two possible definitions for what constitutes a day, and only one of these, the solar day, lasts for exactly 24 hours. A solar day is defined by the apparent movement of the Sun across the sky produced by Earth's rotation. It is the length of time the Sun takes to return to its highest point in the sky from the same point the previous day. The other type of day, the sidereal day, is defined by Earth's rotation relative to the stars. It is the length of time a star takes to return to its highest point in the sky on successive days. A sidereal day is 4 minutes shorter than a solar day.

**APRIL 1, 8:00 P.M.**   **APRIL 8, 8:00 P.M.**   **APRIL 15, 8:00 P.M.**

**SIDEREAL TIME**
The distinctive constellation Orion (see pp.390–391), here pictured as if from 50°N, appears lower in the sky at the same solar time each day, as the daily 4-minute difference between solar and sidereal time mounts up.

**SOLAR AND SIDEREAL DAY**
The disparity between solar and sidereal days results from Earth's orbit and rotation. After rotating once relative to the stars, Earth must rotate a little farther to bring the Sun back to the same point in the sky.

direction of a distant star, against which sidereal time can be measured

Sun

Earth's orbit

noon on first day

Earth's rotation

second noon in solar time

second noon in sidereal time (4 minutes earlier than solar time)

# MEASURING MONTHS

The concept of a month is based on the Moon's orbit around Earth. During each of the Moon's orbits, the angle between Earth, the Moon, and the Sun continuously changes, giving rise to the Moon's phases. The phases cycle through new moon (when the Moon is between Earth and the Sun), crescent, quarter, and gibbous, to full moon (when the Earth lies between the Moon and the Sun). A complete cycle of the Moon's phases takes 29.5 solar days and defines a lunar month. However, Earth's progress around the Sun complicates the expression of a month, just as it confuses the measurement of a day. The Moon in fact takes only 27.3 days to orbit Earth with reference to the background stars. Astronomers call this period a sidereal month. The disparity results because Earth's progress around the Sun alters the angles between the Earth, Sun, and Moon. After one full orbit of Earth (a sidereal month), the Moon must orbit a bit farther to return to its original alignment with Earth and the Sun (a lunar month).

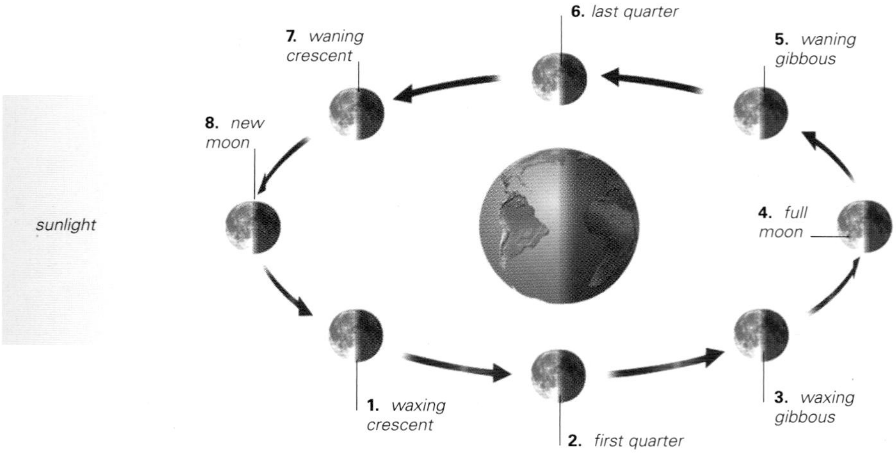

**CHANGING ANGLES**
During each lunar orbit, the angle between Earth, the Moon, and the Sun changes. The part of the Moon's sunlit face seen by an observer on Earth changes in a cyclical fashion.

7. waning crescent
6. last quarter
5. waning gibbous
8. new moon
4. full moon
sunlight
1. waxing crescent
2. first quarter
3. waxing gibbous

INTRODUCTION

## MYTHS AND STORIES
### EVIL PORTENTS

Astronomers have predicted eclipses reliably since about 700 BCE, but that has not stopped doomsayers and astrologers from reading evil omens into these routine celestial events. They have often prophesied disasters associated with eclipses, and although they meet with no more than occasional success, some people listen. The Incas below, for instance, are pictured as awestruck by an eclipse, in a European atlas of 1827. Eclipses may not be useful for predicting the future, but accounts of past eclipses are of great value to today's historians, who can calculate the dates of events with great precision if the historical accounts include records of eclipses.

# LUNAR ECLIPSES

As the Moon orbits the Earth, it occasionally moves into Earth's shadow—an occurrence called a lunar eclipse—or blocks sunlight from reaching a part of Earth's surface—a solar eclipse. Eclipses do not happen every month, because the plane of the Moon's orbit around Earth does not coincide with the plane of Earth's orbit around the Sun. Nevertheless, an eclipse of some kind occurs several times each year. Lunar eclipses are common, occurring two or three times a year, always during a full moon. Astronomers classify lunar eclipses into three different types. In a penumbral eclipse, the Moon passes through Earth's penumbra (part-shadow), leading to only a slight dimming of the Moon. In a partial eclipse, a portion of the Moon passes through Earth's umbra (full shadow), while in a total eclipse the whole Moon passes through the umbra.

### MECHANICS OF A LUNAR ECLIPSE
Earth's shadow consists of the penumbra, within which some sunlight is blocked out, and the umbra, or full shadow. In a total eclipse, the Moon passes through the penumbra, the umbra, and then the penumbra again.

**TOTAL LUNAR ECLIPSE**
This composite photograph shows stages of a total lunar eclipse. The moon appears red at the eclipse's peak (bottom left), because a little red light is bent toward it by refraction in Earth's atmosphere.

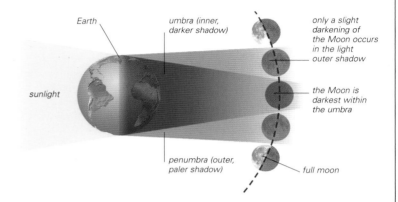

Earth

umbra (inner, darker shadow)

sunlight

penumbra (outer, paler shadow)

only a slight darkening of the Moon occurs in the light outer shadow

the Moon is darkest within the umbra

full moon

# SOLAR ECLIPSES

An eclipse of the Sun occurs when the Moon blocks sunlight from reaching part of the Earth. During a total eclipse, viewers within a strip of Earth's surface, called the path of totality, witness the Sun totally obscured for a few moments by the Moon. Outside this area is a larger region where viewers see the Sun only partly obscured. More common are partial eclipses, which cause no path of totality. A third type of solar eclipse is the annular eclipse, occurring when the Moon is farther from Earth than average, so that its disk is too small to cover the Sun's disk totally. At the peak of an annular eclipse, the Moon looks like a dark disk inside a narrow ring of sunlight. Solar eclipses happen two or three times a year, but total eclipses occur only about once every 18 months. During the period of totality, the Sun's corona (its hot outer atmosphere) becomes visible.

### TOTALITY PATHS
The part of Earth's surface over which the Moon's full shadow will sweep during a total solar eclipse, called the path of totality, can be predicted precisely. Below are the paths for eclipses up to 2030.

**ECLIPSE SEQUENCE**
This multiple exposure photograph depicts more than 20 stages of a total solar eclipse, seen in Mexico in 1991. At the center can be seen the corona around the fully eclipsed Sun.

**BAILY'S BEADS**
At the beginning and end of a total solar eclipse, the Moon's rough, cratered surface breaks a thin slice of Sun into patches of light called "Baily's Beads."

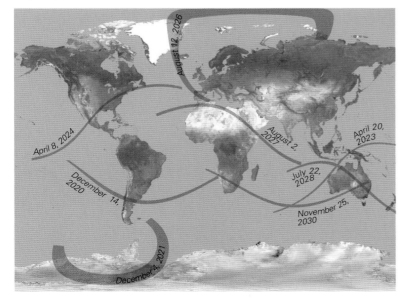

August 12, 2026

August 2, 2027

April 20, 2023

April 8, 2024

December 14, 2020

July 22, 2028

November 25, 2030

December 4, 2021

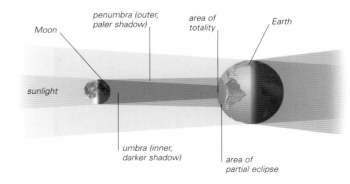

Moon

penumbra (outer, paler shadow)

area of totality

Earth

sunlight

umbra (inner, darker shadow)

area of partial eclipse

**MOON SHADOW**
The shadow cast by the Moon during a total solar eclipse consists of the central umbra (associated with the area of totality) and the penumbra (area of partial eclipse).

# PLANETARY MOTION

THE PLANETS IN THE SOLAR SYSTEM are much closer to Earth than are the stars, so as they orbit the Sun they appear to wander across the starry background. This sky motion is influenced by Earth's own solar orbit, which changes the point of view of Earth-bound observers. The planets closest to Earth move around on the celestial sphere more rapidly than the more distant planets; this is partly due to perspective and partly because the closer a planet is to the Sun, the faster is its orbital speed.

## INFERIOR AND SUPERIOR PLANETS

In terms of their motions in the sky as seen from Earth, the planets are divided into two groups. The inferior planets, Mercury and Venus, are those that orbit closer to the Sun than does Earth. They never move far from the Sun on the celestial sphere—the greatest angle by which the planets stray from the Sun (called their maximum elongation) is 28° for Mercury and 45° for Venus. Because they are close to Earth and orbiting quickly, both planets move rapidly against the background stars. They also display phases, like the Moon's (see p.66), because there is some variation in the angle between Earth, the planet, and the Sun. All the other planets, from Mars outward, are called superior planets. These are not "tied" to the Sun on the celestial sphere, and so can be seen in the middle of the night. Apart from Mars, the superior planets are too far from Earth to display clear phases, and they move slowly on the celestial sphere—the farther they are from the Sun, the slower their movement.

**ALWAYS NEAR THE SUN**
The Moon and Venus appear close together here in the dawn sky. Venus is only ever visible in the eastern sky for up to a few hours before dawn, or in the western sky after dusk—it is never seen in the middle of the night. This is because it orbits closer to the Sun than Earth and so never strays far from the Sun in the sky.

### JOHANNES KEPLER

The German astronomer Johannes Kepler (1571–1630) discovered the laws of planetary motion. His first law states that planets orbit the Sun in elliptical paths. The next states that the closer a planet comes to the Sun, the faster it moves, while his third law describes the link between a planet's distance from the Sun and its orbital period. Newton used Kepler's laws to formulate his theory of gravity.

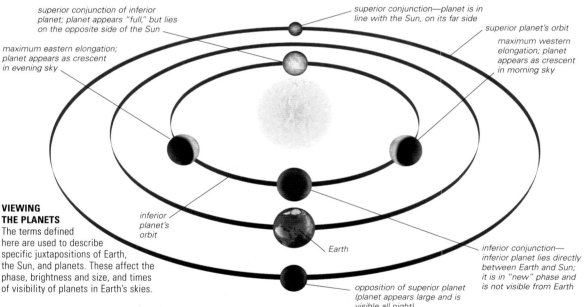

superior conjunction of inferior planet; planet appears "full," but lies on the opposite side of the Sun

maximum eastern elongation; planet appears as crescent in evening sky

superior conjunction—planet is in line with the Sun, on its far side

superior planet's orbit

maximum western elongation; planet appears as crescent in morning sky

**VIEWING THE PLANETS**
The terms defined here are used to describe specific juxtapositions of Earth, the Sun, and planets. These affect the phase, brightness and size, and times of visibility of planets in Earth's skies.

inferior planet's orbit

Earth

inferior conjunction—inferior planet lies directly between Earth and Sun; it is in "new" phase and is not visible from Earth

opposition of superior planet (planet appears large and is visible all night)

## RETROGRADE MOTION

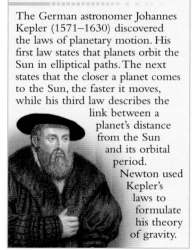

path of Mars across sky

ecliptic plane

Mars's orbit inclined relative to ecliptic plane | Mars | Sun | Earth's orbit | Earth

The planets generally move through the sky from west to east against the background of stars, night by night. However, periodically, a planet moves from east to west for a short time—a phenomenon called retrograde motion. Retrograde motion is an effect of changing perspective. Superior planets such as Mars show retrograde motion when Earth "overtakes" the other planet at opposition (when Earth moves between the superior planet and the Sun). The inferior planets Mercury and Venus show retrograde motion on either side of inferior conjunction. They overtake Earth as they pass between Earth and the Sun.

**ZIGZAG ON THE SKY**
In retrograde motion, a planet may perform a loop or a zigzag on the sky, depending on the angle of its orbit relative to Earth's.

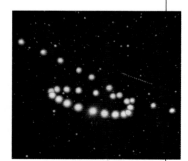

**MARTIAN LOOP-THE-LOOP**
This composite of photographs taken over several months shows a retrograde loop in Mars's motion against the background stars. The additional short dotted line is produced by Uranus.

# ALIGNMENTS IN THE SKY

Because all the planets orbit the Sun roughly in the same plane (see pp.102–103), they never stray from the band in the sky called the zodiac (see p.65). It is not uncommon for several of the planets to be in the same part of the sky at the same time, often arranged roughly in a line. Such events, called planetary conjunctions, are of no deep significance, but can be a spectacular sight. Another type of alignment, called a transit, occurs when an inferior planet comes directly between Earth and the Sun, passing across the Sun's disk. A pair of Venus transits, eight years apart, occurs about once a century or so, while Mercury transits happen about 12 times a century. In earlier times, these transits allowed astronomers to obtain more accurate data on distances in the solar system. A final type of alignment is an occultation—one celestial body passing in front of, and hiding, another. Occultations of one planet by another, such as Venus occulting Jupiter, occur only a few times a century; in contrast, occultations of one or other of the bright planets by the Moon occur 10 or 11 times a year.

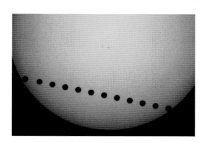

**VENUS'S PATH ACROSS THE SUN'S DISK**
This composite photograph of Venus's 2004 transit spans just over five hours. During this time, astronomers gathered data on the Sun's changing light to use as a model to look for Earth-sized planets orbiting other stars.

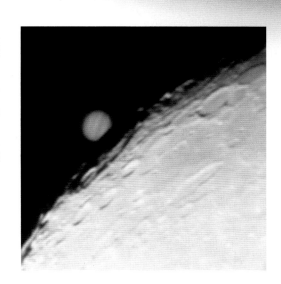

**TRANSIT OF VENUS**
This photograph of the 2004 Venus transit shows our nearest planetary neighbor as a dark circle close to the edge of the Sun's disk. This was the first Venus transit since 1882. Another occurred in 2012, but no more are expected until 2117.

**OCCULTATION OF JUPITER BY THE MOON**
This occultation occurred on January 26, 2002, and was visible above a latitude of 55°N. Here, the planet sinks out of sight beyond the dark far wall of the lunar crater Bailly. Occultations by the Moon tend to run in series, when for a period the planet and Moon wander into alignment as seen from Earth. An occultation then occurs approximately every sidereal month, until eventually the planet and Moon drift out of alignment again.

*Jupiter*

*Saturn*

*Mars*

*Venus*  *Mercury*

**PLANETARY CONJUNCTION, APRIL 2002**
The conjunction shown here, involving all five naked-eye planets, was visible after sunset for several evenings in April 2002. Although the planets appear close, they are separated by tens or hundreds of millions of miles.

## NICOLAUS COPERNICUS

Born in Torun, Poland, Copernicus (1473–1543) studied theology, law, and medicine at university. In 1503, he became the canon of Frauenberg Cathedral. This post provided financial security and left him plenty of time to indulge his passion for astronomy. He described his idea of a Sun-centered universe in his book *On the Revolution of the Heavenly Spheres*, published in the year of his death.

At first, Copernicus's revolutionary new idea made little impact. It was only after the telescopic observations of Galileo Galilei and the discovery of the laws of planetary motion by Johannes Kepler (see panel, opposite) that it was finally accepted.

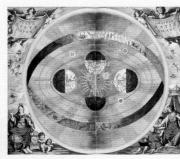

**COPERNICAN MAP**
This map made by Andreas Cellarius demonstrates the Copernican theory of Earth and the other planets circling the Sun, with the zodiac stars beyond.

INTRODUCTION

# STAR MOTION AND PATTERNS

STARS MAY SEEM TO BE FIXED to the celestial sphere, but in fact their positions are changing, albeit very slowly. There are two parts to this motion: a tiny, yearly wobble of a star's position in the sky, called parallax shift; and a continuous directional motion, called proper motion. To record the motion of stars, and properties such as their color and brightness, each star needs a name. Naming systems and catalogs have their roots in the constellations, which were invented to describe the patterns formed by stars in the sky.

## STAR COLORS
Although at first glance they all look white, stars differ in their colors—that is, in the mixture of light wavelengths they emit. This is a long-exposure photograph of the bright stars of Orion, taken while changing the camera's focus. Each star looks white when sharply focused, but when its light is spread out, its true color is revealed.

## EXPLORING SPACE

### HIPPARCOS

Hipparcos was a European Space Agency satellite that between 1989 and 1993 performed surveys of the stars. Its name is short for High Precision Parallax Collecting Satellite and was chosen to honor the Greek astronomer Hipparchus. Its mission has resulted in two catalogs. The Hipparcos catalog records the position, parallax, proper motions, brightness, and color of over 118,000 stars to a high level of precision. The Tycho catalog records over 1 million stars with measurements of lower accuracy.

#### HIPPARCOS SATELLITE
The satellite spun slowly in space, scanning strips of the sky as it rotated. It recorded data for each star about 100 to 150 times.

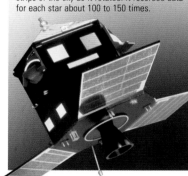

## PARALLAX SHIFT

Parallax shift is an apparent change in the position of a relatively close object against a more distant background as the observer's location changes. When an observer takes two photographs of a nearby star from opposite sides of Earth's orbit around the Sun, the star's position against the background of stars moves slightly. When the observer measures the size of this shift, knowing the diameter of Earth's orbit, she or he can calculate the star's distance using trigonometry. Until recently, this technique was limited to stars within a few hundred light-years of Earth, because the shifts of distant stars were too small to measure accurately. However, by using accurate instruments carried in satellites, much greater precision is possible: those carried in the Hipparcos satellite (see panel, left) have allowed calculation of star distances up to a few thousand light-years from Earth. More recently, a satellite named Gaia has been measuring the distances of stars as far away as the center of our galaxy and beyond.

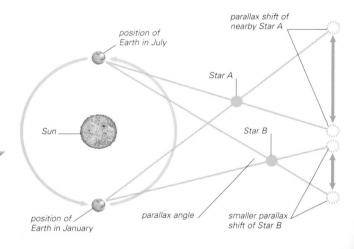

position of Earth in July

Sun

Star A

Star B

parallax shift of nearby Star A

position of Earth in January

parallax angle

smaller parallax shift of Star B

## MEASURING DISTANCE USING PARALLAX
When Star A is observed from opposite sides of Earth's orbit, its apparent shift in position is greater than that of more distant Star B. From the shift, an observer can calculate the parallax angle between the star and the two positions of Earth. The star's distance can be determined from this angle.

## PROPER MOTION OF STARS

All stars in our galaxy are moving at different velocities relative to the solar system, to the galactic center, and to each other. This motion gives rise to an apparent angular movement across the celestial sphere called a star's proper motion—measured in degrees per year. Most stars are so distant that their proper motions are negligible. About 200 have proper motions of more than 1 arc-second a year—or 1 degree of angular movement in 3,600 years. Barnard's star (see p.381) has the fastest proper motion, moving at 10.3 arc-seconds per year. It takes 180 years to travel the diameter of a full moon in the sky. If astronomers know both the proper motion of a star and its distance, they can calculate its transverse velocity relative to Earth—that is, its velocity at right angles to the line of sight from Earth. The other component of a star's velocity relative to Earth is called its radial velocity (its velocity toward or away from Earth), measured by shifts in the star's spectrum (see p.35).

**THE BIG DIPPER IN 100,000 BCE**

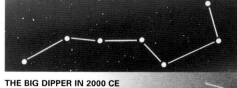

**THE BIG DIPPER IN 2000 CE**

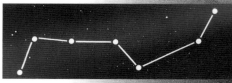

**THE BIG DIPPER IN 100,000 CE**

### CHANGING SHAPE
The shape of the star pattern known as the Big Dipper gradually changes due to the proper motions of its stars. Five stars are moving in unison as a group, but the two stars on the ends are moving independently.

# THE BRIGHTNESS OF STARS

A star's brightness in the sky depends on its distance from Earth and on its intrinsic brightness, which is related to its luminosity (the amount of energy it radiates into space per second, see p.233). To compare how stars would look if they were all at the same distance, astronomers use a measure of intrinsic brightness called the absolute magnitude scale. This scale uses high positive numbers to denote dim stars and negative numbers for the brightest ones. A star's brightness as seen from Earth, on the other hand, is described by its apparent magnitude. Again, the smaller the number of a star's apparent magnitude, the brighter the star. Stars with an apparent magnitude of +6 are only just detectable to the naked eye, whereas the apparent magnitude of most of the 50 brightest stars is between +2 and 0. The four brightest (including the brightest star of all, Sirius) have negative apparent magnitudes.

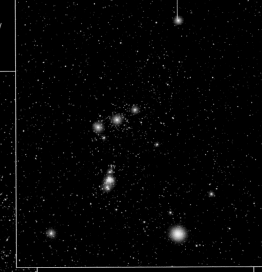

*Betelgeuse*  *Bellatrix*

### INTRINSICALLY BRIGHT STAR
The stars Betelgeuse and Bellatrix mark the shoulders of Orion. Betelgeuse is noticeably brighter (apparent magnitude 0.45) than Bellatrix (1.64), despite being twice as distant. It is a red, high-luminosity supergiant, whereas Bellatrix is a much less luminous giant.

*Alpha Centauri*  *Hadar (Beta Centauri)*

### NEARBY BRIGHT STAR
In the constellation Centaurus, the triple star system Alpha (α) Centauri is a little brighter (apparent magnitude –0.01) than the binary star Hadar, or Beta (β) Centauri (0.61). The reason for Alpha Centauri's brightness is its proximity—it is our closest stellar neighbor. The blue giant stars that make up Hadar are much more luminous than the stars in Alpha Centauri, but they are nearly 100 times farther away.

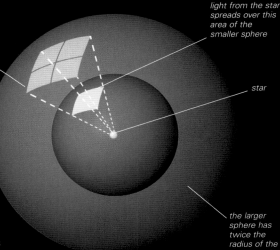

*when the light reaches the larger sphere, it is spread over four times the area (the square of the distance, or 2x2)*

*light from the star spreads over this area of the smaller sphere*

*star*

### THE INVERSE SQUARE RULE
The apparent brightness of a star drops in proportion to the square of its distance from the observer—a rule called the inverse square law. This happens because light energy is spread out over a progressively larger area as it travels away from the star.

*the larger sphere has twice the radius of the smaller sphere*

# CONSTELLATIONS

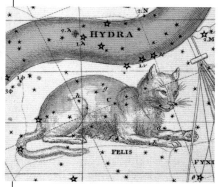

Since ancient times, people have seen imaginary shapes among groups of stars in the night sky. Using lines, they have joined the stars in these groups to form figures called constellations and named these constellations after the shapes they represent. Each constellation has a Latin name, which in most cases is either that of an animal, for example, Leo (the lion); an object, such as Crater (the cup); or a mythological character, such as Hercules. Some constellations, such as Orion (the Hunter), are easy to recognize; others, such as Pisces (the Fishes), are less distinct. Since 1930, an internationally agreed system has divided the celestial sphere into 88 irregular areas, each containing one of these figures. In fact, from an astronomical point of view, the word "constellation" is now applied to the area of the sky containing the figure rather than to the figure itself. All stars inside the boundaries of a constellation area belong to that constellation, even if they are not connected to the stars that produce the constellation figure. Within some constellations are some smaller, distinctive groups of stars known as asterisms; these include Orion's belt (a line of three bright stars in Orion) and the Big Dipper (a group of seven stars in the constellation Ursa Major). A few asterisms cut across constellation boundaries. For example, most of the "Square of Pegasus" asterism is in Pegasus, but one of its corners is in Andromeda.

**LOST CONSTELLATIONS**
Some constellations have proved short-lived. In the 19th century, Felis, the cat, was incorporated into what is now part of the constellation of Hydra. It appeared on several star charts but was not officially adopted.

**LINE-OF-SIGHT EFFECT**
A star pattern such as the Big Dipper in Ursa Major is a two-dimensional view of what may be a widely-scattered sample of stars. The stars might seem to lie in the same plane, but they are at different distances from Earth. If we could view the stars from elsewhere in space, they would form a totally different pattern.

**STAR CHART**
This star chart of Ursa Major (the Great Bear) shows the constellation figure (the pattern of lines joining bright stars) and labels many of the stars, as well as objects such as galaxies, lying within the constellation's boundaries.

constellation borders usually follow lines of right ascension and declination

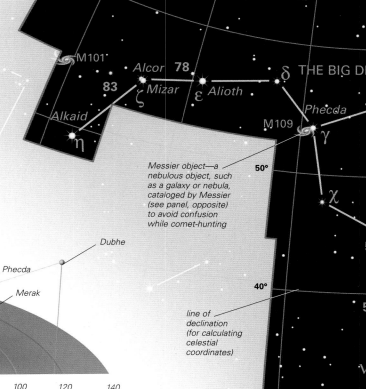

northern border of constellation

Messier object—a nebulous object, such as a galaxy or nebula, cataloged by Messier (see panel, opposite) to avoid confusion while comet-hunting

line of declination (for calculating celestial coordinates)

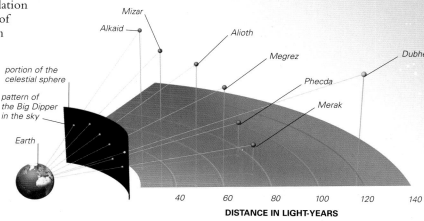

Mizar
Alkaid
Alioth
Megrez
Dubhe
Phecda
Merak

portion of the celestial sphere

pattern of the Big Dipper in the sky

Earth

**DISTANCE IN LIGHT-YEARS**
40   60   80   100   120   140

EXPLORING SPACE

## BAYER'S SYSTEM

Johann Bayer ascribed Greek letters to the stars in a constellation, roughly in order of decreasing brightness. Regulus, the brightest star in the constellation of Leo, was given the name Alpha ($\alpha$) Leonis, the second brightest (Denebola) was called Beta ($\beta$) Leonis, and so on. In some cases, Bayer used other ordering systems. The Big Dipper in Ursa Major is lettered by following the stars from west to east.

**BAYER'S MAP OF URSA MAJOR**
The seven stars of the Big Dipper can be seen in the upper left area of this chart from Bayer's *Uranometria*.

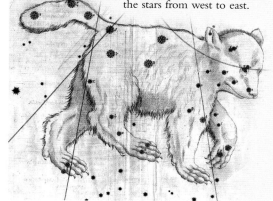

# NAMING THE STARS

Most of the brighter stars in the sky have ancient names of Babylonian, Greek, or Arabic origin. The name Sirius, for example, comes from a Greek word meaning "scorching." The first systematic naming of stars was introduced by Johann Bayer in 1603 (see panel, left, and p.347). Bayer distinguished up to 24 stars in each constellation by labeling them with Greek letters, after which he resorted to using Roman lowercase letters, a to z. In 1712, English astronomer John Flamsteed (1646–1719) introduced another system, in which stars are numbered in order of their right ascension (see p.63) from west to east across their constellation. Stars are usually named by linking their Bayer letter or Flamsteed number with the genitive form (possessive case) of the constellation name— so 56 Cygni denotes the star that is 56th closest to the western edge of the constellation Cygnus. Since the 18th century, numerous further catalogs have identified and numbered many more faint stars, and specialized systems have been devised for cataloging variable, binary, and multiple stars.

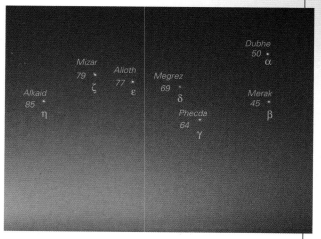

**SYSTEMS OF BAYER AND FLAMSTEED**
This photo of the Big Dipper in Ursa Major shows the ancient name of each star, its Bayer designation, and its Flamsteed number. For example, the star Alkaid can also be called Eta ($\eta$) Ursae Majoris (Bayer), or 85 Ursae Majoris (Flamsteed).

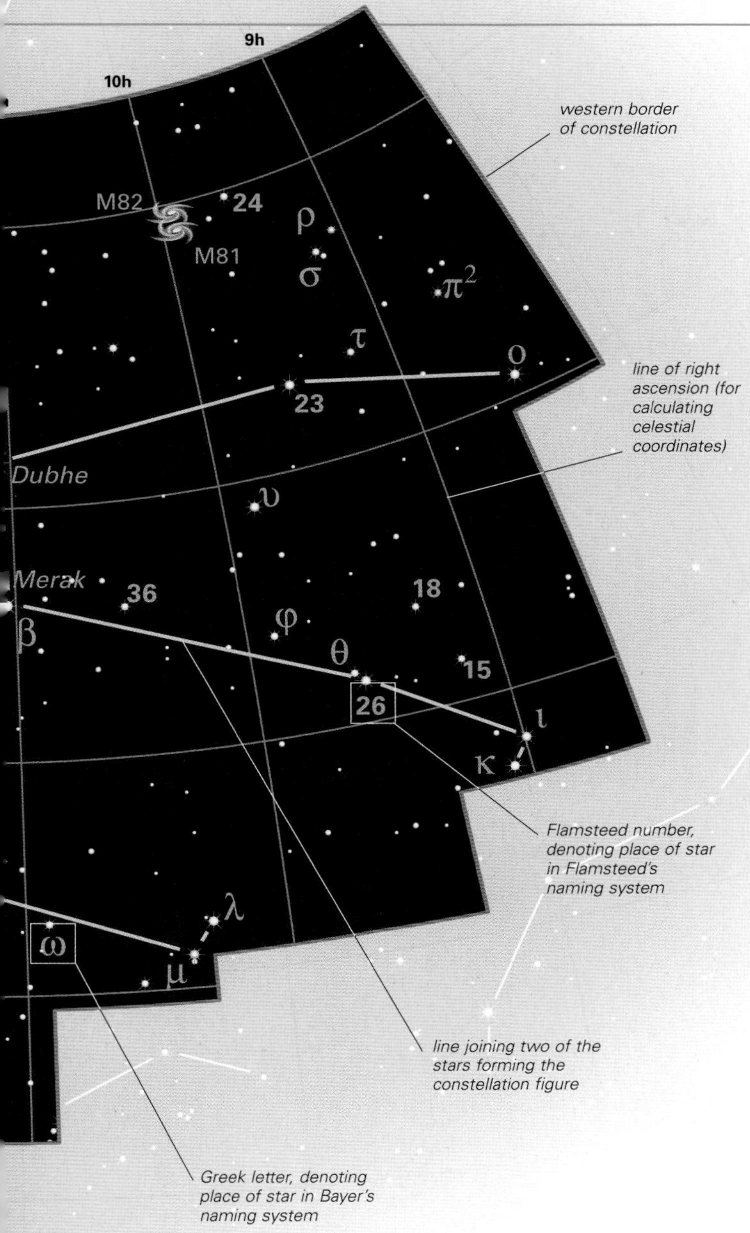

The star chart (left) shows the following labels:

10h, 9h

M82, 24, ρ

M81, σ, π²

τ, ο

23

Dubhe, υ

Merak, 36, φ, 18

β, θ

15

26, ι

κ

λ

ω, μ

*western border of constellation*

*line of right ascension (for calculating celestial coordinates)*

*Flamsteed number, denoting place of star in Flamsteed's naming system*

*line joining two of the stars forming the constellation figure*

*Greek letter, denoting place of star in Bayer's naming system*

# CATALOGS OF NEBULOUS OBJECTS

Besides individual stars, various other types of object, such as star clusters, nebulae, and galaxies, have practically fixed positions on the celestial sphere. Most of these objects appear as no more than hazy blurs in the sky, even through a telescope. The first person to catalog such objects was a French astronomer, Charles Messier (see panel, below), in the 18th century. He compiled a list of 110 hazy objects, though none of these are from the southern polar skies—that is because Messier carried out his observations from Paris, and anything in declination below 40°S was below his horizon. In 1888, a much larger catalog called the *New General Catalog of Nebulae and Star Clusters* (NGC) was published, and this was later expanded by what is called the *Index Catalog* (IC). To this day, the NGC and IC are important catalogs of nebulae, star clusters, and galaxies. Their current versions cover the entire sky and provide data on more than 13,000 objects, all identified by NGC or IC numbers. In addition, several hundred specialized astronomical catalogs are in use, covering different types of objects, parts of the sky, and regions of the electromagnetic spectrum. Most catalogs are now maintained as computer databases accessible over the Internet.

NGC 2841, A SPIRAL GALAXY

NGC 3079, A SPIRAL GALAXY VIEWED EDGE-ON

## NEW GENERAL CATALOG

More than 150 New General Catalog (NGC) objects lie within the constellation Ursa Major. Two are shown here, both spiral galaxies in a region around the Great Bear's forelegs, not far from Theta (θ) Ursae Majoris. NGC 2841 has delicate, tightly wound arms, within which astronomers have recorded many supernovae explosions. NGC 3079 has an active central region, from which rises a lumpy bubble of hot gas, 3,500 light-years wide, driven by star formation.

## CHARLES MESSIER

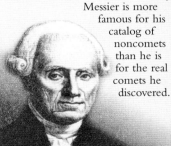

The French comet-hunter Charles Messier (1730–1817) compiled a catalog of 110 nebulous-looking objects in the sky that could be mistaken for comets. Not all of them were discovered by himself— many were spotted by another Frenchman, Pierre Méchain, and yet others had been found years earlier by astronomers such as Edmond Halley. Messier's first true discovery was M3, a globular star cluster in Canes Venatici. Ironically, Messier is more famous for his catalog of noncomets than he is for the real comets he discovered.

## THE MESSIER CATALOG

Messier's catalog includes 57 star clusters, 40 galaxies, 1 supernova remnant (the Crab Nebula), 4 planetary nebulas, 7 diffuse nebulas, and 1 double star. Of these Messier objects, 8 lie in the constellation of Ursa Major, of which 5 are shown here. Each is denoted by the latter M followed by a number. The planetary nebula M97 is also called the Owl Nebula. Galaxies M81 and M82 are neighbors in the sky and can be viewed simultaneously with a good pair of binoculars. M109 lies close to the star Phecda—Gamma (γ) Ursae Majoris—in the Big Dipper.

M81, A SPIRAL GALAXY (SEE P.302)

M82, AN IRREGULAR GALAXY (SEE P.304)

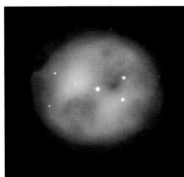
M97, A PLANETARY NEBULA

M108, A SPIRAL GALAXY

M109, A BARRED SPIRAL GALAXY

# LIGHTS IN THE SKY

34–37 Radiation
64–67 Celestial cycles
Naked-eye astronomy 76–77
Earth's atmosphere and weather 127
Comets 214–215
Meteors and meteorites 220–221

AS WELL AS STARS, GALAXIES, NEBULAE, and solar system objects, other phenomena can cause lights to appear in the night sky. Mainly, these originate in light or particles of matter reaching Earth in various indirect ways from the Sun, but some are generated by Earth-bound processes. Amateur stargazers need to be aware of these sources of nocturnal light to avoid confusion with astronomical phenomena.

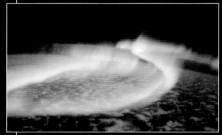

**AURORA FROM THE SPACE SHUTTLE**
This photograph of the aurora australis was taken from the Space Shuttle Discovery during a 1991 mission. A study of the aurora's features was one of the mission tasks.

## AURORAE

The aurora borealis (northern lights) and aurora australis (southern lights) appear when charged particles from the Sun, carried to Earth in the solar wind (see pp.106–107), become trapped by Earth's magnetic field. They are then accelerated into regions above the north and south magnetic poles, where they excite particles of gas in the upper atmosphere, 60–250 miles (100–400 km) above Earth's surface. The appearance and location of aurorae change in response to the solar wind. They are most often visible at high latitudes, toward Earth's magnetic poles, but may be seen at lower latitudes during disturbances in the solar wind, such as after mass ejections from the Sun (see pp.106–107).

**AURORA BOREALIS**
A colorful display of the northern lights is visible here over silhouetted trees near Fairbanks, Alaska. The colors stem from light emission by different atmospheric gases.

## ICE HALOES

Atmospheric haloes are caused by ice crystals high in Earth's atmosphere refracting light. Light either from the Sun or the Moon (that is, reflected sunlight) can cause haloes. The most common halo is a circle of light with a radius of 22° around the Moon or Sun. Also present may be splashes of light, called moon dogs or sun dogs (parhelia), arcs, and circles of light that seem to pass through the Sun or Moon. All these phenomena result from the identical angles between the faces of atmospheric ice crystals. Even if the crystals are not all aligned, they tend to deflect light in some directions more strongly than in others.

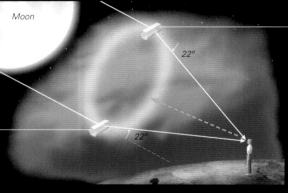

Moon
ice crystal in layer of cirrostratus cloud
22°
crystal's faces act as prism
22°

**OBSERVING A 22° HALO**
This halo is formed when ice crystals in the atmosphere refract light from the Moon to the observer on Earth by an angle of 22°. A light ray is refracted through this angle as it passes through two faces of an ice crystal.

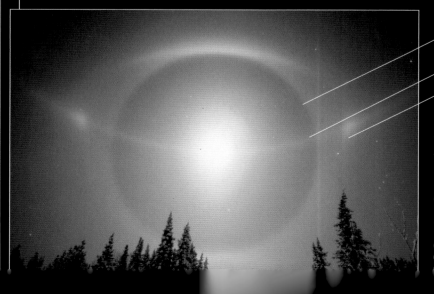

halo
paraselenic circle
moon dog

**HALO AND MOON DOGS**
This photograph taken in Arctic Canada shows several refraction phenomena. The patches of light on either side of the Moon, called moon dogs, are caused by horizontal ice crystals in the atmosphere refracting light. The band of light running through the moon dogs is called a paraselenic circle. Also visible is a circular 22° halo.

## ZODIACAL LIGHT

A faint glow is sometimes visible in the eastern sky before dawn or occasionally in the west after sunset. Called zodiacal light, it is caused by sunlight scattered off interplanetary dust particles in the plane of the solar system—the ecliptic plane (see p. 64). The mixture of wavelengths in the light is the same as that in the Sun's spectrum. A related phenomenon is called the gegenschein (German for "counterglow"). It is sometimes perceivable on a dark night, far from any light pollution, as a spot on the celestial sphere directly opposite the Sun's position in the sky. The dust particles in space responsible for both zodiacal light and gegenschein are thought to be from asteroid collisions and comets and have diameters of about 0.04 in (1 mm).

**THE GEGENSCHEIN**
This faint, circular glow, 10° across, is most often spotted at midnight, in an area above the southern horizon (for Northern Hemisphere viewers).

**SEEING THE ZODIACAL LIGHT**
The zodiacal light is most distinct just before dawn in fall, far from any light pollution. It is near the horizon and forms a rough triangle.

## NOCTILUCENT CLOUDS

Clouds at extremely high altitude (around 50 miles/ 80 km up) in Earth's atmosphere can shine at night by reflecting sunlight long after the Sun has set. These "noctilucent" (night-shining) clouds are seen after sunset or before dawn. It is thought that they consist of small, ice-coated particles that reflect sunlight. Noctilucent clouds are most often seen between latitudes 50° and 65° north and south, from May to August in northern latitudes and November to February in southern latitudes. They may also form at other latitudes and times of year.

**SHINING CLOUDS**
Noctilucent clouds are silvery-blue and usually appear as interwoven streaks. They are only ever seen against a partly lit sky background, the clouds occupying a sunlit portion of Earth's atmosphere.

## MOVING LIGHTS AND FLASHES

Many phenomena can cause moving lights and flashes across the sky. Rapid streaks of light are likely to be meteors or shooting stars—that is, dust particles entering and burning up in the atmosphere. A bigger, but very rare variant is a fireball—simply a larger meteor burning up. Slower-moving, steady, or flashing lights are more likely to be aircraft, satellites, or orbiting spacecraft. Large light flashes are usually electrical discharges connected with lightning storms. In recent years, meteorologists have named two new types of lightning: "red sprites" and "blue jets." Both are electrical discharges between the tops of thunderclouds and the ionosphere above.

### UFO SIGHTINGS

Every year there are reports of unidentified flying objects (UFOs). Most of these can be accounted for by natural phenomena such as brights stars, planets, and meteors, or by man-made objects such as balloons, satellites, and aircraft. After excluding such causes, there are still unexplained cases. It would be unscientific to dismiss the possibility that these UFOs are signs of extraterrestrial visitors without further investigation—just as it would be to accept it before ruling out less exotic explanations.

**FLYING SAUCER?**
This object, suggestive of a flying saucer, is actually a lenticular cloud. Clouds like this are usually formed by vertical air movements around the sides or summits of mountains.

**BLUE JETS**
These cone-shaped discharges are 30–35 miles (50–60 km) high, 6 miles (10 km) wide at the top, and result from lightning in the atmosphere ionizing nitrogen atoms, causing them to glow blue as they reemit light. In the past, blue jets may

**PATH OF THE ISS**
As the International Space Station (ISS) orbits Earth, it is visible from the ground because it reflects sunlight. This photograph of the Space Station was taken using a 60-second camera exposure, which indicates how quickly the spacecraft moves across the night sky.

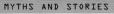

# NAKED-EYE ASTRONOMY

OPTICAL INSTRUMENTS ARE NOT NECESSARY to gain a foothold in astronomy—our ancestors did without them for thousands of years. Today's naked-eye observer, equipped with a little foreknowledge and some basic equipment, can still appreciate the constellations, observe the brightest deep-sky objects, and trace the paths of the Moon and planets in the night sky.

## PREPARING TO STARGAZE

To get the most from stargazing, some preparation is needed. The human eye takes some 20 minutes to adjust to darkness and, as the pupil opens, more detail and fainter objects become visible. Look at a planisphere or monthly sky chart (see pp.426–501) to see what is currently in the sky. A good location is one shielded from street lights and ideally away from their indirect glow. Try to avoid all artificial light—if necessary, use a flashlight with a red filter. Keep a notebook or a prepared report form to record observations, especially if looking for particular phenomena, such as meteors. To see faint stars and deep-sky objects, avoid nights when a bright Moon washes out the sky. Even on a dark, cloudless night, air turbulence can affect the observing quality or "seeing"—the best nights are often those that do not suddenly get colder at sunset.

**SEEING AND TWINKLE**
Variable "seeing" is caused by warm air currents rising from the ground at nightfall. These telescope images of Jupiter show the range of seeing from poor to fine, but seeing also limits the visibility of stars with the naked eye and determines the amount of "twinkle."

**LIGHT POLLUTION**
This composite satellite image shows the extent of artificial lighting on Earth. In industrialized regions, it is almost impossible to find truly dark skies.

**GOOD STREET LIGHTING**
In some countries, non-essential street lights are switched off late at night. Elsewhere, shades are installed to project all the light downward, preventing it from leaking into the sky. Such measures can increase the light on the street, save energy, and preserve the night sky for stargazers.

**PLANISPHERE**
A planisphere is a useful tool for any amateur astronomer. The user rotates the disks so that the time and date markers on the edge match up correctly, and the window reveals a map of the sky at that moment. A single planisphere is useful only for a limited range of latitudes, so be sure to get one with the correct settings.

**THE MOON AND VENUS**
Solar system objects such as the Moon and Venus can be spectacular sights even with the unaided eye. This beautiful twilight pairing was photographed in January 2004.

# MEASUREMENTS ON THE SKY

Distances between objects in the sky are often expressed as degrees of angle. All the way around the horizon measures 360°, while the angle from horizon to zenith (the point directly overhead) is 90°. The Sun and Moon both have an angular diameter of 0.5°, while an outstretched hand can be used to estimate other distances. When studying star charts, bear in mind that 1 hour of right ascension (RA) along the celestial equator is equivalent to 15° of declination (see p.63), but right ascension circles get tighter toward the celestial poles, so at 60°N, an hour's difference in RA is equivalent to only 7.5° of declination.

**FINGER WIDTH**
Held out at arm's length, a typical adult index finger blocks out roughly 1 degree of the sky—enough to cover the Moon twice over.

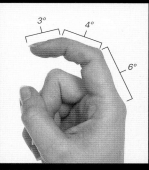

**FINGER JOINTS**
The finger joints provide measures for distances of a few degrees. A side-on fingertip is about 3° wide, the second joint 4°, and the third 6°.

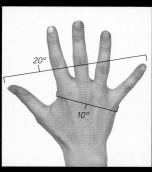

**HAND SPANS**
The hand (not including the thumb) is about 10° across at arm's length, while a stretched hand-span covers 20° of sky.

# STAR-HOPPING

The best way to learn the layout of the night sky is to first find a few bright stars and constellations, then work outward into more obscure areas. Two key regions are the Big Dipper (the brightest seven stars in the constellation Ursa Major, close to the north celestial pole) and the area around the brilliant constellation Orion, including the Winter Triangle (see p.436) on the celestial equator. By following lines between certain stars in these constellations, one can find other stars and begin to learn the sky's overall layout. The Big Dipper is a useful pointer, because two of its stars align with Polaris, the star that marks the north celestial pole. Because the sky seems to revolve around the celestial poles, Polaris is the one fixed point in the northern sky. (There is no bright south Pole Star.) Other useful keystones are the Summer Triangle (see p.466)—comprising the northern stars Vega, Deneb, and Altair—and the Southern Cross (see p.437) and False Cross (see p.443) in the far south.

**STAR HOPS FROM THE BIG DIPPER**
A line through Dubhe and Merak along one side of the Big Dipper points straight to Polaris in one direction and (allowing for the curvature of the sky) toward the bright star Regulus in Leo in the other direction. Following the curve of the Big Dipper's handle, meanwhile, leads to the bright red star Arcturus in Boötes and eventually to Spica in Virgo.

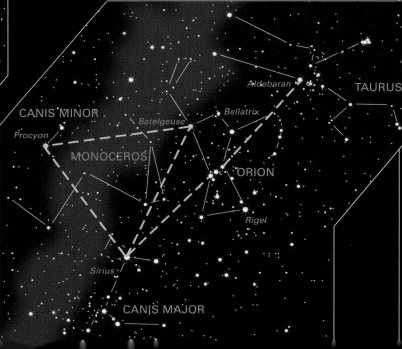

**ORION'S BELT AND THE WINTER TRIANGLE**
The distinctive line of three bright stars forming Orion's belt points in one direction toward the red giant Aldebaran in Taurus and in the other toward Sirius, the brightest star in the sky, in Canis Major. Sirius, Betelgeuse (on Orion's shoulder), and Procyon (in Canis Minor) make up the equilateral Winter Triangle.

**THE MILKY WAY**
The starry band of the Milky Way arches
over the snow-covered cliffs of the Creux
du Van near Neuchâtel, Switzerland, in a
spectacular wide-angle view. The Milky Way is
the plane of our galaxy seen from within—a
mass of distant stars interspersed with dusty,
concealing nebulae and pink patches of
glowing gas where new stars are being
born to join the existing billions.

# BINOCULAR ASTRONOMY

FOR MOST NEWCOMERS to astronomy, the most useful piece of equipment is a pair of binoculars. As well as being easy and comfortable to use, binoculars (unlike telescopes) allow stargazers to see images the right way up. A range of fascinating astronomical objects can be observed through them.

## BINOCULAR CHARACTERISTICS

Binoculars are like a combination of two low-powered telescopes. The two main designs, called Porro-prism and roof-prism, differ in their optics, but either can be useful for astronomy. More important when choosing binoculars are the two main numbers describing their optical qualities; for example, 7×50 or 12×70. The first figure is the magnification. For a beginner, a magnification of 7× or 10× is usually adequate—with a higher magnification, it can be difficult to locate objects in the sky. The second figure is the aperture, or diameter of the objective lenses, measured in millimeters. This number expresses the binoculars' light-gathering power, which is important in seeing faint objects. For night-sky viewing, an aperture of at least 50 mm (2 in) is preferable.

eyepiece

eyepiece focusing ring

prism

main focus ring

objective lens

light enters

**STANDARD BINOCULARS**
These typically have 50-mm (2-in) objective lenses and a magnification of 7× or 10×. This pair has a Porro-prism design.

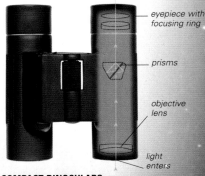

eyepiece with focusing ring

prisms

objective lens

light enters

**COMPACT BINOCULARS**
These are lightweight but their objective lenses are rather small for astronomy. This pair has a roof-prism design.

### BINOCULAR FINDS

Astronomers make some important discoveries using binoculars. Arizona astronomer Peter Collins uses binoculars to search for the stellar outbursts known as novae (see p.282). To make the method effective, he memorizes thousands of star positions. Comets are also frequently first seen by binocular enthusiasts. Japanese astronomer Hyakutake Yuji spotted Comet Hyakutake (see p.215) in 1996 using a pair of giant (25 × 100 mm) binoculars.

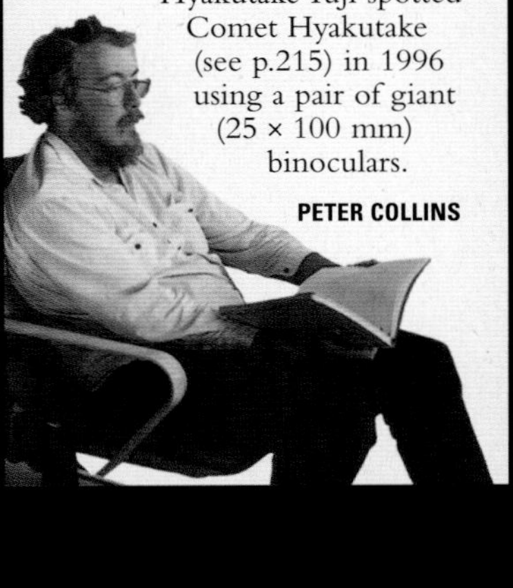

**PETER COLLINS**

**IDYLLIC SKYGAZING**
The modest magnifying power of binoculars is more than enough to reveal many of the sky's most interesting objects. Wilderness camping is a good way to get away from light pollution.

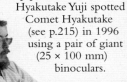

INTRODUCTION

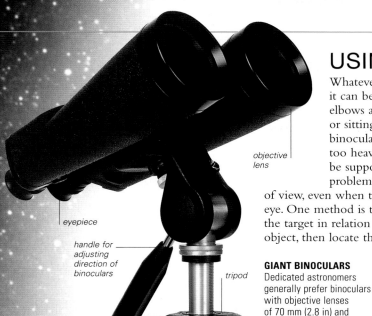

eyepiece

handle for
adjusting
direction of
binoculars

tripod

objective
lens

**GIANT BINOCULARS**
Dedicated astronomers
generally prefer binoculars
with objective lenses
of 70 mm (2.8 in) and
magnifications of 15–20×.

# USING BINOCULARS

Whatever size of binoculars astronomers choose,
it can be difficult to keep them steady. Placing
elbows against something solid, such as a wall,
or sitting down in a lawn chair, can help stop the
binoculars from wobbling. Giant binoculars are
too heavy to hold steady in the hands, so should
be supported on a tripod. Another common
problem is finding the target object in the field
of view, even when the object is visible to the naked
eye. One method is to establish the position of
the target in relation to an easier-to-locate
object, then locate the easier object and
finally navigate to
the target object.
Alternatively, work
upward from a
recognizable feature
on the horizon.

**KEEPING YOUR
BINOCULARS STEADY**
Sitting and placing
the elbows on the
knees can support
the weight of binoculars
and keep them steady.

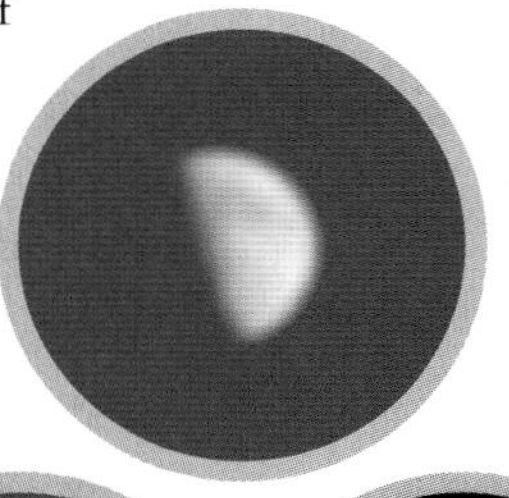

**HOW TO FOCUS A PAIR
OF BINOCULARS**
A pair of binoculars is not
immediately in perfect focus
for every user, since users'
eyesight differs. To fix this,
follow the instructions below.

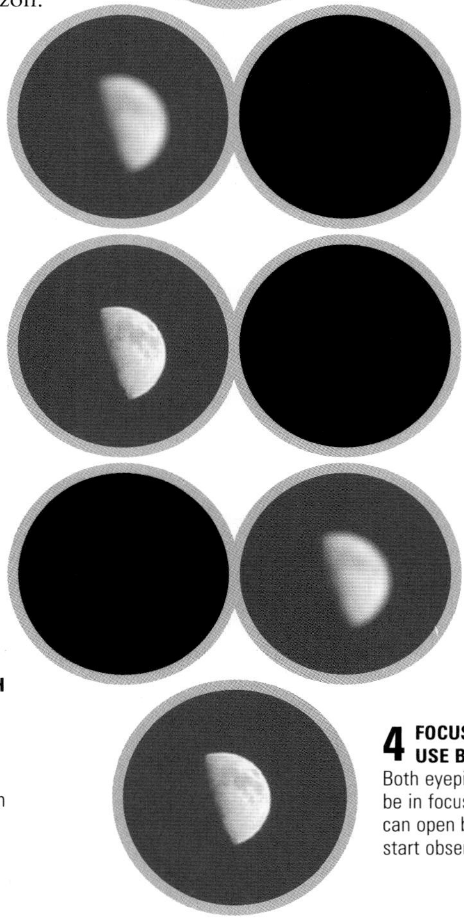

**1 IDENTIFY
FOCUSING RING**
Find which eyepiece
can be rotated to
focus independently
(usually the right).
Look through with
your eye closed on
that side.

**2 FOCUS LEFT
EYEPIECE**
Rotate the binoculars'
main, central focusing
ring, which moves
both eyepieces, until
the left-eyepiece
image comes into
sharp focus.

**3 CLOSE LEFT EYE,
OPEN RIGHT EYE**
Now open only the
other eye (in this
example, the right),
and use the eyepiece
focusing ring to bring
the image into focus.

**4 FOCUS AND THEN
USE BOTH EYES**
Both eyepieces should now
be in focus, so now you
can open both eyes and
start observing.

# BINOCULAR FIELD OF VIEW

The size of the circular area of sky seen through binoculars is called the field
of view and is usually expressed as an angle. The field of view is closely related
to magnification—the higher the magnification, the smaller the field of view.
A typical field of view of a pair of medium-power binoculars (10×) is 6–8°.
This offers a good compromise between adequate magnification and a field
of view wide enough to see most of a large object such
as the Andromeda Galaxy (see pp.312–13). For
viewing larger areas still, lower-power binoculars
(5–7×), with a field of view of at least 9°, are
more suitable. Conversely, for looking at more
compact objects, such as Jupiter
and its moons, binoculars
with higher
magnification, and
a field of view
of 3° or even
less, are better
to use.

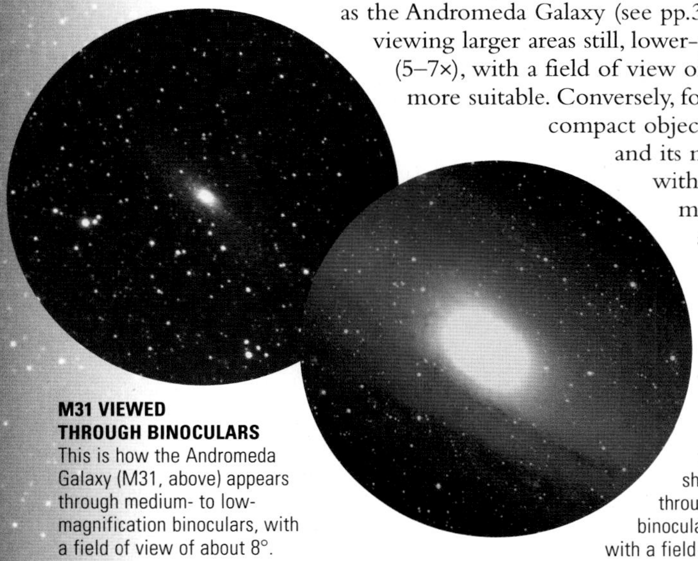

**M31 VIEWED
THROUGH BINOCULARS**
This is how the Andromeda
Galaxy (M31, above) appears
through medium- to low-
magnification binoculars, with
a field of view of about 8°.

**M31 VIEWED THROUGH
A TELESCOPE**
Here the central part of
the Andromeda Galaxy is
shown as you might see it
through very-high-magnification
binoculars, or a small telescope,
with a field of view of about 1.5°.

# BINOCULAR OBJECTS

A striking first object for a novice
binocular user is the Orion Nebula
(see p.241). Other choices might be
the Andromeda Galaxy (above), and
the fabulous star clouds and nebulae
in the Sagittarius and Scorpius
regions of the Milky Way, including
the Lagoon Nebula (see p.243). For
viewers south of 50°N, an excellent
binocular object is the Omega Centauri
star cluster (see p.294). To find these,
all that is needed is some star charts
(see pp.426–501) or an astronomy
app on your phone. Also try
observing the Moon, Jupiter and
its moons, and the phases of Venus.

**THE MILKY WAY**
Shown here is a dense region of the Milky
Way in Sagittarius, as seen through low-
power binoculars with a field of view of 12°.

**ORION NEBULA**
This appears as a blue-green smudge in
Orion, shown here as it appears in medium-
power binoculars with a field of view of 8°.

**THE PLEIADES**
This spectacular star cluster in Taurus is
seen here as it appears through high-power
binoculars with a field of view of about 3°.

INTRODUCTION

# TELESCOPE ASTRONOMY

TELESCOPES ARE THE ULTIMATE optical instruments for astronomy. The simplest spyglass type has changed comparatively little over the centuries, but the most sophisticated amateur instruments now offer the optics and computerized controls once the preserve of professionals.

## EARLY TELESCOPES

The invention of the telescope is usually credited to a Dutch optician named Hans Lippershey (1570–1619). In 1608, Lippershey found that a certain combination of optical lenses mounted at either end of a tube magnified an image—the basis of the refracting telescope. News of this device spread across Europe, and a year later the Italian scientist Galileo Galilei (1564–1642) built telescopes that could magnify up to 30 times. His subsequent observations of the Moon, Sun, and stars helped establish the heliocentric (Sun-centered) theory of the universe proposed by Copernicus (see p.69). In 1668, Isaac Newton developed the reflecting telescope, which used mirrors instead of lenses. There were many advantages: they did not have the optical defects that the refracting instruments had, tubes were shorter, and they could be made with larger apertures. However, the early mirrors were made of metal and tarnished, so they did not catch on initially.

**GALILEO'S SKETCH OF THE MOON**
The Italian astronomer used his telescopes to observe the craters, mountains, and dark lowland areas of the Moon.

*eyepiece lens magnifies image 35 times*

**NEWTON'S TELESCOPE**
Isaac Newton developed this first reflecting telescope, and his design is still used. A mirror at one end of the tube "bounces" light toward the eyepiece at the other end.

*upper tube covered with vellum*

*lower tube made of layers of paper and cardboard*

*screw that holds main mirror in position*

*sphere rotates to point telescope tube in different directions*

**EXPLORING SPACE**

### BEFORE THE TELESCOPE

In the days before telescopes, astronomers used a variety of instruments for measuring the positions of celestial objects. Tycho Brahe (1546–1601), a Danish nobleman, built his own observatory and equipped it with the finest instruments, which included a huge wall-mounted quadrant. German mathematician Johannes Kepler used Brahe's measurements of planetary positions when calculating his laws of planetary motion (see p.68).

## TELESCOPE DESIGNS

A telescope's function is to collect light from distant objects, bring it to a focus, and then magnify it. There are two basic ways of doing this, using either a lens or a concave mirror. A lens refracts, or bends, the light passing through it, directing it to a focal point behind it. A curved mirror reflects light rays back onto converging paths that come to a focus somewhere in front of it. A combination design called a catadioptric telescope is basically a reflector with a thin lens across the front of the tube. Light rays entering a telescope from astronomical objects are near parallel. Once the captured light rays have passed the focus, they begin to diverge again, at which point they are captured by an eyepiece, which returns the rays to parallel directions, magnifying them in the process. Because light rays entering the eyepiece have crossed over as they pass through the focus, the image is usually inverted, which is not generally regarded as a drawback when viewing astronomical objects.

*light enters tube*

*correcting lens*

*finder*

*concave primary mirror*

*hole in primary mirror for light to pass through*

*eyepiece*

*convex secondary mirror*

*equatorial "wedge" mount*

**REFRACTING TELESCOPE**
These telescopes are tubes with a lens, known as the objective, at one end. The lens focuses the incoming light down the tube into an eyepiece at the other end.

*piggyback finderscope*

*refracted light*

*objective lens*

*light enters tube*

*altazimuth fork mount*

*focused light*

*90° eyepiece—a sliding tube allows it to move in and out to focus*

*light enters tube*

*secondary mirror*

*reflected light*

*eyepiece*

*focused light*

**REFLECTING TELESCOPE**
With this design, light falls onto a primary mirror at the base of an open-ended tube. From there it is reflected back up the tube onto a smaller flat mirror, which diverts it into an eyepiece on the side.

*convex primary mirror*

**CATADIOPTRIC TELESCOPE**
In this compact reflector design, a convex secondary mirror directs light to the eyepiece through a hole in the primary mirror. By bouncing the light back on itself, the length of the telescope tube is reduced.

# TELESCOPE MOUNTS

The way a telescope is mounted can greatly affect its performance. The two most common types of mount are the altazimuth and the equatorial. The altazimuth mount allows the instrument to pivot in altitude (up and down) and azimuth (parallel to the horizon). The equatorial mount aligns the telescope's movement with Earth's axis of rotation, so that it can follow the lines of right ascension and declination in the sky (see p.63). Altazimuth mountings are simple to set up, but because objects in the sky are constantly changing their altitude and azimuth, tracking objects requires continued adjustment of both. Equatorial mounts are heavier and take longer to set up but, once aligned to a celestial pole, the observer can follow objects across the sky by turning a single axis.

## ALTAZIMUTH MOUNT
This type of mount is usually light and compact. However, both axes of the telescope must be moved at the same time to track a celestial object—and the higher the telescope's magnification, the faster the object will drift out of the field of view.

movement in altitude
movement in azimuth

movement in right ascension
movement in declination

## EQUATORIAL MOUNT
These mounts are more awkward to set up, but once that is done the observer can track objects just by turning the polar axis. Many equatorial mounts have electric or battery-controlled drive motors that allow for hands-free operation.

## ALTAZIMUTH VARIATIONS
There are two variants of the altazimuth mount. Fork mounts are often used for catadioptric telescopes. Dobsonians are good for large reflectors with wide fields of view and low magnifications.

**FORK MOUNT**          **DOBSONIAN MOUNT**

# APERTURE AND MAGNIFICATION

Two major factors affect an image in a telescope eyepiece—aperture and magnification. The aperture is the diameter of the telescope's primary mirror or objective lens and affects the amount of light it can collect—called its "light grasp." Doubling the aperture quadruples the light grasp. Magnification is dictated by the specification of the telescope's eyepiece. The power of the eyepiece is identified by its focal length—the distance at which it focuses parallel rays of light. The shorter the focal length, the greater the magnification. Objective lenses and primary mirrors also have a focal length, and dividing this measurement by that of the eyepiece gives the combined magnification. An eyepiece can be changed to alter the magnification to suit the observed object.

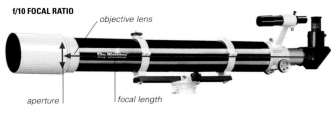

## OBJECTIVE SIZES
The most important specification of a telescope is the diameter of its objective lens. This affects how much light can enter the tube.

**120mm APERTURE**          **66mm APERTURE**

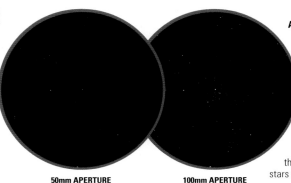
**50mm APERTURE**          **100mm APERTURE**

## APERTURE
These are photographs of the open cluster M35. The image far left was taken through a telescope with a 2 in (50 mm) objective lens; the second image, left, was taken through a 4 in (100 mm) lens. The larger lens has a light grasp four times greater than the smaller one, so the fainter stars can be seen more clearly.

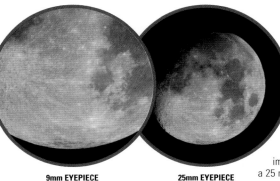
**9mm EYEPIECE**          **25mm EYEPIECE**

## MAGNIFICATION
The shorter the focal length of an eyepiece, the higher its magnifying power but also the smaller its field of view. This can be seen clearly in these two photographs of the Moon. The image far left was taken through a 9 mm eyepiece; the second image was taken through a 25 mm eyepiece.

**f/10 FOCAL RATIO**
objective lens
aperture          focal length

**f/5 FOCAL RATIO**

## VARIATIONS IN FOCAL RATIO
Telescopes with a large focal ratio, such as f/10, above, produce larger images but have smaller fields of view than telescopes with lower focal ratios.

# FOCAL LENGTH AND FOCAL RATIO

After the aperture, the next most important specification of a telescope is its focal length. This is the distance from its primary lens or mirror to the point where the rays of light meet—the focal point. A telescope with a long focal length produces a large but faint image at its focal point, whereas one with a shorter focal length gives a smaller but brighter image. It is easier to make mirrors with short focal lengths than it is lenses, so reflecting telescopes can have shorter tubes for a given aperture. Dividing the focal length of the primary mirror or lens (usually given in millimeters) by the telescope's aperture (also in millimeters) will give its focal ratio, called its "f" number. This ratio can influence the type of celestial object observed. Telescopes with a low focal ratio, around f/5, are best for imaging diffuse objects, such as nebulae or galaxies; those with a focal ratio above f/9 are useful for studying brighter objects, such as the Moon or the planets.

# TELESCOPES FOR BEGINNERS

Choosing a telescope to suit your needs and experience can make a difference to your viewing. Many people start with a basic telescope, possibly on an altazimuth mount (see p.83), such as a Dobsonian, then learn how to find objects in the sky. Others opt for a more advanced computerized go-to telescope that will find celestial objects at the press of a button. These are invaluable for locating hard-to-find objects, and most include a sky-tour that will show the highlights visible at any given time and provide background information. Once an object has been located, the instrument will track it automatically for as long as required. Go-to telescopes may have altazimuth or equatorial mounts—the former are fine for visual observing, the latter are essential for long-exposure photography. Consider, too, whether you need your telescope to be portable. In general, the larger the telescope's aperture the better, but there is no point in having an instrument that you rarely use because it is too big and cumbersome to set up. Most instruments perform well on all subjects. Refracting telescopes tend to be better suited to use in towns, where light pollution can be a problem, while country sites favor reflecting telescopes.

# FINDERS

A telescope's field of view is small, even if using the lowest magnification eyepiece. It is typically only 1°, which is just twice the size of the full Moon in the sky. Simply aiming the telescope can be hit-or-miss. A finder, which is a small refracting telescope that sits on the side of the main instrument, helps you aim your telescope with much greater precision. Almost all telescopes require a finder to help locate objects, or for go-to telescopes to set them up in the first place. There are two main types: finderscopes and red dot finders. Optical finderscopes are useful where there is light pollution because they can show stars not otherwise visible, although red dot finders can be easier to use. Both are mounted on the telescope tube in such a way that they can be adjusted to match the aiming point of the main instrument (see below right). Align the finder by looking at a distant fixed object, ideally in daylight. Never use the Sun, which could blind you. Switch the finder off after using it to avoid a dead battery later.

*finderscope*

*star diagonal rotates image through 90° to give a more comfortable observing position*

*German-type equatorial mount with motor drive for tracking moving objects*

*adjustments to mount enable it to be used at any latitude*

### GO-TO TELESCOPE
This equatorially mounted Schmidt–Cassegrain telescope is a typical computerized instrument. It has a handset for entering the details of target objects and a hand-held controller for adjustments in right ascension and declination. It can also interface with a computer.

*handset for choosing target objects and adjusting telescope position in right ascension and declination*

### FINDERSCOPE
A finderscope magnifies the night sky and gives a field of view of around 5–8°. A crosshair helps center the target in the finderscope. The image through the finderscope is inverted, which can, at first, make finding objects frustrating. Most entry-level telescopes come with a basic finder, but it may be worth upgrading as you progress.

*eyepiece*    *mounting bracket*

**FINDERSCOPE VIEW**

### RED DOT FINDER
A red dot finder indicates where the telescope is pointing by projecting a small red dot onto a piece of transparent glass or plastic. The wider sky remains visible, making it intuitive to use. The brightness of the dot can sometimes be adjusted with a built-in dimmer switch.

*sight*

*alignment adjustment wheel*    *mounting bracket*

**RED DOT FINDER VIEW**

**USING A FINDER**

**1 FIND AN OBJECT**
To align any finder, first select a distant and fixed object in the main instrument using its lowest magnification eyepiece, then center it within the field of view. Clamp the telescope's position.

**2 ALIGN THE FINDER**
Using the adjusters on the finder, bring the same object into the center of the finder's crosshairs or red dot (see far left). You may need to repeat this each time the finder is removed and replaced.

# EYEPIECES

Most telescopes are supplied with one or two eyepieces: one that gives a basic low magnification, or power; and the other providing a higher power. To increase magnification further you need additional eyepieces, but there is also a limit to the power that any telescope can tolerate, often given as twice its aperture in millimeters. For example, the limit of a 130 mm telescope is 260. As the power is increased, the field of view usually decreases, the image dims, any atmospheric turbulence (called the "seeing") is emphasized, and it becomes harder to keep objects within the field of view. One way to increase the power of a set of eyepieces is to place a Barlow lens between the telescope and the eyepiece. This lens typically doubles the power of each eyepiece, giving you a wider range of magnifications from a small set of eyepieces.

**40mm  25mm  9mm  2X BARLOW LENS**

**EYEPIECES**
Telescope eyepieces are available in a range of focal lengths, with the highest figure giving the lowest magnification. The optical design varies, some combining a very wide apparent field of view with a high power.

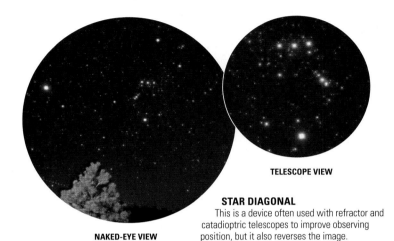

**NAKED-EYE VIEW**

**TELESCOPE VIEW**

**STAR DIAGONAL**
This is a device often used with refractor and catadioptric telescopes to improve observing position, but it also reverses the image.

**ANTI-LIGHT-POLLUTION FILTER**

**FILTERING OUT LIGHT POLLUTION**
Sodium street lamps emit yellow light with a narrow range of wavelengths, making the sky glow orange (above). A light-pollution filter can cut it out while leaving the light from distant stars unaffected (right).

# SOLAR TELESCOPES

The Sun is a fascinating object to observe with constantly changing features, but it is also the most dangerous, because it is so bright that even a momentary view through a telescope can blind the viewer. Specialized filters are available that reduce the brightness of the incoming light, which must be done because light enters the tube rather than at the eyepiece, where the light is focused. Only filters specifically designed for the purpose should be used because other dense material may transmit harmful infrared light. Many of the Sun's most fascinating features are visible only in the deep red hydrogen–alpha wavelength emitted by hydrogen gas. Filters that only transmit this light are very expensive, so even a basic solar telescopes can cost as much as a digital SLR camera. Specialized instruments called solar telescopes are also available. These reveal fascinating detail on the surface of the Sun, as well as the prominences around its edge.

**SOLAR TELESCOPE**

**THE SUN'S SURFACE**
This solar telescope view of the Sun shows granulation and sunspots; bright areas called faculae; prominences at the Sun's edge, or lim; and strandlike filaments seen against the Sun's bright surface.

**STAR PARTY**
Amateur astronomers gather in dark-sky areas at what are often called star parties. Only red lights are allowed, because they interfere less with night vision than lights of any other color.

# SETTING UP A TELESCOPE

THE SKY IS CLEAR, the forecast is good, and your first night of observing lies ahead. However, there is a steep learning curve to negotiate before you can start to see the sky's wonders. Even relatively simple telescopes can magnify objects many dozens of times, so locating apparently obvious bright objects can be surprisingly difficult. The secret to successful observations is to get your bearings before you begin. It may seem obvious, but make sure you know where north and south are—even go-to telescopes may need you to point them in the right direction initially.

## MOUNTING A TELESCOPE

After buying a telescope, it is important to take time to set up its optics, tripod, and mount properly. Careful setup will leave you with a well-aligned and balanced telescope that is a joy to use and that will require minimum tweaking during those precious observing hours. Each telescope is different, so be sure to read the instructions provided before you start or, better still, ask an experienced astronomer to take you through the basics. Below is a brief and general guide to the main points of setting up a typical amateur telescope—a reflector on a motorized equatorial mount. You will probably want to leave your telescope partly set up between observing sessions, so some of the steps will only need to be carried out the first time you use it.

**1 LEVEL TRIPOD**
Set up your tripod on solid, level ground. Use a level to check that the top plate of the tripod is horizontal and adjust the tripod legs as necessary.

**2 ADJUST LEGS**
Avoid extending the sections of the tripod legs to their full extent, because this makes the platform less stable and gives you no latitude for fine adjustment of height later. Double-check that the locks on the legs are secure.

**3 PLACE MOUNT**
Gently position the mount onto the tripod, ensuring that the protrusion on the mount slots into the hole on the tripod.

**4 SECURE MOUNT**
Tighten the mounting screw from beneath the tripod head, making sure it is completely secure.

**5 ATTACH MOTOR DRIVE**
Attach the motor drive to the mount and ensure that the gears of the motor are correctly engaged with those on the mount.

**6 ALIGN NORTH**
If using an equatorial mount, check that the right ascension axis (the long part of the central "T" of the mount) is pointing roughly toward the north (or south) celestial pole, depending on your hemisphere.

**7 ADD COUNTERWEIGHTS**
Slot the counterweights onto the counterweight shaft and use the nut to secure the weights in position. There is usually a safety screw at the end of the shaft that stops the counterweights from sliding off should the main nut fail. Be sure to replace this safety screw after positioning the weights.

**8 MOUNT THE TELESCOPE**
Once the mount is on the tripod, you can mount the telescope tube. Place the tube inside the pair of circular mounting rings (called cradles) and clamp them tight around the tube using the screws.

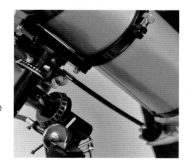

**9 ADD FINE ADJUSTMENT CABLES**
Screw in the fine adjustment cables—these will allow you to make small changes to the right ascension and declination when observing.

**10 ADD THE MOTOR UNIT**
Plug the drive controller into the motor unit, but do not connect it to the power supply.

**11 FIT THE FINDER AND EYEPIECE**
Attach the lowest-magnification eyepiece (the one with the longest focal length) and fit the finder. Align your finder with the main instrument; ideally, do this during the daytime (see p.84).

**12 BALANCE RIGHT ASCENSION (RA)**
Supporting the telescope with one hand, loosen the clamp on the RA axis and disengage the motor drive if necessary. Adjust the position of the counterweight on its shaft until it balances the telescope and the axis turns freely.

**13 BALANCE DECLINATION (DEC)**
Position the mount so that the telescope is out to one side and loosen the declination axis clamp. Support the telescope. Slacken the cradle screws enough to be able to slide the tube back and forth until it is balanced, then tighten them.

**14 POWER UP**
Connect the motor to the power supply, then fine-tune the polar alignment (see below).

# ALIGNING TO THE POLE

To set up an equatorial mount, you need to direct the RA axis, or polar axis, at the celestial pole. How you do this depends on your instrument. With simpler instruments, use the latitude scale usually provided on the mount (see below). On advanced instruments, sight the known position of the pole in the sky (see right). The pole is due south or north, depending on your hemisphere, and at the same angle to the horizon as your geographical latitude. For most observing, approximate alignment by eye is good enough to allow objects to be tracked for many minutes.

**ALIGNING SIMPLE MOUNTS**
Point the polar axis to the north or south depending on hemisphere. Turn the adjuster until the angle on the scale is at your latitude.

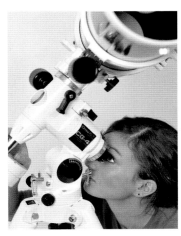

**ALIGNING ADVANCED MOUNTS**
Looking through a Northern Hemisphere polarscope (left), you will see a reticule engraved with several constellations and a circle offset from a crosshair (below). Turn the reticule until the constellations match their positions in the sky, then adjust the whole mount so that Polaris sits in the small circle.

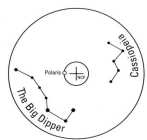

# SETTING UP GO-TO TELESCOPES

Every go-to telescope has a virtual map of the sky in its memory so that once it knows its precise location, the time, and the direction it is pointing to, it can find any celestial object. Encoders on each axis count the number of motor rotations the instrument makes as it "drives" from one object to another. With some simple go-to telescopes (see below), time and location must be input before the instrument is set in its start position, such as leveled and pointed north or south. It then has to be pointed at three bright alignment stars. Depending on the model, these may be chosen from the telescope's catalog or simply any three bright stars or planets. Advanced models are fitted with a GPS (Global Positioning System) receiver that automatically sets the time and location, as well as cameras that locate known bright stars. Whichever type of go-to telescope you have, it is essential that you align the finder with your main instrument before you begin (see p.84).

**1 MOUNT THE TELESCOPE**
Use a level to check that the tripod is level. Gently lower the mount and telescope onto the tripod head and secure it in position.

**2 SET POSITION**
Move the telescope into its start position if required.
For a fork-mounted instrument (shown), this may just mean aligning two arrows; however, an equatorial mount will need polar alignment (see above right).

**3 PREPARE THE TELESCOPE**
Connect the mount to the power pack and switch on the mount. Remove the lens cover.

**4 ENTER START DATA**
Enter the date, time, and location into the handset as prompted. On some go-to telescopes, you select your location from a menu.

**5 ALIGN THE TELESCOPE**
Alignment methods vary between models, but typically, the instrument will choose a bright star and move automatically to where it thinks the star should be. Alternatively, you can choose the first star from the menu.

**6 ADJUST ALIGNMENT**
The first star should be visible in the finder. Center this star using the directional buttons on the handset, then look through the eyepiece to refine its position. Repeat steps 5 and 6 to align two or three more stars as required, then alignment is complete.

**7 SET THE DESTINATION**
The go-to telescope is now ready. To explore the sky, find the name of the object you want to observe (such as Jupiter) in the handset's menus and press "go-to" or "enter." The telescope will then move to center your chosen object in the eyepiece.

# ASTROPHOTOGRAPHY

WITH MODERN TECHNOLOGY, amateur astronomers can now take images that would previously have been possible only from professional observatories. Even compact digital cameras can photograph bright objects, such as the Moon, through a telescope and can capture sky views, such as twilight scenes and constellations.

## BASIC ASTROPHOTOGRAPHY

Almost any camera can be used to take pictures of the night sky, although without a telescope it is limited to recording little more than naked-eye views of the stars, the Moon, bright planets, meteor trails, constellations, and aurorae. The main requirement for basic astrophotography is that the camera can keep the shutter open for long periods—at least several seconds. With long exposures, it is essential to keep the camera steady by mounting it on a tripod. Using a cable release, remote release, or timer to trigger the shutter will also help to avoid shake and blurring of the image.

**FIXED-CAMERA SHOTS**
General sky photography requires exposure times of many seconds with the camera at its most sensitive setting and focused on infinity. Mount the camera on a tripod to hold it steady during the exposure.

**METEORS**
Individual meteors cannot be predicted and so the only way to photograph them is to use long exposures in the hope that one will appear by chance. The field of view of an ordinary camera is ideal, and the exposure time should be as long as possible without the image being saturated by background light. Bright meteors will record as streaks against the background of star trails.

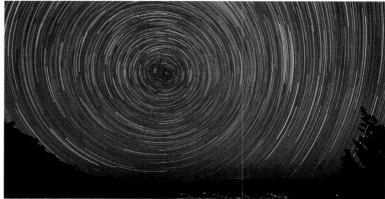

**STAR TRAILS**
During a fixed-camera exposure of more than a few seconds, the stars will trail across the image as they appear to move due to Earth's rotation—in this case, around the celestial pole. In light-polluted areas, take numerous shorter exposures and stack them using image-processing software to avoid an overexposed sky background.

## DIGISCOPING AND PIGGYBACKING

Compact cameras or cell phones can be used to take images directly through a telescope, a technique known as digiscoping. At its simplest, the camera can be mounted on a tripod and pointed down the telescope eyepiece. Alternatively, an adapter can be used to fix the camera to the eyepiece. Attaching the camera on top of a motor-driven equatorially mounted telescope—known as piggybacking—allows long-exposure views of the sky and even deep-sky objects without producing trails on the image. The image recorded is the one captured by the camera, not that seen through the telescope.

**DIGISCOPING IMAGE OF THE MOON**
Excellent images of the Moon can be obtained by digiscoping with even simple cameras. As the Moon is so bright, the exposure time for a Moon picture is similar to that for an ordinary daytime shot.

piggyback-mounted camera with telephoto lens

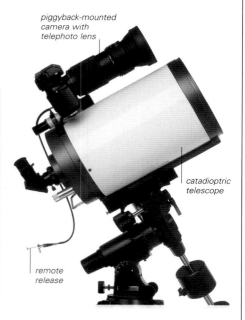

catadioptric telescope

remote release

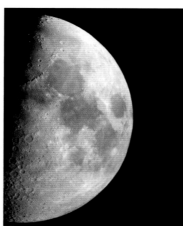

**DIGISCOPING SET-UP**
An adapter enables the camera to be aligned with the telescope eyepiece. Set the camera to manual exposure and use the self-timer to avoid shaking the camera. Experiment with different exposure times for the best results.

**PIGGYBACK SET-UP**
Many motor-driven telescopes have a threaded bolt for piggybacking a camera. If a telephoto lens is mounted on the camera and a long exposure is used, clear images of even deep-sky objects can be obtained.

*catadioptric telescope*

*remote release*

*camera adapter*

*SLR camera*

*equatorial mount with motor drive*

**PRIME FOCUS SET-UP**
This technique uses an adapter to place the camera in the eyepiece position, with or without the eyepiece present. A motor-driven equatorial mount is needed to keep the target object in the field of view.

# PRIME-FOCUS ASTROPHOTOGRAPHY

A telescope is, in effect, a very long telephoto lens, and adapters are available to attach virtually any single-lens reflex (SLR) camera to a telescope, thereby enabling the image produced by the telescope to be recorded. However, the maximum exposure time is often limited by the accuracy of the telescope's drive, which may not be precise enough to prevent star trailing. This problem can be overcome by using many short "sub-exposures" and adding them together with image-processing software to give the equivalent of a single long exposure. The telescope also needs to be kept steady, so a remote release should be used or, if possible, the camera should be operated remotely from a computer.

**PRIME-FOCUS IMAGE OF THE DUMBBELL NEBULA**
Prime-focus imaging is ideal for galaxies and small objects such as planetary nebulae—the Dumbbell Nebula shown here, for example. Exposure times of many minutes are needed for such images. To overcome any drive errors, the technique of sub-exposures (see above) can be used or a device called an autoguider can be fitted to the telescope to monitor the drive rate and make small corrections automatically.

# WEBCAMS AND CCD IMAGING

Digital SLR cameras can produce good astronomical images but many advanced astrophotographers use either webcam-based cameras for planetary imaging or CCD cameras for imaging faint objects that require very long exposures. Planetary imaging is often badly affected by atmospheric turbulence, which typically blurs the view so that it is sharp for only fractions of a second. Webcam-type cameras produce a video stream, taking thousands of images a minute. These images can then be processed by dedicated software that selects and stacks together the best images. For imaging faint objects that require exposures of several hours, cooled CCD cameras produce less electronic noise—and therefore better images—than digital SLRs.

**WEBCAM SET-UP**
A webcam can be used on even small telescopes to image the planets. The webcam slots into the telescope in place of the eyepiece and connects to a computer with a cable. The webcam is then operated from the computer.

*CCD camera*

**CCD IMAGING SET-UP**
Like a webcam, a CCD camera replaces the telescope eyepiece and is connected to a computer. To quickly establish the focus when using a CCD camera, it helps first to focus using a telescope eyepiece with the same focus position as the camera.

# IMAGE PROCESSING

Many images take far longer to process than the original observing time at the telescope, but there are various image-processing programs that can help. For example, software is available for automatically overlaying in register and stacking multiple exposures of the same object. Some cameras (notably CCD cameras) produce monochromatic images but can be used with color filters to produce a series of images that can be combined using stacking software to give a full-color final image. Software can also be used to enhance images by sharpening details, correcting the color balance, altering the brightness, and increasing the contrast. In addition, image-processing software can be used to change individual colors, a technique that is often utilized by professional astronomers to highlight specific features.

**DIGITAL STACKING**
The image (left) of NGC 1977, the Ghost Nebula in Orion, was made through an amateur 12.5 in (300 mm) telescope and is the result of combining four individual 90-minute exposures using dedicated image-stacking software.

**COLOR CONTROL**
This screenshot shows an image of Saturn in Photoshop, software that can be used to enhance or alter an image's features, such as its color. In this image, the brightest ring is composed of ice and needs to be altered to white to show a realistic view of the planet.

# ASTRONOMICAL OBSERVATORIES

SINCE ABOUT THE early 20th century, many new astronomical observatories have been built, housing ever-larger telescopes. Many of these instruments are visible-light telescopes, but with continuing technological advances, telescopes for studying other parts of the electromagnetic spectrum have also been built, such as radio telescopes and gamma-ray telescopes.

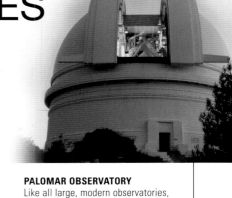

**PALOMAR OBSERVATORY**
Like all large, modern observatories, the Palomar in California, was built at high altitude (5,617 ft/1,712 m) for optimum viewing conditions.

## OBSERVATORY TELESCOPES

Most observatory telescopes are placed away from the air and light pollution of urban areas and at high altitude to minimize atmospheric distortion. The size of a telescope is also important: the larger a telescope's aperture, the greater its light-gathering power. Objective lenses for refractors cannot be made more than about 40 in (1 m) across—the size of the Yerkes refractor (below left)—but single-piece mirrors can be made up to about 26.25 ft (8 m) across—the size of the Gemini South telescope (right). Using segmented mirrors, reflectors can be made even larger. For example, the Gran Telescopio Canarias has a segmented mirror 34 ft (10.4 m) across.

**GEMINI SOUTH TELESCOPE**
Moonlight casts shadows inside the dome of the Gemini South telescope, which sits at an altitude of 8,980 ft (2,737 m) on Cerro Pachón, a mountain in the Chilean Andes. An identical telescope, Gemini North, is situated near the summit of Mauna Kea on the island of Hawaii.

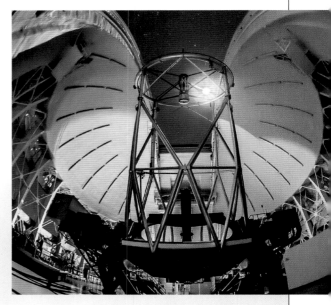

**YERKES REFRACTOR**
Refracting telescopes reached their pinnacle with the 40 in (1 m) instrument at Yerkes Observatory, Wisconsin, shown on the left. Opened in 1897, it remains the largest refractor ever built.

**PARANAL OBSERVATORY**
Situated at an altitude of 8,645 ft (2,635 m) on Cerro Paranal in northern Chile, the Very Large Telescope (VLT) is one of the largest modern telescope arrays, consisting of four 26.9 ft (8.2 m) reflectors. The telescopes operate at visible light and infrared wavelengths and can be used either independently or in combination for greater resolution.

# NEW OPTICAL TECHNOLOGY

In their quest for greater light grasp and sharper images, optical astronomers have utilized innovative new technology, such as mirrors made up of many separate segments. Segmented mirrors can be made much thinner, and hence lighter, than a single large mirror. The segments are usually hexagonal in shape, and each one can be individually controlled to maintain sharp focus as the telescope is moved. Mirrors larger than 26.2 ft (8 m) in diameter are now made in this way, and segmented mirrors up to 128 ft (39 m) wide are planned. Another advance has come from adaptive optics, a technique that removes the blurring effects of the atmosphere and can produce images almost as sharp as those from telescopes in space. This is done by measuring atmospheric distortion using an artificial guide star created by firing a laser beam along the telescope's line of sight. Using these measurements, a flexible secondary mirror (which collects the light from the main mirror) is then deformed to compensate for the distortion.

**SEGMENTED TELESCOPE MIRROR**
The Gran Telescopio Canarias, also known as the GranTeCan or GTC, has a mirror 34 ft (10.4 m) in diameter—the world's largest. Opened in 2009, it is located at the Roque de los Muchachos on La Palma in the Canary Islands. Its mirror (shown left) is composed of 36 hexagonal segments, each of which is 75 in (1.9 m) wide.

**THE LARGE BINOCULAR TELESCOPE**
A novel design for increasing light grasp and resolving power is the Large Binocular Telescope at Mount Graham, Arizona. It consists of two mirrors, each 27.6 ft (8.4 m) in diameter, side by side on the same mount. Together, the two mirrors collect as much light as a single mirror 38.7 ft (11.8 m) across.

**LASER GUIDE STAR**
A powerful beam of orange laser light shoots skyward from one of the components of the Very Large Telescope (VLT) in Chile, creating an artificial guide star 55 miles (90 km) high. The guide star is part of the VLT's adaptive optics system, which helps correct for image distortion caused by atmospheric disturbances.

**EFFECT OF ADAPTIVE OPTICS**
These images of the center of the Galaxy through the Keck II telescope in Hawaii show the effect of adaptive optics. The image on the left was taken without adaptive optics; the much sharper image on the right was taken with the adaptive optics system in operation.

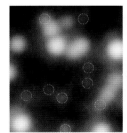

# BEYOND VISIBLE LIGHT

Many celestial objects emit energy outside the visible light spectrum (see pp.36–37), so optical telescopes alone cannot give a complete view. The first non-visible-light telescope was a radio telescope, built in 1937. Radio waves have much longer wavelengths than visible light, so radio telescopes have to be larger to achieve the same resolution. To overcome this restriction, radio dish arrays have been built so that observations from individual dishes can be combined. An example is the Karl G. Jansky Very Large Array near Socorro, New Mexico, which consists of 27 dishes, each 82 ft (25 m) wide, arranged along three arms 13 miles (21 km) long. The largest single radio dish is 1,000 ft (305 m) in diameter at Arecibo, Puerto Rico. Most non-visible-light wavelengths other than radio are blocked by the atmosphere. However, some infrared reaches mountaintops and can be detected by certain telescopes, such as the United Kingdom Infrared Telescope in Hawaii. It is also possible to detect cosmic gamma rays at the Earth's surface. The MAGIC telescope at La Palma in the Canary Islands achieves this by detecting the faint light emitted by particle showers generated by gamma rays.

**GREEN BANK RADIO TELESCOPE**
The world's largest fully steerable radio telescope, at the National Radio Astronomy Observatory at Green Bank, West Virginia, has an elliptical dish 360 x 328 ft (110 x 100 m) across. The dish consists of over 2,000 panels, each of which can be adjusted separately to maintain the shape of the dish as the telescope moves. The secondary reflector (which reflects radio waves from the main dish to the radio detector) is on an arm to avoid obstructing the main dish.

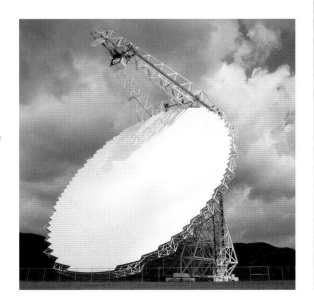

**MILLIMETER ARRAY**
Moonlight illuminates the antennae of the Atacama Large Millimeter Array (ALMA) on the Chajnantor Plateau in Chile. Each of the dishes is 39 ft (12 m) in diameter and observes the sky at millimeter and submillimeter wavelengths, between the infrared and radio parts of the spectrum, detecting objects in nearby star-forming regions to galaxies in the distant universe.

# OBSERVING FROM SPACE

MANY OF THE GREATEST discoveries and most spectacular images of the universe have come from observatories in space. Above Earth's atmosphere, telescopes can see the sky far more clearly than those on the ground, and they can detect wavelengths that the atmosphere blocks.

## VISIBLE AND ULTRAVIOLET LIGHT

Among the first successful space telescopes were those designed to detect ultraviolet light, notably NASA's Orbiting Astronomical Observatory series, launched between 1966 and 1972, and the International Ultraviolet Explorer, which was launched in 1978 and carried a 1.5 ft (0.45 m) telescope. Probably the most famous space telescope is the Hubble Space Telescope (HST), which was launched in 1990 and is still in operation. With a 7.9 ft (2.4 m) telescope designed primarily to detect visible and ultraviolet light, the HST has, among other successes, helped determine the age of the universe and produced evidence for the existence of dark energy. Space telescopes have also advanced more traditional realms of astronomy. For example, a European observatory called Gaia was launched in 2013 to measure the positions, motions, and distances of stars in the Milky Way, building up a three-dimensional picture of our galaxy.

**LAUNCH OF THE HUBBLE SPACE TELESCOPE**
Launched in April 1990 from the Kennedy Space Center, Florida, on board the Space Shuttle Discovery, the Hubble Space Telescope orbits about 380 miles (600 km) above Earth. Initially intended to operate for 10 years, Hubble is still in operation thanks to five servicing missions by astronauts.

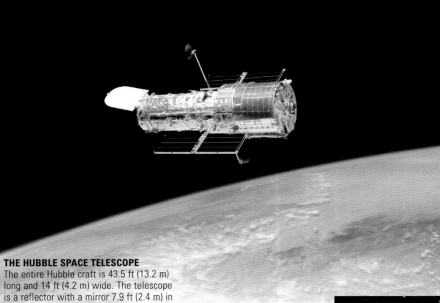

**THE HUBBLE SPACE TELESCOPE**
The entire Hubble craft is 43.5 ft (13.2 m) long and 14 ft (4.2 m) wide. The telescope is a reflector with a mirror 7.9 ft (2.4 m) in diameter. It operates primarily in visible light and ultraviolet, although its coverage also extends into the near-infrared.

**GAIA ASTROMETRY SATELLITE**
Launched at the end of 2013, the European Space Agency's Gaia spacecraft is mapping the precise positions and movements of over a billion stars throughout our galaxy in a mission scheduled to last until 2024.

**HUBBLE DEEP-SKY VIEW**
Thousands of galaxies extending into the remote depths of the universe are visible in this image from the Hubble Space Telescope. Called the Extreme Deep Field, it is the combined result of over 3 weeks of exposure time and covers an area less than a tenth of the width of the full Moon. The most distant galaxies are seen as they were over 13 billion years ago, when the universe was less than 5 percent of its present age. Many of them eventually developed into galaxies similar to our own Milky Way.

# INFRARED AND MICROWAVE

Some infrared and microwave radiation penetrates Earth's atmosphere but to detect the full range requires observing from space. Prominent targets for observation are cool stars and active galaxies, which emit much of their radiation in the infrared. Infrared telescopes also make it possible to see through interstellar dust clouds into regions obscured from optical view, such as the interiors of nebulae and the center of our galaxy. The largest infrared telescope launched so far was the Herschel Space Observatory, which had a mirror 11.5 ft (3.5 m) across and operated from 2009 to 2013. Microwave space telescopes are designed primarily to detect and map the cosmic microwave background radiation, in order to investigate the structure and origin of the universe. The first dedicated microwave space telescopes were the Cosmic Background Explorer, launched in 1989, and the Wilkinson Microwave Anisotropy Probe, launched in 2001. The most recent was the Planck space telescope, which was launched in 2009.

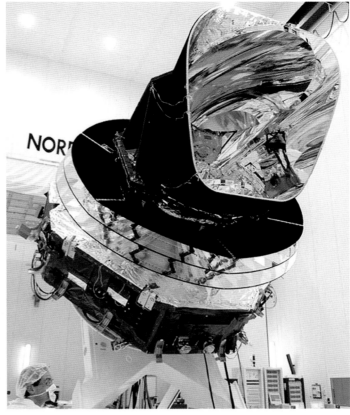

**PLANCK IMAGE OF STAR-FORMATION REGION IN PERSEUS**
This false-color image of a low-activity star-formation region was produced by combining data from Planck at three different microwave wavelengths.

**PLANCK SPACE TELESCOPE**
Shown here being tested before its launch in 2009, the Planck satellite carried a telescope with a 4.9 ft (1.5 m) main mirror to study the cosmic microwave background radiation. During a mission that lasted until 2013, it measured the age of the universe as 13.8 billion years.

# X-RAYS AND GAMMA RAYS

The shortest wavelengths of all, X-rays and gamma rays, are produced by some of the most violent events in the universe, such as supernova explosions. However, like infrared and microwave radiation, X-rays and gamma rays are best studied from space. Major X-ray space observatories include the Chandra X-ray Observatory (see p.37) and XMM-Newton, both launched in 1999, and the NuSTAR, launched in 2012. Notable gamma-ray space telescopes include the Compton Gamma Ray Observatory (see p.37), launched in 1991, and the Fermi Gamma-ray Space Telescope, which was launched in 2008 and carries an instrument designed to study gamma-ray bursts, which are thought to be emitted by the merger of black holes and neutron stars and also by the collapse of massive stars to form black holes.

**X-RAY-EMITTING CLOUD**
This XMM-Newton image shows an X-ray-emitting cloud of ultra-hot gas, at temperatures up to about 90 million °F (50 million °C), around a giant elliptical galaxy.

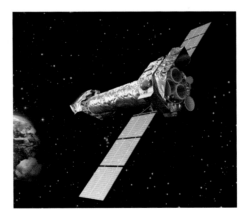

**XMM-NEWTON X-RAY SPACE TELESCOPE**
Launched in 1999, the XMM-Newton contains three X-ray telescopes for the imaging and spectroscopy of X-ray sources. The entire satellite is 33 ft (10 m) long and is in a highly elliptical orbit that, at its most distant, takes the satellite more than 60,000 miles (100,000 km) from Earth.

EXPLORING SPACE

## LAGRANGIAN POINTS

Satellites can be placed in various orbits around the Earth or other celestial objects. Some satellites are placed at specific points called Lagrangian points. These are locations where the orbital motion of a small object (such as a satellite) and the gravitational forces acting on it from larger bodies (such as nearby planets and stars) balance each other. As a result, the small object remains in a fixed position relative to the larger bodies. There are five such points in the Earth–Moon–Sun system.

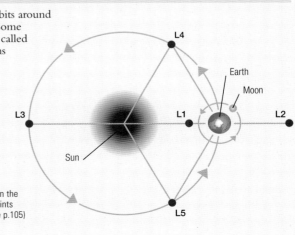

**FIXED ORBITS**
This diagram shows the five Lagrangian points in the Earth–Moon–Sun system. Satellites at these points orbit the Sun, not Earth, and include SOHO (see p.105) at L1, and Herschel, Planck, and Gaia at L2.

# GUIDE TO THE UNIVERSE

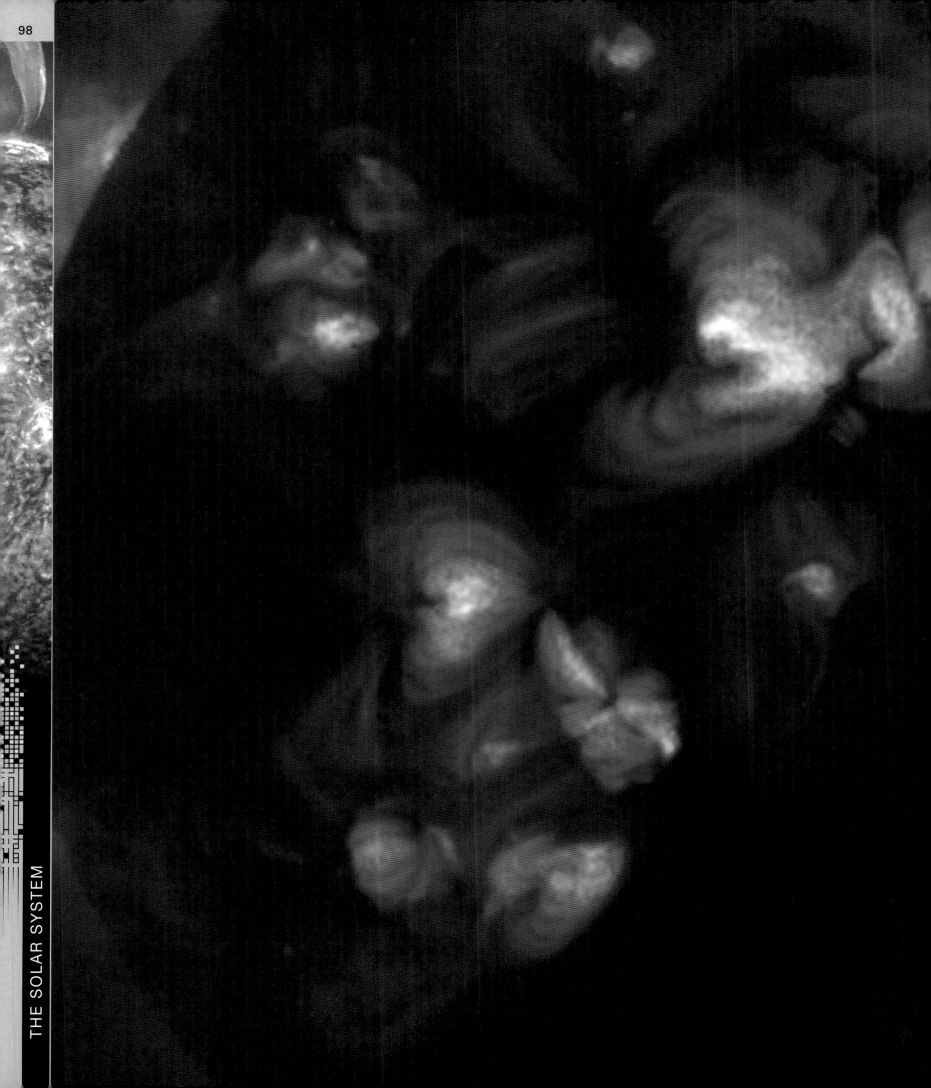

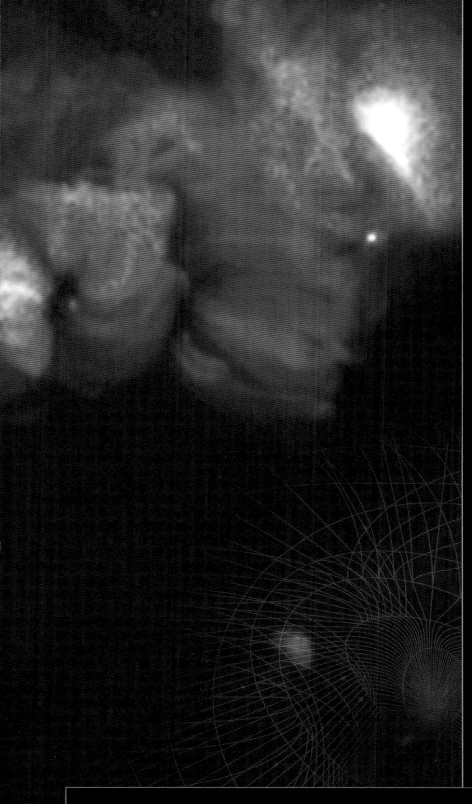

THE SOLAR SYSTEM IS the region of space that falls within the gravitational influence of the Sun, an ordinary yellow star that has shone steadily for almost 5 billion years. After the Sun itself, the most significant objects in the solar system are the planets—a group of assorted rocky, gaseous, and icy worlds that follow independent, roughly circular orbits around their central star. Most of the planets are orbited in turn by moons, while a huge number of smaller lumps of rock and ice also follow their own courses around the Sun—though largely confined in a few relatively crowded zones. Myriad tiny particles flow around all these larger bodies— ranging from fragments of atoms blown out by the Sun to motes of dust and ice left in the wake of comets. Our local corner of the universe has been studied intensively from the time of the first stargazers to the modern era of space probes, yet it is still a source of wonder and surprise.

**SOLAR FLARE**
On the broiling surface of the Sun, a cataclysmic release of magnetic energy triggers a solar flare—a violent outburst of radiation and high-energy particles that will reach Earth within hours.

# THE SOLAR SYSTEM

# THE HISTORY OF THE SOLAR SYSTEM

THE SOLAR SYSTEM IS THOUGHT to have begun forming about 4.6 billion years ago from a gigantic cloud of gas and dust, called the solar nebula. This cloud contained several times the mass of the present-day Sun. Over millions of years, it collapsed into a flat, spinning disk, which had a dense, hot central region. The central part of the disk eventually became the Sun, while the planets and everything else in the solar system formed from a portion of the remaining material.

## THE FORMATION OF THE SOLAR SYSTEM

No one knows for certain what caused the great cloud of gas and dust, the solar nebula from which the solar system formed, to start collapsing. What is certain is that gravity somehow overcame the forces associated with gas pressure that would otherwise have kept it expanded. As it collapsed, the cloud flattened into a pancake-shaped disk with a bulge at its center. Just as an ice skater spins faster as she pulls in her arms, the disk began to rotate faster and faster as it contracted. The central region also became hotter and denser. In the parts of the disk closest to this hot central region, only rocky particles and metals could remain in solid form. Other materials were vaporized. In due course, these rocky and metallic particles gradually came together to form planetesimals (small bodies of rock, up to several miles in diameter) and eventually the inner rocky planets—Mercury, Venus, Earth, and Mars. In the cooler outer regions of the disk, a similar process occurred, but the solid particles that came together to form planetesimals contained large amounts of various ices, such as water, ammonia, and methane ices, as well as rock. These materials were destined eventually to form the cores of the gas-giant planets—Jupiter, Saturn, Uranus, and Neptune.

### 1 SOLAR NEBULA FORMS

The solar nebula started as a huge cloud of cold gas and dust many times larger than our present solar system. Its initial temperature would have been about −382°F (−230°C). From the start, the solar nebula was probably spinning very slowly.

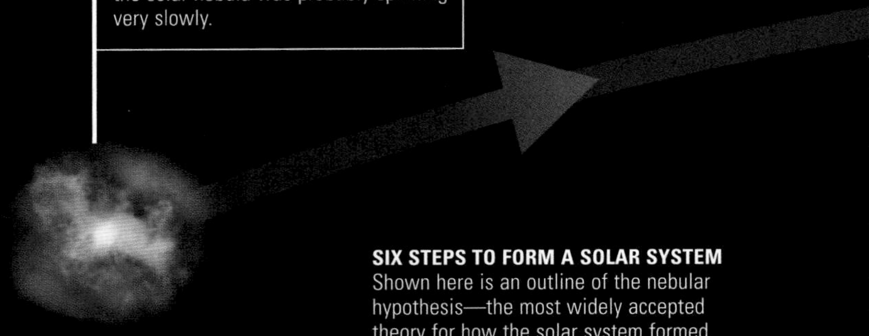

### SIX STEPS TO FORM A SOLAR SYSTEM

Shown here is an outline of the nebular hypothesis—the most widely accepted theory for how the solar system formed. It provides a plausible explanation for many of the basic facts about the solar system. For example, it explains why the orbits of most of the planets lie roughly in the same plane and why the planets all orbit in the same direction.

### PIERRE-SIMON DE LAPLACE

Pierre Laplace (1749–1827) was a French mathematician who developed the nebular hypothesis—the idea, originally proposed by the German philosopher Immanuel Kant, that the solar system originated from the contraction of a huge gaseous nebula. Today, this hypothesis provides the most widely accepted theory for how the solar system formed. Another of Laplace's contributions to science was to analyze the complex forces of gravitational attraction between the planets. He investigated how these might affect the stability of the solar system and concluded that the system is inherently stable.

### 6 REMAINING DEBRIS

Radiation from the Sun blew away most of the remaining gas and other unaccreted material in the planetary solar system. Some of the leftover planetesimals in the outer part of the disk formed the vast and remote Oort Cloud of comets.

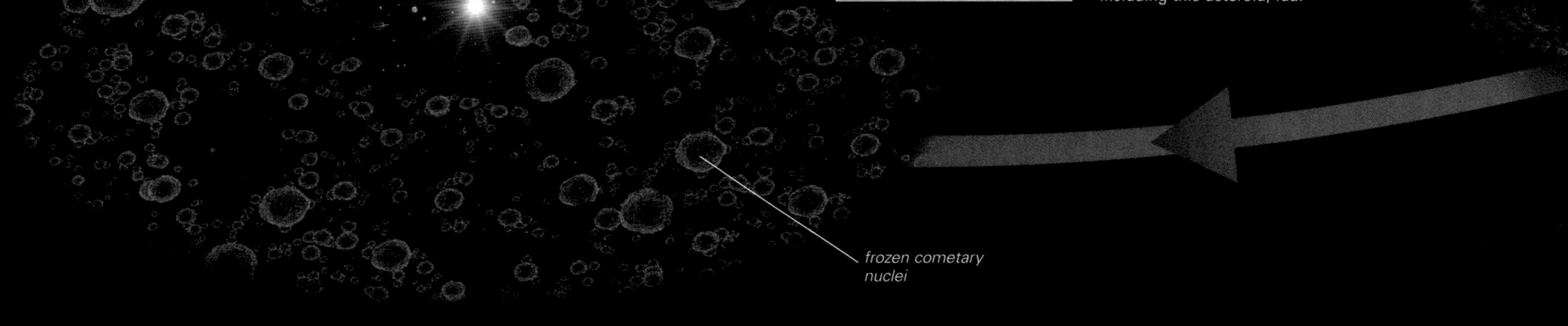

*inner solar system*

### IDA

The ring of planetesimals between Mars and Jupiter failed to form a planet, possibly because of the gravitational influence of Jupiter. Instead, they formed a belt of asteroids, including this asteroid, Ida.

*frozen cometary nuclei*

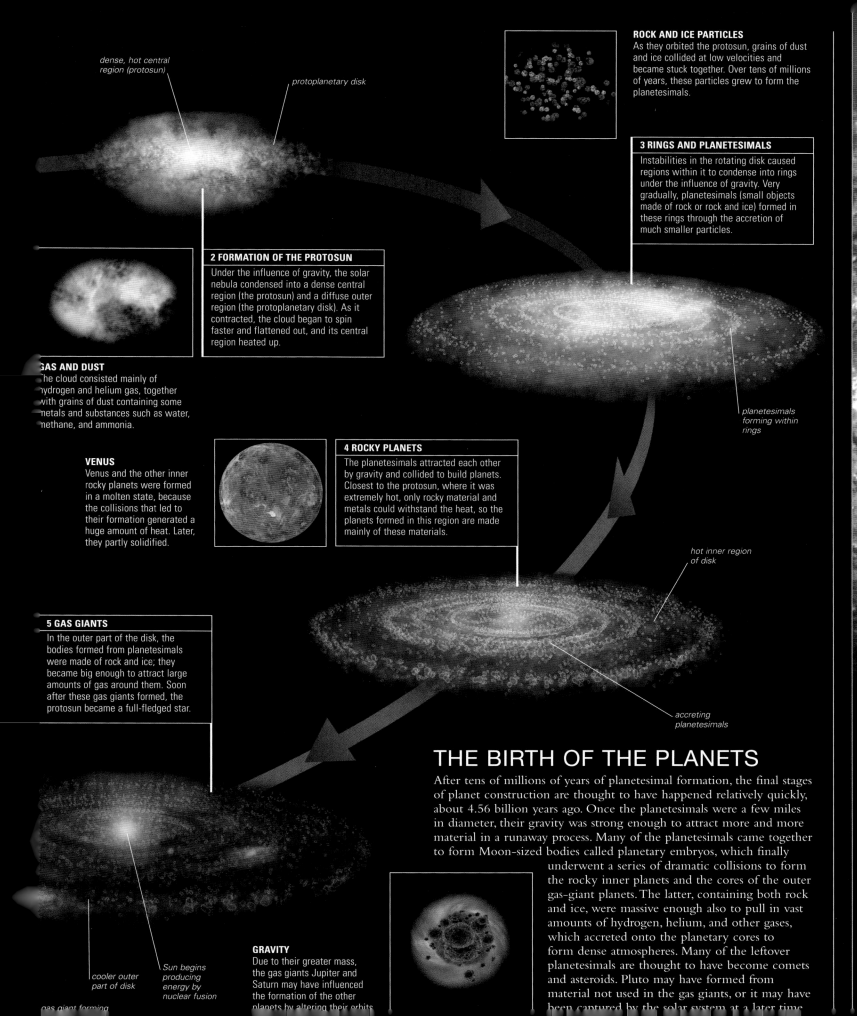

dense, hot central
region (protosun)

protoplanetary disk

**ROCK AND ICE PARTICLES**
As they orbited the protosun, grains of dust
and ice collided at low velocities and
became stuck together. Over tens of millions
of years, these particles grew to form the
planetesimals.

**3 RINGS AND PLANETESIMALS**
Instabilities in the rotating disk caused
regions within it to condense into rings
under the influence of gravity. Very
gradually, planetesimals (small objects
made of rock or rock and ice) formed in
these rings through the accretion of
much smaller particles.

**2 FORMATION OF THE PROTOSUN**
Under the influence of gravity, the solar
nebula condensed into a dense central
region (the protosun) and a diffuse outer
region (the protoplanetary disk). As it
contracted, the cloud began to spin
faster and flattened out, and its central
region heated up.

**GAS AND DUST**
The cloud consisted mainly of
hydrogen and helium gas, together
with grains of dust containing some
metals and substances such as water,
methane, and ammonia.

planetesimals
forming within
rings

**VENUS**
Venus and the other inner
rocky planets were formed
in a molten state, because
the collisions that led to
their formation generated a
huge amount of heat. Later,
they partly solidified.

**4 ROCKY PLANETS**
The planetesimals attracted each other
by gravity and collided to build planets.
Closest to the protosun, where it was
extremely hot, only rocky material and
metals could withstand the heat, so the
planets formed in this region are made
mainly of these materials.

hot inner region
of disk

**5 GAS GIANTS**
In the outer part of the disk, the
bodies formed from planetesimals
were made of rock and ice; they
became big enough to attract large
amounts of gas around them. Soon
after these gas giants formed, the
protosun became a full-fledged star.

accreting
planetesimals

# THE BIRTH OF THE PLANETS

After tens of millions of years of planetesimal formation, the final stages
of planet construction are thought to have happened relatively quickly,
about 4.56 billion years ago. Once the planetesimals were a few miles
in diameter, their gravity was strong enough to attract more and more
material in a runaway process. Many of the planetesimals came together
to form Moon-sized bodies called planetary embryos, which finally
underwent a series of dramatic collisions to form
the rocky inner planets and the cores of the outer
gas–giant planets. The latter, containing both rock
and ice, were massive enough also to pull in vast
amounts of hydrogen, helium, and other gases,
which accreted onto the planetary cores to
form dense atmospheres. Many of the leftover
planetesimals are thought to have become comets
and asteroids. Pluto may have formed from
material not used in the gas giants, or it may have
been captured by the solar system at a later time.

cooler outer
part of disk

Sun begins
producing
energy by
nuclear fusion

**GRAVITY**
Due to their greater mass,
the gas giants Jupiter and
Saturn may have influenced
the formation of the other
planets by altering their orbits.

gas giant forming

# THE FAMILY OF THE SUN

THE SOLAR SYSTEM CONSISTS OF the Sun, eight recognized planets, over 140 moons, and countless small bodies such as asteroids and comets. Its inner region contains the Sun and the rocky planets—Mercury, Venus, Earth, and Mars. Beyond this lies a ring of asteroids, called the Main Belt; the gas giant planets Jupiter, Saturn, Uranus, and Neptune; a region called the Kuiper Belt, which contains icy worlds; and finally a vast cloud of comets. In total, the solar system is about 9.3 trillion miles (15 trillion km) across; the planets occupy a zone extending just 3.25 billion miles (6 billion km) from the Sun.

## URBAIN LE VERRIER

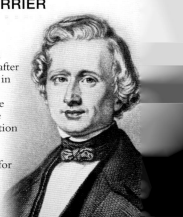

Urbain Le Verrier (1811–1877) was a French mathematician and astronomer who, after studying irregularities in the orbit of Uranus, predicted the existence of the planet Neptune and calculated its position in 1846. He asked the German astronomer Johann Galle to look for Neptune, and within an hour, the planet had been found.

## ORBITS IN THE SOLAR SYSTEM

Most orbits of objects in the solar system have the shapes of ellipses (stretched circles). However, for most of the planets, these ellipses are close to being circular. Only Mercury has an orbit that differs very markedly from being circular. All the planets and nearly all asteroids orbit the Sun in the same direction, which is also the direction in which the Sun spins on its own axis. The orbital period (the time it takes to orbit the Sun) increases with distance from the Sun, from 88 Earth days for Mercury to nearly 250 Earth years for Pluto, following a mathematical relationship first discovered by the German astronomer Johannes Kepler in the early 17th century (see p.68). As well as having longer orbits to complete, the planets farther from the Sun move much more slowly.

**THE SUN**
*The Sun's diameter at the equator is 864,900 miles (1.4 million km), and its equatorial rotation period is about 25 Earth days*

**EARTH**
*Orbits the Sun in 365.26 Earth days at an average distance of 92.9 million miles (149.6 million km)*

**JUPITER**
*Orbits the Sun in 11.86 Earth years at an average distance of 483.4 million miles (778.4 million km)*

**URANUS**
*Orbits the Sun in 84.01 Earth years at an average distance of 1.8 billion miles (2.9 billion km)*

**MERCURY**
*Orbits the Sun in 88 Earth days at an average distance of 36 million miles (57.9 million km)*

**PLANET ORBITS**
All the orbits of the planets, and the Asteroid Belt, lie roughly in a flat plane known as the ecliptic plane. Only Mercury and the dwarf planet Pluto orbit at significant angles to this plane (7.0° and 17.1°, respectively). The planets and their orbits are not shown to scale here.

# THE GAS GIANTS

The four large planets immediately beyond the Asteroid Belt are called the gas giants. These planets have many properties in common. Each has a core composed of rock and ice. This is surrounded by a liquid or semisolid mantle containing hydrogen and helium, or—in the case of Uranus and Neptune—a combination of methane, ammonia, and water ices. Each has a deep, often stormy atmosphere composed mainly of hydrogen and helium. All four have a significant magnetic field, but Jupiter's is exceptional, being 20,000 times stronger than that of Earth. Each of the gas giants is orbited by a large number of moons, several dozen in the case of Jupiter. Finally, all four gas giants have ring systems made of grains of rock or ice. These rings may have been present since the planets formed, or they may be the fragmented remains of moons that were broken up by the gas giants' powerful gravitational fields.

**URANUS AND RINGS**
Uranus has 11 major rings and a blue coloration caused by the presence of methane in its atmosphere. (This is a Hubble infrared image.) Its spin axis is tilted over on its side.

# THE ROCKY PLANETS

The four inner planets of the solar system are also called the rocky planets. They are much smaller than the gas giants, have few or no moons, and have no rings. All four were born in a molten state due to the heat of the collisions that led to their formation. While molten, the materials from which they are made became separated into a metallic core and a rocky mantle and crust. Throughout their later history, all these planets suffered heavy bombardment by meteorites that left craters on their surfaces, although on Earth these craters have largely become hidden by various geological processes. In some other respects, the rocky planets are quite diverse. For example, Venus has a dense atmosphere consisting mainly of carbon dioxide, while Mars has a thin atmosphere composed of the same gas. In contrast, Mercury has virtually no atmosphere and Earth's is rich in nitrogen and oxygen.

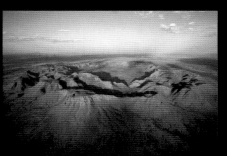

**GOSSES BLUFF CRATER**
This impact crater in a central desert region of Australia resulted from an asteroid 0.6 miles (1 km) wide that smashed into Earth's surface 142 million years ago.

**MAIN BELT**
*Lies between the orbits of Mars and Jupiter and is a source of meteorites; some asteroids orbit the Sun outside the Main Belt*

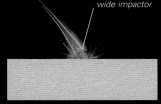

*6-mile (10-km) wide impactor*

**IMPACTOR STRIKES**

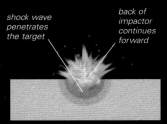

*shock wave penetrates the target*

*back of impactor continues forward*

**EXPLOSION ON IMPACT**

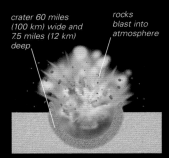

*crater 60 miles (100 km) wide and 7.5 miles (12 km) deep*

*rocks blast into atmosphere*

**CRATER FORMATION**

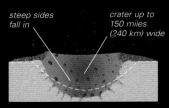

*steep sides fall in*

*crater up to 150 miles (240 km) wide*

**CRATER COLLAPSE**

**DEEP IMPACT**
This sequence shows what typically happens when a 6-mile (10-km) wide projectile hits a rocky planet or moon. The crater formed is much larger than the impactor. The latter usually vaporizes on impact, though some melted or shattered remnants may be left at the site.

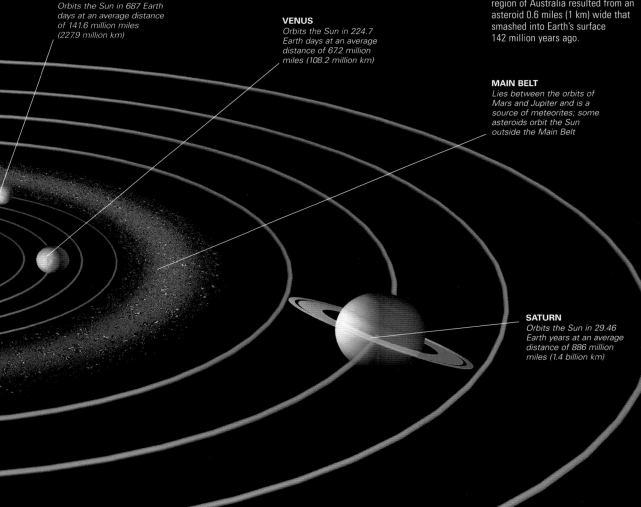

**MARS**
*Orbits the Sun in 687 Earth days at an average distance of 141.6 million miles (227.9 million km)*

**VENUS**
*Orbits the Sun in 224.7 Earth days at an average distance of 67.2 million miles (108.2 million km)*

**SATURN**
*Orbits the Sun in 29.46 Earth years at an average distance of 886 million miles (1.4 billion km)*

**NEPTUNE**
*Orbits the Sun in 164.8 Earth years at an average distance of 2.8 billion miles (4.5 billion km)*

THE SOLAR SYSTEM

# THE SUN

THE SUN IS A 4.6-BILLION-YEAR-OLD main-sequence star. It is a huge sphere of exceedingly hot plasma (ionized gas) containing 750 times the mass of all the solar system's planets put together. In its core, nuclear reactions produce helium from hydrogen and generate colossal amounts of energy. This energy is gradually carried outward until it eventually escapes from the Sun's surface.

## INTERNAL STRUCTURE

The Sun has three internal layers, although there are no sharp boundaries between them. At the center is the core, where temperatures and pressures are extremely high. In the core, nuclear fusion turns the nuclei of hydrogen atoms (protons) into helium nuclei at the rate of about 600 million tons per second. Released as byproducts of the process are energy, in the form of photons of electromagnetic (EM) radiation, and neutrinos (particles with no charge and almost no mass). The EM radiation travels out from the core through a slightly cooler region, the radiative zone. It takes about 1 million years to find its way out of this zone, as the photons are continually absorbed and reemitted by ions in the plasma. Farther out, the energy wells up in a convective zone—where huge flows of rising hot plasma occur next to areas of falling cooler plasma—and is transferred to a surface layer called the photosphere. There, it escapes as heat, light, and other forms of radiation.

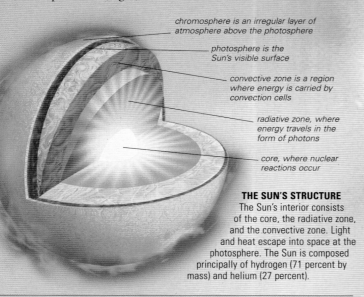

chromosphere is an irregular layer of atmosphere above the photosphere

photosphere is the Sun's visible surface

convective zone is a region where energy is carried by convection cells

radiative zone, where energy travels in the form of photons

core, where nuclear reactions occur

**THE SUN'S STRUCTURE**
The Sun's interior consists of the core, the radiative zone, and the convective zone. Light and heat escape into space at the photosphere. The Sun is composed principally of hydrogen (71 percent by mass) and helium (27 percent).

## SUN PROFILE

**AVERAGE DISTANCE FROM EARTH**
93.0 million miles (149.6 million km)

**SURFACE TEMPERATURE**
9,932°F (5,500°C)

**CORE TEMPERATURE**
27 million °F (15 million °C)

**DIAMETER AT EQUATOR**
864,900 miles (1.4 million km)

**OBSERVATION**
The Sun has an apparent magnitude of −26.7 and should never be observed directly with the naked eye or any optical instrument. It can be observed safely only through special solar filters.

**ROTATION PERIOD (POLAR)**
34 Earth days

**ROTATION PERIOD (EQUATORIAL)**
25 Earth days

**MASS (EARTH = 1)** 333,000

**SIZE COMPARISON**

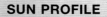

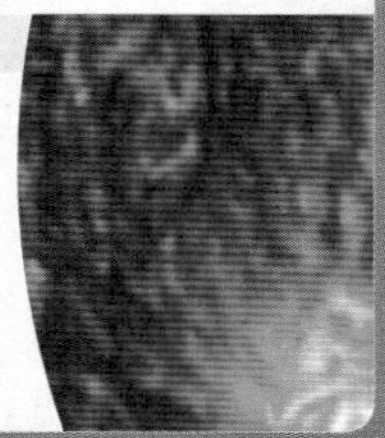

EARTH

THE SUN

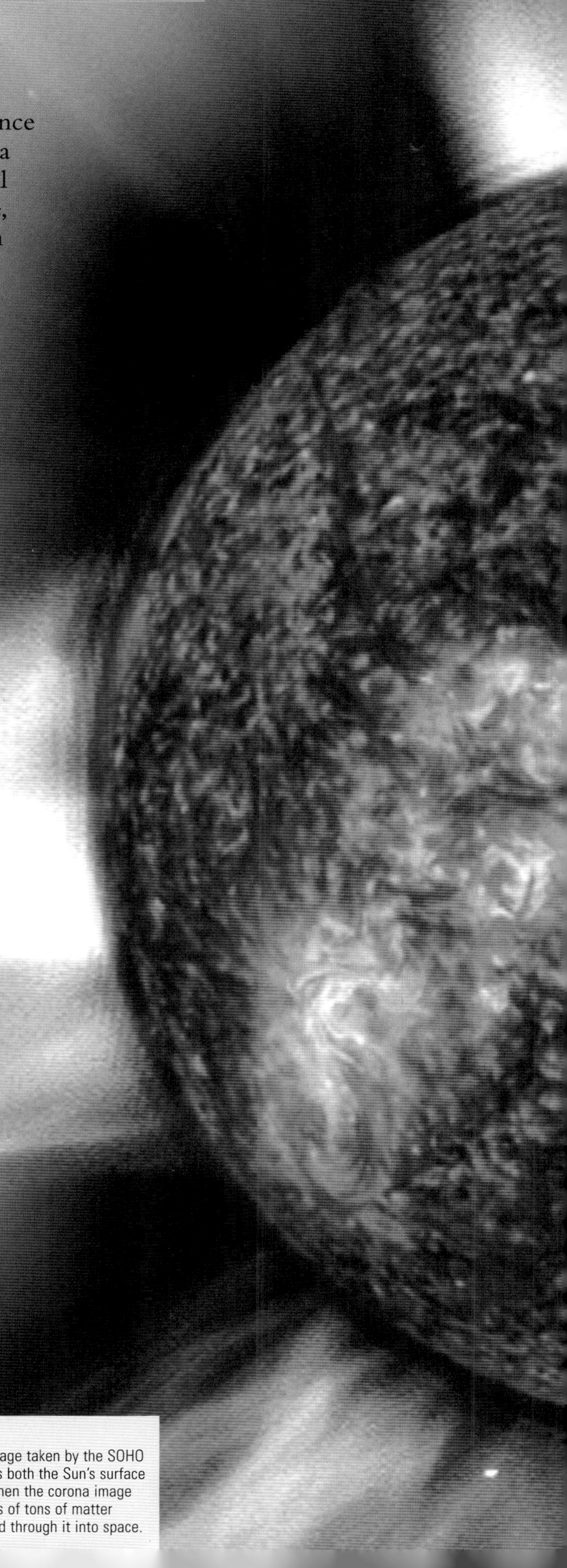

**VIOLENT SUN**
This composite image taken by the SOHO observatory shows both the Sun's surface and its corona. When the corona image was taken, billions of tons of matter were being blasted through it into space.

## STUDYING THE SUN FROM SPACE

Since 1960, a series of space probes and satellites have been launched by NASA and other organizations with the aim of collecting data about the Sun. Some of the most important missions are listed below.

**1960–1968 PIONEERS 5 TO 9 (USA)** These were a series of probes that successfully orbited the Sun and studied the solar wind, solar flares, and the interplanetary magnetic field.

**1974, 1976 HELIOS 1 AND 2 (USA AND GERMANY)** The two Helios probes were put into orbits that involved high-velocity passes close to the Sun's surface. They measured the solar wind and the Sun's magnetic field.

**1980 SOLAR MAXIMUM MISSION (USA)** This studied the Sun at its most active, collecting X-rays, gamma rays, and ultraviolet radiation produced by flares, sunspots, and prominences.

**1990 ULYSSES (USA AND EUROPE)** The first space probe to be sent into an orbit over the Sun's poles, Ulysses has studied the solar wind and the Sun's magnetic field over its polar regions.

**1995 SOHO (USA AND EUROPE)** This solar observatory follows a special "halo" orbit around the Lagrangian point 930,000 miles (1.5 million km) from Earth in the direction of the Sun. SOHO (solar and heliospheric observatory) studies the Sun's interior and events at its surface.

**2006 STEREO (USA)** The Solar Terrestrial Relations Observatory consists of twin spacecraft that observe the Sun from different directions, giving all-around coverage of solar eruptions and the solar wind.

**2010 SDO (USA)** NASA's Solar Dynamics Observatory monitors the Sun to improve our understanding of its activity and to make better predictions of how this activity will affect Earth.

SDO

**2018 PARKER SOLAR PROBE (USA)** Using Venus flybys to slow itself down, Parker will orbit the Sun as close as 3.8 million miles (6.1 million km) to study the Sun's atmosphere.

**2020 SolO (EUROPE AND USA)** From an orbit just inside Earth's orbit to just inside Mercury's orbit, the Solar Orbiter will study the solar wind and the Sun's poles.

**CORONAL MASS EJECTION** *is a bubble of plasma ejected from the Sun into space*

**CORONA** *is hundreds of times hotter than the photosphere*

**GRANULATION** *is the mottling of the surface caused by convection cells*

**FACULAE** *are intensely bright active regions that are associated with the appearance of sunspots*

**PROMINENCE** *is a dense cloud of gas, suspended above the Sun's surface by magnetic field loops, that may persist for days or even weeks*

**SPICULES** *are short-lived jets of gas that are 6,000 miles (10,000 km) long*

# SURFACE

The visible surface of the Sun is called the photosphere. It is a layer of plasma (ionized gas) about 60 miles (100 km) thick and appears granulated or bubbly. The bumps, which are about 600 miles (1,000 km) wide, are the upper surfaces of convection cells that bring hot plasma up from the Sun's interior. Other significant features of the photosphere are sunspots, which are cooler regions that appear dark against their brighter, hotter surroundings. Sunspots and related phenomena, such as solar flares (tremendous explosions on the Sun's surface) and plasma loops, are thought to have a common underlying cause: they are associated with strong magnetic fields or disturbances in these fields. The magnetic fields result from the fact that the Sun is a rotating body that consists largely of electrically charged particles (ions in its plasma). Different parts of the Sun's convective zone rotate at different rates (faster at the equator than the poles), causing the magnetic field lines to become twisted and entangled over time. Sunspots are caused by concentrations of magnetic field lines inhibiting the flow of heat from the interior where they intersect the photosphere. Other types of disturbance are caused by twisted field lines popping out of the Sun's surface, releasing tremendous energy, or by plasma erupting as loops along magnetic field lines. The amount of sunspot and related activity varies from a minimum to a maximum over an 11-year cycle.

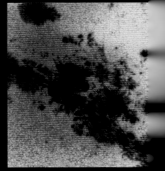

**SUNSPOTS**
Each sunspot has a dark central region, the umbra, and a lighter periphery, the penumbra. Away from the sunspots, the Sun's surface looks granulated. Each granule is the top of a convection cell in the Sun's interior.

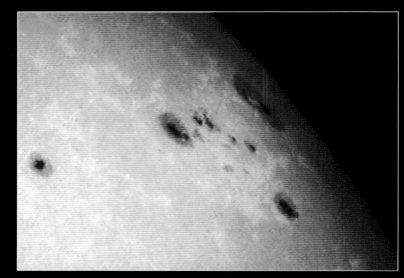

**PHOTOSPHERE**
The base of the photosphere has a temperature of 10,300°F (5,700°C), but its upper layers are cooler and emit less light. Here, the edge of the Sun's disk looks darker because light from it has emanated from these cooler regions.

**SOLAR ACTIVITY**
This ultraviolet image of the Sun was obtained by an instrument onboard the SOHO solar observatory. It shows the Sun's chromosphere (the layer just above the photosphere) and various protuberances, including a huge solar prominence, as well as a number of active regions on the solar surface. The image also shows a coronal mass ejection with a bright central area of ultraviolet emission

**FIRST OBSERVATION OF A SOLAR QUAKE**

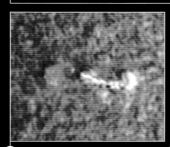

**1** In July 1996, by analyzing data obtained by an instrument on the SOHO observatory, scientists recorded a solar quake for the first time

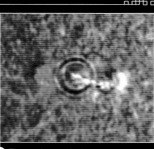

**2** The quake, equivalent to an earthquake of magnitude 11, was caused by a solar flare, visible as the white blob with a "tail" to its left

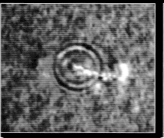

**3** The seismic waves looked like ripples on a pond but were 2 miles (3 km) high and reached a maximum speed of 248,600 mph (400,000 kph)

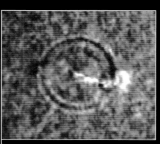

**4** Over the course of an hour, the waves traveled a distance equal to 10 Earth diameters before fading into the fiery background

## JOSEPH VON FRAUNHOFER

A German physicist and optical instrument maker, Joseph von Fraunhofer (1787–1826) is best known for his investigation of dark lines in the Sun's spectrum. Now known as Fraunhofer lines, they correspond to wavelengths of light absorbed by chemical elements in the outer parts of the Sun's atmosphere. Fraunhofer's observations were later used to help determine the composition of the Sun and other stars.

# ATMOSPHERE

As well as forming its visible surface, the photosphere is the lowest layer of the Sun's atmosphere. Above it are three more atmospheric layers. The orangey-red chromosphere lies above the photosphere and is about 1,200 miles (2,000 km) deep. From the bottom to the top, its temperature rises from 8,100°F (4,500°C) to about 36,000°F (20,000°C). The chromosphere contains many flamelike columns of plasma called spicules, each rising up to 6,000 miles (10,000 km) high along local magnetic field lines and lasting for a few minutes. Between the chromosphere and the corona is a thin, irregular layer called the transition region, within which the temperature rises from 36,000°F (20,000°C) to about 1.8 million °F (1 million °C). Scientists are studying this region in an attempt to understand the cause of the temperature increase. The outermost layer of the solar atmosphere, the corona, consists of thin plasma. At a great distance from the Sun, this blends with the solar wind, a stream of charged particles (mainly protons and electrons) flowing away from the Sun across the solar system. The corona is extremely hot, 3.6 million °F (2 million °C), for reasons that are not entirely clear, although magnetic phenomena are believed to be a major cause of the heating. Coronal mass ejections (CMEs) are huge bubbles of plasma, containing billions of tons of material, that are occasionally ejected from the Sun's surface through the corona into space. CMEs can disturb the solar wind, which results in changes to aurorae in Earth's atmosphere (see p.74).

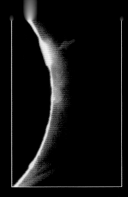

**CHROMOSPHERE**
The Sun's chromosphere is visible here as an irregular, thin red arc adjacent to the much brighter photosphere. Also apparent is a flamelike protuberance from the chromosphere into the corona.

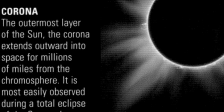

**CORONA**
The outermost layer of the Sun, the corona extends outward into space for millions of miles from the chromosphere. It is most easily observed during a total eclipse of the Sun, as here.

**CORONAL MASS EJECTION**
This image of a coronal mass ejection (top left) was taken by the SOHO solar observatory using a coronagraph—an instrument that blocks direct sunlight by means of an occulter (the central smooth red area in the image). The white circle represents the occulted disk of the Sun.

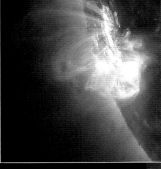

**MAGNETIC ERUPTION**
Hot plasma explodes into the atmosphere, following magnetic field lines. In this TRACE image, colors represent temperature, with blue being the coolest and red the hottest.

**POSTFLARE LOOPS**
These three images of a magnetically active solar region, taken by the TRACE satellite, span a period of 2.5 hours. The loops in the Sun's corona probably followed a solar flare and consist of plasma heated to exceedingly high temperatures along magnetic field lines.

**NORTHERN LIGHTS**
When charged particles from the solar wind reach Earth, they can cause aurorae. This photograph of the aurora borealis was taken in Manitoba, Canada.

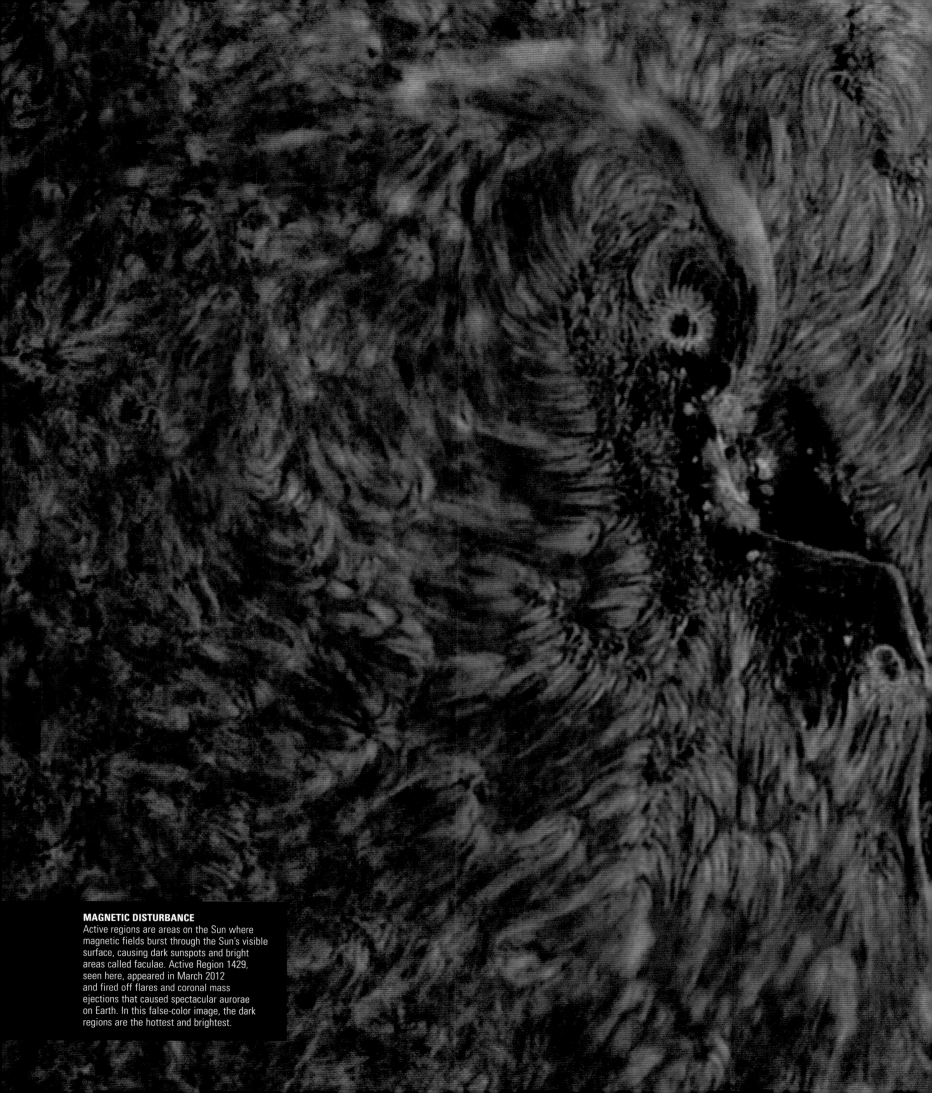

**MAGNETIC DISTURBANCE**
Active regions are areas on the Sun where
magnetic fields burst through the Sun's visible
surface, causing dark sunspots and bright
areas called faculae. Active Region 1429,
seen here, appeared in March 2012
and fired off flares and coronal mass
ejections that caused spectacular aurorae
on Earth. In this false-color image, the dark
regions are the hottest and brightest.

# MERCURY

MERCURY IS THE SMALLEST planet in the solar system, the closest planet to the Sun, and the richest in iron. The surface environment is extremely harsh. There is hardly any shielding atmosphere, and the temperature rises to a blistering 800°F (430°C) during the day, then plummets to an air-freezing −290°F (−180°C) at night. No other planet experiences such a wide range of temperatures. Its surface has been churned up by meteoritic bombardment and is dark and dusty.

## ORBIT

With the exception of Pluto, Mercury has the most eccentric of all the planetary orbits. At perihelion, it is only 28.6 million miles (46 million km) from the Sun, but at aphelion, it is 43.3 million miles (69.8 million km) away. The plane of Mercury's equator coincides with the plane of its orbit. (In other words, its axis of rotation is almost vertical.) This means that the planet has no seasons and that some craters close to the poles never receive any sunlight and are permanently cold. The orbit is inclined at 7° to the plane of the Earth's orbit. Because Mercury orbits inside the Earth's orbit, it displays phases, just like the Moon (see p.62).

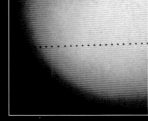

**TRANSIT OF MERCURY**
Mercury passes directly between Earth and the Sun about 13 times a century. This row of dots is a multiple exposure of Mercury's transit across the Sun in 2006.

**SPIN AND ORBIT**
Mercury rotates three times in two orbits. (In other words, there are 3 Mercurian "days" in 2 Mercurian "years.") This unusual spin–orbit coupling means that for an observer standing on Mercury, there would be an interval of 176 Earth days between one sunrise and the next.

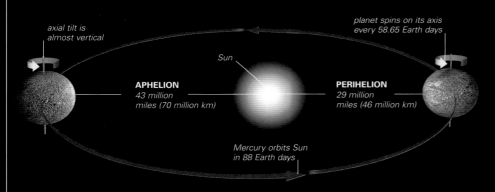

axial tilt is almost vertical

planet spins on its axis every 58.65 Earth days

Sun

**APHELION**
43 million miles (70 million km)

**PERIHELION**
29 million miles (46 million km)

Mercury orbits Sun in 88 Earth days

EXPLORING SPACE

### EINSTEIN AND MERCURY

Mercury's perihelion position moves slightly more than Isaac Newton's theories of motion predict. In the 19th century, it was proposed that a planet (called Vulcan) inside Mercury's orbit produced this effect. In his general theory of relativity of 1915, the German physicist Albert Einstein suggested that space near the Sun was curved and correctly predicted the exact amount by which the perihelion would move.

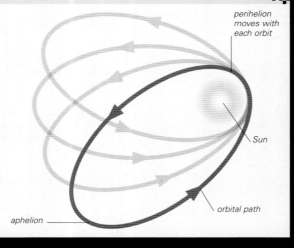

perihelion moves with each orbit

Sun

aphelion

orbital path

**MERCURY'S WOBBLY ORBIT**
Mercury's perihelion advances by about 1.55° every century, which is 0.012° more than is expected given the gravitational influence of nearby planets.

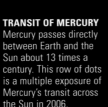

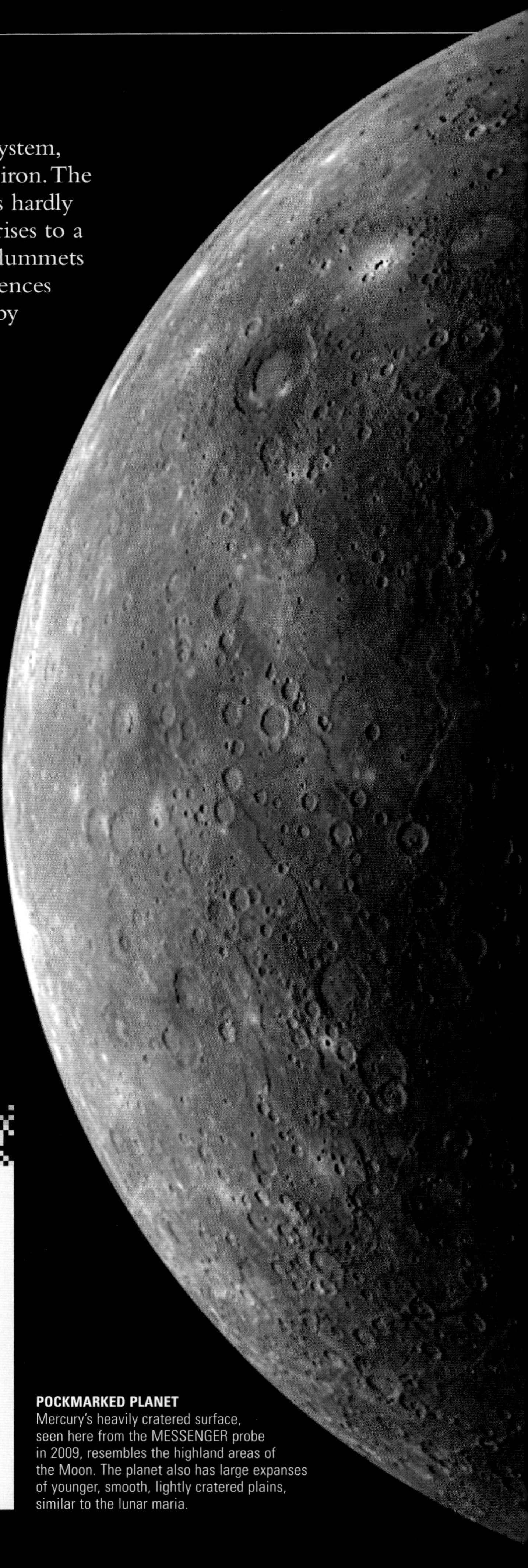

**POCKMARKED PLANET**
Mercury's heavily cratered surface, seen here from the MESSENGER probe in 2009, resembles the highland areas of the Moon. The planet also has large expanses of younger, smooth, lightly cratered plains, similar to the lunar maria.

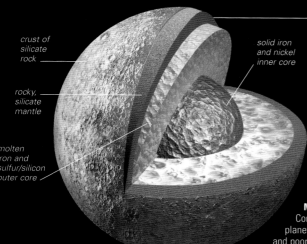

crust of silicate rock

rocky, silicate mantle

molten iron and sulfur/silicon outer core

solid iron and nickel inner core

# STRUCTURE

The very high density of Mercury indicates that it is rich in iron. This iron sank to the center some 4 billion years ago, producing a huge core 2,235 miles (3,600 km) in diameter. There is a possibility that a thin layer of the outer core is still molten. The solid rocky mantle is about 340 miles (550 km) thick and makes up most of the outer 25 percent of the planet. This outer mantle has slowly cooled, and during the last billion years, volcanic eruptions and lava flows have ceased, making the planet tectonically inactive. The mantle and the thin crust mainly consist of the silicate mineral anorthosite, just like the old lunar highlands. There are no iron oxides. Unlike on other planets, it seems that all the iron has gone into the core, which produces a magnetic field with a strength that is about 1 percent of Earth's magnetic field.

**MERCURY INTERIOR**
Compared to the other rocky planets, Mercury is very rich in metals and poor in heat-producing radioactive elements. Its huge iron core is probably solid.

# ATMOSPHERE

Mercury has a very thin temporary atmosphere because the planet's mass is too small for an atmosphere to persist. Mercury is very close to the Sun, so daytime temperatures are extremely high, reaching 800°F (430°C). The escape velocity is less than half that of Earth, so hot, light elements in the atmosphere, such as helium, quickly fly off into space. All the atmospheric gases therefore need constant replenishment. Mercury's atmosphere was analyzed by an ultraviolet spectrometer onboard the Mariner 10 spacecraft in 1974. Oxygen, helium, and hydrogen were detected in this way, and subsequently atmospheric sodium, potassium, and calcium have been detected by Earth-based telescopes. The hydrogen and helium are captured from the solar wind of gas that is constantly escaping from the Sun. The other elements originate from the planet's surface and are intermittently kicked up into the tenuous atmosphere by the impact of ions from Mercury's magnetosphere and micrometeorite particles from the solar system dust cloud. The atmospheric gases are much denser on the cold night side of the planet than on the hot day side, as the molecules have less energy to escape.

oxygen (52%)　　sodium (39%)　　potassium and other gases (1%)

helium (8%)

**ATMOSPHERIC COMPOSITION**
Oxygen is the most abundant gas, followed by sodium and helium. However, loss and regeneration of the gases is continuous, and the atmospheric composition can vary drastically over time.

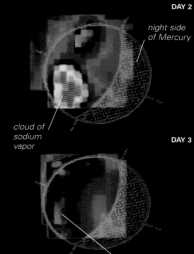

northern hemisphere

direction of sunlight

southern hemisphere

DAY 1

night side of Mercury

cloud of sodium vapor

DAY 2

DAY 3

sodium cloud has disappeared

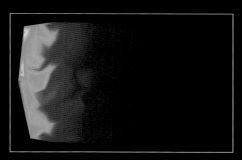

**MERCURY'S SODIUM TAIL**
Pressure exerted by sunlight pushes sodium atoms away from Mercury, forming a "tail" some 25,000 miles (40,000 km) long. Mercury and the Sun are off to the left in this false-colored view of Mercury's sodium tail. Emissions from this tail have previously been observed with Earth-based telescopes, but this image from a spectrometer onboard MESSENGER is the most detailed image yet.

**TEMPORARY ATMOSPHERE**
Thin clouds of sodium suddenly appear over some regions of Mercury and then just as quickly disappear, as seen in these false-color observations made by the Kitt Peak Solar Observatory in the US. The clouds might be produced by meteorite impacts—the freshly cratered surface releases sodium vapor when it is next heated by sunlight. Another possibility is that ionized particles actually hit Mercury's surface and release sodium from the regolith.

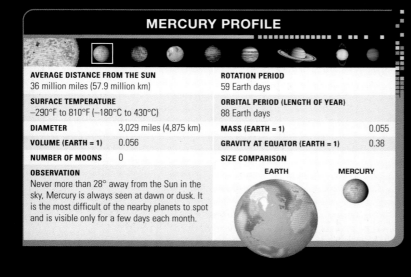

## MERCURY PROFILE

| | |
|---|---|
| **AVERAGE DISTANCE FROM THE SUN**<br>36 million miles (57.9 million km) | **ROTATION PERIOD**<br>59 Earth days |
| **SURFACE TEMPERATURE**<br>−290°F to 810°F (−180°C to 430°C) | **ORBITAL PERIOD (LENGTH OF YEAR)**<br>88 Earth days |
| **DIAMETER** 3,029 miles (4,875 km) | **MASS (EARTH = 1)** 0.055 |
| **VOLUME (EARTH = 1)** 0.056 | **GRAVITY AT EQUATOR (EARTH = 1)** 0.38 |
| **NUMBER OF MOONS** 0 | **SIZE COMPARISON** |

**OBSERVATION**
Never more than 28° away from the Sun in the sky, Mercury is always seen at dawn or dusk. It is the most difficult of the nearby planets to spot and is visible only for a few days each month.

EARTH　　MERCURY

THE SOLAR SYSTEM

# SURFACE FEATURES

Impact craters cover Mercury's visible surface.
As the surface gravity is about twice that of the
Moon, the ejecta blankets are closer to the parent
craters and thicker than those found on the Moon.
Large meteorite impacts have produced multiring
basins. A particularly impressive example is the
Caloris Basin. On the opposite side of the planet
to the basin is a region of strange terrain produced
by earthquakes resulting from the impact (see
right). The craters are interspersed by at least
two generations of flat plains of solidified basaltic
lava, like the lunar maria. Fluid lava oozed gently
out of vents in the crust and pooled in depressions.
Eventually, most of the vents were covered by the
lava. The MESSENGER space probe has photographed
volcanic vents around the perimeter of the Caloris
Basin, which are evidently the source of these lavas.
Mercury's surface also has several ridges, which are
up to 0.6–1.9 miles (1–3 km) high and 310 miles
(500 km) long.

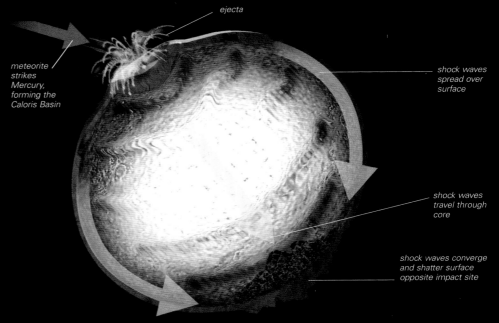

*ejecta*

*meteorite strikes Mercury, forming the Caloris Basin*

*shock waves spread over surface*

*shock waves travel through core*

*shock waves converge and shatter surface opposite impact site*

### SURFACE COMPOSITION

In this false-color mosaic,
yellow represents areas of
the silicate crust that have
been exposed by cratering,
while the blue regions are
younger volcanic rocks.

### IMPACT SHOCK WAVES

A few minutes after the formation
of the Caloris Basin, the shock
waves generated by the impact
came to a focus on the opposite
side of the planet. This caused
a massive upheaval over an area
of 96,500 square miles (250,000
square kilometers), raising ridges
up to 1.1 miles (1.8 km) high
and 3–6 miles (5–10 km) across.
Crater rims were broken into
small hills and depressions.

**CHAOTIC TERRAIN OPPOSITE THE CALORIS BASIN**

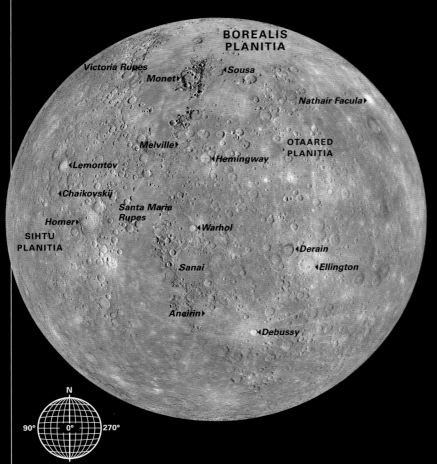

BOREALIS PLANITIA

Victoria Rupes
Monet
Sousa
Nathair Facula
Melville
OTAARED PLANITIA
Hemingway
Lemontov
Chaikovskii
Santa Maria Rupes
Homer
Warhol
SIHTU PLANITIA
Derain
Sanai
Ellington
Aneirin
Debussy

N
90°   0°   270°
S

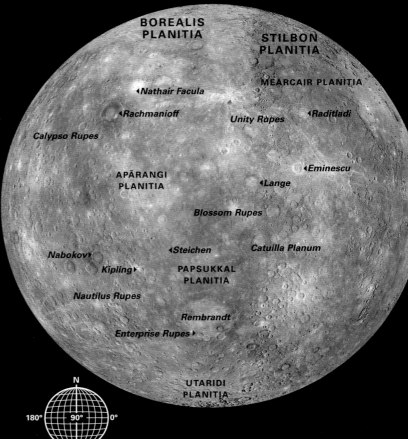

BOREALIS PLANITIA

STILBON PLANITIA

MEARCAIR PLANITIA

Nathair Facula
Rachmaninoff
Unity Rupes
Raditladi
Calypso Rupes
Eminescu
APĀRANGI PLANITIA
Lange
Blossom Rupes
Steichen
Catuilla Planum
Nabokov
Kipling
PAPSUKKAL PLANITIA
Nautilus Rupes
Rembrandt
Enterprise Rupes
UTARIDI PLANITIA

N
180°   90°   0°
S

## GLOBAL CONTRACTION

Escarpments traceable for hundreds of miles across Mercury's surface attest to global contraction that has caused the edges of tracts of crust to become thrust over the adjacent terrain. The biggest scarps are 1–2 miles (2–3 km) high and about 1,000 miles (1,500 km) long and date back billions of years. Except for those near the poles, most are oriented north-south. This suggests that the tidal slowing of the planet's rate of spin might have influenced their origin.

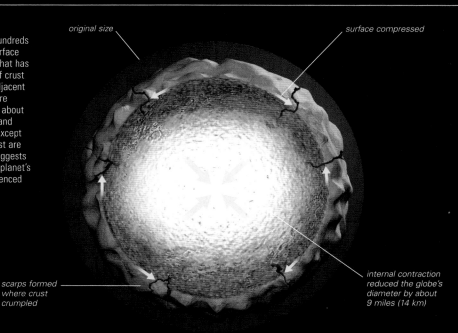

original size

surface compressed

scarps formed where crust crumpled

internal contraction reduced the globe's diameter by about 9 miles (14 km)

## MISSIONS TO MERCURY

Only two spacecraft have visited Mercury, and a third is on the way.

**1974–1975 MARINER 10 (USA)** Mariner 10 was placed in an orbit around the Sun that had exactly twice the period of Mercury's solar orbit. It flew past Mercury three times in 1974–1975, before its fuel supply ran out, but saw only the same half of the planet each time. It discovered the planet's unexpected magnetic field and sent back our first close-up views.

**2011–2015 MESSENGER (USA)** After making three flybys of Mercury, MESSENGER became the first spacecraft to orbit the planet, which it did for more than 4 years before crashing into the surface. It had a more sophisticated suite of instruments than Mariner 10 and carried a ceramic **MESSENGER** sunshade to protect itself from the Sun's heat. To avoid being close to Mercury's hot dayside surface for long periods, MESSENGER was placed into a strongly elliptical orbit.

**2025 BepiColombo (ESA/JAXA)** Launched in 2018, BepiColombo is due to achieve orbit around Mercury in 2025. A larger European orbiter will concentrate on the planet's surface, and a smaller Japanese orbiter will concentrate on its magnetic field.

# GEOGRAPHY

Mercury is an airless and heavily cratered world. Most of its surface consists of vast sheets of lava. The older the lavas, the more scarring has been caused by impact craters. The volcanic eruptions that supplied the lava waned to a trickle about 3 billion years ago, but the rate of impact cratering had waned by then, too, so that the youngest lava surfaces have relatively few craters and are described as "smooth plains." Most plains on Mercury (planitia in Latin) are named after Mercury (the Roman messenger god) in various languages, whereas craters are named after famous artists, authors, and musicians. Raphael, Shakespeare, and Mozart each have a home on this planet. Scarps (rupes in Latin) are named after ships used on voyages of discovery.

**MERCURY MAP**
The entire globe was mapped for the first time by NASA's MESSENGER craft and is shown here in black-and-white images at 90-degree intervals.

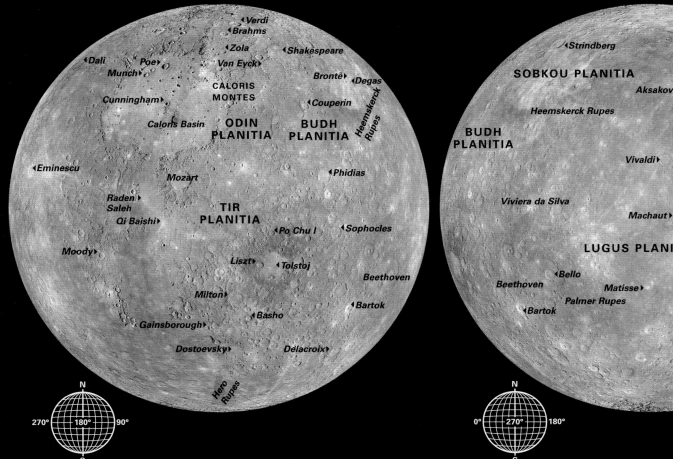

# IMPACT CRATERS

Mercury is covered with impact craters ranging in size from small, bowl-shaped craters to a basin that is a quarter of the diameter of the planet. The relative ages of different surface areas can be deduced because older regions have a greater density of superposed craters and because older craters become degraded by subsequent impacts distributing fragmental material (regolith).

**SOUTH POLE**
The temperature is permanently freezing in Mercury's south polar region because it receives very little sunlight.

## SHAKESPEARE REGION
## Caloris Basin

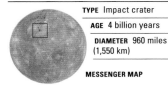

| TYPE | Impact crater |
|---|---|
| AGE | 4 billion years |
| DIAMETER | 960 miles (1,550 km) |

**MESSENGER MAP**

Caloris is the biggest well-preserved basin on Mercury, larger than the state of Texas. Shortly after being formed by a major impact about

3.9 billion years ago, its floor became flooded by lavas that show up as orange in exaggerated-color images, in contrast to the generally darker and bluer terrain outside the basin. Subsequent smaller impacts into the basin penetrated through the lavas, exposing very dark blue material from the original floor of the basin.

There are no "ghost craters" (see Agwo Facula, opposite) on the Caloris floor. This indicates either that the depth of lava flooding was so great that ghost craters are entirely obscured, or that there was not enough time for craters to be made on the floor of the new basin before the lavas were erupted.

Bright orange spots just inside the southern rim of the basin are thought to be material erupted by explosive volcanic eruptions, such as Agwo Facula (see opposite).

The name "Caloris" comes from the Latin for heat, bestowed because the Sun is overhead at this longitude during alternate perihelia, so midday temperatures here are particularly high.

**BASIN FLOOR**
The lavas on the basin floor are wrinkled by ridges and cut by fractures, which reflects a history of lava cooling and subsidence and scarring by impact craters of various sizes.

**IMPACT BASIN**
This exaggerated-color image shows that the lavas covering the basin floor (orange) have a different composition to the surrounding area.

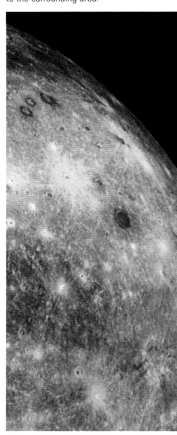

## BEETHOVEN REGION
## Bach Crater

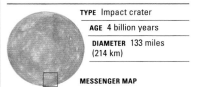

| TYPE | Impact crater |
|---|---|
| AGE | 4 billion years |
| DIAMETER | 133 miles (214 km) |

**MESSENGER MAP**

This two-ringed basin represents an intermediate class of craters, between the slightly smaller ones with central mountainous peaks and larger ones with multiple rings. The prominent inner ring is half the width of the outer, and the overall circularity is impressive. Bach was formed toward the end of the period of heavy bombardment. Lava later flooded the crater, producing the smooth floor.

**TWO-RINGED BASIN**

## SHAKESPEARE REGION
## Degas Crater

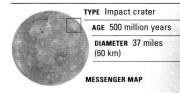

| TYPE | Impact crater |
|---|---|
| AGE | 500 million years |
| DIAMETER | 37 miles (60 km) |

**MESSENGER MAP**

The flat floor of this relatively young crater appears to be covered by a layer of impact melt that cracked as it cooled. The central peak appears bright because volatile substances have migrated there to form hollows (see opposite). Much of the ejecta surrounding the crater's rim is dark, so-called "low reflectance material" excavated by the crater-forming impact and probably originating from a graphite-rich subsurface layer.

**LOW-REFLECTANCE MATERIAL**

## SHAKESPEARE REGION
## Brahms Crater

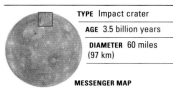

| TYPE | Impact crater |
|---|---|
| AGE | 3.5 billion years |
| DIAMETER | 60 miles (97 km) |

**MESSENGER MAP**

A large, mature, complex crater to the north of the Caloris Basin, Brahms has a prominent central mountainous peak about 12 miles (20 km) across. The walls have slipped inward, forming a series of elaborate, concentric, stairlike terraces and a highly irregular rim. This structure is typical of a crater this size—craters with diameters less than 6 miles (10 km) are bowl-shaped, and craters with diameters

greater than 80 miles (130 km) develop central rings (see Bach, above). Radial hills of ejecta surround Brahms.

**CENTRAL PEAK**
This 2-mile (3-km) high mountain was produced when the subsurface rock rebounded after being struck by an asteroid.

# GEOLOGICAL FEATURES

More than 3 billion years ago, most of Mercury's surface crust erupted as vast sheets of lava like the maria on the Moon but poorer in iron and richer in magnesium, silicon, and various volatile elements. Smaller areas, especially inside some craters, were flooded by lava more recently. There have also been explosive eruptions less than a billion years ago.

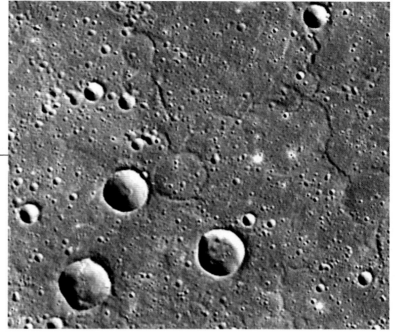

RENOIR REGION

## Discovery Rupes

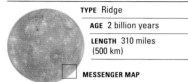

| TYPE | Ridge |
| --- | --- |
| AGE | 2 billion years |
| LENGTH | 310 miles (500 km) |
| | MESSENGER MAP |

This escarpment (*rupes* in Latin), which in places is 1.2 miles (2 km) high, is younger than both the craters and the volcanic plains that it crosses. Discovery runs in a northeast–southwest direction and is 310 miles (500 km) in length. Many similar features, known as "lobate scarps," have been mapped on Mercury. The planet's slow cooling over the past 4 billion years has caused its radius to shrink by about 5 miles (7 km). As a result, adjacent tracts of the rock crust have been thrust over the adjacent terrain.

**DISCOVERY RUPES CUTTING THROUGH IMPACT CRATERS**

## Agwo Facula

CALORIS BASIN FLOOR

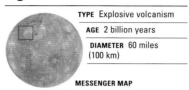

| TYPE | Explosive volcanism |
| --- | --- |
| AGE | 2 billion years |
| DIAMETER | 60 miles (100 km) |
| | MESSENGER MAP |

This orange-colored bright spot (*facula* in Latin) is one of nearly 200 examples on Mercury believed to be deposits resulting from explosive volcanic eruptions. These eruptions blasted out the irregular pits that can usually be seen near their midpoints. Each deposit gets fainter toward its outer edge, which blends into the background. The more powerful the explosion (or series of explosions), the bigger and brighter the deposit. The youngest such explosive eruption deposits on Mercury are probably less than a billion years old. Most eruptions on Mercury produced lava flows, but the less common explosive eruptions are significant

because they can happen only if there is enough gas to expand violently and cause an explosion when molten rock is erupted toward the surface.

**PLAINS VOLCANISM**
This image reveals faint circular traces of "ghost craters" that were completely flooded by lava, creating plains. More obvious craters were formed by later impacts into the lavas.

**COMPOUND VENT**
The 17-mile (27-km) long pit in the middle of Agwo Facula shows it to consist of several overlapping volcanic vents. The deepest exceeds nearly a mile (km) in depth.

## Hollows

RADITLADI REGION

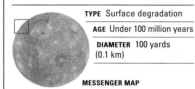

| TYPE | Surface degradation |
| --- | --- |
| AGE | Under 100 million years |
| DIAMETER | 100 yards (0.1 km) |
| | MESSENGER MAP |

The MESSENGER spacecraft revealed finer detail than previously seen on Mercury. One new discovery are features called hollows, which can occur singly but are more often in groups of dozens or hundreds. Hollows are individual patches of surface hundreds of feet (meters) wide, which have been stripped away to a depth of 33–66 ft (10–20 m), to leave steep-sided, flat-bottomed depressions.

Hollows must be young, because no impact craters on Mercury have been made inside them, and some hollows could still be forming today. Their shapes suggest gentle growth, in contrast to the larger and violently excavated volcanic vents

associated with faculae, but it is not known how or why they form. There is no air on Mercury, so it cannot be caused by scouring by wind. A volatile component of the surface material is probably being lost to space, perhaps because of Mercury's searing noontime temperatures, which can exceed 800°F (400°C).

Comparing the surface composition inside and outside hollows in order to work out what is going on is one of the science goals of the BepiColombo mission.

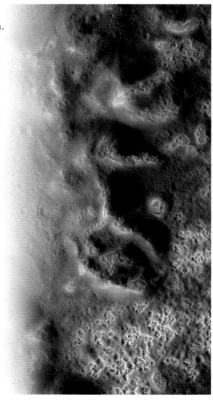

**DETAIL OF HOLLOWS**
This detailed view in exaggerated color covering an area only 10 miles (16 km) wide shows steep-sided, flat-bottomed blue hollows on the red floor and peak-ring inside Raditladi crater.

# VENUS

VENUS IS THE SECOND PLANET FROM THE SUN and Earth's inner neighbor. The two planets are virtually identical in size and composition, but these are very different worlds. An unbroken blanket of dense clouds permanently envelops Venus. Underneath lies a gloomy, lifeless, dry world with a scorching surface, hotter than that of any other planet. Radar has penetrated the clouds and revealed a landscape dominated by volcanism.

## ORBIT

Venus's orbital path is the least elliptical of all the planets. It is almost a perfect circle, so there is little difference between the planet's aphelion and perihelion distances. Venus takes 224.7 Earth days to complete one orbit. As it orbits the Sun, Venus spins extremely slowly on its axis—slower than any other planet. It takes 243 Earth days for just one spin, which means that a Venusian day is longer than a Venusian year. However, the time between one sunrise and the next on Venus is 117 Earth days. This is because the planet is traveling along its orbit as it spins, so any one spot on the surface faces the Sun every 117 Earth days. Venus's slow spin is also in the opposite direction from most other planets. Venus does not have seasons as it moves through its orbit. This is because of its almost circular path and the planet's small axial tilt. Venus's orbit lies inside that of the Earth, and about every 19 months, Venus moves ahead of Earth on its inside track and passes between our planet and the Sun. At this close encounter, Venus is within 100 times the distance to the Moon.

**SPIN AND ORBIT**
Venus is tipped over by 177.4°. This means its spin axis is tilted by just 2.6° from the vertical. As a result, neither of the planet's hemispheres nor poles points markedly toward the Sun during the course of an orbit.

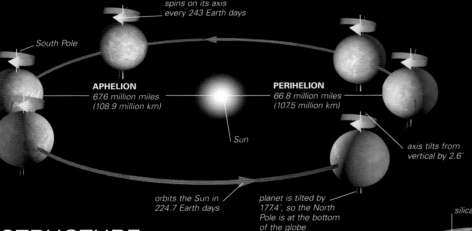

spins on its axis every 243 Earth days

South Pole

**APHELION**
67.6 million miles
(108.9 million km)

**PERIHELION**
66.8 million miles
(107.5 million km)

Sun

axis tilts from vertical by 2.6°

orbits the Sun in 224.7 Earth days

planet is tilted by 177.4°, so the North Pole is at the bottom of the globe

## STRUCTURE

Venus is one of the four terrestrial planets and the most similar of the group to Earth. It is a dense, rocky world just smaller than Earth and with less mass. Venus's Earth-like size and density leads scientists to believe that its internal structure, its core dimensions, and the thickness of its mantle are also similar to Earth's. So Venus's metal core is thought to have a solid inner part and a molten outer part, like Earth's core. In contrast to Earth, Venus has no detectable magnetic field. The planet spins extremely slowly compared to Earth, far too slowly to produce the circulation of the molten core that is needed to generate a magnetic field. Venus's internal heat—generated early in the planet's history and from radioactive decay in the mantle—is lost through the crust by conduction and volcanism. Heat melts the subsurface mantle material, and magma is released onto the surface

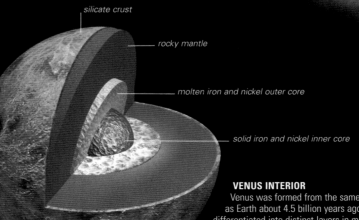

silicate crust

rocky mantle

molten iron and nickel outer core

solid iron and nickel inner core

**VENUS INTERIOR**
Venus was formed from the same material as Earth about 4.5 billion years ago and has differentiated into distinct layers in much the same way. A substantial part of the core has solidified; the exact amount still molten is unknown.

**TERRIBLE BEAUTY**
Venus's thick, reflective clouds cause
the planet to shine brightly, so from
a distance, it looks beguiling and
beautiful, which is why it was
named after the Roman goddess
of love and beauty. Close up,
it is a different story; no
human could survive
on this planet.

## VENUS PROFILE

| | |
|---|---|
| **AVERAGE DISTANCE FROM THE SUN**<br>67.2 million miles (108.2 million km) | **ROTATION PERIOD**<br>243 Earth days |
| **SURFACE TEMPERATURE**<br>867°F (464°C) | **ORBITAL PERIOD (LENGTH OF YEAR)**<br>224.7 Earth days |
| **DIAMETER** 7,521 miles (12,104 km) | **MASS (EARTH = 1)** 0.82 |
| **VOLUME (EARTH = 1)** 0.86 | **GRAVITY AT EQUATOR (EARTH = 1)** 0.9 |
| **NUMBER OF MOONS** 0 | **SIZE COMPARISON** |

**OBSERVATION**
Venus is the brightest planet in Earth's sky and
is surpassed in brightness only by the Sun and
the Moon. Its maximum magnitude is –4.7. It is
seen in the early morning or early evening sky.

SIZE COMPARISON
EARTH          VENUS

*carbon dioxide 96.5%*                    *nitrogen and trace gases 3.5%*

# ATMOSPHERE

Venus's carbon-dioxide-rich
atmosphere stretches up from the
ground for about 50 miles (80 km).
A deck of clouds with three distinct
layers lies within that atmosphere. The lowest layer is the densest
and contains large droplets of sulfuric acid. The middle layer
contains fewer droplets, and the top layer has small droplets. Close
to the planet's surface, the atmosphere moves very slowly and
turns with the planet's spin. Higher up, in the cloudy part of the
atmosphere, fierce winds blow westward. The clouds speed around
Venus once every four Earth days. The clouds reflect the majority
of sunlight reaching Venus back into space, so this is an overcast,
orange-colored world. Venus's equator receives more solar heat than
its polar regions. Yet the surface temperature at the equator and the
poles varies by only a few degrees from 867°F (464°C), as do the
day and night temperatures. The initial difference generates cloudtop
winds that transfer the heat to
the polar regions
in one large
circulation cell.
As a result, Venus
has no weather.

**COMPOSITION OF ATMOSPHERE**
Along with carbon dioxide and
nitrogen, Venus's atmosphere
contains traces of other gases,
such as water vapor, sulfur
dioxide, and argon.

*about 80 percent
of sunlight
reflects away*

*cloud deck
stretches from
about 28 miles
(45 km) to
about 43 miles
(70 km) above
the ground*

*carbon dioxide
in atmosphere
holds in heat*

*reflected light
means cloud
surface is bright
and easy to see*

*thick layers of
sulfuric acid clouds
stop most sunlight
from reaching
the surface*

*infrared radiation is
absorbed by carbon
dioxide and cannot
escape into space*

*20 percent of
sunlight reaches
rocky surface*

**GREENHOUSE EFFECT**
Venus's thick cloud layers trap heat and help produce the planet's
high surface temperature in the same way that glass traps heat in a
greenhouse. Only 20 percent of sunlight reaches the surface. Once
there, it warms up the rock. Heat in the form of infrared radiation is
then released, but it cannot escape and adds to the warming process.

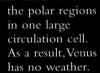

**SOUTH POLAR VORTEX**
This infrared view from the Venus Express
orbiter shows a rotating vortex the size of
Europe above the planet's south pole. We see
deeper into the cloudy atmosphere near the
dark center of this giant storm system. It is
known to have persisted for decades, but it
sometimes splits into two adjacent whorls.

## MISSIONS TO VENUS

Over 20 probes have investigated Venus. The first was the USA's Mariner 2, which made the first successful flyby of a planet in December 1962. Since then, probes have orbited Venus, plunged into its atmosphere, and landed on its scorching-hot surface.

**1967 VENERA 4 (USSR)** Sixteen different Venera probes traveled to Venus between 1961 and 1983. Venera 4 parachuted through Venus's atmosphere in October 1967 and returned the information that it is primarily composed of carbon dioxide.

**VENERA 4**

**1970 VENERA 7 (USSR)** This was the first probe to make a controlled landing on the surface. An instrument capsule landed on the night side and measured the temperature.

**1975 VENERA 9 AND 10 (USSR)** The first image of the surface came from Venera 9. Its lander touched down on October 22, 1975, and returned an image of rocks and soil. Venera 10 did the same 3 days later.

**1978 PIONEER VENUS (USA)** Two Pioneer Venus probes, each with several components, arrived in 1978. An orbiter collected data that was used to make the first global map of Venus, and probes studied the atmosphere.

**1981 VENERA 13 (USSR)** Venera 13 survived on the surface for 2 hours 7 minutes on March 1, 1982. It took the first color images and analyzed a soil sample. Flat slabs of rock and soil can be seen beyond the edge of the probe in the image below.

**1990 MAGELLAN (USA)** Between September 1990 and October 1994, Magellan made four 243-day mapping cycles of Venus. It collected gravity data on the fourth cycle.

**2005 VENUS EXPRESS (EUROPE)** Launched in 2005, the European Space Agency's Venus Express went into a highly elliptical orbit (passing over the planet's poles) in 2006 to monitor its clouds, atmospheric circulation, and magnetic field. The mission lasted until fuel was exhausted in December 2014.

**VENUS EXPRESS**

**2015 AKATSUKI (JAPAN)** Achieved orbit around Venus 5 years late, after its initial orbit insertion burn ended prematurely when the engines unexpectedly cut out.

# GEOLOGICAL FEATURES

Venus could be expected to have global features similar to those on Earth, but it differs in one key respect: it does not have moving plates. This means that its surface tends to move up or down rather than sideways. Yet Venus displays many familiar, Earth-like features formed by a range of tectonic processes, as well as some unfamiliar ones, such as arachnoids (see below). Venus has hundreds of volcanoes, from large, shallow-sloped shield volcanoes such as Maat Mons to small nameless domes. About 85 percent of the planetary surface is low-lying volcanic plains consisting of vast areas of flood lava. There has been volcanic activity as recently as about 500 million years ago, and it is possible that some of the volcanoes could be active. Other features are a result of the crust pulling apart or compressing. There are troughs, rifts, and chasms, as well as mountain belts such as Maxwell Montes, ridges, and rugged highland regions. Venus's highest mountains and biggest volcanoes are comparable in size to the largest on Earth, but overall this planet has less variation in height.

### FRACTURES
This complex network of narrow fractures extends over about 30 miles (50 km) of northwest Aphrodite Terra. It is reminiscent of a river system on Earth, but the angular intersections indicate this is a tectonically formed system of fractures.

### LAVA FLOWS
Solidified lava flows spread out for hundreds of miles in all directions from one of Venus's many volcanoes. The colors represent levels of heat radiation.

### SHIELD VOLCANO
Venus's tallest volcano, Maat Mons, rises to almost 3 miles (5 km) above the surrounding terrain and is 5 miles (8 km) above the planet's mean surface level.

### ARACHNOID
This spiderlike volcanic feature has a central circular depression (or dome) surrounded by a raised rim with radiating ridges and valleys.

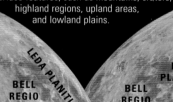

### VENUS MAPS
These four views combine to show the complete surface of Venus. They have been labeled to show surface features, such as mountains, craters, highland regions, upland areas, and lowland plains.

ISHTAR TERRA
LAKSHMI PLANUM
Maxwell Montes
Sacajawea Patera
Sachs Patera
Jeanne Crater
GUINEVERE PLANITIA
Sif Mons ▶ ◀ Gula Mons
Cunitz Crater ▶
EISTLA REGIO
Heng-o ▶ Corona
TINATIN PLANITIA
NAVKA PLANITIA
Danilova Crater ▶
Aglaonice Crater
Saskia Crater ◀
ALPHA REGIO
Stein ▶ Crater
N

LEDA PLANITIA
BELL REGIO
Pavlova ◀ Corona
APHRODITE TERRA

Fortuna Tessera ▲ Cleopatra Crater
LOUHI PLANITIA
LEDA PLANITIA
TELLUS TESSERA
BELL REGIO
Mead Crater ◀
Riley Crater ◀
Hestia Rupes
OVDA REGIO
APHRODITE TERRA
Kuama Chasma
AINO PLANITIA
N

# IMPACT CRATERS

Although many hundreds of impact craters have been identified on Venus, this total is low compared to that for the Moon and Mercury. There were more craters in the past, but they were wiped out by resurfacing due to volcanic activity about 500 million years ago. Venus's craters have some characteristics not seen elsewhere in the solar system, because its dense atmosphere and high temperature affect the incoming impactor and crater ejecta. Ejecta can, for example, be blown by winds and form fluidlike flows. And some potential impactors are too small to reach the surface intact. They break up in the atmosphere and either a resultant shock wave pulverizes the surface or a blanket of fine material formed by the breakup produces a dark halo mark before a crater forms. Wind has also modified the surface, creating wind streaks and what may be sand dunes.

**UNUSUAL CRATER**
This small crater, about 4 miles (6 km) across, has terraced walls and lobes of ejecta radiating out from the rim to give it an unusual starfishlike appearance.

**DARK HALO**
A dark halo surrounds a bright feature that appears to be a cluster of small impacts, ejecta, and debris formed by an impactor that broke up in the atmosphere.

**WIND STREAK**
A 22-mile (35-km) long tail of material has been created on the northeast side of this small volcano by prevailing winds.

**WIND EROSION**
Impact debris thrown 300 miles (500 km) to the northeast of Mead Crater has been modified by surface winds. Wind streaks are visible, but it is not known if these are bright streaks on dark terrain or vice versa.

# GEOGRAPHY

Present maps of Venus are based on data collected by the Magellan probe (see panel, opposite), with additional information from previous missions. The coloring of the maps below and Magellan images is based on the colors recorded by Venera 13 and 14. The orange color is due to the atmosphere filtering out the blue light. The following terminology is used for the surface features: lowland plains are termed planitia; high plains, planum; extensive landmasses, terra; mountain ranges, montes; and mountains or volcanoes, mons. A chasma is a deep, elongated, steep-sided depression. The features are all named after women, both historical and mythological, with the exception of Maxwell Montes (see p.120).

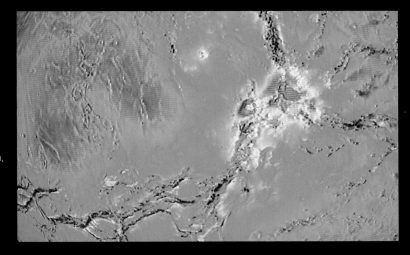

**TOPOGRAPHIC MAP**
This relief map, based on Magellan data, covers the surface about 16° on either side of the equator and between 152 and 217° east. The highest ground is shown as red and the lowest ground as blue. A complex system of troughs snakes through the highland region named Aphrodite Terra.

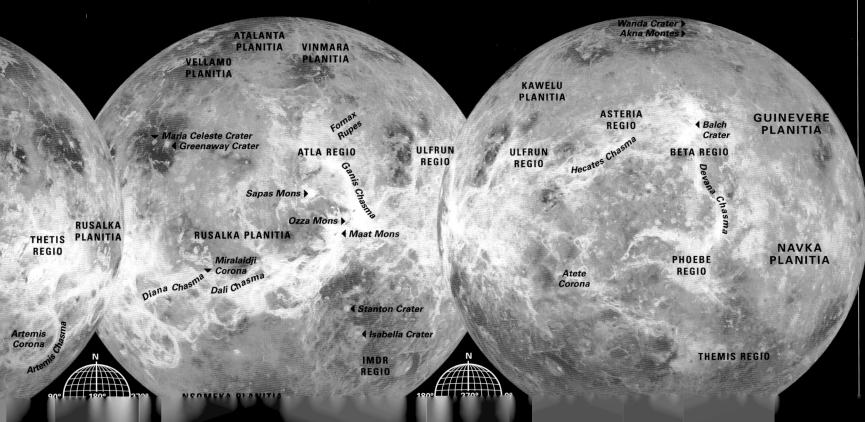

Wanda Crater ▶
Akna Montes ▶

ATALANTA PLANITIA

VINMARA PLANITIA

VELLAMO PLANITIA

KAWELU PLANITIA

ASTERIA REGIO

◀ Balch Crater

GUINEVERE PLANITIA

▽ Maria Celeste Crater
◀ Greenaway Crater

Fornax Rupes

ATLA REGIO

ULFRUN REGIO

ULFRUN REGIO

Hecates Chasma

BETA REGIO

Ganis Chasma

Devana Chasma

Sapas Mons ▶

Ozza Mons ▶

◀ Maat Mons

NAVKA PLANITIA

THETIS REGIO

RUSALKA PLANITIA

RUSALKA PLANITIA

Miralaidji ▼ Corona

PHOEBE REGIO

Diana Chasma

Dali Chasma

Atete Corona

Artemis Corona

Artemis Chasma

◀ Stanton Crater

◀ Isabella Crater

THEMIS REGIO

N

N

90° 180° 270°

NSOMEKA PLANITIA

IMDR REGIO

180° 270° 0°

# GEOLOGICAL FEATURES

Thanks to space-probe exploration, astronomers have a full and detailed picture of Venus's varied landscape. The planet has three main highland regions, termed terra. They are Aphrodite, which dominates the equatorial zone, and Lada and Ishtar. Over 20 smaller, upland areas, termed regio, are found around the planet. Extensive lowland plains, termed planitia, complete the global picture. Volcanic activity is evident across most of the surface, but the volcanoes are not randomly distributed. There are more in the uplands, particularly in Atla and Beta Regio, than in the highlands or plains.

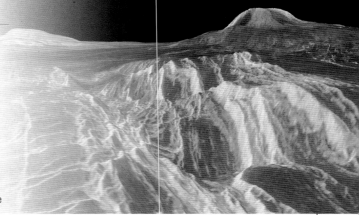

**VOLCANIC TERRAIN**
This view across western Eistla Regio is typical of the Venusian surface. The volcanoes on the skyline are Sif Mons (left) and Gula Mons.

---

### ISHTAR TERRA
## Ishtar Terra

| | |
|---|---|
| **TYPE** | Highland terrain |
| **AGE** | Under 500 million years |
| **LENGTH** | 3,485 miles (5,610 km) |

Ishtar Terra is a large plateau about the size of Australia, which stands 2 miles (3.3 km) above the surrounding lowlands. It is the nearest thing on Venus to the continents on

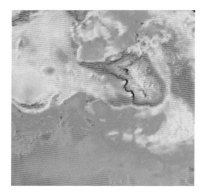

**LAVA CHANNEL**
Running for well over 1,200 miles (2,000 km), this lava channel is unusually long.

Earth. Its western region is the elevated plateau Lakshmi Planum, which is bounded at the northwest by the Akna Montes and the Freyja Montes and to the south by the Danu Montes. The steep-sided Maxwell Montes range makes up the eastern part of Ishtar Terra, along with a deformed area, Fortuna Tessera, to the mountain range's north and east. The plateau was possibly formed as areas of planetary crust were driven together. It is likely that beneath Ishtar there is cooled, thickened crust that is kept up by a rising region of mantle.

**MASSIVE PLATEAU**
This map, color-coded for height as mapped by Magellan, includes the highest point on the planet, Maxwell Montes. Blue represents the lowest elevation and white is the highest.

### ISHTAR TERRA
## Akna Montes

| | |
|---|---|
| **TYPE** | Mountain range |
| **AGE** | Under 500 million years |
| **LENGTH** | 515 miles (830 km) |

Forming the western border of Lakshmi Planum, Akna Montes is a ridge belt that appears to be the result of folding due to northwest–southeast compression. The mountain building is thought to have occurred after the plains formed, as the plains in this region seem to be deformed.

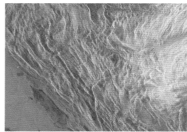

**FOLDING DUE TO COMPRESSION**

### ISHTAR TERRA
## Fortuna Tessera

| | |
|---|---|
| **TYPE** | Ridged terrain |
| **AGE** | Under 500 million years |
| **LENGTH** | 1,739 miles (2,801 km) |

Fortuna Tessera is an area of north–south-trending ridges about 600 miles (1,000 km) wide. The distinctive pattern made by the region's intersecting ridges and grooves led to this type of terrain originally being called parquet terrain, after its resemblance to woodblock flooring, although it is now termed tessera. The image here is a view looking westward across about 155 miles (250 km) of Fortuna Tessera toward the slopes of Maxwell Montes (colored in blue).

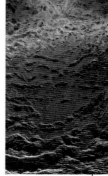

**RIDGES**

---

### ISHTAR TERRA
## Lakshmi Planum

| | |
|---|---|
| **TYPE** | Volcanic plain |
| **AGE** | Under 500 million years |
| **LENGTH** | 1,456 miles (2,345 km) |

The western part of Ishtar Terra consists of Lakshmi Planum. This is a smooth plateau 2.5 miles (4 km) high, formed by extensive volcanic

eruptions. The plateau is encircled by curving mountain belts—the Danu, Akna, Freyja, and Maxwell Montes—and steep escarpments such as Vesta Rupes to its southwest. This massive plain covers an area that is about twice the size of Earth's Tibetan Plateau (see pp.132–133). Two large volcanic features, the Colette Patera and Sacajawea Patera (see opposite), which punctuate the otherwise relatively smooth plain, were identified in Venera 15 and 16 data. Their floors lie over 1.5 miles (2.5 km) below the plateau level. There are just a few planums on Venus, all named after goddesses. Lakshmi is the Indian goddess of love and war.

**LAVA FLOWS**
The eastern Lakshmi region is covered by lava flows. The dark flows are smooth and the light ones are rough in texture. A bright impact crater can be seen on the right.

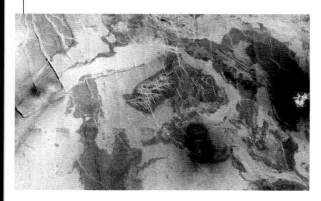

### ISHTAR TERRA
## Maxwell Montes

| | |
|---|---|
| **TYPE** | Mountain range |
| **AGE** | Under 500 million years |
| **LENGTH** | 495 miles (797 km) |

The Maxwell Montes mountain range forms the eastern boundary of Lakshmi Planum. It is the highest point on Venus, rising over 6 miles (10 km) above the surrounding lowlands. In

its higher regions, the ridges, which are 6–12 miles (10–20 km) apart, have a sawtooth pattern. The mountains fall away to Fortuna Tessera to the east. The western side is a complex of grooves and ridges and is particularly steep—Magellan data revealed that the southwestern flank has a slope of 35°. The mountain range was formed by compression and crustal foreshortening, which produced folding and thrust faulting. Venusian mountain ranges are usually named after goddesses, but Maxwell Montes is named after the British physicist James Clerk Maxwell, a pioneer of electromagnetic radiation.

**STEEP SLOPES**
This computer-generated image, looking eastward toward the Maxwell Montes, has been colored to show the presence of iron oxides on the surface.

## ISHTAR TERRA
# Sacajawea Patera

| TYPE | Caldera |
| --- | --- |
| AGE | Under 500 million years |
| DIAMETER | 145 miles (233 km) |

Sacajawea Patera is an elliptically shaped volcanic caldera on Lakshmi Planum. It is thought to have formed when a large underground chamber was drained of magma and then collapsed. The resulting caldera then sagged. The depression is about 0.75 miles (1.2 km) deep and is enclosed by a zone of concentric troughs and scarps that extend up to 60 miles (100 km) in length and are 0.3–2.5 miles (0.5–4 km) apart. They are thought to have formed as the caldera sagged. Sacajawea was a Shoshoni Indian woman, born in 1786, who worked as an interpreter.

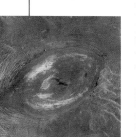

**SAG-CALDERA**
Bright linear scarps extend out from Sacajawea Patera's eastern edge.

## GUINEVERE PLANITIA
# Sachs Patera

| TYPE | Caldera |
| --- | --- |
| AGE | Under 500 million years |
| LENGTH | 40 miles (65 km) |

Sachs Patera is about 420 ft (130 m) deep and is surrounded by scarps spaced 1–3 miles (2–5 km) apart. A second, separately produced arc-shaped set of scarps lies to the north (top in the image below) of the main caldera. Solidified lava flows extend 6–16 miles (10–25 km) to the north and northwest of those scarps.

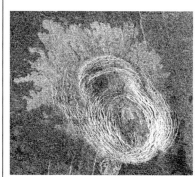

**SCARPS AROUND SACHS PATERA**

## BETA REGIO
# Beta Regio

| TYPE | Volcanic highland |
| --- | --- |
| AGE | Under 500 million years |
| LENGTH | 1,781 miles (2,869 km) |

Beta Regio is a large highland region dominated by Rhea Mons and Theia Mons. Rhea, which lies 500 miles (800 km) to the north of Thea, was originally thought to be a volcano, but Magellan data revealed it to be an uplifted massif cut through by a rift valley, the Devana Chasma (right). Theia Mons is a volcano superimposed onto the rift.

**RHEA AND THEIA MONS**

## BETA REGIO
# Devana Chasma

| TYPE | Fault |
| --- | --- |
| AGE | Under 500 million years |
| LENGTH | 2,860 miles (4,600 km) |

Devana Chasma is a large fault that cuts through Beta Regio (left). This major rift valley extends in a north–south direction and was produced as the planet's crust pulled apart and the surface sank to form a valley floor with steep sides.

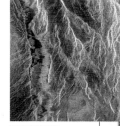

**LANDSLIDE**

It is similar to the Great Rift Valley on Earth (see p.130). Devana Chasma slices through Rhea Mons and Theia Mons. The fault is over 1.2 miles (2 km) deep and about 50 miles (80 km) wide near Rhea Mons. Elsewhere it is broader, as much as 150 miles (240 km) wide. To the south of Theia Mons, it continues to the highland region Phoebe Regio and reaches depths of 3.7 miles (6 km). Faults and grabens cut through and fan out from parts of the rift valley.

## EISTLA REGIO
# Eistla Regio

| TYPE | Volcanic highland |
| --- | --- |
| AGE | Under 500 million years |
| LENGTH | 4,977 miles (8,015 km) |

Eistla Regio is one of Venus's smaller upland areas, which are located in the lower basin land separating the major highland areas. Eistla Regio lies in the equatorial region to the west of the major highland, Aphrodite Terra. It is a series of broad crustal rises, each of which is several thousand miles across. The landscape was seen for the first time in the 1980s, when data collected by the Pioneer Venus Orbiter was used to produce the first accurate topographic map of Venus. Prominent features, such as the volcanoes Sif Mons and Gula Mons (right) and their lava flows, were clearly visible in the west of the region. Eistla Regio was also the first of the equatorial highlands imaged in the 1990s by Magellan, which revealed more detail of the broad volcanic rises and rift zones. An unusual type of volcanic dome unique to Venus is found within Eistla Regio. The domes are circular, flat-topped mounds of lava and so are often referred to as pancake domes. It is believed that when the lava erupted through the surface, it was highly viscous and so didn't flow freely. Cracks and pits in the domes are caused by cooling and withdrawal of lava.

**WESTERN EISTLA REGIO**
Lava flows extending for hundreds of miles fill the foreground of this image. In the distance, Gula Mons (left) and Sif Mons (right) rise above the plain, about 450 miles (730 km) apart.

**PANCAKE DOMES**
The two large, flat pancake domes are each about 37 miles (60 km) across and less than 0.6 miles (1 km) in height.

## EISTLA REGIO
# Gula Mons

| TYPE | Shield volcano |
| --- | --- |
| AGE | Under 500 million years |
| HEIGHT | 2 miles (3 km) |

Gula Mons is the larger of the two volcanoes that dominate the highland rise of western Eistla Regio (left). At its widest, it measures about 250 miles (400 km) across. This shield volcano is encircled by hundreds of miles of lava flows. It does not have a caldera at its summit, but a fracture line 93 miles (150 km) long. The volcano is also at the center of an array of crustal fractures. A particularly prominent one, Guor Linea, is a rift system that extends for at least 600 miles (1,000 km) from the southeast flank.

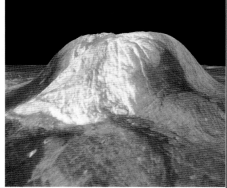

**SOUTHWEST SLOPES OF SUMMIT**

# Sapas Mons

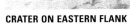

| | |
|---|---|
| **TYPE** | Shield volcano |
| **AGE** | Under 500 million years |
| **HEIGHT** | 1 mile (1.5 km) |

Rising 1 mile (1.5 km) above the surrounding terrain and with a diameter of about 135 miles (217 km), Sapas Mons is one of Venus's shield volcanoes. These are shaped like a shield or inverted plate, with a broad base and gently sloping sides, and are like those found on Earth's Hawaiian Islands. Sapas Mons is located in the Atla Regio, a broad volcanic rise just north of Venus's equator with an average elevation of 2 miles (3 km). The region is believed to have formed as a result of large volumes of molten rock welling up from the planet's interior. It is home to some particularly large shield volcanoes, which are linked by complex systems of fractures. These

**CRATER ON EASTERN FLANK**
Bright lava flows from Sapas Mons have stopped short of an impact crater on the volcano's eastern side. The flows, which are tens of miles long, cover some of the ejecta and therefore are younger than the crater.

include Ozza Mons, which is 4 miles (6 km) high, and the largest Venusian volcano, Maat Mons, which is 5 miles (8 km) high. Sapas Mons is covered in lava flows and grew in size as the layers of lava accumulated. The flows near the summit appear bright in Magellan radar images, suggesting that these are rougher than the dark

flows farther down the volcano's flanks. The flows commonly overlap, and many originate from the flanks rather than the summit. The summit has two mesas with flat to slightly convex tops. Nearby are groups of pits up to 0.6 miles (1 km) wide that are believed to have formed when underground chambers of magma drained away and the surface collapsed. The shield volcanoes are mainly named after goddesses: Sapas was a Phoenician goddess; Ozza, a Persian one; and Maat, an Egyptian.

**DOUBLE SUMMIT**
In this Magellan image of Sapas Mons taken from directly overhead, the two flat-topped mesas, which give the volcano the appearance of a double summit, appear dark against the bright lava flows. The area shown covers about 400 miles (650 km) from top to bottom.

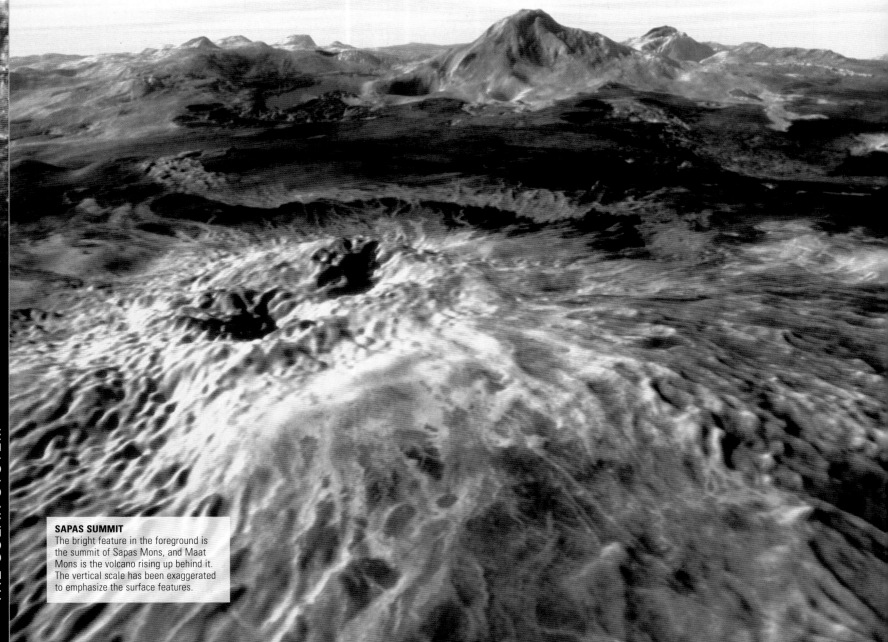

**SAPAS SUMMIT**
The bright feature in the foreground is the summit of Sapas Mons, and Maat Mons is the volcano rising up behind it. The vertical scale has been exaggerated to emphasize the surface features.

## LINEAR RIDGES
Ridges 20–40 miles (30–60 km) long lie along a northern slope of Ovda Regio. Dark lava, or possibly windblown dirt, fills the spaces between the ridges.

# Miralaidji Corona

| TYPE | Corona |
|---|---|
| AGE | Under 500 million years |
| DIAMETER | 186 miles (300 km) |

This large volcanic feature was formed by a plume of magma rising under the Venusian surface. The magma partially melted the crustal rock, which was raised up above the surrounding land to produce the corona, a blisterlike formation with radial faulting. The coronae on Venus range in size from about 30 to 1,600 miles (50 to 2,600 km) across and are circular to elongate in shape. They are named after fertility goddesses. Miralaidji is an Aboriginal fertility goddess.

# Ovda Regio

| TYPE | Highland terrain |
|---|---|
| AGE | Under 500 million years |
| DIAMETER | 3,279 miles (5,280 km) |

Ovda Regio is a highland area in Venus's equatorial region. It forms the western part of Aphrodite Terra, Venus's most extensive highland system, which rises 2 miles (3 km) above the mean surface level. Ovda Regio is one of a handful of highland regions on Venus that displays a type of complex ridge terrain known as tessera, a form of terrain, that was first identified in images taken by

Veneras 15 and 16. Tesserae are raised plateau-shaped regions with chaotic and complex patterns of crisscrossing lines. In places, the planet's crust has been fractured into mile-sized blocks. Elsewhere there are folds, faults, and belts of ridges and grooves hundreds of miles long. These are best seen along Ovda Regio's boundaries, where curving ridges and troughs have developed. There is also evidence that volcanic activity has played its part in the shaping of this

## HIGHLANDS AND LOWLANDS
Tessera ridges run between the Ovda Regio highland (right) and lowland lava flows (left). Some of the highland depressions have been partially filled by smooth material.

landscape. Magma, which may have welled up from the planet's interior, has flowed across part of the region, and ridges formed by compression have filled with lava. Ovda Regio is named after a Marijian (Russian) forest goddess.

**RADIAL FAULTING**

---

# Dali Chasma

| TYPE | Fault |
|---|---|
| AGE | Under 500 million years |
| LENGTH | 1,291 miles (2,077 km) |

The Dali Chasma lies in western Aphrodite Terra. It is a system of canyons and deep troughs coupled with high mountains that makes a broad, curving cut through more than 1,200 miles (2,000 km) of the planet's surface. Along with the Diana Chasma system, it connects the Ovda and Thetis highland regions with the large volcanoes at Atla Regio. The mountain ranges associated with the canyons rise for 2–2.5 miles (3–4 km) above the surrounding terrain. The canyons are 1.2–2.5 miles (2–3 km) deep.

## TROUGHS
In this view along the Dali Chasma, part of the raised rim of the 600-mile (1,000-km) wide Latona Corona is visible on the left.

# Artemis Corona

| TYPE | Corona |
|---|---|
| AGE | Under 500 million years |
| DIAMETER | 1,614 miles (2,600 km) |

Artemis is more than twice as big as the next largest corona on Venus, Heng-o. A nearly circular trough, Artemis Chasma, which has a raised rim, marks its boundary. Within it are complex systems of fractures, volcanic flows, and small volcanoes. Artemis, like other coronae, could have been formed by hot material welling up under the surface. But its large size and the surrounding trough mean that other forces, such as the pulling apart of the crust and surface, were involved.

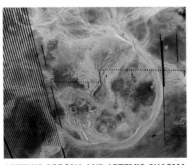

**ARTEMIS CORONA AND ARTEMIS CHASMA**

# Lada Terra

| TYPE | Highland terrain |
|---|---|
| AGE | Under 500 million years |
| LENGTH | 5,350 miles (8,615 km) |

Lada is the second largest of three highland regions on Venus. It is in the south polar region of the planet, largely south of latitude 50°S, and comparatively little is known about it. Lada Terra includes some typical tessera terrain of crisscrossing troughs and ridges. Volcanic activity has also affected the area. Lada includes three large coronae (blisterlike features), called

## LAVA CHANNEL
Part of a 745-mile (1,200-km) long channel, carved through Lada Terra by high-temperature, very fluid lava, runs from west to east across the center of this image.

Quetzalpetlatl, Eithinoha, and Otygen. Lava has flowed over and cut through the northern part of the region. All three terras on Venus are named after goddesses of love: Aphrodite is named after the Greek goddess; Ishtar (see p.120), the Babylonian goddess; and Lada, the Slavic goddess.

## RIDGE BELT
Bright and dark lava flows from the Ammavaru Volcano, which is 200 miles (300 km) to the left of this image, cut across a ridge belt to form a massive pool of lava.

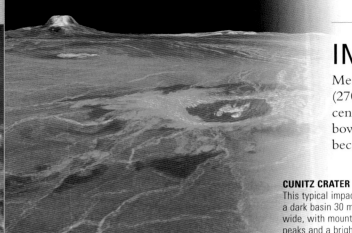

# IMPACT CRATERS

Meteorite impact craters on Venus range in size from 4 miles (7 km) to 168 miles (270 km) across. The largest are multiple-ringed, those of intermediate size have central peaks, and the smaller ones have smooth floors. The smallest of all—simple, bowl-like craters that are common on the Moon and Mars—are scarce on Venus, because the thick atmosphere filters out the small asteroids that would create them. Venusian craters are young and in many cases in pristine condition. The last volcanic resurfacing of Venus could have occurred as recently as 500 million years ago, so its craters must have mostly formed since then, and there has been little geological activity or weathering to affect them. Individual craters on Venus are named after women of note or are given female first names.

**CUNITZ CRATER**
This typical impact crater has a dark basin 30 miles (48 km) wide, with mountainous central peaks and a bright ejecta blanket around it.

## ISHTAR TERRA
# Wanda Crater

**TYPE** Central-peak crater
**AGE** Under 500 million years
**DIAMETER** 13.4 miles (21.6 km)

Wanda Crater is in the northern part of the Akna Montes mountain range. It was mapped first in 1984, by the Venera 15 and 16 missions, and Magellan studied it again a few years later. The crater has a large, rugged peak in the center of its smooth, lava-flooded floor. About one-third of all Venusian craters have such peaks. Material from the mountain ridges seems to have collapsed into the crater's western edge.

**CENTRAL PEAK**

## ISHTAR TERRA
# Cleopatra Crater

**TYPE** Double-ring crater
**AGE** Under 500 million years
**DIAMETER** 65 miles (105 km)

Cleopatra Crater is named after the legendary Egyptian queen. It is located on Maxwell Montes, Venus's highest mountain range and stands out as a smooth, eyelike feature against the rough mountainous terrain. Cleopatra was imaged by the Venera 15 and 16 spacecraft and the Arecibo radio telescope in the mid-1980s. It was one of several circular features that resembled both an impact crater and a volcanic feature. The data of the time revealed a feature of apparently great depth without the rim deposits typical of impact craters. As a result, Cleopatra was classified as a volcanic caldera. However, high-resolution

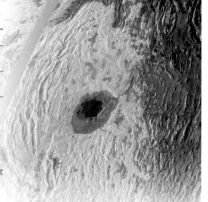

**MYSTERY CRATER**
The dark inner basin, the rim, and the surrounding ejecta revealed in this Magellan image from 1990 convinced astronomers that Cleopatra is an impact crater.

images from Magellan revealed an inner basin and rough ejecta deposits, providing conclusive proof that Cleopatra is an impact crater.

## GUINEVERE PLANITIA
# Jeanne Crater

**TYPE** Central-peak crater
**AGE** Under 500 million years
**DIAMETER** 12 miles (19.4 km)

An asteroid traveling from the southwest smashed obliquely into the Guinevere Planitia and created Jeanne Crater. Ejecta pushed out of the impact basin produced a distinctive triangular shape. Lobes formed to the northwest of the crater as molten material produced by the impact flowed downhill.

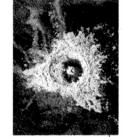

**TRIANGULAR EJECTA**

## BETA REGIO
# Balch Crater

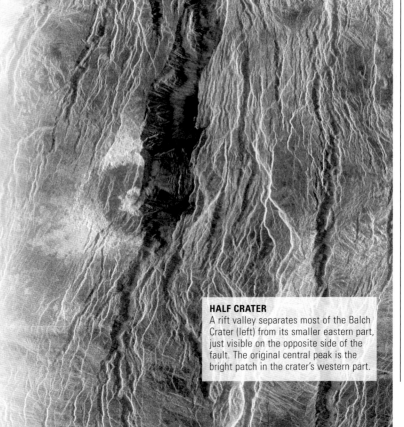

**TYPE** Central-peak crater
**AGE** Under 500 million years
**DIAMETER** 25 miles (40 km)

Most impact craters on Venus have remained unchanged since they were formed and have sharply defined rims. However, a relatively small number have been modified by volcanic eruptions and other kinds of tectonic activity. Balch Crater is one of these. Its circular form was split in two as the land was wrenched apart during the formation of a deep rift valley. The rift, which is up to 12.4 miles (20 km) wide, created a division that runs from north to south through the crater's center. The western half of the crater remains intact, but most of the eastern half was destroyed. A central peak and an ejecta blanket are visible in the western half. The crater was initially called Somerville but is now named after American economist and Nobel laureate Emily Balch.

**HALF CRATER**
A rift valley separates most of the Balch Crater (left) from its smaller eastern part, just visible on the opposite side of the fault. The original central peak is the bright patch in the crater's western part.

## APHRODITE TERRA
# Riley Crater

**TYPE** Central-peak crater
**AGE** Under 500 million years
**DIAMETER** 16 miles (25 km)

Riley Crater, named after 19th-century botanist Margaretta Riley, is one of the few Venusian craters to have been precisely measured. Comparison of images from different angles shows that the 16-mile (25-km) wide crater's floor lies 1,880 ft (580 m) below the surrounding plain, the rim is 2,009 ft (620 m) above it, and the peak is 1,737 ft (536 m) high.

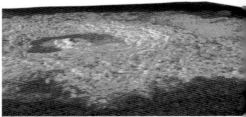

**OBLIQUE VIEW OF RILEY CRATER**

# Mead Crater

| | |
|---|---|
| **TYPE** | Multiringed crater |
| **AGE** | Under 500 million years |
| **DIAMETER** | 168 miles (270 km) |

Mead is the largest impact crater on Venus—although compared to craters on the Moon and Mercury, it is not very large. Mead is a multiple-ringed crater whose inner ring is the rim of the crater basin. This encloses a smooth, flat floor, which hides a possible central peak. The crater floor was flooded at the time of impact as a result of impact melt or by volcanic lava being released from below the surface. This explains why a crater of Mead's size is so shallow; there is a drop of only about 0.6 miles (1 km) between the crater rim and the crater center.

**LARGEST CRATER**
Mead has two distinct rings. Ejecta lies between them and beyond the outer ring. The vertical bands running through the picture are a result of image processing.

# Saskia Crater

| | |
|---|---|
| **TYPE** | Central-peak crater |
| **AGE** | Under 500 million years |
| **DIAMETER** | 23 miles (37.1 km) |

Saskia is a middle-sized crater, and its ejecta pattern is typical for its size. The ejecta blanket extends all the way around the crater's basin, suggesting that the impacting body smashed into the surface at a high angle. The crater has central peaks, formed as the planet's surface recoiled after

**THREE CRATERS**
Saskia lies at the lower left of this 300-mile (500-km) wide segment of Lavinia Planitia. Above it are the Danilova and Aglaonice craters.

being pushed down by the energy released during the impact. The original crater rim has collapsed and formed terraced walls. The incoming object must have been about 1.6 miles (2.5 km) across to produce a crater of this size. Images of Saskia and other craters, such as the similarly

sized Danilova (30.3 miles/48.8 km wide) and Aglaonice (39.6 miles/63.7 km wide), which lie within a few hundred miles of Saskia, have been produced from radar data collected by Magellan. Raw radar images (such as the one above) do not show features as they would appear to the naked eye. Instead, brightness varies according to the smoothness of the surface—rough areas appear light, while smooth ones look dark.

**SASKIA CRATER IN 3-D**
The color in this 3-D perspective view of Saskia is based on the color images of the Venusian surface recorded by the Venera 13 and 14 spacecraft.

# Stein Crater Field

| | |
|---|---|
| **TYPE** | Crater field |
| **AGE** | Under 500 million years |
| **DIAMETER** | 8.7 miles (14 km), 6.8 miles (11 km), and 5.6 miles (9 km) |

Small asteroids heading for Venus's surface can be broken up by the planet's dense atmosphere. The resulting fragments continue heading toward the surface, striking it simultaneously within a relatively small area to form a crater field. The Stein field consists of three small craters. The two smallest ones overlap. Material ejected by all three craters extends mainly to the northeast, suggesting that the fragments struck from the southwest. Material melted by the impacts has formed flow deposits, also lying to the northeast.

**STEIN TRIPLETS**

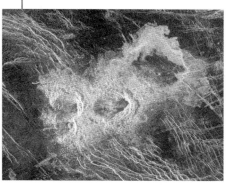

# Addams Crater

| | |
|---|---|
| **TYPE** | Central-peak crater |
| **AGE** | Under 500 million years |
| **DIAMETER** | 54 miles (87 km) |

The large, circular Addams Crater measures almost 55 miles (90 km) across, but it is its long tail that makes this crater unusual. An asteroid hit the ground from the northwest and created a crater basin with an ejecta blanket stretching out beyond about three-quarters of the crater rim. Additionally, impact-melt ejecta and lava extend out from about a third of the rim, creating a mermaid-style

**CRATER AND OUTFLOW**
A 372-mile (600-km) long, radar-bright flow of once-molten debris stretches to the east of Addams Crater.

tail to the east. The molten material flowed downhill for about 370 miles (600 km) from the impact site. The Magellan spacecraft found this area to be radar bright—that is, it bounced back a large portion of the radio waves that Magellan transmitted to it, which suggests it has a rugged surface. Venus's high surface temperature of about 867°F (464°C) allows ejecta to remain molten for a longer time than if it were on Earth. However, once the material cools below about 1,800°F (1,000°C), it becomes so viscous, it stops flowing. The crater is named after the American social reformer Jane Addams.

# Alcott Crater

| | |
|---|---|
| **TYPE** | Degraded crater |
| **AGE** | Under 500 million years |
| **DIAMETER** | 41 miles (66 km) |

Alcott is one of the few craters on Venus that have been modified by volcanic activity not associated with the crater's production. Many craters have floors flooded with lava that came up through the surface as the crater basin was formed. In Alcott's case, lava erupted elsewhere and then flowed over the crater. About half of the crater's rim is still visible, along with ejecta from the original impact lying to the south and east. A channel where lava once flowed touches the southwest edge of the crater.

**MODIFIED BY LAVA**

# EARTH

38–39 Gravity, motion, and orbits
56–57 Life in the universe
64–67 Celestial cycles
68–69 Planetary motion
74–75 Lights in the sky
103 The rocky planets

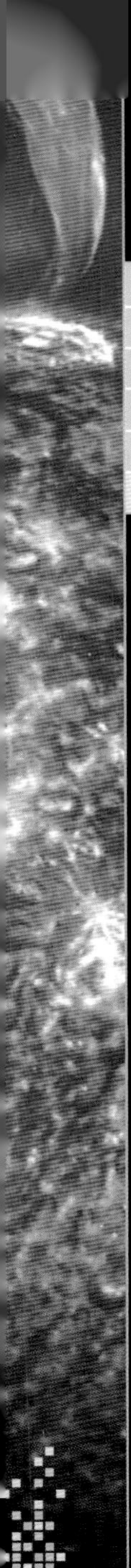

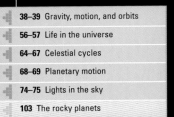

EARTH IS THE THIRD-CLOSEST PLANET to the Sun. The largest of the four rocky planets, it formed approximately 4.56 billion years ago. Earth's internal structure is similar to that of its planetary neighbors, but it is unique in the solar system in that it has abundant liquid water at its surface, an oxygen-rich atmosphere, and is known to support life. Earth's surface is in a state of constant dynamic change as a result of processes occurring within its interior and in its oceans and atmosphere.

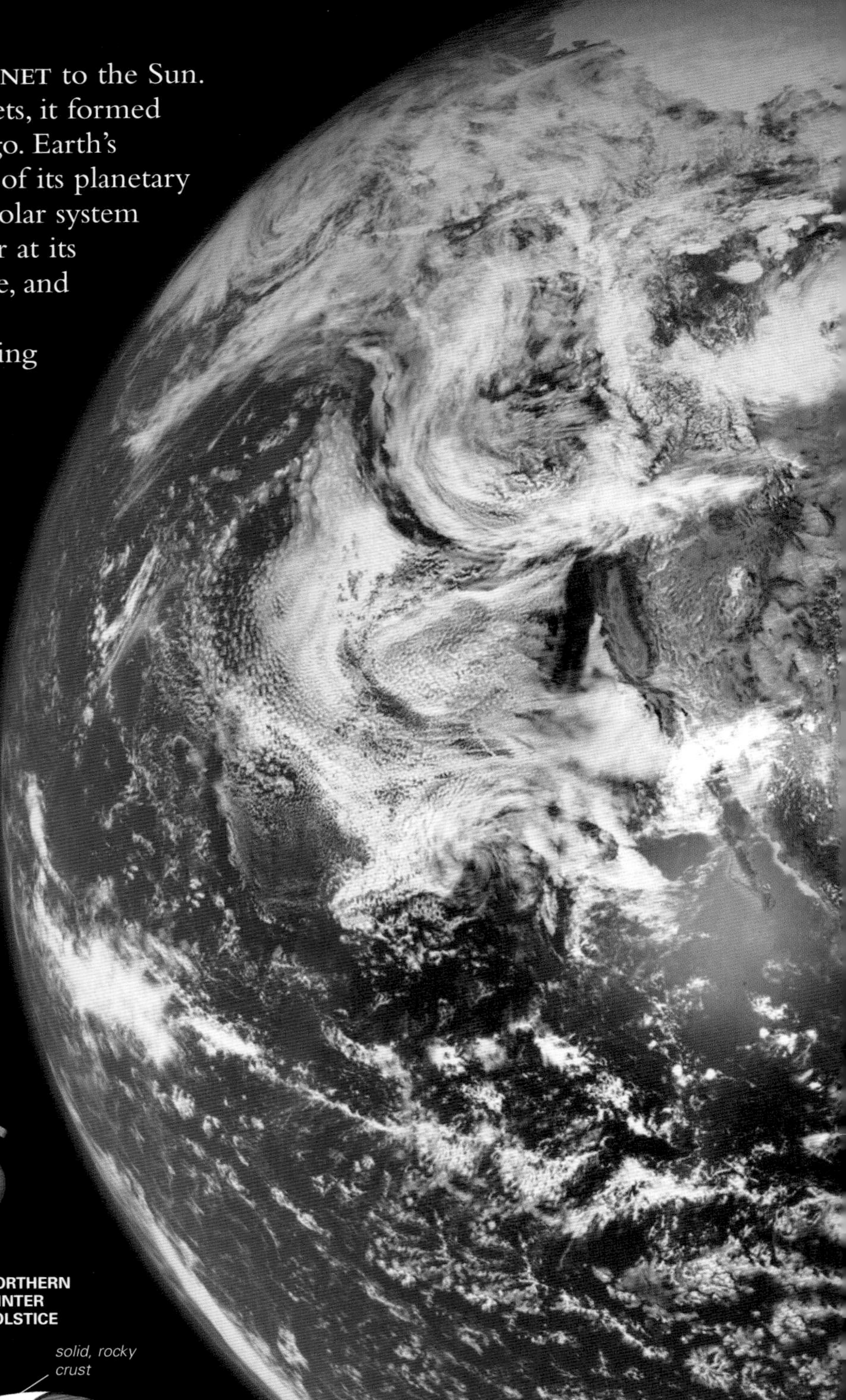

## ORBIT

Earth orbits the Sun at an average speed of 67,000 mph (108,000 kph) in a counterclockwise direction when viewed from above its North Pole. Like the other planets, Earth orbits the Sun along an elliptical path. As a result, about 7 percent more solar radiation currently reaches Earth's surface in January than in July. The plane of Earth's orbit around the Sun is called the ecliptic plane. Earth's spin axis is not perpendicular to this plane but is tilted at an angle of 23.5°. The eccentricity of Earth's elliptical orbit around the Sun (the degree to which it varies from a perfect circle) changes over a cycle of about 100,000 years, and its axial tilt varies over a 42,000-year cycle. Combined with a third cycle—a wobble in the direction in which the spin axis points in space, called precession (see p.64)—these variations are believed to play a part in causing long-term cycles in Earth's climate, such as ice ages.

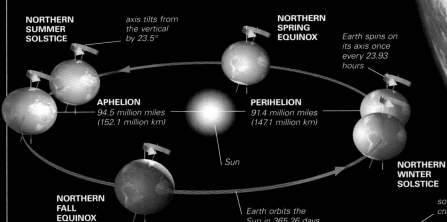

NORTHERN SUMMER SOLSTICE
axis tilts from the vertical by 23.5°
NORTHERN SPRING EQUINOX
Earth spins on its axis once every 23.93 hours
APHELION
94.5 million miles (152.1 million km)
PERIHELION
91.4 million miles (147.1 million km)
Sun
NORTHERN WINTER SOLSTICE
NORTHERN FALL EQUINOX
Earth orbits the Sun in 365.26 days

**SPIN AND ORBIT**
Earth is about 3 percent closer to the Sun at perihelion (in January) than at aphelion (in July). Its axial tilt combined with its elliptical orbit gives rise to seasons (see p.65).

## STRUCTURE

Earth's rotation causes its equatorial regions to bulge out slightly, by about 13 miles (21 km) compared to the poles. Internally, Earth has three main layers. The central core has a diameter of about 4,350 miles (7,000 km) and is made mainly of iron with a small amount of nickel. It has a central solid part, which has a temperature of about 8,500°F (4,700°C), and an outer liquid part. Surrounding the core is the mantle, which contains rocks rich in magnesium and iron and is about 1,700 miles (2,800 km) deep. Earth's crust consists of many different types of rocks and minerals, predominantly silicates, and is differentiated into continental crust and a thinner oceanic crust

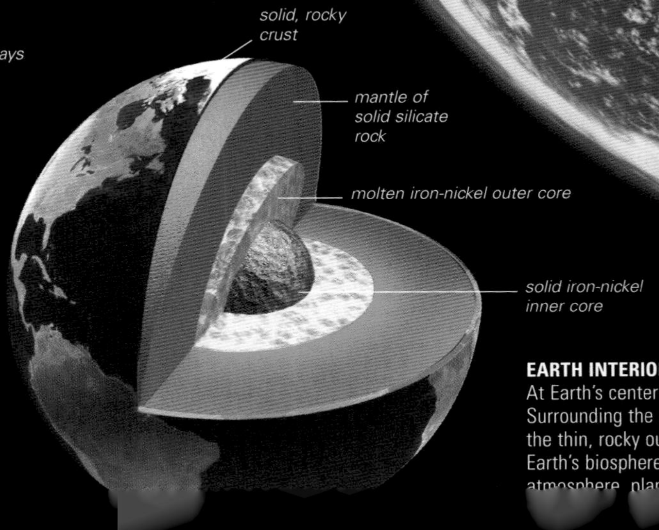

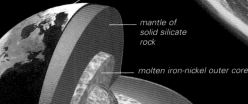

solid, rocky crust
mantle of solid silicate rock
molten iron-nickel outer core
solid iron-nickel inner core

**EARTH INTERIOR**
At Earth's center is a hot, dense core. Surrounding the core are the mantle and the thin, rocky outer crust that supports Earth's biosphere, with its oceans, atmosphere, plants, and animals.

**WATER WORLD**
Viewed from space, what clearly makes Earth unique is the abundance of surface water—in the oceans, lakes, atmosphere, and polar ice caps. The presence of surface water has been a key factor in the development of life on Earth.

## EARTH PROFILE

| AVERAGE DISTANCE FROM THE SUN | ROTATION PERIOD |
|---|---|
| 93.0 million miles (149.6 million km) | 23.93 hours |
| **AVERAGE SURFACE TEMPERATURE** | **ORBITAL PERIOD (LENGTH OF YEAR)** |
| 59°F (15°C) | 365.26 days |
| **DIAMETER** 7,926 miles (12,756 km) | **MASS (EARTH = 1)** 1 |
| **VOLUME (EARTH = 1)** 1 | **GRAVITY AT EQUATOR (EARTH = 1)** 1 |
| **NUMBER OF MOONS** 1 | |

# MAGNETIC FIELD

Earth has a substantial magnetic field, which is thought to be caused by a swirling motion of its liquid metal outer core. This motion is driven by a combination of Earth's rotation and convection currents within the outer core. The magnetic field behaves as though a large bar magnet were present within the Earth, tilted at an angle to its axis of rotation. The lines of the magnetic field converge at two points on Earth's surface called the north and south magnetic poles. The location of these points slowly changes over time. Currently, the north magnetic pole is north of Canada in the Arctic Ocean, while the south magnetic pole is north of eastern Antarctica, in the Southern Ocean. The magnetic field extends into space, forming a protective blanket around the planet by deflecting high-speed streams of charged particles that flow toward Earth in the solar wind (see p.107). A few of the particles escape deflection and become trapped within two regions surrounding Earth called the Van Allen radiation belts (see panel, right). Studies of iron-rich minerals in Earth's crust have shown that at variable time intervals (from less than 100,000 to millions of years), Earth's north and south magnetic poles switch.

## JAMES VAN ALLEN

James Van Allen (1914–2006) is an American physicist who, in the 1950s, designed and built instruments for American satellites. In 1958, a Van Allen-designed instrument carried by the first US satellite, Explorer 1, detected two large, doughnut-shaped belts of radiation around Earth, which carry trapped charged particles. The belts are named after Van Allen.

magnetic axis

magnetic equatorial plane

direction of magnetic force lines

solar wind

bow shock

Van Allen belts

magnetosphere tail

**EARTH'S MAGNETOSPHERE**
The imaginary surface at which Earth's magnetic field first deflects the solar wind is called the bow shock. Behind it is a region of space dominated by the magnetic field, in the sense that the field prevents solar wind particles from entering. Despite its elongated shape, this region is called the magnetosphere.

THE SOLAR SYSTEM

### THERMOSPHERE
The thermosphere extends to over 375 miles (600 km) above the Earth's surface. Temperature rises rapidly in the lower thermosphere due to absorption of solar energy and then increases gradually with altitude, reaching as high as 3,100°F (1,700°C).

### MESOSPHERE
This layer extends up to about 50 miles (80 km). Temperatures fall through the mesosphere to as low as −135°F (−93°C).

### STRATOSPHERE
The stratosphere is a calm layer stretching up to about 30 miles (50 km) above sea level. The temperature rises to 27°F (−3°C) at the top of this layer.

### TROPOSPHERE
This layer extends to 5 miles (8 km) above the poles and 10 miles (16 km) above the equator. It contains 75 percent of the total mass of the atmosphere. Temperatures fall to as low as −62°F (−52°C) at the top.

**HEIGHT ABOVE SEA LEVEL**

| | |
|---|---|
| 81 miles | 130 km |
| 75 miles | 120 km |
| 68 miles | 110 km |
| 62 miles | 100 km |
| 56 miles | 90 km |
| 50 miles | 80 km |
| 43 miles | 70 km |
| 27 miles | 60 km |
| 31 miles | 50 km |
| 25 miles | 40 km |
| 19 miles | 30 km |
| 12 miles | 20 km |
| 6 miles | 10 km |
| sea level | |

aurora

meteor burning up in the atmosphere

ice crystals on meteoric dust

ozone layer absorbs harmful radiation from the Sun

all weather occurs in the lowest level of the atmosphere

# ATMOSPHERE AND WEATHER

Earth is surrounded by the atmosphere, a layer of gases many hundreds of miles thick. This atmosphere is thought to have arisen partly from gases spewed out by ancient volcanoes, although its oxygen content—so vital to most forms of life—was created mainly by plants. Through the effects of gravity, the atmosphere is densest at Earth's surface and rapidly thins with altitude. With increasing altitude, there are also changes in temperature and a progressive drop in atmospheric pressure. For example, at a height of 19 miles (30 km), the pressure is just 1 percent that at sea level. Within the lowest layer of the atmosphere, the troposphere, continual changes occur in temperature, air flow (wind), humidity, and precipitation, known as weather. The basic cause of weather is the fact that Earth absorbs more of the Sun's heat at the equator than the poles. This produces variations in atmospheric pressure, which create wind systems. The winds drive ocean currents and cause masses of air with different temperatures and moisture content to circulate over the planet's surface. Earth's rotation plays a part in causing this atmospheric circulation because of the Coriolis effect (below).

### ATMOSPHERIC LAYERS
The four main layers in Earth's atmosphere are distinguished by different temperature characteristics. No boundary exists at the top of the atmosphere. Its upper regions progressively thin out and merge with space.

initial direction of moving air

deflection to right (Northern Hemisphere)

direction of spin

deflection to left (Southern Hemisphere)

### THE CORIOLIS EFFECT
The Coriolis effect causes deflections of air moving across Earth's surface. It is a consequence of the fact that objects at different latitudes move at different speeds around Earth's spin axis.

nitrogen 78.1%

argon and trace gases 1%

oxygen 20.9%

### COMPOSITION OF ATMOSPHERE
Nitrogen and oxygen make up 99 percent of dry air by volume. About 0.9 percent is argon, and the rest consists of tiny amounts of other gases. The atmosphere also contains variable amounts (up to 4 percent) of water vapor.

# PLATE TECTONICS

Earth's crust and the top part of its mantle are joined in a structure called the lithosphere. This is broken up into several solid structures called plates, which "float" on underlying semimolten regions of the mantle and move relative to each other. Most plates carry both oceanic crust and some thicker continental crust, although a few carry only oceanic crust. The scientific theory concerning the motions of these plates is called plate tectonics, and the phenomena associated with the movements are called tectonic features. Most tectonic features, which include ocean ridges, deep sea trenches, high mountain ranges, and volcanoes, result from processes occurring at plate boundaries. Their nature depends on the type of crust on either side of the boundary and whether the plates are moving toward or away from each other. Tectonic features occurring away from plate boundaries include volcanic island chains, such as the Hawaiian islands. These are caused by magma (molten rock) upwelling from "hot spots" in the mantle, causing a series of volcanoes to form on the overlying plate.

destructive boundary, where tectonic plates converge

plate dragged along by convection current

constructive boundary, where plates diverge and new crust is created

circular motion of convection current

lithospheric tectonic plate

plate in collision descends into mantle

mantle plume rises from lower mantle

upper mantle

lower mantle

outer core

### MOVING PLATES
Earth's plates move relative to each other as a result of convection currents within the mantle. The currents cause parts of the mantle to rise, move sideways, and then sink again, dragging the plates along as they do so.

### TECTONIC PLATES
Earth's surface is broken into seven large plates, such as the Eurasian plate, and many smaller ones, such as the Indian plate. Each continent is embedded in one or more plates.

North American Plate

Eurasian Plate

Pacific Plate

Plate boundary

Indian Plate

Australian Plate

**RAINFOREST**
Forests cover 30 percent of Earth's land surface and range from the cold, dark boreal forest of the far north to the dense rainforests of the humid tropics.

**SANDY DESERT**
Deserts cover about 20 percent of Earth's land surface, but only a small proportion are occupied by sand dunes, like these in the Sahara Desert.

# SURFACE FEATURES

From space, the flatter areas of Earth's land surface (apart from the areas dominated by ice) appear either dark green or various shades of yellow-brown. The green areas are forests and grasslands, which comprise a major component of Earth's biosphere (the planet's life-sustaining regions). The yellow-brown areas are mainly deserts, which have been created over long periods by various weathering and erosional processes. Like the other rocky planets, Earth has suffered many thousands of meteorite impacts over its history (see p.103). But because Earth's surface is so dynamic, the evidence for most of these impacts has disappeared, removed by erosion or covered up by depositional processes.

# WATER

Water is a dominant feature of Earth's surface. Overall, about 97 percent of the water is in oceans (which cover 75 percent of the surface); 2 percent is in ice sheets and glaciers; less than 1 percent is in ground water (underground and in rocks); and the rest is in rivers, lakes, and the atmosphere. The presence of liquid water has been key to the development of life on Earth, and the heat capacity of the oceans has been important in keeping the planet's temperature relatively stable. Liquid water is also responsible for most of the erosion and weathering of Earth's continents, a process unique in the solar system, although it is believed to have occurred on Mars in the past.

## KINGDOMS OF LIFE

Biologists use various systems for classifying living organisms, but the most widely used is the five-kingdom system. This classifies organisms mainly on the basis of their cell structure and method of obtaining nutrients and energy. However, not all scientists accept this system as satisfactory, and some have proposed switching to an eight-kingdom system or one with 30 kingdoms grouped into three superkingdoms.

**ANIMALS**

Animals are multicellular organisms that contain muscles or other contractile structures allowing some method of movement. They acquire nutrients, so they gain energy by ingesting food. Many animals, including mammals, are vertebrates (they possess a backbone), but a far larger number are invertebrates (without a backbone).

**VERTEBRATE**

**PLANTS**

Plants are multicellular organisms that obtain energy from sunlight through the process of photosynthesis. Their cells contain special pigments for absorbing light energy and are enclosed by cell walls made of cellulose.

**FLOWERING PLANT**

**FUNGI**

Fungi acquire nutrients by absorption from other living organisms or dead and decaying organic material. They have no means of locomotion. They range from yeasts (microscopic unicellular organisms) to multicellular forms with large fruiting bodies, such as mushrooms.

**TOADSTOOL**

**PROTISTS**

Protists are microscopic, mainly single-celled organisms whose cells contain nuclei. Some gain energy from sunlight, others ingest food like animals.

**PARAMECIUM**

**MONERANS**

Monerans are the simplest, smallest, most primitive, and most abundant organisms on Earth. The two main groups are bacteria and blue-green algae (cyanobacteria). Monerans are single-celled, but their cells contain no distinct nucleus. Most reproduce by splitting in two.

**MYCOBACTERIUM**

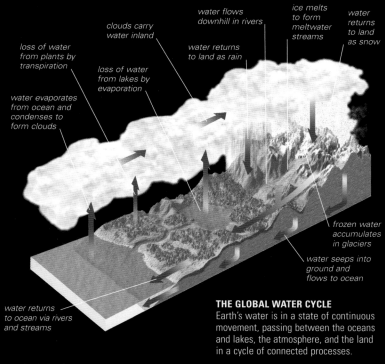

clouds carry water inland

water flows downhill in rivers

ice melts to form meltwater streams

water returns to land as snow

loss of water from plants by transpiration

loss of water from lakes by evaporation

water returns to land as rain

water evaporates from ocean and condenses to form clouds

frozen water accumulates in glaciers

water seeps into ground and flows to ocean

water returns to ocean via rivers and streams

**THE GLOBAL WATER CYCLE**
Earth's water is in a state of continuous movement, passing between the oceans and lakes, the atmosphere, and the land in a cycle of connected processes.

# LIFE ON EARTH

Evidence in ancient rocks points to the presence of simple, bacterialike organisms on Earth some 3.8 billion years ago. However, the prevailing scientific view is that life started on Earth long before that as a result of complex chemical reactions in the oceans or atmosphere. These reactions eventually led to the appearance of a self-replicating and self-repairing molecule, a precursor of DNA (deoxyribonucleic acid). Once life in this rudimentary form had started, processes such as mutation and natural selection inevitably led, over the vast expanses of geological time, to a collection of life forms of increasing diversity and complexity. Life spread from the oceans to the land and to every corner of the planet. Currently, Earth is teeming with life in astonishing abundance and diversity.

# GEOLOGICAL FEATURES

Most of Earth's geological features are associated with plate boundaries. At constructive (or divergent) boundaries, plates move apart and new crust is added. Examples are midocean ridges and the East African Rift. At destructive (or convergent) boundaries, two plates push against each other, producing a range of features, depending on

**THE SAN ANDREAS FAULT**
This fault in California, known for producing earthquakes, marks a transform boundary where two plates push past each other.

the nature of the crust on each plate. Many plate boundaries are associated with an increased frequency of volcanism, earthquakes, or both.

---

**AFRICA** *east* and **ASIA** *southwest*

## East African Rift

**LOCATION** Extending from Mozambique northward through East Africa, the Red Sea, and into Lebanon

**TYPE** Series of rift faults

**LENGTH** 5,300 miles (8,500 km)

The East African Rift provides an example of the geological process of rifting—the stretching and tearing apart of a section of continental crust by a plume of hot magma pushing up

**OL DOINYO LENGAI**
This active volcano in northern Tanzania sits in the middle of the east African part of the East African Rift.

underneath it. Rifting is associated with the development of a constructive plate boundary, which is formed as ascending magma creates new crust and pushes the plates on either side of the rift apart. The main section of the East African Rift runs (in two branches) through eastern Africa. Over tens of millions of years, rifting in this region has caused extensive faulting, the collapse of large chunks of crust, and associated features such as volcanism and a series of lakes in the subsided sections. As rifting continues, it is anticipated that a large area of eastern Africa will eventually split off as a separate island. A northern arm of the rift valley runs up the Red Sea and eventually reaches Lebanon, in the north. This coincides with a divergent boundary that is pushing Arabia away from Africa.

**THE NORTHERN RED SEA**
The Gulf of Aqaba (center right), a branch of the Red Sea, forms part of the northern arm of the East African Rift. The Gulf of Suez (center) is a side branch of the rift.

---

**BLACK SMOKERS**
Hydrothermal vents are underwater geysers located near midocean ridges. The hot water spewed out by some vents, called "black smokers," is discolored by the dark mineral iron sulfide.

**ATLANTIC OCEAN**

## Mid-Atlantic Ridge

**LOCATION** Extending from the Arctic Ocean to the Southern Ocean

**TYPE** Slow-spreading midocean ridge

**LENGTH** 10,000 miles (16,000 km)

The Mid-Atlantic Ridge is the longest mountain chain on Earth and one of its most active volcanic regions, albeit mainly underwater. The ridge sits on top of the Mid-Atlantic Rise, a bulge that runs the length of the Atlantic Ocean floor. Both rise and ridge coincide with plate boundaries that divide the North and South American plates, on the west, from the Eurasian and African plates, on the east. These are constructive plate boundaries, where new ocean crust is formed by magma upwelling from Earth's mantle. As this crust forms, the plates on either side are pushed away from the ridge at a rate of 0.4–4 in (1–10 cm) a year, widening the Atlantic

basin. The discovery in the 1960s of this spreading of the Atlantic sea floor—evidenced by the fact that crustal material near the ridge is younger than that farther away—led to general acceptance of the theory of continental drift. The ridge is a site of extensive earthquake activity and volcanism, along with many seamounts (isolated underwater mountains). Where the volcanoes break the ocean surface, they have formed islands such as Iceland and the Azores.

**SURTSEY**
Between 1963 and 1967, a massive and dramatic submarine eruption, from a section of the Mid-Atlantic Ridge to the south of Iceland, produced the new island of Surtsey.

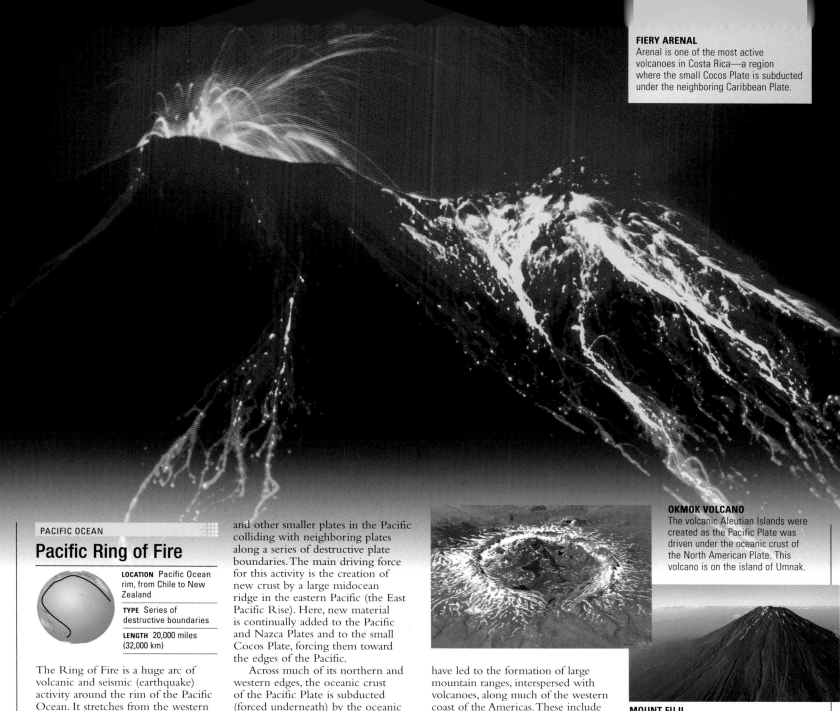

**FIERY ARENAL**
Arenal is one of the most active volcanoes in Costa Rica—a region where the small Cocos Plate is subducted under the neighboring Caribbean Plate.

PACIFIC OCEAN

# Pacific Ring of Fire

**LOCATION** Pacific Ocean rim, from Chile to New Zealand

**TYPE** Series of destructive boundaries

**LENGTH** 20,000 miles (32,000 km)

The Ring of Fire is a huge arc of volcanic and seismic (earthquake) activity around the rim of the Pacific Ocean. It stretches from the western coasts of South America and North America, across the Aleutian Islands of Alaska, and down the eastern edge of Asia, to the northeast of Papua New Guinea, and finally to New Zealand. More than half of the world's active volcanoes above sea level are part of the ring. The Ring of Fire results from the Pacific Plate and other smaller plates in the Pacific colliding with neighboring plates along a series of destructive plate boundaries. The main driving force for this activity is the creation of new crust by a large midocean ridge in the eastern Pacific (the East Pacific Rise). Here, new material is continually added to the Pacific and Nazca Plates and to the small Cocos Plate, forcing them toward the edges of the Pacific.

Across much of its northern and western edges, the oceanic crust of the Pacific Plate is subducted (forced underneath) by the oceanic crust of other plates, forming deep-sea trenches. This predisposes these regions to earthquakes, and the subducted crust also melts deep down to create hot magma, which reaches the surface through volcanoes. The result has been the formation of many highly volcanic island arcs in these regions—examples being the Aleutian Islands, the Kurile Islands, the islands of Japan, and the Mariana Islands.

On the eastern side of the Pacific, the situation is somewhat different. Here, parts of the Pacific, Nazca, and Cocos Plates are being subducted below continental crust. Deep-sea trenches have also formed here, but instead of island arcs, the plate collisions have led to the formation of large mountain ranges, interspersed with volcanoes, along much of the western coast of the Americas. These include the Cascade Range in Washington, home of the active volcano Mount St. Helens, and the Andes in South America, Earth's longest and most active land mountain range.

**MOUNT RUAPEHU**
At the southwest corner of the Ring of Fire is New Zealand. Here, steam rises from the country's tallest volcano, Ruapehu, between eruptions that occurred in 1995 and 1996.

**THE ANDES**
On the western edge of South America, subduction of the Nazca Plate under the South American Plate has created the Andes, another highly active region.

**OKMOK VOLCANO**
The volcanic Aleutian Islands were created as the Pacific Plate was driven under the oceanic crust of the North American Plate. This volcano is on the island of Umnak.

**MOUNT FUJI**
In the northwest Pacific, the subduction of the Pacific Plate under the Eurasian Plate is responsible for creating the islands of Japan, the site of volcanoes such as Mount Fuji, which last erupted in 1707.

# Himalayas

**LOCATION** Running southeast from northern Pakistan and India across Nepal to Bhutan

**TYPE** Continent–continent collision

**LENGTH** 2,400 miles (3,800 km)

The Himalayas are the highest mountain range on Earth, as well as one of the youngest. If the neighboring Karakoram Range is included, the Himalayas contain Earth's 14 highest mountain peaks, each with an altitude of over 5 miles (8 km), including its highest mountain, Mount Everest. These peaks are still being uplifted at the rate of some 20 in (50 cm) per century by the continent–continent collision that originally formed them. However, the mountains are weathered and eroded at almost the same rate, with the debris carried away by great rivers, such as the Ganges and Indus to the south.

**EASTERN HIMALAYAS**
In this satellite view of an eastern region of the Himalayas, which extends into China, the snow-covered high-altitude regions are clearly delineated.

The collision that brought about both the Himalayas and the Tibetan Plateau to its north occurred between 50 and 30 million years ago, when tectonic plate movements caused India—at that time an island continent—to crash into Southeast Asia. For millions of years before the collision, the floor of the ocean between India and Asia (called Tethys) was consumed by subduction under the Eurasian Plate. But once the ocean closed, first the continental margins between India and Asia, and finally the continents themselves, collided. The crust from both was thickened, deformed, and metamorphosed and parts of both continents and the floor of the Tethys Ocean were pushed up to form the Himalayas. Today, because the Himalayas are still rising, earthquakes and accompanying landslides remain a common occurrence.

The mountains form a number of distinct ranges. Traveling northward, from the high plains of the Ganges, the first of these are the Siwalik Hills, a line of gravel deposits carried down from the high mountains. Here, there are subtropical forests of bamboo and other vegetation. Farther north are the Lesser Himalayas, which rise to heights of about 3,000 ft (5,000 m) and are traversed by numerous deep gorges formed by swift-flowing streams. Farthest north are the Great Himalayas, between 20,000 and 29,000 ft (6,000 and 8,800 m) tall and containing the highest peaks. This region is heavily glaciated and contains lakes filled with glacial meltwater.

**MOUNT EVEREST**
At 29,035 ft (8,850 m), Everest is the highest peak on Earth. Satellite studies show it is being uplifted by a fraction of an inch per year.

**TIBETAN RANGE**
The Kailas Range is a central region of the Himalayas, close to the border between Tibet and India. Here, the mountains are viewed from the Tibetan Plateau, which is itself about 3 miles (5 km) above sea level.

**GLACIAL LAKES**
Many of the higher areas of the Himalayas are covered in glaciers and dotted with lakes dammed by glacial moraines. In the left foreground is the Tsho Rolpa Glacier Lake in northeast Nepal, which, at 15,092 ft (4,600 m), is one of the highest lakes on Earth.

# FEATURES FORMED BY WATER

Some of the most obvious and striking features of Earth's surface are large bodies and flows of liquid water, such as oceans, seas, lakes, and rivers. In addition to these, there are landforms caused by the erosional or depositional power of liquid water, which include gorges, river valleys, and coastal features ranging from beaches to eroded headlands. Ice, too, has had a major impact on Earth's appearance. Ice-formed features include existing bodies of ice, such as glaciers and ice sheets, and landforms, such as U-shaped valleys, sculpted by the movement of past glaciers.

**GRAND CANYON**
Carved over millions of years by the Colorado River, the Grand Canyon is Earth's largest gorge.

## NORTH AMERICA *northeast*

# Great Lakes

**LOCATION** Straddling the border of the US and Canada

**TYPE** System of freshwater lakes

**AREA** 94,480 sq. miles (244,767 sq. km)

The Great Lakes of North America are a system of five connected lakes that together form the largest body of fresh water on Earth. The lakes—named, from west to east, Superior, Michigan, Huron, Erie, and Ontario—contain 20 percent of Earth's surface fresh water and drain a basin of approximately 289,900 square miles (751,100 square km). They are connected to each other by short rivers, a strait, and canals and drain into the Atlantic Ocean via the St. Lawrence River. The Great Lakes began to form at the end of the last ice age when glacier-carved

**NIAGARA FALLS**
The greatest drop in water level within the Great Lakes system is at Niagara Falls, between lakes Erie and Ontario. Here, the water plunges a spectacular 167 ft (51 m).

basins were filled with meltwater left by a retreating ice sheet. Originally, several of today's lakes were united in one huge lake, but following post-glacial uplift in the region, they took on their present form about 10,000 years ago. The lake surfaces vary in height above sea level, from 600 ft (183 m) at Lake Superior to 246 ft (75 m) at Lake Ontario. Sprinkled across the lakes are thousands of islands, including Isle Royale on Lake Superior, which is itself big enough to hold several lakes.

**LAKES HURON AND SUPERIOR**
In this photograph taken from a NASA Space Shuttle, the largest lake, Superior, is on the right, and appears partly iced over. Lake Huron is on the left.

## SOUTH AMERICA *north*

# Amazon River

**LOCATION** Flows from the Peruvian Andes, across Brazil to the Atlantic Ocean

**TYPE** River

**LENGTH** 3,995 miles (6,430 km)

The Amazon is the greatest river on Earth, whether measured by the area of the planet's land surface that it drains or by the volume of water that it discharges every year. Overall, the Amazon accounts for nearly 20 percent of all river water discharged into Earth's oceans. The source of the Amazon has been established as a headwater of the Apurímac River, a tributary of the

**MEETING THE ATLANTIC**
The mouth of the Amazon occupies the whole top part of this image, which covers an area of tens of thousands of square miles. Rio Pará, the estuary of a separate major river, the Tocantins, can be seen at bottom.

**BRAIDING**
Over its course, the Amazon frequently braids into channels, creating many temporary islands.

Ucayali, high in the Andes of southern Peru. The Ucayali flows north from this area, turns east, and joins another major tributary, the Marañón River, where it becomes the Amazon proper. The river then meanders for thousands of miles across the Amazon Basin, a vast flat area that contains Earth's largest rainforest, merging with numerous tributaries along the way. Just east of Manaus, at its confluence with the Negro River, the Amazon is already 10 miles (16 km) wide while still 1,000 miles (1,600 km) from the sea. At its mouth, the Amazon discharges into the Atlantic Ocean at the incredible rate of about 200 billion gallons of water (770 billion liters) every hour.

**MEANDERING TRIBUTARY**
The Tigre is a tributary of the Amazon in Peru. Here, it meanders through the Peruvian rainforest, over 1,860 miles (3,000 km) from the Amazon's mouth.

**THE VOLGA DELTA**
The huge triangular delta of the Volga River is visible in the bottom of this image, with the Caspian Sea stretching out beyond it to the south.

ASIA *west*

## Caspian Sea

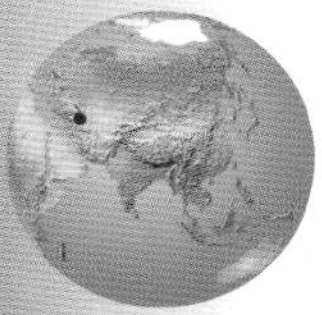

**LOCATION** On the borders of Azerbaijan, Iran, Kazakhstan, Russia, and Turkmenistan

**TYPE** Saline inland sea

**AREA** 143,000 sq. miles (371,000 sq. km)

The Caspian Sea is the largest inland body of water on Earth. It contains salty rather than fresh water, so it can be appropriately described either as a salt lake or as an inland sea. The Caspian was once joined via another inland sea, the Black Sea, to the Mediterranean. However, several million years ago it was cut off from those other seas when water levels fell during an ice age. The sea has no outflow other than by evaporation, but it receives considerable inputs of water from the Volga River (supplying three-quarters of its inflow) and from the Ural, Terek, and several other rivers. Its surface level has changed throughout history in line with discharges from the Volga, which in turn have depended on rainfall levels in the Volga's vast catchment basin in Russia. Today, the Caspian Sea contains about 18,800 cubic miles (78,200 cubic km) of water—about one-third of Earth's inland surface water. Its salinity (saltiness) varies from 1 percent in the north, where the Volga flows in, to about 20 percent in Kara-Bogaz-Bol Bay, a partially cut-off area on its eastern shore.

**OIL EXTRACTION**
Some of Earth's largest oil reserves underlie the Caspian Sea. The greatest concentration of proven reserves and extraction facilities is in its northeastern section.

ANTARCTICA

## Antarctic Ice Sheet

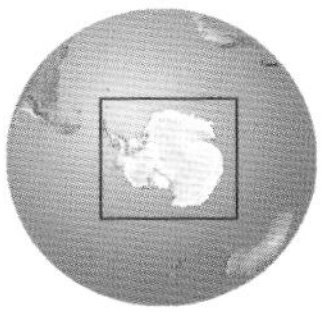

**LOCATION** Covering most of Antarctica

**TYPE** Continental ice sheet

**AREA** 5.3 million sq. miles (13.7 million sq. km)

Earth's largest glacier, the Antarctic Ice Sheet, is an immense mass of ice that covers almost all of the continent of Antarctica and holds over 70 percent of Earth's fresh water. The ice sheet has two distinct parts, separated by a range of mountains called the Transantarctic Range. The West Antarctic Ice Sheet has a maximum ice thickness of 2.2 miles (3.5 km), and its base lies mainly below sea level. The larger East Antarctic Ice Sheet is over 2.8 miles (4.5 km) thick in places with a base above sea level. Both parts of the ice sheet are domed, being slightly higher at their centers and sloping gently down toward their edges. A few areas around the edges of the ice sheets, such as some regions within the Transantarctic Range, are known to be rich sources of meteorites (see pp.220–221). Meteorites continually fall onto the ice sheet and become buried in it. But in a few places, where there is an upward flow of ice and some evaporation, they concentrate again at the surface. There are concerns that the West Antarctic Ice Sheet is shrinking due to global warming. Scientists agree that the West Antarctic Ice Sheet has been showing a general pattern of retreat for over 10,000 years, but think there is a small risk that it will collapse within the next few centuries as a result of climate change.

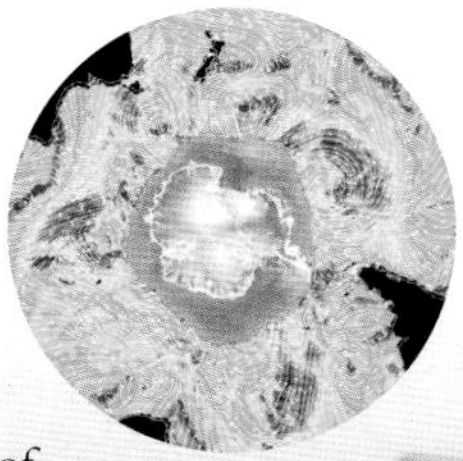

**SATELLITE VIEW**
This radar image shows the whole of Antarctica, with the larger, eastern section of its ice sheet on the left. The gray area around its coast is partly ice shelf and partly sea ice.

**THE BEARDMORE GLACIER**
This huge glacier drains the East Antarctic Ice Sheet into the Ross Ice Shelf. At 260 miles (415 km) in length, it is one of the longest glaciers on Earth.

**THE LARSEN ICE SHELF**
Around the coast of Antarctica, glaciers and ice streams merge to form platforms of floating ice called ice shelves. These are home to large colonies of penguins.

# THE MOON

EVEN THOUGH IT HAS ONLY 1.2 percent of the mass of Earth, the Moon is still the fifth-largest planetary satellite in the solar system. When full, it is the brightest object in our sky after the Sun, and its gravity exerts a strong influence over our planet. However, the Moon is too small to retain a substantial atmosphere, and geological activity has long since ceased, so it is a lifeless, dusty, and dead world. Twelve men have walked on its surface and over 838 lb (380 kg) of lunar rock have been collected, but scientists are still not sure exactly how the Moon formed.

## ORBIT

The Moon has an elliptical orbit around the Earth, so the distance between the two bodies varies. At its closest to Earth (perigee), the Moon is 10 percent closer than when at its farthest point (apogee). The Moon takes 27.32 Earth days to spin on its axis, which is the same time it takes to orbit the Earth. This is known as synchronous rotation (see right) and keeps one side of the Moon permanently facing Earth—although eccentricities in the Moon's orbit called librations allow a few regions of the far side to come into view. Because the Earth is moving around the Sun, the Moon takes 29.53 Earth days to return to the same position relative to the Sun in Earth's sky, completing its cycle of phases (see p.66). This is also the length of a lunar day (the time between successive sunrises on the Moon).

same face always points at Earth

Earth

DAY 1

Moon rotates counterclockwise

direction of Moon's orbit

DAY 8

**SYNCHRONOUS ROTATION**
For each orbit of Earth, the Moon spins once on its axis. As a result, it always keeps the same face toward Earth.

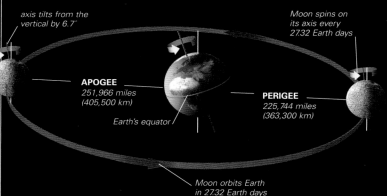

axis tilts from the vertical by 6.7°

Moon spins on its axis every 27.32 Earth days

**APOGEE**
251,966 miles
(405,500 km)

**PERIGEE**
225,744 miles
(363,300 km)

Earth's equator

Moon orbits Earth in 27.32 Earth days

**SPIN AND ORBIT**
The Moon's orbital path is tilted at an angle to Earth's equator, causing its path across the sky to vary in an 18-year cycle. Tidal forces mean that the Moon is slowing down Earth's rotation, while the Moon moves away from the Earth at a rate of about 1 in (3 cm) each year.

## STRUCTURE

The lunar crust is made of calcium-rich, granitelike rock. It is about 30 miles (48 km) thick on the near side and 46 miles (74 km) thick on the far side. Because of the Moon's history of meteorite bombardment, the crust is severely cracked. The cracks extend to a depth of 15 miles (25 km); below that, the crust is completely solid. The Moon's rocky mantle is rich in silicate minerals but poor in metals such as iron. The upper mantle is solid, rigid, and stable. Radioactive decay of minor components of the lunar rock means that the temperature increases with depth. The lower mantle lies about 600 miles (1,000 km) below the crust, and here the rock gradually becomes partially molten. The average density of the Moon indicates that it might have a small iron core. The Apollo missions measured the velocities of shock waves traveling through the Moon, but the results proved inconclusive. Further seismic evidence is needed to confirm the existence of a metallic core.

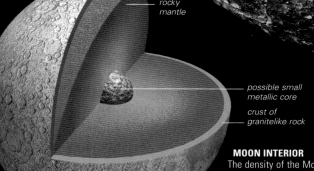

rocky mantle

possible small metallic core

crust of granitelike rock

**MOON INTERIOR**
The density of the Moon is much less than that of the whole Earth but is similar to that of Earth's mantle. It is possible that the Moon is entirely made of solid rock and has no metallic core at all.

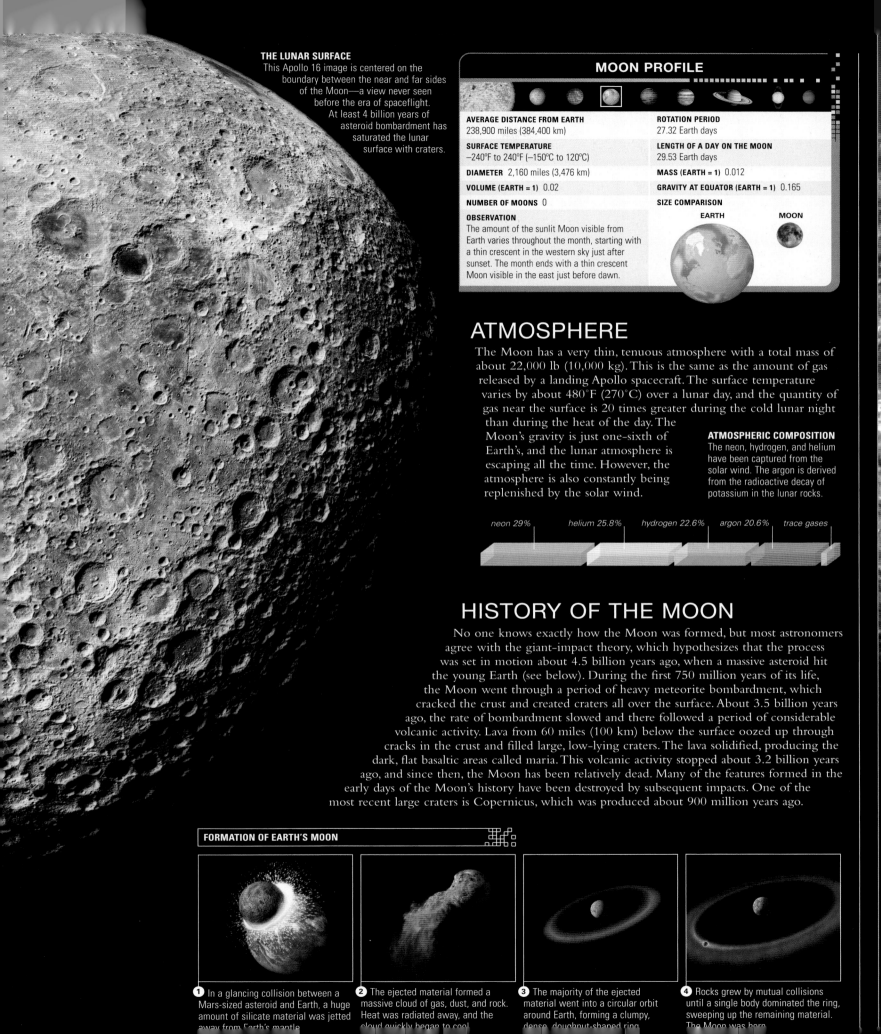

**THE LUNAR SURFACE**
This Apollo 16 image is centered on the boundary between the near and far sides of the Moon—a view never seen before the era of spaceflight. At least 4 billion years of asteroid bombardment has saturated the lunar surface with craters.

## MOON PROFILE

| | |
|---|---|
| **AVERAGE DISTANCE FROM EARTH** 238,900 miles (384,400 km) | **ROTATION PERIOD** 27.32 Earth days |
| **SURFACE TEMPERATURE** −240°F to 240°F (−150°C to 120°C) | **LENGTH OF A DAY ON THE MOON** 29.53 Earth days |
| **DIAMETER** 2,160 miles (3,476 km) | **MASS (EARTH = 1)** 0.012 |
| **VOLUME (EARTH = 1)** 0.02 | **GRAVITY AT EQUATOR (EARTH = 1)** 0.165 |
| **NUMBER OF MOONS** 0 | **SIZE COMPARISON** |

**OBSERVATION**
The amount of the sunlit Moon visible from Earth varies throughout the month, starting with a thin crescent in the western sky just after sunset. The month ends with a thin crescent Moon visible in the east just before dawn.

EARTH　　MOON

## ATMOSPHERE

The Moon has a very thin, tenuous atmosphere with a total mass of about 22,000 lb (10,000 kg). This is the same as the amount of gas released by a landing Apollo spacecraft. The surface temperature varies by about 480°F (270°C) over a lunar day, and the quantity of gas near the surface is 20 times greater during the cold lunar night than during the heat of the day. The Moon's gravity is just one-sixth of Earth's, and the lunar atmosphere is escaping all the time. However, the atmosphere is also constantly being replenished by the solar wind.

**ATMOSPHERIC COMPOSITION**
The neon, hydrogen, and helium have been captured from the solar wind. The argon is derived from the radioactive decay of potassium in the lunar rocks.

neon 29% | helium 25.8% | hydrogen 22.6% | argon 20.6% | trace gases

## HISTORY OF THE MOON

No one knows exactly how the Moon was formed, but most astronomers agree with the giant-impact theory, which hypothesizes that the process was set in motion about 4.5 billion years ago, when a massive asteroid hit the young Earth (see below). During the first 750 million years of its life, the Moon went through a period of heavy meteorite bombardment, which cracked the crust and created craters all over the surface. About 3.5 billion years ago, the rate of bombardment slowed and there followed a period of considerable volcanic activity. Lava from 60 miles (100 km) below the surface oozed up through cracks in the crust and filled large, low-lying craters. The lava solidified, producing the dark, flat basaltic areas called maria. This volcanic activity stopped about 3.2 billion years ago, and since then, the Moon has been relatively dead. Many of the features formed in the early days of the Moon's history have been destroyed by subsequent impacts. One of the most recent large craters is Copernicus, which was produced about 900 million years ago.

### FORMATION OF EARTH'S MOON

**1** In a glancing collision between a Mars-sized asteroid and Earth, a huge amount of silicate material was jetted away from Earth's mantle.

**2** The ejected material formed a massive cloud of gas, dust, and rock. Heat was radiated away, and the cloud quickly began to cool.

**3** The majority of the ejected material went into a circular orbit around Earth, forming a clumpy, dense, doughnut-shaped ring.

**4** Rocks grew by mutual collisions until a single body dominated the ring, sweeping up the remaining material. The Moon was born.

## WEREWOLVES

Many myths and old folk tales attribute strange powers to the Moon. Some say that a full Moon can turn people mad (the origin of the word "lunacy"), and many cultures, from Eurasia to the Americas, share a belief that when the Moon is full, some humans can be transformed into vicious werewolves. The superstition is widespread and ancient—even the Babylonian King Nebuchadnezzar (c. 630–c. 562 BCE) imagined that he had become a werewolf.

# LUNAR INFLUENCES

Although the Moon is much smaller than Earth, its gravity still exerts an influence. The Moon's gravitational attraction is felt most strongly on the side of Earth facing the Moon, and this pulls water in the oceans toward it. Inertia (the tendency of objects with mass to resist forces acting upon them) attempts to keep the water in place, but because the gravitational force is greater, a bulge of water is pulled toward the Moon. On the opposite side of Earth, the water's inertia is stronger than the Moon's gravity, so a second bulge of water is created. As Earth rotates, the bulges sweep over the planet's surface, creating daily changes in sea level called tides. The time of the high tide changes according to the Moon's position in the sky. The height of the tides changes during the lunar cycle, but the actual height also depends on local geography. In shallow coastal bays, the tidal range can be huge.

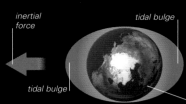

*inertial force*     *tidal bulge*     *gravitational pull of Moon*     *Moon's orbit*

*tidal bulge*

*Earth's spin causes tidal bulges to sweep over surface*

### TIDAL BULGES
Gravitational interaction between Earth and the Moon creates two bulges in Earth's oceans (exaggerated here). As Earth spins on its axis, the bulges of water sweep over the surface, creating tides.

### TIDAL RANGE
The magenta in this satellite image of Morecambe Bay on the northwest coast of England reveals the inlets and mud flats that are left exposed at low tide.

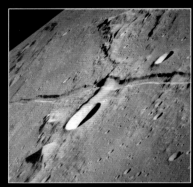

### LAVA TUBE
Over 3 miles (5 km) wide and hundreds of miles long, this rille is a collapsed tubelike structure through which lava once flowed. Moonquakes caused by nearby impacts may have caused the roof to fall in.

# SURFACE FEATURES

The surface of the Moon has been pulverized by meteorites and is covered by a rough, porous blanket of rubble several yards thick. This debris ranges in size from particles of dust to huge lumps of rock dozens of yards across. The soil (or regolith) consists of fine-grained, fragmented bedrock, the size of the grains getting progressively larger with depth. Because there is no wind or rain, the surface material does not move far, and its composition can change considerably from place to place. The thickness also varies—in young mare regions, it is about 16 ft (5 m) thick, but this increases to 32 ft (10 m) in the old highlands. Micrometeorite impacts continuously erode exposed rocks, and they are also damaged by cosmic rays and solar-flare particles. The topmost layer of soil is saturated with hydrogen ions absorbed from the solar wind.

### MOON ROCK
This 6-in (15-cm) wide rock formed as lava from the interior rose to the Moon's surface and solidified. The small holes were formed as gas bubbles escaped.

### TRACKS IN THE SOIL
Lunar Rover tire tracks lead away from the Apollo 15 module "Falcon," nestling near Hadley Rille in 1971. Over a million or more years, they will eventually be erased by meteorite bombardment.

# CRATERS

The vast majority of lunar craters are produced by impacts. Asteroids usually strike the Moon at velocities of about 45,000 mph (72,000 kph). The resulting crater is about 15 times larger than the impacting body. Unless the asteroid nearly skims the surface on entry, the resultant crater is circular. Three types are formed. Those smaller than 6 miles (10 km) across are bowl-shaped, having a depth of around 20 percent of the diameter. Craters between 6 and 90 miles (10–150 km) in diameter have outer walls that have slumped into the initial crater pit. There is often a central mountainous peak produced by the recoil of the underlying stressed rocks. The crater depth is a few miles, and a lot of excavated material falls back into the crater just after the impact. Craters wider than 90 miles (150 km) contain concentric rings of mountains, created as rebounding material rippled out from the center before solidifying. Such craters were so deep that hot magma flooded to the surface and filled the bottom of the crater with lava.

### RAY CRATERS
Material ejected from a crater during an impact is often confined to narrow jets. Where this material hits the surface, it plows up the lunar soil, and this disturbed region then reflects more sunlight than its surroundings. From Earth, these appear as rays. The rays around Tycho Crater (far right) extend for thousands of miles.

### SUNRISE OVER COPERNICUS CRATER

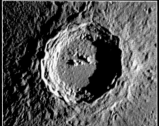

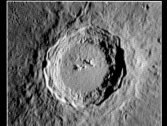

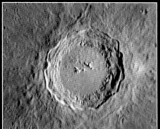

1 Just after dawn in the crater, the low eastern Sun casts long shadows, which emphasize the variation in height between the floor and rim.

2 Halfway through the morning, small shadows enhance the ejecta blanket outside the crater. The temperature inside the crater is rising.

3 At noon, the Sun is overhead, and the scene appears much flatter and washed out. The temperature is now more than 212°F (100°C).

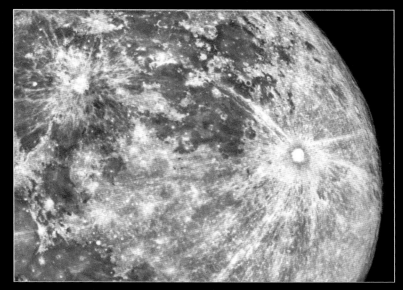

## EUGENE SHOEMAKER

Gene Shoemaker (1928–1997) was an American astrogeologist who studied terrestrial and lunar meteorite impact craters and dreamed of going to the Moon. Addison's disease prevented that. Instead, he taught the Apollo astronauts to be field geologists. In 1969, he joined a team at Palomar, searching for near-Earth asteroids. After Shoemaker died, some of his ashes were carried to the Moon aboard the Lunar Prospector space probe in 1999.

# MAPPING THE MOON

Some ancient Greeks thought that the Moon was like the Earth and that its dark areas were water. This belief continued into the 17th century, when the dark patches were given aquatic names such as mare (sea) and oceanus (ocean) on the first proper maps. Palus Putredinis (the Marsh of Decay) and Sinus Iridum (the Bay of Rainbows) are evocative examples. Italian astronomer Galileo Galilei was the first to realize that the height of surface features could be added to maps by noting how the shadow lengths changed during the lunar day. The first photographic atlas appeared in 1897, but the real leap forward came with the advent of spaceflight. In 1959, the Soviet Union sent the Luna 3 space probe behind the Moon to photograph the far side. NASA's five Lunar Orbiter spacecraft imaged 99 percent of the lunar surface in 1966–1967, paying special attention to potential Apollo landing sites. In the 1990s, the Moon's mineral composition was surveyed by Clementine and Lunar Prospector. Since 2009, Lunar Reconnaissance Orbiter (LRO) has been engaged in a detailed mapping project.

### LUNA 3
On October 7, 1959, the Soviet Union's Luna 3 space probe imaged the far side of the Moon. It had never been seen before.

### LUNAR ORBITER IV
This superb wide-angle image of the half-lit Mare Imbrium was one of 546 images taken by NASA's orbiter on May 11–26, 1967, from a height of about 2,485 miles (4,000 km).

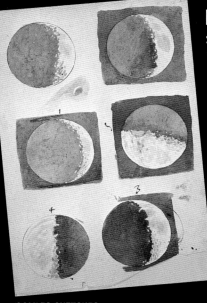

### GALILEO SKETCHES
Galileo's first telescopic observations of the Moon were made on November 30, 1609. The pictures, published in *Sidereus Nuncius* in 1610, emphasized the roughness of the surface.

### SMART-1
During its approach phase, ESA's SMART-1 spacecraft took this image of an illuminated region of the far side, near the lunar north pole, on November 12, 2004, from a distance of about 37,250 miles (60,000 km).

THE SOLAR SYSTEM

# THE NEAR AND FAR SIDES OF THE MOON

The Moon's spin and orbital periods became locked together very early in its existence, when it was much closer to Earth than it is now and the surface was still molten due to the heating produced by massive early impacts. As a result, the Earth's influence has led to noticeable differences in the appearance of the two sides. The far side is on average about 3 miles (5 km) higher with respect to the Moon's center of mass than the near side, and its low-density crust is 16 miles (26 km) thicker. Because the near side is lower, volcanic magma has more easily found its way to the surface here, pouring from volcanic fissures into the low-lying regions of the largest craters and solidifying to form the lunar seas. By contrast, the far side—forever facing outward from Earth—lacks large seas and appears to have suffered heavier bombardment and cratering than the near side.

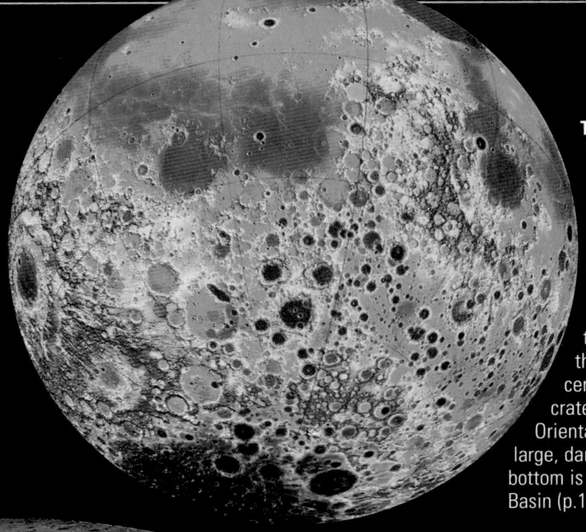

**TOPOGRAPHIC VIEW**
This map of lunar surface relief, color-coded according to height, was produced by the altimeter onboard NASA's Lunar Reconnaissance Orbiter. Red areas are the highest; blue areas the lowest. This view is centered on the bright ray crater Tycho, with the Mare Orientale basin at left. The large, dark blue feature at the bottom is the South Pole–Aitken Basin (p.149).

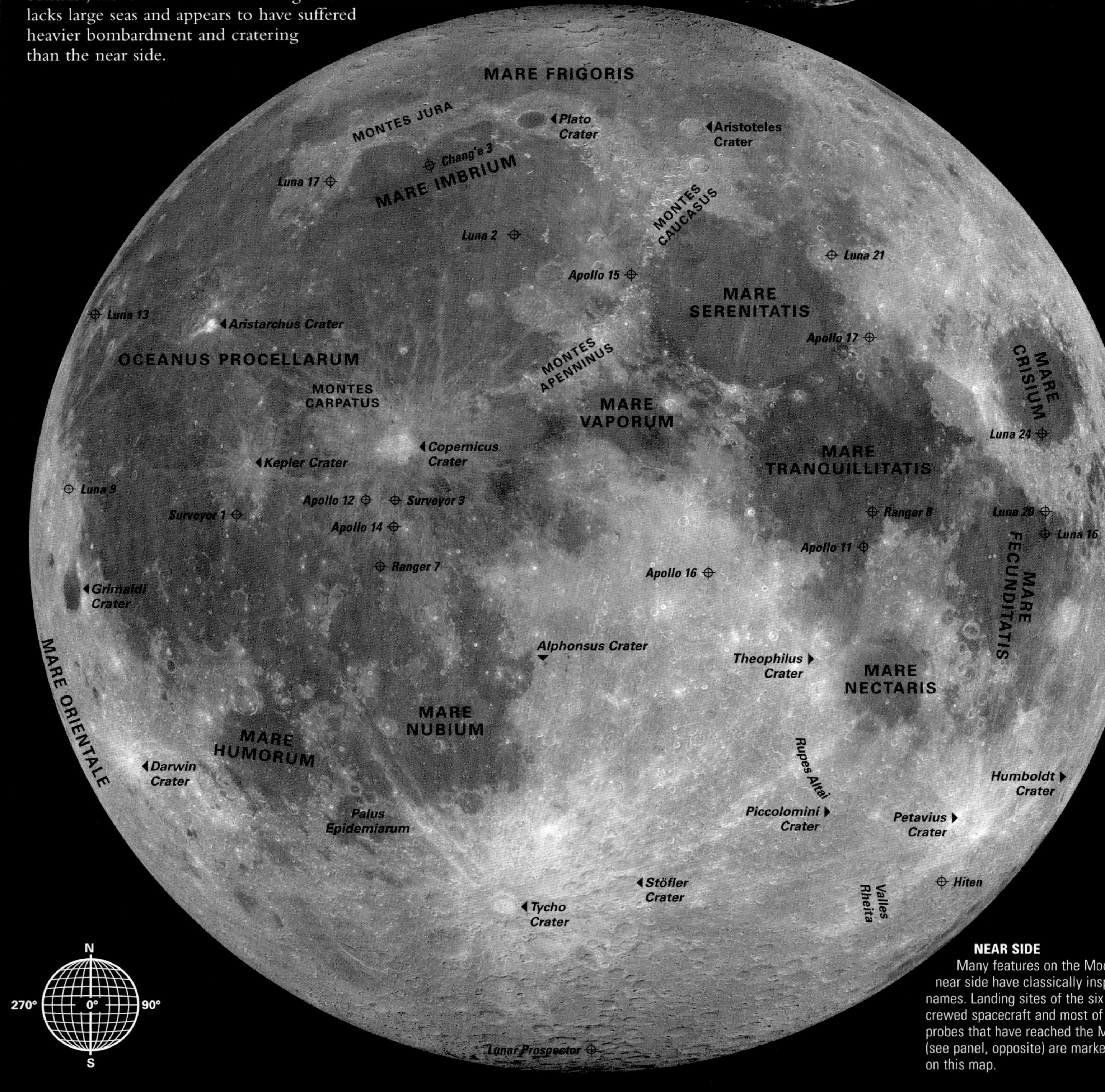

MARE FRIGORIS

MONTES JURA

Plato Crater

Aristoteles Crater

Chang'e 3

Luna 17

MARE IMBRIUM

MONTES CAUCASUS

Luna 2

Luna 21

Apollo 15

MARE SERENITATIS

Luna 13

Apollo 17

Aristarchus Crater

OCEANUS PROCELLARUM

MONTES APENNINUS

MARE CRISIUM

MONTES CARPATUS

MARE VAPORUM

Kepler Crater

Copernicus Crater

Luna 24

MARE TRANQUILLITATIS

Luna 9

Apollo 12

Surveyor 3

Ranger 8

Luna 20

Surveyor 1

Apollo 14

Apollo 11

Luna 16

MARE FECUNDITATIS

Ranger 7

Apollo 16

Grimaldi Crater

Alphonsus Crater

Theophilus Crater

MARE NECTARIS

MARE ORIENTALE

Darwin Crater

MARE HUMORUM

MARE NUBIUM

Rupes Altai

Humboldt Crater

Palus Epidemiarum

Piccolomini Crater

Petavius Crater

Stöfler Crater

Hiten

Tycho Crater

Valles Rheita

N
270°  0°  90°
S

Lunar Prospector

**NEAR SIDE**
Many features on the Moon's near side have classically inspired names. Landing sites of the six crewed spacecraft and most of the probes that have reached the Moon (see panel, opposite) are marked on this map.

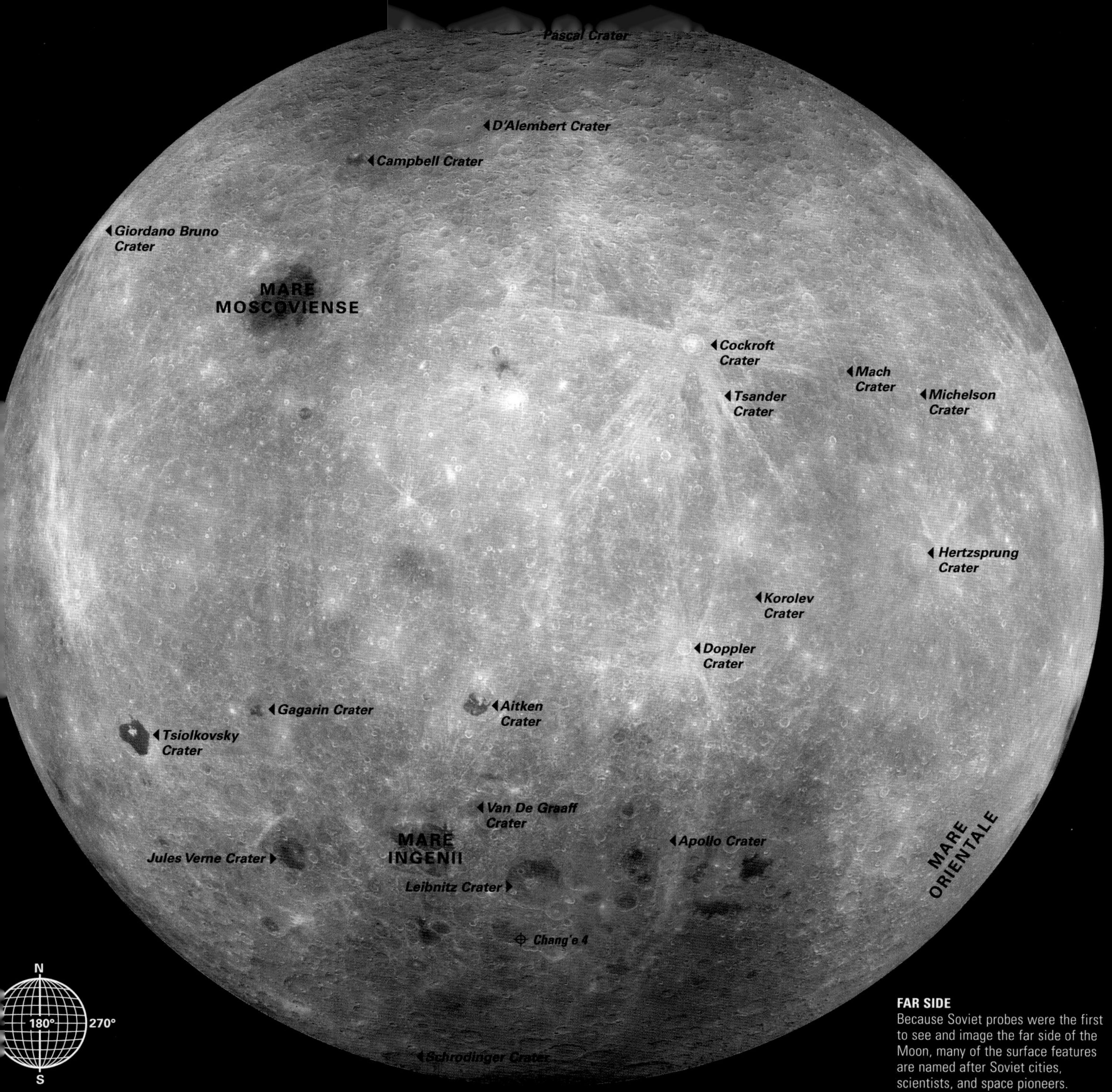

Pascal Crater

D'Alembert Crater

Campbell Crater

Giordano Bruno Crater

MARE MOSCOVIENSE

Cockroft Crater

Mach Crater

Tsander Crater

Michelson Crater

Hertzsprung Crater

Korolev Crater

Doppler Crater

Gagarin Crater

Aitken Crater

Tsiolkovsky Crater

Van De Graaff Crater

MARE INGENII

Jules Verne Crater

Apollo Crater

MARE ORIENTALE

Leibnitz Crater

Chang'e 4

Schrödinger Crater

N
180°    270°
S

**FAR SIDE**
Because Soviet probes were the first to see and image the far side of the Moon, many of the surface features are named after Soviet cities, scientists, and space pioneers.

## SIGNIFICANT LANDINGS ON THE MOON

ween them, automated space probes and human explorers have studied a wide ge of terrains on the near side of the Moon. At first, just crashing a probe into Moon at all was a significant achievement, but by the time of the Apollo sions, landings were targeting particular areas to answer specific questions ut the Moon's geology and history.

| SION | DATE OF ARRIVAL | TYPE | ACHIEVEMENT |
|---|---|---|---|
| 2 (USSR) | September 13, 1959 | Impact | Makes first crash-landing on the Moon |
| ɹer 7 (USA) | July 31, 1964 | Impact | Takes first close-up photos of surface |
| ɹer 8 (USA) | February 20, 1965 | Impact | Takes 7,137 good-quality photos |
| 9 (USSR) | February 3, 1966 | Lander | Makes first soft landing |
| eyor 1 (USA) | June 2, 1966 | Lander | Measures radar reflectivity of surface |
| 13 (USSR) | December 24, 1966 | Lander | Successfully uses mechanical soil probe |
| eyor 3 (USA) | April 20, 1967 | Lander | Images future Apollo 12 landing site |
| ɪlo 11 (USA) | July 20, 1969 | Manned | Lands first astronauts on the Moon |
| ɪo 12 (USA) | November 19, 1969 | Manned | Makes first pinpoint landing |
| 16 (USSR) | September 20, 1970 | Lander | Makes first automated sample return |

| | | | |
|---|---|---|---|
| Luna 17 (USSR) | November 17, 1970 | Rover | Carries first robotic lunar rover |
| Apollo 14 (USA) | February 5, 1971 | Manned | Carries "lunar cart" for sample collection |
| Apollo 15 (USA) | July 30, 1971 | Manned | Carries first manned lunar rovers |
| Luna 20 (USSR) | February 21, 1972 | Lander | Makes automated sample return |
| Apollo 16 (USA) | April 21, 1972 | Manned | Explores central highlands |
| Apollo 17 (USA) | December 11, 1972 | Manned | Makes longest stay on Moon (75 hours) |
| Luna 21 (USSR) | January 15, 1973 | Rover | Explores Posidonius Crater |
| Luna 24 (USSR) | August 14, 1976 | Lander | Returns sample from Mare Crisium |
| Hiten (Japan) | April 10, 1993 | Impact | Crashes into Furnerius region |
| Lunar Prospector (USA) | July 31, 1999 | Impact | Orbiter makes controlled crash near the south pole to look for evidence of water |
| SMART-1 (ESA) | November 14, 2006 | Impact | Simulates a meteor impact with crash |
| Chandrayaan-1 (India) | November 14, 2008 | Impact | Finds evidence of water |
| LCROSS (USA) | October 9, 2009 | Impact | Finds evidence of water |
| Chang'e 3 (China) | December 14, 2013 | Rover | Lander deployed Yutu rover |
| Chang'e 4 (China) | January 3, 2019 | Rover | First landing on the far side; Yutu 2 rover deployed |

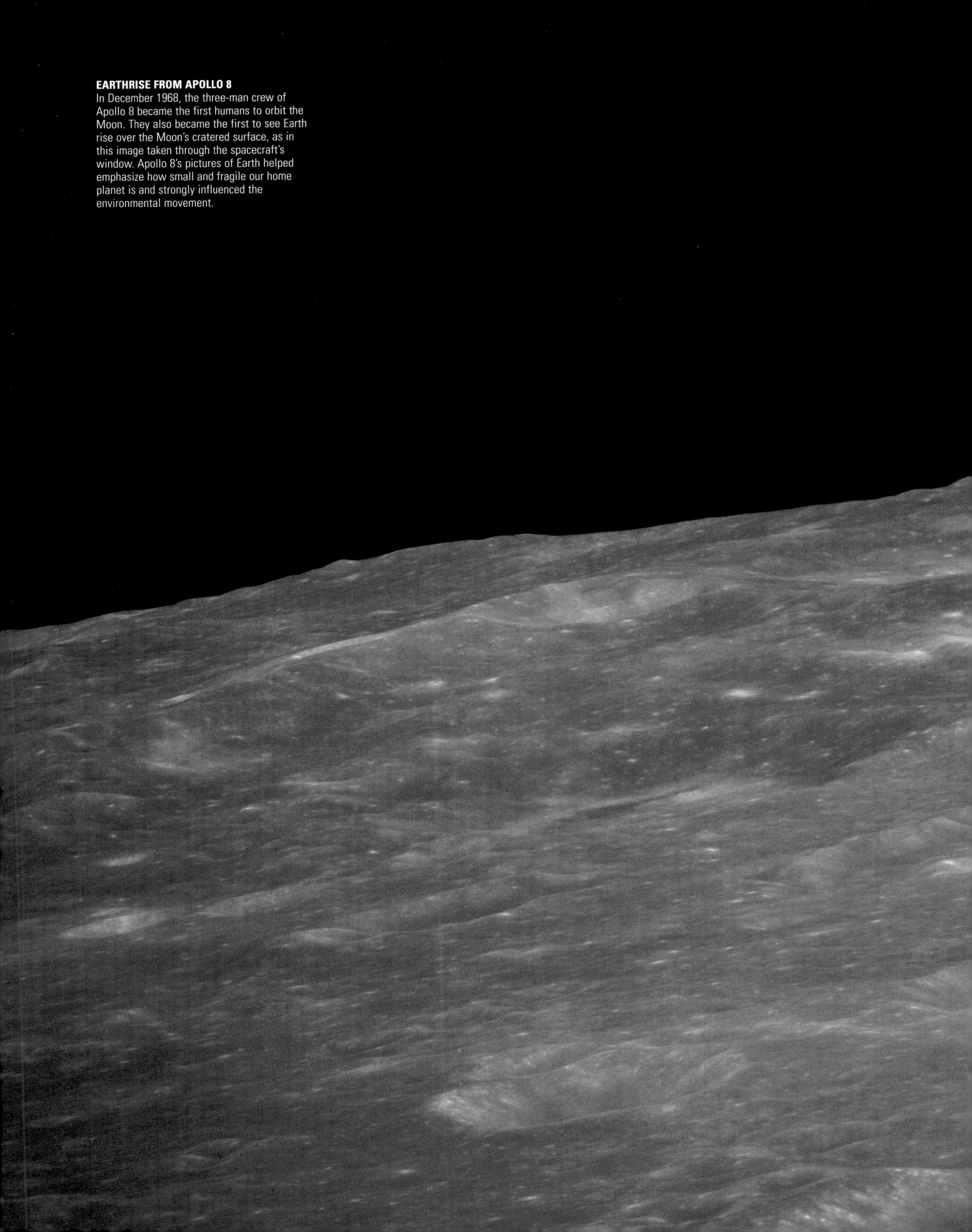

**EARTHRISE FROM APOLLO 8**
In December 1968, the three-man crew of Apollo 8 became the first humans to orbit the Moon. They also became the first to see Earth rise over the Moon's cratered surface, as in this image taken through the spacecraft's window. Apollo 8's pictures of Earth helped emphasize how small and fragile our home planet is and strongly influenced the environmental movement.

# FEATURES OF THE MOON

From afar, the Moon is clearly divided into two types of terrain. There are large, dark plains called maria (Latin for "seas") and also brighter, undulating, heavily cratered highland regions. The whole surface was initially covered with craters, most of which were produced during a time of massive bombardment. The rate at which asteroids have been striking the Moon has decreased over the last 4 billion years. Around 4 billion years ago, the Moon was also volcanically active. Lava rose to the surface through cracks and fissures, filling the lower parts of the large craters to produce the dark plains. The plains reflect only about 4 percent of the sunlight that hits them, whereas the mountains reflect about 11 percent.

**MOSAIC OF THE NORTH POLE**
The lunar north pole is partially hidden from view from Earth and is best imaged by orbiting spacecraft. Galileo took a series of photographs of the region on December 7, 1992, on its way to Jupiter.

## Aristarchus Crater

| | |
|---|---|
| **TYPE** | Impact crater |
| **AGE** | About 300 million years |
| **DIAMETER** | 23 miles (37 km) |

This young crater has a series of nested terraces, which were produced by concentric slices of rock in the wall slipping downward. This both widened the crater and made it considerably shallower, as the initially deep central region was filled with material from the rim. Aristarchus was mapped by the Apollo Infrared Scanning Radiometer. During the night, the temperature in the crater is about 54°F (30°C) higher than that of the surrounding terrain. Young craters contain many large boulders. These take a long time to heat up during the day and also a long time to cool down at night. As time passes, the boulders are broken up by small impacting asteroids, so this thermal difference eventually disappears.

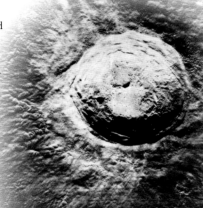

**LUNAR ORBITER 5 IMAGE**
This view of Aristarchus, taken from directly above, underlines the crater's circularity and reveals the extensive surrounding blanket of hummocky ejecta.

## Mare Crisium

| | |
|---|---|
| **TYPE** | Lava-filled impact crater (sea) |
| **AGE** | 3.9 billion years |
| **DIAMETER** | 350 miles (563 km) |

Mare Crisium has an extremely smooth floor, which varies in height by less than 290 ft (90 m). The lava that flooded Crisium had extremely low viscosity and became like a still pond before it solidified. The Soviet Luna 24 probe was the last mission to bring back rock samples from the Moon. In 1976, it returned to Earth with a core of rock weighing 6 oz (170 g), which was collected from Crisium's floor.

**OVAL CRATER**
The Mare Crisium, which can be seen with the naked eye from Earth, is nearly circular in shape. Over 95 percent of lunar craters are completely circular.

## Montes Apenninus

| | |
|---|---|
| **TYPE** | Mountain range |
| **AGE** | 3.9 billion years |
| **LENGTH** | 249 miles (401 km) |

The Lunar Apennine mountains form a ring around the southeastern edge of the Mare Imbrium impact basin. They consist of crustal blocks rising more than 1.9 miles (3 km) above the flat lava plain, pushed up by the shock wave from the Imbrium impact. The mountain chain stretches for some 375 miles (600 km), though its southern end is partially buried beneath lava flows.

**LUNAR MOUNTAINS**
The Apennines lie in the lower right of this Apollo 15 image. The dark area to their left is Palus Putredinis.

## Mare Tranquillitatis

| | |
|---|---|
| **TYPE** | Sea |
| **AGE** | 3.6 billion years |
| **DIAMETER** | 542 miles (873 km) |

The surfaces of lunar maria are much darker than highland rock and are also considerably younger. This means that they are relatively smooth and contain only a few impact craters. Their low reflectivity is due to the chemistry of the very fluid lava that flooded them. The Mare Tranquillitatis (Latin for "Sea of Tranquillity") lies just north of the lunar equator and joins onto the southeast part of the Mare Serenitatis (Sea of Serenity). Together, the two seas form one of the Moon's most prominent features. The basin in which the "sea" formed is very ancient, predating the formation of the Imbrium Basin 3.9 billion years ago. It overlaps with other basins at several points but only flooded with lava about 3.6 billion years ago. The Sea of Tranquillity was famously the landing place of US astronauts Neil Armstrong and Buzz Aldrin on their 1969 Apollo 11 mission.

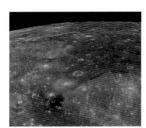

**RICH IN TITANIUM**
This Galileo image has been color-coded according to the titanium content of the rock. The blue Tranquillitatis region is rich in titanium, whereas the orange Serenitatis region at the lower right is titanium-poor.

**BEFORE TOUCHDOWN**
The flat, desolate plain of the Sea of Tranquillity stretches away to the north in this view from the Apollo 11 lunar module taken just before landing.

# Copernicus Crater

| TYPE | Impact crater |
|---|---|
| AGE | 900 million years |
| DIAMETER | 57 miles (91 km) |

This young ray crater has massive terraced walls. The crater floor is below the general level of the surrounding plain and lies 2.3 miles (3.7 km) below the top of the

**LUNAR ORBITER 2 IMAGE**
Copernicus Crater's terraced walls and central peaks were revealed by NASA's second Lunar Orbiter in 1966.

**CRATER CHAINS**
The material excavated by an impact showers down on the surrounding lunar surface, producing long chains of secondary craters.

surrounding walls.
Copernicus is an intermediate-sized crater with high central peaks. These mountains were formed when the rock directly below the crater rebounded after being compressed by the explosion caused by the impacting asteroid. The vicinity of Copernicus is

peppered with secondary craters formed by boulders thrown out during the impact. Fine, light gray rock particles ejected during the crater's formation were collected by the Apollo 12 astronauts near their landing site. Such particles were responsible for forming the rays that surround the crater. The high reflectivity of the rays is due to the ejecta churning up the lunar regolith. (Rough material reflects more light than smooth material.)

# Alphonsus Crater

| TYPE | Impact crater |
|---|---|
| AGE | 4.0 billion years |
| DIAMETER | 80 miles (117 km) |

NASA's Ranger 9 spacecraft was deliberately crash-landed into the Alphonsus Crater on March 24, 1965, taking television pictures as it approached. The crater formed in an impact, but the dark patches and fractures Ranger 9 found on its floor are thought to be a result of volcanic activity—probably explosive eruptions. Because of these features, Alphonsus was considered a possible landing site for later Apollo missions.

**THREE MINUTES BEFORE IMPACT**

---

# Rupes Altai

| TYPE | Cliff |
|---|---|
| AGE | 4.2 billion years |
| LENGTH | 315 miles (507 km) |

Altai is by far the longest cliff on the Moon. It is about 1.1 miles (1.8 km) high. The energy that is released during an impact does more than just excavate a crater and lift material out to form walls and an ejecta blanket. Violent seismic shock waves radiate away from the impact point. An obstacle such as a mountain can halt these waves and the lunar crust may then buckle, forming a long cliff. Altai was created by the Nectaris impact.

**ALTAI ESCARPMENT**
The curving Rupes Altai— 310 miles (500 km) long—runs from top to bottom in this image. The crater at top left of the picture is Piccolomini.

# Humboldt Crater

| TYPE | Impact crater |
|---|---|
| AGE | About 3.8 billion years |
| DIAMETER | 120 miles (189 km) |

This crater is remarkable because its lava-filled floor is crisscrossed with a series of radial and concentric fractures (or rilles). On closer inspection, some look like collapsed tubes through which lava once flowed, and others like rift valleys. Lunar volcanic activity lasted for over 500 million years. Lava would seep up into a crater and then cool, shrink, crack, and sink. It would then be covered by more lava. The final basaltic infill would have many layers.

**CRACKS ON THE FLOOR OF HUMBOLDT**

# Tycho Crater

| TYPE | Impact crater |
|---|---|
| AGE | 100 million years |
| DIAMETER | 52 miles (85 km) |

Lying in the southern highlands, Tycho is one of the most perfect walled craters on the Moon, with a central mountain peak towering 1.8 miles (3 km) above a rough infilled inner region. Surveyor 7 landed on the north rim of Tycho's ejecta blanket in January 1968. About 21,000 photographs were taken, and the soil was chemically analyzed. The highland soil was found to be mainly made of calcium-aluminum silicates, in contrast to the maria material, which is iron-magnesium silicate.

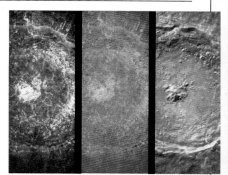

**THREE FILTERED IMAGES OF TYCHO**
The Ultraviolet/Visual camera onboard the Clementine spacecraft was equipped with a series of filters. Differing color combinations revealed the variability in the physical and chemical structure of the crater rock.

**YOUNGEST LARGE LUNAR CRATER?**
Although Tycho is one of the youngest lunar craters (Giordano Bruno may be younger), it still formed in the age of the dinosaurs.

## NEAR SIDE *northern hemisphere*

# Taurus-Littrow Valley

**TYPE** Valley

**AGE** About 3.85 billion years

**LENGTH** 18.6 miles (30 km)

In December 1972, the last crewed mission to the Moon landed in the dark-floored Taurus-Littrow Valley at the edge of the basalt-filled Mare Serenitatis. The range of geological features was impressive, and the Apollo 17 astronauts found three distinct types of rock in the region. One piece of crushed magnesium olivine was 4.6 billion years old and had crystallized directly from the melted shell of the just-formed Moon. The nearby Serenitatis Crater

was produced about 3.9 billion years ago, and much of the basaltic rock dates from that time, when the crater was flooded with lava. The third type of rock was found on the top of nearby hills. This was barium-rich granite and had been ejected from one of the surrounding large craters. Most of the material near the landing site was extremely dark and consisted of cinders and ash ejected billions of years ago from nearby volcanic vents and fissures.

## HIGHLAND MASSIFS

The flat-based Taurus-Littrow Valley can be seen in the center of this image, nestling between the rugged, blocky mountains known prosaically as the North, South, and East Massifs.

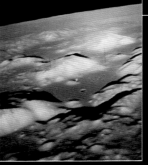

The Taurus-Littrow Valley is surrounded by steep-sided mountains, known as massifs. Moon mountains are different from those found on Earth. On Earth, the crustal plates collide, producing huge mountain ranges like the Alps and Himalayas. These new mountains are subsequently eroded by rain and ice. The Moon's crust is not broken into plates. Nothing moves. All the Moon mountains are produced by impacts, and the mountains around the Taurus-Littrow Valley are the remains of old crater walls. Part of the valley floor just to the north of South

## HARRISON SCHMITT

Harrison "Jack" Schmitt (b. 1935) was born in New Mexico. He studied geology at Caltech and Harvard University. While working for the US Geological Survey, he joined a team instructing astronauts in the art of field geology. In June 1965, Schmitt was selected as a scientist-astronaut by NASA and was later chosen to be the lunar module pilot for Apollo 17. In December 1972, he became the first and only geologist to walk on the Moon. One of the highlights of the Apollo 17 mission was his discovery of orange glass within the lunar rock.

## SHORTY CRATER

Harrison Schmitt stands by the Lunar Rover, parked to the left of the 356-ft (110-m) wide Shorty Crater. Behind Schmitt, 4 miles (6 km) away, is Family Mountain, one of the Taurus-Littrow range named by the astronauts. Near the rover, patches of orange soil can be seen.

Massif was covered with a light mantle of regolith a few yards thick. This had been produced by a rock avalanche, possibly triggered when the area was bombarded by boulders ejected when the nearby Tycho Crater was formed. As the Moon is being continuously bombarded by asteroids, the number of craters per unit area increases with time. There are relatively few craters on the Taurus-Littrow Valley floor, which was taken to indicate that the surface is even younger than the Apollo 12 landing site. One crater in the valley, Shorty, was once thought to be a volcanic vent, but more detailed analysis of its raised rim and central mound indicated that, like millions of other lunar craters, it was produced by an impacting asteroid.

**SPLIT ROCK**
This house-sized boulder was ejected from an impact crater in the Mare Serenitatis and then rolled down into the valley. Scoop marks can be clearly seen where some samples have been taken from its surface.

EXPLORING SPACE

## MOON GLASS

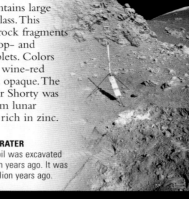

The lunar regolith contains large amounts of volcanic glass. This occurs as glazings on rock fragments and also as tiny teardrop- and dumbbell-shaped droplets. Colors range from green and wine-red through to orange and opaque. The orange glass found near Shorty was typical of high-titanium lunar glasses, but it was also rich in zinc.

**ORANGE SOIL IN SHORTY CRATER**
The glassy orange surface soil was excavated by an impact about 20 million years ago. It was actually formed about 3.6 billion years ago.

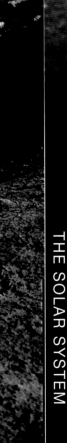

# Pascal Crater

| | |
|---|---|
| **TYPE** | Impact crater |
| **AGE** | About 4.1 billion years |
| **DIAMETER** | 71 miles (115 km) |

This is one of 300 lunar craters named after mathematicians. It honors the Frenchman Blaise Pascal. The image below was taken in 2004, with the camera looking directly down into the crater. The Sun is low in the sky, below the bottom of the picture. The tiny craters around Pascal are bowl-shaped and young, with circular rims much sharper than the older rim of Pascal. The larger crater's rim was initially eroded by slumping and rock slides and is now being worn down further by more recent impacts.

**PASCAL AND ITS YOUNGER NEIGHBORS**

# Tsiolkovsky Crater

| | |
|---|---|
| **TYPE** | Impact crater |
| **AGE** | About 4.2 billion years |
| **DIAMETER** | 123 miles (198 km) |

Only half the size of Mare Crisium, this far-side crater is special because only half the interior basin has been filled with lava. The central peak is also unusually offset from the center of the crater. There have been extensive rock avalanches down the southern rim

**ORBITER 3 IMAGE**
The crest of the rim of Tsiolkovsky Crater runs to the upper right of this image. The diagonal banding to its right is probably the result of a large avalanche down the slope of the rim.

**DARK FLOOR**
If Tsiolkovsky had been formed earlier in lunar history, the volcanic activity would have been greater and more of the crater floor would have been filled with lava.

of the crater. The first images of the lunar far side was obtained in October 1959 by the Soviet spacecraft Luna 3. Resolution was low, but the features that could be seen were nevertheless given names, such as Mare Moscoviense and Sinus Astronautarum. Only a few craters could be made out, including this one. Konstantin Tsiolkovsky was a Russian rocketry pioneer who not only designed a liquid hydrogen/liquid oxygen rocket but also suggested the multistage approach to spaceflight. The crater was penciled in as a possible landing site for one of the post-Apollo 17 missions, which were canceled.

# Van de Graaff Crater

| | |
|---|---|
| **TYPE** | Double impact crater |
| **AGE** | About 3.6 billion years |
| **LENGTH** | 155 miles (250 km) |

Less than 1 percent of the lunar craters are noncircular. Van de Graaff is typical of such irregular craters, which are produced on the rare occasions when the impacting asteroid hits the surface at an angle of less than 4°. Van de Graaff is also special because it is both magnetic and has the highest concentration of natural radiation. Most of the ancient lunar magnetic field decayed away over 3 billion years ago. However, there are still a few magnetic anomalies (magcons), of which Van de Graaff and nearby Aitken are the strongest. Magcons were discovered by small magnetometer subsatellites released by Apollos 15 and 16.

**IRREGULARLY SHAPED CRATER**

# Korolev Crater

| | |
|---|---|
| **TYPE** | Ringed impact crater |
| **AGE** | About 3.7 billion years |
| **DIAMETER** | 250 miles (405 km) |

Sergei Korolev led the Soviet space effort in the 1950s and 1960s and was responsible for the early Sputnik and Vostok spacecraft. He has two craters named after him, one on the Moon and the other on Mars. Korolev is one of only 10 craters on the lunar far side that are more than 125 miles (200 km) across. It is double-ringed and pocked with smaller craters. The outer ring is 252 miles (405 km) in diameter. The inner ring is much less distinct. It is only half the height of the outer ring and its diameter is half that of the outer ring. Together with Hertzsprung and Apollo, Korolev forms a trio of huge ringed formations on the lunar far side. The lunar crust varies in thickness, and it reaches its maximum thickness of 66 miles (107 km) in the region around the Korolev Crater.

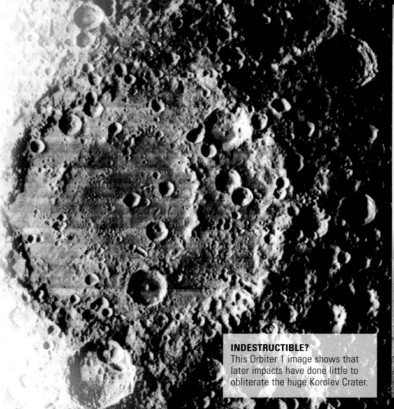

**INDESTRUCTIBLE?**
This Orbiter 1 image shows that later impacts have done little to obliterate the huge Korolev Crater.

# NUCLEAR CRATER

It is very difficult to estimate the relationship between the size of a crater and the size of the asteroid that produced it. Usually the crater is about 20 times bigger. Only in controlled nuclear explosions can an exact relationship between energy release and crater size be established. Sedan Crater in the Nevada Desert (below) is bowl-shaped and 1,200 ft (368 m) across. It was produced by a subsurface nuclear blast equivalent to 100 kilotons of TNT in July 1962. It is very similar to small lunar impact craters such as those within Korolev.

# Mare Orientale

| | |
|---|---|
| **TYPE** | Multiring basin |
| **AGE** | 3.8 billion years |
| **DIAMETER** | 560 miles (900 km) |

This multiring basin is half the size of the near side's Mare Imbrium. It lies on the eastern limb of the far side, and from Earth, the Montes Rook—the innermost eastern portion of the three distinct rings—can be clearly seen. This giant lunar bull's-eye was formed by a massive asteroid, and two theories have been proposed to explain the rings. The first has the impact excavating a deep transient crater. The cracked inner walls of this crater would have been unable to support the weight of surrounding crust, so the rock slumped into the hole, guided by a series of concentric fault systems that account for the rings that remain. Not only was most of the crater filled in, but the break-up of subsurface rock allowed lava from far below the lunar surface to seep up and fill in the central regions. However, the highland crust is about 37 miles (60 km) thick, and rock from

**ROOK AND CORDILLERA MOUNTAINS**
Orientale is surrounded by two huge circular mountain ranges. The outer range is called Montes Cordillera (above right) and the inner one is called Montes Rook (lower left).

below that depth should have been excavated, but this deep rock has not been found. Alternatively, the seismic shocks generated by the massive impact could have briefly turned the surrounding rocks into a fluidized powder. Tsunami-type waves moved out through the pulverized rock but quickly became frozen, resulting in three clearly visible mountain rings.

**LUNAR BULL'S-EYE**
Mare Orientale's concentric rings can be seen in this composite Lunar Reconnaissance Orbiter image.

# South Pole–Aitken Basin

| | |
|---|---|
| **TYPE** | Impact crater |
| **AGE** | 3.9 billion years |
| **DIAMETER** | 1,550 miles (2,500 km) |

The South Pole–Aitken Basin is an immense impact crater lying almost entirely on the far side of the Moon. It stretches from just above the South Pole to beyond the Aitken Crater, which is close to the center of the far side. South Pole–Aitken is a staggering 1,550 miles (2,500 km) in diameter and is over 7.4 miles (12 km) deep. It is one of the largest craters in the solar system and is comparable in size to the largest craters on Mars. Its diameter is about 70 percent of that of the Moon itself. The asteroid that produced it would have been over 60 miles (100 km) across.

Even though the basin was first discovered in 1962, detailed investigation only started when the Galileo spacecraft imaged the Moon in 1992, while on its way to Jupiter. The South Pole–Aitken Basin looked darker than the rest of the far-side highland rocks, indicating that the lower-crustal rocks at the bottom of

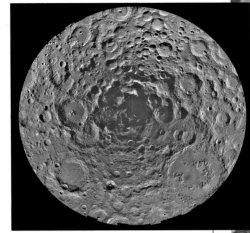

**SOUTH POLE**
The massively cratered, cold lunar South Pole can only be glimpsed tangentially from Earth. NASA's Clementine mission provided the first detailed map of the region in 1994.

the deep crater were richer in iron than normal lunar surface material. Iron oxide and titanium oxide abound. Impact geophysicists are convinced that a normal impact could not have produced a crater this large without digging up large amounts of rock from the mantle that lies below the lunar crust. It may be that the crater was produced by a low-velocity collision, with the impactor coming into the surface at a low angle. Huge amounts of material would have been blasted from the lunar surface and would have moved off around the Moon's orbit. In the subsequent 10 million years, this debris would have collided with the Moon, producing many new craters.

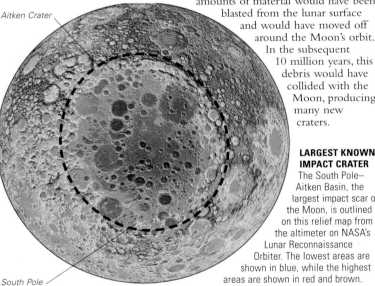

*Aitken Crater*

*South Pole*

**LARGEST KNOWN IMPACT CRATER**
The South Pole–Aitken Basin, the largest impact scar on the Moon, is outlined on this relief map from the altimeter on NASA's Lunar Reconnaissance Orbiter. The lowest areas are shown in blue, while the highest areas are shown in red and brown.

## LOOKING FOR WATER

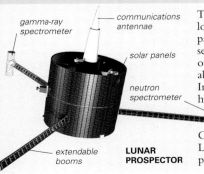

*gamma-ray spectrometer*

*communications antennae*

*solar panels*

*neutron spectrometer*

*extendable booms*

**LUNAR PROSPECTOR**

The South Pole-Aitken is one of the lowest regions on the Moon, and parts of it never see the Sun. Water seeping up from cracks in the mantle or released by an impact will not be able to escape from these "cold traps." In 1998, Lunar Prospector found hydrogen, thought to be from the breakup of water ice, within these traps. Both the Chandrayaan-1 mission (2008) and LCROSS (2009) also indicate the presence of water in this region.

# MARS


38–39 Gravity, motion, and orbits
64–65 Celestial cycles
68–69 Planetary motion
100–101 The history of the solar system
102–103 The family of the Sun


MARS IS THE OUTERMOST of the four rocky planets. Also known as the Red Planet because of its rust-red color, it is named after the Roman god of war. Its varied surface features include deep canyons and the highest volcanoes in the solar system. Although Mars is now a dry planet, a large body of evidence indicates that liquid water once flowed across its surface.

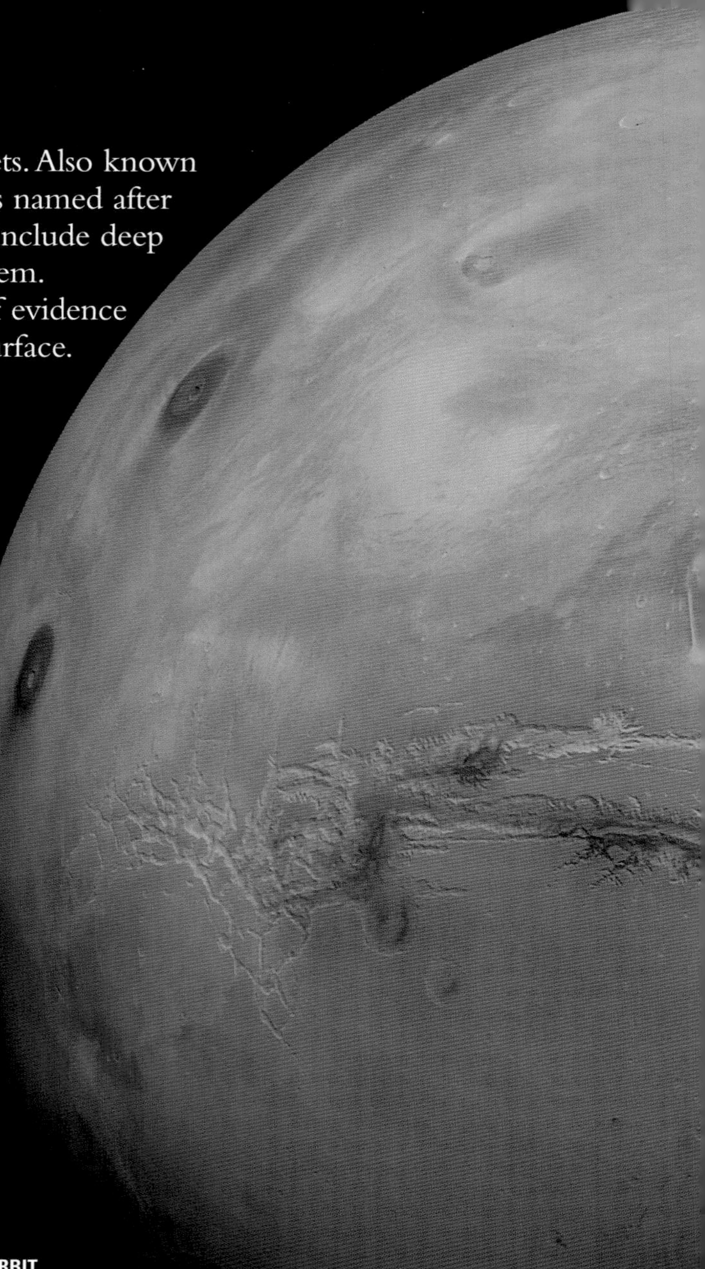

## ORBIT

Mars has an elliptical orbit, so at its closest approach to the Sun (perihelion), it receives 45 percent more solar radiation than at the farthest point (aphelion). This means that the surface temperature can vary from -195°F (-125°C) at the winter pole to 77°F (25°C) during the summer. At 25.2°, the current axial tilt of Mars is similar to that of Earth and, like Earth, Mars experiences changes in seasons as the north pole, and then the south pole, points toward the Sun during the course of its orbit. Throughout its history, Mars's axial tilt has fluctuated greatly due to various factors, including Jupiter's gravitational pull. These fluctuations have caused significant changes in climate. When Mars is heavily tilted, the poles are more exposed to the Sun, causing water ice to vaporize and build up around the colder lower latitudes. At a lesser tilt, water ice becomes concentrated at the colder poles.

direction of sunlight
axis of rotation tilts 60° from vertical
water ice concentrated at cold lower latitudes
south polar region exposed to sunlight is ice-free
60°
water ice still present at equator
equator receives more sunlight than at 60° tilt
45°
water ice builds up at colder north pole
35°
ice-free equator
water ice concentrated around north polar region
25°


**CHANGES IN AXIAL TILT**
Water-ice distribution during a Martian winter in the northern hemisphere varies with the axial tilt. The translucent white areas shown here represent thin ice that melts during the summer, whereas the thick white ice remains.

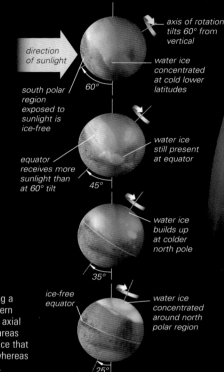

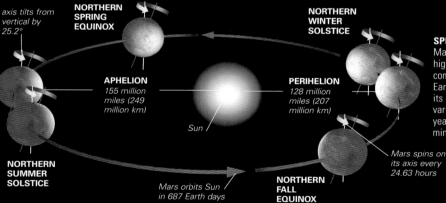

axis tilts from vertical by 25.2°
NORTHERN SPRING EQUINOX
NORTHERN WINTER SOLSTICE
APHELION
155 million miles (249 million km)
PERIHELION
128 million miles (207 million km)
Sun
NORTHERN SUMMER SOLSTICE
Mars orbits Sun in 687 Earth days
NORTHERN FALL EQUINOX
Mars spins on its axis every 24.63 hours


**SPIN AND ORBIT**
Mars's orbit is highly eccentric compared to that of Earth, which means that its distance from the Sun varies more during a Martian year. A Martian day is 42 minutes longer than an Earth day.

## STRUCTURE

Mars is a small planet, about half the size of Earth, and farther away from the Sun. Its size and distance mean that it has cooled more rapidly than Earth, and its once-molten iron core is probably now solid. Its relatively low density compared to the other terrestrial planets indicates that the core may also contain a lighter element, such as sulfur, in the form of iron sulfide. The small core is surrounded by a thick mantle composed of solid silicate rock. The mantle was a source of volcanic activity in the past, but it is now inert. Data gathered by the Mars Global Surveyor spacecraft has revealed that the rocky crust is about 50 miles (80 km) thick in the southern hemisphere, whereas it is only about 22 miles (35 km) thick in the northern hemisphere. Mars has the same total land area as Earth, since it has no liquid water on its surface

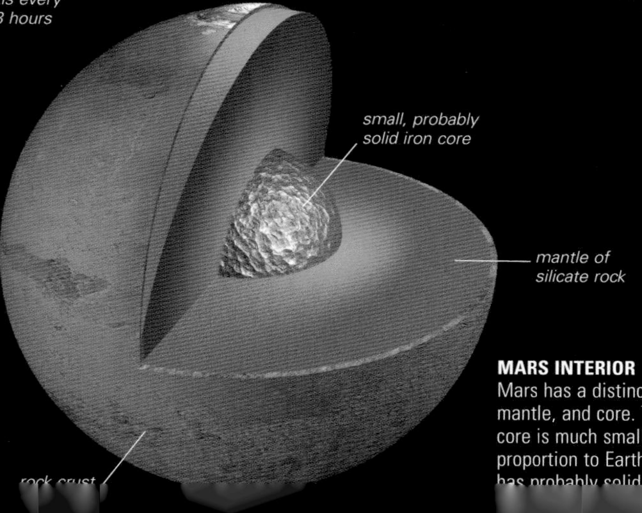

small, probably solid iron core
mantle of silicate rock
rock crust


**MARS INTERIOR**
Mars has a distinct crust, mantle, and core. The core is much smaller in proportion to Earth's and has probably solidified.

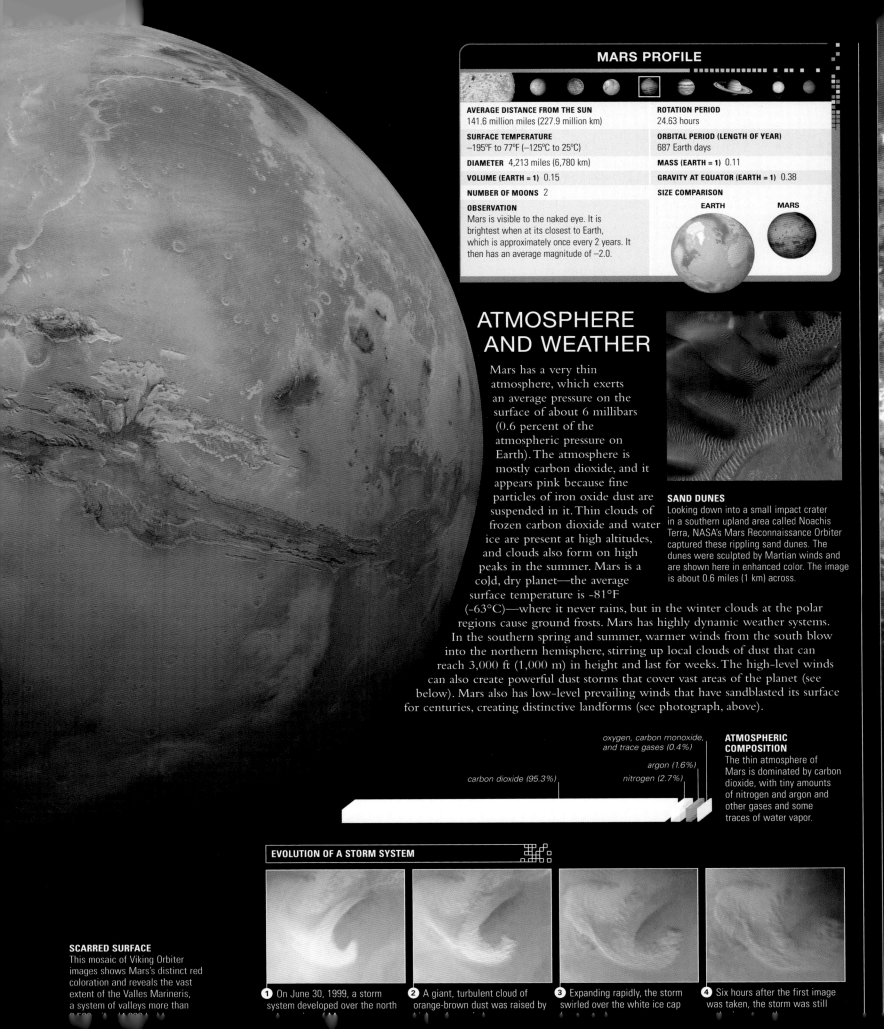

## MARS PROFILE

| AVERAGE DISTANCE FROM THE SUN | ROTATION PERIOD |
|---|---|
| 141.6 million miles (227.9 million km) | 24.63 hours |

| SURFACE TEMPERATURE | ORBITAL PERIOD (LENGTH OF YEAR) |
|---|---|
| –195°F to 77°F (–125°C to 25°C) | 687 Earth days |

| DIAMETER  4,213 miles (6,780 km) | MASS (EARTH = 1)  0.11 |
|---|---|

| VOLUME (EARTH = 1)  0.15 | GRAVITY AT EQUATOR (EARTH = 1)  0.38 |
|---|---|

| NUMBER OF MOONS  2 | SIZE COMPARISON |
|---|---|

**OBSERVATION**
Mars is visible to the naked eye. It is brightest when at its closest to Earth, which is approximately once every 2 years. It then has an average magnitude of –2.0.

SIZE COMPARISON
EARTH      MARS

# ATMOSPHERE AND WEATHER

Mars has a very thin atmosphere, which exerts an average pressure on the surface of about 6 millibars (0.6 percent of the atmospheric pressure on Earth). The atmosphere is mostly carbon dioxide, and it appears pink because fine particles of iron oxide dust are suspended in it. Thin clouds of frozen carbon dioxide and water ice are present at high altitudes, and clouds also form on high peaks in the summer. Mars is a cold, dry planet—the average surface temperature is –81°F (–63°C)—where it never rains, but in the winter clouds at the polar regions cause ground frosts. Mars has highly dynamic weather systems. In the southern spring and summer, warmer winds from the south blow into the northern hemisphere, stirring up local clouds of dust that can reach 3,000 ft (1,000 m) in height and last for weeks. The high-level winds can also create powerful dust storms that cover vast areas of the planet (see below). Mars also has low-level prevailing winds that have sandblasted its surface for centuries, creating distinctive landforms (see photograph, above).

**SAND DUNES**
Looking down into a small impact crater in a southern upland area called Noachis Terra, NASA's Mars Reconnaissance Orbiter captured these rippling sand dunes. The dunes were sculpted by Martian winds and are shown here in enhanced color. The image is about 0.6 miles (1 km) across.

**ATMOSPHERIC COMPOSITION**
The thin atmosphere of Mars is dominated by carbon dioxide, with tiny amounts of nitrogen and argon and other gases and some traces of water vapor.

*oxygen, carbon monoxide, and trace gases (0.4%)*

*argon (1.6%)*

*nitrogen (2.7%)*

*carbon dioxide (95.3%)*

## EVOLUTION OF A STORM SYSTEM

1. On June 30, 1999, a storm system developed over the north

2. A giant, turbulent cloud of orange-brown dust was raised by

3. Expanding rapidly, the storm swirled over the white ice cap

4. Six hours after the first image was taken, the storm was still

**SCARRED SURFACE**
This mosaic of Viking Orbiter images shows Mars's distinct red coloration and reveals the vast extent of the Valles Marineris, a system of valleys more than

## MISSIONS TO MARS

Numerous spacecraft have been sent to Mars since the first missions were undertaken by the USA and the Soviet Union in the 1960s, with varying success due to technical difficulties. A selection of successful missions is described below.

**1976 VIKING 1 AND 2 (USA)** These two craft each consisted of an orbiter and a lander. The orbiters sent back images, while the landers descended to two different sites and sent back analyses of the soil and atmosphere, as well as images.

**1997 MARS PATHFINDER (USA)** his mission sent a stationary lander and a free-ranging robot called Sojourner to the surface of Mars. They landed in an ancient floodplain and sent back pictures and analyses of soil samples.

SOJOURNER

**2003 MARS EXPRESS (EUROPE)** This orbiting spacecraft is imaging the entire surface of Mars, as well as mapping its mineral composition and studying the Martian atmosphere.

**2004 MARS EXPLORATION ROVERS (USA)** Twin rovers Spirit and Opportunity landed on opposite sides of the planet and studied rocks and soil, looking for evidence of how liquid water affected Mars in the past.

MARS ROVER

**2006 MARS RECONNAISSANCE ORBITER (USA)** Looping over the poles of Mars 12 times every Martian day, MRO keeps a constant eye on the red planet's weather and looks for signs of water, past or present, on its surface.

MRO

**2012 CURIOSITY (USA)** About the size of a car, this rover is exploring Gale Crater.

**2016 EXOMARS TRACE GAS ORBITER (EUROPE AND RUSSIA)** This orbiter searches for methane and other minor atmospheric components.

**2018 INSIGHT (USA)** A stationary lander, InSight studies marsquakes and heat flow.

## MARS MAPS

These four views combine to show the complete surface of Mars. They have been labeled to show large-scale features, as well as the landing sites of some of the spacecraft sent to explore its surface.

# SURFACE FEATURES

Mars's surface features have been formed and shaped by meteorite impacts, by the wind (see p.151), and by volcanism and faulting (see Tectonic Features, below). Scientists also believe that water once flowed on and below the surface of Mars (see opposite), carving out features such as valleys and outflow channels. The craters formed during a period of intense meteorite bombardment about 3.9 billion years ago. They are found mainly in the southern hemisphere, which is geologically older than the northern hemisphere, and include the vast Hellas Basin (see p.165), but small craters are found all over Mars. Martian craters are flatter than those on the Moon and show signs of erosion by wind and water; indeed, some have almost been obliterated.

**SPIRIT AT HUSBAND HILL**
The Spirit rover's arm reaches out to investigate a rocky outcrop called Hillary, named after the mountaineer Sir Edmund Hillary, near the summit of Husband Hill.

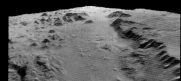

**IMPACT CRATER**
The Herschel impact crater, located in the southern highlands, is about 185 miles (300 km) across. This image has been false-colored to show altitude. The lowest areas are the dark blue floors of smaller craters. The Herschel Crater floor is mostly at 3,240 ft (1,000 m) and the highest parts of the rim (pale pink) are at about 9,720 ft (3,000 m).

# TECTONIC FEATURES

Billions of years ago, when Mars was a young planet, internal adjustments created the large-scale features seen on its surface today. Internal forces created raised areas on the surface, such as the Tharsis Bulge, and stretched and split the surface to create rift valleys, such as the vast Valles Marineris (see pp.158–159). Landslides, wind, and water have since modified the rift valleys. Volcanic activity dates back billions of years and persisted for much of Mars's history. The planet may still be volcanically active today, although no such activity is expected. Lava eruptions of the past formed today's giant volcanoes, including Olympus Mons (see p.157).

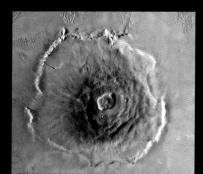

**OLYMPUS MONS**
This mosaic of images of Olympus Mons taken by Viking 1 in 1978 looks deceptively flat—the volcano stands 15 miles (24 km) above the surrounding plain.

**VALLES MARINERIS**
The Valles Marineris is a complex system of canyons that cuts across Mars at an average depth of 5 miles (8 km). If it was on Earth, it would stretch across North America.

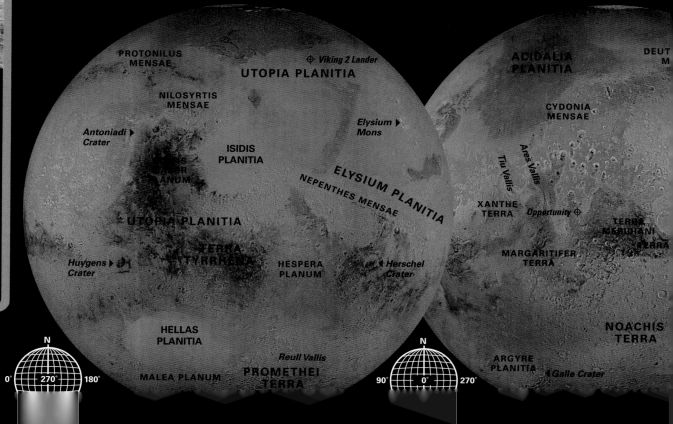

PROTONILUS MENSAE

Viking 2 Lander

UTOPIA PLANITIA

NILOSYRTIS MENSAE

Antoniadi Crater

Elysium Mons

ISIDIS PLANITIA

ELYSIUM PLANITIA

NEPENTHES MENSAE

UTOPIA PLANITIA

TERRA TYRRHENA

Huygens Crater

HESPERA PLANUM

Herschel Crater

HELLAS PLANITIA

N

MALEA PLANUM

Reull Vallis

PROMETHEI TERRA

0° 270° 180°

ACIDALIA PLANITIA

DEUT M

CYDONIA MENSAE

Tiu Vallis

Ares Vallis

XANTHE TERRA

Opportunity

TERRA MERIDIANI TERRA

MARGARITIFER TERRA

NOACHIS TERRA

ARGYRE PLANITIA

Galle Crater

N

90° 0° 270°

# WATER ON MARS

Scientists have been hoping to establish whether water is present on Mars, since this is essential for the development of life. Liquid water is not present, as this is a cold planet, where water can exist only as ice or vapor. The latter can form low-lying mists and fogs and freezes into a thin layer of white water ice on the rocks and soil when the temperature falls. However, dry river beds, valleys, and ancient floodplains bear witness to the presence of large amounts of fast-flowing water on the surface 3–4 billion years ago, when Mars was a warmer, wetter world with a thicker atmosphere. Some of that water remains today in the form of ice, which is present both underground and in the polar ice caps. The ice caps wax and wane with the Martian seasons and are composed of varying amounts of water ice and frozen carbon dioxide.

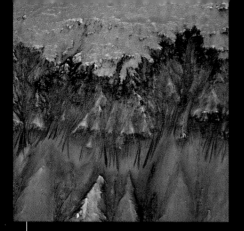

### THE CASE FOR FLOWING WATER
Images from Mars Global Surveyor have been processed to create this 3-D view of seasonally variable features on the inner slopes of the crater Newton, possibly created by running water.

### NORTH POLAR ICE CAP
The north polar cap of Mars, seen by Mars Global Surveyor, is about 600 miles (1,000 km) across and is cut by spiral-shaped troughs. At center right is Chasma Boreale, a valley about the length of Earth's Grand Canyon.

# MOONS

Mars has two small, dark moons named Phobos and Deimos, which were discovered by the American astronomer Asaph Hall in August 1877. Deimos, the smaller of the two, is 9.3 miles (15 km) long and Phobos is 16.6 miles (26.8 km) long. Both are irregular "potato-shaped" rocky bodies and are probably asteroids that were captured in Mars's early history. They both bear the scars of meteorite battering. Deimos orbits Mars at a distance of 14,580 miles (23,500 km). Phobos is only 5,830 miles (9,380 km) from Mars and getting closer; eventually, it will be so close that it will either be torn apart by Mars's gravity field or collide with the planet.

### MOONS' ORBIT
Phobos and Deimos both follow near-circular orbits around Mars, and both exhibit synchronous rotation. From Mars, Phobos rises and sets three times every Martian day.

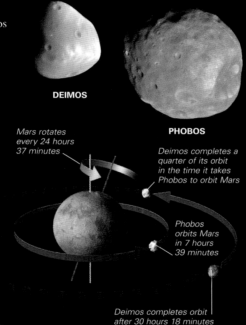

**DEIMOS**

**PHOBOS**

Mars rotates every 24 hours 37 minutes

Deimos completes a quarter of its orbit in the time it takes Phobos to orbit Mars

Phobos orbits Mars in 7 hours 39 minutes

Deimos completes orbit after 30 hours 18 minutes

# GEOGRAPHY

The first reliable maps of Mars were made in the late 19th century, when astronomers drew what they observed through their telescopes. Today's maps are based on data collected by space probes such as Mars Global Surveyor, which obtained 100,000 photos of Mars and completed a survey of the planet, and Mars Express, which is imaging the entire surface. The following terminology is used for the surface features: lowland plains are termed planitia; high plains, planum; extensive landmasses, terra; and mountains or volcanoes, mons. A chasma is a deep, elongated, steep-sided depression, and a labyrinthus is a system of intersecting valleys or canyons. Individual names are allocated depending on the type of feature. Large valleys (vallis) are named after Mars in various languages and small ones after rivers. Large craters are named after past scientists, writers, and others who have studied Mars; smaller craters are named after villages. Other features are named after the nearest albedo feature on the early maps.

US

ALBA PATERA

TEMPE TERRA

ACIDALIA PLANITIA

Mie ▶

ARCADIA PLANITIA

Acheron Fossae

◀ Cassini Crater

BIA RA

◀ Tikhonravo Crater

◀ Olympus Mons

KASEI VALLES

Viking 1 Lander

Belz Crater Mars Pathfinder

CHRYSE PLANITIA

◀ Hecates Tholus

◀ Elysium Mons

◀ Albor Tholus

AMAZONIS PLANITIA

◀ Olympus Mons

HARSIS MONTES

◀ Ascraeus Mons

LUNAE PLANUM

Namedi Vallis

Shalbatana Vallis

Simud Vallis

iaparelli Crater

◀ Pavonis Mons

Curiosity

SIS MONTES

**THE LABYRINTH OF MARS**
Noctis Labyrinthus (which is Latin for "labyrinth of the night") is a complex system of steep-walled canyons at the western end of the giant Valles Marineris rift valley on Mars (see p.156). It is thought to result from faulting, when the giant volcanoes on the Tharsis ridge of Mars caused the crust to bulge in this area. Landslides can be seen on the valley floors.

# GEOLOGICAL FEATURES

Mars has two areas of markedly different terrain. Much of the northern hemisphere is characterized by relatively smooth and low-lying volcanic plains. The older southern landscape is typically cratered highland. The boundary between the two is an imaginary circle tilted by about 30° to the equator. The planet's major geological features are found within a region that extends roughly 30° each side of the equator. It contains Mars's main volcanic center, the Tharsis region, and the Valles Marineris, the vast canyon system that slices across the center of the planet.

**WESTERN FLANK OF OLYMPUS MONS**
Tectonic features on Mars take on familiar forms but are on a much grander scale than those on Earth. This escarpment on the side of Olympus Mons is 4.3 miles (7 km) high.

---

## THARSIS MONTES
# Pavonis Mons

| | |
|---|---|
| **TYPE** | Shield volcano |
| **AGE** | 300 million years |
| **DIAMETER** | 235 miles (375 km) |

A huge bulge in the western hemisphere, commonly known as the Tharsis Bulge, contains volcanoes of various sizes and types, from large shields to smaller domes. Olympus Mons dominates the region. But three other volcanoes, which anywhere else would be considered enormous, are also found here. The three form a line and together make the Tharsis

**CHANNELS**
These deep channels on the volcano's southern flank may have started out as subsurface lava tubes whose roofs collapsed as pits developed over them.

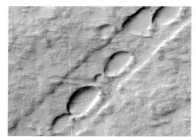

**PIT CHAIN**
A chain of pits lies in a shallow trough on the lower east flank. The pits and trough formed because the ground was either moved apart by tectonic forces or uplifted by molten rock.

Montes mountain range. Pavonis Mons, situated on the equator, is the middle of the three. It is a shield volcano with a broad base and sloping sides and is similar to those found in Hawaii on Earth. The volcano's summit stands 4.3 miles (7 km) above the surrounding plain and has a single caldera within a larger, shallow depression. Hundreds of narrow lava flows are seen to emanate from the rim of the caldera, and others can be traced back to pits situated close by.

**SUMMIT DEPRESSION**
The summit caldera lies within a shallow depression that is almost twice the caldera's size and has faulted sides.

---

## THARSIS MONTES
# Ascraeus Mons

| | |
|---|---|
| **TYPE** | Shield volcano |
| **AGE** | 100 million years |
| **DIAMETER** | 285 miles (460 km) |

Ascraeus Mons is the northernmost of the three Tharsis Montes volcanoes. The three lie on the crest of the Tharsis Bulge and form a line in a southwest–northeast direction. The line marks the position of a major rift zone, long since buried under lava. The three volcanoes grew by the gradual buildup of thousands of individual and successive lava flows that came to the surface through the rift zone. Ascraeus is the tallest of the three, rising over 11 miles (18 km) above the surrounding plain. It has a large number of lines and channels all around the rim of the caldera, showing the paths taken by flowing lava.

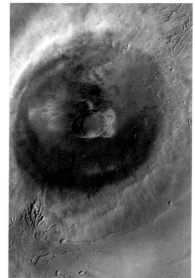

**CALDERA**
The caldera on the summit is made up of eight major depressions and has a nested appearance (center). Its deepest point is over 1.9 miles (3 km) below the rim.

---

## THARSIS MONTES
# Arsia Mons

| | |
|---|---|
| **TYPE** | Shield volcano |
| **AGE** | 700 million years |
| **DIAMETER** | 295 miles (475 km) |

Arsia Mons is second only to the mighty Olympus Mons in terms of volume. It is the southernmost of the three Tharsis Montes volcanoes, and its summit rises more than 5.6 miles (9 km) above the surrounding plain. Like the other two, it has a summit caldera bigger than any known on Earth. Arsia Mons measures 75 miles (120 km) across and is surrounded by arc-shaped faults. Lava flows fan out down the volcano's shallow slopes. The lava is of basaltlike composition and of low viscosity, and the flows are shorter nearer the summit than on the lower flanks.

**CLOUDY SUMMIT**
Water-ice clouds hang over the volcano's summit—a common sight every Martian afternoon in the Tharsis region.

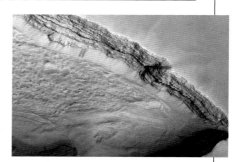

**LAYERED OUTCROP**
This outcrop of layered rock lies in a pit on the volcano's lower west flank. The layers are thought to consist mostly of volcanic rock formed by successive lava flows.

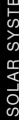

## THARSIS REGION

# Olympus Mons

| | |
|---|---|
| TYPE | Shield volcano |
| AGE | 30 million years |
| DIAMETER | 403 miles (648 km) |

Olympus Mons is unquestionably the largest volcano in the solar system. Its height, of about 15 miles (24 km), makes it the tallest, and its volume is over 50 times that of any shield volcano on Earth. Olympus is one of the giant shield volcanoes of the Tharsis region, which is home to the greatest number of volcanoes on Mars, including the planet's youngest. Volcanoes evolve over long periods of time and can be inactive for hundreds of millions of years. Olympus Mons is considered the youngest of the shield volcanoes. The summit has a complex caldera.

Different areas of its floor are associated with different periods of activity. The largest central area, which is marked by ring-shaped faults, is more recent, at 140 million years old.

The caldera is surrounded by a surface of wide terraces formed by lava flows, crossed by thinner flows. These are encircled by a huge scarp, up to 3.7 miles (6 km) high. Vast plains, termed aureoles, extend from the north and west of the summit like petals from a flower. These regions of gigantic ridges and blocks, whose origins remain unexplained, extend outward for up to 600 miles (1,000 km).

**LAVA FLOWS**
These lava flows and a collapsed lava tube (top left) on the southwest flank have been peppered by tiny impact craters.

### EXPLORING SPACE

## MARTIAN METEORITES

Solidified basaltic lava covers the Tharsis region. Pieces of lava that flowed on the Martian surface as recently as 180 million years ago are now on Earth. Impactors hit Mars and ejected them, and after journeys lasting millions of years, they fell to Earth as meteorites. They include the Shergotty meteorite (right), which landed in Shergahti, India, on August 25, 1865.

**MIGHTY OLYMPUS**
This massive volcano is named after the mountaintop home of the gods and goddesses of Greek mythology. Broad lava-flow terraces surround the caldera at the volcano's summit.

**COMPLEX CALDERA**
In this bird's-eye view of the 32-mile (52-km) wide nested caldera on the summit of Olympus Mons, five roughly circular areas of caldera floor can be seen.

**LANDSLIDE CLIFFS**
Olympus Mons is bounded on all sides by steep cliffs, thought to have been caused by landslides. This close-up taken by Mars Reconnaissance Orbiter shows an area of the cliffs about 0.6 miles (1 km) wide on the northern side of the volcano.

THE SOLAR SYSTEM

# Valles Marineris

**TYPE** Canyon system

**AGE** About 3.5 billion years

**LENGTH** Over 2,500 miles (4,000 km)

Valles Marineris is the largest feature formed by tectonic activity on Mars. It consists of a system of canyons that stretches for over 2,500 miles (4,000 km), is up to 430 miles (700 km) wide, and averages 5 miles (8 km) in depth. The Grand Canyon in Arizona is dwarfed in comparison; it is only about one-tenth as long and one-fifth as deep. Valles Marineris lies just south of the Martian equator, and the system trends, very roughly, west to east. The trend follows a set of fractures that radiates from the Tharsis Bulge at Marineris's western end.

The origins of the system date back a few billion years to when the canyons were formed by faulting. This contrasts with the Grand

Canyon, which is a primarily water-eroded canyon. But water, as well as wind, has played its part in the development of the Marineris system. Buffeting winds, flowing water, and the collapse of unstable walls have all widened and deepened the canyons.

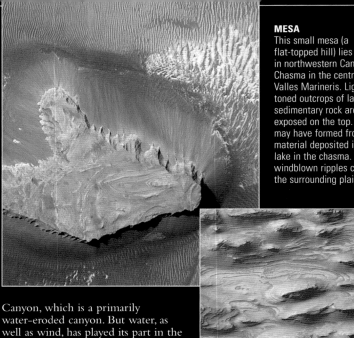

### MESA
This small mesa (a flat-topped hill) lies in northwestern Candor Chasma in the central Valles Marineris. Light-toned outcrops of layered sedimentary rock are exposed on the top. These may have formed from material deposited in a lake in the chasma. Darker windblown ripples cover the surrounding plains.

The Noctis Labyrinthus region marks the western end of the system. This is a roughly triangular area of intersecting rift valleys that form a mazelike arrangement. The eastern end of Valles Marineris is bounded by chaotic terrain of irregular appearance. Here, smaller canyons and depressions give way to outflow canyons. These carried ancient rivers of water out of Marineris toward the lowland region, Chryse Planitia, to the north. This whole area has seen extensive water erosion; millions of cubic miles of material have been removed by water action.

The system's canyons are described as chasma (plural chasmata) and are given identifying names. The main chasma in the western part of the system is Ius. The central complex is made up of three parallel canyons:

### LAYERED DEPOSITS
A detail of the floor of western Candor Chasma shows layered sedimentary rock. Up to 100 layers have been counted, each about 30 ft (10 m) thick. The layers may be made from material deposited in an impact crater before the chasma formed.

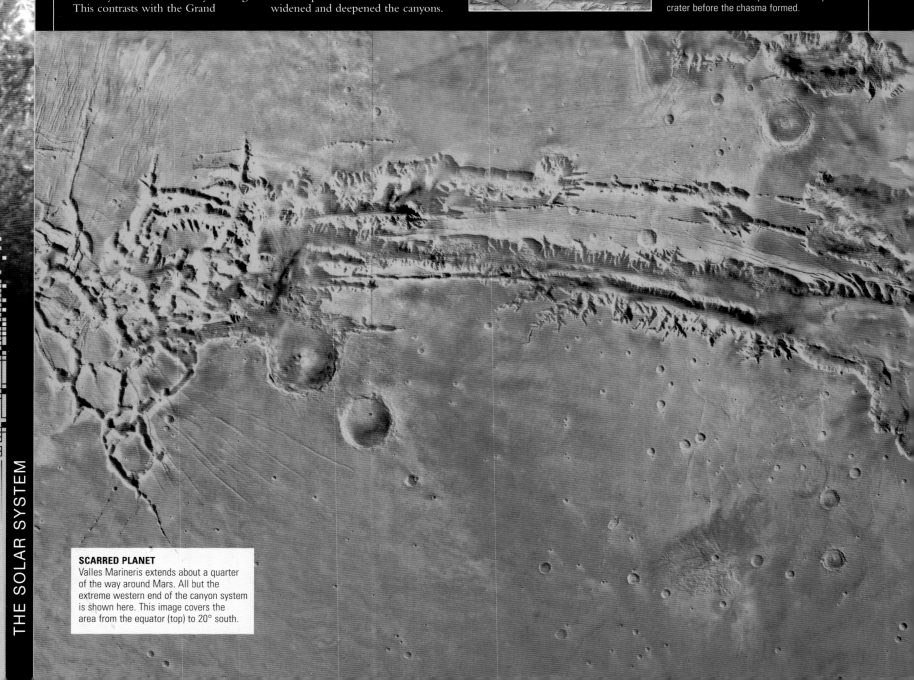

### SCARRED PLANET
Valles Marineris extends about a quarter of the way around Mars. All but the extreme western end of the canyon system is shown here. This image covers the area from the equator (top) to 20° south.

Ophir; Candor; and, to their south, Melas Chasma. The long Coprates Chasma stretches out to the east, where it meets the broader Eos Chasma. The name of the whole system, Valles Marineris, means "Valleys of the Mariner." The Mariner in this case is the Mariner 9 mission that mapped

### LOOKING WEST THROUGH OPHIR CHASMA
Over billions of years, Ophir Chasma has widened as its walls have collapsed and slumped downward, covering the floor with debris.

### EASTERN EOS CHASMA
Water flowed through this broad chasma, out of the Valles Marineris and into a series of valleys and channels.

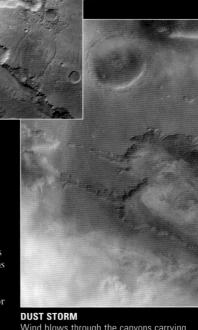

the entire surface of Mars and took the first close-up images of this area. More recent craft, such as Mars Global Surveyor and Mars Express, have provided more detailed coverage. For example, their surveys have revealed layered rock in the canyon walls that could be a profile of the different lava flows that built the plains that the canyons cut through. Rocks on the floor may have formed from windborne dust layers or by deposits in ancient lakes that once filled the canyons.

### DUST STORM
Wind blows through the canyons carrying dust. The pinkish dust cloud at the bottom of this image is moving north across the junction of Ius Chasma and Melas Chasma. The higher, bluish-white clouds are water ice.

### MARS EXPRESS'S STEREO CAMERA

The High Resolution Stereo Camera onboard Mars Express began its 2-year program to map the entire Martian surface in January 2004. Its nine charge-coupled device sensors record data one line at a time. Downward, backward, and forward views are used to build up 3-D images. The Super Resolution Channel provides more detailed information.

Digital Unit includes Camera Control Processor

camera head

instrument frame provides mechanical stability

Super Resolution Channel

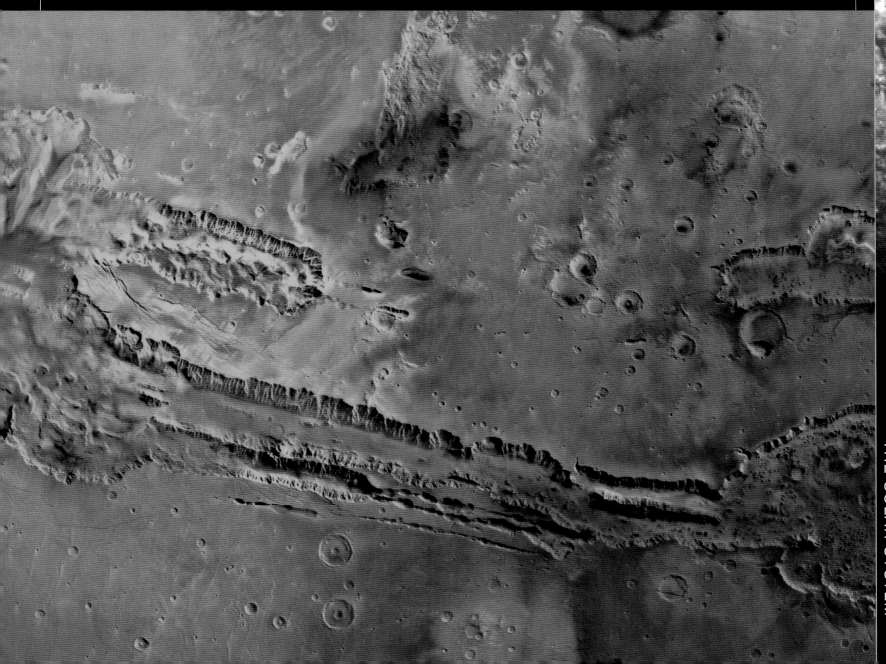

## THARSIS REGION

# Acheron Fossae

| | |
|---|---|
| **TYPE** | Fault system |
| **AGE** | Over 3.5 billion years |
| **LENGTH** | 695 miles (1,120 km) |

Acheron Fossae is a relatively high area that has seen intense tectonic activity in the past. It marks the northern edge of the Tharsis Bulge

**CLIFF FACE**
This is a close-up of the bright, steep slopes of a scarp or cliff. Dark streaks on the cliff face may be formed as the dust mantle that covers the region gives way and produces a dust avalanche.

**BROKEN CRUST**
In this perspective view across the highly deformed area of Acheron Fossae, curved faults trending to the northwest dominate the scene.

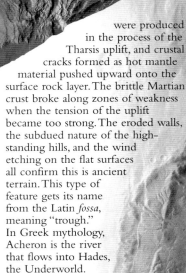

and is located about 600 miles (1,000 km) north of Olympus Mons. Acheron is part of a network of fractures that radiates out from the Tharsis Bulge—a huge region of uplift and volcanic activity. It can be compared to the East African Rift on Earth (see p.130), where continental plates have spread apart. The huge curved faults in the Tharsis Bulge were produced in the process of the Tharsis uplift, and crustal cracks formed as hot mantle material pushed upward onto the surface rock layer. The brittle Martian crust broke along zones of weakness when the tension of the uplift became too strong. The eroded walls, the subdued nature of the high-standing hills, and the wind etching on the flat surfaces all confirm this is ancient terrain. This type of feature gets its name from the Latin *fossa*, meaning "trough." In Greek mythology, Acheron is the river that flows into Hades, the Underworld.

**GRABENS AND HORSTS**
The planetary crust has fallen between parallel faults to form grabens up to 1 mile (1.7 km) deep; remnants of the preexisting heights are termed horsts.

**STRESSED LANDSCAPE**
A fault system cutting across an ancient impact crater is evidence of the stress felt by the Martian crust. The crater floor has since been resurfaced by material from outside the area.

---

## CIMMERIA TERRA

# Apollinaris Patera

| | |
|---|---|
| **TYPE** | Patera volcano |
| **AGE** | 900 million years |
| **DIAMETER** | 184 miles (296 km) |

This is an example of a type of volcano that was first identified on Mars. Known as patera volcanoes, they have very gentle slopes (with angles as low as 0.25°). Apollinaris Patera is one of the largest on the planet, situated on the northern edge of Cimmeria Terra, a few degrees south of the equator. It is the only

**MESAS AND TROUGHS**
A group of mesas was created by pitting and erosion of the surface in an area north of Apollinaris Patera. Windblown dust has filled the troughs between the mesas.

major volcano that is isolated from the two major volcanic regions of Tharsis, to the northeast, and Elysium, to the northwest. Apollinaris is a broad, roughly shield-shaped volcano, reminiscent of an upturned saucer. It is only about 3 miles (5 km) high and has a caldera about 50 miles (80 km) across. It appears to have been formed by both effusive and explosive activity. Lava flows are clearly visible beyond the summit. A cliff surrounding the caldera area is visible on the northern side but has disappeared on the opposite side. It is buried under a fan of material whose surface is marked by broad channels. The fan material could have formed from flowing lava or volcanic rock fragments.

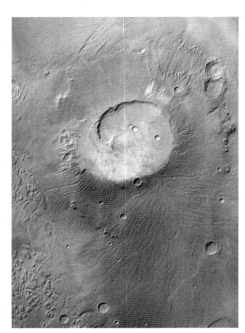

**SPLIT-LEVEL CALDERA**
The caldera has two different floor levels. It is partially hidden here by a patch of blue-white clouds. The summit area is pocked with impact craters.

---

## AONIA TERRA

# Claritas Fossae

| | |
|---|---|
| **TYPE** | Fault system |
| **AGE** | Over 3.5 billion years |
| **LENGTH** | 1,275 miles (2,050 km) |

Claritas Fossae is a series of roughly northwest-to-southeast-trending linear fractures, which forms the southern end of the Tharsis Bulge. It is located south of the equator at the western end of the Valles Marineris. The region is about 95 miles (150 km) wide at its northern end and 340 miles (550 km) wide in the south. Individual fractures range from a few to tens of miles across.

**FAULTS IN CLARITAS FOSSAE**
Running from the volcanic Tharsis Bulge, the linear features in this image of Claritas Fossae coincide with fractures in the Martian crust produced by stretching forces.

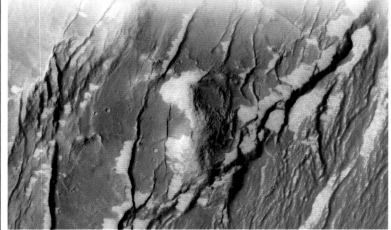

**LAVA BLANKET**
The eastern part of Claritas Fossae (bottom) meets the western part of Solis Planum (top). The lava from Solis has flowed over some of the older fractured terrain of Claritas and surrounds some of the higher ground.

They formed as a result of enormous stresses associated with the formation of the Tharsis Bulge. As the crust pulled apart, blocks of crust dropped between two faults to form features called grabens. Crustal blocks that remained in place or were thrown up are termed horsts. Claritas Fossae separates two volcanic plains: that of Solis Planum to the east and Daedalia Planum to the west.

# FEATURES FORMED BY WATER

Both liquid and solid water have formed and shaped surface features on Mars. Giant channel-like valleys emerge fully formed out of the landscape. Some of these were cut during catastrophic floods, others were formed by water flowing more gradually through networks of river valleys, and others still were carved by glaciers. Some features suggest Mars once had seas, although the evidence is inconclusive. However, any potential rivers and seas have long since vanished, and only water ice remains, most markedly in the two ice plateaus that cap the planet.

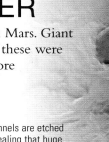

**REULL VALLIS**
Long, wide river channels are etched into the surface, revealing that huge volumes of water flowed across Mars billions of years ago.

## PLANUM BOREUM

# North Polar Region

| | |
|---|---|
| **TYPE** | Polar ice cap |
| **AGE** | Under 2.5 billion years |
| **DIAMETER** | 685 miles (1,100 km) |

Two bright white polar caps stand out against the otherwise dark surface of Mars. The one roughly centered on the north pole is officially named Planum Boreum—the Northern Plain—although it is generally referred to as the North Polar Cap. Both this and its southern counterpart are easy to detect from Earth, but spacecraft have also flown over the poles, allowing monitoring of daily, seasonal, and longer-term change.

The North Polar Cap on Mars is an ice-dominated mound that stands several miles above the surrounding terrain. It consists of a virtually permanent cap of water ice, which is either covered by or free of a deposit of carbon-dioxide ice, depending on the time of the Martian year. The cap is roughly circular but—as is also the case for the South Polar Cap—its bright ice forms a distinctive swirling, loosely spiral pattern when seen from above (see p.163).

The entire region is in darkness for about 6 months during the Martian winter.

**POLAR POLYGONS**
Polygon-shaped structures, similar to those found in Earth's polar regions, pattern parts of Mars's polar landscape. On Earth, they form as a result of stresses induced by repeated freezing and thawing of water.

This is when carbon dioxide in the atmosphere condenses into frost and snow and not only covers the water-ice cap, but also the surrounding region, down to latitudes of about 65° north. When spring turns to summer and the Sun is permanently in the polar sky, its warmth evaporates the carbon dioxide and turns some of the water ice directly into vapor. The polar cap shrinks until just water ice remains.

The cap is not made exclusively of ice but consists of layers of ice and layers of dusty sediment. Frost grains form around small particles of dust during winter dust storms in much the same way that hailstones form on Earth. These cover the ground until the frost is evaporated in the warmer months, leaving a layer of dust. The yards-deep layers take

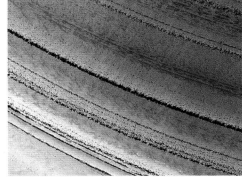

**LAYERED DEPOSITS**
Layers of ice at the Martian north pole attest to past variations in the planet's climate. The layers are exposed at the edge of the ice sheet, which slopes downhill from the bottom to the top in this image. The thickness of the ice is about 0.6 miles (1 km).

millions of years to form, building up at the rate of about 0.04 in (1 mm) per year. A study of these layers will reveal the history of the Martian climate.

**CLIFFS NEAR THE NORTH POLE**
This close-up view of the Martian North Polar Cap shows water ice close to cliffs about 1.2 miles (2 km) high. Dark material in the calderalike structures and dune fields could be volcanic ash.

## UTOPIA PLANITIA
# Utopia Planitia

| TYPE | Lowland plain |
|---|---|
| AGE | 2–3.5 billion years |
| DIAMETER | 2,000 miles (3,200 km) |

Utopia is one of the enormous lava-covered plains of the northern hemisphere. The giant Elysium volcanoes are at its eastern perimeter. From above, it is possible to see that complex albedo patterns, polygonal fractures, and craters mark the vast rolling plain. Down on the surface, the landscape is uniformly flat and rock-strewn. At least that is the view in northeastern Utopia, where Viking 2 landed on September 3, 1976. Angular boulders of basaltic rock cover the landing site, close to Mie Crater. Small holes in the rocks are a result of bursting bubbles of volcanic gas. A thin layer of frost was also seen by the craft, first in mid-1977, when it covered the surface for about 100 Earth days, and then when it built up again in May 1979, 1 Martian year (23 Earth months) later.

These giant polygons are not unique to Utopia; they are also seen on other northern plains such as Acidalia and Elysium (below). They are polygon-shaped chunks of flat land separated by huge cracks, or troughs, and are reminiscent of mud cracks seen in dried-up ponds on Earth. Earth's polygonal patterns are book- to table-sized; the Martian ones are on a far grander scale, the size of a town or small city. The land areas are 3–12 miles (5–20 km) across, and the cracks between them are hundreds of yards wide. Earth's mud cracks form as ground dries through water evaporation. The Martian cracks could have a similar origin. Certainly, Mars has experienced the large-scale floods of water required. And dust-covered cemented rock found by Viking 2 seemed to be held together by salts left behind as briny water vaporized. However, it has also been suggested that the polygons formed in other ways—for example, in cooling lava.

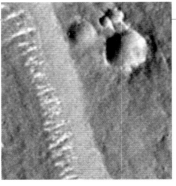

**POLYGON TROUGH FLOOR**
This close-up of one of the huge surface cracks that isolate polygon-shaped areas of land reveals bright, evenly spaced windblown ripples of sediment.

**ICE ON THE ROCKS**
A coating of water ice covers the volcanic rocks and soil at the Viking 2 landing site. The ice layer is incredibly thin, no more than a thousandth of an inch (a fraction of a millimeter) thick.

## LUNAE PLANUM
# Kasei Valles

| TYPE | Outflow channel |
|---|---|
| AGE | 3–3.5 billion years |
| LENGTH | 1,105 miles (1,780 km) |

Kasei, which takes its name from the Japanese word for Mars, is the largest outflow channel. Not only is it long, but parts of its upper reaches are over 125 miles (200 km) across, and in places, it is over 2 miles (3 km) deep. The catastrophic flooding that formed Kasei was greater than any other known flood event on Mars or Earth. Kasei originates in Lunae Planum, directly north of central Valles Marineris (see p.158), then flows across the ridged plain to Chryse Planitia. Along its route lie streamlined islands, isolated as the water flow split and then rejoined.

**PLATEAU EDGE**
This image shows the steep edge of a valley in northern Kasei Valles. The plateau to the left is about 0.8 miles (1.3 km) higher than the floor of the valley, similar to the depth of the Grand Canyon on Earth.

## ELYSIUM PLANITIA
# Elysium Planitia

| TYPE | Lowland plain |
|---|---|
| AGE | Under 2.5 billion years |
| DIAMETER | 1,860 miles (3,000 km) |

The Elysium Planitia is an extensive lava-covered plain just north of the equator. It has been suggested that an area almost directly south of the great volcano Elysium Mons is a dust-covered plain dominated by irregular blocky shapes that look like the rafts of segmented sea ice seen off the coast of Earth's Antarctica. These "ice plateaus" are surrounded by bare rock. They formed when water flooded through a series of fractures in the Martian crust, creating a sea similar in size to Earth's North Sea. As the water froze, floating pack ice broke into rafts. These were later covered in dust from the nearby volcanoes, and this coating protected them. Unprotected ice between the rafts vaporized into the atmosphere, leaving bare rock around the ice plateaus.

**ICE PLATEAUS AND IMPACT CRATERS**
The darker-toned ice plateaus are a few tens of miles across. The relatively small number of impact craters in this area suggests a young surface.

## PROTONILUS MENSAE
# Rock-strewn Glacier

| TYPE | Valley glacier |
|---|---|
| AGE | Under 1 billion years |
| LENGTH | 5 miles (8 km) |

Many glaciers have been mapped across the midlatitudes of Mars. The region known as Protonilus Mensae, which is about 40° north of Mars's equator, hosts some of the nicest examples. They are not as obvious as you might expect, because the bright white ice is hidden below rock debris and dust. However, the flow lineations on their surfaces and the built-up debris (moraine) in front make them convincing evidence of an epoch when accumulated snowfall on Mars allowed glaciers to form and carve out valleys.

**GLACIAL EROSION**
This glacier, flowing out from a region of high ground, has widened and deepened what was probably originally a much narrower valley.

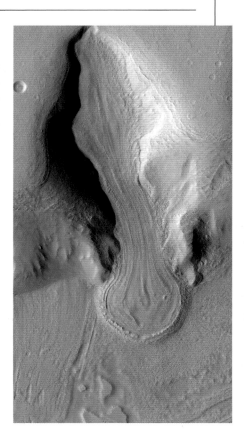

# Meridiani Planum

| | |
|---|---|
| **TYPE** | Highland plain |
| **AGE** | Over 3.5 billion years |
| **DIAMETER** | 680 miles (1,100 km) |

In the westernmost portion of Terra Meridiani and just south of the equator lies the high plain Meridiani Planum. It does not stand out in the global view of Mars but achieved prominence as the landing site and exploration ground for the Opportunity rover. The plain is about 15° due west of Schiaparelli Crater (see p.164). Smaller impact craters pepper the area. They range from Airy, just 26 miles (41 km) across, to much smaller bowl-shaped craters, such as the 72-ft (22-m) wide Eagle Crater where Opportunity landed.

**UNIQUE METEORITE**
This basketball-sized rock has an iron-nickel composition. It is not a Martian rock but a meteorite—the first to be found on a planet other than Earth.

Volcanic basalt is found within the area, but the region is of the greatest interest because it contains ancient layered sedimentary rock that includes the mineral hematite. Some of this mineral, which on Earth almost always forms in liquid water, is exposed and easily found on the surface. The hematite could have been produced from iron-rich lavas, but it is believed that water was involved. This area is dry now but was once soaking wet and could well have been the site of an ancient lake or sea about 3.7 billion years ago. Eroded layered outcrops beyond the landing site support this theory and point to a deep and long-lasting volume of water as large as Earth's Baltic Sea. At this time, Mars must have been a much warmer and wetter place than it is today.

**HEAT SHIELD AND SCORCH MARK**
Subsurface pale dirt was spattered onto the plain when Opportunity landed. The remains of the craft's discarded heat shield are at left and center.

# Reull Vallis

| | |
|---|---|
| **TYPE** | Outflow channel |
| **AGE** | 2–3.5 billion years |
| **LENGTH** | 587 miles (945 km) |

Reull Vallis is one of the larger channels of the southern hemisphere. It extends across the northern part of Promethei Terra, to the east of Hellas Basin (see p.165). Reull is thought to have had a complex evolution because it exhibits the characteristics of all three channel types seen on Mars. In the collapsed region at the southern base of the volcano Hadriaca Patera, for example, it is a fully formed outflow channel. But small tributaries also feed into the main channel, as they would in a runoff channel. And the main channel has the features of a fretted channel—a wide, flat floor and steep walls. Reull Vallis takes its name from the Gaelic word for "planet."

**MERGING CHANNELS**
Reull Vallis (top left) is joined by a tributary, Teviot Vallis (right). The parallel structures in the fretted channel floor were possibly caused by a glacial flow of loose debris mixed with ice.

# South Polar Region

| | |
|---|---|
| **TYPE** | Polar ice cap |
| **AGE** | Under 2.5 billion years |
| **DIAMETER** | 900 miles (1,450 km) |

The South Polar Cap, known formally as Planum Australe (Southern Plain), is an ice-dominated mound a few miles high. It consists of three different parts. First is the bright polar cap that is roughly centered on the South Pole. This is a permanent cap of water ice with a covering of carbon–dioxide ice. Next are the scarps, made primarily of water ice, which fall away from the cap to the surrounding plains. Third, hundreds of square miles of permafrost encircle the region. The permafrost is water ice mixed into the soil and frozen to the hardness of solid rock. The South Polar Cap shrinks and grows with the seasons like the North Polar Cap (see p.161). Yet surprisingly the southern cap does not get warm enough in the summer to lose its carbon–dioxide ice covering. Dust storms that block out the Sun may keep the cap cooler than expected.

carbon-dioxide frost covering

water ice, no carbon dioxide

**SOUTH POLAR CAP**
Carbon-dioxide frost (shown as pink) covers over the water-ice cap (green-blue). Scarps of water ice at the edge of the cap slope toward the surrounding plains.

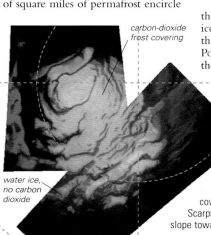

**SPIDERS ON MARS**
These spiderlike features in the south polar region of Mars were cut by dry ice (frozen carbon dioxide) as it turned to gas in the spring. The channels are 3–6 ft (1–2 m) deep.

# IMPACT CRATERS

The Martian surface is scarred by tens of thousands of craters, about 1,000 of which have been given names. They range from simple bowl craters less than 3 miles (5 km) across to basins hundreds of miles wide. The oldest craters are found in the southern hemisphere and have been eroded throughout their lifetimes. Their floors have been filled and their rims degraded, and the craters have become characteristically shallow. Smaller, fresher-looking craters have formed on top of them. The ejecta has been distributed by flowing across the surface rather than being flung through the air.

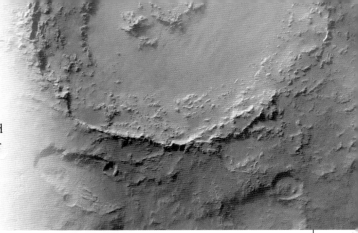

**ANCIENT GEOLOGICAL FEATURE**
Large impact craters such as Hale (right) have had their central peaks and terraced walls continuously eroded for up to 4 billion years.

---

## MERIDIANI PLANUM

## Victoria Crater

| | |
|---|---|
| **TYPE** | Crater |
| **AGE** | Under 100 million years |
| **DIAMETER** | 0.5 miles (800 m) |

Victoria is a small impact crater about two-thirds the size of the Arizona meteorite crater on Earth (see p.221). Victoria's beautifully scalloped edges have been eroded by winds, gradually increasing its diameter, and like many Martian craters, its floor is covered with dunes of wind-blown dust. The

crater was explored by the Mars rover Opportunity over 1 Martian year (or 2 Earth years, from 2006 to 2008). Half of that time was spent driving along part of the crater's rim before it carefully edged down a slope into the interior at an opening called Duck Bay. For the next Earth year, it examined rocky outcrops along the crater's walls with the instruments on its robot arm, finally driving out again to resume its trek across the Martian surface.

**DUNE-FILLED CRATER**
Rippling sand dunes cover the floor of Victoria, as seen in this enhanced-color view from the Mars Reconnaissance Orbiter.

**CAPE ST. VINCENT**
At Cape St. Vincent, a rocky outcrop on the northern rim of Victoria, layers of bedrock are topped by looser material thrown out by the impact that formed Victoria.

## ARABIA TERRA

## Schiaparelli Crater

| | |
|---|---|
| **TYPE** | Large crater |
| **AGE** | About 4 billion years |
| **DIAMETER** | 293 miles (471 km) |

This crater takes its name from the astronomer Giovanni Schiaparelli (see p.220), who spent much of his working life studying Mars. It is a highly circular crater, as are most Martian craters, although a significant number are elliptical—a rarity on the Moon and Mercury. Schiaparelli straddles the equator and is the largest crater in the Arabia Terra. It is an old crater formed by an impacting body when the planet was young and shows signs of degradation. The rim has been smoothed down and in parts is completely missing. Any central peak in the crater has been obliterated. Material has been deposited within the crater, and smaller craters have formed across the whole area. Wind continues to shape the landscape by erosion and by moving surface material.

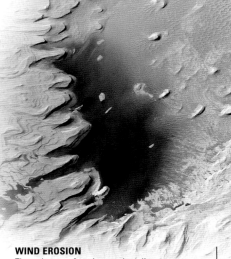

**WIND EROSION**
These layers of ancient rock sediments on the floor of an impact crater lying within the northwestern rim of Schiaparelli have been eroded and exposed by the wind.

**SHALLOW CRATER**
Here, color is used to indicate altitude. The crater floor is at the same height as much of the surrounding terrain. Higher deposits are in green. The degraded rim (yellow) is only about 0.75 miles (1.2 km) above the floor.

---

## TERRA TYRRHENA

## Huygens Crater

| | |
|---|---|
| **TYPE** | Multiringed crater |
| **AGE** | About 4 billion years |
| **DIAMETER** | 292 miles (470 km) |

Huygens is one of the largest impact craters in the heavily cratered southern highlands of Mars. It was formed during the period of intense bombardment within the first 500 million years of the planet's early history. The age of craters such as Huygens is determined by counting the number of craters that overlay their rims. Huygens has a second ring inside its mountainous rim. This has been filled by material carried into the ring. The rim is heavily eroded, and markings on it suggest that surface water has run off it at some

**MESAS**
Smooth-topped hills (mesas) on the crater floor are left behind as a former smooth layer of material is eroded to reveal a more rugged surface.

time. The pattern of markings is reminiscent of dendritic drainage systems on Earth, which from above look like the trunk and branches of a tree. Dark material within this crater's drainage channels was carried either by the draining water or by the wind.

**EASTERN RIM**
In this perspective view across Huygens's eastern rim (foreground) to the surrounding terrain, a branchlike network of drainage channels flows away from the rim, and small, more recently formed craters can be seen.

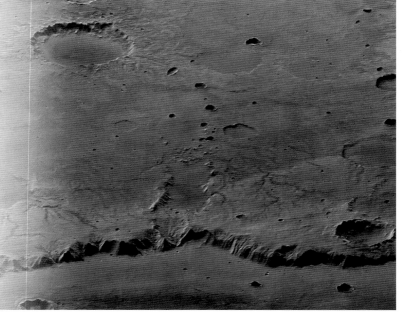

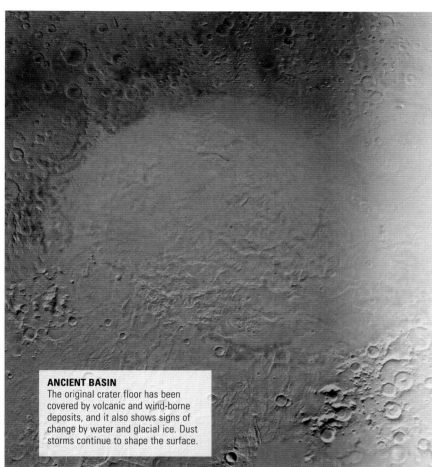

**ANCIENT BASIN**
The original crater floor has been covered by volcanic and wind-borne deposits, and it also shows signs of change by water and glacial ice. Dust storms continue to shape the surface.

# Hellas Planitia

| | |
|---|---|
| **TYPE** | Basin |
| **AGE** | About 4 billion years |
| **DIAMETER** | 1,365 miles (2,200 km) |

The Hellas Basin is the largest impact crater on Mars and one of the largest in the solar system. It is the dominant surface feature in the southern hemisphere. It is not immediately apparent that Hellas is an impact crater. In fact, its official name, Hellas Planitia, indicates that it is a large, low-lying plain. This designation dates from over a century ago, when the Martian surface was observed only through Earth-based telescopes and the true nature of this vast, shallow feature was not known. Hellas is the Greek word for Greece.

Particularly large craters that have been subsequently altered are termed basins. They are analogous to the maria on Earth's Moon. The term basin is also applied to the second-largest Martian crater, Isidis Planitia, and the third-largest,

Argyre Planitia (below). Over the past 3.5–4 billion years, Hellas Basin has had its floor filled by lava and its features changed by wind, water, and fresh crater formation. Despite all of this, some of its original features are still visible. Its overall shape and the remains of its rim can still be seen, as can inward-facing, arc-shaped cliffs lying up to several hundred miles beyond the rim. These are possibly the remnants of multiple rings.

**ROCK OUTCROPS**
Layered sedimentary rocks, which formed long after Hellas, lie in an eroded region northeast of the crater basin. Darker windblown ripples mark the surface.

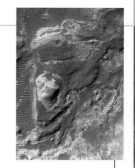

**ERODED RIM**
This perspective view shows the northern rim—the mountain range formed around the crater as the planet's crust was lifted up at the time of impact. Whole portions of the rim are missing to the northeast and southwest.

# Nansen Crater

| | |
|---|---|
| **TYPE** | Large crater |
| **AGE** | About 4 billion years |
| **DIAMETER** | 50 miles (81 km) |

Martian impact craters were first identified in 22 images returned by Mariner 4 in 1966. Nansen Crater was among the first and was named after the Norwegian explorer Fridtjof Nansen. New craters continue to be added to the list as a result of surveys by spacecraft. The Viking orbiter recorded this image of Nansen in 1976. The crater shows signs of erosion; its walls have been nibbled by the wind. Smaller, sharply defined craters have punctuated the surrounding terrain. A more recent crater has formed inside Nansen. Its central dark floor could be volcanic basalt.

**CRATER WITHIN A CRATER**

# Argyre Planitia

| | |
|---|---|
| **TYPE** | Basin |
| **AGE** | About 4 billion years |
| **DIAMETER** | 500 miles (800 km) |

Argyre is the third-largest crater on Mars. Its floor has been flooded by volcanic lava, and it has been heavily eroded by wind and water. It is speculated that in the distant past, water drained into the basin from the south polar ice cap. Channels entering the basin at its southeastern edge and others leading out from its northern edge reveal the water's route. The path cuts through the mountain ranges that define the basin: the Charitum Montes to the south and the Nereidum Montes to the north.

**FROST IN THE SOUTHERN HILLS**
Frost (mainly of carbon dioxide) covers an area of cratered terrain in the Charitum Montes in early June 2003, at which time the south polar frost cap had been retreating southward for about a month.

**CRATER DUNE FIELD**
Argyre's floor and rugged highland rim contain smaller craters. Some of these show signs of erosion. This one lying in the northwestern part of Argyre Basin contains a dark dune field.

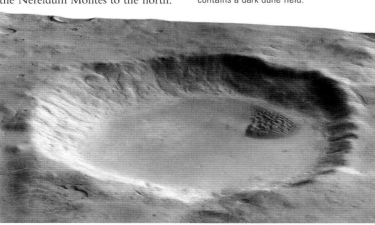

# Lowell Crater

| | |
|---|---|
| **TYPE** | Multiringed crater |
| **AGE** | About 4 billion years |
| **DIAMETER** | 126 miles (203 km) |

Erosion has changed Lowell since its formation early in Mars's history. The edges of both its outer rim and inner ring have been smoothed out, and its fine-grained ejecta soil has been blown around. The crater's appearance continues to undergo long-term changes, but it also changes on a short-term basis. Frost covers the crater's face in the winter months as the frost line extends north from the south polar region.

**LOWELL IN WINTER**

# Endurance Crater

**TYPE** Bowl crater

**AGE** Under 4 billion years

**DIAMETER** 420 ft (130 m)

This small and inconspicuous crater has been explored and investigated to a greater extent than almost any other crater on Mars. In early 2004, it did not even have a name, but by the end of that year, its rim, slopes, and floor had all been imaged and examined by the

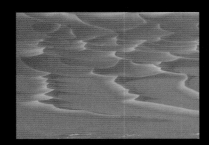

robotic rover Opportunity. The small craft just happened to land within roving distance of this football-field-sized crater when it made its scheduled landing in the Meridiani Planum in Mars's northern hemisphere.

Endurance, named after the ship that carried Irish-born British explorer Ernest Shackleton to the

### SAND DUNES

The center of the crater floor is covered by small sand dunes. The reddish dust has formed flowing tendrils, which are a few inches to a yard or so deep.

Antarctic, is an almost circular crater bounded by a rim of rugged cliffs. Its inner walls slope down to the crater floor, 66–100 ft (20–30 m) below. Layers of bedrock line the crater, some of which are exposed; loose material and sand dunes cover the rest of the floor.

Opportunity spent approximately 6 months exploring Endurance. The rover started by traveling around the southern third of the crater's rim; here, it crossed a region named Karatepe and traveled along the edge of Burns Cliff. It then retraced its route to enter the crater on its southwestern limb.

### BURNS CLIFF

This portion of the crater's southern inner wall is called Burns Cliff. Forty-six Opportunity images taken in November 2004 combine to make this 180° view. The wide-angle camera makes rock walls bulge unrealistically toward the viewer.

Opportunity made its way down the inner slope, examining rocks and soil along its route. It headed toward the crater's center but got less than halfway before doubling back; any farther and it might have gotten stuck in the sandy terrain. It then exited the crater to

### DRAMATIC PANORAMA
This approximately true-color view across Endurance Crater was taken by Opportunity's panoramic camera as the rover perched on the western rim. A dune field lies in the center of the crater.

### WOPMAY ROCK
The 3-ft (1-m) wide rock Wopmay (below) is one of the loose rocks on the crater floor. The image coloring highlights bluish dots in the rock, which are iron-rich spheres. The rover left wheel tracks in the soil (left) as it drove away from Wopmay.

## MARTIAN BLUEBERRIES

Dark, round pebbles nicknamed blueberries were found both within and on the terrain outside Endurance Crater. The name is, however, misleading; the pebbles, which appear bluer than their surroundings, are in fact dark gray. The half-inch blueberries are rich in the mineral hematite, which is also found on Earth. Hematite usually forms in lakes and hot springs on Earth, and this supports the idea that this part of Mars has had a watery past. A second type of round pebble that is lighter-colored and rougher-textured has been nicknamed popcorn.

### EVIDENCE OF WATER
A mixture of blueberries and popcorn lies on top of a rock called Bylot inside Endurance Crater.

move off across the adjoining flat plain, Meridiani Planum.

The exposed layers of rock in walls such as Burns Cliff reveal what lies beneath the Martian surface and what geological processes occurred there in the past. The composition of rocks on the crater floor, including those named Escher, Virginia, and Wopmay, was analyzed and the finer-grained floor material was scrutinized. All the findings suggest that water has affected the rocks both before and after Endurance was formed.

**MOUNT SHARP**

Rovers on the surface of Mars see some dramatic landscapes. This view, recorded by NASA's Curiosity rover in 2015, looks toward the high ground of Mount Sharp, an erosional remnant of layered sedimentary rocks inside Gale Crater. The sandstones in the foreground are about 2 miles (3 km) away. They owe their red color to iron oxides and were probably deposited under water. The whiter outcrop beyond, which lies above the red sandstone, is evidence of subsequent drier conditions.

# ASTEROIDS

ASTEROIDS ARE REMNANTS OF a failed attempt to form a rocky planet that would have been about four times as massive as Earth. They are dry, dusty objects and far too small to have atmospheres. Over 200,000 have been discovered, although over a billion are predicted to exist. The astronomers who discover asteroids have the right to name them.

## ORBITS

Most asteroids are found in a concentration known as the Main Belt, which lies between Mars and Jupiter, about 2.8 times farther from the Sun than Earth. Typically, they take between 4 and 5 years to orbit the Sun. The orbits are slightly elliptical and of low inclination. Even though the asteroids are all orbiting in the same direction, collisions at velocities of a few miles per second often take place. So as time passes, asteroids tend to break up. Some asteroids have been captured into rather strange orbits. The Trojans have the same orbital period as Jupiter and tend to be either 60° in front or 60° behind that planet. Then there are the Amor and Apollo asteroid groups (named after individual asteroids), with paths that cross the orbits of Mars and Earth, respectively. Aten asteroids have such small orbits that they spend most of their time inside Earth's orbit. These three groups are classed as near-Earth asteroids. They can be dangerous, having the potential to hit Earth and cause a great deal of damage. Fortunately, this happens very rarely.

### ASTEROID PATH
To picture stars, the Hubble Space Telescope scans the sky, keeping the stars stationary in the image frame. Asteroids, being much closer than the stars and in orbit around the Sun, form streaked trails (the blue line) during the exposure time.

## STRUCTURES

At the dawn of the solar system, there existed quite a few asteroids nearly as large as Mars. The radioactive decay of elements within the asteroidal rock melted these large bodies and, during their fluid stage, gravity pulled them into a spherical shape before they cooled. Many of these have since been broken up or reshaped by collisions with other asteroids. Smaller asteroids, which cooled more efficiently than larger ones, did not reach melting point and retained a uniform rocky-metallic composition and their original irregular shape. There are three main compositional classes of asteroids. The vast majority are either carbonaceous (C-type) or silicaceous (S-type). The next most populated class is metallic (M-type). These classes correspond to carbonaceous chondrite (stony) meteorites, stony-iron meteorites, and iron meteorites.

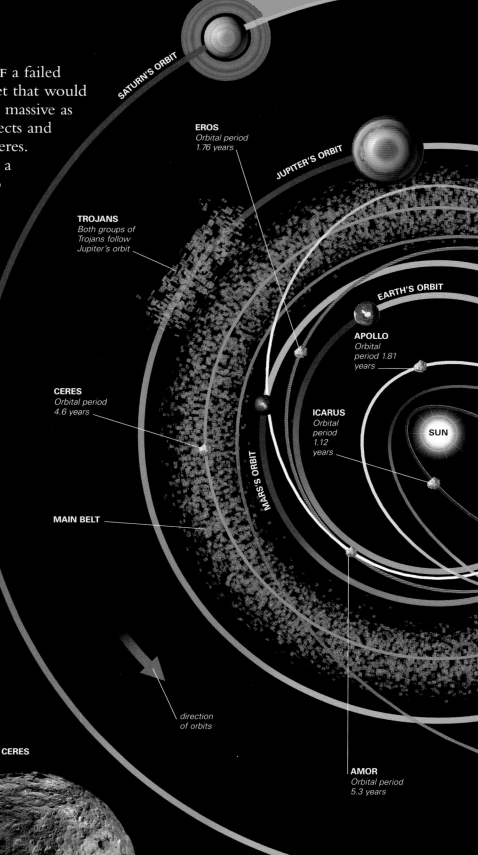

**EROS**
Orbital period 1.76 years

**TROJANS**
Both groups of Trojans follow Jupiter's orbit

SATURN'S ORBIT

JUPITER'S ORBIT

EARTH'S ORBIT

**CERES**
Orbital period 4.6 years

**APOLLO**
Orbital period 1.81 years

**ICARUS**
Orbital period 1.12 years

SUN

MARS'S ORBIT

**MAIN BELT**

direction of orbits

**AMOR**
Orbital period 5.3 years

CERES

VESTA

IDA

### ASTEROID SHAPES AND SIZES
The largest asteroids, such as Ceres and Vesta, are nearly spherical, whereas smaller asteroids, such as Ida, are irregularly shaped. All asteroids have craters on their surfaces, but some areas have been sandblasted and smoothed by a multitude of minor collisions.

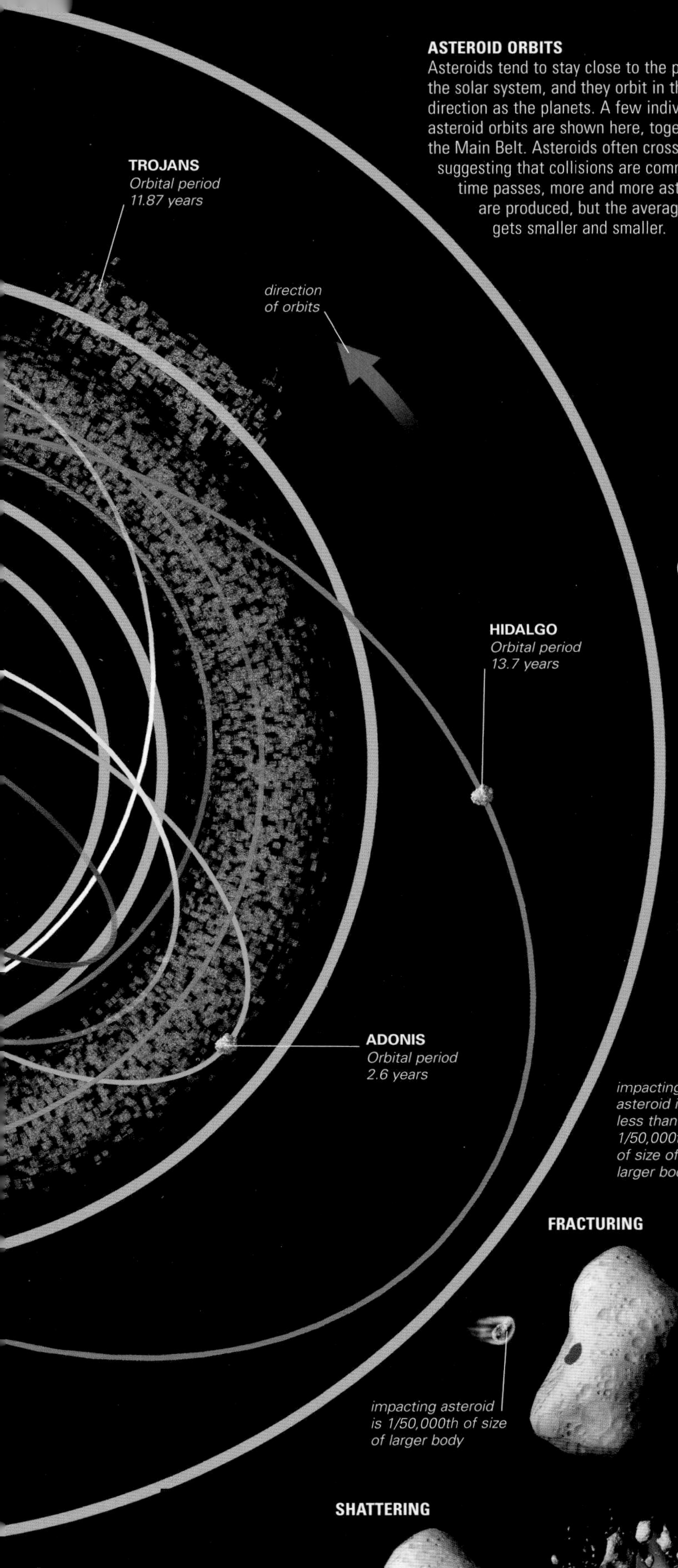

## ASTEROID ORBITS

Asteroids tend to stay close to the plane of the solar system, and they orbit in the same direction as the planets. A few individual asteroid orbits are shown here, together with the Main Belt. Asteroids often cross paths, suggesting that collisions are common. As time passes, more and more asteroids are produced, but the average size gets smaller and smaller.

**TROJANS**
*Orbital period 11.87 years*

*direction of orbits*

**HIDALGO**
*Orbital period 13.7 years*

**ADONIS**
*Orbital period 2.6 years*

## FRANZ XAVER VON ZACH

Franz Xaver von Zach (1754–1832) was a Hungarian baron and the director of the Seeberg Observatory in Germany. He became convinced that there was a missing planet orbiting between Mars and Jupiter. In September 1800, he organized a group of 24 astronomers to help him look. Popularly known as the Celestial Police, they divided the celestial zodiac into 24 parts and started searching but were beaten by a nose by Giuseppe Piazzi's accidental discovery of the asteroid Ceres in 1801. The Police were surprised by how small Ceres was, then were surprised again when more and more asteroids were found in similar orbits.

# COLLISIONS

The effect of a collision between asteroids depends on the sizes of the bodies involved. If a very small asteroid hits a larger one, it will produce a crater on the surface. This crater will be about 10 times the size of the incoming body. Because asteroids are much smaller than planets, the material blasted out of the crater will escape and move off into an independent orbit around the Sun. This orbit will, however, be very similar to that of the impacted asteroid, and there is a good chance that the ejected material will hit the cratered asteroid again. A bigger impactor can break up the asteroid that it hits. But so much energy is used to do this that the resulting fragments cannot escape from the gravitational field, and they will all fall back to form an irregular ball of rubble. Subsequent minor impacts will break up the surface, covering the asteroid in a rocky, dusty layer. A casual observer would not realize that the underlying asteroid is actually in pieces. A large impactor will not only shatter the asteroid, but the fragments will also escape. These will form a family of asteroids that eventually spreads out around the orbit of the original body.

## CRATERING

*impacting asteroid is less than 1/50,000th of size of larger body*

*crater forms*

*asteroid forms dusty ball of rubble*

## FRACTURING

*rocky body fractures*

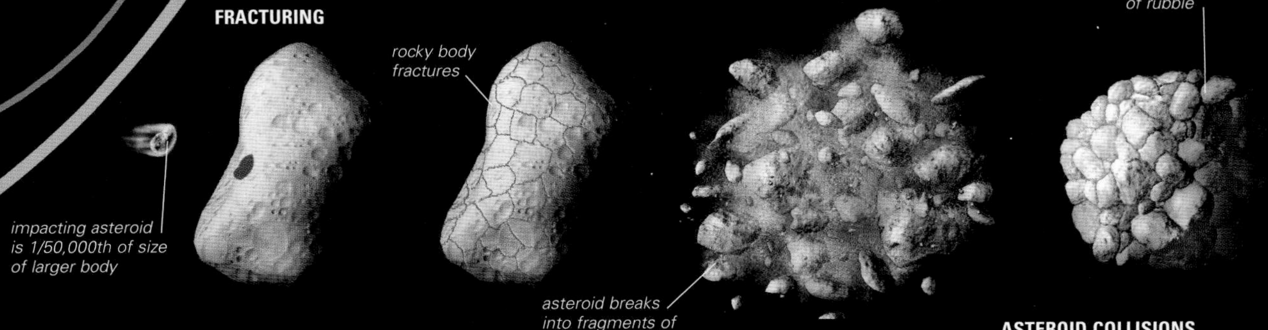

*impacting asteroid is 1/50,000th of size of larger body*

*asteroid breaks into fragments of rock and dust*

## SHATTERING

*impacting asteroid is more than 1/50,000th of size of larger body*

*rocky body shatters into pieces*

*family of asteroids forms*

## ASTEROID COLLISIONS

There are far more small asteroids than large ones. For every asteroid more than 6 miles (10 km) along its longest axis, there are 1,000 longer than 0.6 miles (1 km) and a million longer than 0.06 miles (0.1 km). So cratering is much more common than fracturing, which in turn is much more common than shattering.

# ASTEROIDS

Mainly moving between the orbits of Mars and Jupiter, asteroids are the remnants of a planet-formation process that failed. Today's asteroid belt contains only about 100 asteroids that are larger than 125 miles (200 km) across. But there are 100,000 asteroids greater than about 12.5 miles (20 km) across and a staggering 1 billion that are over 1.25 miles (2 km) along their longest axis. Ceres, the first asteroid to be discovered, in 1801, is now also called a dwarf planet (see p.175). Ceres contains about 25 percent of the mass of all the asteroids combined.

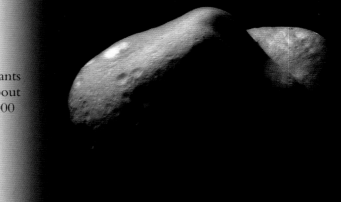

**EROS**
Only asteroids bigger than about 215 miles (350 km) in diameter are spherical. Eros is an irregularly shaped fragment of a much larger body.

---

### MAIN-BELT ASTEROID
## 951 Gaspra

**AVERAGE DISTANCE TO SUN**
206 million miles (331 million km)

**ORBITAL PERIOD** 3.29 years

**ROTATION PERIOD** 7.04 hours

**LENGTH** 11.2 miles (18 km)

**DATE OF DISCOVERY** July 30, 1916

Until 1991, asteroids could be glimpsed only from afar. In October of that year, a much closer view was obtained when the Galileo spacecraft flew within 1,000 miles (1,600 km) of Gaspra, taking 57 color images. Gaspra is a silicate-rich asteroid. The surface is very gray, with some of the recently exposed crater edges being bluish and some of the older, low-lying areas appearing slightly red.

**IRREGULAR SHAPE**

---

### MAIN-BELT ASTEROID
## 5535 Annefrank

**AVERAGE DISTANCE TO SUN** 206 million miles (331 million km)

**ORBITAL PERIOD** 3.29 years

**ROTATION PERIOD** Not known

**LENGTH** 3.7 miles (6 km)

**DATE OF DISCOVERY** March 23, 1942

Annefrank orbits in the inner regions of the Main Belt of asteroids and is a member of the Augusta family. On November 2, 2002, Annefrank was imaged by NASA's Stardust spacecraft as it passed within 2,050 miles (3,300 km) on its way to Comet Wild 2. Annefrank turned out to be twice as large as had been predicted from Earth-based observations. The brightness that is detected from Earth is proportional to the reflectivity multiplied by the surface area, but astronomers had used too high a value for the reflectivity. The asteroid was named after the famous diarist Anne Frank, who died during the Holocaust.

**SURFACE BRIGHTNESS**
This image of Annefrank, taken by the Stardust probe in 2002, shows differences in brightness over the surface of the asteroid. The variations are mainly due to dusty soil layers reflecting different amounts of sunlight in different directions.

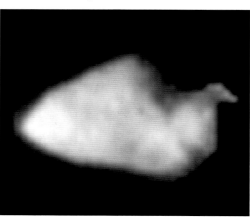

---

### NEAR-EARTH ASTEROID
## 4179 Toutatis

**AVERAGE DISTANCE TO SUN** 234 million miles (376 million km)

**ORBITAL PERIOD** 4.03 years

**ROTATION PERIOD** 5.4 and 7.3 days

**LENGTH** 2.65 miles (4.26 km)

**DATE OF DISCOVERY** January 4, 1989

Toutatis was named after a Celtic god (who, incidentally, appears in the *Asterix* comic books). A typical near-Earth asteroid, it sweeps past the planet nearly every 4 years. In September 2004, it came as close as just four times the distance of the Earth to the Moon. Toutatis is an S-class asteroid, similar to a stony-iron meteorite in composition. It tumbles in space much like a football after a botched pass, spinning around two axes, with periods of 5.4 and 7.3 days.

**RADAR IMAGE**

---

### MAIN-BELT ASTEROID
## 2867 Šteins

**AVERAGE DISTANCE TO SUN**
220 million miles (354 million km)

**ORBITAL PERIOD** 3.63 years

**ROTATION PERIOD** 6.05 hours

**DIAMETER** 4.14 miles (6.67 km)

**DATE OF DISCOVERY** November 4, 1969

Some asteroids are not solid but consist of rock fragments with gaps between. One such example—termed a "rubble pile"—is Šteins, which the Rosetta space probe showed to be shaped like a cut diamond. The impact that produced its largest crater is thought to have fractured the asteroid.

**DIAMOND IN THE SKY**

---

### MAIN-BELT ASTEROID
## 21 Lutetia

**AVERAGE DISTANCE TO SUN** 227 million miles (365 million km)

**ORBITAL PERIOD** 3.80 years

**ROTATION PERIOD** 8.17 hours

**DIAMETER** 75 miles (121 km)

**DATE OF DISCOVERY** November 15, 1852

Lutetia was the second of two asteroids visited by the European Space Agency's Rosetta space probe, the first being 2867 Šteins (see left). At over 60 miles (100 km) across, Lutetia is one of the larger asteroids and is also one of the most dense, suggesting that it contains large amounts of iron and might once have had a molten core. Rosetta's images, taken in July 2010, showed hundreds of craters up to 34 miles (55 km) wide and boulders as large as 1,000 ft (300 m) across on Lutetia's battered surface. Lutetia may have been almost spherical before parts of it were chipped off.

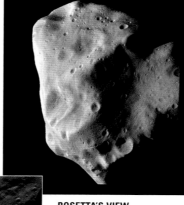

**ROSETTA'S VIEW**
Lutetia was photographed by Rosetta at its closest approach of 1,970 miles (3,170 km) in the top image. Craters and grooves on the asteroid's surface are visible in the close-up (left).

Some of its craters have been partly or completely buried by landslides set off by the vibrations from later impacts. Lutetia seems to be a link between small "rubble pile" asteroids and terrestrial planets such as Earth.

---

### MAIN-BELT ASTEROID
## 253 Mathilde

**AVERAGE DISTANCE TO SUN** 246 million miles (396 million km)

**ORBITAL PERIOD** 4.31 years

**ROTATION PERIOD** About 418 hours

**LENGTH** 41 miles (66 km)

**DATE OF DISCOVERY** November 12, 1885

Mathilde was visited by the NEAR Shoemaker space probe in 1997 but, because it spins very slowly, only about half of the surface was imaged. It is a primitive carbonaceous asteroid with a density much lower than that of most rocks, suggesting that it is full of holes. Mathilde is probably a compacted pile of rubble.

**WEDGE-SHAPED CRATER**

## MAIN-BELT ASTEROID

# 243 Ida

| | |
|---|---|
| **AVERAGE DISTANCE TO SUN** | 266 million miles (428 million km) |
| **ORBITAL PERIOD** | 4.84 years |
| **ROTATION PERIOD** | 4.63 hours |
| **LENGTH** | 37 miles (60 km) |
| **DATE OF DISCOVERY** | September 29, 1884 |

Ida was one of 119 asteroids discovered by the Austrian astronomer Johann Palisa, who, together with Max Wolf of Heidelberg, Germany, was a pioneer in the use of photography to produce star maps and hunt for minor planets (another name for asteroids). Ida is a member of the Koronis

family. Asteroidal families were discovered by the Japanese astronomer Hirayama Kiyotsugu in 1918. He found that there were groups of asteroids with very similar orbital parameters. The individual members were strung out on one orbit and formed a stream of minor bodies in the inner solar system (see p.170). Koronis is the most prominent member of Ida's family.

Ida is famous because the Galileo spacecraft imaged it in detail as it flew within 6,800 miles (11,000 km) during August 1993, on its way to Jupiter. Because Ida makes a complete rotation every 4 hours 36 minutes, Galileo was able to image most of the surface during the flyby. Ida was originally thought to be an S-type

## DACTYL

At just 1 mile (1.6 km) long, Dactyl is tiny. Its orbit around Ida is nearly circular, with a radius of about 56 miles (90 km) and an orbital period of about 27 hours.

asteroid like Gaspra (see opposite), but observations revealed that its density is too low and it is more likely to be a C-type asteroid. It has about five times more craters per unit area than Gaspra, indicating that its surface is considerably older. The most exciting outcome of the Galileo flyby was the discovery that Ida has its own moon, Dactyl. This binary system is thought to have been formed during the asteroid collision and breakup that created the Koronis family.

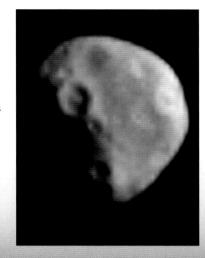

### IDA AND ITS MOON
Dactyl was the first asteroid satellite to be discovered. Ann Harch, a Galileo mission member, noticed it when examining images that had been stored on the spacecraft when it passed Ida 6 months earlier.

MAIN-BELT ASTEROID

# 4 Vesta

| | |
|---|---|
| **AVERAGE DISTANCE TO SUN** | 219 million miles (353 million km) |
| **ORBITAL PERIOD** | 3.63 years |
| **ROTATION PERIOD** | 5.34 hours |
| **DIAMETER** | 330 miles (530 km) |
| **DATE OF DISCOVERY** | March 29, 1807 |

Vesta is one of the largest asteroids, with a surface that reflects, on average, 42 percent of the incoming light. This makes it the brightest asteroid in the night sky and the only one that is visible to the unaided eye. Most asteroids of Vesta's size are expected to be nearly spherical, but Vesta's shape has been distorted by a massive impact at the south pole that created a huge basin called Rheasilvia, which measures some 300 miles (500 km) across—almost as wide as Vesta itself. Like many

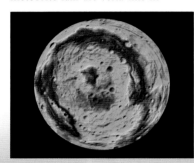

large craters on the Moon, Rheasilvia has a mountainous central peak. It overlaps an older and slightly smaller crater, which is 230 miles (375 km) wide. Some of the fragments of Vesta's crust produced by the cratering process are still trailing Vesta on similar orbits. Others have hit Earth and been recognized as strange meteorites with a composition similar to the igneous rock basalt. Six percent of Earth's recent meteorite falls are Vesta-like in

their mineralogical makeup. The composition is similar to the lava that spews out of Hawaiian volcanoes. In its early life, Vesta melted and then resolidified, with denser material sinking to the center. It now has a layered structure like the rocky planets, with a low-density crust lying above layers of pyroxene and olivine and an iron core. Vesta is thought to be the only remaining differentiated asteroid in the Main Belt. It is one of the densest asteroids known, with a density similar to that of Mars. In July 2011, NASA's Dawn spacecraft went into orbit around Vesta, beginning a year-long study of the asteroid. Vesta's

### SOUTH-POLAR IMPACT BASIN
A giant impact basin called Rheasilvia at the south pole of Vesta is seen here in a false-color image taken by NASA's Dawn spacecraft. It is 300 miles (500 km) across and has a central peak (shown in red) and a rim that is 9 miles (15 km) high.

### METEORITE
This meteorite, which landed in Western Australia in October 1960, originated from Vesta.

surface turned out to be old and heavily cratered, with grooves running around the equator, possibly fractures from the south-polar impact. One of the asteroid's most distinctive features was a chain of three craters nicknamed the Snowman, seen in the image below. From top to bottom, the individual craters are called Minucia, which is 13 miles (21.5 km) wide; Calpurnia, 30 miles (50 km) wide; and Marcia, 36 miles (58 km) wide. The Dawn spacecraft's instruments are studying the surface composition of Vesta, with the objective of matching meteorites to specific areas on the asteroid. One likely source is the south-polar mountain, which rises 13 miles (22 km) within the Rheasilvia impact basin.

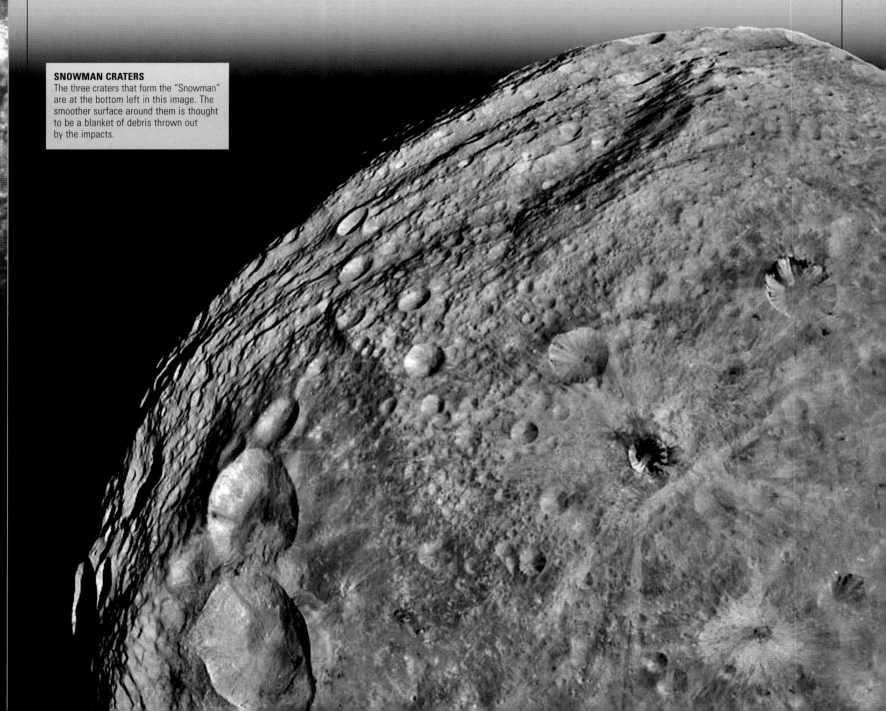

### SNOWMAN CRATERS
The three craters that form the "Snowman" are at the bottom left in this image. The smoother surface around them is thought to be a blanket of debris thrown out by the impacts.

## MAIN-BELT ASTEROID

# 1 Ceres

| | |
|---|---|
| **AVERAGE DISTANCE TO SUN** 257 million miles (414 million km) | |
| **ORBITAL PERIOD** 4.60 years | |
| **ROTATION PERIOD** 9.08 hours | |
| **DIAMETER** 590 miles (950 km) | |
| **DATE OF DISCOVERY** January 1, 1801 | |

Ceres was discovered by accident in 1801 by Giuseppe Piazzi, the director of the Palermo Observatory in Italy, while he was compiling a catalog of fixed stars (see panel, right). One of the "stars" had moved during the night, and this turned out to be the first

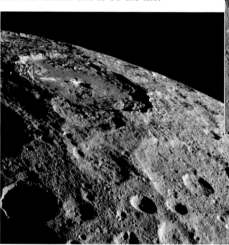

known asteroid, Ceres. Some 100 years before, Johannes Kepler (see p.68) had suspected that there was a "missing" planet in the gap between the orbits of Mars and Jupiter (see pp.170–171). By 1800, some of Europe's leading astronomers had started to look for objects in this gap, and Piazzi made the first discovery. About a year later, a German doctor and astronomer, Heinrich Olbers, was observing the path of Ceres in an attempt to produce a more accurate estimate of its orbital parameters when he discovered a second asteroid. This

**LANDSCAPES**
Ceres landscape varies across its surface. A 57-mile (92-km) wide crater dominates the view on the left. Bright spots on its floor are probably salts left after a briny liquid oozed up from below. The 22-mile (35-km) wide view on the right contains a series of extensional fractures.

was given the number 2 and named Pallas. The orbits of Ceres and Pallas were found to cross, and Olbers concluded that they were fragments of a planet that had broken up. As the century progressed, more and more asteroids were discovered that were smaller and fainter than the first two. Using spectroscopes to analyze the light reflected from their surfaces, astronomers found that asteroids had different colors due to different compositions, and this led to the establishment of a classification system (see p.170). Ceres was classified as a carbonaceous, or C-type, asteroid.

In 2006, the new category of dwarf planets was introduced to describe objects that are rounded in shape but have not swept their orbits clear of other bodies (see p.209). Ceres was placed in this category. However, it remains the largest member of the asteroid belt, so it can be said to have a dual identity as both a dwarf planet and an asteroid. NASA's Dawn probe collected data from orbit about Ceres between April 2015 and November 2018, after previously studying Vesta (see opposite).

see p.68
see pp.170–171
see p.170
see p.209

## EXPLORING SPACE

# PIAZZI'S TELESCOPE

Known as the Palermo Circle, this telescope was made between 1787 and 1789 by Jesse Ramsden of London, England, the greatest European instrument maker of the 18th century. Its lens has an aperture of 3 in (75 mm). The circular altitude scale and the horizontal azimuth scale are both read using microscopes. In its day, it was the most southerly European telescope, being on top of the Royal Palace in Palermo, Sicily. While measuring star positions with this telescope, Piazzi discovered the first asteroid, Ceres.

## NEAR-EARTH ASTEROID

# 25143 Itokawa

| | |
|---|---|
| **AVERAGE DISTANCE TO SUN** 123 million miles (198 million km) | |
| **ORBITAL PERIOD** 1.52 years | |
| **ROTATION PERIOD** 12.1 hours | |
| **LENGTH** 0.34 miles (0.54 km) | |
| **DATE OF DISCOVERY** September 26, 1998 | |

Itokawa is a small, irregularly shaped asteroid of the type known as a "rubble pile," meaning that it is not a solid, coherent body. The Japan Aerospace Exploration Agency (JAXA) chose it as the target for a sample-collection mission called

Hayabusa, which the agency launched in May 2003 (see panel, below). At the time of launch, the asteroid bore only the reference number 1998SF36, but it was named in honor of Hideo Itokawa (1912–1999), known as the father of Japanese rocketry, while the probe was on its way.

The Hayabusa probe reached Itokawa in September 2005 and spent 2 months surveying the asteroid before attempting a landing. The asteroid's picklelike shape, 1,110 ft (540 m) long and 690 ft (210 m) across at its narrowest, gives it the appearance of two separate masses stuck together. Astronomers think that Itokawa was once much bigger, possibly up to 12 miles (20 km)

**ELONGATED ASTEROID**
This image of Itokawa was taken by the Hayabusa probe, which landed in the smooth area near the middle to collect dust samples. The asteroid lacks obvious impact craters, in contrast to most others visited by spacecraft.

across. Large impacts broke the asteroid into smaller pieces, which then gently reassembled to form the low-density rubble-pile structure that we see today.

Rocks that are up to 165 ft (50 m) wide are dotted over Itokawa's surface, while the narrow neck near the middle is smoother and covered with dust. In November 2005, the Hayabusa space probe gently touched down on the smooth part. It collected numerous microscopic particles of the asteroid's dust, which it brought back to Earth. The sample return capsule blazed through the Earth's atmosphere to land at Woomera, Australia, in June 2010. The capsule was retrieved and opened under sterile conditions in a laboratory in Japan. The asteroid's surface rocks proved to be rich in the mineral olivine, similar to common types of chondrite meteorites.

**ROUGH SURFACE**
Large rocks can be seen strewn over Itokawa's surface in this close-up. These rocks are probably fragments of an earlier breakup of the asteroid that have since collected together again.

## EXPLORING SPACE

# THE HAYABUSA MISSION

Hayabusa was a Japanese probe sent to rendezvous with and bring back samples from the near-Earth asteroid Itokawa. Hayabusa is the Japanese name for the peregrine falcon. It was intended to swoop down on the asteroid like a hawk, to take samples, and then return them to Earth. It spent 30 minutes on the asteroid's surface before taking off again with dust samples.

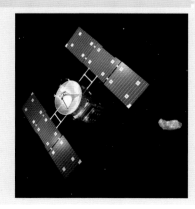

**ASTEROID APPROACH**
An artist's impression shows the Japanese space probe Hayabusa approaching asteroid Itokawa in November 2005.

# 433 Eros

| AVERAGE DISTANCE TO SUN | 136 million miles (218 million km) |
| --- | --- |
| ORBITAL PERIOD | 1.76 years |
| ROTATION PERIOD | 5.27 hours |
| LENGTH | 19.25 miles (31 km) |
| DATE OF DISCOVERY | August 13, 1898 |

Lying in near-Earth orbit, outside the Main Belt, Eros is usually closer to the Sun than to Mars (see p.170). Its orbit also brings it close to Earth; at the last close approach, in 1975, Eros came within 14 million miles (22 million km) of the planet. The orbit is unstable, and Eros has a 1-in-10 chance of hitting either Earth or Mars in the next million years. In 1960, Eros was detected by radar, and infrared measurements taken in the 1970s indicated that the surface was not just bare rock but was covered by a thermally insulating blanket of dust and rock fragments. Eros was the first asteroid to be orbited by a spacecraft and the first to be landed on. It was chosen for close study because it is big and nearby.

On February 14, 2000, the Near Earth Asteroid Rendezvous (NEAR) spacecraft (renamed NEAR Shoemaker in March 2000) went into orbit around Eros. It landed 363 days later. About 160,000 images were recorded. They revealed an irregularly shaped body that had heavily cratered 2-billion-year-old areas lying next to relatively smooth regions. Even though the gravitational

**CLOSING IN**
NEAR Shoemaker took this image from a height of 3,770 ft (1,150 m) shortly before it touched down on Eros.

field is very small, several thousand boulders larger than 50 ft (15 m) across have fallen back to the surface after being ejected by impacts, and some surface dust has rolled down the slopes to form sand dunes a few yards (meters) deep. Laser measurements of the NEAR–Eros distance as the spacecraft orbited have not only produced an accurate map of the asteroid's shape but also indicated that the interior is nearly uniform, with a density about the same as that of Earth's crust. Eros is not a pile of rubble like Mathilde; it is a single solid lump of rock. The spacecraft's gamma-ray spectrometer worked for 2 weeks after touchdown. Eros was found to be silicate-rich and highly reflective.

**ROCK AND REGOLITH**
Some of the rocks and regolith on Eros's surface have a red coloring. The longer their exposure to minor impacts and the solar wind, the redder they appear.

**COMPUTER MODELS**
The gravity on Eros is about 1/2,000th of that on Earth but varies by nearly a factor of two from place to place. The colors in this image represent the rate at which a rock would roll downhill. It would roll fastest in the red areas, and it wouldn't move at all in the blue areas.

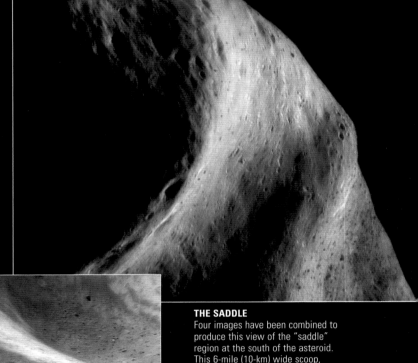

**THE SADDLE**
Four images have been combined to produce this view of the "saddle" region at the south of the asteroid. This 6-mile (10-km) wide scoop, which has been named Himeros, is relatively boulder-free, unlike the region at the lower right of the frame.

**MOSAIC IMAGE OF BENNU**
This global view of Bennu was created using 12 images collected from a range of 15 miles (24 km). The "spinning top" shape is clear. The giant boulder at the lower right is about 180 ft (55 m) across.

# 101955 Bennu

| AVERAGE DISTANCE TO SUN | 105 million miles (169 million km) |
| --- | --- |
| ORBITAL PERIOD | 1.20 years |
| ROTATION PERIOD | 4.30 hours |
| DIAMETER | 0.30 miles (0.49 km) |
| DATE OF DISCOVERY | September 11, 1999 |

Bennu is the target of NASA's OSIRIS-REx mission, which arrived in orbit around it in December 2018 and is expected to return samples to Earth in 2023. Previously known only by the provisional designation 1999 RQ$_{36}$, Bennu's name was chosen out of more than 8,000 entries as the winner of a school "name that asteroid" contest after it had been selected as the target for OSIRIS-REx.

**CLOSELY PACKED BOULDERS**
This view from a range of only 3 miles (5 km) shows jagged boulders of many sizes, some darker than others. The largest here is about 24 ft (7 m) across.

Its shape resembles that of a spinning top, which results from the effect of its rapid rotation on its unconsolidated interior. Despite being rocky, its overall density is only slightly greater than that of water, so it is almost certainly a "rubble pile" with spaces between its internal components. The rubble-pile structure is similar to that of Itokawa (see p.175), but Bennu is rotating about twice as fast and its rotation period is decreasing by about 1 second every 100 years. Its exposure to heat from the Sun while it rotates causes uneven emissions of thermal radiation from its surface, which accelerates its rotation speed. This is a phenomenon known as the YORP (Yarkovsky-O'Keefe-Radzievskii-Paddack) effect.

The way it reflects sunlight suggests that Bennu's composition is similar to carbonaceous chondrite asteroids, but that won't be confirmed until the samples are returned and analyzed.

Bennu's orbit crosses the Earth's, and it is estimated to have a 1-in-2,700 chance of colliding with the Earth between 2175 and 2199.

**EJECTING PARTICLES**
Several particle ejection events have been seen at Bennu. It is not known whether these are caused by meteoroid impacts, thermal stress fracturing, or release of water vapor.

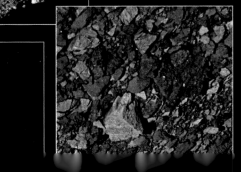

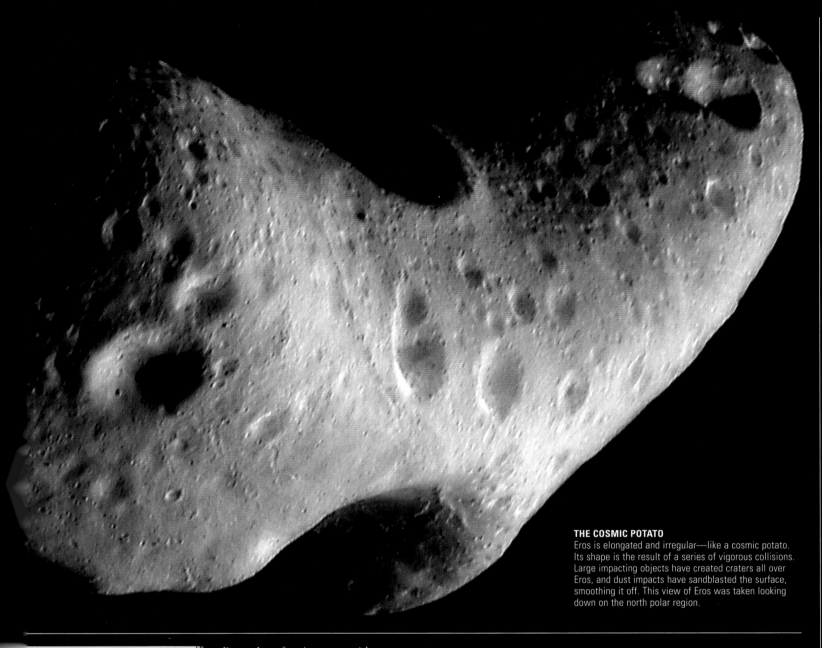

### THE COSMIC POTATO

Eros is elongated and irregular—like a cosmic potato. Its shape is the result of a series of vigorous collisions. Large impacting objects have created craters all over Eros, and dust impacts have sandblasted the surface, smoothing it off. This view of Eros was taken looking down on the north polar region.

# 162173 Ryugu

**AVERAGE DISTANCE TO SUN** 111 million miles 178 million km)

**ORBITAL PERIOD** 1.30 years

**ROTATION PERIOD** 7.63 hours

**DIAMETER** 0.54 miles (0.87 km)

**DATE OF DISCOVERY** May 10, 1999

Ryugu is the target of the Japanese probe Hayabusa 2, which arrived in orbit around it in June 2018. After collecting samples, Hayabusa 2 left for Earth in November 2019 and is expected to return samples to Earth in 2020.

Ryugu has about twice the diameter of Bennu (see opposite) and also is likely to be a carbonaceous asteroid. It has a low density and a "spinning top" shape, which is controlled by its rapid rotation. Interestingly, both Ryugu and Bennu rotate in a retrograde sense, meaning that they spin clockwise about their axes while orbiting counterclockwise around the Sun.

Ryugu's surface is strewn with large boulders, separated by smoother areas. Those are not so dusty as they might appear; seen up close, they consist mostly of angular rock fragments. Hayabusa 2 deployed four small hopping rovers to collect surface material. Some of the samples were freshly exposed from below the surface in a 33-ft (10-m) crater made by using a free-flying gun called the "Small Carry-on Impactor" or SCI to fire a copper projectile at the surface from a range of about 1,641 ft (500 m).

Ryugu's orbit crosses the Earth's, but it poses less of a hazard than Bennu because currently it is predicted not to come closer than about a quarter of the distance between the Moon and the Earth.

### RYUGU

Strewn with angular boulders of all scales, Ryugu's "spinning top" shape is clearly seen. The ridge along its equator bears at least one impact crater whose indistinct form may be symptomatic of this asteroid's rubbly interior.

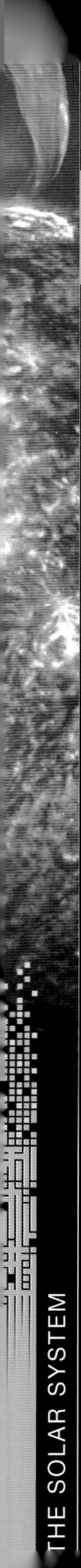

# JUPITER

**JUPITER IS THE LARGEST AND MOST MASSIVE** of all the planets. It has almost 2.5 times the mass of the other seven planets combined, and over 1,300 Earths could fit inside it. Jupiter bears the name of the most important of all the Roman gods (known as Zeus in Greek mythology). The planet has the largest family of moons in the solar system, its members named after Jupiter's lovers, descendants, and attendants.

## ORBIT

Jupiter is the fifth planet from the Sun. It lies approximately five times as far away as Earth, but its distance from the Sun is not constant. Its orbit is elliptical and there is a difference of 47.3 million miles (76.1 million km) between its aphelion and perihelion distances. Jupiter's spin axis tilts by 3.1°, and this means that neither of the planet's hemispheres point markedly toward or away from the Sun as it moves around its orbit. Consequently, Jupiter does not have obvious seasons. The planet spins quickly around its axis more quickly than any other planet. Its rapid spin throws material in its equatorial region outward. The result is a bulging equator and a slightly squashed appearance.

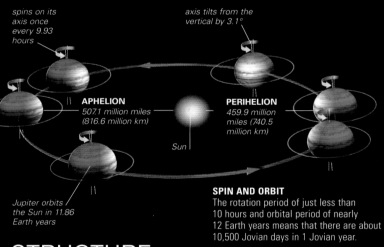

spins on its axis once every 9.93 hours

axis tilts from the vertical by 3.1°

**APHELION**
507.1 million miles (816.6 million km)

Sun

**PERIHELION**
459.9 million miles (740.5 million km)

Jupiter orbits the Sun in 11.86 Earth years

**SPIN AND ORBIT**
The rotation period of just less than 10 hours and orbital period of nearly 12 Earth years means that there are about 10,500 Jovian days in 1 Jovian year.

## STRUCTURE

Although it is the most massive planet (318 times the mass of the Earth), Jupiter's great size means that its density is low. Its composition is more like the Sun's than any other planet in the solar system. Jupiter's hydrogen and helium is in a gaseous form in the outer part of the planet, where the temperature is about –166°F (–110°C). Closer to the center, the pressure, density, and temperature increase. The state of the hydrogen and helium changes accordingly. By about 4,350 miles (7,000 km) deep, at about 3,600°F (2,000°C), hydrogen acts more like a liquid than a gas. By 8,700 miles (14,000 km), at about 9,000°F (5,000°C), hydrogen has compacted to metallic hydrogen and acts like a molten metal. Deep inside, at a depth of about 37,260 miles (60,000 km), is a solid core of rock, metal, and hydrogen compounds. The core is small compared to Jupiter's great size but is about 10 times the mass of Earth.

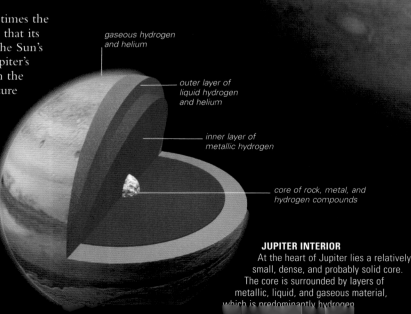

gaseous hydrogen and helium

outer layer of liquid hydrogen and helium

inner layer of metallic hydrogen

core of rock, metal, and hydrogen compounds

**JUPITER INTERIOR**
At the heart of Jupiter lies a relatively small, dense, and probably solid core. The core is surrounded by layers of metallic, liquid, and gaseous material, which is predominantly hydrogen.

**GAS GIANT**
Jupiter's surface is not solid. Each light or dark band and every big or small swirl or spot is a part of the planet's cloudy atmosphere.

## JUPITER PROFILE

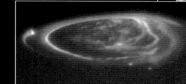

| AVERAGE DISTANCE FROM THE SUN | ROTATION PERIOD |
|---|---|
| 483.6 million miles (778.3 million km) | 9.93 hours |

| CLOUD-TOP TEMPERATURE | ORBITAL PERIOD (LENGTH OF YEAR) |
|---|---|
| −160°F (−110°C) | 11.86 Earth years |

| DIAMETER 88,846 miles (142,984 km) | MASS (EARTH = 1) 318 |
|---|---|

| VOLUME (EARTH = 1) 1,321 | CLOUD-TOP GRAVITY (EARTH = 1) 2.53 |
|---|---|

| NUMBER OF MOONS 64 | SIZE COMPARISON |
|---|---|

**OBSERVATION**
Jupiter is bright and easy to spot. It has a maximum magnitude of −2.9. Even at its faintest, it is brighter than Sirius, the brightest star in the sky. Jupiter is best seen at opposition, which occurs once every 13 months.

SIZE COMPARISON
EARTH        JUPITER

# MAGNETIC FIELD

Jupiter has a magnetic field—it is as if the planet had a large bar magnet deep inside it. The field is generated by electric currents within the thick layer of metallic hydrogen, and the axis joining the magnetic poles is tilted at about 11° to the spin axis. The field is stronger than that of any other planet. It is about 20,000 times stronger than Earth's magnetic field and has great influence on the volume of space surrounding Jupiter. Solar wind particles (see p.107) streaming from the Sun plow into the field. They are slowed down and rerouted to spiral along the field's magnetic lines of force. Some particles enter Jupiter's upper atmosphere around its magnetic poles. They collide with the atmospheric gases, which radiate and produce aurorae. Other charged particles (plasma) are trapped and form a disklike sheet around Jupiter's magnetic equator. Electric currents flow within this sheet. High-energy particles are trapped and form radiation belts, similar to but much more intense than the Van Allen belts around Earth (see p.125). Jupiter's magnetic field is shaped by the solar wind, forming a vast region called the magnetosphere. Its size varies with changes in pressure of the solar wind, but the tail is thought to have a length of about 370 million miles (600 million km).

**AURORA**
This striking electric-blue aurora centered on Jupiter's north magnetic pole was imaged by the Hubble Space Telescope in 1998.

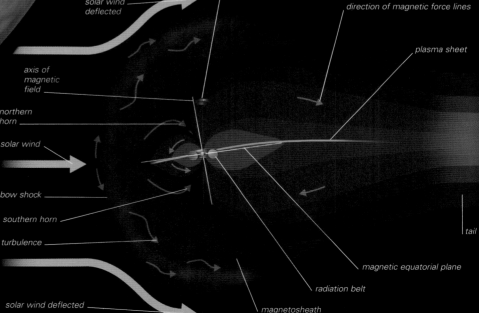

axis of rotation
solar wind deflected
direction of magnetic force lines
plasma sheet
axis of magnetic field
northern horn
solar wind
bow shock
southern horn
turbulence
tail
magnetic equatorial plane
radiation belt
solar wind deflected
magnetosheath

**JUPITER'S MAGNETOSPHERE**
Jupiter's magnetosphere—the bubblelike region around Jupiter dominated by the planet's magnetic field—is enormous; it is a thousand times the volume of the Sun, and its tail stretches away from the planet as far as Saturn's orbit. This slice through the magnetosphere reveals its structure.

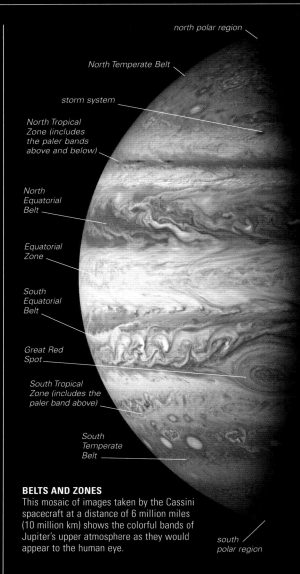

- north polar region
- North Temperate Belt
- storm system
- North Tropical Zone (includes the paler bands above and below)
- North Equatorial Belt
- Equatorial Zone
- South Equatorial Belt
- Great Red Spot
- South Tropical Zone (includes the paler band above)
- South Temperate Belt
- south polar region

**BELTS AND ZONES**
This mosaic of images taken by the Cassini spacecraft at a distance of 6 million miles (10 million km) shows the colorful bands of Jupiter's upper atmosphere as they would appear to the human eye.

# ATMOSPHERE

Jupiter's atmosphere is dominated by hydrogen, with helium being the next most common gas. The rest is made up of simple hydrogen compounds—such as methane, ammonia, and water—and more complex ones such as ethane, acetylene, and propane. It is these compounds that condense to form the different-colored clouds of the upper atmosphere and help give Jupiter its distinctive banded appearance. The temperature of the atmosphere increases toward the planet's interior. As gases condense at different temperatures, different types of clouds form at specific altitudes. All the while, the gas in Jupiter's equatorial region is heated by the Sun, and this rises and moves toward the polar regions. Cooler air flows from the polar regions at a lower altitude to take its place, creating in effect a large circulation cell. This hemisphere-wide circulation transfer would be straightforward if Jupiter were stationary. It is not—it rotates, and speedily at that, and a force known as the Coriolis effect (see p.126) deflects the north–south flow into an east–west flow. As a result, the large circulation cell is split into many smaller cells of rising and falling air. These are seen on Jupiter's surface as alternating bands of color. Jupiter's white bands of cool rising air are called zones. The red-brown bands of warmer falling air are known as belts.

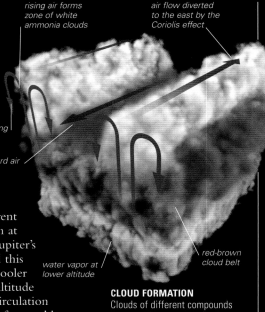

- rising air forms zone of white ammonia clouds
- air flow diverted to the east by the Coriolis effect
- descending cooler air
- westward air flow
- water vapor at lower altitude
- red-brown cloud belt

**CLOUD FORMATION**
Clouds of different compounds form at different altitudes in the atmosphere. Convection currents move the mixture of gas upward. Water is first to reach the altitude where it is cool enough to condense to form clouds. Higher up, where it is cooler, red-brown ammonium hydrosulfide clouds form, and highest of all, where it is coolest, are the white ammonia clouds.

- hydrogen (89.8%)
- helium with traces of methane and ammonia (10.2%)

**COMPOSITION OF ATMOSPHERE**
Hydrogen dominates, but it is the trace compounds that color Jupiter's upper atmosphere.

# MOONS

Jupiter has over 79 known moons, over two-thirds of which have been discovered since January 2000. Only 50 of the moons have been given names, and several have yet to have their orbit confirmed. The recent discoveries are typically irregularly shaped rocky bodies a few miles across and are thought to be captured asteroids. By contrast, Jupiter's four largest moons are spherical bodies that were formed at the same time as Jupiter. Collectively known as the Galilean Moons (see p.182), they were the first moons to be discovered after Earth's Moon. As they orbit Jupiter, passing between it and the Sun, their shadows sweep across the planetary surface; from within the shadow, the Sun appears eclipsed. A triple eclipse happens just once or twice a decade.

**TRIPLE ECLIPSE**
Three shadows were cast onto Jupiter's surface on March 28, 2004, as its three largest moons passed between the planet and the Sun. Io is the white circle in the center, with its shadow to its left. Ganymede is the blue circle at upper right, and its shadow lies on Jupiter's left edge. Callisto's shadow is on the upper-right edge, but the moon itself is out of view, to the right of the planet.

**JUPITER'S MOONS**

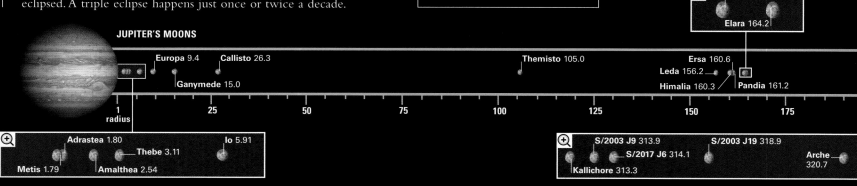

- Lysithea 163.9
- Elara 164.2
- Europa 9.4
- Callisto 26.3
- Ganymede 15.0
- Themisto 105.0
- Ersa 160.6
- Leda 156.2
- Himalia 160.3
- Pandia 161.2
- 1 radius
- 25 50 75 100 125 150 175
- Adrastea 1.80
- Thebe 3.11
- Io 5.91
- Metis 1.79
- Amalthea 2.54
- S/2003 J9 313.9
- S/2017 J6 314.1
- S/2003 J19 318.9
- Arche 320.7
- Kallichore 313.3

Scale in radii of Jupiter
1 radius = 44,397 miles (71,492 km)

Moons are not to scale and increase in size for magnification purposes only

# WEATHER

Jupiter has no notable seasons, and the planet's temperature is virtually uniform. Its polar regions have temperatures similar to those of its equatorial regions because of internal heating. Jupiter radiates about 1.7 times more heat than it absorbs from the Sun. The excess is infrared heat left from when the planet was formed. Most of Jupiter's weather occurs in the part of its atmosphere that contains its distinct white and red-brown cloud layers and is dominated by clouds, winds, and storms. The rising warm air and descending cool air within the atmosphere produce winds that are channeled around the planet, both to the east and west, by Jupiter's fast spin. The wind speed changes with latitude; winds within the equatorial region are particularly strong and reach speeds in excess of 250 mph (400 kph). The solar and infrared heat, the wind, and Jupiter's spin combine to produce regions of turbulent motion, including circular and oval cloud structures, which are giant storms. The smallest of these storms are like the largest hurricanes on Earth. They can be relatively short-lived and last for just days at a time, but others endure for years. Jupiter's most prominent feature, the Great Red Spot, is an enormous high-pressure storm that may have first been sighted from Earth over 340 years ago.

**THE GREAT RED SPOT**
This giant storm, which is bigger than Earth, is constantly changing its size, shape, and color. It rotates counterclockwise every 6 to 7 days.

## THE POLES

Jupiter's poles had only ever been seen at an angle until NASA's Juno probe arrived in polar orbit in 2016. The familiar belt and zone pattern is replaced by spiral storms, known as cyclones. In 2016, the south polar cyclone was surrounded by a polygonal pattern of five circumpolar cyclones (of which three are in sunlight in the accompanying image), whereas the north pole had a pattern of eight circumpolar cyclones. By late 2019, a sixth southern circumpolar cyclone had appeared.

**JUPITER'S SOUTH POLE**
This was the view on December 16, 2017, from Juno 65,000 miles (104,000 km) above the polar cloudtops.

# RINGS

Jupiter's ring system was revealed for the first time in an image taken by Voyager 1 in 1979. It is a thin, faint system made of dust-sized particles knocked off Jupiter's four inner moons. The system consists of three parts. The main ring is flat and is about 4,350 miles (7,000 km) wide and less than 18 miles (30 km) thick. Outside this is the flat gossamer ring, which is 528,000 miles (850,000 km) wide and stretches beyond Amalthea to Thebe's orbit. On the inside edge of the main ring is the 12,400-mile (20,000-km) thick doughnut-shaped halo. Its tiny dust grains reach down to Jupiter's cloudtops.

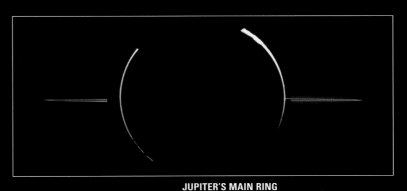

**JUPITER'S MAIN RING**
Jupiter's main ring was imaged by Galileo with the Sun behind the planet. From this position, small particles within the ring and in Jupiter's upper atmosphere stand out. The halo and gossamer ring are revealed only if the main ring is overexposed.

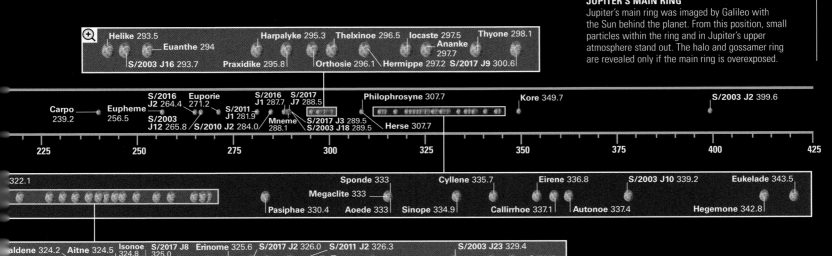

# JUPITER'S MOONS

Jupiter's moons fall into three categories: the four inner moons; the four large Galilean moons; and the rest, the small outer moons. The inner and Galilean moons orbit in the usual direction—that is, the same direction as Jupiter's spin (counterclockwise viewed from above the north pole). Most of the outer moons travel in the opposite direction, suggesting that they originated from an asteroid that fragmented after it was captured by Jupiter's gravitational field.

**SO NEAR AND YET SO FAR**
Io, one of the largest of Jupiter's 64 moons, appears close to its planet, but the two are almost three times the diameter of Jupiter apart.

## Europa

**DISTANCE FROM JUPITER** 416,630 miles (670,900 km)

**ORBITAL PERIOD** 3.55 Earth days

**DIAMETER** 1,939 miles (3,122 km)

Europa is an ice-covered ball of rock, which has been studied for about 400 years but whose intriguing nature was only fully revealed once the Galileo space probe started its study in 1996. The probe was named after the Italian scientist Galileo Galilei, who observed Europa, along with the three other moons that collectively bear his name, in January 1610, from Padua, Italy. The German astronomer Simon Marius (1573–1624) is believed to

**DAYTIME TEMPERATURE**
Infrared observations reveal heat radiation from Europa's surface at midday. Temperatures at the equator (shown here as yellow) reach about −225°F (−140°C). Farther away from the equator, the surface temperatures are even lower.

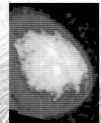

---

## Metis

**DISTANCE FROM JUPITER** 79,460 miles (127,960 km)

**ORBITAL PERIOD** 6 hours 58 minutes

**DIAMETER** 25 miles (40 km)

Metis, the closest moon to Jupiter, is irregular in shape and lies within the planet's main ring. It was discovered on March 4, 1979, by the Voyager 1 probe. Metis is named after the first wife of Zeus, who was swallowed by him when she became pregnant.

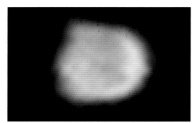

**JUPITER'S CLOSEST MOON**

---

## Adrastea

**DISTANCE FROM JUPITER** 80,100 miles (128,980 km)

**ORBITAL PERIOD** 7 hours 9 minutes

**LENGTH** 16 miles (26 km)

The small, irregularly shaped Adrastea is the second moon out from Jupiter and lies within its main ring. For each

**ADRASTEA**

orbit of Jupiter, Adrastea spins once on its axis, so the same side of the moon always faces the planet. This synchronous rotation is also exhibited by Adrastea's three closest neighbors, Metis, Amalthea, and Thebe. Adrastea was discovered by Voyager 2 in July 1979, and is named after a nymph of Crete into whose care, according to Greek mythology, the infant Zeus was entrusted.

---

## Thebe

**DISTANCE FROM JUPITER** 137,800 miles (221,900 km)

**ORBITAL PERIOD** 16 hours 5 minutes

**LENGTH** 68 miles (110 km)

The most distant of the inner moons, Thebe is named after an Egyptian king's daughter who was a granddaughter of Io. The moon, which was discovered on March 5, 1979, by Voyager 1, lies within the outer part of the Gossamer Ring (see p.181).

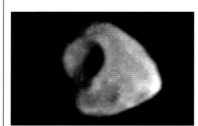

**THEBE SHOWING IMPACT CRATER**

---

## Amalthea

**DISTANCE FROM JUPITER** 112,590 miles (181,300 km)

**ORBITAL PERIOD** 11 hours 46 minutes

**LENGTH** 163 miles (262 km)

The largest of Jupiter's inner moons and the third from the planet, Amalthea is named after the nurse of newborn Zeus. The irregularly shaped moon lies within the Gossamer Ring and is believed to be a source of ring material. Meteoroids from outside the Jovian system collide with Amalthea and the other inner moons, chipping off flecks of dust, which then become part of the ring system. Amalthea's unexpected discovery on September 9, 1892, over 280 years after the four much larger Galilean moons had been discovered was headline news.

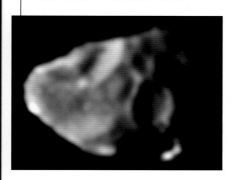

---

**BARNARD'S TELESCOPE**
Amalthea was the last of Jupiter's moons to be discovered by direct visual observation (as opposed to photography). Its discoverer, American Edward Barnard, used a 36-in (91-cm) refractor telescope, which is now preserved at the Lick Observatory in California.

**BATTERED SURFACE**
The circular feature in this image is Pan, which, with a diameter of about 56 miles (90 km), is the largest impact crater on Amalthea. The bright spot below Pan is associated with another, smaller crater, Gaea (bottom).

---

## Themisto

**DISTANCE FROM JUPITER** 4.66 million miles (7.5 million km)

**ORBITAL PERIOD** 130 Earth days

**DIAMETER** 5 miles (8 km)

In November 2000, astronomers at the Mauna Kea Observatory, Hawaii, carried out a systematic search for new moons and identified 11 small moons. Observations recorded on subsequent nights revealed that one of the 11, since named Themisto, was a moon that had been discovered by American astronomer Charles Kowal on September 30, 1975, but then lost.

**THEMISTO REDISCOVERED**
This digital image is one of a series that shows Themisto (highlighted) and its changing position against the background stars, which led to its rediscovery in November 2000.

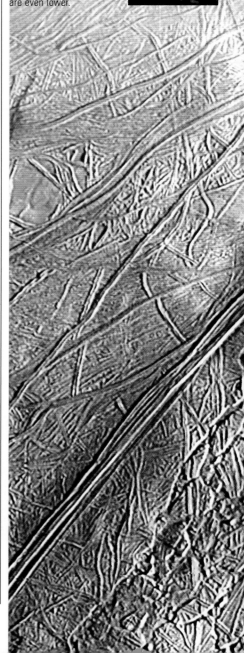

have observed the moons first, but it was Galileo who published his findings and brought the moons to the attention of the scientific and wider community.

Jupiter's fourth-largest satellite is a fascinating world. It is a little smaller than Earth's Moon but much brighter, since its icy surface reflects five times as much light. A liquid ocean may lie below Europa's water-ice crust, which is just tens of miles thick. This watery layer, which is estimated to be 50–105 miles (80–170 km) deep and to contain more liquid than Earth's oceans combined, could be a haven for life. Below lies a rocky mantle surrounding a metallic core.

**OVERHEAD VIEW**

**TERRAIN MODEL**

**PWYLL CRATER**
This three-dimensional model of the 16-mile (26-km) wide Pwyll Crater (above) was made by combining images (see example, left) taken from different angles and then applying color. Unusually, the crater floor (blue) is the same height as the moon's surface, and the central peak (red) is much higher than the crater's rim.

fractures in crust

Pwyll Crater

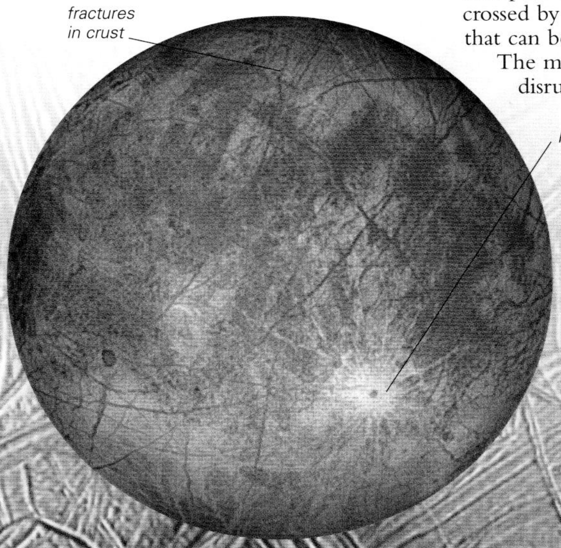

**IMAGE OF THE FAR SIDE**
This is how the far side of Europa would appear to the human eye. Bright plains in the polar areas (top and bottom) sandwich a darker, disrupted region of the crust.

The surface appears to be geologically young and consists of smooth ice plains, disrupted terrain, and regions criss-crossed by dark linear structures that can be thousands of miles long. The mottled appearance of the disrupted terrain comes from

crust that has broken up and floated into new positions. Round or oblong, city-sized dark spots freckle the surface. Known as lenticulae, these form as large globules of warm and slushy ice push up from underneath and briefly melt the surface ice. Exactly how the dark lines formed is unclear, but volcanically heated water and ice and other kinds of tectonic activity were involved. Tidal forces fractured the crust, and liquid or icy water erupted through the crack to freeze almost instantly on the surface.

In Greek myth, Europa was the girl who was seduced by Zeus in the form of a white bull and carried off to Crete.

## KEEP EUROPA CLEAN

After a 6-year journey from Earth, the Galileo space probe spent 8 years studying the Jovian system and made 11 close flybys of Europa. The decision was made to destroy the probe because NASA wanted to avoid an impact with Europa and the potential contamination of its subsurface ocean, which could possibly harbor life. With little fuel left, Galileo was put on a collision course with Jupiter. The probe disintegrated in the planet's atmosphere on September 21, 2003.

**GALILEO**

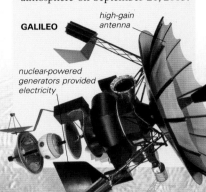

high-gain antenna

nuclear-powered generators provided electricity

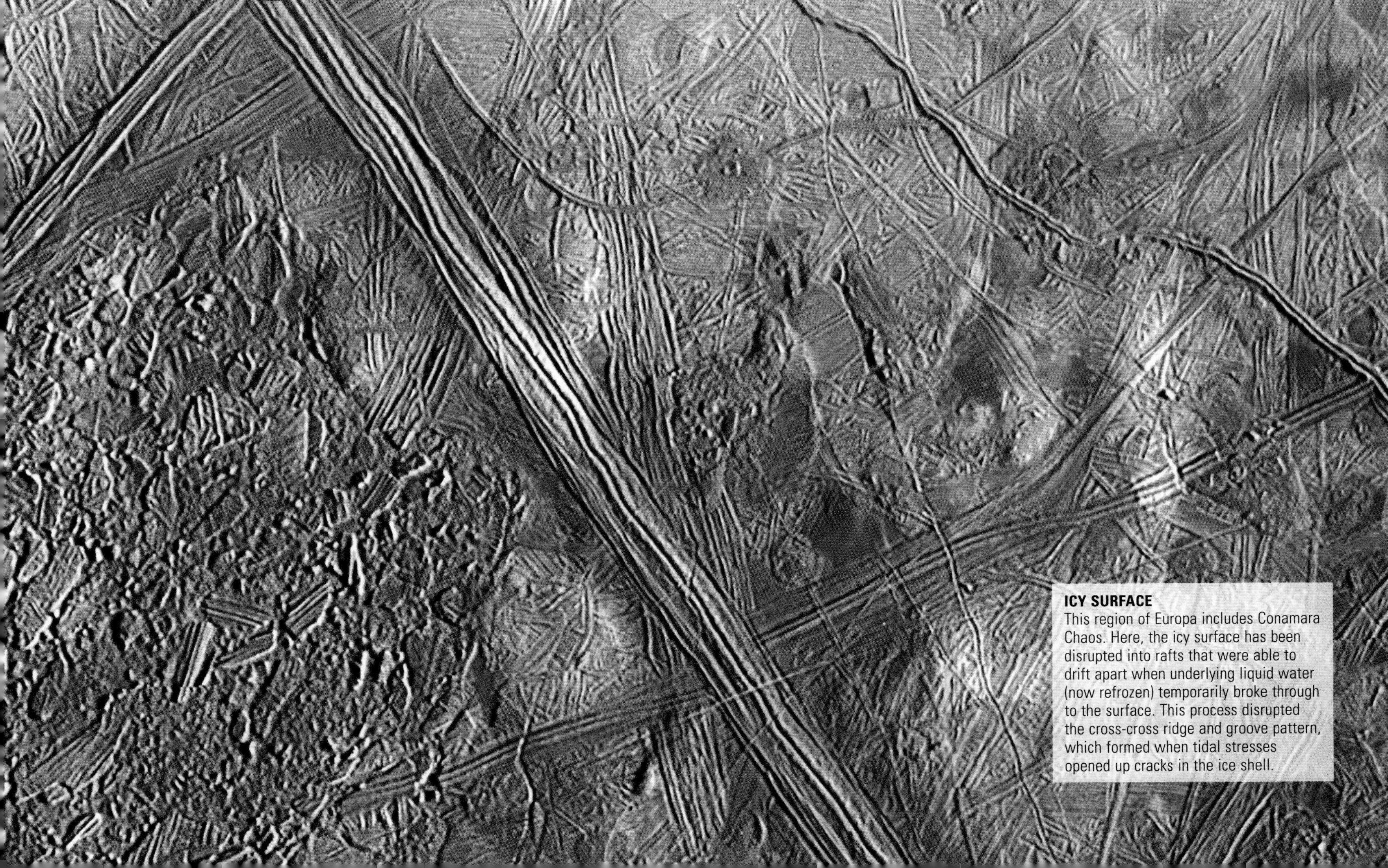

**ICY SURFACE**
This region of Europa includes Conamara Chaos. Here, the icy surface has been disrupted into rafts that were able to drift apart when underlying liquid water (now refrozen) temporarily broke through to the surface. This process disrupted the cross-cross ridge and groove pattern, which formed when tidal stresses opened up cracks in the ice shell.

THE SOLAR SYSTEM

# Io

**DISTANCE FROM JUPITER** 261,800 miles (421,600 km)

**ORBITAL PERIOD** 1.77 Earth days

**DIAMETER** 2,262 miles (3,643 km)

Io is a little larger and denser than Earth's Moon and orbits Jupiter at a distance only slightly greater than the Moon's from Earth. But there the similarities end. Io is a highly colored world of volcanic pits, calderas and vents, lava flows, and high-reaching plumes. The moon's nature was revealed first by the two Voyager probes and then more fully explored by the Galileo mission. Prior to Voyager 1's arrival in March 1979, scientists expected to find a cold, impact-cratered moon. Instead, it found the most volcanic body in the solar system.

Io has a thin silicate crust that surrounds a molten silicate layer. Below this lies a comparatively large iron-rich core that extends about halfway to the surface. Io orbits Jupiter quickly, every 42.5 hours or so. As it orbits, it is subjected to the strong gravitational pull of Jupiter on one side and the lesser pull of Europa on the other. Io's surface flexes as a consequence of the varying strength and direction of the pull it experiences. The flexing is accompanied by friction, which produces the heat that keeps part of Io's interior molten. It is this material that erupts through the surface and constantly renews it.

The evidence of such volcanism is seen all over Io. Over 80 major active volcanic sites and more than 300 vents have been identified. Features known as plumes are also found at the surface; these fast-moving and long-lived columns of cold gas and frost grains are more like geysers than volcanic explosions. They are created as superheated sulfur dioxide shoots through fractures in Io's crust. The material in the plumes falls slowly back to the surface as snow and leaves circular or oval frost deposits. Plume material also spreads into space surrounding Io and supplies a doughnut-shaped body of material that has formed along Io's orbital path. Temperatures

**JUPITERSHINE**
Sunlight reflected off Jupiter illuminates Io's western side. The eastern side is in shadow except for a burst of light beyond the limb where the plume of the volcano Prometheus is lit. The yellowish sky is produced by sodium atoms surrounding Io scattering the sunlight.

ring of sulfur-dioxide snow

Culann Patera

Tohil Mons

**VOLCANIC ACTIVITY**
In this color-enhanced Galileo image, the dark spots on Io's surface are active volcanic centers. The dark, eruptive area of Prometheus at center left is encircled by a pale yellow ring of sulfur-dioxide snow deposited by the volcano's plume.

at the volcanic hot spots can be over 2,240°F (1,230°C), the highest surface temperatures in the solar system outside the Sun. Elsewhere, the surface is cold, reaching just –244°F (–153°C).

Simon Marius (see p.182) suggested the names of the Galilean moons. Io is named after one of Zeus's loves, whom he changed into a cow to hide her from his jealous wife. Hera was not fooled and sent a gadfly to torment Io forever. Other surface features are named after people and places from the Io myth or from Dante's *Inferno*, or after fire, sun, volcano, and thunder gods, goddesses, and heroes.

**TOHIL MONS**
Nonvolcanic mountains are also found on Io. Here, the sunlit peak of the 185-mile (300-km) wide Tohil Mons rises 3.4 miles (5.4 km) above Io's surface.

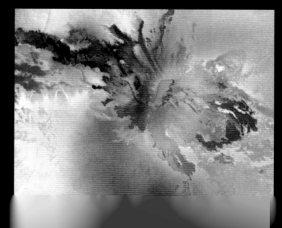

**CULANN PATERA**
Colorful lava flows stream away from the irregularly shaped green-floored volcanic crater of Culann Patera (right of center). The reasons for the varied colors are uncertain. The diffuse red material is thought to be a compound of sulfur deposited from a plume of gas. The green deposits may be formed when sulfur-rich material coats warm silicate lava.

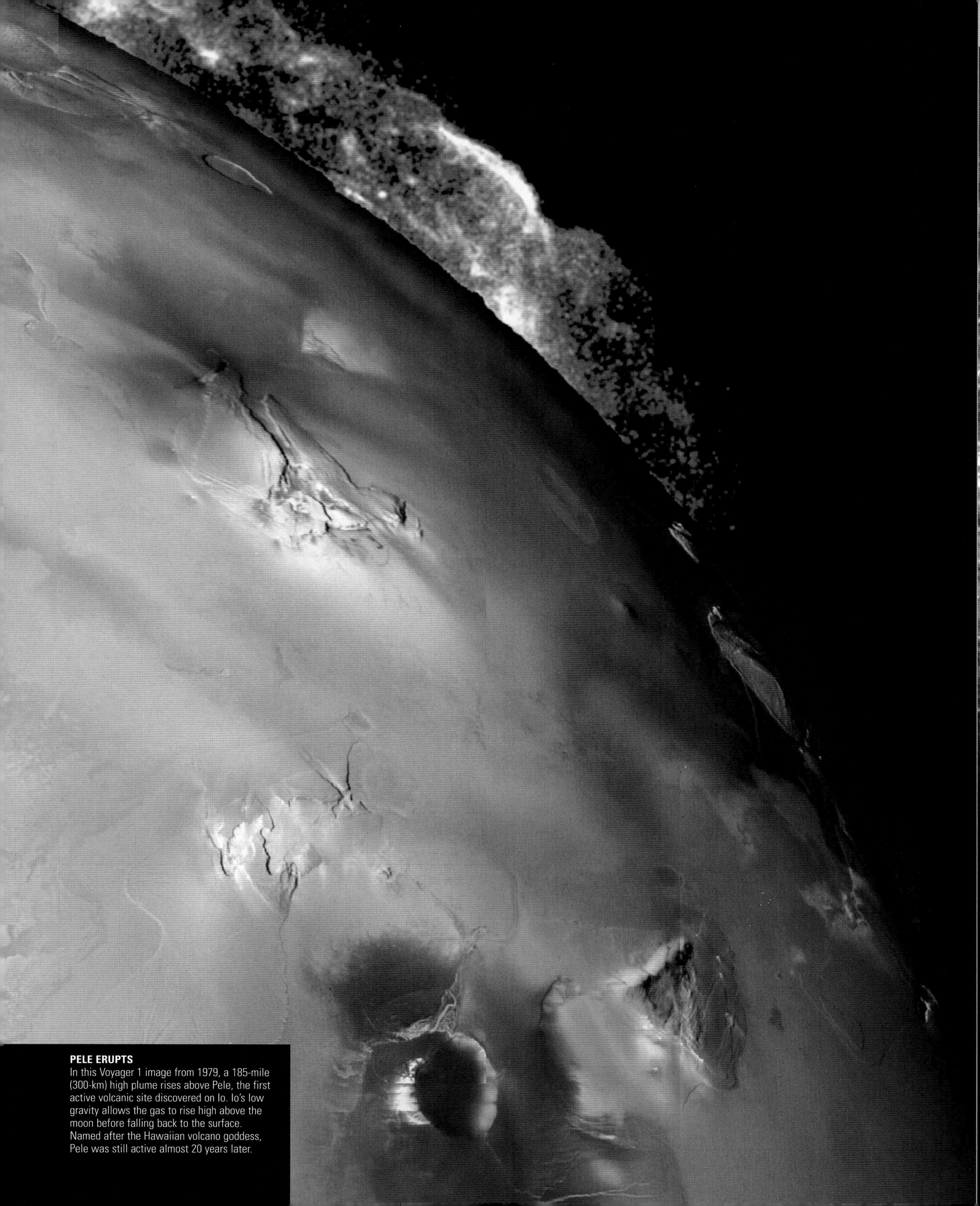

**PELE ERUPTS**
In this Voyager 1 image from 1979, a 185-mile (300-km) high plume rises above Pele, the first active volcanic site discovered on Io. Io's low gravity allows the gas to rise high above the moon before falling back to the surface. Named after the Hawaiian volcano goddess, Pele was still active almost 20 years later.

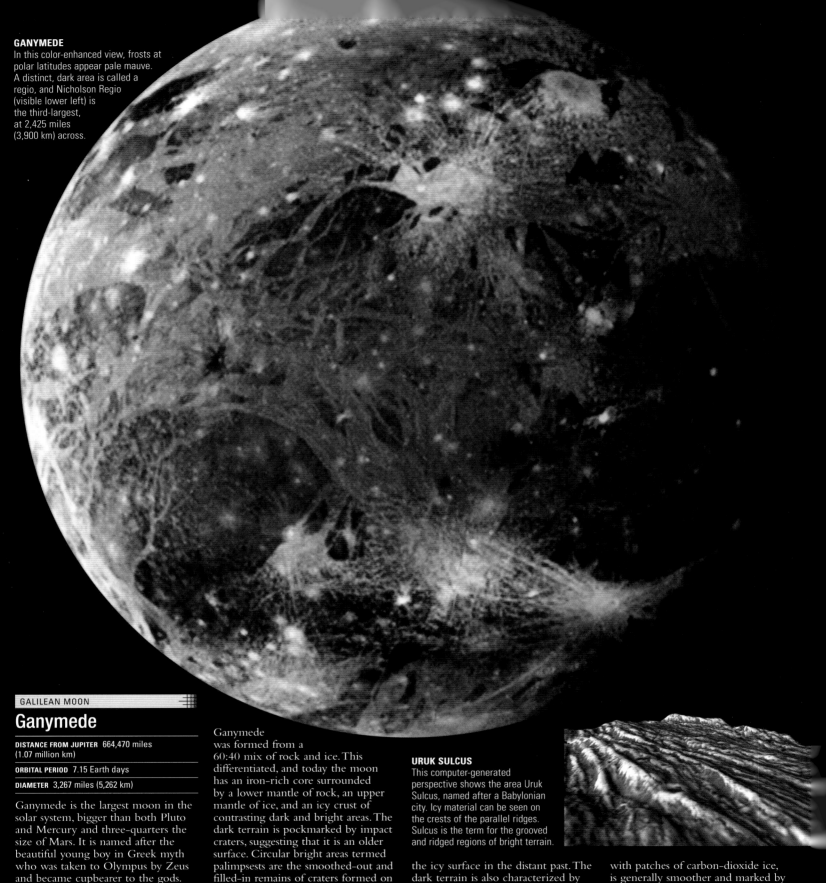

**GANYMEDE**

In this color-enhanced view, frosts at polar latitudes appear pale mauve. A distinct, dark area is called a regio, and Nicholson Regio (visible lower left) is the third-largest, at 2,425 miles (3,900 km) across.

# Ganymede

**DISTANCE FROM JUPITER** 664,470 miles (1.07 million km)

**ORBITAL PERIOD** 7.15 Earth days

**DIAMETER** 3,267 miles (5,262 km)

Ganymede is the largest moon in the solar system, bigger than both Pluto and Mercury and three-quarters the size of Mars. It is named after the beautiful young boy in Greek myth who was taken to Olympus by Zeus and became cupbearer to the gods.

Ganymede was formed from a 60:40 mix of rock and ice. This differentiated, and today the moon has an iron-rich core surrounded by a lower mantle of rock, an upper mantle of ice, and an icy crust of contrasting dark and bright areas. The dark terrain is pockmarked by impact craters, suggesting that it is an older surface. Circular bright areas termed palimpsests are the smoothed-out and filled-in remains of craters formed on

**INFRARED MAPPING**

The infrared image on the left, taken by Galileo, locates surface water ice— the brighter the shading, the greater the amount. The colors of the right-hand image indicate the location of minerals (red) and the size of ice grains (shades of blue).

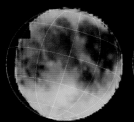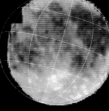

**URUK SULCUS**

This computer-generated perspective shows the area Uruk Sulcus, named after a Babylonian city. Icy material can be seen on the crests of the parallel ridges. Sulcus is the term for the grooved and ridged regions of bright terrain.

the icy surface in the distant past. The dark terrain is also characterized by long depressions about 4 miles (7 km) wide, called furrows. These may have formed as subsurface ice flowed into recently formed craters and material dragged across the surface created the bow-shaped troughs.

The bright terrain, which is rich in water ice

with patches of carbon-dioxide ice, is generally smoother and marked by fewer craters. It is crisscrossed by ridges and grooves produced by the tectonic stretching of the moon's surface.

**SIPPAR SULCUS**

This depression within Sippar Sulcus appears to be an old caldera (a volcano's collapsed underground reservoir) containing frozen lava.

GALILEAN MOON

# Callisto

**DISTANCE FROM JUPITER** 1.17 million miles
(1.88 million km)

**ORBITAL PERIOD** 16.69 Earth days

**DIAMETER** 2,994 miles (4,821 km)

The most distant, second-largest, and darkest of the Galilean moons, Callisto is still brighter than Earth's Moon since its surface contains ice that reflects sunlight. Callisto has undergone little internal change since its formation. Its original mix of rock and ice is only partly differentiated, so the moon is rockier toward its center and icier

*dark areas lack ice*

*ice on crater rim and floor shines brightly*

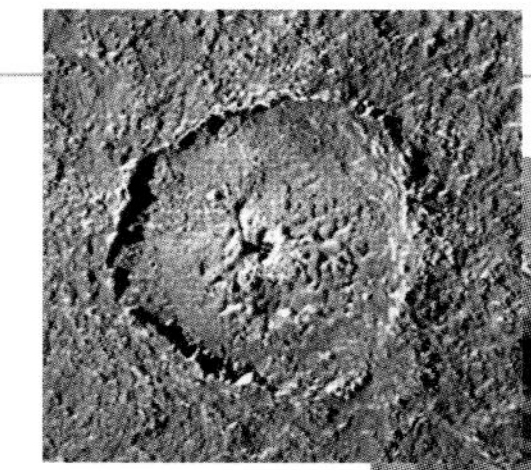

### TINDR CRATER
The partial collapse of the rim of this 47-mile (76-km) wide crater and its pitted floor is probably the result of erosion by ice.

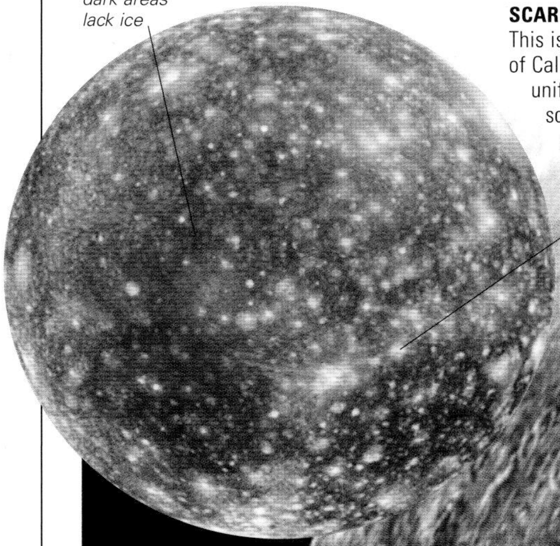

toward its crust. The surface, scarred by craters and multiringed structures created by meteorite impacts, bears few signs of geological activity. Callisto does not appear to have been shaped by plate tectonics or cryovolcanism, where ice behaves like volcanic lava, although the ice has eroded the rock in places, causing crater rims to be worn away and sometimes collapse.

### SCARRED SURFACE
This is the only complete global color image of Callisto obtained by Galileo. The surface is uniformly cratered, and the bright impact scars are easily visible against its otherwise dark, smooth surface.

The craters are named after heroes and heroines of northern myths, and the large, ringed features, such as the Valhalla Basin (see below), after homes of the gods or heroes. About 1,600 miles (2,600 km) across, the Valhalla Basin was probably formed by a large meteorite strike early in Callisto's history, which fractured the moon's cold, brittle crust, allowing ice that was previously below the surface to flood the impact site.

### VALHALLA REGION
This photograph of part of the Valhalla Basin, lit by sunlight streaming in from the left, shows a 6-mile (10-km) wide fault scarp, part of Valhalla's ring system. The smallest craters visible are about 510 ft (155 m) across.

## CALLISTO

Callisto was a beautiful follower of the huntress Artemis, who was seduced by Zeus and bore him a son. According to one myth, Zeus's jealous wife, Hera, turned Callisto into a bear. One day, Callisto came across her son, Arcas, now grown. Fearful for his life, Arcas was only stopped from killing Callisto by Zeus, who raised a whirlwind that carried the pair up into the sky. Callisto became the constellation Ursa Major and Arcas formed Boötes.

### MULTIRING BASIN
The multiringed Valhalla Basin dominates Callisto's surface. The bright, ice-covered central zone is about 370 miles (600 km) across. It is surrounded by rings, which are troughs about 30 miles (50 km) apart.

# SATURN

**SATURN IS THE SECOND-LARGEST PLANET** and the sixth from the Sun—it is the most distant planet normally visible to the naked eye. A huge ball of gas and liquid, Saturn has a bulging equator and an internal energy source. With a composition dominated by hydrogen, it is the least dense of all the planets. A spectacular system of rings encircles the planet itself, and it also has a large family of moons.

## ORBIT

Saturn takes 29.46 Earth years to complete one orbit of the Sun. It is tilted to its orbital plane by 26.7°, a little more than Earth's axial tilt. This means that as Saturn moves along its orbit, the north and south poles take turns pointing toward the Sun. The changing orientation of Saturn to the Sun is seen from Earth by the apparent opening and closing of the planet's ring system. The rings are seen edge-on, for example, at the start of an orbital period. Then an increasing portion of the rings is seen from above as the North Pole tips toward the Sun. The rings slowly close up and disappear from view as the North Pole starts to tip away until, 14.73 Earth years (half an orbit) later, they appear edge-on again. Now the South Pole tips sunward and the rings are seen increasingly from below. They close up once again as the South Pole turns away, until they are seen edge-on once more as the orbit is completed. The strength of the Sun at Saturn is only about 1 percent of that received on Earth, but it is enough to generate seasonal smog. Saturn is at perihelion at the time the South Pole is facing the Sun.

**NORTHERN SUMMER SOLSTICE**

*spins on its axis every 10.66 hours*

**NORTHERN SPRING EQUINOX**

*axis tilts from the vertical by 26.7°*

**NORTHERN WINTER SOLSTICE**

**APHELION**
*938 million miles (1.51 billion km)*

*Sun*

**PERIHELION**
*838 million miles (1.35 billion km)*

*Saturn orbits the Sun in 29.46 Earth years*

**NORTHERN FALL EQUINOX**

**SPIN AND ORBIT**
Saturn spins on its axis as it orbits. The rapid spin flings material outward, with the result that Saturn is about 10 percent wider at its equator than its poles. Its bulging equator is bigger than that of any other planet.

## STRUCTURE

Saturn's mass is only 95 times that of Earth, yet 764 Earths could fit inside it. This is because Saturn is composed mainly of the lightest elements, hydrogen and helium, which are in both gaseous and liquid states. Saturn is the least dense of all the planets. If it were possible to put Saturn in an ocean of water, it would float. The planet has no discernible surface—its outer layer is gaseous atmosphere. Inside the planet, pressure and temperature increase with depth and the hydrogen and helium molecules are forced closer and closer together until they become fluid. Deeper still, the atoms are stripped of their electrons and act as a liquid metal. Electric currents within this region generate a magnetic field 71 percent the strength of Earth's (see p.127).

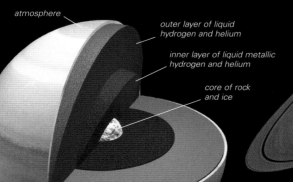

*atmosphere*

*outer layer of liquid hydrogen and helium*

*inner layer of liquid metallic hydrogen and helium*

*core of rock and ice*

**SATURN'S INTERIOR**
A thin, gaseous atmosphere surrounds a vast shell of liquid hydrogen and helium. The central core is about 10–20 times the mass of Earth.

**RINGLEADER**
Girdled by its bright system of rings, Saturn has a hazy, muted appearance in this Cassini image, which shows the planet in its natural colors. A number of Saturn's moons are also visible in this image.

# ATMOSPHERE

Saturn's atmosphere forms the planet's visible surface.
It is seen as a pale-yellow cloud deck with muted bands
of various shades, which lie parallel to the planet's equator.
Its upper clouds have a temperature of about −220°F
(−140°C). The atmospheric temperature decreases with
height, and as different compounds condense into liquid
droplets at different temperatures, clouds of different
composition form at different levels. Saturn is believed to
have three cloud layers. The highest, visible layer is made of
ammonia ice crystals; beneath this lies a layer of ammonium
hydrosulfide; water-ice clouds, so far unseen, form the lowest
layer. The upper atmosphere absorbs ultraviolet light, and the
temperature rises here, leading to the production of a thin
layer of smoggy haze; it is this layer that gives the planet its
indistinct, muted appearance. The smog builds up on the
hemisphere that is tilted toward the Sun. Saturn radiates almost
twice the amount of energy it receives from the Sun. The extra
heat is generated by helium rain droplets within the planet's
metallic shell. These convert motion energy to heat energy as
they fall toward the planet's
center. The heat is transported
through the lower atmosphere
and, along with the planet's
rotation, generates Saturn's winds.

**COMPOSITION OF ATMOSPHERE**
The trace gases include methane,
ammonia, and ethane. It is not known
which elements or compounds color the
atmosphere's clouds and spots.

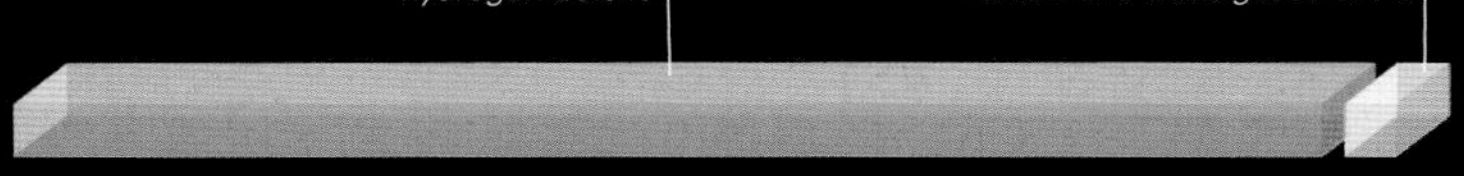

*hydrogen 96.3%*      *helium and trace gases 3.7%*

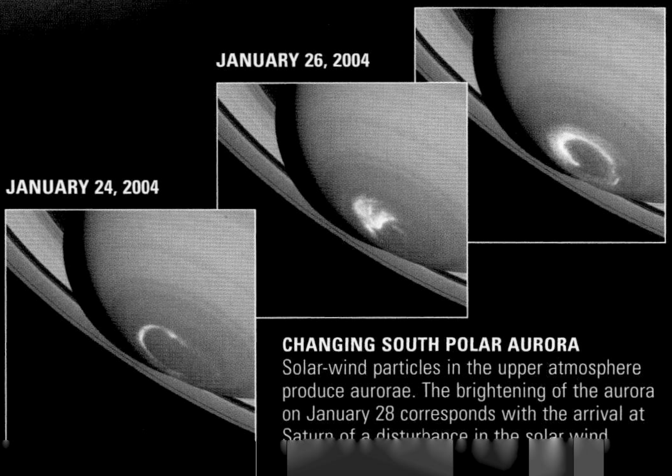

**JANUARY 28, 2004**

**JANUARY 26, 2004**

**JANUARY 24, 2004**

### CHANGING SOUTH POLAR AURORA
Solar-wind particles in the upper atmosphere
produce aurorae. The brightening of the aurora
on January 28 corresponds with the arrival at
Saturn of a disturbance in the solar wind.

## SATURN PROFILE

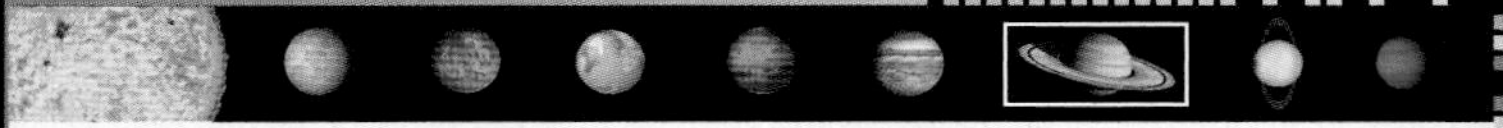

**AVERAGE DISTANCE FROM THE SUN**
888 million miles (1.43 billion km)

**CLOUDTOP TEMPERATURE**
−220°F (−140°C)

**DIAMETER** 74,898 miles (120,536 km)

**VOLUME (EARTH = 1)** 763.59

**NUMBER OF MOONS** 82

**OBSERVATION**
Saturn is visible to the naked eye for about
10 months of the year. It appears like a star
and takes about 2.5 years to pass though
one zodiacal constellation. A telescope is
needed to make out the ring system.

**ROTATION PERIOD**
10.66 hours

**ORBITAL PERIOD (LENGTH OF YEAR)**
29.46 Earth years

**MASS (EARTH = 1)** 95

**GRAVITY AT CLOUDTOPS (EARTH = 1)** 1.07

**SIZE COMPARISON**

EARTH          SATURN

# WEATHER

Giant upper-atmosphere storms composed of white ammonia ice can be seen from Earth when they rise through the haze. They occur once every 30 years or so, when it is midsummer in the northern hemisphere, but there is no accepted explanation for the storms yet. The last of these "Great White Spots" was discovered on September 25, 1990. It spread around the planet, almost encircling the equatorial region over about a month. Smaller, different-colored oval spots and ribbonlike features have been observed on a more regular basis. In 2004, Cassini revealed a region then dominated by storm activity, nicknamed "storm alley." Wind speed and direction on the planet are determined by tracking storms and clouds. Saturn's dominant winds blow eastward, in the same direction as the planet's spin. Near the equator, they reach 1,200 mph (1,800 kph).

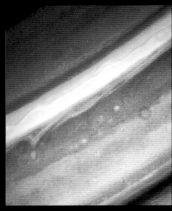

**BANDS AND SPOTS**
Bands of clouds, spots, and ribbonlike features move across Saturn's visible surface. The spots look small but can be thousands of miles across.

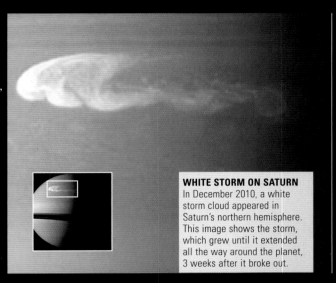

**WHITE STORM ON SATURN**
In December 2010, a white storm cloud appeared in Saturn's northern hemisphere. This image shows the storm, which grew until it extended all the way around the planet, 3 weeks after it broke out.

— C ring —                                                                 — B ring —

# MOONS

Saturn has over 80 known moons. Most of these have been discovered since 1980, through exploration by the Voyager and Cassini probes and by improved Earth-based observing techniques. Future observations are expected to confirm the presence of more moons. Titan is the largest and was the first to be discovered, in 1655. It is a unique moon, being the only one in the solar system to have a substantial atmosphere. Saturn's moons are mixes of rock and water ice. Some have ancient, cratered surfaces, and others show signs of resurfacing by tectonics or ice volcanoes. The moons are mostly named after mythological giants. The first to be discovered were named after the Titans, the brothers and sisters of Cronus (Saturn) in Greek mythology. More recent discoveries have Gallic, Inuit, and Norse names.

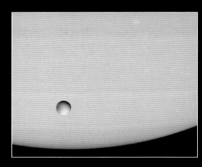

**DIONE**
Cassini produced this image of the moon Dione against the backdrop of Saturn's clouds in December 2005. This true-color view reveals the variations in brightness of the moon's icy surface. Dione is Saturn's fourth-largest moon.

**SATURN'S MOONS**

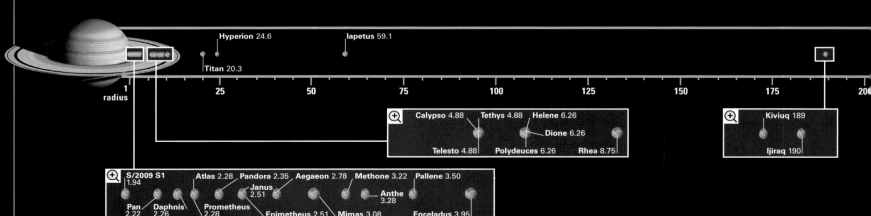

Hyperion 24.6    Iapetus 59.1
Titan 20.3

1 radius    25    50    75    100    125    150    175    200

Calypso 4.88    Tethys 4.88    Helene 6.26
Dione 6.26
Telesto 4.88    Polydeuces 6.26    Rhea 8.75

Kiviuq 189
Ijiraq 190

S/2009 S1 1.94    Atlas 2.28    Pandora 2.35    Aegaeon 2.78    Methone 3.22    Pallene 3.50
Janus 2.51    Anthe 3.28
Pan 2.22    Daphnis 2.26    Prometheus 2.28    Epimetheus 2.51    Mimas 3.08    Enceladus 3.95

# RINGS

Saturn's visible rings are the most extensive, massive, and spectacular in the solar system. From Earth, they appear as a band of material whose appearance changes according to Saturn's position. The rings are, in fact, collections of separate pieces of dirty water ice following individual orbits around Saturn. The pieces range from dust grains to boulders several yards across. They are very reflective, so the rings are bright and easy to see. Individual rings are identified by letters, allocated in order of discovery. The readily seen rings are the C, B, and A rings. These are bounded by others made of tiny particles that are almost transparent. The thin F ring, the broader G ring, and the diffuse E ring lie outside the main rings. The D ring, inside C, completes the system. The rings change slowly over time, and moons orbiting within the system shepherd particles into rings and maintain gaps such as the Encke gap. Far beyond the visible system is a huge, doughnut-shaped ring. Almost impossible to see, it was discovered in 2009 by the infrared glow of its cool and sparse dust.

**MAIN RING SYSTEM**
This mosaic of six images shows the main rings in natural color and reveals the ringlets within the Cassini Division. The distance from the inner edge of C to the F ring is about 40,500 miles (65,000 km).

**COMPOSITIONAL DIFFERENCES**
In this ultraviolet image of the outer portion of the C ring (left) and inner B ring (right), red indicates the presence of dirty particles and turquoise indicates purer ice particles.

Encke gap

F ring

Cassini Division

A ring

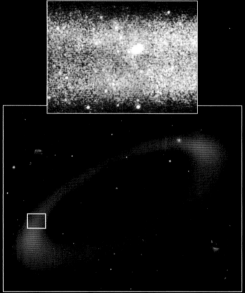

**PROMETHEUS AND THE F RING**
Saturn's innermost moons orbit within the ring system and interact with it. Some act as shepherd moons, confining particles within specific rings. Prometheus (just below the rings in this image) and Pandora work in this way on either side of the F ring.

**PANDORA**
The small shepherd moon Pandora orbits just beyond the F ring. It is visible as a white dot in this view taken by Cassini on February 18, 2005.

**INVISIBLE RINGS**
Saturn's largest ring is invisible to the eye. It consists of dust and was discovered at infrared wavelengths by the Spitzer Space Telescope. Above is an artist's representation of the ring, which lies between about 3.7 and 11.2 million miles (6 million and 18 million km) from Saturn. At the top is a Spitzer image of part of the ring. The ring is tilted about 27 degrees from Saturn's main ring plane.

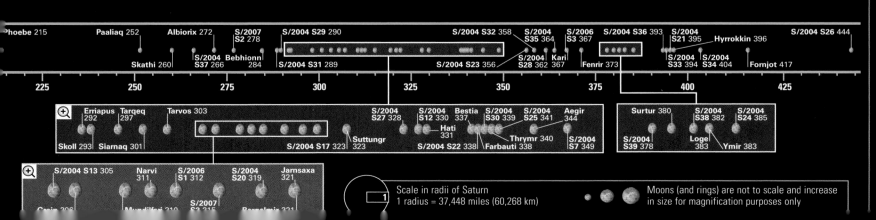

Phoebe 215 · Paaliaq 252 · Albiorix 272 · S/2007 S2 278 · S/2004 S29 290 · S/2004 S32 358 · S/2004 S35 364 · S/2006 S3 367 · S/2004 S36 393 · S/2004 S21 395 · Hyrrokkin 396 · S/2004 S26 444

Skathi 260 · S/2004 S37 266 · Bebhionn 284 · S/2004 S31 289 · S/2004 S23 356 · S/2004 S28 362 · Kari 367 · Fenrir 373 · S/2004 S33 394 · S/2004 S34 404 · Fornjot 417

225 · 250 · 275 · 300 · 325 · 350 · 375 · 400 · 425

Erriapus 292 · Tarqeq 297 · Tarvos 303 · S/2004 S27 328 · S/2004 S12 330 · Bestia 337 · S/2004 S30 339 · S/2004 S25 341 · Aegir 344 · Surtur 380 · S/2004 S38 382 · S/2004 S24 385

Skoll 293 · Siarnaq 301 · Suttungr 323 · S/2004 S17 323 · Hati 331 · S/2004 S22 338 · Farbauti 338 · Thrymr 340 · S/2004 S7 349 · S/2004 S39 378 · Logei 383 · Ymir 383

S/2004 S13 305 · Narvi 311 · S/2006 S1 312 · S/2004 S20 319 · Jarnsaxa 321

Scale in radii of Saturn
1 radius = 37,448 miles (60,268 km)

Moons (and rings) are not to scale and increase in size for magnification purposes only

# SATURN'S MOONS

The moons of Saturn are divided into three groups. The first consists of the major moons, which are large and spherical. The second group, the inner moons, are smaller and irregularly shaped. Members of both these groups orbit within or outside the ring system. The third set of moons lies far beyond the other two—the most distant orbit over 15 million miles (25 million km) from Saturn. These irregularly shaped moons are tiny, just a few miles to tens of miles across. They have inclined orbits, which suggests that they are captured objects. From Earth, Saturn's moons appear as little more than disks of light, but Voyager and Cassini revealed many of them as worlds in their own right.

**DWARFED BY SATURN**
Saturn's moons, such as Tethys (top) and Dione (below), are not only small compared to their host planet but, with the exception of Titan, they are all smaller than Earth's Moon.

---

## Prometheus

**DISTANCE FROM SATURN** 86,539 miles (139,353 km)

**ORBITAL PERIOD** 0.61 Earth days

**LENGTH** 84 miles (136 km)

Prometheus is a small, elongated moon orbiting just inside the multi-stranded F ring. Along with Pandora, it is a "shepherd" of the F ring.

Cassini images of Prometheus and the F ring show them to be linked by a fine thread of material, produced as Prometheus pulls particles out of the ring. The moon's long axis points toward Saturn.

**SHEPHERD MOON**

---

## Janus

**DISTANCE FROM SATURN** 94,120 miles (151,472 km)

**ORBITAL PERIOD** 0.69 Earth days

**LENGTH** 126 miles (203 km)

Heavily cratered and irregularly shaped, Janus orbits Saturn just beyond the F ring and only 30 miles (50 km) farther away than its co-orbital moon, Epimetheus. Its existence was first reported in December 1966, and it was named after the Roman god Janus, who could look forward and back at the same time. Yet it was only confirmed as a moon in February 1980, after Voyager 1 data had been studied.

**BEYOND THE F RING**

---

## Epimetheus

**DISTANCE FROM SATURN** 94,089 miles (151,422 km)

**ORBITAL PERIOD** 0.69 Earth days

**LENGTH** 81 miles (130 km)

Occasionally, moons orbit a planet within about 30 miles (50 km) of each other. They are described as co-orbital since they virtually share an orbit. The two moons Epimetheus and Janus

**CO-ORBITAL MOON**
Epimetheus orbits against the backdrop of Saturn's rings, which are seen nearly edge-on in this view taken by Cassini's narrow-angle camera on February 18, 2005.

(below left), which orbit just beyond the F ring, are such a pair. They swap orbits every 4 years, taking turns being slightly closer to Saturn. Epimetheus is a lumpy moon just 17 miles (28 km) longer than it is wide or deep, and it is one of 16 moons that lie within the ring system. Epimetheus is in synchronous rotation—that is, it keeps the same face toward Saturn at all times because its rotation and orbital periods are the same. As it orbits Saturn, it works as a shepherd moon, confining the ring particles within the F ring. Prometheus (left) works

**BATTERED SURFACE**
Epimetheus (left) and its co-orbital moon, Janus, are believed to be the remnants of a larger object that was broken apart by an impact.

the same way on the inner side of the ring. The existence of Epimetheus was suspected in 1967, but was not confirmed until February 26, 1980. It was one of eight moons discovered in Voyager data that year. The moon is named after a Titan, the family of giants in Greek mythology who once ruled the Earth. Prometheus was one of Epimetheus's five brothers.

---

## Methone

**DISTANCE FROM SATURN** 120,550 miles (194,000 km)

**ORBITAL PERIOD** 1 Earth day

**DIAMETER** 1.65 miles (3 km)

Two small moons orbiting between the major moons Mimas and Enceladus were discovered in 2004 in data collected by the Cassini probe. As with all such discoveries, the moons were initially identified by numerical designations (S/2004 S1 and S/2004 S2). The two moons are now known as Methone and

Pallene. They were not discovered by chance, but were identified in images taken as part of a search for new moons within this region around Saturn. The contrast of the images was enhanced to increase visibility. Methone was visited by the Cassini spacecraft in 2012, which gave scientists a better look at the egg-shaped moon.

**SMOOTH SATELLITE**
In 2012, the Cassini spacecraft made two approaches to Methone, getting as close as 1,200 miles (1,900 km) to the moon. Cassini images show Methone's surface to be relatively smooth and free from visible craters.

THE SOLAR SYSTEM

MAJOR MOON

# Mimas

**DISTANCE FROM SATURN** 115,208 miles
(185,520 km)

**ORBITAL PERIOD** 0.94 Earth days

**DIAMETER** 246 miles (396 km)

Mimas is the first of the major moons out from Saturn, and it orbits the planet in the outer part of the ring system. It is in synchronous rotation, so the same side of the moon always faces the planet in the same way that one side of the Moon always faces Earth. Mimas is a round moon, but it is not a perfect sphere—this icy object is about 19 miles (30 km) longer than it is wide and deep. Its surface is covered in deep, bowl-shaped impact craters. Many of those greater than about 12 miles (20 km) across have central peaks. One crater, Herschel, dwarfs the rest and is the moon's most prominent feature. It is about 80 miles (130 km) wide, almost 6 miles (10 km) deep, and has a prominent central peak. If the impacting body that formed the crater had been much bigger, it might have smashed the moon apart. The crater is named after the astronomer William Herschel, who discovered Mimas on July 18, 1789. It was the sixth of Saturn's moons to be discovered and the first of two discovered by Herschel. Mimas is named after a Titan (see p.190).

**EMPHASIZING DIFFERENCES**
False color highlights slight differences in surface composition on Mimas—for example, the bluish terrain near the crater Herschel, possibly caused by impact ejecta, and greenish terrain elsewhere.

**GIANT CRATER**
The crater Herschel lies on the moon's leading hemisphere (the side that points in the direction in which it is moving) and is about one-third of the diameter of Mimas itself. The impact that formed Herschel must have come close to shattering the moon.

**TRUE BLUE**
Mimas drifts against the backdrop of Saturn's northern hemisphere in this true-color view. Scattering of sunlight in the relatively cloud-free area gives the planet a bluish hue. The dark lines cutting across the atmosphere are shadows cast by Saturn's rings.

## MAJOR MOON
# Enceladus

**DISTANCE FROM SATURN** 147,898 miles
(238,020 km)

**ORBITAL PERIOD** 1.37 Earth days

**DIAMETER** 313 miles (504 km)

Enceladus orbits within the broad
E ring of Saturn. Its orbit lies within
the densest part of the ring, which
suggests that Enceladus could be
supplying the ring with material.
The moon is in synchronous rotation
with Saturn. The frosty surface of
Enceladus is highly reflective and
makes this moon particularly bright,
the brightest in the solar system.
The surface terrain suggests that
this frigid moon has experienced
a long history of tectonic activity
and resurfacing. The extent of the
geological change is surprising in

**WATER JETS**
Ice and water vapor spray out from
so-called tiger stripes near the south
pole of Enceladus, as shown here in
this image from the Cassini probe.

such a small world—Mimas (see
p.193) is about the same size but is
inactive. Craters are concentrated in
some regions, and elsewhere there
are grooves, fractures, and ridges.
Images processed to accentuate color
differences have revealed previously
unseen detail. The blue color seen in
some fracture walls could be due to
the exposure of solid ice or because
the composition or size of
particles in the buried ice
is different from that on
the surface. Enceladus was
discovered by William
Herschel on August 28, 1789.

**BAGHDAD SULCUS**
This is a close-up of part of Baghdad
Sulcus, the longest of several
linear structures popularly termed
tiger stripes in the south polar
region of Enceladus.

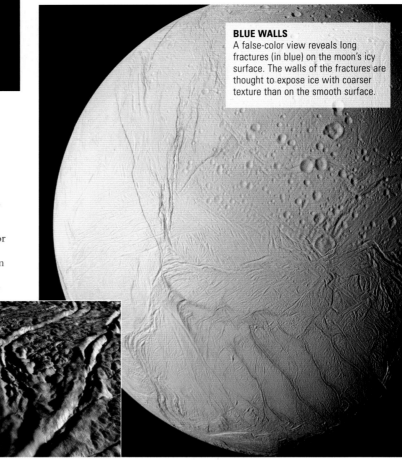

**BLUE WALLS**
A false-color view reveals long
fractures (in blue) on the moon's icy
surface. The walls of the fractures are
thought to expose ice with coarser
texture than on the smooth surface.

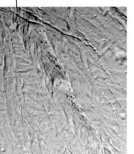

**SMOOTH PLAINS**
This region of
smooth plains has a
band of chevron-
shaped features
running across its
center, cut across
at the top by a
system of crevasses.

---

## INNER MOON
# Telesto

**DISTANCE FROM SATURN** 183,093 miles
(294,660 km)

**ORBITAL PERIOD** 1.89 Earth days

**LENGTH** 20 miles (32.5 km)

Telesto shares an orbit within the
E ring with two other moons:
Calypso, which is about the same
size as Telesto, and the much larger
Tethys (right). Telesto moves along
the orbit 60° ahead of Tethys, and
Calypso follows 60° behind Tethys.
The positions taken on the orbit by
the two smaller moons are called the
Lagrange points. In these positions,
the two small moons can maintain
a stable orbit balanced between the
gravitational pull of Saturn and that
of Tethys. Telesto and Calypso were
discovered in 1980—Calypso by
Earth-based observation and Telesto
in Voyager images. The probe revealed
two irregularly shaped moons.

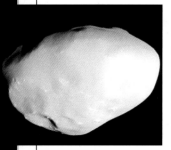

**SMOOTH
MOON**
Telesto's surface
appears less
cratered than
most of Saturn's
other moons in
this image taken
by Cassini from
a distance of
only 9,000 miles
(14,500 km).

---

**ICY MOON**
The pale, icy disk of Tethys was imaged by
Cassini on January 18, 2005, as it orbited
below Saturn's south polar region, where
fierce storms were raging.

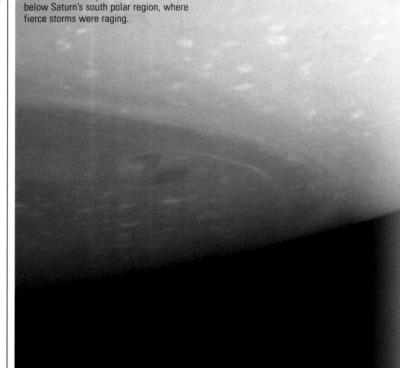

---

## MAJOR MOON
# Tethys

**DISTANCE FROM SATURN** 183,093 miles
(294,660 km)

**ORBITAL PERIOD** 1.89 Earth days

**DIAMETER** 660 miles (1,062 km)

The Italian–French astronomer
Giovanni Cassini discovered Tethys
on March 21, 1684. Nearly 300 years
later, it was discovered that Tethys
shares its orbit with two far smaller
moons: Telesto (left) and Calypso.
Its surface shows that Tethys has
undergone tectonic change and
resurfacing. Two features stand out.
A 248-mile (400-km) wide impact
crater called Odysseus dominates the
leading hemisphere. Large but shallow,
its original bowl shape has been
flattened by ice flows. The second
large feature is the Ithaca Chasma
on the side of Tethys facing Saturn.
This vast canyon system extends
across half of the moon. It may have
been formed by tensional fracturing
as a result of the impact that produced
the Odysseus Crater, or when Tethys's
interior froze and the moon expanded
in size and stretched its surface.

**ITHACA CHASMA**
This canyon system,
which is up to
2.5 miles (4 km)
deep, runs from
the lower left of the
prominent Telemachus
Crater (top right).

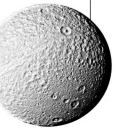

## MAJOR MOON

# Dione

**DISTANCE FROM SATURN** 234,505 miles (377,400 km)

**ORBITAL PERIOD** 2.74 Earth days

**DIAMETER** 698 miles (1,123 km)

Dione is the most distant moon within Saturn's ring system, but it is not alone in the outer reaches of the E ring. Two other moons, Helene and Polydeuces, follow the same orbit; Helene is ahead of Dione by 60° and Polydeuces follows 60° behind. Helene was discovered in March 1980; Polydeuces was discovered in Cassini data some 24 years later, just after the probe arrived at Saturn. Giovanni Cassini discovered Dione in 1684, on the same day that he discovered Tethys (opposite). Dione has a higher proportion of rock in its rock–ice mix than most of the other moons (only Titan has more), so it

### IMPACT CRATERS
The well-defined central peaks of Dione's largest craters are visible in this Voyager image. Dido Crater lies just left of center, with Romulus and Remus just above it and Aeneas Crater near the upper limb.

is the second-densest of Saturn's moons. The terrain displays evidence of tectonic activity and resurfacing. There are ridges, faults, valleys, and depressions. There are also craters, which are more densely distributed in some regions than others—Dione's leading face, for example, has more than the trailing face. The largest crater is over 124 miles (200 km) across. Dione also has bright streaks on its surface. These wispy features are composed of narrow, bright, icy lines.

### DIONE'S FAR SIDE
Impact craters scar the surface of the side of Dione that is permanently turned away from Saturn because it is in synchronous rotation. Areas of wispy terrain are visible on the left of this image.

### CLIFFS OF ICE
A Cassini close-up of Dione's wispy terrain reveals that it is formed from lines of ice cliffs created by tectonic fractures rather than deposits of ice and frost as was previously thought.

## MAJOR MOON

# Rhea

**DISTANCE FROM SATURN** 327,487 miles (527,040 km)

**ORBITAL PERIOD** 4.52 Earth days

**DIAMETER** 949 miles (1,527 km)

Vast sweeps of ancient cratered terrain cover large parts of Rhea. At first glance, the landscape resembles that seen on Earth's Moon, although Rhea's surface is bright ice. There is some evidence of resurfacing, although not as much as expected for such a large moon. Rhea is Saturn's second-largest moon, but other, smaller moons, such as its inner neighbors Dione and Tethys, show more resurfacing. It is thought that

Rhea froze early in its history and became frigid. Its ice would then have behaved like hard rock. Rhea's craters, for example, are freshly preserved in its icy crust. The craters on other icy moons, such as Jupiter's Callisto (see p.187), have collapsed in the soft, icy crust. Rhea is the first of Saturn's moons to lie beyond the ring system. It is named after the Titan Rhea, who was the mother of Zeus in Greek mythology.

### ANCIENT SURFACE
Two large impact basins (top center) are visible in this enhanced-color image of Rhea's heavily cratered surface. The great age of these basins is indicated by the many smaller craters upon them.

### HEAVILY CRATERED
Rhea's icy surface is heavily cratered, suggesting that it dates back to the period immediately following the formation of the planets. This image shows the region around the moon's North Pole. The largest craters are several miles deep.

### FRESH ICE
An enhanced-color view of the surface of Rhea shows blue patches of freshly uncovered ice. The ice is thought to have been exposed when debris in orbit around Rhea struck the surface along the equator.

## MAJOR MOON

# Titan

**DISTANCE FROM SATURN** 758,073 miles
(1.22 million km)

**ORBITAL PERIOD** 15.95 Earth days

**DIAMETER** 3,200 miles (5,150 km)

Titan was discovered in 1655 by the Dutch scientist Christiaan Huygens. It is the second-largest moon in the solar system after Jupiter's Ganymede (see p.186) and is by far the largest of Saturn's moons. This Mercury-sized body is also one of the most fascinating. A veil of smoggy haze shrouds the moon and permanently obscures the world below. Titan is intriguing, not least because the

chemistry of the atmosphere has similarities to that of the young Earth before life began. The first chance to see the surface and test the atmosphere came in 2005, when Cassini turned its attention to Titan, and Huygens plunged through the atmosphere to the surface (see panel, right).

The nitrogen-rich atmosphere extends for hundreds of miles above Titan. Layers of yellow-orange, smoglike haze high within it are the result of chemical reactions triggered by ultraviolet light. Methane clouds form much closer to the surface. These rain methane onto Titan, where it forms rivers and lakes. It then evaporates and forms clouds, and the cycle, which is reminiscent of the water cycle on Earth, continues.

Titan is the densest of Saturn's moons: it is a 50:50 mix of rock and water ice with a surface temperature of –292°F (–180°C). It is a gloomy world because the smog blocks 90 percent of the incident sunlight. Cassini revealed

that its surface is shaped by Earth-like processes—tectonics, erosion, and winds—and perhaps ice volcanism. No liquid methane was detected on the initial flybys, but drainage channels and dark elliptical regions, thought to be evaporated lakes, showed where fluids had been. Linear features nicknamed "cat scratches" were also identified.

### TITAN'S ATMOSPHERE

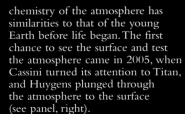

Infrared and ultraviolet data combined reveals aspects of the atmosphere. Areas where methane absorbs light appear orange and green. The high atmosphere is blue.

### ORANGE AND PURPLE HAZE
The upper atmosphere consists of separate layers of haze; up to 12 have been detected in this ultraviolet, true-color image of Titan's night-side limb.

## THE HUYGENS PROBE

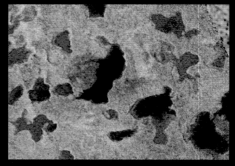

The European probe Huygens traveled to Titan onboard NASA's Cassini spacecraft. Once there, it separated from the larger craft and parachuted into Titan's haze. During its 2.5-hour descent, Huygens tested the atmosphere, measured the speed of the buffeting winds, and took images of the moon's surface. An instrument recorded the first surface touch on January 14, 2005, sending back evidence of a thin, hard crust with softer material beneath.

### HUYGENS AND CASSINI
The shield-shaped Huygens probe (right) is attached to Cassini's frame in preparation for the launch from Cape Canaveral, Florida, in October 1997.

### POLAR LAKES
Lakes of liquid methane and ethane have formed at Titan's north pole. The lakes are shown in blue. The largest are bigger than the Great Lakes but much shallower.

### ICY SURFACE
The surface of Titan is shown in this image taken by the Huygens lander in 2005. The pebbles in the foreground are up to about 6 in (15 cm) across and are thought to be composed of frozen water.

### TITAN REVEALED
Infrared observations that cut through Titan's clouds reveal bright highlands; dark, dune-filled regions; and, to the left and center, a large impact crater.

**DARK COATING**
Iapetus, the two-toned moon of Saturn, is seen in this close-up from the space probe Cassini. One side of Iapetus consists of bright ice, while the other is covered with a dark coating.

## MAJOR MOON

# Iapetus

**DISTANCE FROM SATURN** 2.21 million miles (3.56 million km)

**ORBITAL PERIOD** 79.33 Earth days

**DIAMETER** 913 miles (1,469 km)

Most of Saturn's inner and major moons orbit in the equatorial plane (also the plane of the rings). Iapetus is an exception, its orbit being inclined by 14.72° to the equatorial plane. Other moons follow orbits with greater inclination, but these are the much smaller, outer moons. Iapetus is Saturn's most distant major moon. It is also in synchronous rotation.

Iapetus was discovered by Italian astronomer Giovanni Cassini while he was working from Paris on October 25, 1671. He noticed that Iapetus has a naturally dark leading hemisphere and a bright trailing hemisphere. The dark region is called Cassini Regio and is covered in material as dark as coal, in contrast to the icy surface

**SURFACE COMPOSITION**
False colors represent Iapetus's vastly different surface compositions. Bright blue signifies an area rich in water ice, dark brown indicates a substance rich in organic material, and the yellow region is composed of a mix of ice and organic chemicals.

**LANDSLIDE IN CASSINI REGIO**
Land has collapsed down a 9-mile (15-km) high scarp, which marks the edge of a huge impact crater, into a smaller crater. The long distance traveled by the material along the floor indicates that it could be fine-grained.

on the bright side. Although the Cassini probe revealed more of the moon's heavily cratered surface, the origin of the dark material remains a mystery. It has been suggested that the material erupted from the moon's interior or that it is ejecta from impacts on a more distant moon, such as Phoebe (right). A unique feature revealed by Cassini has provided another mystery. It is not known whether a 800-mile (1,300-km) long ridge that coincides almost exactly with the moon's equator is a folded mountain belt or material that erupted through a crack in the surface.

## INNER MOON

# Hyperion

**DISTANCE FROM SATURN** 919,620 miles (1.48 million km)

**ORBITAL PERIOD** 21.28 Earth days

**LENGTH** 224 miles (360 km)

Nothing about Hyperion is typical. First, it is an irregularly shaped moon with an average width of about 174 miles (280 km). This makes it one of the largest nonspherical bodies in the solar system. Second, it follows an elliptical orbit just beyond Saturn's largest moon, Titan (opposite). And, as it orbits, it rotates chaotically: its rotation axis wobbles, and the moon appears to tumble as it travels. Its

**STRANGE CRATERS**
Hyperion has a strange, spongy appearance resulting from its low density and weak gravity.

## OUTER MOON

# Phoebe

**DISTANCE FROM SATURN** 8.05 million miles (12.95 million km)

**ORBITAL PERIOD** 550 Earth days

**DIAMETER** 132 miles (213 km)

Phoebe was discovered in 1898 and, until 2000, was thought to be Saturn's only outer moon. Many others are now known to exist. Phoebe has a long orbital period and follows a highly inclined orbit, a characteristic of the outer moons. Phoebe's orbit is inclined by 175.3°, so it travels in a retrograde manner (backward

surface is cratered, and there are segments of cliff faces. Its shape and scarred surface suggest Hyperion could be a fragment of a once-larger object that was broken by a major impact. Even Hyperion's discovery was unusual. Astronomers in the US and England found it independently and within 2 days of each other in September 1848.

compared to most moons). Half the outer moons orbit this way. Phoebe is by far the largest outer moon; the others are, at most, only 12 miles (20 km) across. From Cassini images, it appears to be an ice-rich body coated with a thin layer of dark material.

**CRATER FLOOR**
Debris covers the floor of this impact crater. The streaks inside the crater indicate where loose ejecta has slid down toward the center.

**BLASTED PHOEBE**
Phoebe's impact craters were revealed by Cassini in June 2004. They are named after the Argonauts in Greek mythology. The largest, Jason, is about 60 miles (100 km) across.

Erginus
Jason
Iphitus
Euphemus
Butes
Eurydamas
Canthus
Oileus

**SATURN FROM ABOVE**
This image recorded by Cassini in 2016 shows Saturn as it can never be seen from Earth, looking obliquely down across its north pole. At this time, Saturn's northern hemisphere midsummer was fast approaching, so its north pole was fully illuminated. In addition to the more familiar linear belts of clouds girdling the planet, this view shows a hexagon-shaped jet stream encircling the pole. The complexity of the rings is apparent, and Saturn's shadow falls across them to the left, extending to just beyond the Cassini division (see p.190).

# URANUS

URANUS IS THE THIRD-LARGEST planet and lies twice as far from the Sun as its neighbor Saturn. It is pale blue and featureless, with a sparse ring system and an extensive family of moons. The planet is tipped on its side, so from Earth the moons and rings appear to encircle it from top to bottom. Uranus was the first planet to be discovered by telescope, but little was known about it until the Voyager 2 spacecraft flew past in January 1986.

**PALE BLUE DISK**
Voyager 2 images have been combined to show the southern hemisphere of Uranus as it would appear to a human onboard the spacecraft.

## ORBIT

Uranus takes 84 Earth years to complete one orbit around the Sun. Its axis of rotation is tipped over by 98°, and the planet moves along the orbital path on its side. Uranus's spin is retrograde, spinning in the opposite direction of most planets. The planet would not have always been like this. Its sideways stance is probably the result of a collision with a planet-sized body when Uranus was young. Each of the poles points to the Sun for 21 years at a time, during the periods centered on the solstices. This means that while one pole experiences a long period of continuous sunlight, the other experiences a similar period of complete darkness. The strength of the sunlight received by the planet is 0.25 percent of that on Earth. When Voyager encountered Uranus in 1986, its south pole was pointing almost directly at the Sun. Uranus's equator then became increasingly edge-on to the Sun. After 2007, it has progressively turned away, and the north pole will face the Sun in 2030.

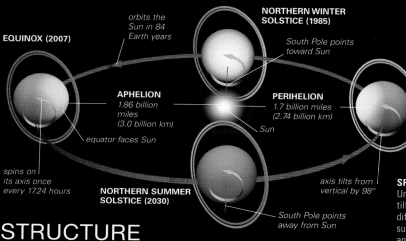

EQUINOX (2007)

orbits the Sun in 84 Earth years

**NORTHERN WINTER SOLSTICE (1985)**

South Pole points toward Sun

APHELION
1.86 billion miles
(3.0 billion km)

PERIHELION
1.7 billion miles
(2.74 billion km)

Sun

EQUINOX
(1965)

equator faces Sun

spins on its axis once every 17.24 hours

**NORTHERN SUMMER SOLSTICE (2030)**

axis tilts from vertical by 98°

South Pole points away from Sun

## STRUCTURE

Uranus is big. It is four times the size of Earth and could contain 63 Earths inside it, yet it has only 14.5 times the mass of Earth. So the material it is made of must be less dense than that of Earth. Uranus is too massive for its main ingredient to be hydrogen, which is the main constituent of the bigger planets, Saturn and Jupiter. It is made mainly of water, methane, and ammonia ices, which are surrounded by a gaseous layer. Electric currents within its icy layer are believed to generate the planet's magnetic field, which is offset by 58.6° from Uranus's spin axis.

**SPIN AND ORBIT**
Uranus's long orbit and its extreme tilt combine to produce long seasonal differences. Each pole experiences summer when pointing toward the Sun and winter when it is pointing away. At such times, the pole is in the middle of Uranus's disk when viewed from Earth. At the equinoxes, the equator and rings are edge-on to the Sun.

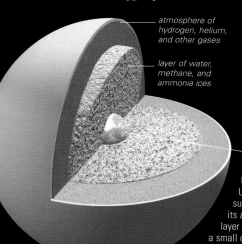

atmosphere of hydrogen, helium, and other gases

layer of water, methane, and ammonia ices

core of rock and possibly ice

**URANUS INTERIOR**
Uranus does not have a solid surface. The visible surface is its atmosphere. Below this lies a layer of water and ices, which surrounds a small core of rock and possibly ice.

### URANUS PROFILE

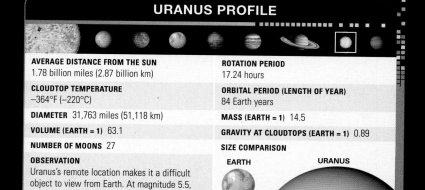

| AVERAGE DISTANCE FROM THE SUN | ROTATION PERIOD |
|---|---|
| 1.78 billion miles (2.87 billion km) | 17.24 hours |

| CLOUDTOP TEMPERATURE | ORBITAL PERIOD (LENGTH OF YEAR) |
|---|---|
| −364°F (−220°C) | 84 Earth years |

| DIAMETER 31,763 miles (51,118 km) | MASS (EARTH = 1) 14.5 |
| VOLUME (EARTH = 1) 63.1 | GRAVITY AT CLOUDTOPS (EARTH = 1) 0.89 |
| NUMBER OF MOONS 27 | SIZE COMPARISON |

**OBSERVATION**
Uranus's remote location makes it a difficult object to view from Earth. At magnitude 5.5, it is just visible to the naked eye and looks like a star. There is no perceptible change in brightness when Uranus is at opposition.

EARTH          URANUS

# ATMOSPHERE AND WEATHER

Uranus's blue color is a result of the absorption of the incoming sunlight's red wavelengths by methane-ice clouds within the planet's cold atmosphere. The cloudtop temperature of −364°F (−220°C) appears to be fairly uniform across the planet. The action of ultraviolet sunlight on the methane produces haze particles, and these hide the lower atmosphere, making Uranus appear calm. The planet is, however, actively changing. The Voyager 2 data revealed the movement of ammonia and water clouds around Uranus, carried by wind and the planet's rotation. It also revealed that Uranus radiates about the same amount of energy as it receives from the Sun and has no significant internal heat to drive a complex weather system. More recently, observations made using ground-based telescopes have also made it possible for astronomers to track changes in Uranus's atmosphere.

**CLOUDS**
This Keck II telescope infrared image has been processed to show vertical structure. The highest clouds appear white; mid-level ones, bright blue; and the lowest clouds, darker blue. As a by-product, the rings are colored red.

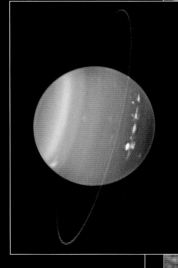

**COMPOSITION OF ATMOSPHERE**
The atmosphere is made mainly of hydrogen, which extends beyond the visible cloudtops and forms a corona around Uranus.

hydrogen 82.5%  helium 15.2%  methane 2.3%

# RINGS AND MOONS

Uranus has 11 rings that together extend out from 7,700 to 15,900 miles (12,400–25,600 km) from the planet. The rings are so widely separated and so narrow that the system has more gap than ring. All but the inner and outer rings are between 0.6 and 8 miles (1 and 13 km) wide, and all are less than 9 miles (15 km) high. They are made of charcoal-dark pieces of carbon-rich material measuring from a few inches to possibly a few yards across, plus dust particles. The first five rings were discovered in 1977 (see panel, right). The rings do not lie quite in the equatorial plane, nor are they circular or of uniform width. This is probably due to the gravitational influence of small, nearby moons. One of these, Cordelia, lies within the ring system. Uranus has 27 moons. The five major moons were discovered using Earth-based telescopes. Smaller ones have been found since the mid-1980s, through analysis of Voyager 2 data or by using today's improved observing techniques. More discoveries are expected.

**FALSE-COLOR VIEW OF THE RINGS**
Nine of Uranus's rings are visible in this Voyager 2 image. The faint pastel lines are due to image enhancement. The brightest, colorless ring (far right) is the outermost ring, epsilon. To its left are five rings in shades of blue-green, then three in off-white.

## RINGS DISCOVERED

In March 1977, astronomers onboard the Kuiper Airborne Observatory, an adapted high-flying aircraft, were preparing to observe a rare occultation of a star by Uranus in order to measure the planet's diameter. Before the star was covered by the planet's disk, it blinked on and off five times. A second set of blinks was recorded after the star appeared from behind the planet. Rings around Uranus had blocked out the star's light.

**KUIPER AIRBORNE OBSERVATORY**
Astronomers and technicians operate an infrared telescope, which looks out to space through an open door in the side of the aircraft.

**URANUS'S MOONS**

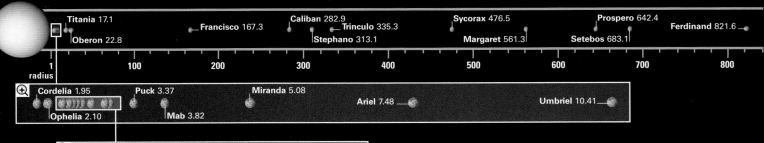

Titania 17.1
Oberon 22.8
Francisco 167.3
Caliban 282.9
Trinculo 335.3
Stephano 313.1
Sycorax 476.5
Margaret 561.3
Prospero 642.4
Setebos 683.1
Ferdinand 821.6

1 radius   100   200   300   400   500   600   700   800

Cordelia 1.95
Ophelia 2.10
Puck 3.37
Mab 3.82
Miranda 5.08
Ariel 7.48
Umbriel 10.41

Desdemona 2.45   Juliet 2.52   Belinda 2.94   Perdita 2.99
Bianca 2.32   Rosalind 2.74
Cressida 2.42   Portia 2.59   Cupid 2.93

**Scale in radii of Uranus**
1 radius = 15,872 miles (25,559 km)

Moons (not to scale) increase in size for magnification purposes

THE SOLAR SYSTEM

**THE VIEW FROM EARTH**
Some of the 27 moons that orbit Uranus
can be seen in this infrared image, which
was taken by the Hubble Space
Telescope in 1998.

# URANUS'S MOONS

Uranus's moons can be divided into three groups. Moving out from Uranus, they are: the small inner satellites; the five major moons, which orbit in a regular manner; and the small outer moons, many of which follow retrograde orbits. Much of what is known about the moons, and the only close-up views, came from the Voyager 2 flyby in 1985–1986. This revealed the major moons to be dark, dense rocky bodies with icy surfaces, featuring impact craters, fractures, and volcanic water-ice flows. The moons are named after characters in the plays of the English dramatist William Shakespeare or in the verse of the English poet Alexander Pope.

---

## INNER MOON
### Cordelia

**DISTANCE FROM URANUS** 30,910 miles (49,770 km)

**ORBITAL PERIOD** 0.34 Earth days

**DIAMETER** 25 miles (40 km)

Cordelia is the innermost and one of the smallest of Uranus's moons. A team of Voyager 2 astronomers discovered it on January 20, 1986. Cordelia was one of 10 moons that were discovered in the weeks between December 30, 1985, and January 23, 1986, as the Voyager 2 spacecraft flew by Uranus and transmitted images

**SHEPHERD MOON**
Cordelia is the innermost of two shepherd moons lying on either side of Uranus's bright outer ring.

back to Earth. Astronomers had expected to find some more moons in orbit around Uranus. In particular, it was expected that pairs of shepherd moons—moons that are positioned on either side of a ring and keep the ring's constituent particles in place— would be found. Surprisingly, just one pair, that of Cordelia and Ophelia, was discovered. Cordelia takes its name from the daughter of Lear in Shakespeare's *King Lear*.

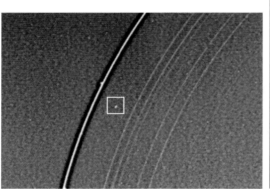

---

## INNER MOON
### Ophelia

**DISTANCE FROM URANUS** 33,400 miles (53,790 km)

**ORBITAL PERIOD** 0.38 Earth days

**DIAMETER** 26 miles (42 km)

Ophelia is one of a pair of moons that orbit either side of Uranus's outer ring, the epsilon ring. It was discovered at the same time as its partner, Cordelia, on January 20, 1986. The two are small—not much bigger than the particles that make up the thin, narrow ring. The moon is named after the heroine in Shakespeare's *Hamlet*.

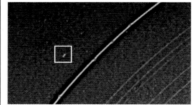

**OPHELIA LIES OUTSIDE THE EPSILON RING**

---

## INNER MOON
### Puck

**DISTANCE FROM URANUS** 53,410 miles (86,010 km)

**ORBITAL PERIOD** 0.76 Earth days

**DIAMETER** 101 miles (162 km)

Puck was discovered on December 30, 1985, and was the first of the 10 small moons to be found in the Voyager 2 data. It is the second-farthest inner moon from Uranus and was

**CRATERED MOON**

discovered as the probe approached the planet. There was time to calculate that an image could be recorded on January 24, the day of closest approach. The image (above) revealed an almost circular moon with craters. The largest crater (upper right) is named Lob, after a British Puck-like sprite.

---

## MAJOR MOON
### Miranda

**DISTANCE FROM URANUS** 80,350 miles (129,390 km)

**ORBITAL PERIOD** 1.41 Earth days

**DIAMETER** 300 miles (480 km)

Miranda is the smallest and innermost of Uranus's five major moons and was discovered by Dutch-born American astronomer Gerard Kuiper on February 16, 1948. When all five major moons were seen in close-up for the first time, on January 24, 1986, it was Miranda that gave astronomers the biggest surprise. As Voyager 2 passed within 19,870 miles (32,000 km) of its surface, the probe revealed a bizarre-looking world where various surface features butt up against one another in a seemingly

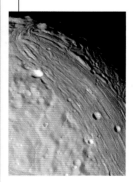

**GEOLOGICAL MIX**
On the left lies an ancient terrain of rolling hills and degraded craters; to the right is a younger terrain of valleys and ridges.

**FULL DISK**
The complex terrain of the bright, chevron-shaped Inverness Corona stands out in this south polar view of Miranda.

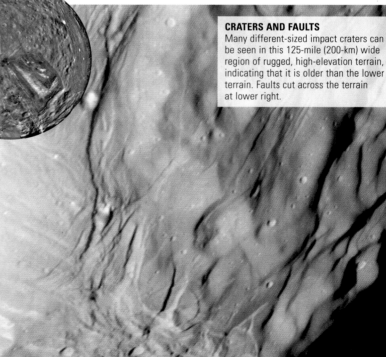

unnatural way. One explanation for this strange appearance is that Miranda experienced a catastrophic collision in its past. The moon shattered into pieces and then reassembled in the disjointed way seen today. An alternative theory says that the moon's evolution was halted before it could be completed. Soon after its formation, dense, rocky material began to sink and lighter material, such as water ice, rose to the surface. This process then stopped because the necessary internal heat had disappeared. The surface clearly has different types of terrain from different time periods.

**CRATERS AND FAULTS**
Many different-sized impact craters can be seen in this 125-mile (200-km) wide region of rugged, high-elevation terrain, indicating that it is older than the lower terrain. Faults cut across the terrain at lower right.

## MAJOR MOON
# Ariel

DISTANCE FROM URANUS 118,620 miles (191,020 km)

ORBITAL PERIOD 2.52 Earth days

DIAMETER 722 miles (1,162 km)

Ariel and Umbriel (below) were both discovered on October 24, 1851, by the English brewer and astronomer William Lassell (see p.207). Ariel is named after a spirit in Shakespeare's play *The Tempest*. Of the four largest moons, this is the brightest, with the youngest surface. It has impact craters,

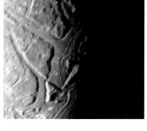

**COMPLEX TERRAIN**
The long, broad valley faults in Ariel's southern hemisphere are filled with deposits and are more sparsely cratered than the surrounding terrain.

**VOYAGER 2 MOSAIC**
Four Voyager 2 images were combined to produce this view of Ariel. Kachina Chasmata slices across the top and the Domovoy Crater is on the left, below the center. Below and to its right is the 30-mile (50-km) wide Melusine Crater, which is surrounded by bright ejecta.

but these are relatively small—many are just 3–6 miles (5–10 km) wide. Domovoy, at 44 miles (71 km) across, is one of the largest. The sites of any older, larger craters that Ariel once had have been resurfaced. Long faults that formed when Ariel's crust expanded cut across the moon to a depth of 6 miles (10 km). One fault, Kachina Chasmata, is 386 miles (622 km) long. The floors of such valleys are covered in icy deposits that seeped to the surface from below.

## MAJOR MOON
# Umbriel

DISTANCE FROM URANUS 140,530 miles (226,300 km)

ORBITAL PERIOD 4.14 Earth days

DIAMETER 726 miles (1,169 km)

Umbriel is the darkest of Uranus's major moons, reflecting only 16 percent of the light striking its surface. It is just slightly larger than Ariel, a fact confirmed by the Voyager 2 data. Previous observations had led astronomers to believe that Umbriel was much smaller. This was because of the difficulty in

observing such a small, distant moon that reflects little light. Voyager 2 revealed a world covered in craters, many of which are tens of miles across. Unlike Ariel, Umbriel appears to have no bright, young ray craters, which means that its surface is older. There is no indication that it has been changed by internal activity. Umbriel's one bright feature, Wunda, is classified as a crater, although its nature is unknown.

**SOUTHERN HEMISPHERE**
Umbriel is almost uniformly covered by impact craters. Its one bright feature, the 81-mile (131-km) wide Wunda at the top of this image, is unfortunately virtually hidden from view.

## MAJOR MOON
# Titania

DISTANCE FROM URANUS 270,700 miles (435,910 km)

ORBITAL PERIOD 8.7 Earth days

DIAMETER 979 miles (1,578 km)

At a little less than half the size of the Moon, Titania is Uranus's largest moon. This rocky world has a gray, icy surface that is covered by impact craters. Icy material ejected when the craters formed reflects the light and stands out on Titania's surface. Large cracks are also visible and are an indication of an active interior. Some of these cut across the craters and appear to be the moon's most recent geological features. They were

**FULL DISK**
At top right is Titania's largest crater, Gertrude, which is 202 miles (326 km) across. Below it, the Messina Chasmata cuts across the moon.

probably caused by the expansion of water freezing under the crust. There are also smooth regions with few craters that may have been formed by the extrusion of ice and rock. Titania was discovered by the German-born astronomer William Herschel on January 11, 1787, using his homemade 20-ft (6-m) telescope in his backyard in England.

## MYTHS AND STORIES
# QUEEN OF THE FAIRIES

Titania and Oberon are the king and queen of the fairies in William Shakespeare's play *A Midsummer Night's Dream*. After a disagreement, Oberon squeezes flower juice into Titania's eyes as she sleeps so that on awakening she will fall in love with the next person she sees. Titania wakes and falls in love with Bottom, the weaver (seen here in a movie still from 1999), who has been given an ass's head by the impish sprite Puck.

## MAJOR MOON
# Oberon

DISTANCE FROM URANUS 362,370 miles (583,520 km)

ORBITAL PERIOD 13.46 Earth days

DIAMETER 946 miles (1,523 km)

Oberon was the first Uranian moon to be discovered—William Herschel observed it before spotting Titania. It has an icy surface pockmarked by ancient impact craters. There are several large craters surrounded by bright ejecta rays. Hamlet, which is just below center in the Voyager 2 image below, has a diameter of 184 miles (296 km). Its floor is partially covered by dark material, and it has a bright central peak. A 4-mile (6-km) high mountain protrudes from the lower left limb of the moon.

**ICY SURFACE**

## OUTER MOON
# Caliban

DISTANCE FROM URANUS 4.5 million miles (7.2 million km)

ORBITAL PERIOD 579.5 Earth days

DIAMETER 60 miles (96 km)

Caliban and another small moon, Sycorax, were discovered in September 1997. Both moons follow retrograde and highly inclined orbits. Sycorax is the more distant of the two, at 7.6 million miles (12.2 million km) from Uranus. They were the first of Uranus's irregular moons to be discovered and are believed to be icy asteroids that were captured soon after the planet's formation.

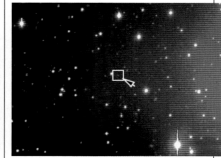

**CALIBAN DISCOVERED**
Caliban lies within the square outline in this image, which was taken using the Hale telescope at Mount Palomar, California. The glow on the right is from Uranus, and the bright dots are background stars.

# NEPTUNE

NEPTUNE IS THE SMALLEST and the coldest of the four gas giants, as well as the most distant from the Sun. It was discovered in 1846, and just one spacecraft, Voyager 2, has been to investigate this remote world. When the probe flew by in 1989, it provided the first close-up view of Neptune and revealed that it is the windiest planet in the solar system. Voyager 2 also found a set of rings encircling Neptune, as well as six new moons.

## ORBIT

Neptune takes 164.8 Earth years to orbit the Sun, which means that it has completed only one circuit since its discovery in 1846. The planet is tilted to its orbital plane by 28.3°, and as it progresses on its orbit, the north and south poles point sunward in turn. Neptune is about 30 times farther from the Sun than Earth, and at this distance the Sun is 900 times dimmer. Yet this remote, cold world is still affected by the Sun's heat and light and apparently undergoes seasonal change. Ground-based and Hubble Space Telescope observations show that the southern hemisphere has grown brighter since 1980, and this—as well as an observed increase in the amount, width, and brightness of banded cloud features—has been taken as an indication of seasonal change. However, a longer period of observations is needed to be sure that this seasonal model is correct. The change is slow and the seasons are long. The southern hemisphere is currently in the middle of summer. Once this is over, it is expected to move through fall, into a colder winter. Then, after 40 years of spring and a gradual increase in temperature and brightness, it will experience summer once more.

**SPIN AND ORBIT**
Neptune's orbit is elliptical, but less so than most planets. Only Venus has a more circular orbit. This means there is no marked difference between Neptune's aphelion and perihelion distances.

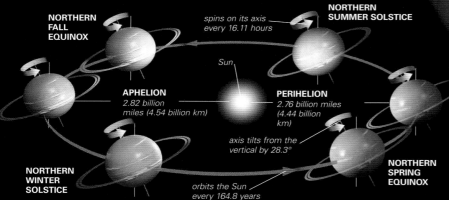

NORTHERN FALL EQUINOX

spins on its axis every 16.11 hours

NORTHERN SUMMER SOLSTICE

Sun

APHELION
2.82 billion miles (4.54 billion km)

PERIHELION
2.76 billion miles (4.44 billion km)

axis tilts from the vertical by 28.3°

NORTHERN WINTER SOLSTICE

orbits the Sun every 164.8 years

NORTHERN SPRING EQUINOX

## STRUCTURE

Neptune is very similar in size and structure to Uranus, and neither planet has a discernible solid surface. Like its inner neighbor, Neptune is too massive in relation to its size to be composed mainly of hydrogen. Only about 15 percent of the planet's mass is hydrogen. Its main ingredient is a mix of water, ammonia, and methane ices that makes up the planet's biggest layer. Neptune's magnetic field, which is tilted by 46.8° to the spin axis, originates in this layer. Above it lies the atmosphere. This is a shallow, hydrogen-rich layer that also contains helium and methane gas. Below the layer of water and ices, there is a small core of rock and possibly ice. The boundaries between the layers are not clearly defined. The planet rotates quickly on its axis, taking 16.11 hours for one spin, and as a result Neptune has an equatorial bulge. Its polar diameter is 527 miles (848 km) less than its equatorial diameter

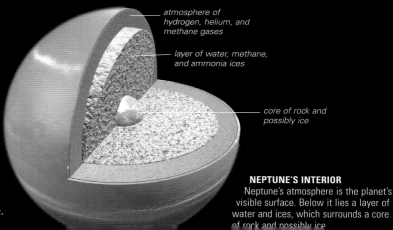

atmosphere of hydrogen, helium, and methane gases

layer of water, methane, and ammonia ices

core of rock and possibly ice

**NEPTUNE'S INTERIOR**
Neptune's atmosphere is the planet's visible surface. Below it lies a layer of water and ices, which surrounds a core of rock and possibly ice

**THE BLUE PLANET**
This image of Neptune, which was taken by Voyager 2 on August 19, 1989, reveals the planet's dynamic atmosphere. The Great Dark Spot, which is almost as big as Earth, lies in the center of the planet's disk. A little dark spot and, just above it, the fast-moving cloud feature named the Scooter, are visible on the west limb. A band of cloud stretches across the northern polar region.

## NEPTUNE PROFILE

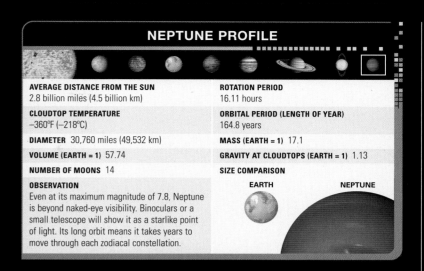

| AVERAGE DISTANCE FROM THE SUN | ROTATION PERIOD |
|---|---|
| 2.8 billion miles (4.5 billion km) | 16.11 hours |
| **CLOUDTOP TEMPERATURE** | **ORBITAL PERIOD (LENGTH OF YEAR)** |
| –360°F (–218°C) | 164.8 years |
| **DIAMETER** 30,760 miles (49,532 km) | **MASS (EARTH = 1)** 17.1 |
| **VOLUME (EARTH = 1)** 57.74 | **GRAVITY AT CLOUDTOPS (EARTH = 1)** 1.13 |
| **NUMBER OF MOONS** 14 | **SIZE COMPARISON** |

**OBSERVATION**
Even at its maximum magnitude of 7.8, Neptune is beyond naked-eye visibility. Binoculars or a small telescope will show it as a starlike point of light. Its long orbit means it takes years to move through each zodiacal constellation.

SIZE COMPARISON: EARTH — NEPTUNE

# ATMOSPHERE AND WEATHER

Neptune is a perplexing place. For a planet so far from the Sun, it has a surprisingly dynamic atmosphere that exhibits colossal storms and super-fast winds. The heat Neptune receives from the Sun is not enough to drive its weather. The atmosphere may be warmed from below by Neptune's internal heat source, and this is the trigger for larger-scale atmospheric changes. The white bands that encircle the planet are cloud cover, produced when the heated atmosphere rises and then condenses, forming clouds. The winds are most ferocious in the equatorial regions, where they blow westward and reach a staggering 1,340 mph (2,160 kph). Gigantic, dark, stormlike features accompanied by bright, high-altitude clouds appear and then disappear. One, the Great Dark Spot, was seen by Voyager 2 in 1989. When the Hubble Space Telescope looked for the storm in 1996, it had disappeared.

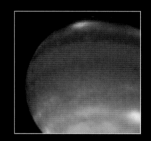

**CLOUDS OVER NEPTUNE**
Neptune's atmosphere lies in bands, which are parallel to the equator. The bright patches are high-altitude clouds, floating above the blue methane layer.

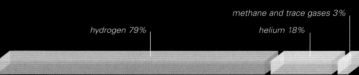

hydrogen 79%    helium 18%    methane and trace gases 3%

**COMPOSITION OF ATMOSPHERE**
Neptune's atmosphere is made mostly of hydrogen. But it is the methane that gives the planet its deep blue color, absorbing red light and reflecting blue.

# RINGS AND MOONS

The first indication that Neptune has a ring system came in the 1980s, when stars were seen to blink on and off near the planet's disk. Intriguingly, Neptune seemed to have ring arcs. The mystery was solved when Voyager 2 discovered that Neptune has a ring system with an outer ring so thinly populated that it does not dim starlight but contains three dense regions that do. Neptune has five sparse yet complete rings; moving in from the outer Adams ring, they are Arago, Lassell, Le Verrier, and Galle. A sixth, unnamed partial ring lies within Adams. The rings are made of tiny pieces of unknown composition, which together would make a body just a few miles across. The material is believed to have come from nearby moons. Five of Neptune's 14 moons are within the ring system. It is one of the moons, Galatea, that prevents the arc material from spreading uniformly around the Adams ring. Only one of the 14, Triton, is of notable size. Triton and Nereid were discovered before the days of space probes. Six small moons have been discovered since 2002, and more will probably be found.

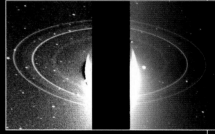

**THE RINGS OF NEPTUNE**
Two Voyager 2 images placed together reveal Neptune's ring system. The two bright rings are Adams and Le Verrier. The faint Galle ring is innermost, and the diffuse band, Lassell, is visible between the two bright ones.

**NEPTUNE'S MOONS**

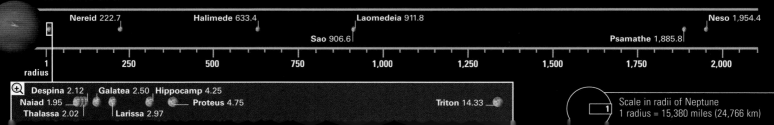

Nereid 222.7    Halimede 633.4    Laomedeia 911.8    Neso 1,954.4
Sao 906.6    Psamathe 1,885.8

1 radius    250    500    750    1,000    1,250    1,500    1,750    2,000

Despina 2.12    Galatea 2.50    Hippocamp 4.25
Naiad 1.95    Proteus 4.75    Triton 14.33
Thalassa 2.02    Larissa 2.97

Scale in radii of Neptune
1 radius = 15,380 miles (24,766 km)

# NEPTUNE'S MOONS

Neptune has only one major moon: Triton. All its other satellites are small and can be described as inner or outer moons depending on whether they are closer to or farther from Neptune than Triton. Six of the seven inner moons were discovered by analysis of Voyager 2 data in 1989. The moons are named after characters associated with the Roman god of the sea, Neptune, or his Greek counterpart, Poseidon.

**NEPTUNE AND TRITON**
This image of the crescent moon of Triton below the crescent of Neptune was captured by Voyager 2 as it flew away from the planet.

---

## INNER MOON
## Larissa

| | |
|---|---|
| **DISTANCE FROM NEPTUNE** | 45,617 miles (73,458 km) |
| **ORBITAL PERIOD** | 0.55 Earth days |
| **LENGTH** | 134 miles (216 km) |

Larissa is the fifth moon from Neptune, lying outside the ring system. The moon was first spotted from Earth in 1981, but astronomers eventually decided that it was a ring arc circling Neptune. In late July 1989, a Voyager 2 team of astronomers confirmed that it is, in fact, an irregularly shaped, cratered moon. It was named after a lover of Poseidon.

**IRREGULARLY SHAPED MOON**

---

## INNER MOON
## Proteus

| | |
|---|---|
| **DISTANCE FROM NEPTUNE** | 73,059 miles (117,647 km) |
| **ORBITAL PERIOD** | 1.12 Earth days |
| **LENGTH** | 273 miles (440 km) |

The most distant of the inner moons from Neptune, Proteus is also the largest of the seven—with the exception of S/2004 N1, their size increases with distance. It has an almost equatorial orbit, moving around Neptune in less than 27 hours. Its visible surface has extensive cratering, but just one major feature stands out: a large, almost circular depression 158 miles (255 km) across, with a rugged floor.
Proteus was the first of the inner moons to be found by Voyager 2 scientists. It was detected in mid-June 1989, within 2 months of the probe's closest approach to Neptune, enabling the observation sequence to be changed. The images later recorded by Voyager 2 revealed a gray, irregular, but roughly spheroid moon that reflects 6 percent of the sunlight hitting it. The moon was later named Proteus after a Greek sea god.

rim of circular depression

cratered surface

**TWO VIEWS**
The first image of Proteus (far right) shows the moon half-lit. The second was taken closer in. (The black dots are a processing artifact.)

---

## OUTER MOON
## Nereid

| | |
|---|---|
| **DISTANCE FROM NEPTUNE** | 3.4 million miles (5.5 million km) |
| **ORBITAL PERIOD** | 360.1 Earth days |
| **DIAMETER** | 211 miles (340 km) |

Nereid was discovered on May 1, 1949, by the Dutch-born astronomer Gerard Kuiper while working at the McDonald Observatory in Texas. Little is still known about this moon. Voyager 2 flew by at a distance of 2.9 million miles (4.7 million km) in 1989 and could take only a low-resolution image. Nereid's outstanding characteristic is its highly eccentric and inclined orbit, which takes the moon out as far as about 5.9 million miles (9.5 million km) from Neptune and to within just 507,500 miles (817,200 km) at its closest approach.

**BEST VIEW**
Voyager 2 revealed Nereid to be a dark moon, reflecting only 14 percent of the sunlight it receives.

---

## OUTER MOON
## Halimede

| | |
|---|---|
| **DISTANCE FROM NEPTUNE** | 9.7 million miles (15.7 million km) |
| **ORBITAL PERIOD** | 1,874.8 Earth days |
| **DIAMETER** | 30 miles (48 km) |

Halimede was discovered by an international team of astronomers who were carrying out a systematic search for new Neptunian moons. Their task was not easy because moons as small and as distant as Halimede are extremely difficult to detect. Halimede follows a highly inclined and elliptical orbit. The origin of the irregular outer moons, which now number five, is unknown. More may be found, since these moons could be the result of an ancient collision between a former moon and a passing body such as a Kuiper Belt object.

### EXPLORING SPACE
## LOOKING FOR NEW MOONS

A team of astronomers announced the discovery of three new moons, including Halimede, on January 13, 2003. They had taken multiple images of the sky around Neptune from two sites in Hawaii and Chile. The images were combined to boost the signal of faint objects. The new moons showed up as points of light against the background of stars, which appeared as streaks of light.

**MAUNA KEA OBSERVATORY, HAWAII**
The Canada-France-Hawaii Telescope used in the search is at Mauna Kea. The other site was the Cerro Tololo Inter-American Observatory in Chile.

---

## MAJOR MOON
## Triton

| | |
|---|---|
| **DISTANCE FROM NEPTUNE** | 220,306 miles (354,760 km) |
| **ORBITAL PERIOD** | 5.88 Earth days |
| **DIAMETER** | 1,681 miles (2,707 km) |

Triton was the first of Neptune's moons to be discovered, just 17 days after the discovery of the planet was announced. William Lassell (see panel, right) used the coordinates published in *The Times* to locate Neptune in early October 1846. On October 10, he found its biggest moon using the 24-in (61-cm) reflecting telescope at his observatory in Liverpool, England. The moon was named Triton after the sea-god son of Poseidon. The Voyager 2 flyby nearly 143 years later revealed most of what is now known about this icy world.

**SMOOTH PLAIN**
The 185-mile (300-km) wide Ruach Planitia is in the cantaloupe terrain. It may be an old impact crater that has been filled in.

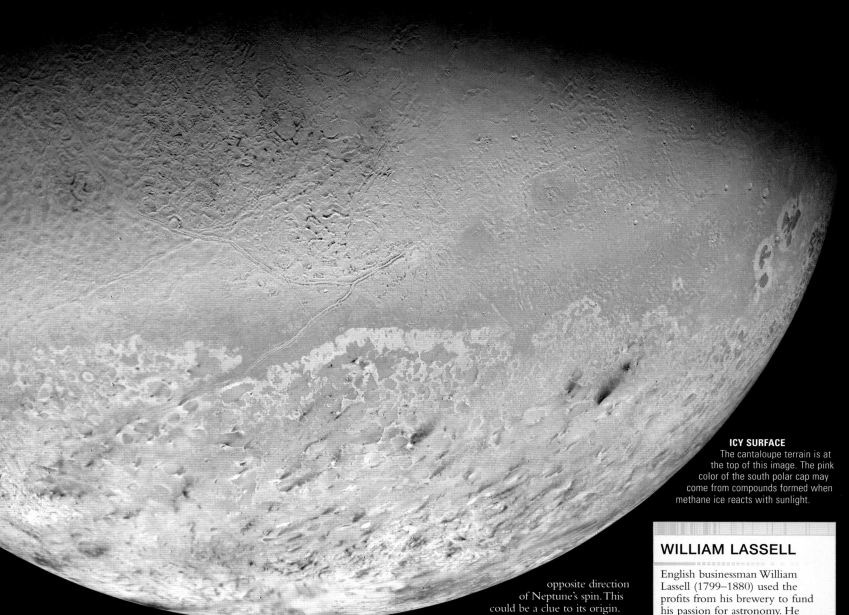

### ICY SURFACE
The cantaloupe terrain is at the top of this image. The pink color of the south polar cap may come from compounds formed when methane ice reacts with sunlight.

## WILLIAM LASSELL

English businessman William Lassell (1799–1880) used the profits from his brewery to fund his passion for astronomy. He designed and built large reflecting telescopes that were the finest of the day. He made his observations first from his home in Liverpool, England, then from the island of Malta. In addition to Triton, Lassell discovered the Uranian moons Ariel and Umbriel and Saturn's moon Hyperion.

Triton is by far the largest of Neptune's moons and is bigger than Pluto. It follows a circular orbit and exhibits synchronous rotation, so the same side always faces Neptune. Peculiarly for such a large moon, Triton is in retrograde motion, traveling in the opposite direction of Neptune's spin. This could be a clue to its origin. Triton may have formed elsewhere in the solar system and been captured by Neptune. The moon's mix of two parts rock to one part ice is differentiated into a rocky core, a possibly liquid mantle, and an ice crust. Its geologically young, icy surface has few craters and displays a range of features. An area of linear grooves, ridges, and circular depressions is nicknamed the cantaloupe after its resemblance to a melon's skin. Dark patches mark the south polar region. These form when solar heat turns subsurface nitrogen ice into gas. This erupts through surface vents in geyserlike plumes, which carry dark, possibly carbonaceous dust into the atmosphere before depositing it on Triton's surface.

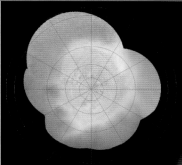

### POLAR PROJECTION
Triton's south polar region is seen head-on in this image. A band of bluish material extends out from the central polar cap into the equatorial region. It is probably fresh nitrogen frost or snow redistributed by the wind.

mottled crust

### SOUTHERN HEMISPHERE
Three Voyager 2 images were combined to produce this almost full-disk image of Triton. From a distance, the southern-hemisphere terrain appears mottled.

# PLUTO AND ITS MOONS

DISCOVERED BY US astronomer Clyde Tombaugh in 1930 (see p.209), Pluto was once classified as a planet in its own right. This outrider of the solar system was reclassified as a dwarf planet in 2006 and is acknowledged as the first Kuiper Belt Object (KBO) to be discovered. It orbits closer to the Sun than other large KBOs but is probably not exceptional other than having been seen at close quarters by the New Horizons mission (see p.211) in 2015.

## ORBIT

Pluto's orbit is inclined and eccentric. Its 248-year orbit ranges between about 2.7 billion miles (4.4 billion km) and 4.6 billion miles (7.4 billion km) from the Sun, meaning that Pluto sometimes lies closer in than Neptune (most recently between 1979 and 1999). However, the pronounced tilt of Pluto's orbit (at an angle of 17.1° to the ecliptic), combined with the fact that it completes precisely two orbits in the time it takes Neptune to complete exactly three orbits, makes close encounters between the two impossible.

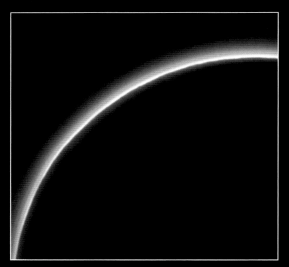

**BLUE SKY**
Looking back toward the Sun 3.5 hours after it had passed Pluto, from a range of about 120,000 miles (200,000 km) beyond, the New Horizons probe could see sunlight scattered and made blue by the photochemical haze in Pluto's atmosphere.

## ATMOSPHERE

The existence of Pluto's atmosphere was confirmed in 1988, when the planet occulted a star and the starlight was cut off much more slowly than would have happened with an airless body. It is mostly nitrogen mixed with less than 1 percent of methane and a smaller proportion of carbon monoxide. The surface pressure of the atmosphere was found to be about 100,000 times less than the Earth's atmospheric pressure by New Horizons in 2015, and will probably decrease as Pluto's orbit carries it farther from the Sun, allowing more of its atmosphere to freeze onto the ground. The surface temperature is -351°F to -387°F (-213°C to -233°C), but the atmospheric temperature increases with height to a maximum of about -261°F (-163°C) at a height of about 19 miles (30 km) because of the absorption of sunlight by methane. This process leads to methane molecules linking together into hydrocarbon chains constituting smog particles, which slowly settle to the ground and make Pluto's atmosphere hazy.

## STRUCTURE

Pluto is similar to a slightly smaller and less massive version of Neptune's largest moon, Triton (see p.207). Pluto's crust is mostly water ice, which is very strong and rigid under Pluto's low-temperature surface conditions. Pluto's density indicates that it must have a substantial rocky core. Radioactive heat generation within the core may be sufficient to melt the overlying ice to maintain an internal ocean of liquid water. The surface is covered mostly by nitrogen ice, with traces of frozen methane and carbon monoxide, too. Dark red discoloration of parts of the surface is probably caused by hydrocarbon haze particles that have settled to the ground.

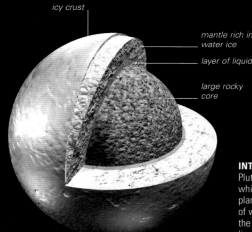

*icy crust*
*mantle rich in water ice*
*layer of liquid water*
*large rocky core*

**INTERIOR OF PLUTO**
Pluto is thought to consist of a rocky core, which makes up about 70 percent of the planet's diameter, surrounded by a mantle of water ice and a thin, icy crust. Heat from the core may help sustain a thin layer of liquid water between the core and mantle.

## CLOSE-UP OF PLUTO

This 200-mile (300-km) wide view looks across Pluto's Tenzing Montes. These are 2–4-mile (3–6-km) high blocks of ice near the edge of the much flatter nitrogen ice that occupies Sputnik Planitia.

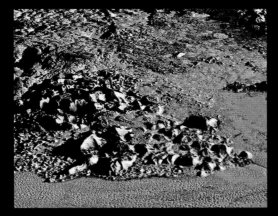

# MOONS

The presence of Pluto's largest moon, Charon, was detected in 1978. Since then, four smaller moons have been discovered orbiting Pluto. Charon is by far the largest of Pluto's moons and is about 750 miles (1,200 km) across. Its mass is about 15 percent that of Pluto. The size and mass of Charon makes it the largest known satellite in relation to its parent body. Charon and Pluto are in a synchronous orbit, meaning they keep the same face toward each other at all times. Both spin on their axes every 6.38 Earth days, and Charon orbits Pluto once in the same period. Because they are so similar in mass, their common center of mass (called their barycenter) about which they each orbit does not lie inside Pluto, but in open space between the two globes.

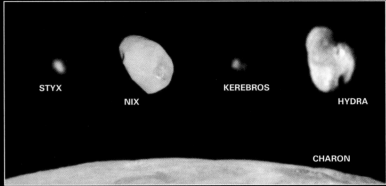

STYX

NIX

KEREBROS

HYDRA

CHARON

## PLUTO AND ITS MOONS

Pluto has five known moons—Charon, Styx, Nix, Kerebros, and Hydra—which are shown here to scale. Charon is by far the largest, with a diameter of about 750 miles (1,200 km). Hydra is about 70 miles (115 km) across, and Nix is about 55 miles (90 km) across. Kerebros and Styx, which were found in 2011 and 2012, are much smaller than the other three moons.

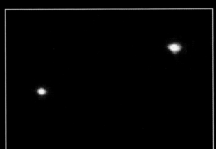

## GROUND-BASED IMAGE

This image of the Pluto–Charon system was taken by one of the 26.9-ft (8.2-m) telescopes at Paranal Observatory, Chile. Charon was discovered in 1978 by James Christy of the US Naval Observatory, Arizona, who noticed that Pluto's image became elongated periodically. He realized that this was because Pluto has a moon. Charon orbits Pluto at a distance of 10,890 miles (17,530 km).

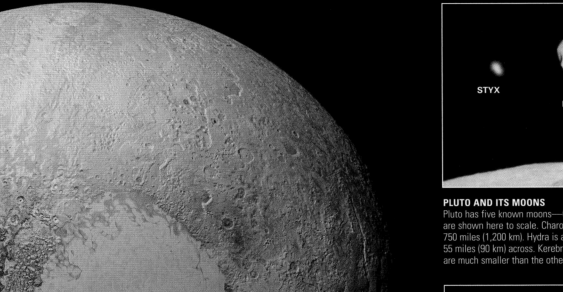

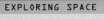

## PLUTO AND CHARON

Pluto (below) and Charon (above) are shown here to scale in a composite of New Horizons images. Charon has no atmosphere, and its dramatically fractured surface is dominated by water ice, apart from the red-stained area near the north pole that might be contamination from Pluto. Pluto itself has a fascinatingly varied surface. The bright patch near the middle of the globe, named Sputnik Planitia, is a low-lying region occupied by nitrogen ice that is weak enough to flow and convect even in Pluto's extreme cold. It may have escaped from Pluto's interior to occupy the scar left by a major impact. Elsewhere, the more ancient icy crust is marked by smaller impact craters.

EXPLORING SPACE

## SEARCHING FOR A PLANET

Pluto was discovered as the result of a deliberate hunt for a "Planet X," which in the early 20th century was thought to affect the orbits of Uranus and Neptune. American astronomer Clyde Tombaugh began his attempt to find this planet at the Lowell Observatory, Arizona, in 1929. His method involved photographing the same area of sky a few days apart and comparing the two images to look for any objects that had moved. On January 23, 1930, Tombaugh took a long exposure of the Delta Geminorum region. On January 29, he imaged the area again, and one "star" in his plates (indicated by the red arrow) had moved. He had discovered Pluto. It later became clear that Pluto was too small to be Planet X, and astronomers today realize there is no need for a Planet X in our models of the solar system.

# THE KUIPER BELT AND THE OORT CLOUD

BEYOND THE ORBITS of the giant planets, the solar system is surrounded by billions of small worlds made mostly of various kinds of ice. The inner ones orbit within the Classical Kuiper Belt and are known as Kuiper Belt Objects (KBOs), while others beyond that occur in the Scattered Disk. Some of these icy bodies are the size of small planets, and one—Pluto—was originally classified as a planet in its own right. Beyond the Kuiper Belt lies an enormous halo of smaller icy bodies known as the Oort Cloud. Believed to contain trillions of objects, the Oort Cloud is the source of many of the comets that visit the inner solar system.

Scattered Disk    Kuiper Cliff    Classical Kuiper Belt

**LOCATION OF THE KUIPER BELT**
The Kuiper Belt extends out from the orbit of Neptune to about 9.3 million miles (15 billion km) from the Sun. It has two subregions: the Classical Kuiper Belt, extending out to about 4.7 billion miles (7.5 billion km), and the Scattered Disk, stretching from the Classical Belt to the edge of the entire Kuiper Belt.

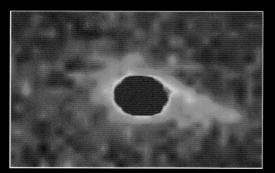

**CHIRON**
Discovered in 1977, Chiron is the prototype of a group of icy bodies following orbits around those of Saturn and Uranus. Known as Centaurs, they are thought to be Scattered Disk Objects that have been pulled inward by interactions with Neptune's gravity. They may go on to become short-period comets.

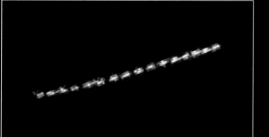

**QUAOAR**
The multiple exposures in this image show Quaoar moving across the sky. Discovered in 2002, it has an estimated diameter of around 730 miles (1,170 km), about half that of Pluto, but Quaoar is much denser than Pluto, indicating that it contains more rock than ice.

## DWARF PLANETS

On its discovery in 1930, Pluto was designated as the ninth planet from the Sun, though at the time there was no formal definition of the term "planet." The subsequent discovery of several Pluto-sized objects in the Kuiper Belt made it clear that Pluto was merely the first Kuiper Belt Object to be discovered. The realization that the Kuiper Belt is populated by numerous bodies, ranging from Pluto-sized objects down to smaller sizes, meant that any definition of a planet could not be based on size alone because there is no natural break in the size distribution. In 2006, the International Astronomical Union (IAU) made a formal definition of the term "planet," stipulating three essential criteria: it must orbit the Sun, it must be massive enough for its own gravity to have pulled it into a round shape, and it must have cleared the neighborhood around its orbit. Pluto fails the third criterion because it shares its orbital space with several objects of comparable size and also crosses the orbit of the vastly more massive Neptune. Pluto was therefore no longer officially a planet. Although this decision was controversial, it is clear that if Pluto had been discovered in 2010 instead of 1930, it would never have been called a planet in the first place. However, the IAU coined a new term, dwarf planet, to denote a body that fails only the third criterion for planet definition. Thus Pluto, Eris, Haumea, Makemake, and probably a few other large KBOs are dwarf planets, as is the largest asteroid, Ceres (see p.175).

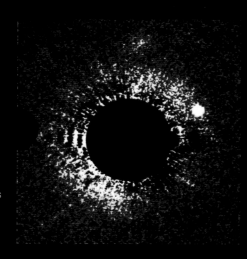

**EXTRASOLAR DEBRIS DISK**
Several Kuiper Belt-like structures have been found around other stars that are thought to be debris left over from the processes of planet formation. The disk around the billion-year-old HD 53143 (shown right), a cool star about 60 light-years from Earth, stretches to around 10.2 billion miles (16.5 billion km) from its central star— roughly the same diameter as our Kuiper Belt and Scattered Disk.

### GERARD KUIPER

Gerard Kuiper (1905–1973) was one of the most influential planetary scientists of the 20th century. After studying astronomy at the University of Leiden in the Netherlands, he moved to the United States in 1933. He founded the Lunar and Planetary Institute at Tucson, Arizona, in 1960, and later worked on early planetary probes. He discovered the moons Miranda and Nereid and was also the first to identify carbon dioxide in the atmosphere of Mars. In 1951, he proposed the existence of what we now call the Kuiper Belt, although he believed that its existence had been a short-lived phase of the early solar system.

*Sun*    *Uranus's orbit*    *Neptune's orbit*    *Pluto's orbit*

# THE KUIPER BELT AND ITS CONSTITUENTS

The Kuiper Belt is a broad ring of objects that begins around the orbit of Neptune and extends out to roughly 9.3 billion miles (15 billion km) from the Sun. The possibility of such a belt was initially put forward in 1930, soon after the discovery of Pluto (see pp.208–209). The first theoretical models for how such a belt could have formed were proposed in 1943 by British astronomer Kenneth Edgeworth and in 1951 by Gerard Kuiper (see opposite). For this reason, the belt is sometimes known as the Edgeworth–Kuiper Belt, or EKB. However, the belt remained purely theoretical until 1992, when astronomers identified a small body with a diameter of about 100 miles (160 km), now known as 1992 QB1. This was the first confirmation that there were other objects in addition to Pluto in the space beyond Neptune, and since then, about a thousand more such objects have been discovered. The Kuiper Belt as a whole can be split into an inner zone called the Classical Kuiper Belt and an outer zone called the Scattered Disk. The Classical Kuiper Belt extends out to about 4.7 billion miles (7.5 billion km) from the Sun and is relatively densely populated with objects that have roughly circular orbits. The drop in density at its outer edge is known as the Kuiper Cliff. Beyond this is the Scattered Disk, which is relatively sparsely populated with objects that have more eccentric and tilted orbits.

## NEW HORIZONS

This NASA mission was the first (and so far only) spacecraft to visit any object more remote than Neptune. It was launched in January 2006, and flew past Pluto and its moons in July 2015. Its subsequent trajectory was adjusted to pass close to a newly discovered small Kuiper Belt Object temporally designated as 2014MU 69, which later received the permanent name 486958 Arrokoth. New Horizons made a successful flyby of this on January 1, 2019, passing within 2,200 miles (3,500 km) at a speed of 32,000 mph (51,500 kph). The spacecraft remains healthy, and its radioactive power source (consisting of plutonium-238) should last well into the 2030s. The search is on to locate an even more distant KBO that is close enough to New Horizons' trajectory that there is enough fuel remaining to adjust the spacecraft's course to achieve another close flyby.

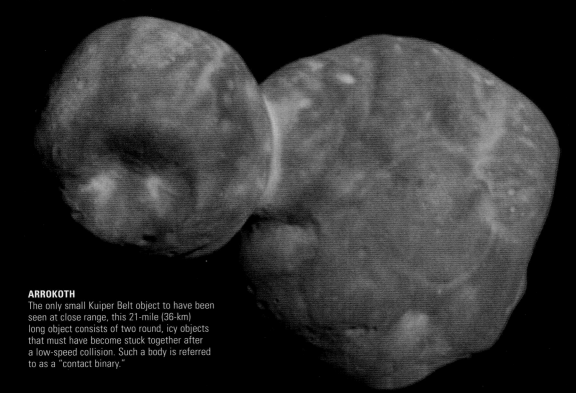

**ARROKOTH**
The only small Kuiper Belt object to have been seen at close range, this 21-mile (36-km) long object consists of two round, icy objects that must have become stuck together after a low-speed collision. Such a body is referred to as a "contact binary."

# CLASSICAL KUIPER BELT OBJECTS

The objects of the Classical Kuiper Belt, often called KBOs, form several distinct groups that have different compositions and probably originate from different parts of the solar system. One distinction is between "cold" and "hot" KBOs. Despite their name, these groups are identified not by differences in their surface temperatures but by the shape and tilt of their orbits. Cold KBOs have relatively circular orbits with shallow tilts. They also have reddish surfaces, indicating the presence of methane ice. Hot KBOs, such as Makemake, follow more eccentric and tilted orbits and have bluish-white surfaces. Cold KBOs are thought to have originated in roughly the same region where they currently orbit, while hot KBOs probably originated closer to the Sun than they are now. A third group, known as Plutinos, occupy stable orbits in a 2:3 resonance with Neptune. (That is, they orbit the Sun twice for every three orbits of Neptune.) This configuration protects them from Neptune's gravitational influence and ensures their orbits remain stable. However, the Plutinos, which include Haumea and Pluto itself, are not considered to be Classical KBOs by some astronomers.

**MAKEMAKE**
Discovered in 2005, Makemake has an estimated diameter of 845–920 miles (1,360–1,480 km), about two-thirds the size of Pluto. With a temperature of only about -405°F (-243°C), Makemake's surface is covered with methane, ethane, and possibly nitrogen ices.

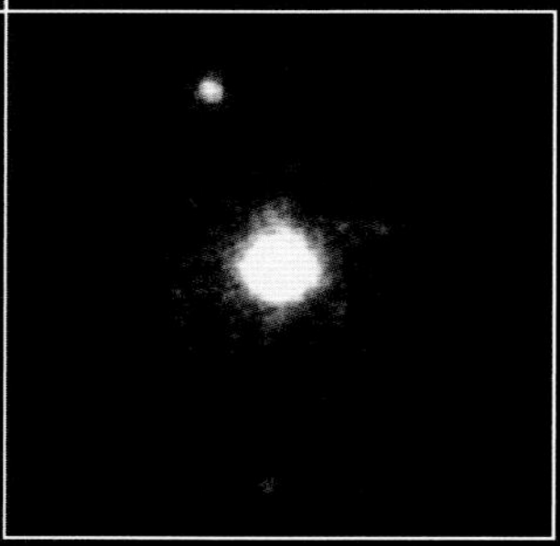

**HAUMEA**
With a long axis of about 1,218 miles (1,960 km) and a short axis only half this length, Haumea is unusually elongated for a KBO. It also has a very short rotational period, spinning on its axis once every 4 hours. It was discovered in 2004.

# SCATTERED DISK OBJECTS

Beyond the Classical Kuiper Belt is another distinct group of objects, known as Scattered Disk Objects (SDOs). These SDOs move around the Sun in eccentric, often highly tilted orbits that sometimes cross the Classical Belt but also venture much farther out, to 9.3 billion miles (15 billion km) from the Sun or more. They are thought to have originated closer to the Sun and been ejected outward by the gravitational influence of the outer planets. SDOs are still affected by Neptune's gravity, and the Scattered Disk is thought to be the source of Centaur objects, such as Chiron, as well as some comets. The largest known SDO is Eris, discovered in 2005. Eris has a moon of its own, Dysnomia, whose orbit demonstrates that Eris has about 28 percent more mass than Pluto, despite being probably slightly smaller. Astronomers faced a choice of either promoting Eris to an official 10th planet of the solar system or demoting Pluto, since it was now revealed to be not outstanding within the KBO in terms of either mass or size. They chose the latter option and introduced a new category—dwarf planet—for objects that have planetlike features but lack sufficient gravity to clear their neighboring region of other objects (see p.210).

**OORT CLOUD**
The Oort Cloud is thought to consist of two distinct regions: a spherical, sparsely populated outer cloud and a doughnut-shaped inner cloud. Comets in the more densely populated inner cloud are frequently ejected into the outer cloud and help keep it replenished.

Sun

Kuiper Belt

typical elongated orbit of long-period comet

few comets lie in the region between the inner and outer Oort Cloud

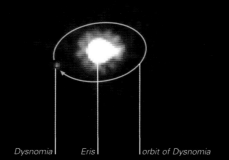

Dysnomia    Eris    orbit of Dysnomia

**ERIS AND DYSNOMIA**
On September 10, 2005, astronomers using the 32.8-ft (10-m) Keck Telescope in Hawaii discovered that Eris has a moon (seen to the left of Eris), now named Dysnomia. This moon orbits Eris about once every 16 days. Together, Eris and Dysnomia move around the Sun in an eccentric orbit that lasts 557 years.

# THE OORT CLOUD

Surrounding the solar system beyond the Kuiper Belt lies an enormous cloud of long-period comets known as the Oort Cloud. Its outer reaches extend to almost a light-year from the Sun, and it is thought to contain trillions of objects with a total mass of roughly five Earths. The Oort Cloud is impossible to observe directly, although there is strong evidence for it from the orbits of comets that pass through the inner solar system. Its existence was first suggested by Estonian astronomer Ernst Öpik in 1932, but it was also proposed independently by Jan Oort (see panel, right) in 1950. The comets in the Oort Cloud are thought to have originated much closer to the Sun, in the region where the giant planets now orbit. However, early in the solar system's history, as the giant planets migrated toward their current positions, close encounters with these planets pushed enormous numbers of the comets into highly elliptical orbits. In the outer solar system, these comets were only weakly bound by the Sun's gravity, so tidal forces from other stars and the Milky Way itself were able to act on them, gradually "circularizing" their orbits. Today, similar tidal effects occasionally knock comets out of the Oort Cloud toward the Sun. However, according to other theories, some Oort Cloud comets might have begun their lives in orbit around other stars and were later captured by the Sun's gravity.

comet's orbit takes it to the edge of the Oort Cloud

comet orbiting close to the plane of the solar system

inner cloud

outer cloud

## JAN HENDRIK OORT

Jan Oort (1900–1992) was born in Franeker, the Netherlands, and spent most of his career at the University of Leiden, also in the Netherlands. Oort is mostly remembered for his idea that the solar system is surrounded by the vast symmetrical cloud of comets that was named after him. He is also famous as a pioneer of radio astronomy, for discovering the rotation of the Milky Way and estimating its distance and the direction of its center from Earth, and for discovering evidence that the universe contains "missing mass" (now known as dark matter).

**LONG-PERIOD COMET**
Long-period comets, such as Hyakutake (left), typically approach the inner solar system from all directions and at high speeds, indicating that they come from a spherical region that surrounds the Sun at a vast distance—the Oort Cloud.

**DISTANT OBJECT**
When discovered on November 14, 2003, Sedna was nearly 90 times farther from the Sun than Earth, making it the most distant solar system object then observed.

# SEDNA

In 2003, astronomers searching for objects in the region of space beyond Neptune discovered a world about 8.4 billion miles (13.5 billion km) from the Sun but moving in an eccentric orbit that takes it out to a maximum distance of 87.1 billion miles (140.2 billion km). This was the most distant object yet found in the solar system. With an estimated surface temperature of –436°F (–260°C), it was also the coldest solar system body, so it was named after the Inuit goddess of the Arctic Ocean, Sedna. Some astronomers believe that Sedna could offer our first glimpse of an object from the inner Oort Cloud. However, its orbit is unusual even for an object from the innermost part of the Cloud, suggesting that it must have been disrupted in the past. Sedna's orbit is too remote for it to have been influenced by Neptune, but other possible explanations include disruption by the gravity of other stars or even by the influence of a large, as-yet-undiscovered planet far beyond Neptune.

Sedna takes approximately 11,400 Earth years to orbit the Sun

**SEDNA'S ORBIT**
Sedna takes about 11,400 years to complete one orbit around the Sun. It will be at perihelion in 2076, about 7.1 billion miles (11.4 billion km) from the Sun, but spends most of its orbit between the Scattered Disk and the inner Oort Cloud.

Pluto

Sun

Kuiper Belt

Sedna's closest approach to the Sun (perihelion) will be in 2076

Sedna will reach aphelion (farthest point from the Sun) in about 7776

**RED BODY**
Sedna's diameter is estimated at 750–1,000 miles (1,200–1,600 km), and it has a dark red surface, as shown in this artist's impression.

# COMETS

COMETS PRODUCE A STRIKING celestial spectacle when they enter the inner solar system. Their small nuclei become surrounded by a bright cloud, or coma, of dust and gas about 60,000 miles (100,000 km) across. Large comets that get close to the Sun also produce long, glowing tails that can extend many tens of millions of miles into space and are bright enough to be seen in Earth's sky.

## ORBITS

Cometary orbits divide into two classes. Short-period comets orbit the Sun in the same direction as the planets. Most have orbital periods of about 7 years and get no farther from the Sun than Jupiter. Short-period comets were captured into the inner solar system by the gravitational influence of Jupiter. If they remain in these small orbits, they will decay quickly. Some, however, will be ejected by Jupiter into much larger orbits and then possibly recaptured. Intermediate- and long-period comets have orbital periods greater than 20 years (see pp.216–217). Their orbital planes are inclined at random to the plane of the solar system. Many of these comets travel huge distances into the interstellar regions. Most of the recorded comets get close to the Sun, where they develop comae and tails and can be easily discovered. There are vast numbers of comets on more distant orbits that are too faint to be found.

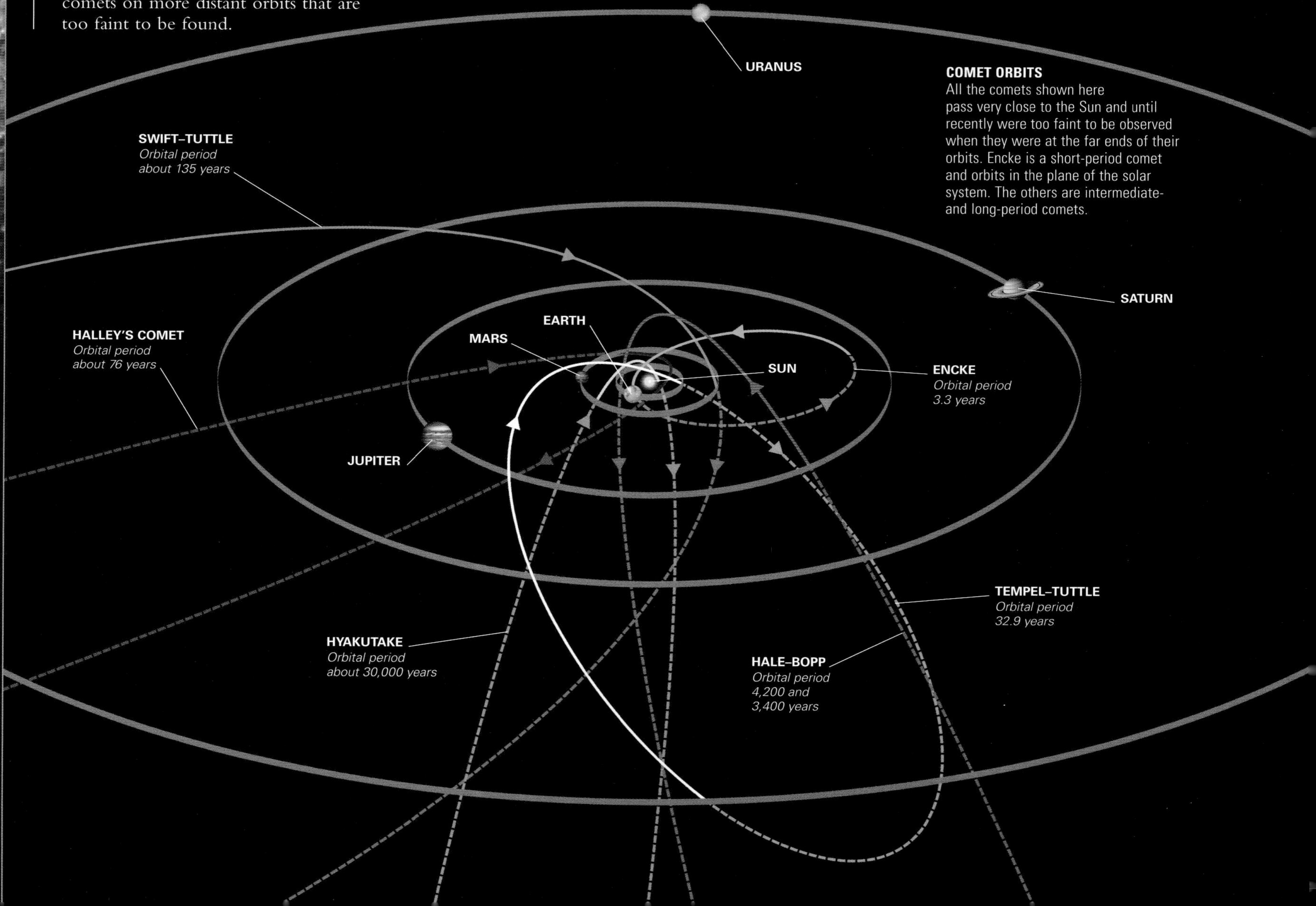

**SPECTACULAR SIGHT**
Most comets are only discovered when they are bright enough to glow in Earth's sky. Comet Hale–Bopp was discovered in late July 1997 and could be seen for several weeks afterward. It will return to Earth's sky again around 5400 CE.

URANUS

**COMET ORBITS**
All the comets shown here pass very close to the Sun and until recently were too faint to be observed when they were at the far ends of their orbits. Encke is a short-period comet and orbits in the plane of the solar system. The others are intermediate- and long-period comets.

**SWIFT–TUTTLE**
*Orbital period about 135 years*

SATURN

**HALLEY'S COMET**
*Orbital period about 76 years*

EARTH

MARS

SUN

**ENCKE**
*Orbital period 3.3 years*

JUPITER

**TEMPEL–TUTTLE**
*Orbital period 32.9 years*

**HYAKUTAKE**
*Orbital period about 30,000 years*

**HALE–BOPP**
*Orbital period 4,200 and 3,400 years*

# STRUCTURES

The fount of all cometary activity is a low-density, fragile, irregularly shaped, small nucleus that resembles a "dirty snowball." The dirt is silicate rock in the form of small dust particles. The snow is mainly composed of water, but about 1 in 20 molecules are more exotic, being carbon dioxide, carbon monoxide, methane, ammonia, or more complex organic compounds. The nucleus is covered by a thin, dusty layer, which is composed of cometary material that has lost snow from between its cracks and crevices. The snow is converted directly from the solid into the gaseous state by the high level of solar radiation the comet receives when it is close to the Sun.

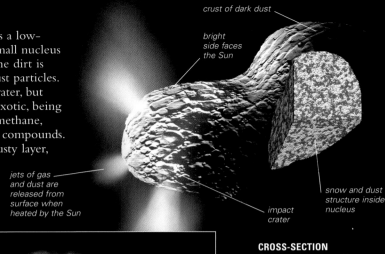

crust of dark dust

bright side faces the Sun

jets of gas and dust are released from surface when heated by the Sun

impact crater

snow and dust structure inside nucleus

## NUCLEUS
The central part of Comet Borrelly's elongated nucleus has a smooth terrain, but the more "mottled" regions consist of steep-sided hills that are separated by pits and troughs.

## CROSS-SECTION
The nucleus has a uniform structure consisting of many smaller "dirty snowballs." The surface dust layer is only a few inches thick and appears dark because it reflects little light. The strength of the whole structure is negligible. Not only do tidal forces pull comets apart, but many simply fragment at random.

## FRED WHIPPLE
Fred Whipple (1906–2004) was an astronomy professor at Harvard University and the director of the Smithsonian Astrophysical Observatory from 1955 to 1973. In 1951, he introduced the "dirty snowball" model of the cometary nucleus, in which the snowball spins. As the Sun heated one side, its heat was slowly transmitted down to the underlying snows, which eventually turn straight to gas. This resulted in a jet force along the cometary orbit which either accelerated or decelerated the nucleus depending on the direction of its spin.

# LIFE CYCLES

A comet spends the vast majority of its life in a dormant, deep-freeze state. Activity is triggered by an increase in temperature. When the comet gets closer to the Sun than the outer part of the Main Belt (see p.170), frozen carbon dioxide and carbon monoxide in the nucleus start to sublime—that is, they pass directly from the solid to the gaseous state. Once the comet is inside the orbit of Mars, it is hot enough for water to join in the activity. The nucleus quickly surrounds itself with an expanding spherical cloud of gas and dust, called a coma. The coma is at its maximum size when the comet is closest to the Sun. A comet that passes through the inner solar system will lose the equivalent of a 6-ft (2-m) thick layer from its surface. The comet moving away from the Sun is thus smaller than it was on its approach. Mass is lost every time a comet passes perihelion. Borrelly, for example, orbits the Sun about every 7 years. If it stays on the same orbit, its 2-mile (3.2-km) wide nucleus will be reduced to nothing in about 6,000 years. Comets are transient members of the inner solar system. They are soon dissipated by solar radiation. Large cometary dust particles form a meteoroid stream around the orbit. Gas molecules and small particles of dust are just blown away from the Sun and join the galactic disk.

## CRATER CHAIN
This 120-mile (200-km) long chain of impact craters, named Enki Catena, is on Ganymede, the largest of Jupiter's moons. It is likely that Ganymede was struck by 12 or so fragments of a comet that had just been pulled apart by tidal forces as it passed too close to Jupiter.

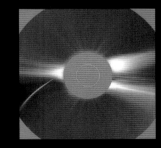

## COMET SOHO-6
Large numbers of sungrazing comets have been discovered by the SOHO satellite. Here, Soho-6 is seen as an orange streak, at left, approaching the masked Sun.

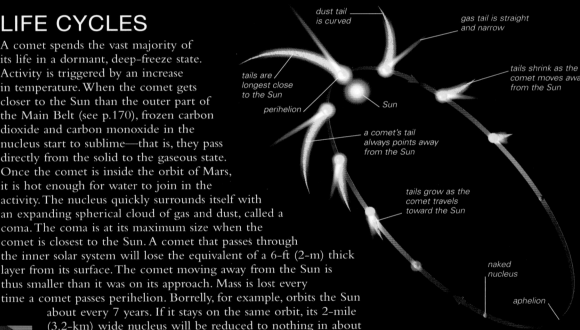

dust tail is curved

gas tail is straight and narrow

tails are longest close to the Sun

tails shrink as the comet moves away from the Sun

perihelion

Sun

a comet's tail always points away from the Sun

tails grow as the comet travels toward the Sun

naked nucleus

aphelion

## COMETARY TAILS
As a comet nears the Sun, it develops two tails. The curved tail is formed of dust that is pushed away by solar radiation. The straight tail consists of ionized gas that has been blown away from the coma by the solar wind.

## HALLEY'S TAIL
These 14 images of Halley's Comet were taken between April 26 and June 11, 1910, around the time it passed perihelion. An impressive tail was produced and dissipated in just 7 weeks of its 76-year orbit.

THE SOLAR SYSTEM

# COMETS

There are billions of comets at the edge of the solar system, but very few have been observed, since they are bright enough to be seen in Earth's sky only when they travel into the inner solar system and approach the Sun. Nearly 2,000 comets have been recorded and their orbits calculated so far. About 200 of the cataloged comets are periodic, having orbital periods of less than 20 years (short period) or between 20 and 200 years (intermediate period). Most, but not all, comets are named after their discoverers.

**COMET HALE–BOPP**
Caught in the evening sky above Germany in 1997, Hale–Bopp, one of the brightest comets of the 20th century, clearly has two tails.

---

## INTERMEDIATE-PERIOD COMET

# Ikeya–Seki

**CLOSEST APPROACH TO THE SUN** 290,000 miles (470,000 km)

**ORBITAL PERIOD** 184 years

**FIRST RECORDED** September 8, 1965

This comet is named after the two amateur Japanese comet hunters, Ikeya Kaoru and Seki Tsutomu, who discovered it independently (and within 5 minutes of each other) in 1965. On October 21, 1965, as it passed perihelion, the comet was so bright that it was visible in the noon sky only 2 degrees from the Sun. Tidal forces then caused the nucleus to split into three parts. Ikeya–Seki faded quickly as it moved away from the Sun, but the tail grew until it extended over 60 degrees across the sky. At this stage, it was 121 million miles (195 million km) from the Sun.

### SUNGRAZER

Ikeya–Seki is a sungrazer and passed within just 290,000 miles (470,000 km) of the Sun's surface in 1965. It is one of over 1,000 comets in the Kreutz sungrazer family.

## LONG-PERIOD COMET

# Great Comet of 1680

**CLOSEST APPROACH TO THE SUN** 580,000 miles (940,000 km)

**ORBITAL PERIOD** 9,400 years

**FIRST RECORDED** November 14, 1680

This comet has two great claims to fame. It was the first comet to be discovered by telescope and the first to have a known orbit. Some 70 years after the telescope was invented, the German astronomer Gottfried Kirch found the comet by accident when observing the Moon in 1680. The orbit was calculated by the English

mathematician Isaac Newton using his new theory of universal gravity, and the results were published in his masterpiece *Principia* in 1687. The comet is a sungrazer and was seen twice: first, as a morning phenomenon, when it was approaching the Sun; and subsequently in the evenings, when it was receding. Newton was the first to realize that these apparitions were of the same comet. The English physicist Robert Hooke noticed a stream of light issuing from the nucleus. This was the first description of jets of material emanating from active areas.

**GREAT COMETS**
Great comets, such as this 1680 comet, are extremely bright and can be very startling when they appear.

## COMET ORBIT

Isaac Newton made observations of the Great Comet of 1680. At the time, a conventional view held that comets traveled in straight lines, passing through the solar system only once. Based on his observations, Newton realized that he had seen a comet traveling around the Sun on a parabolic curve. In 1687, in the *Principia*, he used his study of comets and other phenomena to confirm his law of universal gravitation. He also showed how to calculate a comet's orbit from three accurate observations of its position. Using Newton's laws, Edmond Halley successfully predicted the return of the comet named after him.

**NEWTON'S ORBIT SKETCH**

---

## INTERMEDIATE-PERIOD COMET

# Swift–Tuttle

**CLOSEST APPROACH TO THE SUN** 88 million miles (143 million km)

**ORBITAL PERIOD** About 135 years

**FIRST RECORDED** July 16, 1862

After Swift–Tuttle's discovery in 1862, calculations of its orbit established the relationship between comets and meteoroid streams. Every August, the Earth passes through a stream of dust particles that produces the Perseid meteor shower, named after the constellation from which the shooting stars appear to be emanating.

### PERSEIDS

It takes about 2 weeks for the Earth to pass through this meteoroid stream. The peak rate is on August 12, at about 50 visible meteors per hour.

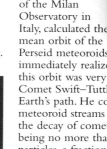

**PERIODIC COMET**
Swift–Tuttle was discovered independently by American astronomers Lewis Swift and Horace Tuttle in 1862. This optical image was taken in 1992, when the comet approached the Sun once again.

In 1866, Giovanni Schiaparelli (see p.220), the director of the Milan Observatory in Italy, calculated the mean orbit of the Perseid meteoroids. He immediately realized that this orbit was very similar to that of Comet Swift–Tuttle, which intersects Earth's path. He concluded that meteoroid streams were produced by the decay of comets, the meteoroids being no more than cometary dust particles, a fraction of a gram in mass, hitting the Earth's upper atmosphere at velocities of about 134,000 mph (216,000 kph). About the same number of Perseid meteors are seen each year, so the dust must be evenly spread around the cometary orbit. This uniformity takes a long time to come about. Swift–Tuttle must have passed the Sun on the same orbit a few hundred times to produce this effect. Comets are decaying, but they have to pass through the inner solar system a thousand times or so before they are whittled down to nothing.

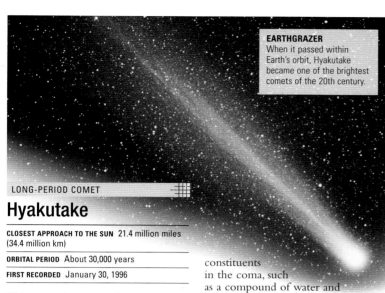

**EARTHGRAZER**
When it passed within Earth's orbit, Hyakutake became one of the brightest comets of the 20th century.

**HALLEY AGAINST A STAR FIELD**
This photograph was taken from Australia on March 11, 1986, 3 days before the comet was visited by the Giotto spacecraft.

LONG-PERIOD COMET

# Hyakutake

**CLOSEST APPROACH TO THE SUN** 21.4 million miles (34.4 million km)

**ORBITAL PERIOD** About 30,000 years

**FIRST RECORDED** January 30, 1996

This comet became a Great Comet, not (like Hale–Bopp) because the nucleus was big, but because on March 24, 1996, it came within a mere 9 million miles (15 million km) of Earth. It was discovered by the Japanese amateur astronomer Hyakutake Yuji using only a pair of high-powered binoculars. The comet became so bright that large radio-telescope spectrometers could detect minor

**TELESCOPIC VIEW**
In March and April 1996, superb short-exposure photographs of Hyakutake could be obtained using only large telephoto lenses or small telescopes.

constituents in the coma, such as a compound of water and deuterium (HDO) and methanol ($CH_3OH$). Hyakutake was the first comet to be observed to emit X-rays. Subsequently, it was found that other comets are also sources of X-rays, the rays being produced when electrons in the coma are captured by ions in the solar wind. On May 1, 1996, the Ulysses spacecraft detected Hyakutake's gas tail when 355 million miles (570 million km) from the nucleus. This is the longest comet tail to be detected to date. Sections of Hyakutake's gas tail have disconnected due to interactions between magnetic fields in the solar wind and the tail.

SHORT-PERIOD COMET

# Encke

**CLOSEST APPROACH TO THE SUN** 32 million miles (51 million km)

**ORBITAL PERIOD** 3.3 years

**FIRST RECORDED** January 17, 1786

Comet Encke was "discovered" in 1786 (by the French astronomer Pierre Méchain), in 1795 (by the German-born astronomer Caroline Herschel), and in 1805 and 1818–1819 (by the French astronomer Jean Louis Pons). These comets were found to be the same only after orbital calculations in 1819 by the German astronomer Johann Encke, who then predicted its

return in 1822. Comet Encke is unusual in that, like Halley's Comet, it is not named after its discoverer. It has the shortest period of any known comet and has been seen returning to the Sun at intervals of 3.3 years since then. The orbit is also shrinking in size, as Encke comes back to perihelion about 2.5 hours sooner than it should. Some astronomers have suggested that this is due to the comet plowing through a resistive medium in the solar system. But other comets have returned later than predicted, and the time error has varied from one orbit to the next. Astronomers have realized that the changing orbits were caused by the "jet effect" of gas escaping from the comet's nucleus. The comet receives a push from the expanding gases and, depending on its direction of spin in relation to its orbit, it is either accelerated or decelerated.

**COMA**
Not all comets have tails. Some, such as Encke, often only have a dense spherical envelope of gas and dust around the nucleus called the coma. The density of the gas decreases as it flows away from the nucleus. Cometary comae have no boundaries; they just fade away.

INTERMEDIATE-PERIOD COMET

# Halley's Comet

**CLOSEST APPROACH TO THE SUN** 55 million miles (88 million km)

**ORBITAL PERIOD** About 76 years

**FIRST RECORDED** 240 BCE

In 1696, Edmond Halley, England's second Astronomer Royal, reported to the Royal Society in London that comets that had been recorded in 1531, 1607, and 1682 had very similar orbits. He concluded that this was the same comet returning to the inner solar system about every 76 years, moving under the influence of the newly discovered solar gravitational force. What is more, Halley predicted that the comet would return in 1758. Halley's Comet was the first periodic comet to be discovered. This indicated that at least some comets were permanent members of the solar system.

Orbital analysis has revealed that Halley's Comet has been recorded 30 times, the first known sighting being in Chinese historical diaries of 240 BCE. The last appearance, in 1986, was 30 years after the start of the space age, and five spacecraft visited the comet. The most productive was ESA's Giotto mission. This flew to within 370 miles (600 km) of the nucleus and took the first-ever pictures. Giotto proved that cometary

nuclei are large, potato-shaped dirty snowballs and that the majority of the snow is water ice. Halley was about 93 million miles (150 million km) from the Sun when Giotto encountered it. Only about 10 percent of the surface was actively emitting gas and dust at the time. On average, a comet loses a surface layer about 6.5 ft (2 m) deep every time it passes through the inner solar system. At this rate, Halley's Comet will survive for about another 200,000 years.

**NUCLEUS**
Giotto revealed that Halley's nucleus is 9.5 miles (15.3 km) long. The brightest parts of this image are jets of dust streaming toward the Sun.

MYTHS AND STORIES

## CELESTIAL OMEN

Some superstitious people regard comets as portents of death and disaster. Before Edmond Halley's work, all comets were unexpected. They were often compared to flaming swords. England's King Harold II was worried by the appearance of Halley's Comet in 1066. But what was a bad omen for him was a good sign for the Norman Duke William, who conquered Harold at Hastings.

**BAYEUX TAPESTRY**
This crewel embroidery beautifully depicts the coma and tail of Halley's Comet (top left), as seen in 1066. It looks like a primitive rocket spewing out flames.

## LONG-PERIOD COMET

# Hale–Bopp

**CLOSEST APPROACH TO THE SUN** 85 million miles (137 million km)

**ORBITAL PERIOD** 2,530 and 4,200 years

**FIRST RECORDED** July 23, 1995

Comet Hale–Bopp was discovered independently and accidentally by the American amateur astronomers Alan Hale and Thomas Bopp, who were looking at Messier objects in the clear skies of the western US (it was close to M70). Later, after the orbit had been calculated, Hale–Bopp was found to be at a distance of over 600 million miles (1 billion km). This is between the orbits of Jupiter and Saturn and is an almost unprecedented distance for the discovery of a nonperiodic comet. The orbit showed that it had been to the inner solar system before, some 4,200 years ago, but because it passed close to Jupiter a few months after discovery, it will return again in about 2,510 years. Hale–Bopp passed perihelion on April 1, 1997. It was one of the brightest comets of the century—not, like Hyakutake, because it came very close to Earth, but simply because it had a huge nucleus, about 22 miles (35 km) across.

**TWIN TAILS**
The two tails of Comet Hale–Bopp shine brightly over the Little Ajo Mountains in Arizona shortly after sunset in 1997.

## SHORT-PERIOD COMET

# Borrelly

**CLOSEST APPROACH TO THE SUN** 126 million miles (203 million km)

**ORBITAL PERIOD** 6.86 years

**FIRST RECORDED** December 28, 1904

The flyby of NASA's Deep Space 1 mission on September 22, 2001, revealed that this periodic comet has a nucleus shaped like a bowling pin, about 5 miles (8 km) long. Reflecting on average only 3 percent of the sunlight that hits it, Borrelly has the darkest known surface in the inner solar system. Any ice in the nucleus is hidden below the hot and dry, mottled, sooty black surface.

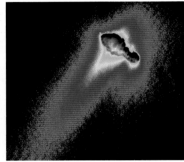

**DEEP SPACE 1 IMAGE**
The production of the jets of gas and dust emanating from Borrelly's nucleus is eroding the surface. There is a possibility that the nucleus will split in two in the future.

## SHORT-PERIOD COMET

# 67P/Churyomov– Gerasimenko

**CLOSEST APPROACH TO THE SUN** 322 million miles (518 million km)

**ORBITAL PERIOD** 6.44 years

**FIRST RECORDED** September 11, 1969

This was the target of ESA's Rosetta mission, which orbited it from September 2014 until September 2016, and deployed a partially successful lander (Philae) to its surface in November 2014. The twin lobed shape of the nucleus suggests that it grew by a gentle collisional merger between two bodies. Rugged areas are icy "bedrock," whereas smoother areas are dusty, with some icy boulders.

**COMET'S ATMOSPHERE**
This Rosetta image, 5 months before perihelion, shows jets of gas and dust escaping from the comet's surface as it became warmed by the Sun.

## SHORT-PERIOD COMET

# Wild 2

**CLOSEST APPROACH TO THE SUN** 147 million miles (236 million km)

**ORBITAL PERIOD** 6.39 years

**FIRST RECORDED** January 6, 1978

Wild 2 is a relatively fresh comet that was brought into an orbit in the inner solar system as recently as September 1974, when it had a close encounter with Jupiter. It is too faint to be seen with the naked eye, since its nucleus is only 3.4 miles (5.5 km) long. Wild 2's present path around the Sun takes it very close to the orbits of both Mars and Jupiter. It may oscillate between its present orbit and an orbit with a period of about 30 years that brings it only as close as Jupiter. Wild 2 was

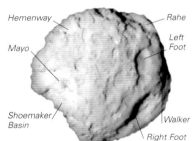

Hemenway — Rahe
Mayo — Left Foot
Shoemaker Basin — Walker
Right Foot

**CLOSE-UP OF NUCLEUS**
The surface of the nucleus is covered by steep-walled depressions hundreds of yards (meters) deep. They are mostly named after famous cometary scientists.

chosen for NASA's Stardust mission (see panel, below) because the spacecraft could fly by at the relatively low speed of 13,600 mph (21,900 kph), capturing comet dust on the way.

## SHORT-PERIOD COMET

# Shoemaker–Levy 9

**ORBITAL DISTANCE FROM JUPITER** 56,000 miles (90,000 km)

**ORBITAL PERIOD AROUND JUPITER** 2.03 years

**FIRST RECORDED** March 25, 1993

Unlike normal comets, this one was discovered in orbit around Jupiter by the American astronomers Gene and Carolyn Shoemaker and David Levy. Even more remarkably, it was in 22 pieces, having been ripped apart on July 7, 1992, when it passed close to Jupiter. These fragments subsequently crashed into the atmosphere in Jupiter's southern hemisphere in July 1994 (see p.181). Observatories all over the world and the Hubble Space Telescope witnessed the sequence of events. The nucleus was originally just over 0.6 miles (1 km) across and was most likely captured by Jupiter in the 1920s.

**SHATTERED NUCLEUS**
This false-color Hubble Space Telescope image show strung-out fragments of the nucleus, each surrounded by its own coma, 2 months before they impacted onto Jupiter.

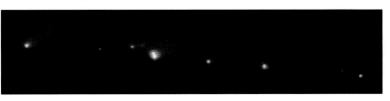

## CAROLYN SHOEMAKER

After taking up astronomy at the age of 51 after her three children had grown up, Carolyn Shoemaker (b. 1929) has now discovered over 800 asteroids and 32 comets. She uses the 18-in (46-cm) Schmidt wide-angle telescope at the Palomar Observatory in California. Her patience and attention to detail are vital when it comes to inspecting photographic plates that are taken about an hour apart and then studied stereoscopically. Typically, 100 hours of searching are required for each comet discovery. Carolyn was married to Gene Shoemaker (see p.139).

EXPLORING SPACE

## THE STARDUST MISSION

The Stardust spacecraft flew by Wild 2 on January 2, 2004. It has captured both interstellar dust and dust blown away from the comet's nucleus. Aerogel placed on an extended tennis-racket-shaped collector was used to capture the particles without heating them up or changing their physical characteristics. The craft returned to Earth in 2006, and the collector, stowed in a canister, parachuted to safety in the desert in Utah.

**AEROGEL**
Although it has a ghostly appearance, aerogel is solid. It is a silicon-based sponge-like foam, 1,000 times less dense than glass.

# Tempel 1

**CLOSEST APPROACH TO THE SUN** 140 million miles
(226 million km)

**ORBITAL PERIOD** 5.52 years

**FIRST RECORDED** April 3, 1867

This comet was first discovered in 1867 by the German astronomer Wilhelm Tempel, but after two reappearances, it disappeared because its orbit had been changed by close approaches to Jupiter. Following calculations by British astronomer Brian Marsden in 1963, the comet was rediscovered, and it has been followed ever since as it orbits between Mars and Jupiter.

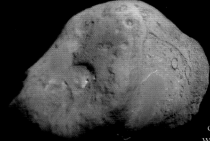

**TEMPLE NUCLEUS**
The potato-shaped nucleus of Comet Tempel 1 was photographed by the Deep Impact probe in July 2005. The impactor hit between the craters at center right.

**DEEP IMPACT**
A fountain of dust, shown in false color, sprays off the nucleus of Comet Tempel 1. This image was taken on July 4, 2005, about 50 minutes after the comet's nucleus was hit by the impactor released by Deep Impact.

## EXPLORING SPACE

### STUDYING COMETS

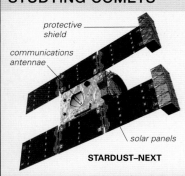

protective shield

communications antennae

solar panels

**STARDUST–NEXT**

Space-probe exploration of comets began at the last return of Halley's Comet, in 1986. Since then, probes have brought back samples of dust (Stardust-NExT, left) and hit the nucleus (Deep Impact). The next step was to orbit and land on a comet's nucleus—the mission of the European probe Rosetta. Comet probes are fitted with shields to protect them from the fast-moving specks of dust from the comets.

To find out what lies beneath the dusty crust of a comet's nucleus, NASA launched an ambitious mission to Tempel 1 in 2005. Called the Deep Impact probe, its aim was to punch a crater in the crust and uncover the subsurface ice, which is thought to have survived unchanged since the formation of the solar system. As the probe approached the nucleus of Tempel 1 in July 2005, it released a 820-lb (370-kg) copper impactor into its path. This collided with the nucleus at a speed of over 22,800 mph (36,000 kph), spraying out a fountain of dust and gas. Because the dust was so fine, about the same size as the particles in talcum powder, it appeared very bright—the comet temporarily brightened tenfold as a result, but it was still not visible to the naked eye. Deep Impact observed the ejected material to determine its composition. Most of the gas was steam (water vapor) and carbon dioxide at an initial temperature of over 1,340°F (720°C), resulting from the heat of the impact. There was so much dust that the crater formed by the impactor was hidden from view.

As Deep Impact flew past the comet, it took detailed images of the nucleus, which turned out to be shaped like a potato. It measured about 4¾ miles (7.5 km) long and 3 miles (5 km) across and rotated every 41 hours. It was very different from the nuclei of other comets that have been seen close up, such as Wild 2 and Borrelly. Surface features were visible, including a plateau fringed by a 66-ft (20-m) cliff (possibly the result of a landslide), and two apparent impact craters, each about 1,000 ft (300 m) wide. The impactor hit the nucleus between these craters.

Stardust, the craft that had collected dust samples from Comet Wild 2 (see p.218), was later sent to photograph the crater produced by Deep Impact. Renamed Stardust-NExT (for New Exploration of Tempel 1), it arrived in 2011, but saw little. It seems that the scar had been covered by dust that fell back onto it.

---

# Hartley 2

**CLOSEST APPROACH TO THE SUN** 98 million miles
(158 million km)

**ORBITAL PERIOD** 6.47 years

**FIRST RECORDED** March 15, 1986

This comet was discovered by British astronomer Malcolm Hartley while he was working at the Schmidt Telescope Unit at Siding Spring Observatory, Australia, in 1986. The nucleus of Comet Hartley 2 has been observed up close in a flyby from the Deep Impact probe.

Following its encounter with Comet Tempel 1 (see above), the Deep Impact probe was sent to take a closer look at Hartley 2. After a journey of 5 years, it arrived near the comet in November 2010 and flew past it at a distance of just under 435 miles (700 km). The spacecraft was not carrying a second impactor, so it could not hit the nucleus. Instead, research concentrated on the comet's appearance and composition.

Deep Impact's observations revealed that Hartley 2's nucleus, at only about 1.2 miles (2 km) long, was the smallest ever visited by a space probe. The comet is peanut-shaped, with two lobes that are connected by a smoother neck only about ¼ mile (0.4 km) wide. Jets of carbon-dioxide gas shoot out from the two lobes at either end of the nucleus, while water vapor is released from the middle. The levels of gas production also vary as the nucleus rotates over a period of

**SPECTACULAR JETS**
Huge jets of gas and dust spew from the elongated nucleus of Comet Hartley 2 as seen in this image from the Deep Impact probe. The image was taken on November 4, 2010, when the spacecraft was at it closest to the comet.

about 18 hours. In addition, for the first time with any comet, the nucleus was seen to be shedding lumps of ice that ranged in size from golf balls to basketballs. Investigation by the Deep Impact probe also revealed larger blocks of ice up to 260 ft (80 m) high on the lobes of nucleus.

Although the spacecraft retained the name Deep Impact for this encounter, its extended space mission has been renamed EPOXI. This name comes from a combination of two acronyms: EPOCh, which stands for Extrasolar Planet Observation and Characterization, since its instruments observed a number of stars for evidence of transits by orbiting planets; and DIXI, short for Deep Impact Extended Investigation.

---

# Lovejoy

**CLOSEST APPROACH TO THE SUN** 515,000 miles
(829,000 km)

**ORBITAL PERIOD** 565 years

**FIRST RECORDED** November 27, 2011

This sungrazer comet was discovered by an Australian amateur astronomer, Terry Lovejoy, less than 3 weeks before its closet approach to the Sun. Sungrazers are comets that skim so close to the Sun, they either evaporate in the intense heat or they crash into its surface. They are usually seen only by telescopes onboard satellites that monitor the region around the Sun, such as the Solar and Heliospheric Observatory (SOHO, see p.105).

In December 2011, Comet Lovejoy not only defied predictions by surviving passing so close to the Sun but emerged to become a brilliant object that could be seen from Earth. On December 16, 2011,

satellites including the Solar Dynamics Observatory (SDO) watched as the comet passed the Sun at a distance of just over 82,000 miles (130,000 km). Over the following days, observers in the Southern Hemisphere were astounded as the comet moved away from the Sun, becoming visible in the morning skies, and grew a long, featherlike tail. Astronauts on the International Space Station got a particularly good view. Sungrazer comets are thought to be fragments of a much larger comet that broke up long ago, possibly in the 12th century. The pieces have continued to orbit the Sun, disintegrating further as they do so. They are also known Kreutz sungrazers, because they were first studied in the 19th century by German astronomer Heinrich Kreutz. Comet Lovejoy has a calculated orbital period of 565 years.

**SURVIVING THE SUN**
Comet Lovejoy (circled) emerges from its passage through the Sun's inner corona as seen by NASA's Solar Dynamics Observatory.

**SPACE STATION VIEW**
This view from the International Space Station shows Comet Lovejoy's tail extending upward from the horizon. The bands beneath are part of Earth's atmosphere.

# METEORS AND METEORITES

POPULARLY KNOWN AS SHOOTING STARS, meteors are linear trails of light-radiating material produced in Earth's upper atmosphere by the impact of often small, dusty fragments of comets or asteroids called meteoroids. About 1 million visible meteors are produced each day. If the meteor is not completely destroyed by the atmosphere, it will hit the ground and is then called a meteorite. If the meteorite is very large, a crater will be formed by the impact.

## METEOROIDS

Most of the dusty meteoroids responsible for visual meteors come from the decaying surfaces of cometary nuclei. When a comet is close to the Sun, its surface becomes hot, and snow just below the surface is converted into gas. This gas escapes and breaks up the surface of the friable, dusty nucleus and blows small dust particles away from the comet. These dusty meteoroids have velocities that are slightly different from that of their parent comet. This causes them to have slightly different orbits, and as time passes, they form a stream of particles all around the original orbit of the comet. This stream is fed by new meteoroids every time the parent comet swings past the Sun. The inner solar system is full of these streams. Dense streams are produced by large comets that get close to the Sun. Streams with relatively few meteoroids are formed by smaller and more distant comets. As the Earth orbits the Sun, it continually passes in and out of these streams, colliding with some of the meteoroids that they contain. Names are given to some meteor showers that occur at fixed times of year, such as the Leonids (right).

**CHELYABINSK FIREBALL**
This is the vapor trail left when a house-sized stony asteroid streaked across the predawn sky in Russia's southern Ural region on February 15, 2013. It broke apart before reaching the ground, but many fragments were recovered.

**LEONID METEOR SHOWER**
Leonid meteors are seen around November 17 every year and are so called because they appear to pour out of the constellation of Leo. Every 33 years, the shower strengthens into a veritable storm. The woodcut on the right was carved by the Swiss artist Karl Jauslin in 1888; it represents the maximum activity of the 1833 Leonids.

## METEORITES

### GIOVANNI SCHIAPARELLI

Giovanni Schiaparelli (1835–1910) was an Italian astronomer who worked at the Brera Observatory in Milan and has two claims to fame. In 1866, he calculated the orbits of the Leonid and Perseid meteoroids and realized that they were similar to the orbits of comets Tempel–Tuttle and Swift–Tuttle, respectively. He concluded that cometary decay produced meteoroid streams. In the late 1870s, he went on to map Mars's surface.

Small extraterrestrial bodies that hit the Earth's atmosphere are completely destroyed during the production of the associated meteor. If, however, the impacting body has a mass of between about 70 lb (30 kg) and 10,000 tons, only the surface layers are lost during atmospheric entry, and the atmosphere slows down the incoming body until it eventually reaches a "free-fall" velocity of just over 90 mph (150 kph). The central remnant then hits the ground. The fraction of the incoming body that survives depends on its initial velocity and composition. Meteorites are referred to as "falls" if they are seen to enter and are then picked up just afterward. Those that are discovered some time later are called "finds." Meteorites are classified as one of three compositional types.

**STONY**
This is by far the most common type of meteorite, comprising 93.3 percent of all falls. They are subdivided into chondrites and achondrites.

**STONY-IRON**
The rarest meteorites—just 1.3 percent of meteorite falls—are a mixture of stone and iron-nickel alloy, similar to the composition of the ...

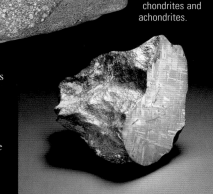

**IRON**
Iron meteorites make up 5.4 percent of all falls. They are composed mainly of iron-nickel alloy (consisting of 5–10 percent nickel by weight) and ...

# METEORITE IMPACTS

Earth's atmosphere shields the surface from the vast majority of incoming extraterrestrial bodies. The typical impact velocity at the top of the atmosphere is about 45,000 mph (72,000 kph), and the leading surface of the meteoroid quickly heats up and starts boiling as a result of hitting air molecules at this speed. Usually, the body is so small that it boils away completely. Parts of medium-sized bodies survive to fall as meteorites. However, a very large body, having a mass greater than about 100,000 tons, is hardly affected by the atmosphere. It punches through the gas like a bullet through tissue paper, energetically slamming into Earth's surface and gouging out a circular crater that is typically 20 times larger than its own size (see p.103). The enormous energy generated ensures that most of the impactor is vaporized in the process, and seismic shocks and blast waves are produced. The resulting huge earthquake will topple any trees for many miles around. The surrounding atmosphere reaches furnace temperatures, causing widespread fires. A tsunami will be produced if the impact is in the ocean. An impact crater greater than 12 miles (20 km) in diameter is produced on Earth about once every 500,000 years.

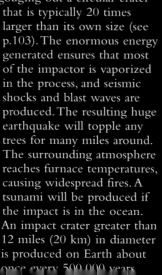

**MOLDAVITE (GREEN GLASS)**

**DISK-SHAPED TEKTITE**

**IMPACTITES**
These half-inch, glassy bodies are formed when Earth's rock melts or shatters due to the heat and pressure of an impact.

**IMPACT CRATER**
About 50,000 years ago, an iron meteorite hit this desert region in Arizona. The resulting crater, called Meteor Crater, is 0.75 miles (1.2 km) wide and 550 ft (170 m) deep. Ejecta produced by the impact can be seen as hummocky deposits lying beyond the crater rim.

# METEORITES

Meteorites are mainly pieces of asteroids that have fallen to Earth from space, but a few very rare meteorites have come from the surface of Mars and the Moon. Some meteorites are made up of the primitive material that originally formed rocky planets. These give researchers a glimpse of the conditions at the dawn of the solar system. Others are fragments of bodies that have differentiated into metallic cores and rocky surfaces, providing an indirect opportunity to study the deep interior of a rocky planet. Meteorites are named after the place where they landed.

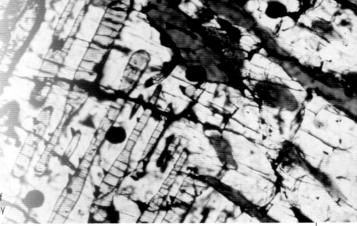

**METEORITE CROSS-SECTION**
By shining polarized light through thin sections of chondrites (a type of stony meteorite), scientists can study their crystalline structure.

---

## NORTH AMERICA north

### Tagish Lake

| | |
|---|---|
| **LOCATION** | British Columbia, Canada |
| **TYPE** | Stony |
| **MASS** | About 2.2 lb (1 kg) |
| **DATE OF DISCOVERY** | 2000 |

Over 500 fragments of this meteorite rained down onto the frozen surface of Tagish Lake on January 18, 2000. The meteorite was dark red and rich in carbon. Analysis showed that it was extremely primitive, containing many unaltered stellar dust grains that had been part of the cloud of material that formed the Sun and the planets.

**FRAGMENT ENCASED IN ICE**

## NORTH AMERICA southwest

### Canyon Diablo

| | |
|---|---|
| **LOCATION** | Arizona |
| **TYPE** | Iron |
| **MASS** | 30 tons |
| **DATE OF DISCOVERY** | 1891 |

Many pieces of this meteorite, ranging from minute fragments to chunks weighing about 1,100 lb (500 kg), have been found near Meteor Crater in Arizona. Much more is thought to be buried under one of the crater rims. If a Canyon Diablo meteorite is sawn in half and then one of the faces is polished and etched with acid, a characteristic surface pattern appears.

*nodule of iron sulfide*

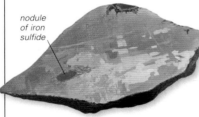

**ACID-ETCHED, POLISHED CROSS-SECTION**

## NORTH AMERICA south

### Allende

| | |
|---|---|
| **LOCATION** | Chihuahua, Mexico |
| **TYPE** | Stony |
| **MASS** | 2 tons |
| **DATE OF DISCOVERY** | 1969 |

On February 8, 1969, a fireball was seen streaking across the sky above Mexico. It exploded, and a shower of stones fell over an area of about 60 square miles (150 square km). Two tons of material were speedily collected

**CHONDRULE**
This thin, magnified section of an Allende meteorite shows one of many spherical, pea-sized chondrules that are locked in the stony matrix. Chondrules are droplets of silicate rock that have cooled extremely rapidly from a molten state.

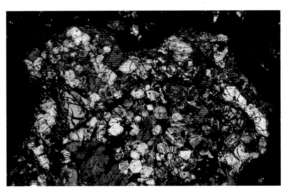

and distributed among the scientific community. Allende was found to be a very rare type of primitive meteorite. Previously, only gram-sized amounts of this meteorite type were known. Because such large samples of Allende were available, destructive analysis was possible. The white calcium- and aluminum-rich crystals were separated from the surrounding rock. They were found to contain the decay products of radioactive aluminum-26, indicating that these crystals were formed in the outer shells of stars that exploded as supernovae and were subsequently incorporated into planetary material.

---

## EUROPE west

### Glatton

| | |
|---|---|
| **LOCATION** | Cambridgeshire, UK |
| **TYPE** | Stony |
| **MASS** | 27 oz (767 g) |
| **DATE OF DISCOVERY** | 1991 |

On May 5, 1991, while planting out a bed of onions just before Sunday lunch, retired English civil servant Arthur Pettifor heard a loud whining noise. Noticing one of the conifers in his hedge waving around, he got up and looked in the bottom of the hedge. He spotted a small stone that was lukewarm to the touch. If Pettifor had not been gardening, the stony meteorite would never have been found.

**LUCKY FIND**

## EUROPE west

### Ensisheim

| | |
|---|---|
| **LOCATION** | Alsace, France |
| **TYPE** | Stony |
| **MASS** | 280 lb (127 kg) |
| **DATE OF DISCOVERY** | 1492 |

This large stone is the oldest meteorite fall that can be positively dated. It was carefully preserved by being hung from the roof of the parish church of Ensisheim, Alsace. This veneration was due to the fall being regarded by the Holy Roman Emperor Maximilian as a favorable omen for the success of his war with France and his efforts to repel Turkish

**METEORITE FRAGMENT**
This highly valuable 17.6-lb (8-kg) sample of the Ensisheim meteorite is kept at the Museum of Paris, France.

**MEDIEVAL WOODCUT**
The woodcut at the top of this medieval manuscript shows the meteorite falling near Ensisheim after producing a brilliant fireball in the sky on November 16, 1492.

invasions. Initially, Ensisheim was thought to be a "thunderstone," a rock ejected from a nearby volcano and subsequently struck by lightning. In the early 19th century, it was chemically analyzed and found to contain 2.3 percent nickel. This is very rare in rocks on Earth, and theories of an extraterrestrial origin started to proliferate.

## AFRICA north

### Nakhla

| | |
|---|---|
| **LOCATION** | Alexandria, Egypt |
| **TYPE** | Stony |
| **MASS** | 88 lb (40 kg) |
| **DATE OF DISCOVERY** | 1911 |

On June 28, 1911, about 40 stones landed near Alexandria, the largest weighing 4 lb (1.8 kg). Nakhla is a volcanic, lavalike rock that formed 1,200 million years ago. It is one of over 16 meteorites that have been blasted from the surface of Mars and, after many millions of years in space, fallen to Earth.

*black, glassy fusion crust formed during fall*

**MARTIAN METEORITE**

## AFRICA southwest

# Hoba West

| | |
|---|---|
| **LOCATION** | Grootfontein, Namibia |
| **TYPE** | Iron |
| **MASS** | 66 tons |
| **DATE OF DISCOVERY** | 1920 |

The largest meteorite to have been found on Earth, Hoba West measures 8.9 x 8.9 x 3 ft (2.7 x 2.7 x 0.9 m). It consists of 84 percent iron and 16 percent nickel. Hoba West has never been moved from where it landed. In the past, enterprising individuals tried to recover this valuable lump of "scrap" metal. To protect it from damage and sample-taking, the Namibian Government has declared it to be a national monument. Hoba West represents the maximum mass that the Earth's

atmosphere can slow down to a free-fall velocity. If its parent meteoroid had been much bigger or the trajectory of the fall steeper, the impact with the ground would have been much faster. This would have led to the destruction of most of the meteorite and the production of a crater in the Earth's surface. Large lumps of surface iron, such as Hoba West, are hard to overlook.

**RUSTING AWAY**
The Hoba West meteorite weighed about 66 tons when it was discovered, but it has started to rust away and today weighs less than 60 tons.

**LARGEST KNOWN METEORITE**
A team of scientists from Kings College, London, UK, poses on top of Hoba West in the 1920s. Standing second from the left is Dr. L. J. Spencer, who became Keeper of Minerals at the British Museum, London, in the 1930s.

---

## AFRICA south

# Cold Bokkeveld

| | |
|---|---|
| **LOCATION** | Western Cape, South Africa |
| **TYPE** | Stony |
| **MASS** | About 8.8 lb (4 kg) |
| **DATE OF DISCOVERY** | 1838 |

This meteorite is a perfect example of a stony chondrite, a class of primitive meteorite that makes up almost 90 percent of those found so far. They consist of silicate, metallic, and sulfide minerals and are thought to represent the material from which the Earth was formed. They contain tiny, spherical chondrules cemented into a rocky matrix. These rocky droplets solidified extremely quickly from a starting temperature of at least 2,600°F (1,400°C). Chondrules contain a mixture of imperfect crystals and glass. Cold Bokkeveld is carbonaceous, which means that it contains compounds of carbon, hydrogen, oxygen, and nitrogen. These are the main constituents of living cells. Carbonaceous chondrites thus contain the building blocks of life.

**WATER FROM STONE**
This tiny chondrule is surrounded by a water-rich matrix (shown in black). Cold Bokkeveld contains about 10 percent water by mass, which would be released if it was heated.

---

## AUSTRALIA west

# Mundrabilla

| | |
|---|---|
| **LOCATION** | Nullarbor Plain, Western Australia |
| **TYPE** | Iron |
| **MASS** | About 18 tons |
| **DATE OF DISCOVERY** | 1911 |

Mundrabilla is on the Trans-Australian railroad line in a featureless desert. Three small irons were found there in 1911 and 1918. Renewed interest in 1966 led to the discovery of two meteorites weighing 5 and 11 tons. Mundrabilla took many millions of years to solidify, and it offers a rare chance to investigate the formation of alloys at low gravity. A 100-lb (45-kg) core of one of the meteorites (below) is undergoing computer X-ray analysis by NASA.

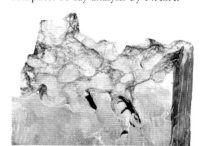

**UNDER INVESTIGATION**

---

## ANTARCTICA

# ALH 81005

| | |
|---|---|
| **LOCATION** | Allan Hills, Antarctica |
| **TYPE** | Stony |
| **MASS** | 1.1 oz (31.4 g) |
| **DATE OF DISCOVERY** | 1982 |

ALH 81005 is a lunar meteorite. About 36 have been discovered, a mere 0.08 percent of the present total. The cosmic-ray damage they have suffered indicates that they have been blasted from the surface of the Moon by a meteorite impact in the last 20 million years. The main mineral is anorthite (calcium aluminum silicate), which is very rare in asteroids. The composition of these stony meteorites is very similar to that of the lunar-highland rocks brought back to Earth by the Apollo astronauts.

**MOON ROCK**
This golf-ball-sized rock was found by the US Antarctic Search for Meteorites program in 1982. It was the first meteorite to be recognized as being of lunar origin.

anorthite

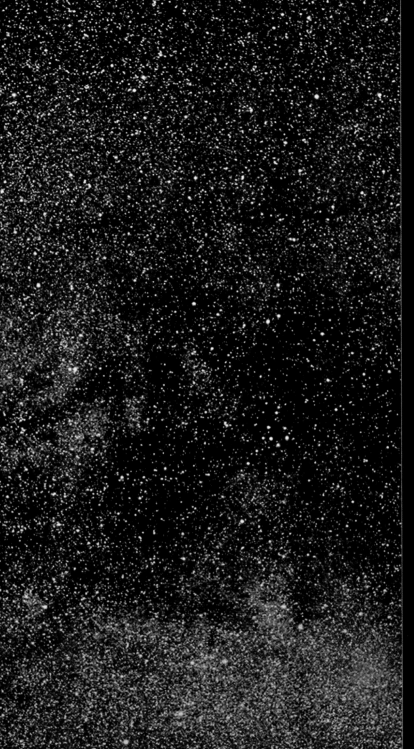

> *"A broad and ample road, whose dust is gold,*
> *And pavement stars, as stars to thee appear*
> *Seen in the galaxy, that milky way*
> *Which nightly as a circling zone thou seest*
> *Powder'd with stars."*
>
> John Milton

THE SOLAR SYSTEM is part of a vast collection of stars, gas, and dust called the Milky Way Galaxy. Galaxies can take various forms, but the Milky Way is a spiral. The Sun and its system of planets lie halfway from the center, on the edge of one of the spiral arms. For thousands of years, humans have pondered the significance of the pale white band that stretches through the sky. This Milky Way is the light from millions of stars that lie in the disk of the galaxy. Within the Milky Way lie stars at every stage of creation, from the immense clouds of interstellar material that contain the building material of stars to the exotic stellar black holes, neutron stars, and white dwarfs that are the end points of a star's life. Most of the Milky Way's visible mass consists of stellar material, but about 90 percent of its total mass is made up of invisible "dark matter," which is a mystery yet to be explained.

**GLOWING PATHWAY**
From Earth, the Milky Way presents a glowing pathway of stars and gas vaulting across the night sky. The billions of stars that make up the Milky Way are arranged in a great spiral disk, and from our position, halfway from its center, we view the disk end-on.

# THE MILKY WAY

# THE MILKY WAY

THE SUN IS ONE STAR of around 200 billion that make up the Milky Way, a relatively large spiral galaxy (see p.302) that started to form around 13.5 billion years ago. From our position inside the Milky Way, it appears as a bright band of stars stretching across the night sky.

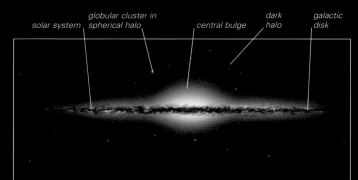

**BAND OF STARS**
As we look out along the disk of the Milky Way from our position within, we see a bright band of thousands of stars that has captured humankind's imagination throughout history.

## THE GEOGRAPHY OF THE MILKY WAY

At the very center of the Milky Way lies a black hole with a mass of about 4 million solar masses. This core or nucleus of the galaxy is surrounded by a bulge of stars that grows denser closer to the center. This forms an ellipsoid of about 15,000 by 6,000 light-years, the longest dimension lying along the plane of the Milky Way. Lying in the plane is the disk containing most of the galaxy's stellar materials. Young stars etch out a spiral pattern, and it is thought that they radiate out from a bar. Surrounding the bulge and disk is a spherical halo in which lie some 200 globular clusters, and this in turn may be surrounded by a dark halo, the corona.

**MILKY WAY GALAXY**
The Milky Way has a diameter of about 180,000 light-years and a thickness of about 2,000 light-years. The Sun lies about 26,000 light-years from the center.

solar system    globular cluster in spherical halo    central bulge    dark halo    galactic disk

Galactic core containing older stars    younger, OB stars    dense molecular clouds    globular cluster containing older stars    nebula, consisting of ionized gas    interstellar gas and dust

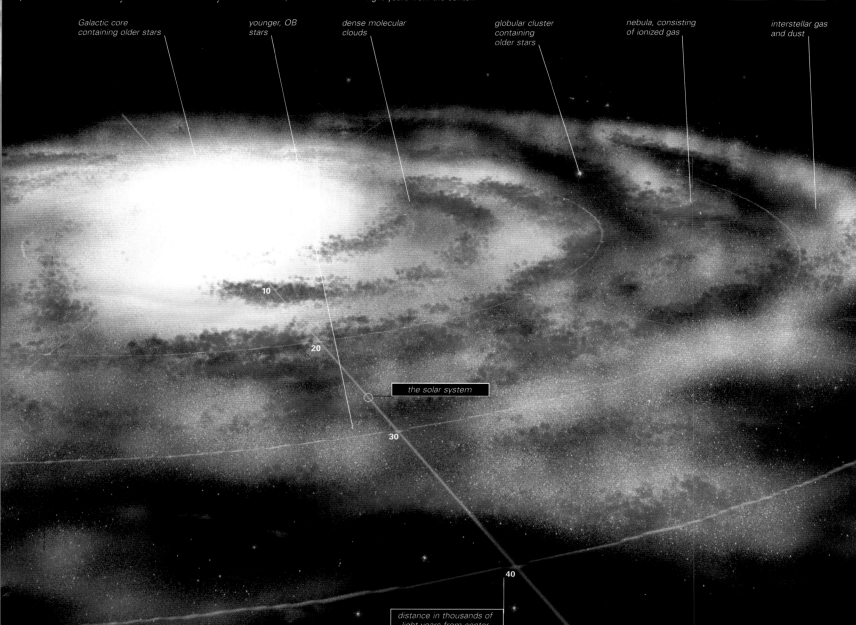

10

20

the solar system

30

40

distance in thousands of light-years from center

180°

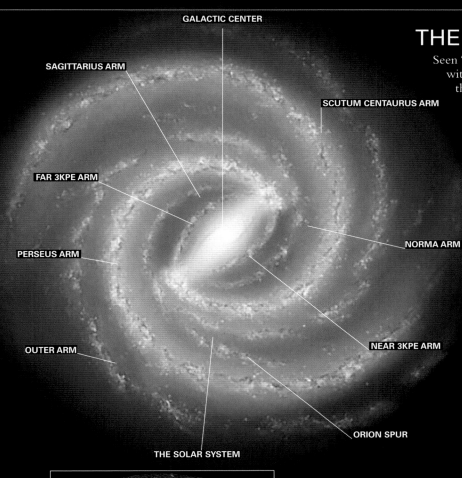

GALACTIC CENTER

SAGITTARIUS ARM

SCUTUM CENTAURUS ARM

FAR 3KPE ARM

NORMA ARM

PERSEUS ARM

OUTER ARM

NEAR 3KPE ARM

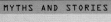

ORION SPUR

THE SOLAR SYSTEM

# THE SPIRAL ARMS

Seen "face-on," the Milky Way would look like a huge catherine wheel, with the majority of its light coming from the arms spiraling out from the ends of a bar of stars that extends from the central bulge. In fact, the material in the spiral arms is generally only slightly denser than the matter in the rest of the disk. It is only because the stars that lie within them are younger, and therefore brighter, that the pattern in spiral galaxies shows up. Two mechanisms are thought to create the Milky Way's spiral structure. Density waves, probably caused by gravitational attraction from other galaxies, ripple out through the disk, creating waves of slightly denser material and triggering star formation (see pp.238–239). By the time the stars have become bright enough to etch out the spiral pattern, the density waves have moved on through the disk, starting more episodes of star formation and leaving the young stars to age and fade. High-mass stars eventually explode as supernovae, sending out blast waves that also pass through the star-making material, triggering further star formation.

### SPINNING GALAXY

The galaxy rotates differentially—the closer objects are to the center, the less time they take to complete an orbit. The Sun travels around the galactic center at about 500,000 mph (800,000 kph), taking around 225 million years to make one orbit.

## HEAVENLY MILK

There are many myths involving the formation of the Milky Way. In Greek mythology, Hercules was the illegitimate son of Zeus and a mortal woman, Alcmene. It was said that when Zeus's wife, while suckling Hercules, heard he was the son of Alcmene, she pulled her breast away and her milk flowed among the stars.

**MYTH IN ART**
*The Origin of the Milky Way* (c.1575) by Jacopo Tintoretto was inspired by the Greek myth.

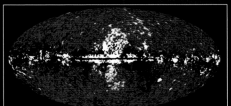

### GAMMA-RAY BUBBLES
Two huge bubbles of gamma-ray-emitting gas fill the space above and below the Milky Way's hub. They may be remnants of past activity in the central black hole.

# STELLAR POPULATIONS

Stars are broadly classified into two groups, called populations, based on age and chemical content. Population I consists of the youngest stars, which tend to be richer in heavy elements. These elements are primarily produced by stars, and Population I stars are created from materials shed by existing stars. In the Milky Way, the majority of Population I stars lie in the galactic disk, where there is an abundance of star-making material. Population II stars are older, metal-poor stars, existing primarily in the halo, but also in the bulge. Most are found within globular clusters, where all star-making materials have been used up and no new star formation is taking place.

### STAR MOTION
The stars in the bulge have the highest orbital rates. They can travel thousands of light-years above and below the plane of the Milky Way. Within the disk, stars stay mainly in the plane of the galaxy as they orbit the galactic center. Stars in the halo plunge through the disk, reaching distances many thousands of light-years above and below it.

*bulge-star orbit*

*halo-star orbit*

*disk-star orbit*

### MAPPING THE MILKY WAY
The Milky Way's structure is defined by its major arms, each named after the constellation in which it is most prominent—the brightest arm is that in Sagittarius, beyond which lies the galactic nucleus. The solar system lies near the inner edge of the Orion Arm. All the arms lie in a plane defined by the galactic disk. The nucleus forms a bulge at its center, and globular clusters orbit above and below it in the halo region.

# THE INTERSTELLAR MEDIUM

The interstellar medium, permeating the space between the stars, consists mainly of hydrogen in various states, together with dust grains. It constitutes about 10 percent of the mass of the Milky Way and is concentrated in the galactic disk. It is not distributed uniformly: there are clouds of denser material, where star formation takes place, and regions where material has been shed by stars, interspersed with areas of very low density. Within the interstellar medium, there is a wide range of temperatures. In the cooler regions, at around −440°F (−260°C), hydrogen exists as clouds of molecules. These cold molecular clouds contain molecules other than hydrogen, and star formation occurs where such clouds collapse. There are also clouds of neutral hydrogen (HI regions) with temperatures ranging from −280°F (−170°C) to 1,340°F (730°C), and areas of ionized hydrogen heated by stars (HII regions) with temperatures around 18,000°F (10,000°C). Dust grains contribute about 1 percent of the galactic mass and are found throughout the medium. They are mostly small, solid grains 0.01 to 0.1 micrometers in diameter, consisting of carbon, silicates (compounds of silicon and oxygen), or iron, with mantles of water and ammonia ice or, in the cooler clouds, possibly solid carbon dioxide.

**NONUNIFORM MEDIUM**
As this image of the Cygnus Loop supernova remnant (see p.269) shows, material in the interstellar medium is very uneven. The blast wave from the supernova explosion is still expanding through the interstellar matter. Where it hits denser areas and slows down, atoms in the medium become excited and emit optical and ultraviolet light.

**INVISIBLE COSMIC RAYS**
Cosmic rays travel throughout the Milky Way. These are highly energetic particles that spiral along magnetic field lines. Cosmic rays are primarily ions and electrons and are an important part of the interstellar medium, producing a pressure comparable to that of the interstellar gas.

**STARS**
Stars are an important factor in the composition of the interstellar medium since they enrich the medium with heavy, metallic elements. A supernova explosion, the death of a massive star (see p.266), is the only mechanism that produces elements heavier than iron.

**DARK NEBULAE**
Dark nebulae are cool clouds composed of dust and the molecular form of hydrogen. They are only observed optically when silhouetted against a brighter background as they absorb light and reradiate the energy in infrared wavelengths. Stars are formed when dark nebulae collapse.

**MAGNETIC FIELDS**
Galactic magnetic fields are weak fields that appear to lie in the plane of the Milky Way, increasing in strength toward the center. They are aligned with the spiral arms but are distorted locally by events such as the collapse of molecular clouds and supernovae.

**BETWEEN THE STARS**
Contrary to early popular belief, the space between stars is not empty. The interstellar medium is fundamental in the process of star formation and galaxy evolution. Temperature defines the material's appearance and the processes occurring within it.

**DUST CLOUDS**
Young stars are often surrounded by massive disks of dust. These disks are believed to be the material from which solar systems are formed. There are often high levels of dust around stars in the later stages of their lives as they lose material to the interstellar medium.

**REFLECTION NEBULAE**
Material surrounding young stars contains dust grains that scatter starlight. In these nebulae, the density of the dust is sufficient to produce a noticeable optical effect. The nebulae appear blue because the shorter-wavelength, bluer light is scattered more efficiently.

**EMISSION NEBULAE**
When the interstellar medium is heated by stars, the hydrogen is ionized, producing a so-called HII region. The electrons freed by the ionization process are continually absorbed and reemitted, producing the red coloring observed in emission nebulae.

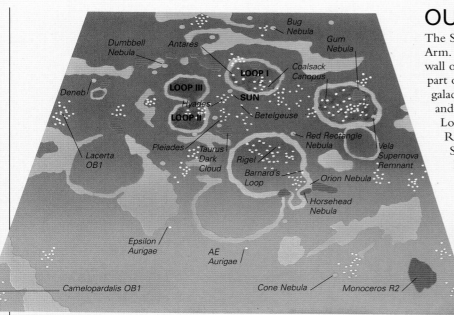

**REGIONAL MAP**
This schematic representation of the solar system's local neighborhood maps out a section of the Milky Way's Orion Arm about 5,000 light-years across. The Sun is located just above center. Hydrogen gas clouds are marked in brown, molecular clouds are in red, and interstellar bubbles are colored green. Nebulae are shown in pink, while star clusters and giant stars are featured in white.

# OUR LOCAL NEIGHBORHOOD

The Sun lies in one of the less-dense regions of the Milky Way's Orion Arm. It sits in a "bubble" of hot, ionized hydrogen gas bounded by a wall of colder and denser neutral hydrogen gas. The Local Bubble is part of a tubelike chimney that extends through the disk into the galactic halo. The largest local coherent structure, detected by radio- and X-rays, is known as Loop I. This is believed to be part of the Local Bubble impacting into a molecular cloud known as the Aquila Rift. Two other expanding bubbles, Loops II and III, lie nearby. The Sun is traveling through material flowing out from the young stars known as the Scorpius–Centaurus Association, toward the Local Interstellar Cloud, a mass of dense interstellar gas.

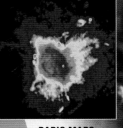

**LOCAL BUBBLE**
A map of local space from NASA's Interstellar Boundary Explorer (IBEX) satellite shows the Sun's motion through tenuous clouds of gas whose own motions through space are indicated by blue arrows. In general, the gas is blowing away from the Scorpius-Centaurus Association of young stars.

# THE GALACTIC CENTER

Dense layers of dust and gas obscure the center of the Milky Way from us in optical wavelengths. However, the brightest radio source in the sky is located toward the Galactic Center in the constellation of Sagittarius. This source, known as Sagittarius A, consists of two parts. Sagittarius A East is believed to be a bubble of ionized gas, possibly a supernova remnant. Sagittarius A West is a cloud of hot gas, and embedded within it is a very strong and compact radio source, called Sagittarius A★ (Sgr A★). Sgr A★ appears to have no orbital motion and therefore probably lies at the very center of the Milky Way. It has a radius of less than 1.4 billion miles (2.2 billion km)—smaller than that of Saturn's orbit—and orbital motions of the gas clouds around it indicate that it surrounds a supermassive black hole of about 3 million solar masses. Centered on Sgr A★ is a three-pronged mini-spiral of hot gas, about 10 light-years in diameter, and surrounding this is a disk of cooler gas and dust called the Circumnuclear Disk.

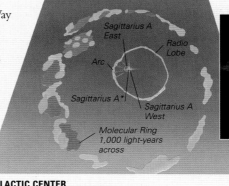

**GALACTIC CENTER**
Surrounding Sagittarius A, the Radio Lobe is a region of magnetized gas including an arc of twisted gas filaments. Farther out, the expanding Molecular Ring consists of a series of huge molecular clouds (red) and an association of hydrogen clouds (brown) and nebulae (pink). The two smaller gas disks around Sagittarius A cannot be seen at this scale.

**RADIO MAPS**
Radio maps of Sagittarius A show a spiral pattern of hot, ionized gas that appears to be falling into the very center of the Milky Way. Situated at the middle of the maps is the point source Sagittarius A★, thought to be a supermassive black hole at the very heart of the Milky Way.

# J. C. KAPTEYN

Dutch astronomer Jacobus Cornelius Kapteyn (1851–1922) was fascinated by the structure of the Milky Way. Studying at the University of Groningen, he used photography to plot star densities. He arrived at a lens-shaped galaxy with the Sun near its center. Although his positioning of the Sun was incorrect, many subsequent studies of the structure of the Milky Way stemmed from his work.

**GLOBULAR CLUSTER**
Like bees around a honey pot, the stars of a globular cluster swarm in a compact sphere. Containing up to a million (mostly Population II) stars, most of these clusters are found in the Milky Way's halo.

# THE EDGES OF THE MILKY WAY

Surrounding the disk and central bulge of the Milky Way is the spherical halo, stretching out to a diameter of more than 100,000 light-years. Compared to the density of the disk and the bulge, the density of the halo is very low, and it decreases as it extends away from the disk. Throughout the halo are about 200 globular clusters (see pp.288–289), spherical concentrations of older, Population II stars (see p.227). Individual Population II stars also exist in the halo. These halo stars orbit the galactic center in paths that take them far from the galactic disk, and because they do not follow the motion of the majority of the stars in the disk, their relative motion to the Sun is high. For this reason, they are sometimes called high-velocity stars. Calculations of the mass of the Milky Way suggest that 90 percent consists of mysterious dark matter (see p.27). Some of this may be composed of objects with low luminosities, such as brown dwarfs and black holes, but most is believed to be composed of exotic particles, the nature of which have yet to be discovered. The halo extends into the corona, which reaches out to encompass the Magellanic Clouds (see pp.310–311), the Milky Way's nearby neighbors in space.

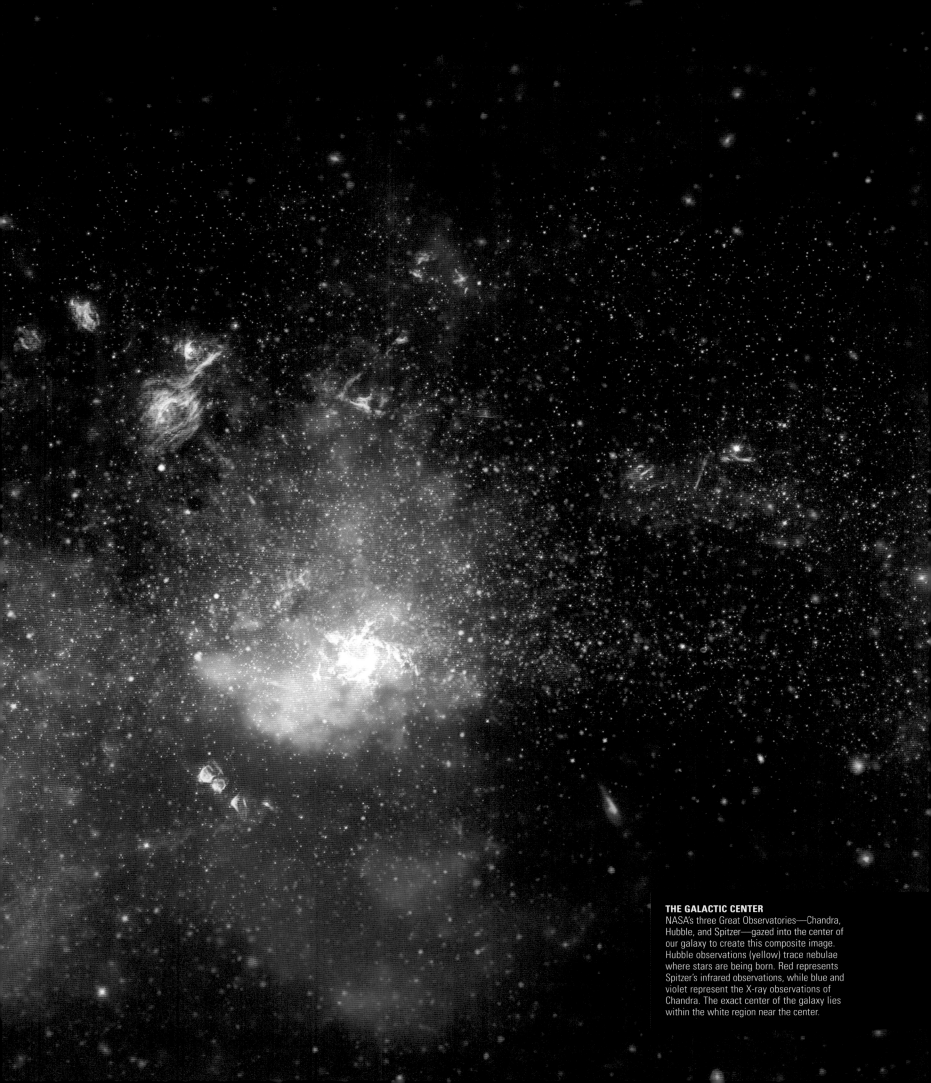

**THE GALACTIC CENTER**
NASA's three Great Observatories—Chandra, Hubble, and Spitzer—gazed into the center of our galaxy to create this composite image. Hubble observations (yellow) trace nebulae where stars are being born. Red represents Spitzer's infrared observations, while blue and violet represent the X-ray observations of Chandra. The exact center of the galaxy lies within the white region near the center.

# STARS

STARS ARE MASSIVE gaseous bodies that generate energy by nuclear reactions and shine because of this energy source. The mass of a star determines its properties—such as luminosity, temperature, and size—and its evolution over time. Throughout its life, a star achieves equilibrium by balancing its internal pressure against gravity.

## WHAT IS A STAR?

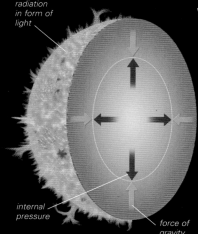

radiation in form of light

internal pressure

force of gravity

**PRESSURE BALANCE**
The state and behavior of any star, at any stage in its evolution, are dictated by the balance between its internal pressure and its gravitational force.

A collapsing cloud of interstellar matter becomes a star when the pressure and temperature at its center become so high that nuclear reactions start (see pp.238–239). A star converts the hydrogen in its core into helium, releasing energy that escapes through the star's body and radiates out into space. The pressure of the escaping energy would blow the star apart if it were not for the force of gravity acting in opposition. When these forces are in equilibrium, the star is stable, but a shift in the balance will change the star's state. Stars fall within a relatively narrow mass range, since nuclear reactions cannot be sustained below about 0.08 solar masses and in excess of about 100 solar masses stars become unstable. A star's life cycle, as well as its potential age, is directly linked to its mass. High-mass stars burn their fuel at higher rates and live much shorter lives than low-mass stars.

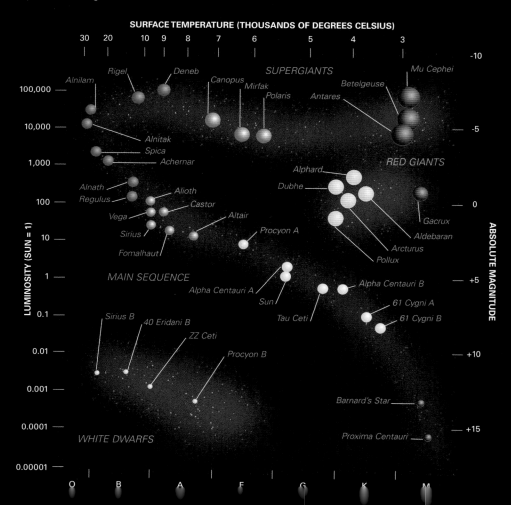

**SURFACE TEMPERATURE (THOUSANDS OF DEGREES CELSIUS)**

30  20  10  9  8  7  6  5  4  3

SUPERGIANTS

Alnilam · Rigel · Deneb · Canopus · Mirfak · Polaris · Betelgeuse · Mu Cephei · Antares

Alnitak · Spica · Achernar

RED GIANTS

Alphard · Dubhe · Gacrux · Aldebaran · Arcturus · Pollux

Alnath · Regulus · Alioth · Castor · Altair · Procyon A

Vega · Sirius · Fomalhaut

MAIN SEQUENCE

Alpha Centauri A · Sun · Tau Ceti · Alpha Centauri B · 61 Cygni A · 61 Cygni B

Sirius B · 40 Eridani B · ZZ Ceti · Procyon B

Barnard's Star

WHITE DWARFS

Proxima Centauri

LUMINOSITY (SUN = 1): 100,000 · 10,000 · 1,000 · 100 · 10 · 1 · 0.1 · 0.01 · 0.001 · 0.0001 · 0.00001

ABSOLUTE MAGNITUDE: -10 · -5 · 0 · +5 · +10 · +15

O  B  A  F  G  K  M

## THE H–R DIAGRAM

Named after the Danish and American astronomers Ejnar Hertzsprung and Henry Russell, the Hertzsprung–Russell (H–R) diagram graphically illustrates the relationship between the luminosity, surface temperature, and radius of stars. The astronomers' independent studies had revealed that a star's color and spectral type are indications of its temperature. When the temperature of stars was plotted against their luminosity, it was noticed that stars did not fall randomly, but tended to be grouped. Most stars lie on the main sequence, a curved diagonal band stretching across the diagram. Star radius increases diagonally from bottom left to top right. Protostars evolve onto the main sequence as they reduce in radius and increase in temperature. On the main sequence, stars remain at their most stable before evolving into red giants or supergiants, moving to the right of the diagram as their radius increases and their temperature falls. White dwarfs are at the bottom left with small radii and high temperatures.

**IMPORTANT DIAGRAM**
The H–R diagram is the most important diagram in astronomy. It illustrates the state of a star throughout its life. Distinct groupings represent different stellar stages, and few stars are found outside these groups, since they spend little time migrating.

## STELLAR SPECTRAL TYPES

| TYPE | PROMINENT SPECTRAL LINES | COLOR | | AVERAGE TEMPERATURE | EXAMPLE |
|------|--------------------------|-------|---|---------------------|---------|
| O | $He^+$, He, H, $O^{2+}$, $N^{2+}$, $C^{2+}$, $Si^{3+}$ | Blue | | 80,000°F (45,000°C) | Gamma Velorum (p.253) |
| B | He, H, $C^+$, $O^+$, $N^+$, $Fe^{2+}$, $Mg^{2+}$ | Bluish white | | 55,000°F (30,000°C) | Rigel (p.281) |
| A | H, ionized metals | White | | 22,000°F (12,000°C) | Sirius (p.252) |
| F | H, $Ca^+$, $Ti^+$, $Fe^+$ | Yellowish white | | 14,000°F (8,000°C) | Procyon (p.284) |
| G | $Ca^+$, Fe, Ti, Mg, H, some molecular bands | Yellow | | 12,000°F (6,500°C) | The Sun (pp.104–107) |
| K | $Ca^+$, H, molecular bands | Orange | | 9,000°F (5,000°C) | Aldebaran (p.256) |
| M | TiO, Ca, molecular bands | Red | | 6,500°F (3,500°C) | Betelgeuse (p.256) |

# STELLAR CLASSIFICATION

Stars are classified by group, according to the characteristics of their spectra. If the light from a star is split into a spectrum, dark absorption and bright emission lines are seen (see p.35). The positions of these lines indicate what elements exist in the photosphere of the star, and the strengths of the lines give an indication of its temperature. The classification system has seven main spectral types, running from the hottest O stars to the coolest M stars. Each spectral type is further divided into 10 subclasses denoted by a number from 0 to 9. Stars are also divided into luminosity classes, denoted by a Roman numeral, which indicates the type of star and its position on the H–R diagram. For example, class V is for main-sequence stars and class II for bright giants, while dim dwarfs are class VI. In addition to the main spectral types, there are classes for stars that show unusual properties, such as the carbon stars (C class). A small letter after the spectral class can also indicate a special property—for example, "v" means variable.

**CONTRASTING SUPERGIANTS**
Both Betelgeuse (above) and Rigel (left) are supergiants, but they are at opposite ends of the stellar spectrum. Betelgeuse (see p.256) is a cool, red star in its later stages, while Rigel is a hot, blue, relatively young star (see p.281).

**MAIN-SEQUENCE STAR**
Shown here in a false-color image, the Sun is a yellowish main-sequence star with a surface temperature of 9,900°F (5,500°C) and spectral type G2, class V.

# LUMINOSITY

The luminosity of a star is its brightness, defined as the total energy it radiates per second. It can be calculated over all wavelengths—the bolometric luminosity—or at particular wavelengths. Measuring the brightness of a star as it appears in the night sky gives its apparent magnitude, but this does not take account of its distance from Earth. Stars that are located at vastly different distances from Earth can have the same apparent magnitudes if the farther star is sufficiently bright (see p.71). Once a star's distance is known, its absolute magnitude can be determined. This is its intrinsic brightness, and from this its luminosity can be determined. Stellar luminosities are generally expressed as factors of the Sun's luminosity. There is a very large range of stellar luminosities, from less than one ten-thousandth to about a million times that of the Sun. If stars are of the same chemical composition, their luminosities are dependent on their mass. Apart from highly evolved stars, they generally obey a consistent mass–luminosity relation, which means that if a star's luminosity is known, its mass can be determined.

**DENEB AND VEGA**
Although Deneb (bottom) and its neighbor Vega (top) are similar in apparent brightness, Deneb is about 300 times more distant. If Deneb were moved to Vega's distance of only 25 light-years from Earth, it would appear to be as bright as a crescent moon.

## CECILIA PAYNE-GAPOSCHKIN

Born and educated in England, Cecilia Helena Payne (1900–1979) married fellow astronomer Sergei Gaposchkin. Initially studying at Cambridge University, England, Payne-Gaposchkin was one of the first astronomy graduates to enter Harvard College Observatory. She studied the spectra of stars and suggested in her doctoral thesis that the different strengths of absorption lines in stellar spectra were a result of temperature differences rather than chemical content. She also suggested that hydrogen was the most abundant element in stars. Her ideas were initially dismissed but finally accepted in 1929.

**HARVARD PROFESSOR**
Cecilia Payne-Gaposchkin was the first woman to become a full professor at Harvard.

# THE LIFE CYCLES OF STARS

STARS FORM WHEN clouds of interstellar gas collapse under the influence of gravity (see pp.238–239). During their lifetimes, stars pass through a series of stages, with the sequence and timing depending crucially on the mass of the star. As a star passes through these stages, different elements are created, again depending on the star's mass. When stars have completed their development, they shed their material back into the interstellar medium, enriching the matter from which future generations of stars will form.

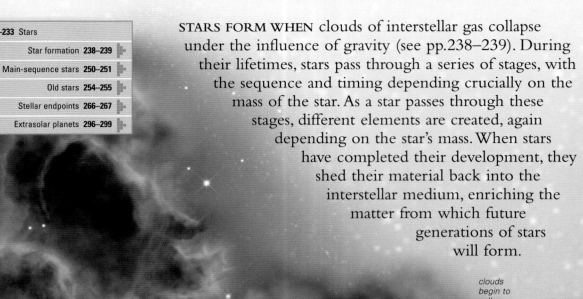

**LIFE STAGES**
The environs of the nebula NGC 3603 display most stellar life stages, from "pregnant" dark nebulae and pillars of hydrogen, to a cluster of young stars, to a red star nearing its end.

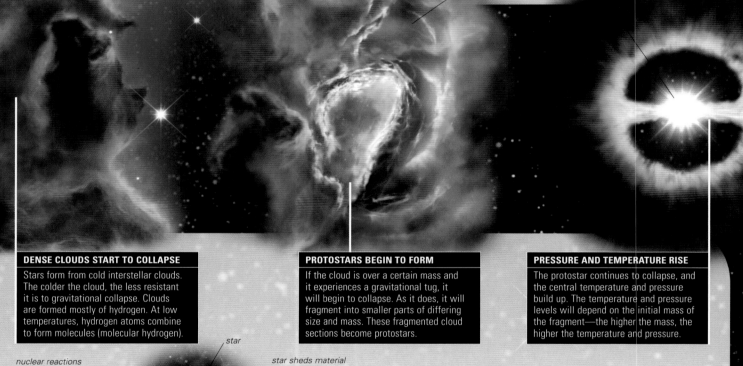

clouds begin to collapse

protostar

shroud of gas and dust

**DENSE CLOUDS START TO COLLAPSE**
Stars form from cold interstellar clouds. The colder the cloud, the less resistant it is to gravitational collapse. Clouds are formed mostly of hydrogen. At low temperatures, hydrogen atoms combine to form molecules (molecular hydrogen).

**PROTOSTARS BEGIN TO FORM**
If the cloud is over a certain mass and it experiences a gravitational tug, it will begin to collapse. As it does, it will fragment into smaller parts of differing size and mass. These fragmented cloud sections become protostars.

**PRESSURE AND TEMPERATURE RISE**
The protostar continues to collapse, and the central temperature and pressure build up. The temperature and pressure levels will depend on the initial mass of the fragment—the higher the mass, the higher the temperature and pressure.

nuclear reactions in star produce heavier elements

star

star sheds material during the course of its life

## STAR-MAKING RECIPE

The basic ingredients of stars are found in cold clouds made mostly of hydrogen molecules. The early stages of star formation are initiated by gravity, which can be exerted by the tug of a passing object, a supernova shock wave, or the compression of one of the Milky Way's density waves. If the cloud has sufficient mass, it will collapse into a protostar, which contracts until nuclear reactions start in its core. At this point, a star is born. During its lifetime, a star will convert hydrogen to helium and a series of heavier materials, depending on its mass. These materials are gradually

stars forming

clouds condense to form stars

mass loss

molecular cloud

gas and dust particles shed by stars join with interstellar material in gigantic molecular clouds

**ONGOING CYCLE**
Stars form from material shed by previous generations of stars, and the death of massive stars can trigger the birth of others.

lost to the interstellar medium until the star has used up most of its fuel and begins to collapse. For a high-mass star, this will result in a supernova that scatters much of the remaining material into space.

**BROWN DWARF**
In protostars less than 0.08 solar masses, the pressure and temperature at the core do not get high enough for nuclear reactions to begin. These protostars become brown dwarfs.

# STELLAR EVOLUTION

If they are of a sufficient mass, new stars will go onto the main sequence, where they will remain for most of their lives. When the hydrogen fuel in their cores is exhausted, they will evolve off the main sequence to become red giants or supergiants. Mass dictates what path stars will follow in their maturity. When a star expands as it burns fuel in its atmosphere or collapses after using up its fuel, it crosses a region to the right of the main sequence on the Hertzsprung–Russell (H–R) diagram (see p.232) known as an instability strip. The more massive the star, the more times it will expand and contract. High-mass stars explode as supernovae in the supergiant region of the H–R diagram, while low-mass stars cross back over the main-sequence band as they collapse to form white dwarfs. Being small and hot, white dwarfs appear in the bottom left of the H–R diagram. As they cool, they move to the right, eventually cooling to become black dwarfs. Neutron stars and black holes do not appear on the H–R diagram because they do not fit the mass–luminosity relationship that it represents.

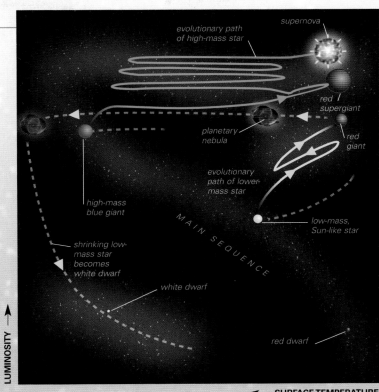

strong stellar winds

### STELLAR MATURITY
The paths of mature stars on their journey toward death can be traced on the H–R diagram. Stars expand off to the right as they get larger and cooler to become red giants or supergiants. They travel back leftward as they collapse, after burning fuel in their atmospheres.

evolutionary path of high-mass star · supernova · red supergiant · planetary nebula · red giant · evolutionary path of lower-mass star · high-mass blue giant · low-mass, Sun-like star · shrinking low-mass star becomes white dwarf · MAIN SEQUENCE · white dwarf · red dwarf · LUMINOSITY ← · ← SURFACE TEMPERATURE

### STELLAR ADOLESCENCE
The gas that contracts to make a protostar starts to rotate slowly and speeds up as it is pulled inward, creating a disk of stellar material. Before joining the main sequence, the protostar exhibits unstable behavior such as rapid rotation and strong winds.

### STAR REACHES MAIN SEQUENCE
For protostars with a mass of more than about 0.08 solar masses, the pressure and temperature within become high enough for nuclear reactions to start. The pressure balances gravity, and the protostar becomes a star.

circumstellar disk

### FORMATION OF ORBITING PLANETS
Once a star is on the main sequence and stable, any disk of remaining material will start to cool. As it cools, elements condense out and begin to stick together. The larger clumps attract the smaller ones until conglomerations are planet-sized.

# FORMATION OF A PLANETARY SYSTEM

Most young stars, unless they are in a close binary system, are surrounded by the remnants of the material from which they have formed. Rotation and stellar winds often shape the material into a flattened disk around the equatorial radius. Initially, the disk of material is hot, but as the star settles down onto the main sequence, it begins to cool. As it cools, different elements condense out, depending on the disk's temperature. Elements can exist in different states throughout the disk. Moving out from the star, temperatures fall, so water, for example, will exist as ice far away from the star and steam close to the star. Tiny condensing particles gradually stick together and grow larger. The ones that grow fastest will gravitationally attract others, becoming larger still, though in the dynamic early stages they may be broken back into pieces by collisions with other growing particles. Eventually, as the disk cools down, it becomes a calmer environment, and some particles will grow large enough to be classed as planetesimals—embryonic planets. Remnants of the original disk that do not form planets become asteroids or comets, depending on their distance from the parent star. Atmospheres are formed by gas attracted from the circumstellar disk, from gases erupting from the planets, or from bombarding comets.

### CIRCUMSTELLAR DISK
Within the circumstellar disk of AB Aurigae, knots of material may be in the early stages of planet formation. This swirling disk of stellar material is about 30 times the size of the solar system.

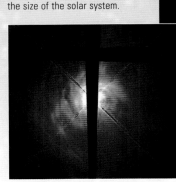

# FROM MATURITY TO OLD AGE

When a star has finished burning hydrogen in its core, it will start burning its outer layers in a series of concentric shells. The star will expand as the source of heat moves outward and its outer layers cool. Stars with very low mass will eventually fade and cool; Sun-like stars will evolve into red giants; and high-mass stars will become supergiants. Once a star has used up all its available nuclear fuel, it will deflate, because there is no longer any power source to replace the energy lost from its surface. As it collapses, if it has enough mass, its helium core starts to burn and change into carbon. Once the fuel in its core is used up again, helium-shell burning begins in the star's atmosphere and the star expands. In very massive stars, this process is repeated until iron is produced. When a Sun-like star has used up all of its fuel, it will lose its outer atmosphere in a spectacular planetary nebula and collapse to become a white dwarf. A high-mass star will explode as a supernova and leave behind a neutron star or black hole.

*star expands as hydrogen-shell burning occurs*

*star starts to collapse as hydrogen is used up*

## LOW-MASS STAR

Once a star with a mass less than half that of the Sun has used up the hydrogen in its core, it will convert the hydrogen in its atmosphere to helium and collapse, just as in higher-mass stars. However, low-mass stars do not have enough mass for the temperature and pressure at its core to get high enough for helium burning to occur. These stars will just gradually fade as they cool.

## STAR NOW ON THE MAIN SEQUENCE

Stars spend the greatest proportion of their lives on the main sequence. The more massive the star, the shorter the period of time it will spend on the main sequence, since larger stars burn their fuel at a faster rate than smaller ones.

## SUN-LIKE STAR

When a Sun-like star exhausts the hydrogen in its core, hydrogen-shell burning begins and it becomes a red giant, often losing its outer layers to produce a planetary nebula. It eventually collapses, and the temperature and pressure at its core initiate helium-core burning. The star again expands as helium-shell burning occurs before finally collapsing to become a white dwarf that gradually fades to black.

*star becomes a red giant as hydrogen-shell burning starts*

## MOSTLY MAIN-SEQUENCE STARS

About 90 percent of the visible stars in a typical view of the night sky are on the main sequence. This corresponds with the fact that most stars spend 90 percent of their life on the main sequence.

## HIGH-MASS STAR

The higher the mass of the star, the more times it will expand and contract—its mass dictates the temperature of the core each time it contracts. Different elements are produced at each stage. If the star is massive enough, an iron core is formed, but elements heavier than iron cannot be formed within stellar cores. They are formed in supernova explosions that leave behind neutron stars or black holes.

*supergiant star produces heavier elements through nuclear reactions*

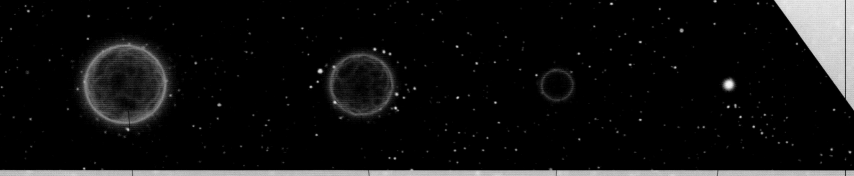

**OLD RED GIANTS**
Red giants and supergiants appear very distinctive in the sky since they are noticeably red. As they are so large, they are also very luminous, which makes them easy to detect.

star continues to collapse as no helium burning occurs

only gas pressure counterbalances gravity

small, dim star, gradually fades

star eventually becomes a small, dim black dwarf

red giant

star collapses after burning its helium shell to become a white dwarf

white dwarf will fade over time to become a black dwarf

red giant's outer layers start to form planetary nebula

planetary nebula

neutron stars are extremely compact and dense, composed mainly of neutrons

star explodes as a supernova, producing elements heavier than iron

black holes are objects so dense that even light cannot escape

**COLLAPSING STAR**
After undergoing its red giant or supergiant stage, the stellar remnant will collapse. If its mass is over 1.4 solar masses, it will collapse to become a neutron star. If the remnant is above about 3 solar masses, it will collapse to become a black hole.

THE MILKY WAY

# STAR FORMATION

STARS ARE FORMED by the gravitational collapse of cool, dense interstellar clouds. These clouds are composed mainly of molecular hydrogen (see p.228). A cloud has to be of a certain mass for gravitational collapse to occur, and a trigger is needed for the collapse to start, as the clouds are held up by their own internal pressure. Larger clouds fragment as they collapse, forming sibling protostars that initially lie close together—some so close that they are gravitationally bound. The material heats up as it collapses until, in some clouds, the temperature and pressure at their centers become so great that nuclear fusion begins and a star is born.

**STAR-FORMING REGION**
In the nebula RCW 120, in the southern Milky Way, an expanding bubble of ionized gas is causing the surrounding material to collapse into dense clumps, in which new stars will be born.

## STELLAR NURSERIES

As well as being among the most beautiful objects in the universe, star-forming nebulae contain a combination of raw materials that makes star birth possible. These clouds of hydrogen molecules, helium, and dust can be massive systems hundreds of light-years across or smaller individual clouds, known as Bok globules. Although they may lie undisturbed for millions of years, disturbances can trigger these nebulae to collapse and fragment into smaller clouds from which stars are formed. Remnants from the star-forming nebulae will surround the stars, and the stellar winds produced by the new stars can, in turn, cause these remnants to collapse. If the clouds are part of a larger complex, this can become a great stellar nursery. Massive stars have relatively short lives, and they can be born, live, and die as a supernova while their less-massive siblings are still forming. The shock wave from the supernova may plow through nearby interstellar matter, triggering yet more star birth.

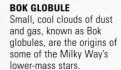

**BOK GLOBULE**
Small, cool clouds of dust and gas, known as Bok globules, are the origins of some of the Milky Way's lower-mass stars.

Bok globule

stellar EGGS

**STELLAR EGGS**
Within the evaporating gaseous globules (EGGS) of the Eagle Nebula, interstellar material is collapsing to form stars.

**FORMATION IN ACTION**
This spectacular vista captures the outstretched wings of the so-called Seagull Nebula, a star-forming region in the constellation of Monoceros. Young, bright stars are still surrounded by glowing nebulosity, while dark clouds of denser gas and dust silhouetted against the glowing background mark locations where star birth is still continuing.

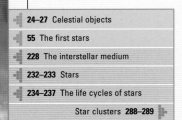

# TRIGGERS TO STAR FORMATION

Clouds of interstellar material need a trigger to start them collapsing, since they are held up by their own pressure and that of internal magnetic fields. Such a trigger might be as simple as the gravitational tug from a passing star, or it might be a shock wave caused by the blast from a supernova or the collision of two or more galaxies. In spiral galaxies such as the Milky Way, density waves move through the dust and gas in the galactic disk (see p.227). As the waves pass, they temporarily increase the local density of interstellar material, causing it to collapse. Once the waves have passed, their shape can be picked out by the trails of bright young stars.

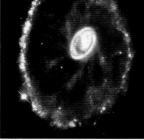

**GALACTIC COLLISIONS**
A ring of stars is created when two galaxies collide. Here, shock waves have rippled out, triggering star formation in the interstellar material.

**FROM OLD TO NEW**
Shock waves and material from a supernova blast spread out through the interstellar medium, triggering new star formation.

## STAR CLUSTERS

When they have formed from the fragmentation of a single collapsing molecular cloud, young stars are often clustered together. Many stars are formed so close to their neighbors that they are gravitationally bound, and some are even close enough to transfer material. It is unusual for a star not to be in a multiple system such as a binary pair (see pp.274–275), and in this respect, the Sun is uncommon. Stars within a cluster usually have a similar chemical composition, although, since successive generations of stars may be produced by a single nebula, clusters may contain stars of different ages (see pp.288–289). Remnants of dust and gas from the initial cloud will linger, and the dust grains often reflect the starlight, predominantly in the shorter blue wavelengths.

Thus, young star clusters are often surrounded by distinctive blue reflection nebulae. Young stars are hot and bright, and any nearby interstellar material will be heated by new stars' heat, producing red emission nebulae. Stars' individual motions will eventually cause a young star cluster to dissipate, though multiple stellar systems may remain gravitationally bound and may move through a galaxy together.

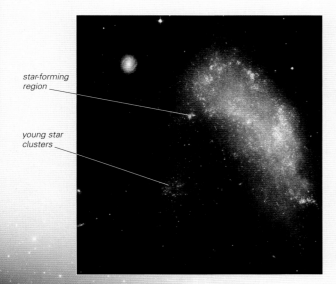

star-forming region

young star clusters

**VIOLENT STAR FORMATION**
Young star clusters (blue) and star-forming regions (pink) abound in NGC 1427A. As the galaxy's gas collides with the intergalactic medium through which the galaxy is traveling, the resulting pressure triggers violent but stunning star-cluster formation.

## TOWARD THE MAIN SEQUENCE

As collapsing fragments of nebulae continue to shrink, their matter coalesces and contracts to form protostars. These stellar fledglings release a great deal of energy as they continue to collapse under their own gravity. However, they are not easily seen because they are generally surrounded by the remnants of the cloud from which they formed. The heat and pressure generated within protostars acts against the gravity of their mass, opposing the collapse. Eventually, matter at the centers of the protostars gets so hot and dense that nuclear fusion starts and a star is born. At this stage, stars are very unstable. They lose mass by expelling strong stellar winds, which are often directed in two opposing jets channeled by a disk of dust and gas that forms around their equators. Gradually, the balance between gravity and pressure begins to equalize and the stars settle down on to the main sequence (see pp.234–237).

### J. L. E. DREYER

Danish-Irish astronomer John Louis Emil Dreyer (1852–1926) compiled the New General Catalog of Nebulae and Clusters of Stars, from which nebulae and galaxies get their NGC number. At the time of compilation, it was not known if all the nebulous objects were within the Milky Way. Dreyer studied the proper motions of many and concluded the "spiral nebulae," now known to be spiral galaxies, were likely to be more distant objects.

polar gas jets

accretion disk

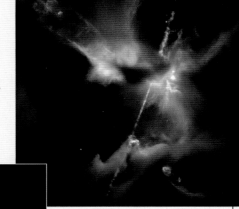

**ADOLESCENT STAR**
As young stars gather more material onto their equators, they eject excess matter from their pole (left). This can produce twin beams of glowing gas known as Herbig-Haro objects. HH 24 (above) is one spectacular example.

# STAR-FORMING NEBULAE

Star formation can be seen throughout the Milky Way, but it is principally evident in the spiral arms and toward the galactic center, where there is an abundance of star-making ingredients: dust and gas. In these regions, the interstellar matter is dense enough for molecular clouds to exist. These clouds are cold and appear as dark nebulae that are visible only when framed against a brighter background. When stars are born, these clouds are illuminated from within to become emission nebulae, some of the most beautiful sights in the Milky Way.

**STELLAR NURSERY**
Bright young stars within the Omega Nebula, M17, light up the nebula from which they were born.

## DARK NEBULA

# BHR 71

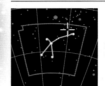

**CATALOG NUMBER**
BHR 71

**DISTANCE FROM SUN**
600 light-years

**MUSCA**

The small dark nebula BHR 71 is called a Bok globule (see p.238) and has a diameter of about 1 light-year. Within the dark molecular cloud are two sources of infrared and radio rays believed to be very close embryonic stars: HH 320 and HH 321, both losing vast amounts of material as they collapse. HH 320 has the strongest outflow, and it is probably surrounded by a massive disk of previously ejected stellar material. Although not optically visible, HH 320 has 10 times the luminosity of the Sun. BHR 71 and its protostars offer a rare opportunity for the study of star-formation processes.

**BINARY FORMATION**
Jets from BHR 71's newly forming binary star system have created the filamentary structure seen in this composite image made from four separate images.

## DARK NEBULA

# Horsehead Nebula

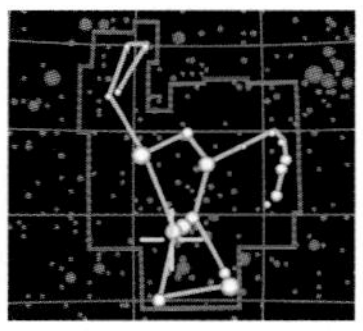

**CATALOG NUMBER**
Barnard 33

**DISTANCE FROM SUN**
1,500 light-years

**ORION**

One of the most beautiful and well-known astronomical sights, the Horsehead Nebula can be located in the night sky just south of the bright star Zeta (ζ) Orionis, the left star of the three in Orion's belt (see pp.390–391). The nebula is an extremely dense, cold, dark cloud of gas and dust, silhouetted against the bright, active nebula IC 434. It is about 16 light-years across and has a total mass about 300 times that of the Sun. The Horsehead shape is sculpted out of dense interstellar material by the radiation from the hot young star Sigma (σ) Orionis. Within the dark cloud from which the Horsehead rears is a scattering of young stars in the process of forming. The streaks that extend through the bright area above the Horsehead are probably caused by magnetic fields within the nebula.

**DARK KNIGHT**
One of the most photographed objects in the night sky, this dark nebula resembles the head of a sea horse or a knight on a chessboard. Its unusual shape was first discovered on a photographic plate in 1888.

**NEW STARS**
At the top of this image are the Trapezium stars forming within the Orion Nebula. Also visible, toward the bottom left-hand corner, is a line of shock waves created by material outflowing from the embryonic stars at a speed of 450,000 mph (720,000 kph).

## EMISSION NEBULA
# Orion Nebula

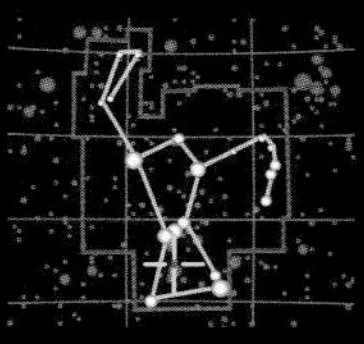

**CATALOG NUMBERS**
M42, NGC 1976

**DISTANCE FROM SUN**
1,500 light-years

**MAGNITUDE** 4

**ORION**

The most famous and the brightest nebula in the night sky, the Orion Nebula is easily visible with the naked eye as a diffuse, reddish patch below Orion's belt (see pp.390–391). It is also the closest emission nebula to Earth and has been extensively studied. The nebula spans about 30 light-years and has an apparent diameter four times that of a full moon. However, it is a small part of a much larger molecular cloud system known as OMC-1, which has a diameter of several hundred light-years. The Orion Nebula sits at the edge of OMC-1, which stretches as far as the Horsehead Nebula (opposite). The nebula glows with the ultraviolet radiation of the new stars forming within it. Many of these stars have been shown to have protoplanetary disks surrounding them. The principal stars whose radiation is ionizing the cloud of dust and gas belong to the Trapezium star cluster (see p.391), located at the heart of the nebula. At about 30,000 years old, the Trapezium is one of the youngest clusters known. It is a quadruple star system consisting of hot OB stars (see pp.232–233). In 1967, an extended dusty region was discovered directly behind the Orion Nebula. Known as the Kleinmann–Low Nebula, it has strong sources of infrared radiation embedded within it. These sources are believed to be protostars and newly formed stars.

## EXPLORING SPACE
# FIRST PHOTOGRAPH

A pioneer of astrophotography, the American scientist Henry Draper (1837–1882) took the first photo of a nebula in September 1870 after he turned his camera to the Orion Nebula, the brightest one in the sky. Although his photograph was relatively crude, 12 years later, he used an 11-in (28-cm) photographic refractor to obtain a much-improved image. The Orion Nebula has since been photographed probably more times than any other nebula.

**THE GREAT ORION NEBULA**
This view was captured with the VISTA telescope in Chile. It is an infrared image, revealing newborn stars within the nebula's dusty interior.

THE MILKY WAY

## DARK NEBULA

# Cone Nebula

**CATALOG NUMBER**
NGC 2264

**DISTANCE FROM SUN**
2,500 light-years

**MAGNITUDE** 3.9

**MONOCEROS**

Discovered by William Herschel in 1785, the Cone Nebula is a dark nebula located at the edge of an immense, turbulent star-forming region. This conical pillar of dust and gas is more than 7 light-years long and at its "top" is 2.5 light-years across. The Cone Nebula is closely associated with the star cluster NGC 2264, commonly known as the Christmas Tree Cluster. This cluster,

which spans a distance of 50 light-years, is made up of at least 250 stars, and it is the light from some of its newborn stars that allows the Cone Nebula to be seen in silhouette. The Cone Nebula is located at the top of the Christmas Tree Cluster, pointing downward to the bottom of the tree. At the opposite end, the 5th-magnitude star S Mon marks the left of the base of the tree (see below left). Jets of stellar material thrown out by newly forming stars have been detected within the star cluster. These Herbig-Haro objects also help shape the material in the surrounding nebula. One explanation for the shape of the Cone Nebula suggests that it was formed by stellar wind particles from an energetic source blowing past a Bok globule at the head of the cone. Buried in the dust and gas near

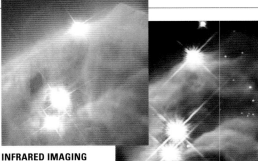

**INFRARED IMAGING**
Unseen in an optical image (left), a remarkable infrared view of the tip of the Cone Nebula (right) reveals, to the right of the image, a clutch of faint newborn stars.

the top of the Cone is a massive star known as NGC 2264 IRS, which is surrounded by six smaller Sun-like stars. It is thought that the outflow of stellar material during the early years of this massive star triggered the formation of the surrounding six and also helped to sculpt the shape of the Cone Nebula itself. None of these stars are visible with optical telescopes. Infrared observations have revealed further embryonic stars embedded in the nebulosity (above), making this one of the most active star-forming regions in this area of the Milky Way.

**CHRISTMAS TREE CLUSTER**
The stars of the open cluster NGC 2264 can be seen in this image resembling an upside-down Christmas tree, with the Core Nebula (boxed) at the apex of the tree.

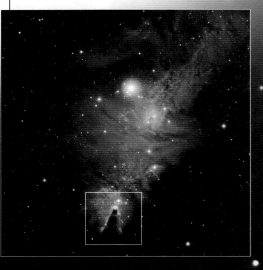

**TOWER OF RESISTANCE**
Born in immense clouds of dust and gas, the great tower of the Cone Nebula is a slightly denser region of material that has resisted erosion by radiation from its neighboring stars.

## EMISSION NEBULA

# IC 1396

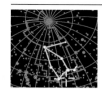

**CATALOG NUMBER**
IC 1396

**DISTANCE FROM SUN**
3,000 light-years

**CEPHEUS**

Occupying an area hundreds of light-years across, the IC 1396 complex contains one of the largest emission nebulae close enough to be observed in detail. It has an apparent diameter in the night sky 10 times that of a full moon. The mass of the nebula is estimated to be an immense 12,000 times the mass of the Sun, mainly consisting of hydrogen and helium in various forms. HD 206267, a massive, young blue star at the center of the region, produces most of the radiation that illuminates the nebula's interstellar material. Observations have shown that ionized clouds form a rough ring around this star at distances between 80 and 130 light-years. These clouds are the remains of the molecular cloud that originally gave birth to HD 206267 and its siblings, which compose the star cluster known as Tr37. Tracts of cool, dark material lie farther away from HD 206267. Among the most dramatic of these

**GIGANTIC STELLAR NURSERY**
The immense IC 1396 complex of emission nebulae, dark nebulae, and a young star cluster is shown here in a composite image. Mu Cephei is located at the center, and the Elephant's Trunk Nebula is boxed.

structures is one commonly known as the Elephant's Trunk Nebula. Research suggests that some of this material has been blown away from the star by strong stellar winds, causing the material to form elongated structures such as the Elephant's Trunk. Some of these structures stretch radially away from HD 206267 for up to 20 light-years. Within IC 1396 lies Mu (μ) Cephei, also known as Herschel's Garnet Star. One of the largest and brightest stars known, Mu Cephei is a red supergiant emitting 350,000 times the power of the Sun.

**ELEPHANT'S TRUNK**
The Elephant's Trunk Nebula is sculpted from a huge cloud of interstellar material in which star formation may take place in the future.

## EMISSION NEBULA

# DR6

**CATALOG NUMBER**
DR6

**DISTANCE FROM SUN**
4,000 light-years

**CYGNUS**

Strong stellar winds from about 10 young stars at the center of this unusual nebula have created cavities within its interstellar material, making it resemble a human skull. The nebula has a diameter of about 15 light-years, and the "nose," where the stars that have sculpted the nebula are located, is about 3.5 light-years across. The central group of stars is very young, having formed less than 100,000 years ago. The picture below is a composite of four infrared images.

**GALACTIC GHOUL**

**TWISTS OF GAS**
Creative chaos is revealed within the vast Lagoon Nebula, as radiation and strong winds from newborn stars interact with surrounding clouds of interstellar dust and gas.

**DARK GLOBULES**
A Hubble Space Telescope image shows knots of dense gas and dust shaped by stellar winds from the nebula's newborn stars. Curtainlike streaks of dark and light show where the nebula is being eroded by the pressure of radiation.

## EMISSION NEBULA

# Lagoon Nebula

**CATALOG NUMBERS**
M8, NGC 6523

**DISTANCE FROM SUN**
5,200 light-years

**MAGNITUDE** 6

**SAGITTARIUS**

The Lagoon Nebula is a productive star-forming region situated within rich, conspicuous fields of interstellar matter. Covering an apparent diameter of more than three full moons, the Lagoon Nebula is so large and luminous that it is visible to the naked eye. The region contains young star clusters, distinctive Bok globules (see p.238), and very energetic star-forming regions. There are also many examples of twisted-rope structures thought to have been created by hot stellar winds colliding with cooler dust clouds. The bright center of the Lagoon Nebula is illuminated by the energy of several very hot young stars, including the 6th-magnitude 9 Sagittarii and the 9th-magnitude Herschel 36. Also found in the brightest region is the famous Hourglass Nebula (see p.263). The open cluster NGC 6530 (to the left of center in the main image) contains 50 to 100 stars that are only a few million years old. Clearly visible across the Lagoon Nebula are dark Bok globules.

## EMISSION NEBULA

# Eagle Nebula

**CATALOG NUMBER**
IC 4703

**DISTANCE FROM SUN**
7,000 light-years

**MAGNITUDE** 6

**SERPENS**

Observations of the Eagle Nebula have introduced new ideas into the theory of star formation. Lying in one of the dense spiral arms of the Milky Way, this is an immense stellar nursery where young stars flourish, new stars are being created, and the material and triggers exist for future star formation. In optical wavelengths, this region is dominated by the light from the bright young star cluster M16. This cluster was discovered by the Swiss astronomer Philippe Loys de Chéseaux in around 1745, but it was nearly 20 years later that the surrounding nebula, from which the star cluster had formed, was discovered by Charles Messier (see p.73). The star cluster itself is only about 5 million years old and has a diameter of about 15 light-years. The Eagle Nebula is much larger than the star cluster, with a diameter of about 70 light-years.

In 1995, the Hubble Space Telescope imaged features within the nebula that are commonly known as the Pillars of Creation (see panel, below). These famous pillars are towers of dense material that have resisted evaporation by radiation from local young stars. However, the stars' ultraviolet radiation is gradually boiling their surfaces away through a process called photoevaporation. Because the towers themselves are not of a consistent density, the continuing photoevaporation has caused some of the smaller nodules, known as evaporating gaseous globules (EGGs), to become detached from the main gas towers. At this point, these dense stellar nurseries cease to accrue more material, and any embryonic star

within has its upper mass limit fixed. It is thought that this type of star formation inhibits the formation of accretion disks around the stars, which are believed to be the material from which planets are formed. These detailed images of the Pillars of Creation were the first to suggest this process of star creation. The Eagle Nebula also contains many Bok globules, regions where future star formation is probably occurring.

**HUGE STELLAR NURSERY**
This wide-field image shows the immensity of the Eagle Nebula, with the three Pillars of Creation located near the center. This huge cloud of gas lies in the galaxy's Sagittarius–Carina arm, toward the galactic center.

(see p.73)

---

## THE PILLARS OF CREATION

This image, taken by the Hubble Space Telescope in 1995, has become one of the most famous and iconic astronomical images. Revealing for the first time in dramatic detail a previously unsuspected process of star formation, it captured the public's imagination and inspired a new interest in astronomy. The image's aesthetic appeal and the sense of wonder it inspires have led to it being displayed on posters, in magazines, and even on stamps.

**STELLAR CLOSE-UP**
These spectacular pillars of dust and gas are several light-years long but represent only a small section of the Eagle Nebula.

**TWISTED PILLARS**
The three Pillars of Creation are shown twisting through a rich star field in this composite infrared image. Not all these stars are in the Eagle Nebula—some lie far behind and others lie in front.

## EMISSION NEBULA

# IC 2944

**CATALOG NUMBER**
IC 2944

**DISTANCE FROM SUN**
5,900 light-years

**MAGNITUDE** 4.5

**CENTAURUS**

Between the constellations Crux and Centaurus lies the bright, busy star-forming nebula IC 2944. This nebula is made up of dust and gas that is illuminated by a loose cluster of massive young stars. IC 2944 is perhaps best known for the many Bok globules that are viewed in silhouette against its backdrop. Bok globules are thought to be cool, opaque regions of molecular material that will eventually collapse to form stars. However, studies of the globules in IC 2944 have

### THACKERAY'S GLOBULES
The Bok globules in IC 2944 were first observed in 1950 by South African astronomer A. D. Thackeray. This globule has recently been shown to be two overlapping clouds.

revealed that the material of which they are composed is in constant motion. This may be caused by radiation from the loose cluster of massive young stars embedded in IC 2944. The stars' ultraviolet radiation is gradually eroding the globules, and it is possible that this could prevent them from collapsing to form stars. In addition to radiation, the stars also emit strong stellar winds that send out material at high velocities, causing heating and erosion of interstellar material. The largest Bok globule in IC 2944 (below) is about 1.4 light-years across, with a mass about 15 times that of the Sun.

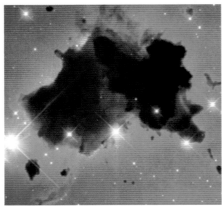

## EMISSION NEBULA

# Sharpless 29

**CATALOG NUMBER**
SH 2-29

**DISTANCE FROM SUN**
4,100 light-years

**MAGNITUDE** 8

**SAGITTARIUS**

Lying close to the brighter and more famous Lagoon nebula in Sagittarius, Sharpless 29 is a compact region of star formation where the turbulence caused by radiation from newborn stars can be seen in detail. Stars within the nebula are around 2 million years old, and their fierce ultraviolet light excites nearby hydrogen gas to create a series of glowing halos. Meanwhile, radiation and stellar winds combine to blow bubbles within the nebula and ripple its silhouetted dust lanes.

### BINARY BUBBLE
Near the center of this image, a newborn binary system has blown a large cavity in the nebula, creating a bright shock front on its upper edge.

## EMISSION NEBULA

# Trifid Nebula

**CATALOG NUMBER**
M20

**DISTANCE FROM SUN**
7,600 light-years

**MAGNITUDE** 6.3

**SAGITTARIUS**

This emission nebula is one of the youngest yet discovered. It was first called the Trifid Nebula by English astronomer John Herschel because of its three-lobed appearance when seen through his 18th-century telescope. The nebula is a region of interstellar dust and gas being illuminated by stars forming within it. It spans a distance of around 50 light-years. The young star cluster at its center, NGC 6514, was formed only about 100,000 years ago. The Trifid's lobes, the brightest of which is actually a multiple system, are created by dark filaments lying in and around the bright nebula. The whole area is surrounded by a blue reflection nebula, particularly conspicuous in the upper part, where dust particles disperse light.

### HEART OF THE TRIFID
The main image, spanning about 20 light-years, reveals details of the NGC 6514 star cluster and the filaments of dust weaving through the Trifid Nebula. A wider view (above) shows the full breadth of the nebula.

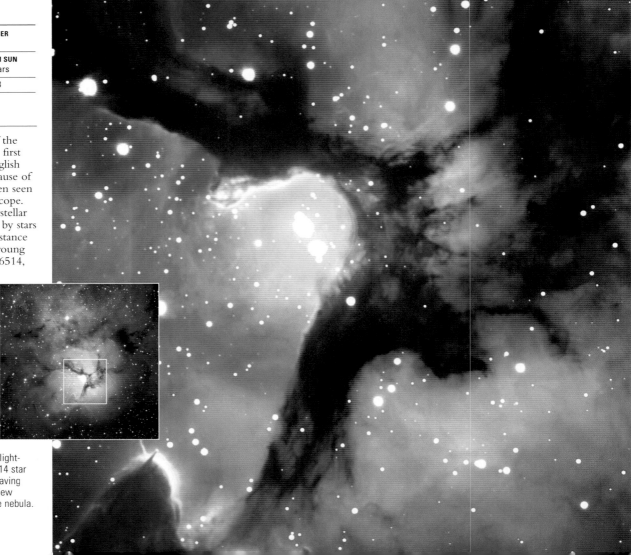

## EMISSION NEBULA

# Carina Nebula

**CARINA**

| | |
|---|---|
| **CATALOG NUMBER** | NGC 3372 |
| **DISTANCE FROM SUN** | 8,000 light-years |
| **MAGNITUDE** | 1 |

Also known as the Eta (η) Carinae Nebula, this is one of the largest and brightest nebulae to be discovered. It has a diameter of more than 200 light-years, stretching up to 300 light-years if its fainter outer filaments are included. Within its heart, and heating up its dust and gas, is an interesting zoo of young stars. These include examples of the most massive stars known, with a spectral type of O3 (see pp.232–233). This type of star was first discovered in the Carina Nebula, and the nebula remains the closest location of O3 stars to Earth. Also within the Carina Nebula are three Wolf–Rayet stars with spectral type WN (see pp.254–255). These stars are believed to be evolved O3 stars with very large rates of mass ejection. One of the best-known features within the Carina Nebula is the blue supergiant star Eta (η) Carinae (see p.262), embedded within part of the nebula known as the Keyhole Nebula. Recent observations made with infrared

**PROBING THE NEBULA**
An infrared image reveals the stars lying within the nebula's dense dust and gas. The open clusters Trumpler 14 and Trumpler 16 are visible to the left and top of the image.

telescopes reveal that portions of the Carina Nebula are moving at very high speeds—up to 522,000 mph (828,000 kph)—in varying directions. Collisions of interstellar clouds at these speeds heat material to such high temperatures that it emits high-energy X-rays, and the entire Carina Nebula is a source of extended X-ray emission. The movement of these clouds of material is thought to be due to the strong stellar winds emitted by the massive stars within, bombarding the surrounding material and accelerating it to its high velocities.

**ERODING TOWER**
A tower of cool hydrogen gas and dust 3 light-years long extends from the Carina Nebula in this false-color Hubble image. The tower is being eroded by the energy from hot, young stars nearby.

## EMISSION NEBULA

# RCW 49

**CARINA**

| | |
|---|---|
| **CATALOG NUMBERS** | RCW 49, GUM 29 |
| **DISTANCE FROM SUN** | 14,000 light-years |

One of the most productive regions of star formation to have been found in the Milky Way, RCW 49 spans a distance of about 350 light-years. It is thought that over 2,200 stars reside within RCW 49, but because of the nebula's dense areas of dust and gas, the stars are hidden from view at optical wavelengths of light. However, the infrared telescope onboard the Spitzer spacecraft (see panel, right) has recently revealed the presence of up to 300 newly formed stars. Stars have been observed at every stage of their early evolution in this area, making it a remarkable source of data for studying star formation and development. One surprising preliminary observation suggests that most of the stars have accretion disks around them. This is a far higher ratio than would usually be expected. Detailed observations of two of the disks reveal that they are composed of exactly what is required in a planet-forming system. These are the farthest and faintest potential planet-forming disks ever observed. This discovery supports the theory that planet-forming disks are a natural part of a star's evolution. It also suggests that solar systems like our own are probably not rare in the Milky Way (see pp.296–299).

**COSMIC CONSTRUCTION**
This false-color image, composed of four separate images taken in different infrared wavelengths, reveals more than 300 newborn stars scattered throughout the RCW 49 nebula. The oldest stars of the nebula appear in the center in blue, gas filaments appear in green, and dusty tendrils are shown in pink.

## EXPLORING SPACE

# SPITZER TELESCOPE

Launched in August 2003, the Spitzer Telescope is one of the largest infrared telescopes put into orbit. It has been very successful in probing the dense dust and gas that lies in the interstellar medium and has revealed features and details within star-forming clouds that have never been seen before. As Spitzer observes in infrared, its instruments are cooled almost to absolute zero to ensure that their own heat does not interfere with the observations. A solar shield protects the telescope from the Sun.

**INSIDE SPITZER**
The Spitzer craft has a 34-in (85-cm) telescope and three super-cooled processing instruments.

**THE MILKY WAY**

**THE CARINA NEBULA**
This panoramic infrared image of the Carina Nebula, taken using the Very Large Telescope (see p.90), reveals countless newborn stars and complex radiation-sculpted features embedded within the maelstrom of gas and dust. The nebula contains at least a dozen stars that are 50 to 100 times the mass of the Sun, including the supergiant Eta Carinae (see p.254), an unstable star near the very end of its life.

# MAIN-SEQUENCE STARS

MAIN–SEQUENCE STARS are those that convert hydrogen into helium in their cores by nuclear reactions. Stars spend a high proportion of their lives on the main sequence, during which time they are very stable. The higher the mass of the star, the less time it spends on the main sequence, as nuclear reactions occur faster in higher–mass stars.

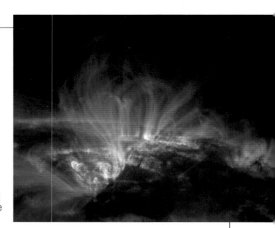

**STAR FLARES**
The Sun's photosphere radiates huge amounts of energy as solar flares contribute to the solar wind.

## STAR ENERGY

The cores of main-sequence stars initially consist mainly of hydrogen. When the temperature and pressure become high enough, the hydrogen is converted into helium by nuclear reactions. For stars of less than about 1.5 solar masses, this is done by means of a process called the proton–proton chain reaction (the pp chain). For stars of more than about 1.5 solar masses and with core temperatures of more than about 36 million °F (65 million °C), carbon, nitrogen, and oxygen are used as catalysts in a process called the carbon cycle (CNO cycle). When hydrogen is converted to helium, a tiny amount of energy is released as gamma rays, which gradually permeate their way out through the photosphere (the Sun's visible surface). The huge amounts of energy radiated by main-sequence stars are due to the immense masses of hydrogen they contain. In the core of the Sun, 600 million tons of hydrogen are converted into helium every second.

**MASSIVE STAR**
Achernar, or Alpha (α) Eridani, the ninth-brightest star in the sky, is a blue main-sequence star of about 6 to 8 solar masses. Main-sequence stars of this size convert hydrogen to helium through a process called the carbon cycle.

## STELLAR STRUCTURE

Energy, in the form of gamma rays, is released in the nuclear reactions occurring within stellar cores. This energy can be transported outward by two processes: convection and radiation. In convection, hot material rises to cooler zones, expanding and cooling, then sinks back to hotter levels, just like water being boiled in a saucepan. In the radiation process, photons are continually absorbed and reemitted. They can be emitted in any direction, and sometimes travel back into the central core. They follow a path called a "random walk" but gradually diffuse outward, losing energy as they do so. Their energy matches the temperature of the surrounding material, so they start as gamma rays, but at the Sun's surface (the photosphere), they appear in the visible part of the electromagnetic spectrum.

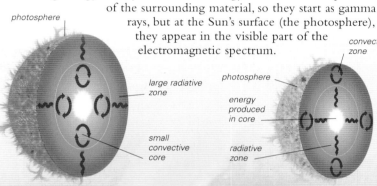

*photosphere*

*large radiative zone*

*small convective core*

*convective zone*

*photosphere*

*energy produced in core*

*radiative zone*

**HIGH-MASS STAR**
Stars with a mass greater than 1.5 solar masses produce energy through the CNO cycle. They have convective cores and a large radiative zone reaching to the photosphere.

**LOW-MASS STAR**
In stars with a mass smaller than 1.5 solar masses, the pp chain dominates and a large, inner radiative zone reaches out to a smaller convection zone near the star's photosphere.

**ERUPTIVE SURFACE**
Main-sequence stars, such as the Sun, appear smooth in optical light. In reality though, their photospheres are extremely turbulent, with huge prominences of material constrained by magnetic fields.

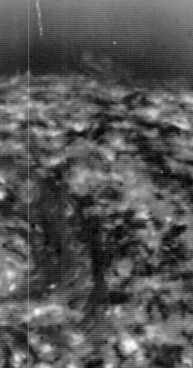

# ROTATION AND MAGNETISM

The pressures and temperatures within stars mean they are composed of plasma (see p.30). Within this ionized matter, negatively charged electrons travel free from the positively charged ions. This has a profound effect on magnetic fields, since charged particles do not cross magnetic field lines easily. Magnetic field lines can dictate the movement of stellar material, but the movement of the plasma can also affect magnetic fields. All stars rotate, and some spin so fast they bulge out at the equator and are very flattened at the poles. As stars rotate, magnetic field lines are carried around by the plasma. This "winds up" the magnetic field and creates pockets of intense magnetic flux where field lines are brought close together. The movement of stellar material and the transfer of heat is restricted in these areas, so they are appreciably cooler than the surrounding material. As they are cooler, they appear dark against the rest of the photosphere. Dark star spots on the surface of stars are areas of intense activity, because the buildup of heat around them can suddenly be released as flares.

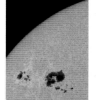
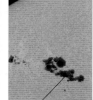

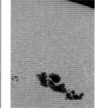
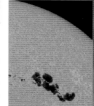

sun-spot group has rotated from previous position

region of equal area to the Earth

**SOLAR ROTATION**
As the Sun rotates, sun-spot groups are observed traveling across its disk. Main-sequence stars rotate differentially, with material at the equator rotating faster than that at the poles. On the Sun, sunspots closer to the equator travel across the solar disk more rapidly.

## ARTHUR EDDINGTON

British astronomer Arthur Stanley Eddington (1882–1944) studied the internal structure of stars and derived a mass–luminosity relationship for main-sequence stars. In 1926, he published *The Internal Constitution of Stars*, in which he suggested that nuclear reactions were the power source of stars. While working at the Royal Greenwich Observatory, Eddington led two expeditions to view total solar eclipses and in 1919 provided evidence for the theory of general relativity. Eddington also calculated the abundance of hydrogen within stars and developed a model for Cepheid variable pulsation (see p.282). He became Plumian professor of astronomy at Cambridge in 1913 and director of the Cambridge Observatory in 1914. He was knighted in 1930.

# THE MAIN SEQUENCE

A star enters the main sequence when it starts to burn hydrogen in its core. As soon as the nuclear reactions instigating this process begin, it is said to be at age zero on the main sequence. A star's life on the main sequence is very stable, with the pressure from the nuclear reactions in its core being balanced by its gravity trying to compress all of its mass into the center. A star will spend most of its life on the main sequence, and consequently about 90 percent of the stars observed in the sky are main-sequence stars. A star's time on the main sequence is dependent on its mass. The more massive the star, the hotter and denser its core and the faster it will convert hydrogen into helium. The Sun is a relatively small main-sequence star and will be on the main sequence for about 10 billion years. A 10-solar-mass star will be on the main sequence for only 10 million years. While on the main sequence, a star will conform to the mass–luminosity relation, which means that the absolute magnitude or luminosity of a star will give an indication of its mass. As it converts hydrogen into helium, a star's chemical composition and internal structure will change and it will move slightly to the right of its zero-age position on the H–R diagram (below). As soon as the hydrogen in the core is depleted and hydrogen burning in the atmosphere begins, the star leaves the main sequence (see p.236).

**DIAGONAL PATH**
The main sequence is a diagonal curving path of stars on the Hertzsprung–Russell diagram, a simplified version of which is shown here (see also p.232). The curve runs from bottom right (low mass and cool) to top left (massive and hot). Each star has a "zero-age" position (a point on the curve indicating its mass and temperature). It hardly strays at all from this position during its time on the main sequence.

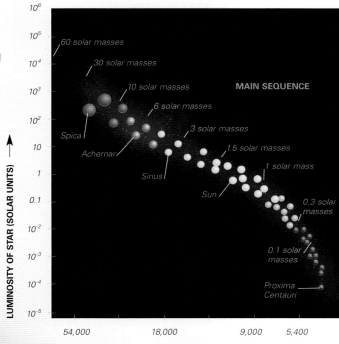

MAIN SEQUENCE

60 solar masses
30 solar masses
10 solar masses
6 solar masses
3 solar masses
Spica
1.5 solar masses
Achernar
1 solar mass
Sirius
Sun
0.3 solar masses
0.1 solar masses
Proxima Centauri

LUMINOSITY OF STAR (SOLAR UNITS)

$10^6$
$10^5$
$10^4$
$10^3$
$10^2$
$10$
$1$
$0.1$
$10^{-2}$
$10^{-3}$
$10^{-4}$
$10^{-5}$

54,000    18,000    9,000    5,400

← SURFACE TEMPERATURE OF STAR (°F)

# MAIN-SEQUENCE STARS

During a star's life, it passes through many phases, but most of its time will be spent on the main sequence. This means that the chances of seeing any star are greatest during its main-sequence lifetime. In fact, about 90 percent of all observed stars are on the main sequence. Although main-sequence stars are spread throughout the Milky Way, they appear predominantly in its plane and central bulge.

**PROMINENT STARS**
Known as the Pointers, Alpha and Beta Centauri are prominent main-sequence stars guiding the way to the Southern Cross.

---

## Proxima Centauri

| | |
|---|---|
| **DISTANCE FROM SUN** | 4.2 light-years |
| **MAGNITUDE** | 11.05 |
| **SPECTRAL TYPE** | M |

**CENTAURUS**

The closest star to the Sun, Proxima Centauri is far too faint to be seen with the naked eye—as a result, it was only discovered in 1915. A red dwarf star with just one-tenth the mass of the Sun, it is thought to be an outlying member of the Alpha Centauri system (see right), orbiting the central stellar pair at a distance of 10,000 au, with a period of at least 1 million years. Proxima is also a flare star—despite its faintness, it periodically emits huge bursts of radiation that see it brighten by a whole magnitude (see pp.282–283). Even during these eruptions, however, it remains 18,000 times dimmer than the Sun—despite becoming an intense source of low-energy X-rays and high-energy ultraviolet. In 2016, tiny shifts in the wavelength of Proxima's light (see p.297) revealed the presence of a roughly Earth-sized planet orbiting in Proxima's habitable zone where water might survive on the surface.

**FLARING DWARF**
Despite its general faintness, Proxima's harsh radiation outbursts make life on its orbiting planet extremely unlikely.

---

## Alpha Centauri

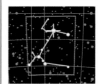

| | |
|---|---|
| **DISTANCE FROM SUN** | 4.3 light-years |
| **MAGNITUDES** | 0.0 and 1.3 |
| **SPECTRAL TYPES** | G and K |

**CENTAURUS**

The two stars of Alpha Centauri—also known as Rigil Kentaurus—orbit each other every 79.9 years. They are very close and, in some images (below), are distinguishable only by seeing two sets of diffraction spikes. Alpha Centauri A is the brighter and more massive, at 1.57 times the luminosity and 1.1 times the mass of the Sun. Alpha Centauri B is both less massive and less luminous than the Sun.

**ALPHA CENTAURI A AND B**

---

## Sirius A

| | |
|---|---|
| **DISTANCE FROM SUN** | 8.6 light-years |
| **MAGNITUDE** | -1.46 |
| **SPECTRAL TYPE** | A |

**CANIS MAJOR**

The brightest star in the night sky, Sirius is the ninth-closest star to Earth. It is a binary star, with Sirius A being a main-sequence star and its companion a white dwarf. Sirius A has twice the mass of the Sun and is 23 times as luminous. Recent observations suggest that it may have a stellar wind—the first spectral type A star to show evidence of one.

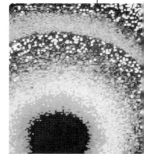

**SCORCHING STAR**
A false-color image shows the diffraction pattern of Sirius, the brightest star in the sky. Its name is from the Greek for "scorching."

---

## 61 Cygni

| | |
|---|---|
| **DISTANCE FROM SUN** | 11.4 light-years |
| **MAGNITUDES** | 5.2 and 6.1 |
| **SPECTRAL TYPE** | K |

**CYGNUS**

61 Cygni is a binary system of two main-sequence stars that orbit each other every 653 years. It is believed that 61 Cygni has at least one massive planet and possibly as many as three. In 1838, German astronomer Friedrich Bessel became the first to measure the distance of a star from Earth accurately, when he calculated 61 Cygni's annual parallax (see p.70). He chose 61 Cygni because, at that time, it was the star with the largest known proper motion.

**FAST STAR**

---

## Altair

| | |
|---|---|
| **DISTANCE FROM SUN** | 16.8 light-years |
| **MAGNITUDE** | 0.77 |
| **SPECTRAL TYPE** | A |

**AQUILA**

One of the three stars of the Summer Triangle, Altair is the 12th-brightest star in the sky. With a diameter about 1.6 times that of the Sun, it rotates once every 6.5 hours. This puts its equatorial spin rate at about 559,000 mph (900,000 kph), which causes distortion of its overall shape. This distortion is such that the star becomes wider at the equator and flattened at the poles, and estimates have suggested that its equatorial diameter is as much as double its polar diameter. It has a surface temperature of about 17,000°F (9,500°C) and a high rate of proper motion through the Milky Way.

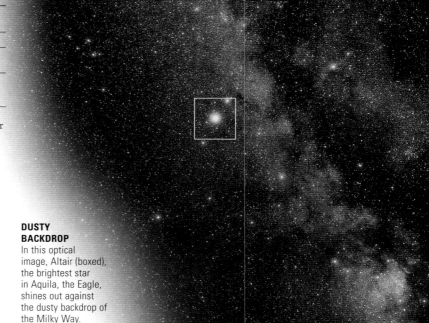

**DUSTY BACKDROP**
In this optical image, Altair (boxed), the brightest star in Aquila, the Eagle, shines out against the dusty backdrop of the Milky Way.

## WHITE STAR
# Fomalhaut

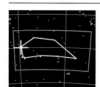

| | |
|---|---|
| **DISTANCE FROM SUN** | 25.1 light-years |
| **MAGNITUDE** | 1.16 |
| **SPECTRAL TYPE** | A |

**PISCIS AUSTRINUS**

The brightest star in Piscis Austrinus, Fomalhaut is the 18th-brightest star in the sky. It has a surface temperature of about 15,000°F (8,500°C), with a luminosity 16 times that of the Sun. In 1983, the infrared telescope IRAS revealed that it was a source of greater infrared radiation than expected. Further observations revealed that

**GLOWING RING**
This image from the ALMA radio telescope reveals Fomalhaut's sharply defined outer dust ring, which is thought to be kept in line by unseen planets.

the infrared radiation is being emitted by a ring of dust particles—with a diameter over twice that of the solar system—around Fomalhaut. In 2008, astronomers announced Hubble images that revealed a gas-giant-sized object near the outer edge of this ring, around 10.7 billion miles (17.2 billion km) from the star. However, later observations have raised doubts over whether "Fomalhaut b" is a true planet or a temporary accumulation of rubble within the debris ring.

**DISTINCTIVE STAR**
Fomalhaut, the "mouth of the fish," is the most distinctive star in the constellation Piscis Austrinus.

## WHITE STAR
# Vega

| | |
|---|---|
| **DISTANCE FROM SUN** | 25.3 light-years |
| **MAGNITUDE** | 0.03 |
| **SPECTRAL TYPE** | A |

**LYRA**

Also known as Alpha (α) Lyrae, Vega is the fifth-brightest star in the sky. Along with Altair (opposite) and Deneb, it makes up the Summer Triangle. Vega has a mass of about 2.5 solar masses, a luminosity 54 times that of the Sun, and a surface temperature of about 16,500°F (9,300°C). Around 12,000 years ago, it was the north Pole Star, and it will be so again in about 14,000 years. In 1983, the infrared satellite IRAS revealed that it is surrounded by a disk of dusty material that is possibly the precursor to a planetary system. Vega is the ultimate "standard" star used to calibrate the spectral range and apparent magnitude of stars in optical astronomy (see p.233).

**BRIGHT BEACON**
The brightest star in the northern summer sky, Vega takes its name from an Arabic word meaning "swooping eagle."

## YELLOW-WHITE STAR
# Porrima

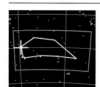

| | |
|---|---|
| **DISTANCE FROM SUN** | 38 light-years |
| **MAGNITUDE** | 2.74 |
| **SPECTRAL TYPE** | F |

**VIRGO**

**THE PORRIMA PAIR**

Porrima, also known as Gamma (γ) Virginis, is a binary system made up of two almost identical stars, both about 1.5 times the mass of the Sun. Their surface temperatures are around 13,000°F (7,000°C), and they appear creamy white in amateur telescopes. Their luminosities are each about four times that of the Sun. They orbit each other in a very elliptical path that takes around 170 years to complete.

## BLUE-WHITE STAR
# Regulus

| | |
|---|---|
| **DISTANCE FROM SUN** | 78 light-years |
| **MAGNITUDE** | 1.35 |
| **SPECTRAL TYPE** | B |

**LEO**

The brightest star in the constellation Leo, Regulus just makes it into the top 25 brightest stars as seen from Earth. *Regulus* is a Latin word meaning "little king." The star is situated at the base of the distinctive sickle asterism (shaped like a reversed question mark) in the constellation. It lies very close to the ecliptic (see pp.62–65) and is often occulted by the Moon (right). Regulus is a triple system. The brightest component is a blue-white

main-sequence star about 3.5 times the mass of the Sun and with a diameter also around 3.5 times that of the Sun. It has a surface temperature of about 22,000°F (12,000°C) and shines at about 140 times the brightness of the Sun. It is also an emitter of high levels of ultraviolet radiation. Regulus has a companion binary star composed of an orange dwarf and a red dwarf separated by about 9 billion miles (14 billion km). These dwarf components orbit each other over a period of about 1,000 years, and they in turn orbit the main star once every 130,000 years.

**REGULUS OCCULTED**
Poised at the top-left curve of the Moon, Regulus is about to be occulted as the Moon passes in front of it. Occultations can help astronomers determine the diameters of large stars and ascertain whether they are binary systems. Occultations by the Moon can also reveal details about the Moon's surface features.

## BLUE STAR
# Gamma Velorum

| | |
|---|---|
| **DISTANCE FROM SUN** | 840 light-years |
| **MAGNITUDE** | 1.8 |
| **SPECTRAL TYPES** | O and WR |

**VELA**

This blue star is also sometimes known as Regor—"Roger" spelled backward—in honor of the astronaut Roger Chaffee, who died in a fire during a routine test onboard the Apollo 1 spacecraft in 1967. Gamma (γ) Velorum is a complex star system dominated by a blue subgiant poised to evolve off the main sequence. Its evolution has been affected by being in a very close binary orbit with a star that is now a Wolf–Rayet star. They lie as close as Earth does to the Sun and orbit each other every 78.5 days. The Wolf–Rayet star is now the less massive component of the close binary, but probably started as the more massive and evolved much more rapidly. The subgiant has around 30 times the mass of the Sun, with a surface temperature of 60,000°F (35,000°C) and a luminosity around 200,000 times that of the Sun. There are also two other components to the system lying much farther away, one of which is a hot B-type star (see pp.232–233) at a distance of about 0.16 light-years.

# OLD STARS

232–233 Stars
234–237 The life cycles of stars
Stellar end points 266–267
The role of black holes 307

OLD STARS INCLUDE low-mass main-sequence stars that came into existence billions of years ago and also some high-mass stars that will explode as supernovae after existing for less than a million years. Some of the most beautiful sights in the Milky Way are old stars undergoing their death throes.

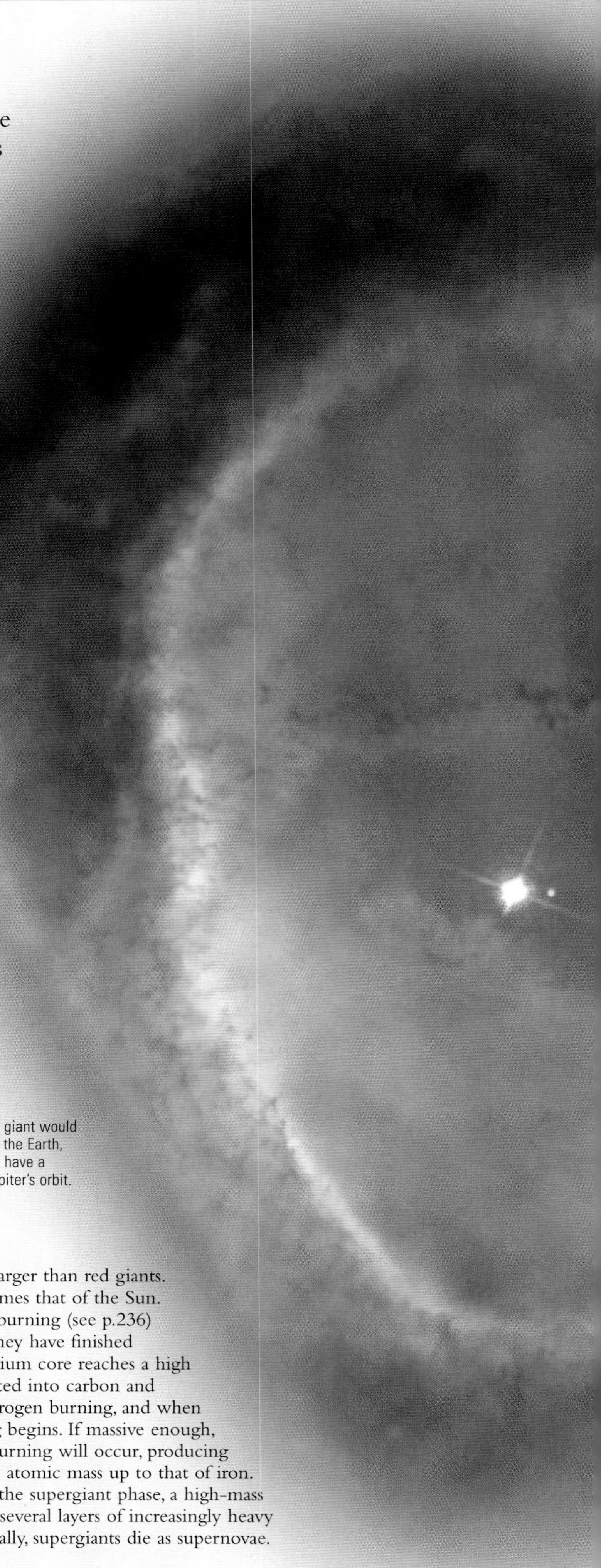

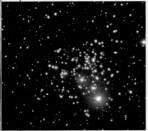

**EVOLVED STARS**
It is easy to pick out the evolved red giant stars in this image of the ancient star cluster NGC 2266.

## RED GIANTS

When a star has depleted the hydrogen in its core, it will start to burn the hydrogen in a shell surrounding the core. This shell gradually moves outward through the atmosphere of the star as fuel is used up. The expanding source of radiation heats the outer atmosphere, which expands and then cools. The result is a large star with a relatively low surface temperature. It remains luminous because of its huge size, though some red giants are hidden from view by extensive dust clouds.

Red giants have surface temperatures of 3,600–7,200°F (2,000–4,000°C) and radii 10–100 times that of the Sun. Because they are so large, gravity does not have much effect on their outer layers and they can lose a great deal of mass to the interstellar medium, either by stellar winds or in the form of planetary nebulae. Red giants are often variable stars; their outer layers pulsate, causing changes in luminosity (see p.282).

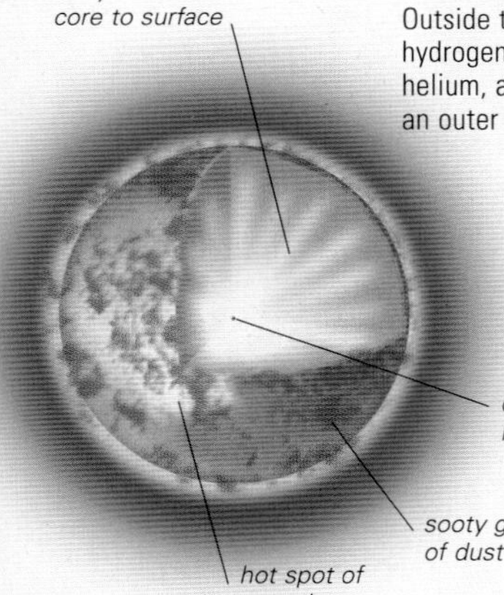

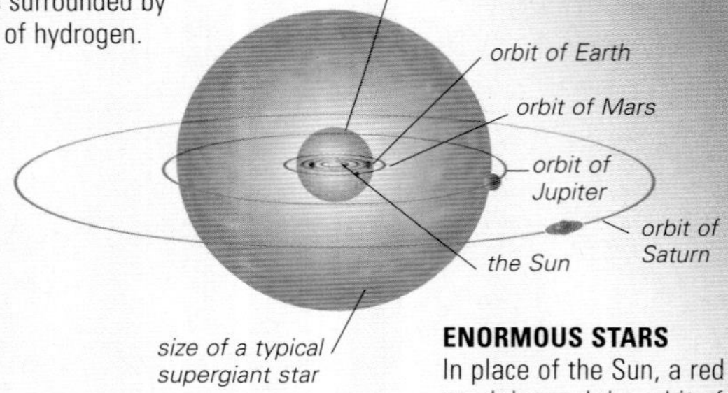

convection cells carry heat from core to surface
core of helium
sooty grains of dust
hot spot of escaping gas

**INSIDE A RED GIANT**
A red giant's helium core is contained by an inert helium shell. Outside this zone, a shell of hydrogen is being converted into helium, and this is surrounded by an outer envelope of hydrogen.

size of a large red giant star
orbit of Earth
orbit of Mars
orbit of Jupiter
orbit of Saturn
the Sun
size of a typical supergiant star

**ENORMOUS STARS**
In place of the Sun, a red giant would reach beyond the orbit of the Earth, while a supergiant would have a radius reaching out to Jupiter's orbit.

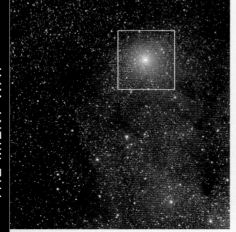

**GARNET STAR**
One of the largest stars visible in the night sky, Mu Cephei or the Garnet Star is a red supergiant with a radius greater than that of Jupiter's orbit.

## SUPERGIANTS

Stars of very high mass expand to become even larger than red giants. Red supergiants can have radii several hundred times that of the Sun. Just like red giants, they undergo hydrogen-shell burning (see p.236) and leave the main sequence (see p.232). When they have finished hydrogen-shell burning, they collapse and the helium core reaches a high enough temperature for the helium to be converted into carbon and oxygen. Helium-core burning is briefer than hydrogen burning, and when the helium core is depleted, helium-shell burning begins. If massive enough, further nuclear burning will occur, producing elements with an atomic mass up to that of iron. Near the end of the supergiant phase, a high-mass star will develop several layers of increasingly heavy elements. Eventually, supergiants die as supernovae.

# HELIUM FLASHING

Once hydrogen burning has produced a core of helium, if its temperature reaches higher than about 180 million °F (100 million °C), the helium will be fused together to form carbon. In stars of around 2 to 3 solar masses, helium burning can start in an explosive process called a helium flash. As the core collapses after hydrogen burning, it temporarily arrives at a dormant or "degenerate" state as the collapse is halted by the pressure between the helium's electrons. The temperature continues to rise, but the dormant core does not change in pressure, so it does not expand and cool. The rising temperature causes the helium to burn at an increasing rate, causing a "flash" that rids the core of the degenerate electrons. In higher-mass stars, the temperature rises high enough for helium fusion to begin before the core becomes degenerate.

**INSTABILITY STRIPS**
Many red giants and supergiants are pulsating variable stars that appear in regions of the H–R diagram (right, see also p.232) called instability strips. Three types of variable star are shown here.

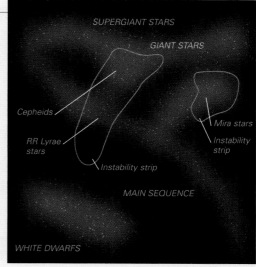

# WOLF–RAYET STARS

Massive stars, of about 10 solar masses, that have strong, broad emission lines in their spectra (see p.35) but few absorption lines are named Wolf–Rayet stars, after Charles Wolf and Georges Rayet (see p.264), who discovered them in 1867. They are hot, luminous stars whose strong stellar winds have blown away their outer atmospheres, revealing the stars' inner layers. They are broadly classified as WN, WC, and WO stars, depending on their spectra. The emission lines of WN stars are dominated by hydrogen and nitrogen, those of WC stars by carbon and helium, and those of WO by oxygen as well as carbon and helium. More than half of the known Wolf–Rayet stars are members of binary systems (see pp.274–275) with O or B stars as companions. It is believed that the Wolf–Rayet star was originally the more massive partner but lost its outer envelope to the companion star.

**GREAT ILLUMINATION**
A Wolf–Rayet star illuminates the heart of N44C, a nebula of glowing hydrogen gas surrounding young stars in the Large Magellanic Cloud.

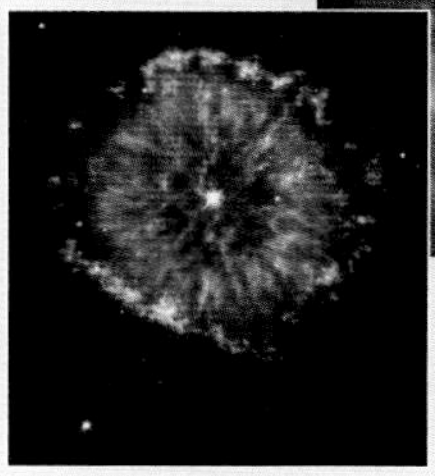

**STRONG WINDS**
The planetary nebula NGC 6751 may have a Wolf–Rayet star at its center. Its strong winds created the elaborate filaments.

# PLANETARY NEBULAE

Planetary nebulae are heated halos of material shed by dying stars. They were termed planetary nebulae by William Herschel in 1785 because of their disklike appearance through 18th-century telescopes. Planetary nebulae include some of the most stunning sights in the universe, contorted into various shapes by magnetic fields and the orbital motion of binary systems (see pp.274–275). They are composed of low-density gas thrown off by low-mass stars in the red-giant phase of their lives, and this gas is heated by the ultraviolet radiation given off by the hot inner cores of the dying stars. This stage of a star's life is relatively short. Eventually, the planetary nebula will disperse back into the interstellar medium, enriching the material there with the elements that have been produced by its parent star. These elements include hydrogen, nitrogen, and oxygen. At one time, the oxygen identified in the emission spectra of planetary nebulae (see p.35) was regarded as a new element called nebulium. It was later realized that "forbidden" emission lines of oxygen were present—forbidden because under usual conditions on Earth, they are very unlikely to occur. The central stars of planetary nebulae are among the hottest stars known. They are the contracting cores of red giants evolving into white dwarfs. Some planetary nebulae have been observed surrounding the resulting white dwarf. Current studies of planetary nebulae are revealing new facts about the late evolution of red giants and the manner of mass loss from these aging stars.

**BUTTERFLY NEBULA**
The Hubble 5 nebula is a prime example of a "butterfly" or bi-polar nebula, created by the funneling of expanding gas.

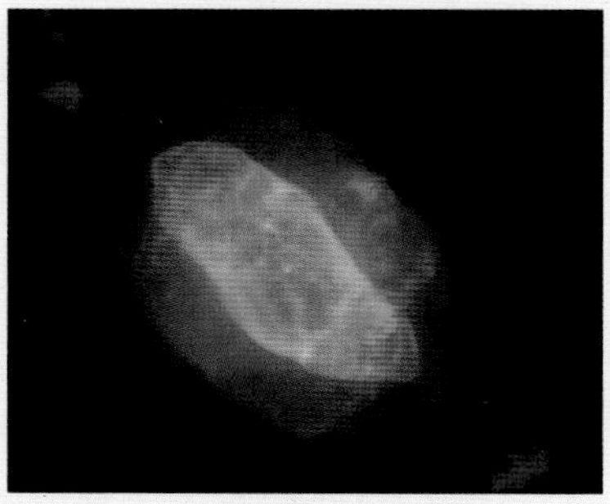

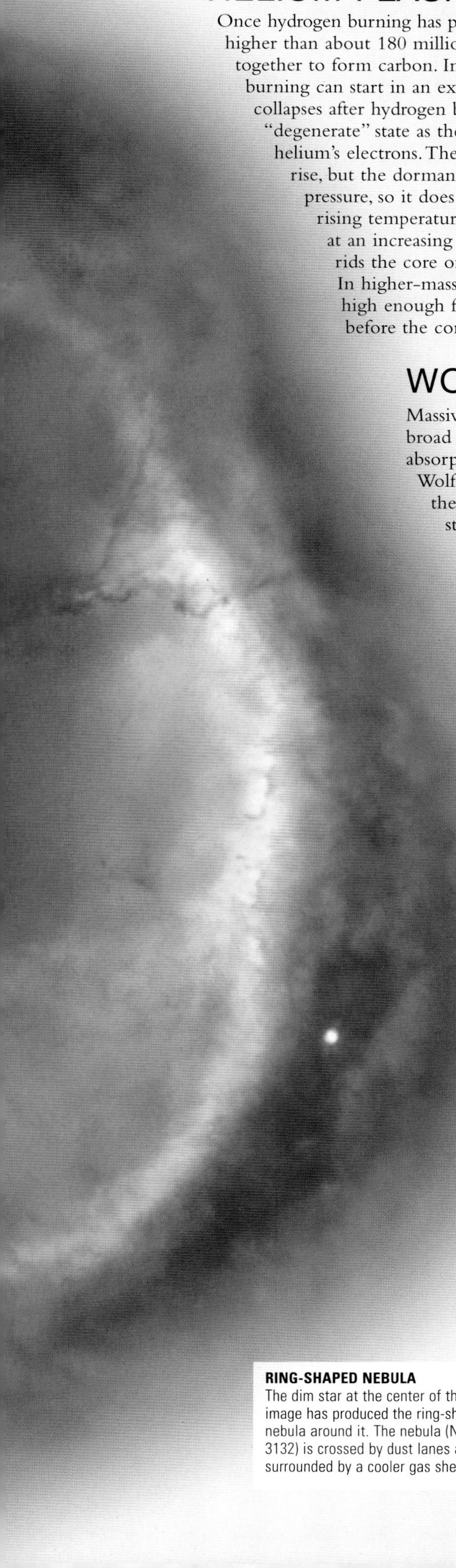

**RING-SHAPED NEBULA**
The dim star at the center of this image has produced the ring-shaped nebula around it. The nebula (NGC 3132) is crossed by dust lanes and surrounded by a cooler gas shell.

**UNUSUAL NEBULA**
The Saturn Nebula was shaped by early ejected material confining subsequent stellar winds into jets.

THE MILKY WAY

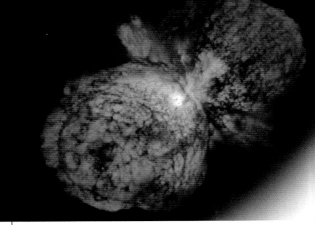

# OLD STARS

Some of the most visible and familiar bodies in the sky are stars that are approaching the ends of their lives or are experiencing their final death throes. In Wolf–Rayet stars and planetary nebulae, these old stars also present some of the most dramatic events and most beautiful sights in the universe. Although different types of old stars exist throughout the Milky Way, the oldest are situated far out in the galactic halo (see pp.226–229) or within the globular clusters (see pp.288–295). Some of these stars are nearly as old as the universe itself.

**DYING STAR**
Eta Carinae is a large, extremely old, and unstable star ejecting material into the interstellar medium. It could explode as a supernova at any time.

## RED GIANT
# Aldebaran

| DISTANCE FROM SUN | 67 light-years |
| --- | --- |
| MAGNITUDE | 0.85 |
| SPECTRAL TYPE | K5 |

**TAURUS**

Also known as Alpha (α) Tauri, Aldebaran is the brightest star in the constellation Taurus and the 13th-brightest star in the sky. Its surface temperature of only 6,740°F (3,730°C) makes it glow a dull red that can easily be seen by the naked eye. Aldebaran's diameter is about 45 times that of the Sun and, in place of the Sun, it would extend halfway to the orbit of Mercury. The star appears to be part of the Hyades cluster (see p.290), but this is a line-of-sight effect, with Aldebaran lying about 40 light-years closer to the Sun. This elderly star is a slow rotator, taking 2 years for each rotation, and an irregular variable, pulsating erratically. It has at least two faint stellar companions. Its name is derived from the Arabic *Al Dabaran*, meaning "the Follower," because it rises after

**BULL'S EYE**
The red tinge of Aldebaran makes it very distinctive against the whiter stars of the Hyades cluster. It is often depicted as the eye of the bull in the constellation Taurus.

the prominent Pleiades star cluster and pursues it across the sky. Aldebaran was one of the Royal Stars or Guardians of the Sky of ancient Persian astronomers and marked the coming of spring.

## RED SUPERGIANT
# Betelgeuse

| DISTANCE FROM SUN | 500 light-years |
| --- | --- |
| MAGNITUDE | 0.5 |
| SPECTRAL TYPE | M2 |

**ORION**

The right shoulder of the hunter, Orion, is marked by this distinctive, bright red star. Betelgeuse, or Alpha (α) Orionis, is a massive supergiant and the first star after the Sun to have its size reliably determined. Its diameter is more than twice that of the orbit of Mars, or about 500 times that of the Sun, and because of its huge size, it is about 14,000 times brighter. Betelgeuse is the 10th-brightest star in the sky, although as it pulsates, its brightness varies over a period of about 6 years.

**SURFACE SPOTS**
The infrared image of Betelgeuse above shows bright surface spots that could be convection cells. The infrared image at left shows gas and dust shed by the star, which has been masked by a black disk so that the gas and dust are visible.

It is a strong emitter of infrared radiation, which is produced by three concentric shells of material ejected by the star over its lifetime. It is slowly using up its remaining fuel and one day will probably explode as a supernova.

## RED SUPERGIANT
# Antares

| DISTANCE FROM SUN | 520 light-years |
| --- | --- |
| MAGNITUDE | 0.96 |
| SPECTRAL TYPE | M1.5 |

**SCORPIUS**

Antares or Alpha (α) Scorpii is the 15th-brightest star in the sky. Estimates of its diameter range from 280 to 700 times that of the Sun. It is about 15 times more massive than the Sun and shines 10,000 times brighter. This elderly star pulsates irregularly and has a binary companion that orbits in a period of about 1,000 years. This companion lies close enough to be affected by Antares' stellar wind and is a hot radio source. When viewed through an optical telescope, this blue companion looks green because of the color contrast with red Antares.

**A RIVAL OF MARS**
The glowing Antares (bottom right) looks a lot like Mars, the red planet. Its name derives from the Greek for "rival of Mars" (or *anti Ares*).

## RED GIANT
# TT Cygni

| DISTANCE FROM SUN | 1,500 light-years |
| --- | --- |
| MAGNITUDE | 7.55 |
| SPECTRAL TYPE | G |

**CYGNUS**

With a high ratio of carbon to oxygen in its surface layers, TT Cygni is known as a carbon star. The carbon, produced during helium burning, has been dredged up from inside the star. An outer shell, about half a light-year across, was emitted about 6,000 years before the star was as it appears to us now.

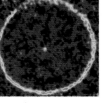

**CARBON RING**
This false-color image shows a shell of carbon monoxide surrounding the carbon star TT Cygni.

# Helix Nebula

| CATALOG NUMBER |
| --- |
| NGC 7293 |
| **DISTANCE FROM SUN** |
| Up to 650 light-years |
| **MAGNITUDE** 6.5 |

**AQUARIUS**

The Helix Nebula is the closest planetary nebula to the Sun, but its actual distance is uncertain, and estimates vary from 85 to 650 light-years. It is called the Helix Nebula because, from Earth, the outer gases of the star expelled into space give the impression that we are looking down the length of a helix. One of the largest known planetary nebulae, its main rings are about 1.5 light-years in diameter and span an apparent distance of more than half the width of a full moon. Its outer halo extends up to twice this distance. The dying star at the center of the nebula is destined to become a white dwarf, and as it continues to use up all its energy, it will continue to expel material into the interstellar medium. The Helix Nebula presents an impressive example of the final stage that stars like our Sun will experience before collapsing for the last time. It was first discovered by the German astronomer Karl Ludwig Harding in around 1824, and its size and proximity mean that it has been extensively observed and imaged.

**GLOWING HALO**
Rings of expelled material glow red in the light produced by nitrogen and hydrogen atoms when they are energized by ultraviolet radiation.

Detailed images made of the inner edge of the ring of material surrounding the central star have shown "droplets" of cooler gas twice the diameter of our solar system, radiating outward for billions of miles. These were probably formed when a fast-moving shell of gas, expelled by the dying star, collided with slower-moving material thrown off thousands of years before.

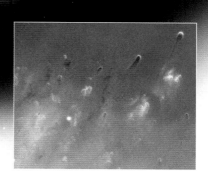

**COMETLIKE KNOTS**
Resembling comets, these tadpole-shaped gaseous knots are several billion miles across. They lie like spokes in a wheel along the inner edge of the ring of ejected gas surrounding the central star.

---

# Ring Nebula

| CATALOG NUMBER |
| --- |
| M57 |
| **DISTANCE FROM SUN** |
| 2,000 light-years |
| **MAGNITUDE** 8.8 |

**LYRA**

One of the best known planetary nebulae, the Ring Nebula was discovered in 1779 by French astronomer Antoine Darquier de Pellepoix. When seen through a small telescope, it appears larger than the planet Jupiter. Its central star, a planet-sized white dwarf of only about 15th magnitude, was not discovered until 1800, when it was found by German astronomer Friedrich von Hahn. There has been a great deal of discussion about the true shape of the Ring Nebula. Although it appears like a flattened ring, some astronomers believe the stellar material has been expelled in a spherical shell that only looks like a ring because we view it through a thicker layer at its edges. Others believe it is a torus (shaped like a ring doughnut), which would look similar to the Dumbbell Nebula if viewed side-on, or that it is cylindrical or tubelike. The nebula appears to be about 1 light-year in diameter, but it has an outer halo of material that extends for more than 2 light-years. This is possibly a remnant of the central star's stellar winds before the nebula itself was ejected. The nebula is lit by fluorescence caused by the large amount of ultraviolet radiation emitted by the central star. The rate of the ring's expansion indicates that the nebula started to form about 20,000 years before it was as it appears to us now.

**COMPLEX AND COLORFUL**
This unique composite image brings out details of the Ring Nebula, including a surrounding many-lobed hydrogen cloud and the blue glow of helium energized by radiation from the white dwarf at its center.

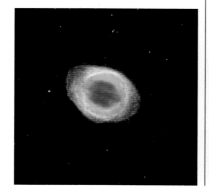

---

# Twin Jet Nebula

| CATALOG NUMBER |
| --- |
| M2–9 |
| **DISTANCE FROM SUN** |
| 2,100 light-years |
| **MAGNITUDE** 14.7 |

**OPHIUCHUS**

The Twin Jet Nebula is one of the most striking examples of a butterfly or bipolar planetary nebula. It is believed that the star at the center of this nebula is actually a binary pair with a period of about 100 years, whose complex interplay has affected the shape of the resulting planetary nebula. The gravitational interaction between the stars has pulled stellar material around them into a dense disk with a diameter about 10 times that of Pluto's orbit. About 1,200 years before this happened, one of the stars had an outburst, ejecting material in a strong stellar wind. This rammed into the disk, which acted like a nozzle, deflecting the material in perpendicular directions, forming the two lobes stretching out into space. This is very similar to the process that takes place in jet propulsion engines. Studies have suggested that the nebula's size has increased steadily with time and that the material is flowing outward at up to 450,000 mph (720,000 kph).

**BUTTERFLY WINGS**
This false-color image reveals some of the elements present in the nebula's lobes—red indicates sulfur, while blue highlights oxygen.

## PLANETARY NEBULA

# Red Rectangle Nebula

**CATALOG NUMBER**
HD 44179

**DISTANCE FROM SUN**
2,300 light-years

**MAGNITUDE** 9.02

**MONOCEROS**

Nature does not often create rectangles, so astronomers were surprised to observe this planetary nebula's unusual shape. The shape of the Red Rectangle Nebula is created by a pair of stars orbiting so close to each other that they experience gravitational interactions. This close binary star has created a dense disk of material around itself, which has restricted the direction of further outflows. This has caused subsequently ejected material to be expelled in expanding cone shapes perpendicular to the disk. Our view of the Red Rectangle is from the side, at right angles to these cones.

**COMPLEX STRUCTURE**
One of the most unusual celestial bodies in the Milky Way, the Red Rectangle Nebula has a distinctive shape that reflects an extremely complex inner structure.

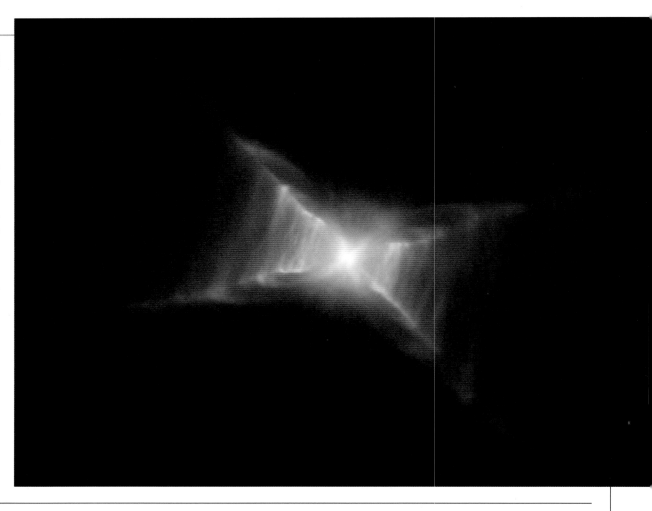

## PLANETARY NEBULA

# Cat's Eye Nebula

**CATALOG NUMBER**
NGC 6543

**DISTANCE FROM SUN**
3,000 light-years

**MAGNITUDE** 9.8

**DRACO**

The Cat's Eye Nebula is one of the most complex of all planetary nebulae. It is thought that its intricate structures may be produced either by the interactions of a close binary system or by the recurring magnetic activity of a solitary central star. At 3,000 light-years away, it is too far even for the Hubble Space Telescope to resolve its central star. The "eye" of the nebula is estimated to be more than half a light-year in diameter, with a much larger outer halo stretching into the interstellar medium. Although models of planetary nebulae once assumed a continuous outflow of stellar material, this nebula contains concentric rings that are the edges of bubbles of stellar material ejected at intervals. Eleven of these bubbles have been identified, possibly ejected at intervals of 1,500 years. The Cat's Eye also contains jets of high-speed gas, as well as bow waves created when the gas slammed into slower-moving, previously ejected material.

**WAVES AND SYMMETRIES**
A composite picture (above) shows emission from nitrogen atoms as red and oxygen atoms as green and blue shades, thus revealing successive waves of expelled stellar material. The nebula's symmetrical properties are further revealed by a false-color image processed to highlight its ring structure (right).

## PLANETARY NEBULA

# Egg Nebula

**CATALOG NUMBER**
CRL 2688

**DISTANCE FROM SUN**
3,000 light-years

**MAGNITUDE** 14

**CYGNUS**

The Egg Nebula's central star, which was a red giant until a few hundred years ago, is hidden by a dense cocoon of dust (visible in the image below as the dark band of material across the middle of the nebula). The material shed by the dying star is expanding at the rate of 45,000 mph (72,000 kph). Distinct arcs of material suggest a varying density throughout the nebula. The light from the central star shines like searchlights through the thinner parts of its cocoon and reflects off dust particles in the outer layers of the nebula.

**BRIGHT SEARCHLIGHTS**

## PLANETARY NEBULA

# Ant Nebula

**CATALOG NUMBER**
Menzel 3

**DISTANCE FROM SUN**
4,500 light-years

**MAGNITUDE** 13.8

**NORMA**

There are two main theories about what has caused the unusual shape of this planetary nebula. Either the central star is a close binary, its interacting gravitational forces shaping

the outflowing gas, or it is a single spinning star whose magnetic field is directing the material it has ejected. The expelled stellar material is traveling at around 2.25 million mph (3.6 million kph) and impacting into the surrounding slower-moving medium; the lobes of the nebula stretch to a distance of more than 1.5 light-years. Observations of the Ant Nebula may reveal the future of our own star, because its central star appears to be very similar to the Sun.

**HEAD AND THORAX**
Even through a small telescope, this planetary nebula resembles the head and thorax of a common garden ant.

## PLANETARY NEBULA

# Crescent Nebula

**CATALOG NUMBER**
NGC 6888

**DISTANCE FROM SUN**
4,700 light-years

**MAGNITUDE** 7.44

**CYGNUS**

The central star of the Crescent Nebula is a Wolf–Rayet star. Only about 4.5 million years after its formation (one-thousandth the age of the Sun), this massive star expanded to become a red giant and ejected its outer layers at about 22,000 mph (35,000 kph). Two hundred thousand years later, the intense radiation from the exposed, hot inner layer of the star began pushing gas away at speeds in excess of 2.8 million mph

**GASEOUS COCOON**
This composite image of the Crescent Nebula shows a compact semicircle of dense material surrounding a presupernova star (center). The Crescent spans a distance of about 3 light-years.

(4.5 million kph). This strong stellar wind expelled material equivalent to the Sun's mass every 10,000 years, forming a series of dense, concentric shells that are visible today. Typical of emission nebulae, the radiation from the hot central star excites the stellar material, principally hydrogen, causing it to shine in the red part of the spectrum. It is thought that the nebula's central star will probably explode as a supernova in about 100,000 years.

## WOLF–RAYET STAR

# WR 104

**DISTANCE FROM SUN**
4,800 light-years

**MAGNITUDE** 13.54

**SPECTRAL TYPE**
WCvar+

**SAGITTARIUS**

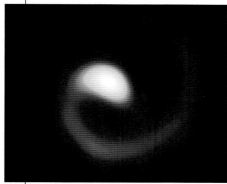

**STELLAR SPIRAL**

Like water from a cosmic lawn sprinkler, dust streaming from this rotating star system creates a pinwheel pattern. Since Wolf–Rayet stars are so hot that any dust they emit is usually vaporized, it is surprising that WR 104 has dust streaming away from it in this obvious spiral pattern. One theory is that this is a binary system, with each star emitting a strong stellar wind. Where these winds meet, there is a "shock front" that compresses the outflowing material, creating a denser, slightly cooler environment in which dust can exist. The orbital motion of the two stars then causes the spiral shape.

## PLANETARY NEBULA

# Eskimo Nebula

**CATALOG NUMBER**
NGC 2392

**DISTANCE FROM SUN**
5,000 light-years

**MAGNITUDE**
10.11

**GEMINI**

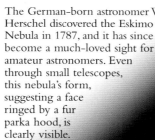

The German-born astronomer William Herschel discovered the Eskimo Nebula in 1787, and it has since become a much-loved sight for amateur astronomers. Even through small telescopes, this nebula's form, suggesting a face ringed by a fur parka hood, is clearly visible.

Hubble Space Telescope images reveal a complex structure featuring an inner nebula and an outer halo. The inner nebula consists of material ejected from the central star in two elliptical lobes around 10,000 years before the star was as we now see it. Each lobe is about 1 light-year long and about half a light-year wide and contains filaments of dense matter. Astronomers think that a ring of dense material around the star's equator, ejected during its red-giant phase, helped

create the nebula's "face." The surrounding "hood" contains unusual orange filaments, each about 1 light-year long, streaming away from the central star at up to 75,000 mph (120,000 kph). One explanation for these is that they were created when a fast-moving outflow from the central star impacted into slower-moving, previously ejected material.

**HOODED NEBULA**
In the center of this image, the apparent "face" of the Eskimo consists of one bubble of ejected material lying in front of the other, with the central star visible in the middle.

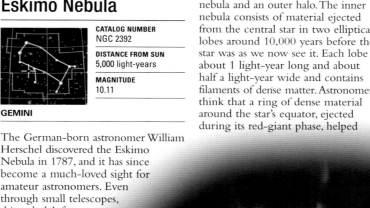

PLANETARY NEBULA

# Bug Nebula

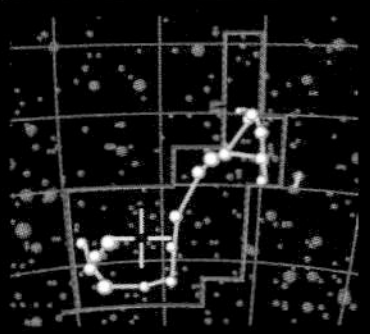

**CATALOG NUMBER**
NGC 6302

**DISTANCE FROM SUN**
4,000 light-years

**MAGNITUDE** 7.1

**SCORPIUS**

First discovered in 1826 by Scottish astronomer James Dunlop, then rediscovered in the late 19th century by the great American astronomer E. E. Barnard, the Bug Nebula is one of the brightest planetary nebulae. The central star is thought to have an extremely high temperature, and its intense ultraviolet radiation lights up the surrounding stellar material. The star itself is not visible at optical wavelengths because it is hidden by a blanket of dust. It is believed that the central star ejected a ring of dark material about 10,000 years before it was as we see it now, but astronomers cannot explain why it was not destroyed long ago by the star's ultraviolet emissions. The composition of the surrounding material is also surprising because it contains carbonates, which usually form when carbon dioxide dissolves in liquid water. Although ice exists in the nebula, along with hydrocarbons and iron, there is no evidence of liquid water.

**COLORFUL BUG**
The Bug Nebula is the ejected outer layers of a dying star that was once about five times the mass of the Sun. Ultraviolet radiation from the intensely hot central star is making the cast-off material glow.

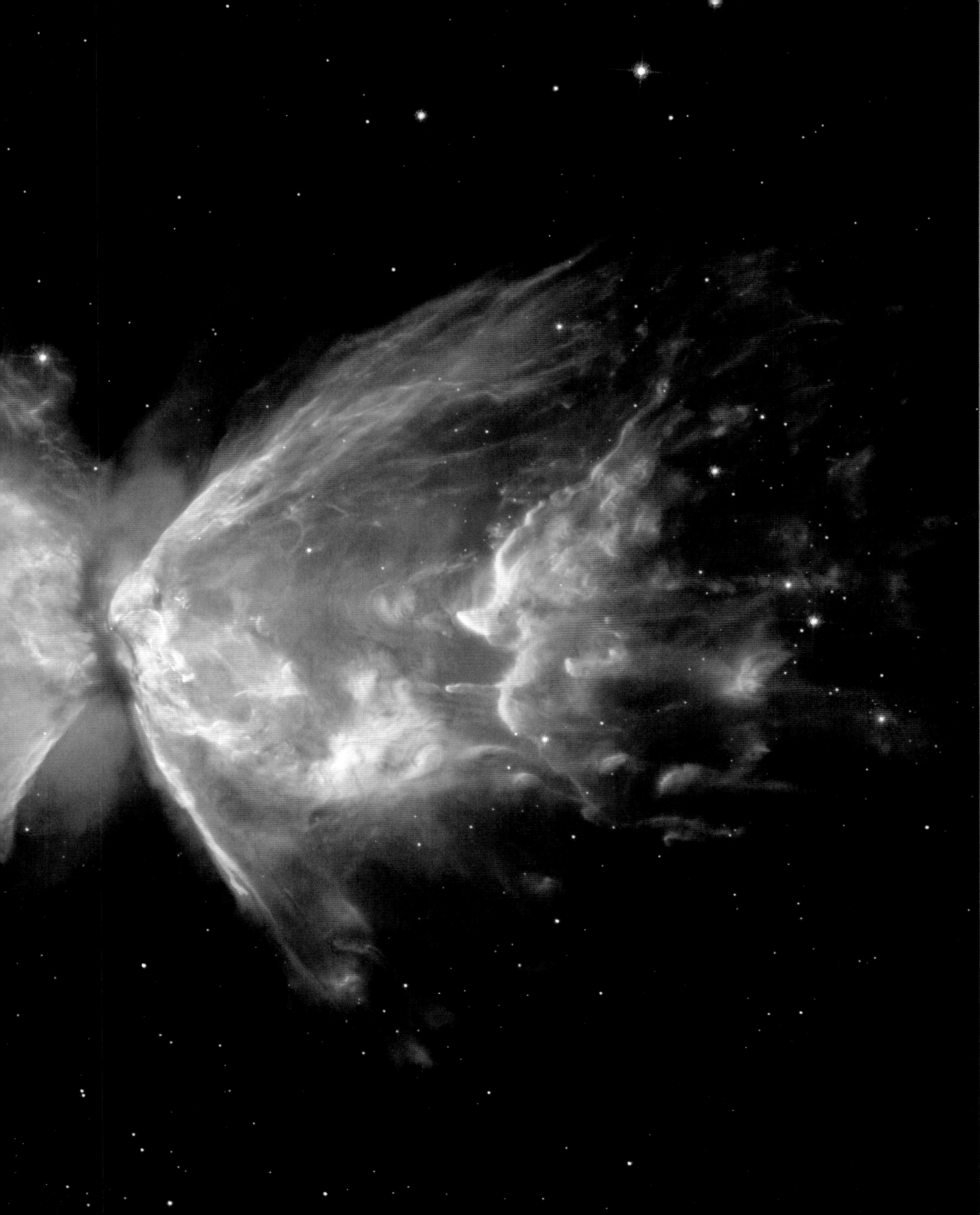

PLANETARY NEBULA

# Calabash Nebula

**CATALOG NUMBER**
OH231.8+4.2

**DISTANCE FROM SUN**
5,000 light-years

**MAGNITUDE** 9.47

**PUPPIS**

One of the most dynamic planetary nebulae, the Calabash Nebula's central star is expelling gas at a speed of 435,000 mph (700,000 kph). The fast-moving material is being channeled into streamers on one side and into a jet on the other. The jet of material appears to be striking denser, slower-moving material, creating shock waves. Radio observations have revealed an unusually large amount of sulfur in the gas around the star, which may have been produced by the shock waves. This planetary nebula is in the earliest stages of formation and has offered astronomers the chance to observe the kind of processes that led to the creation of more established planetary nebulae elsewhere in the Milky Way.

**ROTTEN EGG NEBULA**
The Calabash Nebula is popularly called the Rotten Egg Nebula because it contains a lot of sulfur, which smells like rotten eggs. The outflows of expelled gas show up bright yellow-orange in the center of this picture.

PLANETARY NEBULA

# Southern Crab Nebula

**CATALOG NUMBER**
Hen 2-104

**DISTANCE FROM SUN**
10,000 light-years

**MAGNITUDE** 14.20

**CENTAURUS**

This beautiful nebula has a complex "symbiotic" binary star system at its center—a Mira-type pulsating red giant and a white dwarf orbiting so closely that the dwarf's gravity is pulling away gas from the giant's upper atmosphere. The result is a nebula that looks relatively similar to other double-lobed planetaries but is formed by a somewhat different mechanism. A disk of hot material spiraling onto the white dwarf acts as a constraint around the middle of the nebula, while a relatively slow stellar wind of material escaping from the red giant interacts with the much faster wind from the white dwarf to drive the expanding lobes of gas above and below. Although these lobes are expanding in size, the inner and outer structures that can currently be seen both have a similar age.

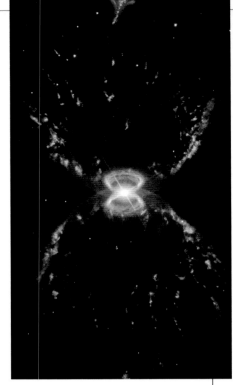

**LOBES AND JETS**
The Southern Crab's outer lobes glow in cometlike streaks where they plow into surrounding interstellar material. Similarly, narrow jets escaping from above and below the white dwarf's poles only becomes visible where they encounter this material.

BLUE SUPERGIANT

# Eta Carinae

**DISTANCE FROM SUN**
8,000 light-years

**MAGNITUDE** 6

**SPECTRAL TYPE** B0

**CARINA**

Long thought to be a single massive star, Eta Carinae is in fact a binary system of two huge stars, one of which has about 120 times the mass of the Sun and shines more than 5 million times as bright. A vast dumbbell of ejected stellar material, which is expanding at a rate of about 1.2 million mph (2 million kph), hides the stars from direct view. Eta Carinae experiences two types of irregular eruptions—gradual brightening of one to two magnitudes (see pp.232–233) that can last for a few years, and shorter-lived intense outbursts that are accompanied by the ejection of more than a solar mass of material. As a result, Eta Carinae has varied in brightness from as faint as magnitude 7 to as bright as –1 since it was first cataloged by the English astronomer Edmond Halley in 1677. The last major eruption, in 1841, saw its brightness rival Sirius and released the two lobes of material that now surround the system. Eta Carinae survived that outburst and currently hovers around sixth magnitude, but it will probably eventually destroy itself in a supernova.

**HOMUNCULUS NEBULA**
A Chandra X-ray image reveals super-hot gas created where high-energy stellar winds from the central stars energize the expanding gas cloud, known as the Homunculus.

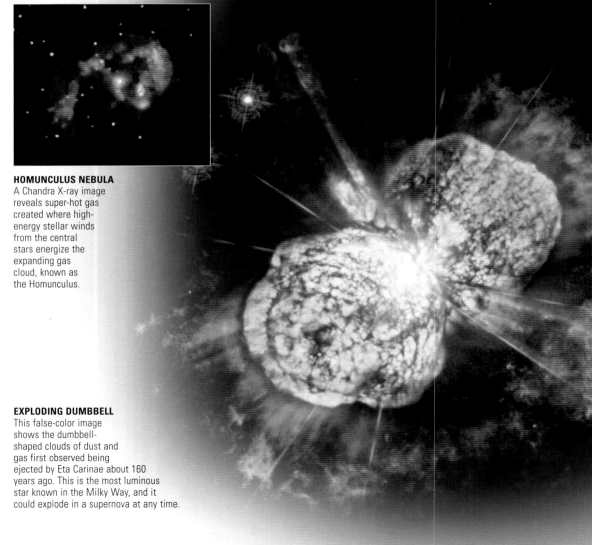

**EXPLODING DUMBBELL**
This false-color image shows the dumbbell-shaped clouds of dust and gas first observed being ejected by Eta Carinae about 160 years ago. This is the most luminous star known in the Milky Way, and it could explode in a supernova at any time.

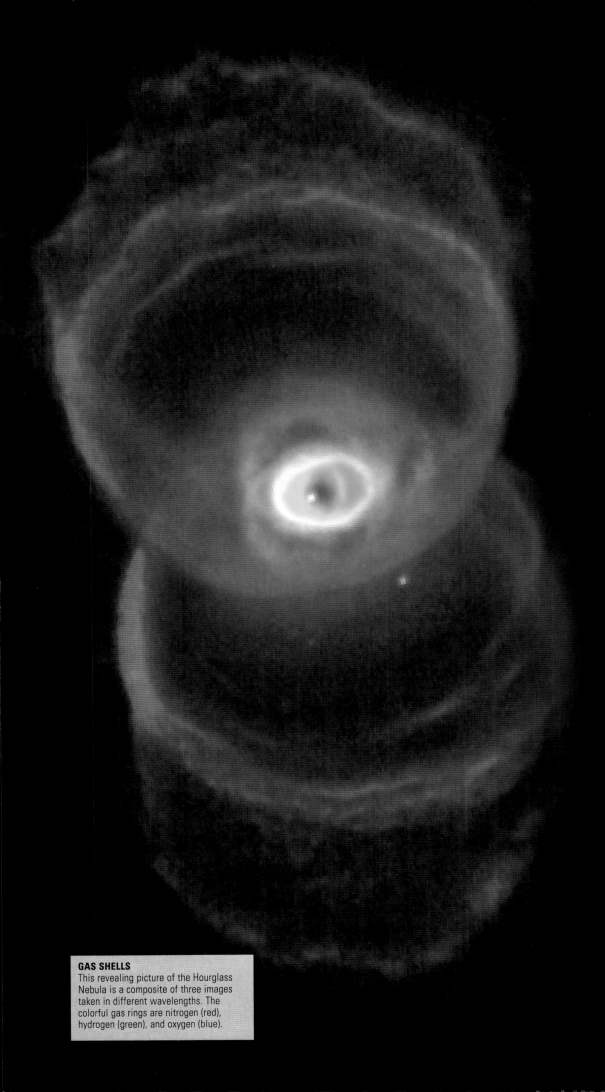

# Hourglass Nebula

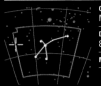

**CATALOG NUMBER**
MyCn18

**DISTANCE FROM SUN**
8,000 light-years

**MAGNITUDE** 11.8

**MUSCA**

The distinctive shape of the stunning Hourglass Nebula has caused much debate over its formation among astronomers. One suggestion is that as the aging, intermediate-mass star started to expand into a red giant, the escaping gas and dust accumulated first as a belt around the star's equator. As the volume of escaping gas continued to grow, the belt constricted the star's midsection, forcing the increasingly fast-moving gas into an hourglass shape. Other astronomers argue that the central star has a massive, heavy-element core that produces a strong magnetic field. In this scenario, the shape is a result of the ejected material being constrained by the magnetic field. Yet another suggestion is that the central star is in fact a binary and one of the pair is a white dwarf. A disk of dense material is produced around its middle by the gravitational interactions between the two components, which pinches in the "waist" of the expanding nebula. However, other features of the Hourglass Nebula have so far defied explanation. Astronomers have observed a second hourglass-shaped nebula within the larger one but, unusually, neither is positioned symmetrically around the central star. Two rings of material seen around the "eye" of the hourglass, perpendicular to one another, are the subjects of continuing studies.

## NEBULA IN ACTION

The beautiful images of the Hourglass Nebula captured by the Hubble Space Telescope have revealed details within planetary nebulae that have revolutionized the study of these elusive but beautiful objects, especially in regard to the creation of nonspherical planetary nebulae. These fascinating nebulae are observed in many varied shapes, and an equally large number of hypotheses have been suggested to account for them. The life of a planetary nebula is a mere blink of an eye when compared to the lifetime of a star, but it is a very important stage. When a star is evolving off the main sequence, it loses huge quantities of its material and thus enriches the interstellar medium in elements heavier than helium, which can then be recycled to form other celestial objects.

**GAS SHELLS**
This revealing picture of the Hourglass Nebula is a composite of three images taken in different wavelengths. The colorful gas rings are nitrogen (red), hydrogen (green), and oxygen (blue).

## WOLF–RAYET STAR

# WR 7

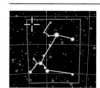

| DISTANCE FROM SUN | 15,000 light-years |
| --- | --- |
| MAGNITUDE | 11.4 |
| SPECTRAL TYPE | WN5 |

**CANIS MAJOR**

The emission nebula NGC 2359, which has a diameter of around 30 light-years, has been produced by an extremely hot Wolf–Rayet star, visible at its center. This star, designated WR 7, has a surface temperature of between 54,000°F (30,000°C) and 90,000°F (50,000°C)—6 to 10 times as hot as the Sun. It is also highly unstable, ejecting stellar material into the interstellar medium at speeds approaching 4.5 million mph (7.2 million kph). Even though it is a massive star of around 10 solar masses, it is losing about the equivalent of the mass of the Sun every thousand years. With this level of mass loss, Wolf–Rayet stars like WR 7 are unable to exist in this stage of their life for long and are therefore rarely observed: only about 550 such stars are known in the Milky Way. Material

**THOR'S HELMET**
The popular name for the nebula surrounding WR 7 is Thor's Helmet, because it looks like a helmet with wings (above). The nebulae surrounding Wolf–Rayet stars are sometimes called bubble nebulae, and WR 7 lies at the center of the nebula's main bubble of hot gas. (The star is above and to the right of center in the image to the right.)

from the star has been ejected in an even, spherical manner, producing a bubble of material. This bubble has been further shaped by interactions with the surrounding interstellar medium. WR 7 is unusual because it lies at the edge of a dense, warm molecular cloud, and the asymmetrical shape of the outer parts of the surrounding nebula is due to "bow shocks," produced when fast stellar winds hit denser, static material.

**STELLAR FIREBALL**
WR 124 can be seen as a glowing body at the center of a huge, chaotic fireball. The fiery nebula surrounding the star consists of vast arcs of glowing gas violently expanding outward into space.

## WOLF–RAYET STAR

# WR 124

| DISTANCE FROM SUN | 15,000 light-years |
| --- | --- |
| MAGNITUDE | 11.04 |
| SPECTRAL TYPE | WN |

**SAGITTARIUS**

With a surface temperature of around 90,000°F (50,000°C), WR 124 is one of the hottest known Wolf–Rayet stars. This massive, unstable star is blowing itself apart—its material is traveling at up to 90,000 mph (150,000 kph). The observed state of M1-67, the relatively young nebula surrounding WR 124, is only 10,000 years old, and it contains clumps of material with masses about 30 times that of Earth and diameters of 90 billion miles (150 billion km).

## PLANETARY NEBULA

# Stingray Nebula

| CATALOG NUMBER | Hen-1357 |
| --- | --- |
| DISTANCE FROM SUN | 18,000 light-years |
| MAGNITUDE | 10.75 |

**ARA**

The Stingray Nebula is the youngest known planetary nebula. Observations made in the 1970s revealed that the dying star at the center of the nebula was not hot enough to cause the surrounding gases to glow. By the 1990s, further observations had shown that the central star had rapidly heated up as it entered the final stages of its life, causing the nebula to shine. This afforded astronomers a remarkable opportunity to observe the star in an exceedingly brief phase of its evolution. Because of its young age, the Stingray Nebula is one-tenth the size of most planetary nebulae, with a diameter only about 130 times that of the solar system. A ring of ionized oxygen surrounds the central star, and bubbles of gas billow out in opposite directions above and below the ring. Material traveling rapidly outward from the central star has opened holes

**GRACEFUL SYMMETRY**
The graceful, symmetrical shape of this very young planetary nebula gives it its popular name. In this enhanced true-color image, the Stingray Nebula's central star has a companion star just visible above it to the left.

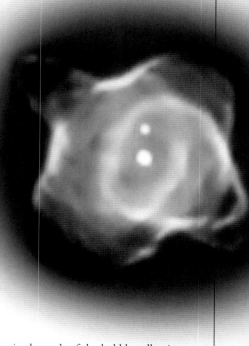

in the ends of the bubbles, allowing streams of gas to escape in opposite directions. On the outer edges of the nebula, the central star's winds crash into the walls of the gas bubbles, generating shock waves and heat that cause the gas to glow brightly.

## CHARLES WOLF AND GEORGES RAYET

French astronomers Charles Wolf (1827–1918) and Georges Rayet (1839–1906) co-discovered the type of unusual, hot stars that now bear their name. In 1867, they used the Paris Observatory's 16-in (40-cm) Foucault telescope to discover three stars whose spectra were dominated by broad emission lines rather than the usual narrow absorption lines (see pp.254–255). Today, over 500 Wolf–Rayet stars are known in our galaxy. Rayet later became Director of the Bordeaux Observatory.

**GEORGES RAYET**

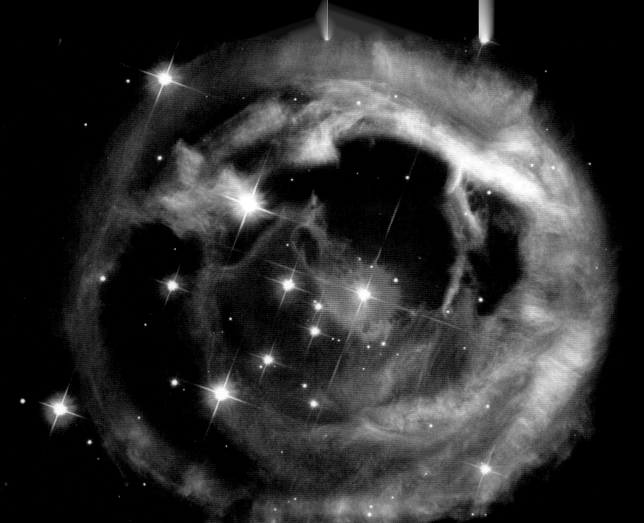

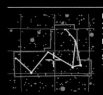

| DISTANCE FROM SUN | |
|---|---|
| 20,000 light-years | |
| MAGNITUDE | 10 |
| SPECTRAL TYPE | K |

**MONOCEROS**

Discovered on January 6, 2002, by an amateur astronomer, V838 Monocerotis is one of the most interesting stars. Its precise nature is not yet fully understood, but astronomers believe its recent evolution has moved it off the main sequence to become a red supergiant. While this phase would usually take hundreds or thousands of years, here it has happened in a matter of months. Its first viewed outburst, in January 2002, was followed a month later by a second in which it brightened from magnitude 15.6 to 6.7 in a single day—an increase of several thousand times. Finally, in March 2002, it brightened from magnitude 9 to 7.5 over just a few days. The energy emitted in the outbursts caused previously ejected shells of material to brighten and become visible.

**LIGHT ECHOES**
Light echoes from recent outbursts illuminate the ghostly shells of ejected material around the enigmatic star V838 Monocerotis (seen glowing red).

---

## BLUE SUPERGIANT

# Sher 25

| DISTANCE FROM SUN | |
|---|---|
| 20,000 light-years | |
| MAGNITUDE | 12.2 |
| SPECTRAL TYPE | |
| B1.5 | |

**CARINA**

This blue supergiant is poised to explode as a supernova, possibly within the next few thousand years. The prediction of its apparent closeness to death has been based on observations that reveal striking similarities between Sher 25 and Sk-69 202, the progenitor star of the supernova that occurred in the Large Magellanic Cloud in 1987 (now known as SN 1987A, see p.310). Sher 25 lies at the center of a clumpy ring of ejected material, and additional material from the star is escaping perpendicular to this ring. This has caused the ejected stellar material to form an hourglass-shaped nebula with Sher 25 lying at its middle. The ring and nebula are similar to those observed around Sk-69 202 before that blue supergiant exploded. Spectroscopy reveals that the nebula

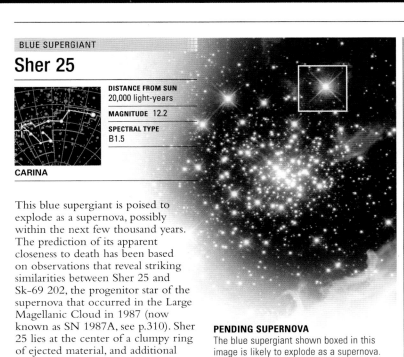

**PENDING SUPERNOVA**
The blue supergiant shown boxed in this image is likely to explode as a supernova. The open cluster of bright white stars and the surrounding nebula are known as NGC 3603.

surrounding Sher 25 is rich in nitrogen, indicating that it has passed through a red supergiant phase, again displaying an evolutionary path similar to that of the star Sk-69 202.

---

## BLUE VARIABLE

# Pistol Star

| DISTANCE FROM SUN | |
|---|---|
| 25,000 light-years | |
| SPECTRAL TYPE | LBV |

**SAGITTARIUS**

One of the most luminous stars ever discovered is located at the center of the Pistol Nebula and is known as a luminous blue variable. The Pistol Star emits around 10 million times more light than the Sun, unleashing as much energy in 6 seconds as the Sun does in 1 year. When it first formed, it contained at least 100 solar masses of material, but a stellar wind billions of times more powerful than the Sun's, along with huge eruptions, have seen it shed at least 10 times the Sun's mass. The biggest eruptions occurred about 4,000 and 6,000 years before its presently seen state. In the Sun's position, the star would fill the diameter of Earth's orbit. Despite its size and luminosity, the star is obscured at visible wavelengths by the ejected material that has formed the pistol-shaped nebula surrounding it.

**VAST NEBULA**
Seen in infrared light, the Pistol Nebula glows a brightly. The nebula is 4 light-years across and would nearly span the distance from the Sun to Proxima Centauri, the closest star to the solar system.

# STELLAR END POINTS

THE FORM A STAR TAKES in the ultimate stage of its life is called a stellar end point. Such end points include some of the most exotic objects in the Milky Way. The fate of a star is dictated by its mass, with lower-mass stars becoming white dwarfs and the highest-mass stars becoming black holes from which not even light can escape. Between these are neutron stars, including spinning pulsars.

**WHITE DWARFS IN NGC 6791**
The faint stars inside the squares in this image are white dwarfs in the globular cluster NGC 6791. Too faint to be seen from the ground, the stars were captured here by the Hubble Space Telescope.

## WHITE DWARFS

Once a star has used up all of its fuel through nuclear fusion, the stellar remnant will collapse, as it cannot maintain enough internal pressure to counteract its gravity. Stars of less than about 8 solar masses will lose up to 90 percent of their material in stellar winds and by creating planetary nebulae (see p.255). If the remnants of these stars have less than 1.4 solar masses (the Chandrasekhar limit), they will become white dwarfs. White dwarfs are supported by what is known as electron degeneracy pressure, created by the repulsion between electrons in their core material. More massive stars collapse to the smallest diameters and highest densities. The first white dwarf to be discovered, Sirius B (see p.268), has a mass similar to that of the Sun but a radius only twice that of the Earth. Although they have surface temperatures of around 180,000°F (100,000°C) at first, white dwarfs fade over periods of hundreds of millions of years, eventually becoming cold black dwarfs.

**MORGUE OF STARS**
Spanning a distance of 900 light-years, this mosaic of X-ray images of the center of the Milky Way reveals hundreds of white-dwarf stars, neutron stars, and black holes. They are all embedded in a hot, incandescent fog of interstellar gas. The supermassive black hole at the center of the galaxy is located inside the central bright white patch.

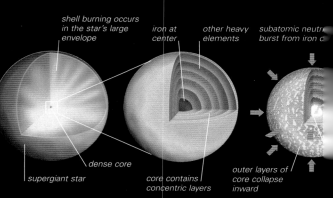

## SUPERNOVAE

Massive stars die spectacularly, blasting their outer layers off into space in type II supernovae explosions. A type I supernova is a type of variable star (see p.283). When a star of more than about 10 solar masses reaches the end of its hydrogen-burning stage, it will eventually produce an iron core. Initially, this core is held up by its internal pressure, but when it reaches a mass greater than 1.4 solar masses (the Chandrasekhar limit), it starts to collapse, forming an extremely dense core almost entirely made of neutrons. Supernova detonation occurs when the outer layers of the star, which have continued to implode, impact on the rigid core and rebound back into space at speeds of up to 45 million mph (70 million kph). This releases massive amounts of energy, creating a great rise in luminosity that may last for several months, before fading. A supernova remnant consisting of the debris will become a nebula.

shell burning occurs in the star's large envelope · iron at center · other heavy elements · subatomic neutr... burst from iron c...

supergiant star · dense core · core contains concentric layers · outer layers of core collapse inward

**COLLAPSING STAR**
As a massive star collapses, elements heavier than helium are produced in a series of shell-burning layers. Elements heavier than iron cannot be produced in this way, and an iron core may collapse to produce a neutron star.

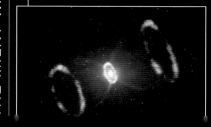

**ASYMMETRIC EXPLOSION**
The uneven shape of hot gas clouds expanding from Supernova 1987A indicates extreme turbulence during the initial explosion.

# NEUTRON STARS

Neutron stars are one of the byproducts of type II supernovae explosions. During an explosion, the outer layers of a star are blown off, leaving an extremely dense, compact star consisting predominantly of neutrons with a smaller amount of electrons and protons. Neutron stars have a mass between 0.1 and 3 solar masses. Beyond this limit, a star will collapse further to become a black hole (below). As the neutron star forms, the magnetic field of the parent star becomes concentrated and grows in strength. Similarly, the original rotation of the star increases in speed as the star collapses. Neutron stars are characterized by their strong magnetic fields and rapid rotation. Over time, their rotation slows as they lose energy. However, some neutron stars show a temporary rise in rotation rate, possibly due to tremors (known as starquakes), in their thin, crystalline outer crusts. Neutron stars that emit directed pulses of radiation at regular intervals are known as pulsars (below).

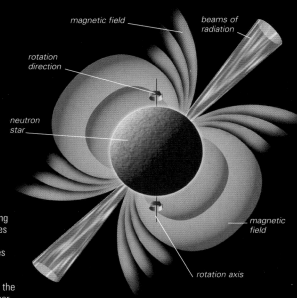

magnetic field

beams of radiation

rotation direction

neutron star

magnetic field

rotation axis

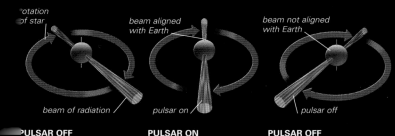

rotation of star

beam aligned with Earth

beam not aligned with Earth

beam of radiation

pulsar on

pulsar off

**PULSAR OFF**

**PULSAR ON**

**PULSAR OFF**

### HOW PULSARS WORK
Charged particles spiral along the star's magnetic field lines and produce a beam of radiation. If the beam passes across the field of Earth, it can be detected as a pulse. Depending on the energy of the radiation, this can be in either the radio or X-ray part of the electromagnetic spectrum.

# BLACK HOLES

If the remnant of a supernova explosion is greater than about 3 solar masses, there is no mechanism that can stop it from collapsing. It becomes so small and dense that its resulting gravitational pull is great enough to stop even radiation, including visible light, from escaping. Stellar-mass black holes, as such objects are known, can be detected only by the effect they have on objects around them. Light from far-off objects can be bent around a black hole as it acts as a gravitational lens, while the movement of nearby objects can be affected by a black hole's strong gravitational field (see pp.42–43). If a stellar-mass black hole is a member of a close binary system (see pp.274–275), the material from its companion star will be pulled toward it by its immense gravity. Matter will not fall directly onto the black hole due to its rotational motion. Instead, it will first be pulled into a accretion disk around the black hole. Matter impacts onto this disk, creating hot spots that can be detected by the radiation they emit. As matter in the disk gradually spirals into the black hole, friction will heat up the gas and radiation is emitted, predominantly in

### BLACK HOLE
Here, the gas from a companion star is drawn into a black hole via an accretion disk. When the gas crosses a limit called the event horizon, the gravitational field has become so strong that light cannot escape, and it disappears from view.

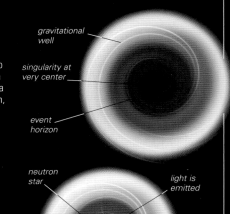

gravitational well

singularity at very center

event horizon

### NEUTRON STAR
The gas drawn from a companion star approaches a neutron star in the same manner. However, when the gas strikes the solid surface of the neutron star, light is emitted and

neutron star

light is emitted

gravitational

# STELLAR END POINTS

Stars end their lives in a variety of ways, but many are difficult or impossible to observe. It is thought that unobserved dead stars contribute significantly to the Milky Way's mysterious missing mass (see pp.226–229). Often, black holes and small white dwarfs can be observed only by the effect they have on nearby objects, and neutron stars are visible only in gamma-ray wavelengths. However, some stellar end points and their remnants, such as supernovae, are among the galaxy's most spectacular sights.

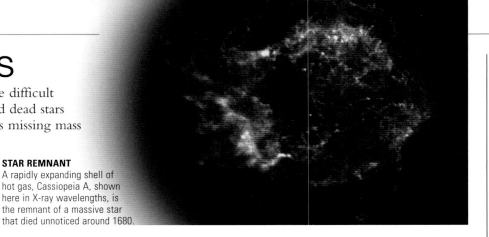

**STAR REMNANT**
A rapidly expanding shell of hot gas, Cassiopeia A, shown here in X-ray wavelengths, is the remnant of a massive star that died unnoticed around 1680.

---

## WHITE DWARF
## Sirius B

| | |
|---|---|
| **CATALOG NUMBER** | HD 48915 B |
| **DISTANCE FROM SUN** | 8.6 light-years |
| **MAGNITUDE** | 8.5 |

**CANIS MAJOR**

This was the first white dwarf to be discovered. First observed in 1862, it was found to be a stellar remnant when its spectrum was analyzed in 1915. Although "Sirius A, its companion, is the brightest star in the sky, Sirius B appears brighter in X-ray images (such as the one below). Sirius B's diameter is only 90 percent that of Earth, but since its mass is equal to that of the Sun, its gravity is 400,000 times that on Earth.

**CLOSE COMPANIONS**

---

## NEUTRON STAR
## RX J1856.5-3754

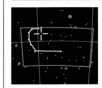

| | |
|---|---|
| **CATALOG NUMBER** | 1ES 1853-37.9 |
| **DISTANCE FROM SUN** | 200–400 light-years |
| **MAGNITUDE** | 26 |

**CORONA AUSTRALIS**

This lone star is one of the closest known neutron stars to Earth. Discussions are ongoing as to its true distance, but estimates vary from 200 to 400 light-years. There is also much speculation about its age. Some astronomers believe it is an old neutron star emitting X-rays because it is accreting material onto its surface from the surrounding interstellar medium. Others believe it is a young neutron star emitting X-rays as it cools. It is possible that it formed about 1 million years ago, when a massive star in a close binary system exploded. It is traveling through the interstellar medium at about 240,000 mph (390,000 kph). RX J1856.5-3754 is moving away from a group of young stars in the constellation of Scorpius. Also moving away from this group of stars is the ultra-hot

blue star now known as Zeta (ζ) Ophiuchi. It is possible that RX J1856.5-3754 is the remnant of Zeta Ophiuchi's original binary companion. As the closest neutron star, it is being extensively studied, but its diminutive size makes it difficult for astronomers to obtain conclusive results. Estimates of the diameter of RX J1856.5-3754 vary from 6 miles (10 km) to 20 miles (30 km). This puts it very close to the theoretical limit of how small a neutron star can be, challenging some models of their internal structure. Its X-ray emissions suggest it has a surface temperature of around 1,000,000°F (600,000°C). Its visual magnitude of only 26 means that this star is 100 million times fainter than an object on the limit of naked-eye visibility.

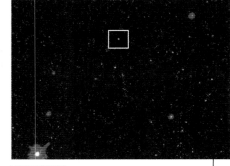

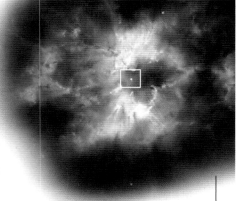

**RARE VIEWS**
Taken in 1997, a Hubble image (above) offered astronomers an unusual glimpse of a neutron star in visible light. The star's movement through the interstellar medium has produced a cone-shaped nebula, visible in a later image (below).

---

## NEUTRON STAR
## Geminga Pulsar

| | |
|---|---|
| **CATALOG NUMBER** | SN 437 |
| **DISTANCE FROM SUN** | 500 light-years |
| **MAGNITUDE** | 25.5 |

**GEMINI**

Discovered in 1972, the Geminga Pulsar, a pulsating neutron star, is the second-brightest source of high-energy gamma rays known in the Milky Way. Its name is a contraction of "Gemini gamma-ray source"; it is also an expression, in the Milanese dialect, meaning "It's not there," because only recently has this object been observed in wavelengths other than gamma rays. Variations in the pulsar's period of luminosity (see pp.280–281) have suggested that it may have a

**GAMMA RAY SOURCE**
The Geminga Pulsar shines bright in an image taken through a gamma-ray telescope. Gamma-ray photons are blocked from Earth's surface by the atmosphere.

companion planet, but they may also be due to irregularities in the star's rotation. Geminga is believed to be the remnant of a supernova that took place about 300,000 years earlier in the star's life. It is traveling through space at almost 15,000 mph (25,000 kph), at the head of a shock wave 2 billion miles (3.2 billion km) long.

---

## WHITE DWARF
## NGC 2440 nucleus

| | |
|---|---|
| **CATALOG NUMBER** | HD 62166 |
| **DISTANCE FROM SUN** | 3,600 light-years |
| **MAGNITUDE** | 11 |

**PUPPIS**

The central star of the planetary nebula NGC 2440 has one of the highest surface temperatures of all known white dwarfs. This stellar remnant has a surface temperature of around 360,000°F (200,000°C)—40 times hotter than that of the Sun. This also makes it intrinsically very bright, with a luminosity more than 250 times that of the Sun. The complex structure of the surrounding nebula has led some astronomers to believe that there have been periodic ejections of material

**INNER LIGHT**
Energy from the extremely hot surface of NGC 2440's central white dwarf makes this beautiful and delicate-looking planetary nebula fluoresce.

from the dying central star. The structure of the nebula also suggests that the material was ejected in various directions during each episode.

# The Cygnus Loop

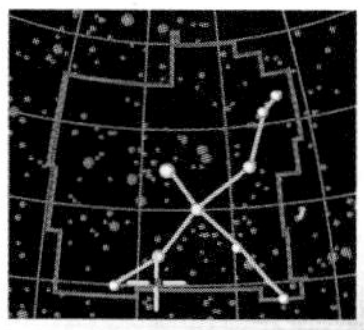

| | |
|---|---|
| **CATALOG NUMBER** | NGC 6960/95 |
| **DISTANCE FROM SUN** | 2,600 light-years |
| **MAGNITUDE** | 11 |

**CYGNUS**

the Veil Nebula, and because it is so large, the Cygnus Loop has been cataloged using many different reference numbers. The supernova remnant is some 80 light-years long and sprawls 3.5 degrees across the sky—about seven full moons across. It shines in the light generated by shock waves

The Cygnus Loop is the remnant of a dying star that blew itself up in a supernova. Estimates of how long ago in the star's lifetime this event occurred vary from 5,000 to 15,000 years. The most prominent parts of the nebula seen in visible light are often called produced as stellar material from the supernova hits material in the interstellar medium. Observations of this stellar laboratory have revealed an inconsistent composition and structure of the interstellar medium, as well as that of the supernova remnant.

**GLOWING FILAMENTS**
Filaments of shocked interstellar gas glow in the light emitted by excited hydrogen atoms. This side-on view shows a small portion of the Cygnus Loop moving upward at about 380,000 mph (612,000 kph).

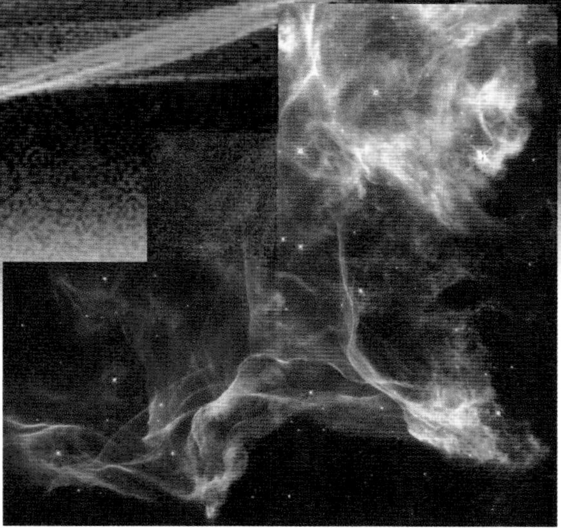

**COLORFUL GASES**
This composite image of a section of the Cygnus Loop reveals the presence of different kinds of atoms excited by shock waves: oxygen (blue), sulfur (red), and hydrogen (green).

# Vela Supernova

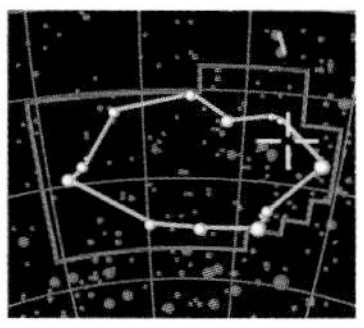

| | |
|---|---|
| **CATALOG NUMBER** | NGC 2736 |
| **DISTANCE FROM SUN** | 6,000 light-years |
| **MAGNITUDE** | 12 |

**VELA**

The Vela Supernova Remnant is the brightest object in the sky at gamma-ray wavelengths. It is estimated that the star that produced it exploded between 5,000 and 11,000 years previously, and that its final explosion would have rivaled the Moon as the brightest object in the night sky. The star that died has become a pulsar, a rapidly spinning neutron star, which rotates about 11 times each second. The Vela Pulsar is about 12 miles (19 km) in diameter and was only the second pulsar to be discovered optically, the optical flashes being observed in 1977. As with other pulsars, the rotation rate of the Vela Pulsar is gradually slowing down. Since 1967, it has suffered several brief glitches where its rotation rate has temporarily increased before continuing to slow.

**EXPANDING SHELL**
This optical photograph of the Vela Supernova Remnant reveals details of the 100-light-year-wide expanding gas bubble against more distant nebulosity.

**DYNAMIC JET**
A sequence of X-ray images from the Chandra satellite reveals changes in the jet of high-energy particles emitted by the Vela Pulsar. The jet remains tightly structured for more than half a light-year before losing coherence due to contact with surrounding interstellar medium.

**JUNE 2010**

**JULY 2010**

**AUGUST 2010**

**SEPTEMBER 2010**

THE MILKY WAY

**BRIGHT REMNANT**
The conspicuous Crab Nebula is shown here in a composite image captured at five different wavelengths—radio emission (red), infrared (yellow), visible light (green), ultraviolet (blue), and X-rays (purple).

SUPERNOVA REMNANT

# Crab Nebula

**CATALOG NUMBERS**
M1, NGC 1952

**DISTANCE FROM SUN**
6,500 light-years

**MAGNITUDE** 8.4

**TAURUS**

In the summer of 1054, during the Sung dynasty, Chinese astronomers recorded that a star, in the present-day constellation Taurus, had suddenly become as bright as the full moon. They described it as a reddish-white "guest star" and observed it over a period of 2 years as it slowly faded. Their records show it was visible in daylight for more than 3 weeks. They had witnessed a supernova, and the stellar material flung off in this cataclysmic explosion now shines as the wispy filaments of the Crab Nebula. This nebula is the very first object, and the only supernova remnant, to be listed by Charles Messier (see p.73) in his famous catalog. The nebula is easily visible in binoculars and small telescopes. It spans a distance of about 10 light-years with a magnitude of between 8 and 9.

The remains of the original star have become a spinning neutron star, a pulsar, rotating at about 30 times per second.

**HUBBLE'S ANALYSIS**
This visible-light mosaic highlights elements expelled in the supernova explosion. Blue indicates neutral oxygen; green indicates ionized sulfur; and red indicates super-hot, doubly ionized oxygen.

**CORE OF THE NEBULA**
This Hubble close-up reveals the Crab Pulsar (upper-right of the two stars at the center), surrounded by the glowing blue haze emitted by electrons traveling close to the speed of light in its powerful magnetic field.

The pulsar (known as PSR 0531+21) is observable optically and in radio, X-ray, and gamma-ray wavelengths because the beams it generates happen to be directed toward Earth during part of its revolution. It was discovered in 1967, but had been known previously as a powerful emitter of radio waves and X-rays. It was the first pulsar to be identified optically and is of 16th magnitude. It is estimated to have a diameter of only about 6 miles (10 km) but a mass greater than the Sun's. Its energy output is more than 750,000 times that of the Sun. Its rotation is decreasing by about 36.4 nanoseconds every day, which means that over 2,500 years from its presently observed state, its rotation period will have doubled (see pp.282–283). The loss of rotational energy is being translated into energy, which is heating the surrounding Crab Nebula.

As the most easily observable supernova remnant, the Crab Nebula has been extensively studied. Detailed observations show that the material within the central portion of the nebula changes within a time scale of only a few weeks. Wispy features, each about a light-year across, have been observed streaming away from the pulsar at half the speed of light. These are created by an equatorial wind emitted by the pulsar (see left). They brighten and then fade as they move away from the pulsar and expand out into the main body of the nebula. The most dynamic feature within the center is the point where one of the polar jets from the pulsar crashes into the surrounding previously ejected material, forming a shock front. The shape and position of this feature have been observed to change over very short time scales.

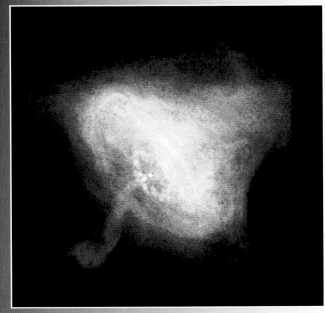

**PULSAR CLOSE-UP**
This X-ray image of the central region of the Crab Nebula shows its pulsar as a white dot near the center. Jets of matter stream away from the poles of the rapidly rotating pulsar, and energetic particles from its equator plow into the surrounding nebula.

## NEUTRON STAR

# PSR B1620-26

| CATALOG NUMBER |
| --- |
| PSR B1620-26 |
| **DISTANCE FROM SUN** |
| 7,000 light-years |
| **MAGNITUDE** 21.3 |

**SCORPIUS**

Situated in the globular cluster M4, the pulsar PSR B1620-26 rotates more than 90 times per second and has a mass of about 1.3 solar masses. It has a white-dwarf companion (boxed in the image below). A third companion is thought to be a planet twice the mass of Jupiter (see pp.296–299). This planet is named Methuselah, as it may be up to 13 billion years old.

**WHITE-DWARF COMPANION**

## BLACK HOLE

# GRO J1655-40

| CATALOG NUMBER |
| --- |
| V* V1033 Sco |
| **DISTANCE FROM SUN** |
| 6,000–9,000 light-years |
| **MAGNITUDE** 17 |

**SCORPIUS**

Discovered in 1994, as a source of unusual X-ray emissions, this black hole produces outbursts in which jets of material are ejected at speeds close to the speed of light. In addition to this, the gas surrounding GRO J1655-40 displays an unusual flicker (at a rate of 450 times per second) that can be explained as a rapidly rotating black hole. This is only the second object of this type to have been found in the Milky Way. It has been suggested that a subgiant star is orbiting the black hole, which is six to seven times the mass of the Sun. Their orbits are thought to be inclined at 70 degrees to each other, causing partial eclipses. Mass has been pulled off the subgiant star by the gravitational interaction from the black hole and formed a disk of material around the system. This system has been dubbed a mini-quasar because of its similarity to active galactic nuclei (AGNs) (see pp.306–309).

## BLACK HOLE

# Cygnus X-1

| CATALOG NUMBER |
| --- |
| HDE 226868 |
| **DISTANCE FROM SUN** |
| 8,200 light-years |
| **MAGNITUDE** 8.95 |

**CYGNUS**

This X-ray source was one of the first to be discovered and is one of the strongest X-ray sources in the sky. The X-ray emissions from Cygnus X-1 flicker at a rate of 1,000 times per second. In 1971, astronomers observed a radio

source at the same position in the sky and also identified an optical object, the blue supergiant star HDE 226868. This star has a mass of 20–30 solar masses and is visible through binoculars. It is in a 5.6-day orbit with Cygnus X-1, which has a mass of about six solar masses. Further observations have shown that the black hole is slowly pulling material from its companion supergiant and increasing its own mass. Cygnus X-1 was the first object to be identified as a stellar-mass black hole.

**ELUSIVE BLACK HOLE**
Cygnus X-1 is located close to the red emission nebula Sh2-101, within the rich Cygnus Star Cloud (below). A negative optical image helps to pinpoint its companion, HDE 226868.

## SUPERNOVA

# Tycho's Supernova

| CATALOG NUMBER |
| --- |
| SN 1572 |
| **DISTANCE FROM SUN** |
| 7,500 light-years |
| **MAXIMUM MAGNITUDE** |
| –3.5 |

**CASSIOPEIA**

In 1572, Tycho Brahe (see panel, below) observed a supernova in the constellation Cassiopeia and recorded its brightness changes in exceptional detail. It brightened to around –3.5—

## TYCHO BRAHE

The leading astronomer of his day, Tycho Brahe (1546–1601) founded a great observatory in Uraniborg, Denmark, and spent years making detailed observations of planetary movements and the positions of the stars. Johannes Kepler became his assistant, and Tycho's work was to give the empirical basis for Kepler's laws of planetary motion (see p.68).

**RADIO ENERGY**
A radio image of Tycho's Supernova shows areas of low (blue), medium (green), and high (red) energy. A shock wave produced by the expanding debris is shown by the pale blue circular arcs on the outer rim.

as bright in the sky as Venus—before fading over a period of 6 months. This brilliant new object was to help astronomers reject the idea that the heavens were immutable. The remnant from this supernova is still expanding and has a current diameter estimated at nearly 20 light-years. Its stellar material is estimated to be traveling at 14.5–18 million mph (21.5–27 million kph), which is the highest expansion rate observed for any supernova remnant. No strong central point source is detected in the remnant, which suggests that Tycho was a Type Ia supernova. The model for this type of supernova is the destruction of a white dwarf when infalling matter from a companion star increases its mass beyond the Chandrasekhar limit (see pp.266–267). This concurs with the recent discovery of what astronomers think is the burned-out star from the heart of the supernova. The star was discovered because it is moving at three times the speed of other objects in the region. At the edge of the remnant is a shock wave heating the stellar material to 36 million °F (20 million °C); the interior gas is much cooler, at 18 million °F (10 million °C).

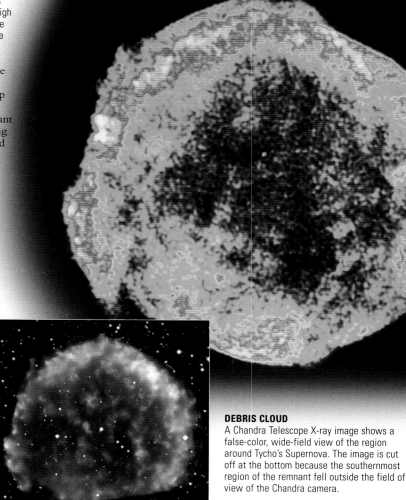

**DEBRIS CLOUD**
A Chandra Telescope X-ray image shows a false-color, wide-field view of the region around Tycho's Supernova. The image is cut off at the bottom because the southernmost region of the remnant fell outside the field of view of the Chandra camera.

## SUPERNOVA

# Kepler's Star

| CATALOG NUMBER |
| --- |
| SN 1604 |
| **DISTANCE FROM SUN** |
| 13,000 light-years |
| **MAXIMUM MAGNITUDE** |
| –2.5 |

**OPHIUCHUS**

The last supernova explosion in the Milky Way to be observed is named after Johannes Kepler, who witnessed it in October 1604. This previously unremarkable star reached a magnitude of –2.5 and remained visible to the naked eye for more than a year. Its position is now marked by a strong radio source and, in optical light, by a wispy supernova remnant, generally known as Kepler's Star. Observations have revealed that the supernova remnant has a diameter of about 14 light-years and that the material within it is expanding at 4.5 million mph (7.2 million kph). Kepler's Star has been imaged by three of NASA's great observatories: the Hubble Space Telescope, the Spitzer Space Telescope, and the Chandra X-Ray Observatory.

**VISIBLE WISPS**
In this optical image, the supernova remnant appears as a faint ring of gas filaments. Having been expelled by the original explosion, this stellar material becomes heated and glows as it plows through the interstellar medium.

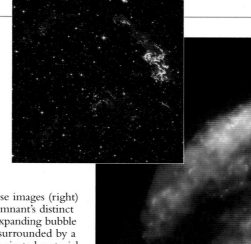

A combination of these images (right) has highlighted the remnant's distinct features. It shows an expanding bubble of iron-rich material surrounded by a shock wave, created as ejected material slams into the interstellar medium. This shock wave, shown in yellow, can also be seen optically (above). The red color is produced by microscopic dust particles, which have been heated by the shock wave. The blue and green regions represent locations of hot gas: blue indicates high-energy X-rays and the highest temperatures; green represents lower-energy X-rays.

**COMBINED IMAGE**
A composite picture made using images from three separate telescopes offers a view ranging from X-ray through to infrared.

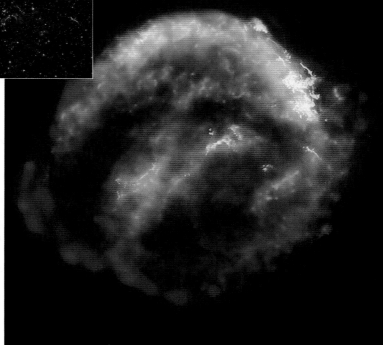

## SUPERNOVA

# Cassiopeia A

| CATALOG NUMBER |
| --- |
| SN 1680 |
| **DISTANCE FROM SUN** |
| 10,000 light-years |
| **MAXIMUM MAGNITUDE** 6 |

**CASSIOPEIA**

An intense radio source, Cassiopeia A is the remnant of a supernova explosion that occurred in the middle of the 17th century. The fact that no reports of the original explosion have been found suggests it may have been of unusually low optical luminosity. Today, Cassiopeia A is the strongest discrete low-frequency radio source in the sky (after the Sun). The radio waves are produced by electrons spiraling in a strong magnetic field. Cassiopeia A is about 10 light-years in diameter and is expanding at a rate of about 5 million mph (8 million kph).

**COLOR-CODED IMAGE**
This Hubble Space Telescope image of Cassiopeia A's cooling filaments and knots has been color-coded to help astronomers understand the chemical processes involved in the recycling of stellar material.

**SHOCK WAVES**
This false-color image combines X-ray data from Chandra (blue and green) with near- and far-infrared images from the Hubble and Spitzer telescopes (yellow and red).

## BLACK HOLE

# MACHO 96

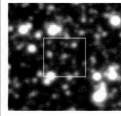

| CATALOG NUMBER |
| --- |
| MACHO 96 |
| **DISTANCE FROM SUN** |
| Up to 100,000 light-years |

**SAGITTARIUS**

Although we cannot see black holes, we can detect their presence by measuring their effects on objects around them. The existence of the black hole named MACHO 96 is inferred from the observed brightening of a star lying beyond the black hole caused by a process called lensing (see p.327). Through this process, the black hole's mass bends the light from the star in the same way as a lens does. The distant star is temporarily magnified, and we see a brief and subtle brightening in the star's output. The dark lensing object MACHO 96

has been calculated to be a six-solar-mass black hole that is moving independently among other stars. The chances of observing such a lensing event are estimated to be extremely slim. Therefore, astronomers monitor millions of stars every night, using computers to analyze the brightness of the stellar images captured by advanced camera systems. So far, fewer than 20 events have been detected looking toward the Large Magellanic Cloud, a nearby galaxy (see pp.310–311). MACHO 96 was initially detected by the MACHO Alert System in 1996 and subsequently monitored by the Global Microlensing Alert Network. However, it was only by studying images taken by the Hubble Space Telescope that astronomers could identify the lensed star and determine its true brightness (see below). Observations have suggested that the distant star may be a close binary system, but astronomers are still debating whether the lensing object lies in the Milky Way's Galactic Halo or in the Large Magellanic Cloud.

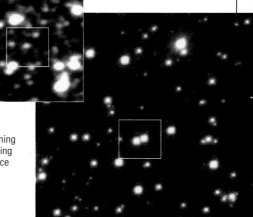

**PASSING BLACK HOLE**
Two ground-based images of a crowded star field (above) show the slight brightening of a star caused by the gravitational lensing of the passing MACHO 96. A Hubble Space Telescope image of the same area (right) resolves the star and allows its true brightness to be determined.

# MULTIPLE STARS

A MULTIPLE STAR IS A SYSTEM of two or more stars bound together by gravity. Systems with two stars are called binary or double stars. Although at first sight only a few stars appear to be multiple, it is estimated that they may account for over 60 percent of stars in the Milky Way. Binary stars orbit each other at a great variety of distances, with orbital periods ranging from a few hours to millions of years. Multiple stars allow astronomers to determine stellar masses and diameters and give them insights into stellar evolution.

## BINARIES AND BEYOND

Although there are many millions of multiple systems within the Milky Way, not all of them consist of just two stars in mutual orbit. What may appear to be a double or binary star can often reveal itself to be a more complex system of three or more stars. A simple binary system consists of two stars orbiting each other. If the stars are of similar mass, they orbit around a common center of gravity, located between them. If one of the stars is much more massive than the other, the common center of gravity may be located inside the massive star. The more massive star then merely exhibits a wobble, while the secondary star appears to take on all the orbital motion. However, multiple systems may have a greater orbital complexity, with multiple centers of gravity. For example, a quadruple system may have two pairs of stars orbiting each other, while the individual stars within the pair are also in mutual orbit.

**BRIGHT BINARY**
One of the brightest stars in our sky, Alpha (α) Centauri is also a striking double star, with two sunlike components that orbit each other in just under 80 years.

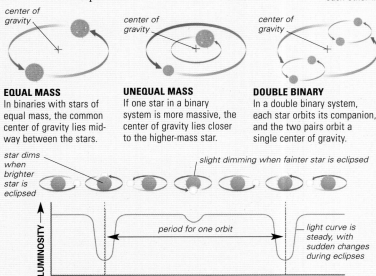

**EQUAL MASS**
In binaries with stars of equal mass, the common center of gravity lies mid-way between the stars.

**UNEQUAL MASS**
If one star in a binary system is more massive, the center of gravity lies closer to the higher-mass star.

**DOUBLE BINARY**
In a double binary system, each star orbits its companion, and the two pairs orbit a single center of gravity.

center of gravity

star dims when brighter star is eclipsed

slight dimming when fainter star is eclipsed

LUMINOSITY

period for one orbit

light curve is steady, with sudden changes during eclipses

TIME

## DETECTING BINARIES

Astronomers detect binary stars in a variety of ways. Line-of-sight binaries consist of stars that appear in the sky to be related but, in fact, are not physically associated. These are usually identified by determining the distances to the individual stars. Visual binaries are detected when the naked eye or magnification splits the stars and shows each one separately. Measurements of each star's position over time allow astronomers to compute their orbit. Although they cannot be separated with a telescope, astrometric binaries are detected when an unseen companion causes a star to wobble periodically through its gravitational influence. Spectroscopy can also be used to identify binary stars, when a star's spectrum appears doubled up and actually consists of the combined spectrums of two stars orbiting each other. These systems are known as spectroscopic binaries. The apparent magnitude of a binary star may show periodic fluctuations, caused by the stars eclipsing each other. Such stars are known as eclipsing binaries.

**ECLIPSING BINARIES**
Eclipsing binary stars are detected by variations in a star's magnitude. These variations occur when stars periodically pass in front of each other during orbit.

## CO-EVOLUTION

Like all stars, those within a multiple system evolve. A binary system can start out as two main-sequence stars with a mutual, regular orbit and predictable eclipses. However, over millions of years, the stars progress through their evolutionary stages, which may result in a binary system with two stars of completely different characteristics. One example is the Sirius system (see p.252). The evolution of one star within a system can change the behavior of the whole system. For example, should a star expand and become a red giant, the expansion can bring the evolving star to interact with its companion star. This leads to mass transfer, and if the companion has itself evolved into a white dwarf, the result can be a cataclysmic explosion (see p.283). Stellar evolution can thus convert a stable binary system into a scene of immense violence.

material is being transferred continuously

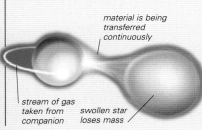

stream of gas taken from companion

swollen star loses mass

**INTERACTING BINARIES**
The stars in some binary systems are located so close together that material can pass between them. Here, one of the stars has swollen and is spilling gas onto the other.

## EXTREME BINARIES

Many binary systems exhibit perfectly regular behavior, with the stars orbiting each other for millions of years with no dramatic changes. However, other binary systems, particularly those that have undergone evolutionary changes, may exhibit much more extreme behavior. One example is a contact binary system, in which the two stars are touching each other. In this case, a massive star transfers material to the secondary star at a faster rate than the secondary star can absorb. This results in the material forming a common envelope that surrounds both stars. The envelope then creates frictional drag, causing the stars' orbital periods to change. In this way, a binary system with a wide separation and an orbital period of about a decade may be converted into a rapid system with the stars orbiting in a matter of hours. Other binary systems seem to operate at the extremes of physics. The discovery in 1974 of a binary pulsar system opened up a new field of observation in gravitational physics. A strong source of gravitational waves, binary pulsars are very regular and precise systems.

**HUB OF STARS**
One of the most famous multiple star systems, Theta (θ) Orionis, or the Trapezium (top left of image), is the middle star in the sword of Orion (see pp.390–391). Its four brightest stars are easily separated with a telescope, but it is made up of a total of at least 10 stars.

# MULTIPLE STARS

Most of the stars in the Milky Way are members of either binary or multiple systems—single stars like the Sun are more unusual. These systems vary from distant pairs in slow, centuries-long orbits around a common center of mass to tightly bound groups that orbit each other in days

**TRAPEZIUM**
The multiple star known as Theta Orionis, or the Trapezium, is a system containing at least 10 individual stars.

and may even distort each other's shape. Most multiples are so close together that we know about them only from their spectra. They also vary widely in size and color—stars of any age and type can be members of a multiple star system.

## TRIPLE STAR
### Omicron Eridani

| | |
|---|---|
| DISTANCE FROM SUN | 16 light-years |
| MAGNITUDE | 9.5 |
| SPECTRAL TYPE | DA |

**ERIDANUS**

Originally Omicron (ο) Eridani was classed as a double star, Omicron-1 Eridani and Omicron-2 Eridani. Nineteenth-century observations revealed that the system is actually three stars, now called 40 Eridani A, B, and C. A is a main-sequence orange-red dwarf, and C is a faint red dwarf. However, it is 40 Eridani B that is the gem. This young white dwarf is the brightest white dwarf visible through a small telescope.

**TRIPLE SYSTEM**

## SEXTUPLE SYSTEM
### Castor

| | |
|---|---|
| DISTANCE FROM SUN | 51 light-years |
| MAGNITUDE | 1.6 |
| SPECTRAL TYPE | A2 |

**GEMINI**

Easily visible to the naked eye, Castor appears to be an ordinary A-type star. However, a telescope reveals that Castor is in fact a pair of bright A-type stars, Castor A and Castor B, with a fainter third companion, Castor C. Spectrographic analysis shows that both the A and B components of Castor are themselves double stars.

Castor A consists of two stars in a very close 9.2-day orbit, while Castor B's components orbit each other in a rapid 2.9 days. The faint Castor C star is also a double—a pair of faint red-dwarf stars orbiting each other with a period of only 20 hours. Castor is therefore a sextuplet star: a double-double-double.

**DOUBLE-DOUBLE-DOUBLE**
Castor (boxed) and its neighbor Pollux are the two brightest stars in Gemini (below). Only when viewed through a telescope are the individual stars, Castor A and Castor B, separated (right).

## QUADRUPLE STAR
### Mizar and Alcor

| | |
|---|---|
| DISTANCE FROM SUN | 81 light-years |
| MAGNITUDE | 2 |
| SPECTRAL TYPE | A2 |

**URSA MAJOR**

Although Mizar and Alcor are a famous naked-eye double, easily visible in the handle of the Big Dipper and known since ancient times as the horse and rider, it is still unknown whether or not they are a genuine double. Mizar itself is a double star—the first double star to be discovered. Spectrography reveals, however, that Mizar is a double-double star—that is, two double stars in orbit around each other.

**MIZAR THROUGH A TELESCOPE**

## QUADRUPLE STAR
### Algol

| | |
|---|---|
| DISTANCE FROM SUN | 93 light-years |
| MAGNITUDE | 2.1 |
| SPECTRAL TYPE | B8 |

**PERSEUS**

Algol, or Beta (β) Persei, appears to the naked eye as a single star. However, exactly every 2.867 days, the star's brightness drops by 70 percent for a few hours—a variation that was discovered as early as 1667. This variation is caused by Algol's being eclipsed by a faint giant star Algol B, which is larger than the bright primary Algol A.

**ECLIPSING BINARY**

## QUADRUPLE STAR
### Epsilon Lyrae

| | |
|---|---|
| DISTANCE FROM SUN | 160 light-years |
| MAGNITUDE | 3.9 |
| SPECTRAL TYPE | A4 |

**LYRA**

Epsilon (ε) Lyrae is visible as a double star on a clear, dark night, but closer observation reveals that, in fact, each star is itself a double. Unlike other double-double systems, Epsilon Lyrae is within reach of amateur astronomers—its four component stars can each be seen through a telescope, and spectroscopy is not needed to detect their presence (see p.274). The two bright stars visible to the naked eye, Epsilon-1 and Epsilon-2, are widely separated, with an orbital period of millions of years. The components of each pair

**ISOLATED PAIRS**
This double-double system is easily separated into its four components through a telescope. Although the stars in each pair are strongly bound to one another, the link between the pairs is tenuous.

orbit much more closely, with a period of about 1,000 years. Epsilon-1 and Epsilon-2 are so far apart, they are hardly bound by gravity at all, and eventually Epsilon Lyrae will become two separate star systems.

THE MILKY WAY

## DOUBLE STAR

# Zeta Boötis

| | |
|---|---|
| **DISTANCE FROM SUN** | 180 light-years |
| **MAGNITUDE** | 3.8 |
| **SPECTRAL TYPE** | A3 |

**BOOTES**

Zeta (ζ) Boötis would appear to be a standard double star—two A-type stars orbiting each other with a period of about 123 years. However, anomalies in calculations of its mass have suggested that there is something strange about the Zeta Boötis system. The answer lies in a highly elongated orbit, in which the stars range from 130–5,900 million miles (210–9,500 million km) apart. At their closest, they are almost as close as the Sun and Earth, and no telescope can visually split them. The Zeta Boötis system is about 40 times as luminous as the Sun, with about four times its mass, and has a temperature of about 15,700°F (8,700°C).

**ENHANCED IMAGES**
When the components of Zeta Boötis are at their farthest apart, image-processing software can be used to separate them and even split their spectra.

## DOUBLE STAR

# Izar

| | |
|---|---|
| **DISTANCE FROM SUN** | 210 light-years |
| **MAGNITUDE** | 2.4 |
| **SPECTRAL TYPE** | A0 |

**BOOTES**

Izar, or Epsilon (ε) Boötis, is one of the best double stars in the sky. Its stars exhibit a striking color contrast—an orange giant close to a white dwarf—and it was given the name Pulcherrima, "most beautiful," by its discoverer, German-born Friedrich Struve. The dwarf star is about twice the size of the Sun, while the orange giant is about 34 times the size. The dwarf and giant orbit each other with a period of more than 1,000 years. This double star is not particularly astronomically unusual but is well known to amateur astronomers for its visual splendor.

**DWARF AND GIANT**

## QUADRUPLE STAR

# Almach

| | |
|---|---|
| **DISTANCE FROM SUN** | 355 light-years |
| **MAGNITUDE** | 2.3 |
| **SPECTRAL TYPE** | K3 |

**ANDROMEDA**

Almach, or Gamma (γ) Andromedae, is well known to amateur astronomers as being a fine example of a double star with contrasting colors. The brighter star is yellow-orange, and the fainter star is blue; through a telescope, the two colors enhance

## DOUBLE STAR

# M40

| | |
|---|---|
| **DISTANCE FROM SUN** | 1,900 and 550 light-years |
| **MAGNITUDE** | 8.4 |
| **SPECTRAL TYPE** | G0 |

**URSA MAJOR**

Some multiple stars are famous for their beauty, and others are famous for the dramatic astrophysics played out within the system. In the case of M40, neither applies. When compiling his well-known catalog of star clusters and nebulae, Charles Messier

**ALMACH**

each other. The brighter star is a giant K-type star, while the fainter star is itself a double star, consisting of two hot, white main-sequence stars in a mutual orbit, with a period of about 60 years. It is difficult to split these two stars visually, but spectroscopic analysis reveals that one of them is also a double star in turn, making Almach a quadruple system.

**OPTICAL PAIR**

(see p.73) observed two stars close to each other in the night sky and erroneously included them. The two stars are nothing more than an optical double—that is, they happen to lie on the same line of sight. Modern distance measurements have shown that they are not truly associated with each other. M40 is therefore a double that achieves fame through error.

## QUADRUPLE STAR

# Alcyone

| | |
|---|---|
| **DISTANCE FROM SUN** | 368 light-years |
| **MAGNITUDE** | 2.9 |
| **SPECTRAL TYPE** | B7 |

**TAURUS**

Alcyone, one of the sisters of the Pleiades (see p.291), is a bright giant star of spectral type B that shines about 1,500 times more brightly than the Sun. Orbiting around Alcyone are three stars forming a compact system: 24 Tau (magnitude 6.3) and V647 Tau (magnitude 8.3) are both A-type stars, while HD 23608 (magnitude 8.7) is an F-type star. V647 Tau is a variable of the Delta Scuti type. The system of three stars orbits Alcyone at a distance

of a few billion miles. Alcyone is unusual in that it rotates at high speed. This has caused it to throw gas off at its equator, which forms a light-emitting disk. It is classified as a Be star (see p.285), similar to Gamma (γ) Cassiopeiae.

**SEASONAL SIGNAL**
Alcyone is the brightest star in the Pleiades Cluster (see p.291). Its appearance over the eastern horizon signals the start of fall in the Northern Hemisphere.

**NORTH POLE STAR**
Polaris may appear motionless, but a
long-exposure photograph reveals it is
slightly offset from the celestial pole.
Polaris's movement is marked by the
bright arc just left of center.

## DOUBLE STAR

# Polaris

| | |
|---|---|
| **DISTANCE FROM SUN** | 430 light-years |
| **MAGNITUDE** | 2 |
| **SPECTRAL TYPE** | F7 |

**URSA MINOR**

Polaris is famous as the current north Pole Star and consequently is known to every observer of the northern sky (see panel, below). However, it is also an interesting system in terms of its component stars. Polaris is a double star consisting of Polaris A, a supergiant, and Polaris B, a main-sequence star. The two stars can be separated through a modest amateur telescope, and Polaris B was first detected by William Herschel in 1780. The distance between them has been estimated at more than 190 billion miles (300 billion km). Polaris A is more than 1,800 times more luminous than the Sun and is also a Cepheid variable with a period of just under 4 days (see p.282). The radial velocity, or line-of-sight motion, of Polaris has been accurately measured (see p.70) and found to vary regularly with a period of 30.5 years. This indicates that Polaris is also an astrometric binary—that is, the presence of an unseen companion is detected by the movement it induces in the primary star (see p.274). The companion, which was seen for the first time in a Hubble Space Telescope image in 2005, orbits Polaris with a 30.5-year period but is so faint that it has no effect on Polaris's spectrum.

### DISTINCTIVE STAR
One of the best-known stars in the northern sky, Polaris lies just away from the celestial pole, in the tail of Ursa Minor, the Little Bear. This telescope view reveals the faint companion, Polaris B, but a second smaller companion remains invisible.

## EXPLORING SPACE

## CELESTIAL SIGNPOST

Polaris has long been regarded as the most important star in the northern sky. Because it is located almost directly overhead at the north pole, it has long been used, just like a compass, to locate north. By calculating the relative angle of Polaris above the horizon, travelers by land and sea have also used Polaris to establish approximate latitudinal positions on the Earth's surface. The status accorded to Polaris by disparate cultures is reflected in their myths. In Norse mythology, Polaris was the jewel on the head of the spike that the gods stuck through the universe. The Mongols called Polaris the Golden Peg that held the world together. In ancient China, Polaris was known as Tou Mu, the goddess of the North Star.

### IN THE LITTLE BEAR'S TAIL
In Arabic mythology, Polaris was an evil star who killed the great warrior of the sky. The dead warrior was said to lie in the tail of the little bear, a constellation that also represented a funeral bier.

DOUBLE STAR

# 15 Monocerotis

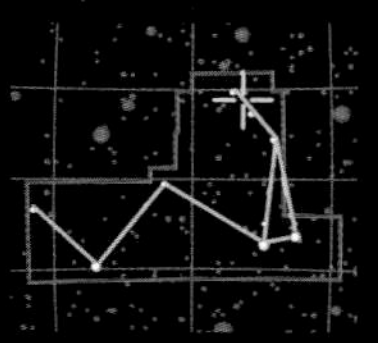

| DISTANCE FROM SUN | |
|---|---|
| 1,020 light-years | |
| MAGNITUDE | 4.7 |
| SPECTRAL TYPE | O7 |

**MONOCEROS**

15 Monocerotis (15 Mon), also known as S Monocerotis, is an O-type binary system located within the open cluster NGC 2264. It is a blue supergiant star—young, massive, and about 8,500 times more luminous than the Sun. It is also a variable star, exhibiting a small (0.4 magnitude) change in brightness. 15 Mon is responsible for illuminating the Cone Nebula (see p.242) and consequently is an easy target for amateur astronomers. 15 Mon is an astrometric and spectroscopic binary—that is, its companion star is detected through observations of the motion of 15 Mon, and also through spectroscopic analysis of 15 Mon's starlight (see p.274). The companion orbits 15 Mon with a period of 24 years, and recent studies using the Hubble Space Telescope show that the closest approach between the stars occurred in 1996. It has been suggested that 15 Mon is a multiple system, with three other bright giants nearby, but there is no evidence that the other giants are associated with 15 Mon.

**BRIGHTEST STAR**
The brightest star in the open star cluster NGC 2264, 15 Monocerotis sits in close visual proximity to the Cone Nebula (see p.242).

**BRILLIANT ILLUMINATION**
Even the most powerful telescopes cannot separate the two stars of 15 Monocerotis visually. The brilliant blue star lights up the emission nebula that surrounds it.

## TRIPLE STAR
# Beta Monocerotis

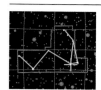

| DISTANCE FROM SUN | 700 light-years |
|---|---|
| MAGNITUDE | 5.4 |
| SPECTRAL TYPE | B2 |

**MONOCEROS**

Beta (β) Monocerotis is a triple star system with components A, B, and C. The BC pair orbits each other with a period of about 4,000 years, and A orbits the BC pair with a period of about 14,000 years. The system is unusual because the three stars are so similar. All are hot, blue–white B-type stars, each more than 1,000 times as luminous and six times as massive as the Sun. All three stars also exhibit the same rotation speed and have circumstellar disks.

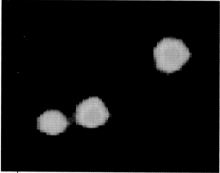

**COMPUTER-ENHANCED OPTICAL IMAGE**

## TRIPLE STAR
# Rigel

| DISTANCE FROM SUN | 860 light-years |
|---|---|
| MAGNITUDE | 0.1 |
| SPECTRAL TYPE | B8 |

**ORION**

Rigel is a blue supergiant star shining 40,000 times more brightly than the Sun and has a faint close companion, Rigel B. The luminosity of Rigel makes observation of the companion difficult. Rigel B has been discovered to be

a double star itself. It consists of two faint B-type main-sequence stars, Rigel B and Rigel C, separated by about 2.5 billion miles (4 billion km) and orbiting each other in an almost circular orbit. By contrast, the separation between the bright supergiant and the BC pair is over 190 billion miles (300 billion km).

### BRILLIANT GIANT
Rigel is the brightest star in the constellation Orion and the 7th-brightest star in the night sky. Rigel B and C, its companion stars, are obscured by Rigel's great luminosity.

## DOUBLE STAR
# Beta Lyrae

| DISTANCE FROM SUN | 880 light-years |
|---|---|
| MAGNITUDE | 3.5 |
| SPECTRAL TYPE | B7 |

**LYRA**

Beta (β) Lyrae, or Sheliak, is the prototype of a class of eclipsing binary stars known as Beta Lyrae stars or EB variables (see p.274). The brightness of the system varies by about one magnitude every 12 days 22 hours and is easily visible to the naked eye. Beta Lyrae's component stars are contact binaries and are so close together that they are greatly distorted by their mutual attraction. Material pouring out of the stars is forming a thick accretion disk.

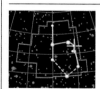

**CLOSE BINARY**

## QUINTUPLE STAR
# Sigma Orionis

| DISTANCE FROM SUN | 1,150 light-years |
|---|---|
| MAGNITUDE | 3.8 |
| SPECTRAL TYPE | O9 |

**ORION**

Sigma (σ) Orionis is a quintuple system containing four bright, easily visible stars and one fainter component, with the brightest being a close double. The two main stars, A and B, are more than 30,000 times as luminous as the Sun and have a combined mass more than 30 times greater than the Sun. The AB pair is one of the more massive binary systems in the Milky Way. It is in a stable orbit, but the C, D, and E stars are not, and gravitational forces may well throw them out of the system in the future.

**SIGMA ORIONIS'S FOUR BRIGHT STARS**

## QUADRUPLE STAR
# Theta Orionis

| DISTANCE FROM SUN | 1,800 light-years |
|---|---|
| MAGNITUDE | 4.7 |
| SPECTRAL TYPE | B |

**ORION**

Theta (θ) Orionis, perhaps better known as the Trapezium, appears to the naked eye to be a single star but is revealed by any telescope to be a quadruple system. Theta Orionis provides much of the ultraviolet radiation that illuminates the Orion Nebula (see p.241). All four stars are hot O- and B-type stars, the largest being Theta-1 C, with 40 times the

mass of the Sun; about 200,000 times its luminosity; and a temperature of 72,000°F (40,000°C). Theta-1 C is the hottest star visible to the naked eye. Theta-1 A is an eclipsing double star with an additional companion, Theta-1 D is a double star, and Theta-1 B is an eclipsing binary star with a companion double (making it quadruple in itself). The multiple star forms the core of an open cluster containing hundreds of young stars.

### THE TRAPEZIUM GROUP
The stars of Theta Orionis light up the center of the Orion Nebula. A false-color image (below) helps define the four main stars in the system.

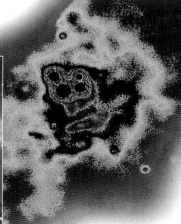

## DOUBLE STAR
# Epsilon Aurigae

| DISTANCE FROM SUN | 2,040 light-years |
|---|---|
| MAGNITUDE | 3 |
| SPECTRAL TYPE | A8 |

**AURIGA**

The hot giant Epsilon (ε) Aurigae, or Almaaz, is an eclipsing binary star. Unusually, its eclipse lasts for 2 years, suggesting that the system is huge. The giant star is being eclipsed by something far bigger than itself, but exactly what is uncertain. One theory is that Epsilon Aurigae's companion is an unseen binary system surrounded by a huge, dusty ring, through which the bright star shines through this ring during an eclipse.

### DISTANT BINARY
Dwarfed in this image by its celestial neighbor Capella, Epsilon Aurigae experiences its strange eclipses every 27 years.

**THE MILKY WAY**

# VARIABLE STARS

ALTHOUGH AT FIRST SIGHT the stars in the night sky seem to be unchanging, many thousands of stars change their brightness over periods ranging from a few days to decades. True, or intrinsic, variable stars vary in brightness due to physical changes within the star. Others, such as eclipsing binaries (see p.274), only appear to vary because they have orbiting companions.

**PROTOTYPE**
Mira is one of the most famous variable stars in the Milky Way (see p.285). It is a long-period, pulsating star that has given its name to one of the main types of variable stars.

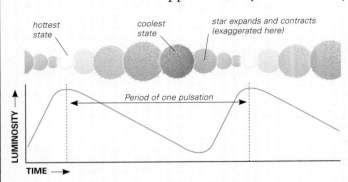

hottest state

coolest state

star expands and contracts (exaggerated here)

Period of one pulsation

LUMINOSITY →

TIME →

**LIGHT CURVE**
The light curve of a Cepheid variable shows the regular variation in luminosity during a period of one pulsation.

## PULSATING VARIABLES

Pulsating variable stars are intrinsic variables that undergo repetitive expansion and contraction of their outer layers. A pulsating star is constantly trying to reach equilibrium between the inward gravitational force and the outward radiation and gas pressure. This causes the star's brightness to vary. In many types of pulsating stars, including Cepheids (see p.286), a star's period of variation is related to its luminosity. Knowledge of the star's luminosity, coupled with its apparent magnitude, enables astronomers to calculate a star's distance. Pulsating variables are therefore a useful tool for determining distances to faraway objects such as other galaxies.

## NOVAE

A nova is a binary system consisting of a giant star that is being orbited by a smaller white dwarf. The giant star has grown so large that its outer material is no longer gravitationally bound to the star and instead falls onto the white-dwarf companion. Eventually, this gain in material triggers a thermonuclear explosion on the surface of the white dwarf, which brightens it by many magnitudes, increasing its energy output by a factor of a million or more. The surface gases of the white dwarf are in a "degenerate" state, and they do not obey the normal gas laws. Usually a gas explosion will cause the gas to expand, thereby reducing the explosion and finishing it. However, the degenerate gases of a white dwarf do not expand, and the explosion turns into a runaway event that does not finish until the fuel is exhausted. Prior to this, the binary system would be invisible to the naked eye, and the nova outburst would then bring the system into visibility as a "new"—in Latin, *nova*—star.

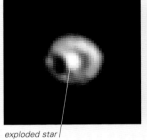

exploded star

hot bubble of gas

**CATACLYSMIC BINARY**
The most widely studied nova, Nova Cygni 1992, was witnessed exploding in 1992 (see p.287). Its magnitude rose by such a degree that at its brightest the nova was visible to the naked eye.

**LIGHT ECHOES**
In 2002, the star V838 Monocerotis (see p.265) emitted an outburst of light that echoed off the surrounding dust. The star has been imaged several times since. At first thought to be a nova, it might in fact be a new type of eruptive star.

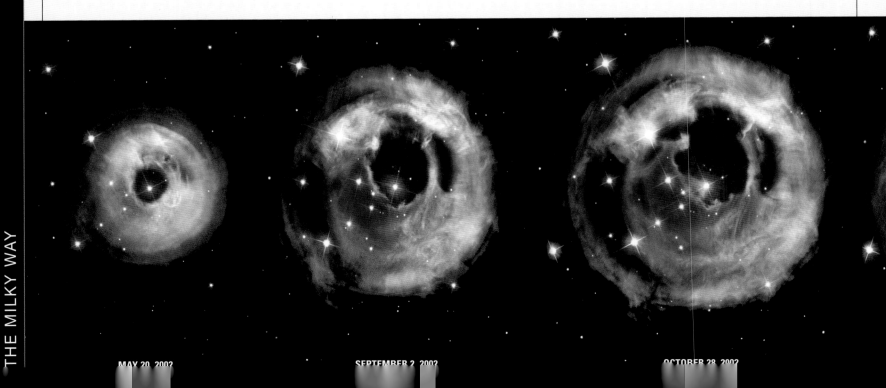

MAY 20, 2002

SEPTEMBER 2, 2002

OCTOBER 28, 2002

# TYPE I SUPERNOVAE

As in a nova (opposite), the source of a type I supernova is a binary system consisting of a giant star and a white dwarf. In type I supernovae, rather than triggering a nova, the material transfer onto the white dwarf continues to increase the mass of the star until it collapses and then explodes, destroying itself. The class of type I supernovae is subdivided depending on which chemical elements are present in the supernova's spectrum. In type Ia supernovae, the core of the white dwarf reaches a critical density, triggering the fusion of carbon and oxygen. This fusion is unconstrained and results in a massive explosion, with an associated leap in luminosity and the ejection of matter into interstellar space. According to theory, all type Ia supernovae have identical luminosities. This means that the distance to a supernova of this type can be determined by comparing its intrinsic luminosity with its apparent brightness.

large companion star

material being pulled from companion star

white dwarf

**POWERFUL SUPERNOVA**
This white-dwarf star pulls gas from a larger companion. Its mass rises until it can no longer support itself and it collapses in a huge explosion.

Supernova 1994D

**DISTANT SUPERNOVA**
Like other type Ia supernovae, 1994D—seen in the outskirts of the distant galaxy NGC 4526—has an intrinsic brightness that allows its distance to be known.

# BIZARRE VARIABLES

Many variable stars exhibit magnitude variations that are regular and are easily explained by eclipsing or by a pulsation mechanism occurring in a star's outer layers. However, there are other variable stars whose magnitude variations seem to defy explanation. One example is Epsilon (ε) Aurigae or Almaaz, a giant star with eclipses that last for 2 years, far longer than expected for a normal eclipsing system. As Almaaz is itself huge, whatever is eclipsing it must be even larger, but in the absence of decisive observations, astronomers can only theorize. One theory is that there is an unseen companion star or stars surrounded by a large dust ring, and it is the extended dust ring that eclipses Almaaz. Another bizarre variable is R Coronae Borealis (R CrB). This star can suddenly drop eight magnitudes, a large range that cannot be explained by physical changes within the star's structure (see p.287). The variation cannot be due to an eclipse, since the drop in magnitude is irregular and not periodic. Some astronomers have suggested that an orbiting dust cloud is responsible, but the more popular theory is that R CrB is ejecting material from its surface and that this ejected material blocks the light from the star before being blown away. Although the majority of variable stars are well understood, even to the extent that they can be used as reliable distance indicators, there are many individual stars that require further study before they reveal their secrets.

**MYSTERIOUS STAR**
One of the strangest stars known to astronomers, Epsilon Aurigae is a giant star that is being eclipsed by something even bigger than itself. One theory is that it is being orbited by a large, dusty disk surrounding a binary companion.

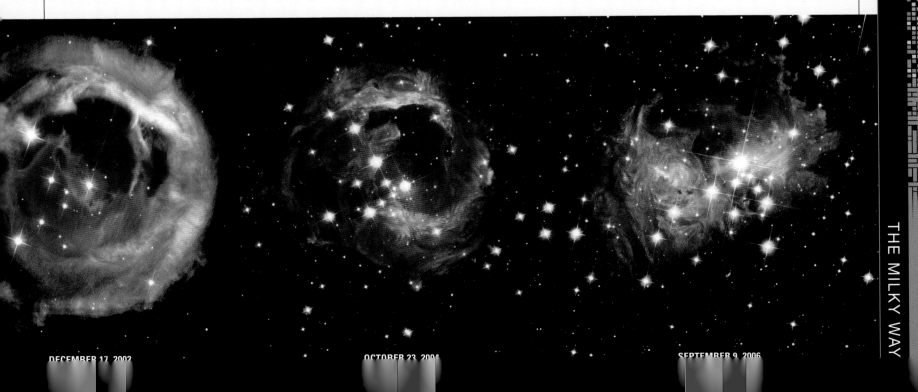

DECEMBER 17, 2002

OCTOBER 23, 2004

SEPTEMBER 9, 2006

# VARIABLE STARS

More than 30,000 variable stars are known within the Milky Way, and it is likely that there are many thousands more waiting to be discovered. Variable star research is a fundamental and vital branch of astronomy, as it provides information about stellar masses, temperatures, structure, and evolution. Variable stars often have periods ranging from years to decades, and professional astronomers do not have the resources to continuously monitor such stars. Consequently, amateur astronomers play a key role within this field, submitting thousands of observations into an international database.

**IRREGULAR VARIABLE**
The brightness of the variable star, Gamma (γ) Cassiopeiae, changes irregularly and unpredictably by up to two magnitudes.

---

## ROTATING VARIABLE

# Procyon

| | |
|---|---|
| **DISTANCE FROM SUN** | 11.4 light-years |
| **MAGNITUDE** | 0.34 |
| **SPECTRAL TYPE** | F5 |

**CANIS MINOR**

Procyon is only about seven times as luminous as the Sun but appears bright in the sky due to its proximity

**CONSPICUOUS VARIABLE**
Seven times more luminous than the Sun, Procyon is the eighth-brightest star in the night sky.

to Earth. Procyon has a companion, Procyon B, a white-dwarf star about the same size as Earth. Procyon shows small changes in magnitude caused by surface features, such as star spots, passing in and out of view as the star rotates. This type of variation classifies Procyon as a BY Draconis-type variable. In addition to surface changes, the tiny, brighter companion also increases the apparent brightness of Procyon when it passes in front of the star as seen from Earth.

---

## ERUPTIVE VARIABLE

# U Geminorum

| | |
|---|---|
| **DISTANCE FROM SUN** | 250 light-years |
| **MAGNITUDE** | 8.8 |
| **SPECTRAL TYPE** | B |

**GEMINI**

The prototype cataclysmic variable star, U Geminorum, is a close binary system consisting of a red main-sequence star orbiting and eclipsing a white dwarf and its accretion disk. Material falls from the main-sequence star onto the disk, causing localized heating and rapid increases in brightness of three to five magnitudes.

**PROTOTYPE**
U Geminorum lends its name to a type of irregular variable star that displays sudden increases in brightness.

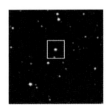

---

## INDETERMINATE VARIABLE

# Tabby's Star

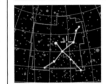

| | |
|---|---|
| **DISTANCE FROM SUN** | 1,470 light-years |
| **MAGNITUDE** | 11.7 |
| **SPECTRAL TYPE** | F3V |

**CYGNUS**

The planet-hunting satellite Kepler revealed this faint star (named after codiscoverer Tabetha S. Boyajian) to be one of the most intriguing variables in the sky. Tabby's Star is prone to unpredictable dips that reduce its brightness by up to 22 percent. These dips are too large to be caused by a transiting planet or eclipsing star, but spectroscopic studies show no signs of dust or planet-forming material in orbit. A precise cycle to the eclipses has also proved hard to pin down. One hypothesis is that Tabby's Star is orbited by a swarm of small compact bodies, while others suggest some new type of pulsation mechanism in the star itself.

---

## ECLIPSING BINARY

# Eta Geminorum

| | |
|---|---|
| **DISTANCE FROM SUN** | 349 light-years |
| **MAGNITUDE** | 3.3 |
| **SPECTRAL TYPE** | M3 |

**GEMINI**

Commonly known as Propus, Eta (η) Geminorum is a red giant star, and its red coloring is very apparent through binoculars. It is a semiregular variable star with a 0.6-magnitude variation—ranging between magnitudes 3.3 and 3.9—over 234 days. Propus is also a spectroscopic eclipsing binary, having a cool spectral-type-B companion star orbiting it with a period of 8.2 years and at a distance of about 625 million miles (1 billion km). Propus is therefore eclipsed every 8.2 years and is a target for amateur variable-star observers. A second star orbits at a greater distance, with a period of 700 years, but with no eclipsing. Although Propus is a cool star, with

**CHANCE ALIGNMENT**
In Earth's skies, Eta Geminorum lies close to the far more distant supernova remnant IC 433.

a temperature of about 6,500°F (3,600°C), it is more than 2,000 times as luminous as the Sun. Its temperature and luminosity suggest that it is 130 times larger than the Sun. This agrees with optical measurements, but there is some uncertainty—the star has different sizes when observed at different wavelengths due to dark bands of titanium oxide in its spectrum. This uncertainty in measuring Propus's size is typical of large, cool stars. Propus is evolving—observations by amateur

astronomers show that its average brightness has increased by 0.1 magnitude over the last decade. It has a dead helium core and is slowly entering a new phase: it is destined to become a Mira variable (see opposite).

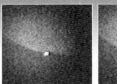   

**ETA GEMINORUM OCCULTED**
In an event that takes less than one-thirtieth of a second (above), Propus is occulted by the Moon (see p.71). In an optical image (top), Propus is pictured alongside the much more distant supernova remnant IC 443.

THE MILKY WAY

ECLIPSING BINARY

## Alpha Herculis

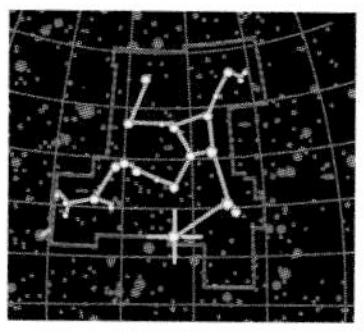

**DISTANCE FROM SUN** 382 light-years

**MAGNITUDE** 3

**SPECTRAL TYPE** M5

**HERCULES**

Alpha (α) Herculis, or Ras Algethi, is a cool red supergiant star that varies in brightness by almost one magnitude over a period of about 128 days. It is a complex star system with a much smaller companion that is itself a binary, consisting of a giant and a Sun-like star. There is a strong wind of stellar material blowing from the star, which reaches and engulfs its companions. Alpha Herculis is wider than the orbit of Mars. The outer atmosphere of the supergiant is slowly being removed, and the star will eventually become a white dwarf.

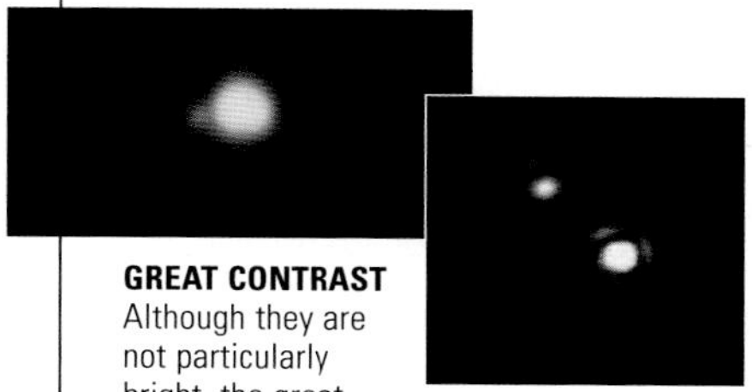

**GREAT CONTRAST**
Although they are not particularly bright, the great contrast in size and color of the stars that make up Alpha Herculis allow them to be separated easily through a telescope.

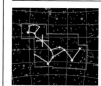

PULSATING VARIABLE

## Mira

**DISTANCE FROM SUN** 418 light-years

**MAGNITUDE** 3

**SPECTRAL TYPE** M7

**CETUS**

Omicron (o) Ceti, better known as Mira, is among the best known of all variable stars. At its brightest, it reaches 2nd magnitude, and at its faintest, it drops to 10th—far too faint for the naked eye. It undergoes this variation with a period of 330 days. Therefore, an observer can find Mira when it is at its brightest and over a period of time watch it completely disappear. Although Mira is one of the coolest stars visible in the sky, with a temperature of just 3,600°F (2,000°C), it is at least 15,000 times more luminous than the Sun. Internal changes in the star have left it so distended that the Hubble Space Telescope has revealed that it is not perfectly spherical (right). The variation in Mira's magnitude is caused by pulsations that cause temperature changes and therefore changes in the star's luminosity. Furthermore, Mira is shedding material from its outer layers in the form of a stellar wind. In the future, Mira will lose its outer structure and be left as a small white dwarf. In this way, Mira represents the future of the Sun.

EXPLORING SPACE

### WONDERFUL MIRA

When German astronomer David Fabricius discovered Mira in 1596, it was the first long-period variable star to be recognized. In 1642, Johannes Hevelius named the star Mira, meaning "wonderful." It has become the most famous long-period pulsating variable in the sky and one of the most popular stars for amateur astronomers. The American Association of Variable Star Observers has received more than 50,000 observations of Mira by over 1,600 observers.

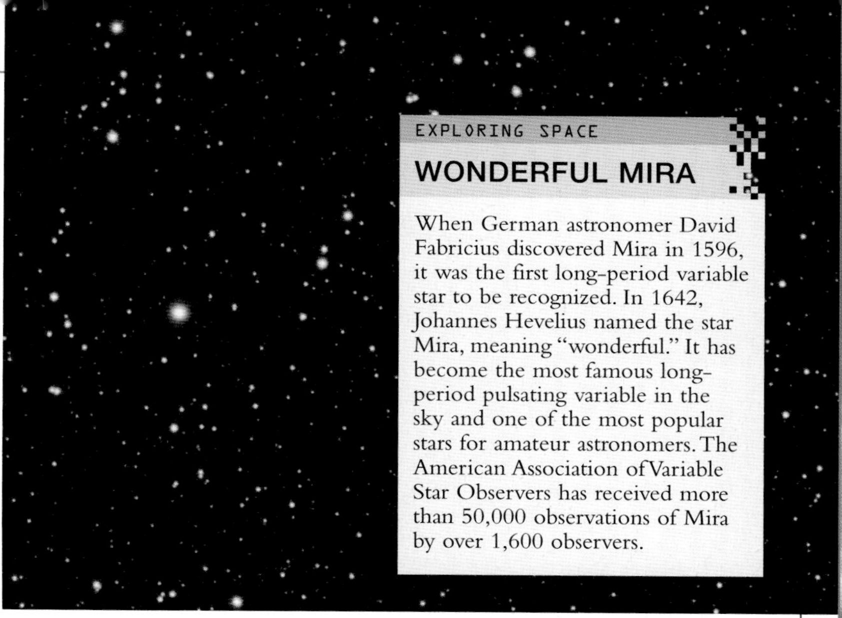

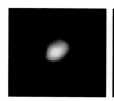

**THE ORIGINAL MIRA**
Easily recognized in the night sky, Mira lends its name to a type of long-period variable, of which thousands are known.

**DISTORTED SHAPE**
Enhancement of Hubble's Mira images reveals the star's asymmetrical atmosphere in visible (left) and ultraviolet light (right).

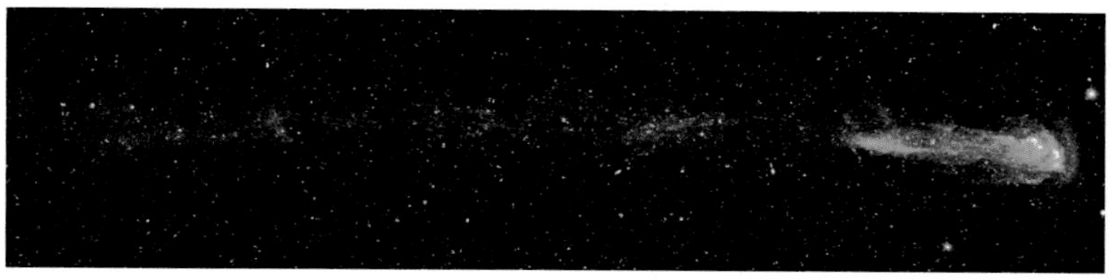

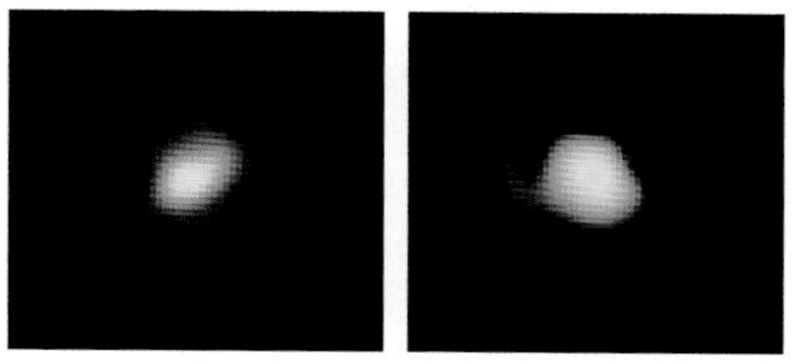

**GLOWING TAIL**
Mira is shedding gas as it moves through space, producing a tail 13 light-years long that shows up at ultraviolet wavelengths, as seen in this image from NASA's Galaxy Evolution Explorer.

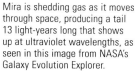

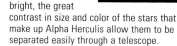

IRREGULAR VARIABLE

## Gamma Cassiopeiae

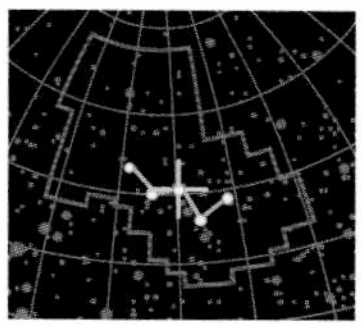

**DISTANCE FROM SUN** 613 light-years

**MAGNITUDE** 2.4

**SPECTRAL TYPE** B0

**CASSIOPEIA**

A hot blue star with a surface temperature of 45,000°F (25,000°C), Gamma (γ) Cassiopeiae is about 70,000 times more luminous than the Sun. It is a variable star with unpredictable changes in magnitude. Astronomers have observed it as bright as 1st magnitude and as faint as 3rd magnitude. It may have been fainter in ancient times, which might explain its lack of a common name. Gamma Cassiopeiae is a Be star (see panel, right), rotating at more than 625,000 mph (1 million kph) at its equator and shedding material from its surface. The ejected material forms a surrounding disk, and it is the disk that makes varying and unpredictable emissions. Gamma Cassiopeiae may also be transferring material to an undiscovered dense companion star.

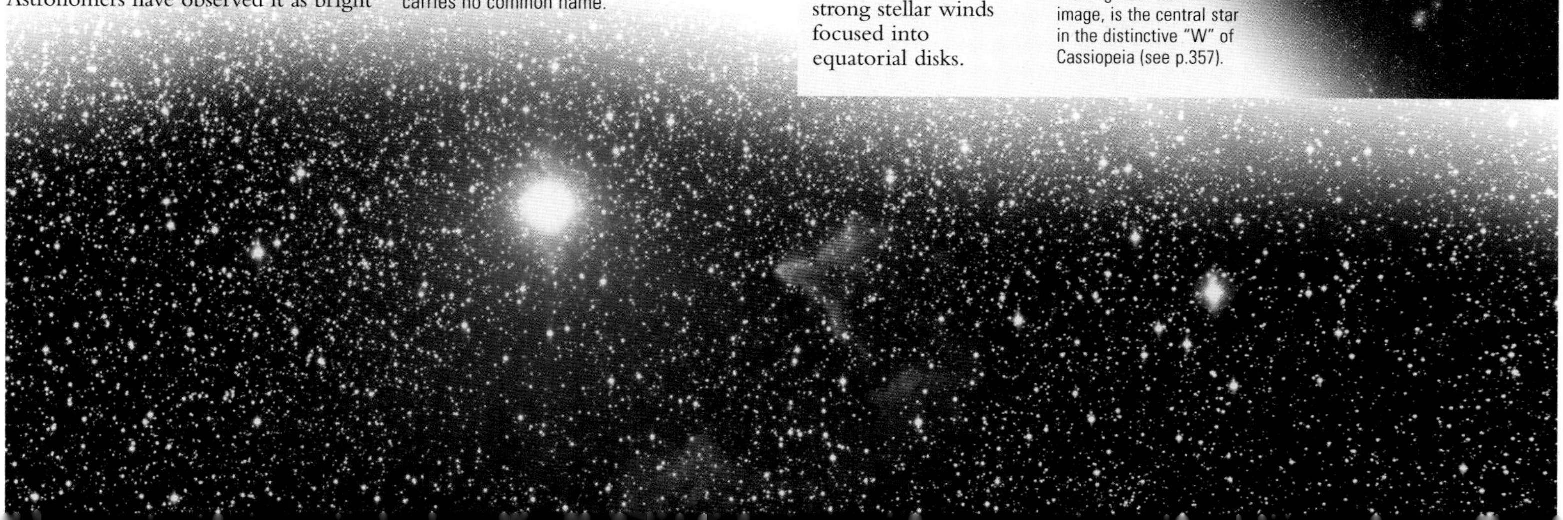

**NAMELESS STAR**
Pictured here with the red-colored emission nebula IC 63, Gamma Cassiopeiae is among the most prominent stars in the sky that carries no common name.

EXPLORING SPACE

### THE FIRST "BE" STAR

When, in 1866, Father Angelo Secchi, director of the Vatican Observatory and scientific advisor to Pope Pius IX, studied the spectrum of Gamma Cassiopeiae, he discovered that the star emitted light at particular wavelengths associated with hydrogen emission (see p.35). He is therefore credited with the discovery of the first Be star—a star of spectral type B but with "e" for emission. Be stars are characterized by their high rotation speeds, high surface temperatures, and strong stellar winds focused into equatorial disks.

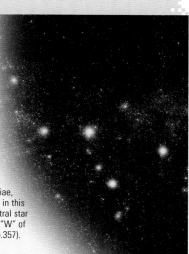

**CENTRAL STAR**
Gamma Cassiopeiae, the brightest star in this image, is the central star in the distinctive "W" of Cassiopeia (see p.357).

## PULSATING VARIABLE

# W Virginis

| | |
|---|---|
| DISTANCE FROM SUN | 10,000 light-years |
| MAGNITUDE | 9.6 |
| SPECTRAL TYPE | F0 |

**VIRGO**

W Virginis lends its name to a class of variable stars that are similar to Cepheid variables (see p.282) and are also known as Population II Cepheids. W Virginis is a pulsating yellow giant star. The outer layers of its atmosphere expand and contract with a period of 17.27 days. The period has lengthened over the last 100 years of observation. The pulsation causes a 1.2-magnitude variation, as the star doubles its size during the cycle. As a Population II star (see p.227), W Virginis is among the oldest stars in the Milky Way.

**W VIRGINIS**
W Virginis is located high above the galactic plane in the diffuse halo of old stars that surrounds the Milky Way (see pp.226–229). Like other W Virginis variables, it is an old Population II star, on average lower in mass and magnitude than a Cepheid variable.

## PULSATING VARIABLE

# RR Lyrae

| | |
|---|---|
| DISTANCE FROM SUN | 744 light-years |
| MAGNITUDE | 7.1 |
| SPECTRAL TYPE | F5 |

**LYRA**

RR Lyrae is the brightest member of the class of variables that takes its name. RR Lyrae stars are similar to Cepheid variables (see p.282) but are less luminous and tend to have shorter periods—ranging from about 5 hours to just over a day. RR Lyrae's period is 0.567 days, and its magnitude varies between 7.06 and 8.12. By comparing the luminosity of RR Lyrae variables with their apparent magnitude, a good distance determination can be made. In this way, RR Lyrae variable stars are important tools for calculating astronomical distances.

### BRIGHT VARIABLE
RR Lyrae has an average luminosity 40 times that of the Sun and a surface temperature of about 12,000°F (6,700°C). RR Lyrae stars are often found in globular clusters, and they are sometimes referred to as cluster variables.

## PULSATING VARIABLE

# Delta Cephei

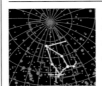

| | |
|---|---|
| DISTANCE FROM SUN | 982 light-years |
| MAGNITUDE | 4 |
| SPECTRAL TYPE | F5 |

**CEPHEUS**

Delta (δ) Cephei is the prototype of the Cepheid class of variable stars (see p.282), and to astronomers it is one of the most famous stars in the sky. Its magnitude variation, from 3.48 to 4.37, is visible to the naked eye, and its short period of 5 days, 8 hours, and 37.5 minutes makes it a popular target for amateur observers. Its position in the sky makes it easy to find, and it is close to two comparison stars with magnitudes at the ends of Delta Cephei's range. Delta Cephei is a supergiant with a spectral type that varies between F5 and G2.

**BOW SHOCK**
Strong stellar winds blowing out from Delta Cephei excite interstellar gas and dust in this infrared image.

## EXPLORING SPACE

# THE CEPHEID PROTOTYPE

In 1921, Henrietta Leavitt (1868–1921), an astronomer based at the Harvard Observatory, discovered a strong link between the period and luminosity of a group of stars later known as Cepheid variables (see p.282), of which Delta Cephei was the prototype. This correlation provided astronomers with a new way of measuring distances in space. In 1923, Edwin Hubble used it to prove that the Andromeda Galaxy is situated outside the Milky Way. Since then, Cepheids have provided more useful information about the universe than any other star type.

**HENRIETTA LEAVITT**

## PULSATING VARIABLE

# Zeta Geminorum

| | |
|---|---|
| DISTANCE FROM SUN | 1,168 light-years |
| MAGNITUDE | 4 |
| SPECTRAL TYPE | G0 |

**GEMINI**

Also known as Mekbuda, Zeta (ζ) Geminorum is a yellow supergiant about 3,000 times as luminous as the Sun. It is one of the easiest Cepheid variable stars to observe (see p.282). Zeta Geminorum, like all Cepheids, is unstable and pulsates, changing its temperature, size, and spectral type. It has a period of 10.2 days and a magnitude that varies from 3.6 to 4.2. Its period is shortening at the rate of about 3 seconds per year. Zeta Geminorum is also a binary star with a faint, magnitude-10.5 companion.

## PULSATING VARIABLE

# Eta Aquilae

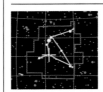

| | |
|---|---|
| DISTANCE FROM SUN | 1,400 light-years |
| MAGNITUDE | 3.9 |
| SPECTRAL TYPE | F6 |

**AQUILA**

Eta (η) Aquilae is a yellow supergiant star with a luminosity 3,000 times that of the Sun. It is one of the brightest Cepheid variables in the night sky (see p.282), and also one of the first to be discovered. Eta Aquilae varies in magnitude from 3.5 to 4.4 over a period of 7.2 days. Such a brightness variation is easily detectable with the naked eye. The magnitude range, by coincidence, is the same as the prototype of the class, Delta Cephei (above). Over this period, Eta Aquilae also varies in spectral type between G2 and F6.

## NOVA

# T Coronae Borealis

| | |
|---|---|
| DISTANCE FROM SUN | 2,025 light-years |
| MAGNITUDE | 11 |
| SPECTRAL TYPE | M3 |

**CORONA BOREALIS**

T Coronae Borealis (T CrB), also known as the Blaze Star, is a recurrent nova (see p.282). It has displayed two major outbursts, one witnessed in 1866 and the other in 1946. T CrB's usual apparent magnitude is 10.8, but during outbursts, it has reached 2nd or 3rd magnitude. T CrB is a spectroscopic binary consisting of a red giant of spectral type M3 and a smaller blue-white dwarf. T CrB is usually about 50 times as luminous as the Sun, but during outbursts, it becomes more than 200,000 times as luminous. In between the outbursts,

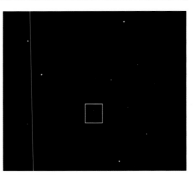

**BLAZE STAR**
Although T Coronae Borealis cannot usually be seen without a telescope, during eruptions, it has "blazed" bright enough to be seen with the naked eye.

stellar dust and gas from the outer layers of the red giant are drawn onto the white dwarf. Eventually, the total mass of the white dwarf reaches a critical level, causing the outer layers of the white dwarf to explode violently. After the explosion, the two stars return to normal, to repeat the process many years later.

# Mu Cephei

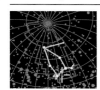

| DISTANCE FROM SUN | |
|---|---|
| 5,258 light-years | |
| MAGNITUDE | 4 |
| SPECTRAL TYPE | M2 |

**CEPHEUS**

Mu (μ) Cephei, or the Garnet Star, is also known as Herschel's Garnet Star, after the pioneering German-born astronomer William Herschel, who first described its distinctive red color and noted its resemblance to the precious stone garnet. Mu Cephei is one of the most luminous stars in the Milky Way, outshining the Sun by a factor of more than 200,000. A red supergiant, it is also one of the largest stars that can be seen with the naked eye. Its great size means that if placed in the Sun's position at the center of the solar system, its outer layers would fall between Jupiter and Saturn. As with most large supergiants, Mu Cephei is an unstable star, expanding and contracting in diameter with a corresponding variation in magnitude. It is classed as a semiregular supergiant variable with a spectral type of M2 and a magnitude varying between 3.43 and 5.1. It has two periods of variation (730 and 4,400 days) overlaid on one another. The pulsations of Mu Cephei, caused by internal absorption and release of energy, have thrown off the outer layers of the star's atmosphere, creating concentric shells of dust and gas around it. Observations have also shown that Mu Cephei is surrounded by a sphere of water vapor. Mu Cephei probably started its life as a star of around 20 solar masses. Typically for such a high-mass star, it has evolved very rapidly, and we are seeing it as it hurtles headlong toward the end of its short life. One day soon (on an astronomical timescale), Mu Cephei will erupt in a cataclysmic supernova, after which only the core will remain, ending its days as a neutron star or black hole.

**VARIABLE JEWEL**
Known for its distinctive color, Mu Cephei, or the Garnet Star, is the bright reddish-orange star at the top left of the image. It is pictured above the red emission nebula IC 1396 (see p.243).

# RS Ophiuchi

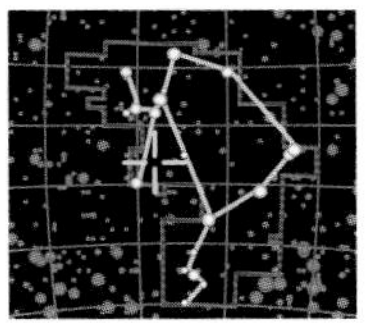

| DISTANCE FROM SUN | |
|---|---|
| 2,000–5,000 light-years | |
| MAGNITUDE | 12.5 |
| SPECTRAL TYPE | M2 |

**OPHIUCHUS**

RS Ophiuchi is a recurrent nova (see p.282), having been witnessed erupting in 1898, 1933, 1958, 1967, 1985, and—most recently—in 2006. During its periods of normality, it shines at magnitude 12.5, but during outbursts, it has reached 4th magnitude. While RS Ophiuchi is usually invisible to the naked eye, during outbursts, it can be seen in the night sky without a telescope. RS Ophiuchi is classed as a cataclysmic variable—a binary system consisting of a giant star shedding material and a dwarf companion receiving the material. Eventually, a thermonuclear explosion is triggered on the surface of the dwarf, resulting in the ejection of a shell and an increase in brightness. RS Ophiuchi is constantly monitored by amateur astronomers, and the American Association of Variable Star Observers has more than 30,000 observations in its database.

# R Coronae Borealis

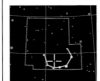

| DISTANCE FROM SUN | |
|---|---|
| 6,037 light-years | |
| MAGNITUDE | 5.9 |
| SPECTRAL TYPE | G0 |

**CORONA BOREALIS**

The prototype of a class of variable stars, R Corona Borealis (R CrB) drops in magnitude from 5.9 to 14.4 at irregular intervals. There are two theories for this variation. One is that an orbiting dust cloud obscures R CrB when it passes in front of the star. The other is that R CrB ejects material, which obscures the light the star emits, before being blown away.

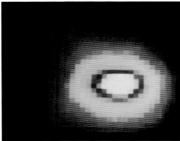

**FALSE-COLOR INFRARED IMAGE**

# Nova Cygni 1992

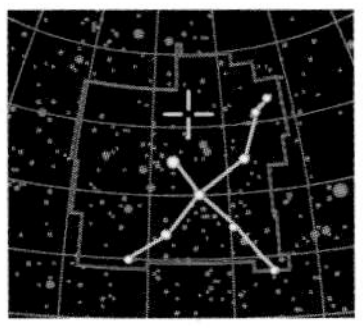

| DISTANCE FROM SUN | |
|---|---|
| 10,430 light-years | |
| MAGNITUDE | 4.3 |
| SPECTRAL TYPE | Q |

**CYGNUS**

Nova Cygni 1992, a cataclysmic binary, was discovered on the night of February 18–19, 1992, shining with a magnitude of 7.2 at a location where there should have been no such star. The discoverer, Peter Collins (see p.80), alerted astronomical authorities, and soon a whole range of instruments—both ground-based and space-borne (including the Voyager spacecraft)—were observing it at a variety of wavelengths. Over the next few days, the nova continued to brighten to magnitude 4.3, making Nova Cygni 1992 not only the first nova to be observed so extensively, but also the first to be thoroughly observed before it had reached its peak. The nova eruption was the result of material falling from one star onto a white dwarf, triggering an explosion and the ejection of a shell of material. The Hubble Space Telescope observed the system in 1993, detecting the ring thrown out by the binary system and also an unusual barlike structure across the middle of the ring, the origin of which is unknown.

**BRIGHT NOVA**
Nova Cygni 1992 was one of the brightest nova to be witnessed erupting in recent history. Targeted by some of the most powerful telescopes in the world, it could be seen at its brightest with the naked eye.

**HOT BUBBLE**
This Hubble Space Telescope photograph reveals the irregularly shaped bubble of hot stellar material blasted into space by the eruption of Nova Cygni 1992.

# STAR CLUSTERS

ALTHOUGH THE STARS in our night sky appear to live out their lives in isolation, many millions of stars reside in groups called open and globular clusters. Open clusters are young and often the site of new star creation, whereas globulars are ancient, dense cities of stars, some of which contain as many stars as a small galaxy.

## OPEN STAR CLUSTERS

Open clusters are made up of "sibling" stars of similar age formed from the same nebulous clouds of interstellar gas and dust. This often results in stars within an open cluster having the same chemical composition. However, an open cluster's stars can exhibit a wide range of masses due to variations within the original nebula and other influences during their formation. Open clusters reside within the galactic plane and often remain associated with the nebulous clouds from which they were produced. Open clusters do not hold onto their stars for long—as they orbit the center of the galaxy, they lose their members over a period of hundreds of millions of years. More than 2,000 open clusters have been discovered within the Milky Way, representing perhaps only 1 percent of the total population.

**YOUNG OPEN CLUSTER**
Spanning an area in the sky larger than a full moon, M39 is a large but sparsely populated open cluster. It contains about 30 loosely bound stars, each around 300 million years old and therefore much younger than the Sun.

**LARGEST CLUSTER**
The largest globular cluster in the Milky Way, Omega (ω) Centauri probably contains more than 10 million stars. This makes it larger than some small galaxies.

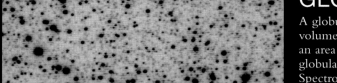

## GLOBULAR CLUSTERS

A globular cluster is a massive group of stars bound together by gravity within a spherical volume. Globular clusters can contain between 10,000 and several million stars, all within an area often less than 200 light-years across. As in open clusters, the stars within a globular cluster all have the same origin, and thus similar ages and chemical compositions. Spectroscopic studies of the starlight from globulars reveal that their stars are very old—older than most of the stars currently within the disk of the Milky Way. Analysis of their properties also reveals that they are about the same age, implying that they all formed together over a short period of time. Estimates of their ages vary, but they are thought to be over 10 billion years old. More than 150 globular clusters have been discovered in the Milky Way. Although a few are found in its central bulge, most are located in the halo. The chemistry of globular clusters shows that they represent the remnants of the early stages of the formation of the Milky Way and possibly formed even before the Milky Way had a disk. Four globulars may have originally been part of a dwarf galaxy that has been absorbed into the Milky Way. Globular clusters are made up of Population II stars (see p.227), which have their own independent orbits. These orbits are highly elliptical and can take the globulars out to distances of hundreds of thousands of light-years from the center of the Milky Way. Globular clusters are not unique to the Milky Way, and some galaxies have more globular clusters than our own.

**DENSE CLUSTER**
An image of part of the Omega (ω) Centauri globular cluster, captured in red light, reveals a great swarm of tightly bound stars. Omega Centauri is one of the densest and most populated globular clusters known within the Milky Way or beyond.

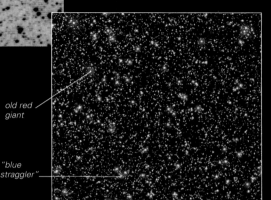

old red giant

"blue straggler"

**BLUE STRAGGLERS**
In the central region of the Omega Centauri cluster, among its old red stars, are seen a few young blue stars. These stars, called "blue stragglers," are thought to have been created by densely packed stars colliding.

## CLUSTER EVOLUTION

Star clusters, whether open or globular, are not static through time. Over millions of years, the clusters change physically and the stars within them age and die. However, there are major differences between the evolution of globular clusters and open clusters. An open cluster starts its life with a set of stars of similar chemical composition and age. Over hundreds of millions of years, it loses its members, either due to death of the stars or losing them to the gravitational tugs of other stars within the Milky Way. However, an open cluster can continue to manufacture stars from the original nebulous cloud from which it formed. Because of this, open clusters often contain stars of different ages at various stages of evolution. A globular cluster is more tightly bound and less likely to lose its stars. It also spends most of its time away from the disk of the galaxy, avoiding interactions. In this way, its structure is preserved for thousands of millions of years—far longer than open clusters. Similarly, once a globular cluster has formed, the original gas and dust is ejected and the cluster is then unable to form new stars. As the stars within a globular cluster age and die, so the cluster itself ages and dies.

spiral arm

**CLUSTER DISTRIBUTION**
The difference in the distributions of open and globular clusters within the Milky Way reflects their differences in age and orbit. Open clusters, formed from relatively young, Population I stars, are located within the Milky Way's rotating disk. Globular clusters, made up of Population II stars, have independent orbits mostly located out in the Milky Way's halo.

halo

central bulge

globular clusters

open clusters

**EVOLVED CLUSTER**
At about 1 billion years old, NGC 2266 is a relatively old and well-evolved open cluster. Many of its stars, clearly seen here, have reached the red-giant stage of their life cycle, while young blue stars are also present.

# STAR CLUSTERS

More than 2,000 open clusters have been cataloged in the Milky Way. About half contain fewer than 100 stars, but the largest have more than 1,000. Open clusters are asymmetrical and range in size from 5 to 75 light-years across. By contrast, globular clusters contain up to a million stars, spread symmetrically across several hundred light-years. Only about 150 globular clusters are known in the Milky Way and, unlike open clusters, which are found mainly in the galaxy's spiral arms, most are scattered around the periphery.

**MASSIVE CLUSTER**
Omega Centauri is a prime example of a globular star cluster. It contains more than 10 million old stars and has a mass of 5 million solar masses.

---

OPEN CLUSTER

## Hyades

**CATALOG NUMBER**
MEL 25

**DISTANCE FROM SUN**
150 light-years

**MAGNITUDE** 0.5

**TAURUS**

The Hyades is one of the closest open clusters to Earth and has been recognized since ancient times. The brightest of the cluster's 200 stars form a V-shape in the sky, clearly visible to the naked eye. The cluster's central group is about 10 light-years

in diameter, and its outlying members span up to 80 light-years. Most of the stars in this cluster are of spectral classes G and K (see pp.232–233) and are average in size, with temperatures comparable to that of the Sun. The brightest star in the field of the Hyades, the red giant Aldebaran (see p.256), is not a member of the cluster and is much closer to Earth. The cluster's stars all move in a common direction, toward a point east of Betelgeuse in Orion (see p.256). Studies of the movement of the stars of the Hyades show that they have a common origin with the Beehive Cluster (see below). The Hyades is thought to be about 790 million years old, and this age matches that of the Beehive Cluster. The parallel movement of stars in the Hyades has allowed their distance to be measured using the moving cluster method for stellar distances (see pp.232–233).

**PROMINENT CLUSTER**
First recorded by Homer around 750 BCE, the Hyades is one of the few star clusters visible to the naked eye. Aldebaran, the bright red giant in this image, is not part of the cluster but is 90 light-years closer to Earth.

---

**OPEN CLUSTER**

## Butterfly Cluster

**CATALOG NUMBERS**
M6, NGC 6405

**DISTANCE FROM SUN**
2,000 light-years

**MAGNITUDE** 5.3

**SCORPIUS**

The Butterfly Cluster, located toward the center of the Milky Way, is about 12 light-years across and has an estimated age of 100 million years. In the night sky, the cluster occupies an area the size of a full moon, and, to some, it resembles the shape of a butterfly. The cluster is made up of about 80 stars, most of them very hot, blue main-sequence stars with spectral types B4 and B5 (see pp.232–233). The brightest star in the cluster, BM Scorpii, is an orange supergiant star that is also a semi-regular variable (see pp.282–283). At its brightest, this star is visible to the naked eye; at its faintest, binoculars are needed. The Butterfly Cluster displays a striking contrast between the blue main-sequence stars and the orange supergiant.

**SKY SPECTACLE**
The Butterfly Cluster is one of the largest and brightest open star clusters in the Milky Way. It can best be seen with binoculars in a dark sky and can be located within the constellation Scorpius.

---

**OPEN CLUSTER**

## Beehive Cluster

**CATALOG NUMBER**
M44

**DISTANCE FROM SUN**
577 light-years

**MAGNITUDE** 3.7

**CANCER**

The Beehive Cluster, also known as Praesepe, is easily visible to the naked eye. The cluster contains over 350 stars spread across 10 light-years, but most of them can be seen only with a large telescope. It is thought to be about 730 million years old. Age, distance, and motion measurements suggest that the Beehive Cluster most likely originated in the same star-forming nebula as the Hyades (above).

**CELESTIAL BEEHIVE**

---

**OPEN CLUSTER**

## M93

**CATALOG NUMBERS**
M93, NGC 2447

**DISTANCE FROM SUN**
3,600 light-years

**MAGNITUDE** 6

**PUPPIS**

M93 is a bright open cluster and, at about 25 light-years across, is relatively small. It lies in the southern sky, close to the galactic equator. The cluster consists of about 80 stars, but only a few of the stars—blue giants of spectral type B9 (see pp.232–233)—provide most of the cluster's light. At about 100 million years old, M93 is young in astronomical terms.

**SOUTHERN CLUSTER**

---

**OPEN CLUSTER**

## M52

**CATALOG NUMBERS**
M52, NGC 7654

**DISTANCE FROM SUN**
3,000–7,000 light-years

**MAGNITUDE** 7.5

**CASSIOPEIA**

An open cluster of about 200 stars, M52 lies against a rich Milky Way background. It was first cataloged in 1774 by Charles Messier (see p.73). The distance to the cluster is uncertain, with estimates ranging from 3,000 to 7,000 light-years. The uncertainty is due to high interstellar absorption that affects the light from the cluster during its journey to Earth. The uncertain distance also means that the cluster's size is unknown, but midrange estimates give a size of about 20 light-years across. The age of the cluster is calculated to be about 35 million years. The brightest stars in M52 have magnitudes of only 7.7 and

**CLUSTER AND NEBULA**
This image, stretching more than twice the diameter of a full moon, captures the open cluster M52 (top left) and the glowing Bubble Nebula (bottom right).

8.2, and with an overall magnitude of 7.5, the cluster is too faint to be seen with the naked eye. However, through binoculars, the cluster can be viewed as a faint nebulous patch, while a small telescope reveals a rich, compressed cluster of stars.

# Pleiades

| CATALOG NUMBER |
|---|
| NGC 1435 |
| DISTANCE FROM SUN |
| 380 light-years |
| MAGNITUDE 4.17 |

**TAURUS**

The Pleiades, also known as the Seven Sisters, is the best-known open cluster in the sky and has been recognized since ancient times (see panel, right). The cluster is easily visible to the naked eye, but although most people

### CLUSTERS IN TAURUS
The two best-known clusters, the Pleiades (boxed) and the Hyades (opposite), both lie in Taurus. However, the Pleiades is more than 200 light-years more distant.

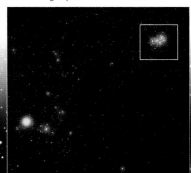

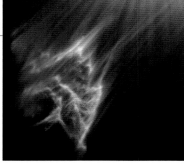

### GHOSTLY NEBULA
This haunting image shows an interstellar cloud caught in the process of destruction by strong radiation from the star Merope in the Pleiades. The cloud is called IC 349 or Barnard's Merope Nebula.

can see seven stars in the Pleiades, the seventh can often be elusive. Nine stars can be seen on a very dark and clear night. The nine brightest stars are known as the father, Atlas; the mother, Pleione; and the sisters Alcyone, Maia, Asterope, Taygeta, Celaeno, Merope, and Electra. Small telescopes and binoculars reveal many more stars, and larger telescopes show that the cluster, in fact, contains hundreds of stars. The Pleiades is about 100 million years old and will remain a cluster for only

another 250 million years or so, by which time it will have broken up into separate isolated stars. The stars of the Pleiades are blue giants of spectral class B (see pp.232–233) and are hotter and more luminous than the Sun. Long-exposure photography reveals that the Pleiades stars are embedded in clouds of interstellar dust. The clouds are illuminated by radiation from the stars, and they glow as reflection nebulae (see p.228). Although most gas and dust surrounding star clusters represents the material that gave birth to the stars, here the clouds are merely moving through the cluster. The clouds are traveling relative to the Pleiades at 25,000 mph (40,000 kph) and will eventually pass through the cluster and travel into deep space, where they will once again become dark and invisible.

### GLOWING NEBULOSITY
The stars of the Pleiades are surrounded by clouds of dusty material that are reflecting the blue light of the stars. However, the stars were not produced from this material, which seems simply to be passing by.

## BRONZE AGE CLUSTER

The Nebra Disk is perhaps the oldest semirealistic depiction of the night sky. It was discovered in 1999, near the German town of Nebra, and other artifacts found nearby have allowed it to be dated to about 1600 BCE. The disk depicts a crescent moon, a full moon, randomly placed stars, and a star cluster likely to be the Pleiades. Although its authenticity remains uncertain, the Nebra Disk may be proof that European Bronze Age cultures had a more sophisticated appreciation of the night sky than had previously been accepted.

**ANCIENT PLEIADES**
The cluster of seven gold dots (above and right of center) has been interpreted as the Pleiades as it appeared 3,600 years ago.

**M9 GLOBULAR CLUSTER**
Globular clusters are swarms of very old stars that were born long before the Sun. Most are concentrated toward the center of the Milky Way, including this one, M9, some 25,000 light-years away. M9 is estimated to contain a quarter of a million stars. In this image from the Hubble Space Telescope, hot blue stars and cooler red giants can be identified by their colors.

## GLOBULAR CLUSTER
# M4

| CATALOG NUMBERS |
| --- |
| M4, NGC 6121 |
| DISTANCE FROM SUN |
| 7,000 light-years |
| MAGNITUDE 7.1 |

**SCORPIUS**

M4 is one of the closest globular clusters to Earth and can be seen with the naked eye on a dark, clear night. The cluster has a diameter of about 70 light-years and contains more than 100,000 stars, but about half the cluster's mass resides within 8 light-years of its center. The Hubble Space Telescope has revealed a planet within M4, with about twice the mass of Jupiter, orbiting a white dwarf star. The planet is estimated to be 13 billion years old.

**DENSE CENTER**

## OPEN CLUSTER
# Jewel Box

| CATALOG NUMBER |
| --- |
| NGC 4755 |
| DISTANCE FROM SUN |
| 8,150 light-years |
| MAGNITUDE 4.2 |

**CRUX**

**GLITTERING JEWELS**

The Jewel Box, also known as the Kappa Crucis Cluster, is an open cluster of about 100 stars and is about 20 light-years across. At less than 10 million years old, it is one of the youngest open clusters known. The three brightest stars are blue giants, while the fourth-brightest star is a red supergiant. The different colors are very apparent in photographs of the cluster, hence its popular name. Lying within the constellation Crux, the Jewel Box is visible only to observers in the Southern Hemisphere.

## GLOBULAR CLUSTER
# 47 Tucanae

| CATALOG NUMBER |
| --- |
| NGC 104 |
| DISTANCE FROM SUN |
| 13,400 light-years |
| MAGNITUDE 4.9 |

**TUCANA**

47 Tucanae is so named because it was originally cataloged as a star—the 47th in order of right ascension in the constellation Tucana. In reality, it is the second-largest and second-brightest globular cluster in the sky, containing several million stars—enough to make a small galaxy. These stars are spread over an area about 120 light-years across, and the cluster's central region is so crowded that there is a high rate of stellar collisions. As a globular cluster ages, the stars within it also age, but 47 Tucanae is home to a number of blue stragglers—stars that are too blue and too massive to still be there if they were original members of the cluster. Astronomers have determined that it is the stellar collisions within the cluster that cause the formation of these blue stragglers.

**SOUTHERN SPECTACLE**
An optical image (top) captures 47 Tucanae and the Small Magellanic Cloud (see p.311), a satellite galaxy of the Milky Way. A close-up of 47 Tucanae (boxed) reveals one of the most spectacular globular clusters in the sky.

## GLOBULAR CLUSTER
# Omega Centauri

| CATALOG NUMBER |
| --- |
| NGC 5139 |
| DISTANCE FROM SUN |
| 17,000 light-years |
| MAGNITUDE 5.33 |

**CENTAURUS**

Omega Centauri is the largest globular cluster in the Milky Way—up to 10 times as massive as other globular clusters. Containing more than 10 million stars and having a width of 150 light-years, Omega Centauri is as massive as some small galaxies. To the naked eye, it appears as a fuzzy star, but a small telescope starts to resolve its individual stars. Studies of the cluster's stellar population have revealed that Omega Centauri is one of the oldest objects in the Milky Way—almost as old as the universe itself—and that there have been several episodes of star formation within the cluster. This is unusual for a globular cluster, and one explanation for this is that Omega Centauri may once have been a dwarf galaxy that collided with our own. It would have had about 1,000 times its current mass, but the Milky Way would have ripped it apart, leaving Omega Centauri as the remnant core.

**GIGANTIC GLOBULAR**
Easily the biggest of all known globular clusters in the Milky Way, Omega Centauri has a mass of more than 5 million solar masses. The stars in this globular cluster are generally older, redder, and less massive than the Sun.

## GLOBULAR CLUSTER
# NGC 3201

| CATALOG NUMBER |
| --- |
| NGC 3201 |
| DISTANCE FROM SUN |
| 15,000 light-years |
| MAGNITUDE 8.2 |

**VELA**

The globular cluster NGC 3201 contains many bright red giant stars, which give the cluster an overall reddish appearance. The cluster lies close to the galactic plane, so its appearance is further reddened by interstellar absorption. With a visual magnitude of only 8.2, the cluster is too faint to be seen with the naked eye. NGC 3201 is less condensed than most globular clusters, and several observers have suggested that some of the stars appear in short, curved rays, like jets of water from a fountain.

**RED-TINGED CLUSTER**

## GLOBULAR CLUSTER
# M12

**CATALOG NUMBERS**
M12, NGC 6128

**DISTANCE FROM SUN**
16,000–18,000 light-years

**MAGNITUDE** 7.7

**OPHIUCHUS**

Discovered by Charles Messier (see p.73) in 1764, M12 was one of the first globular clusters to be recognized. M12 is at the very limit of naked-eye visibility and therefore best viewed with a telescope. The cluster contains many bright stars and is condensed toward the center. Its stars are spread across a distance of about 70 light-years, making it less compact than most. Because of this, M12 was originally regarded as an intermediate form of cluster, between open clusters and globular clusters, before the two types were recognized as being fundamentally different.

**EARLY DISCOVERY**

## GLOBULAR CLUSTER
# NGC 4833

**CATALOG NUMBER**
NGC 4833

**DISTANCE FROM SUN**
17,000 light-years

**MAGNITUDE** 7.8

**MUSCA**

NGC 4833 is a small globular cluster in the southern constellation Musca and therefore is not visible to most observers in the Northern Hemisphere. It was discovered by Nicolas Louis de Lacaille (see p.422) during his 1751–1752 journey to South Africa. Although NGC 4833 is too faint to see with the naked eye, it is easily visible through a small telescope. However, because the cluster is rich and compact, even a moderate amateur telescope fails to resolve its stars fully. The center of the cluster is only slightly more dense than its surroundings, and consequently the cluster lacks the gravitational pull needed to hold onto its stars and many have already left the cluster. NGC 4833 is located below the galactic plane behind a dusty region. The dust absorbs light from the cluster and causes

its starlight to redden. Because of this reddening, astronomers studying this globular cluster have had to correct the apparent magnitudes of the various stars being studied. The technique used is applied to all globulars lying near the galactic plane. The cluster contains at least 13 confirmed RR Lyrae variable stars (see pp.282–283), which have helped astronomers to estimate the cluster's age at about 13 billion years.

**COMPACT CLUSTER**
NGC 4833 was first recorded by Nicolas Louis de Lacaille in 1752 as resembling a comet. However, with modern, high-powered telescopes, it is seen as a well-resolved and compact cluster with a scattering of outlying stars.

**DISTANT GLOBULAR**

## GLOBULAR CLUSTER
# M14

**CATALOG NUMBERS**
M14, NGC 6402

**DISTANCE FROM SUN**
23,000–30,000 light-years

**MAGNITUDE** 8.3

**OPHIUCHUS**

The globular cluster known as M14 has a diameter of about 100 light-years and contains several hundred thousand stars. Because of its considerable distance, it is too faint to be seen with the naked eye and, although binoculars or a small telescope will reveal the cluster, a larger instrument is needed to resolve individual stars. Many amateur observers mistakenly identify this object as an elliptical galaxy. In 1938, M14 was home to the first nova photographed in a globular cluster. However, subsequent searches with some of the world's most powerful telescopes have failed to find either the nova star or any of its remnants.

## GLOBULAR CLUSTER
# M107

**CATALOG NUMBERS**
M107, NGC 6171

**DISTANCE FROM SUN**
27,000 light-years

**MAGNITUDE** 8.9

**OPHIUCHUS**

A relatively "open" globular cluster lying close to the galactic plane, M107 is too faint to be seen with the unaided eye. Observations through large telescopes have revealed that the cluster contains dark regions of interstellar dust that obscure some of its stars. This is quite unusual in globular clusters. M107 spans a distance of about 50 light-years.

**LOOSE CLUSTER**

## GLOBULAR CLUSTER
# M68

**CATALOG NUMBERS**
M68, NGC 4590

**DISTANCE FROM SUN**
33,000–44,000 light-years

**MAGNITUDE** 9.7

**HYDRA**

M68 is a globular cluster that is visible only through telescopes. It appears as a small patch when viewed with binoculars, but small telescopes can reveal its constituent stars and its densely populated center. The cluster has a diameter of about 105 light-years, and its orbit around the center of the Milky Way means that it is approaching the solar system at about 250,000 mph (400,000 kph). Although many variable stars (see pp.282–283) have been detected within the cluster—more than 40 to date, including RR Lyrae stars—the distance to M68 is still uncertain.

**DENSE BALL**

## GLOBULAR CLUSTER
# M15

**CATALOG NUMBERS**
M15, NGC 7078

**DISTANCE FROM SUN**
35,000–45,000 light-years

**MAGNITUDE** 6.4

**PEGASUS**

At the limit of naked-eye visibility, M15 is one of the densest globular clusters in the Milky Way. The cluster has a diameter of about 175 light-years, but, as the center of the cluster has collapsed in on itself, half of its mass is located within its 1-light-year-wide superdense core. M15 also contains nine pulsars, remnants of ancient supernova explosions (see pp.266–267) from

**PACKED CORE**
At its core, this globular cluster has the highest concentration of stars in the Milky Way outside the galactic center.

the time when the cluster was young. Unusually, two of these pulsating neutron stars form a contact binary pair (see p.274).

**TRUE COLORS**
The brightest stars in M15 are red giants, with surface temperatures lower than the Sun's. Most of the fainter stars are hotter, giving them a bluish-white tint.

# EXTRASOLAR PLANETS

THE SUN IS NOT the only star with a planetary system. More than 4,000 planets have so far been found orbiting other stars, with the list growing rapidly every year. Extrasolar planets have been detected around stars of a range of types and ages, suggesting that planet formation is a robust process and that planetary systems are commonplace.

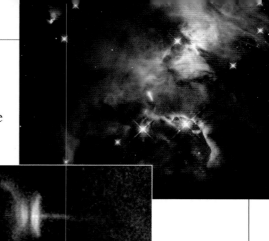

**TELLTALE SHADOWS**
Planet-forming disks around young stars can be seen in direct silhouette (right) or when they cast twin lighthouselike shadows across nearby nebulae (top right).

## PLANET-FORMING DISKS

Some of the first evidence leading to the detection of extrasolar planets, or exoplanets, was the discovery of flattened disks of material around some young stars. This fit the standard theory of planetary-system formation (originally put forward to explain the origins of the solar system), in which planets form from a disk of dust and gas rotating around a star. Some such circumstellar disks—also called debris disks—are symmetrical, suggesting that they are in their early stages, before planet formation. Others are distorted or have gaps or other structural features that suggest that planets have formed and are disturbing material in the disks. For example, the disk around the bright young star Fomalhaut has a distinctive bright outer ring of concentrated material that has probably been "shepherded" into place by the influence of nearby planets. Dusty disks are also found around mature stars. Vega (see p.253) is surrounded by an extensive dust disk, which is fully revealed only at infrared wavelengths. This fine dust is thought to be the debris from a large and relatively recent collision between Pluto-sized bodies orbiting the star at a distance of 8 billion miles (13 billion km). Irregularities in Vega's debris disk also suggest the presence of at least one planet.

**YOUNG DISK**
The protoplanetary disk around the star AS 209 appears perfect—the star is so young that no planets have yet had time to form.

**DEVELOPING SPIRAL**
The debris disk around the young star Elias 2-27 has formed a spiral pattern with concentrations of matter out of which planets may soon emerge. The structure is created by density waves similar to those seen in spiral galaxies (see p.303).

**FOMALHAUT'S COMPLEX DISK**
This image combines a Hubble Space Telescope view of Fomalhaut (blue) with a radio map from the ALMA array (orange) to show scattered dust around the ring, with the main concentration of around 140 AU from the central star.

# DETECTING EXTRASOLAR PLANETS

As extrasolar planets are invariably much smaller and dimmer than their parent stars, detecting them presents many challenges. Only about 30 or so have been found by direct imaging, which involves first blocking out the light from the parent star. All other exoplanet discoveries have been made by indirect methods. The most productive so far has been the Doppler spectroscopy or radial velocity method, which is based on the use of a sensitive instrument called a spectrograph. It relies on the fact that as an exoplanet orbits its parent star, its gravitational pull produces a tiny "wobble" in the star's movement relative to Earth. A second indirect approach proving increasingly productive is the transit method, which involves looking for repeated slight dips in the brightness of a star as a planet passes in front of it. One advantage of this method is that it reveals the planet's diameter. Several other indirect detection methods have also been employed, with varying success. These include gravitational microlensing (detecting variations in the lensing effect of a star's gravitational field, caused by a planet orbiting the star) and the pulsar timing method, which detects exoplanets orbiting pulsars from slight anomalies in the timing of the pulsars' radio pulses.

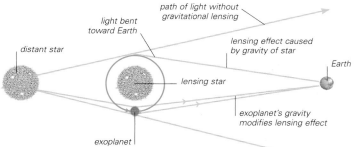

exoplanet · star · planet tracks across face of star · dip in star's light curve

BRIGHTNESS · TIME

## THE TRANSIT METHOD
This approach involves observing repeated transits of a planet in front of its parent star. Each transit causes a slight dip in the star's brightness—of the order of 0.01 percent for an Earth-size planet.

## DOPPLER SPECTROSCOPY
An exoplanet's orbit causes a "wobble" in the motion of its parent star. As a result, light waves coming from the star appear to be alternately slightly lengthened (red-shifted) and shortened (blue-shifted)—a phenomenon that can be detected by sensitive spectrographs (see p.33).

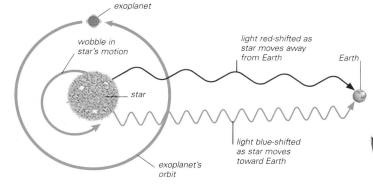

exoplanet · wobble in star's motion · light red-shifted as star moves away from Earth · Earth · star · light blue-shifted as star moves toward Earth · exoplanet's orbit

## DIRECT IMAGING
This composite image was made in 2004 with a telescope located at the Paranal Observatory in Chile. It shows a brown dwarf star (here appearing bright white) known as 2M1207 and a smaller red companion, thought to be a hot gas-giant planet. This red object is the first extrasolar planet ever to have been directly imaged.

path of light without gravitational lensing · light bent toward Earth · distant star · lensing effect caused by gravity of star · Earth · lensing star · exoplanet's gravity modifies lensing effect · exoplanet

## GRAVITATIONAL MICROLENSING
The gravitational field of a star acts like a lens that can bend light rays coming from a distant background star, thus magnifying that star as seen from Earth. The presence of an exoplanet orbiting the lensing star produces detectable variations in the degree of magnification, or lensing effect, over time.

# MYSTERIOUS PULSAR PLANETS

The very first exoplanets to be discovered, in 1992, were not in solar systems around Sun-like stars, but instead orbited pulsar PSR B1257+12—the rapidly spinning burned-out remnant of a once-massive star (see p.267). So-called "pulsar planets" can be identified because a pulsar's rate of rotation is constant over very long periods and can be measured with great precision. When the gravity of one or more planets pulls on the pulsar, it can create slow, regular changes in the pulsation rate. Three planets were discovered orbiting PSR B1257+12 (claims of a fourth are now disproved), the innermost of which—just twice the mass of the Moon—is the least massive planet known. The discovery was unexpected, since pulsars are formed during supernova explosions that would destroy any systems in orbit around them. Astronomers therefore think that pulsar planets are a second generation formed from debris left in orbit once the explosion subsides. Planets orbiting three more pulsars have since been confirmed, with several others proposed. In 2006, the Spitzer Space Telescope even discovered an apparently recent protoplanetary disk orbiting around a 100,000-year-old supernova.

THE MILKY WAY

# GAS GIANTS

The first exoplanets to be discovered around Sun-like stars were all giants with masses from about that of Neptune to much larger than Jupiter, with short-period orbits. While this reflects the fact that planets of this type are easiest to find with the radial velocity method (and many giants at greater distances have since been discovered), the existence of these so-called "hot Jupiters" in large numbers was a surprise. According to current theories, such giant planets must have formed farther out from their stars and spiraled inward to their current orbits. Temperatures in the atmospheres of hot Jupiters frequently exceed 2,000°F (1,100°C), causing them to slowly evaporate into space. More recently, several possible "Chthonian" planets—the bare solid cores of former hot Jupiters—have also been identified.

## KAPPA ANDROMEDAE B

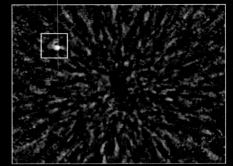

This rare view shows the massive "super-Jupiter" planet orbiting Kappa Andromedae, a hot young star some 170 light-years from Earth (whose direct light is blocked in this image). Kappa Andromedae b has 12.8 times the mass of Jupiter, putting it right on the borderline between a giant planet and a brown dwarf "failed star."

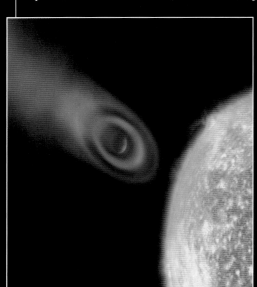

## ESCAPING ATMOSPHERE

The "hot Jupiter"-type exoplanet HD 209458b orbits close-in on its parent star (seen here in an artist's impression). In 2003–2004, astronomers discovered an extended ellipsoidal envelope containing hydrogen and other gases evaporating from the planet. It is thought this type of atmosphere loss may be common to all "hot Jupiters." When all their volatile materials have evaporated away, an exposed rocky or metallic core may survive as a "Chthonian planet."

## 51 PEGASI B

This artist's impression shows the first exoplanet discovered around a Sun-like star—initially nicknamed Bellerophon but now formally named Dimidium. Lying some 50 light-years from Earth, 51 Pegasi B has a mass at least half of Jupiter's, and orbits its star at an average 1/20th of the Earth–Sun distance, making it the prototype of the "hot Jupiter" class of giant planets.

## EXOPLANET TEMPERATURES MAP

This map produced by the Spitzer Space Telescope shows temperature variation across the surface of the "hot Jupiter"-type exoplanet HD 189733b. One side of the planet always faces its parent star. The hottest area is slightly displaced from the point on the planet exactly facing the star—evidence that fierce winds operate in its atmosphere.

# PLANETARY SYSTEMS

Multiple exoplanets have been observed orbiting a number of relatively close stars. The first system of this type to be identified, in 1999, was found orbiting Upsilon (υ) Andromedae A, a Sun-like star located approximately 44 light-years away. The system is now known to include at least four planets, all thought to be comparable in size to Jupiter. HR 8799, a young main-sequence star located 129 light-years from Earth, also has at least four high-mass planets orbiting it. This quartet of giants, which have been directly imaged, orbit inside a large debris disk that surrounds the star—at orbital radii that are two to three times those of the four gas giants orbiting our own Sun. This is surpassed by the star 55 Cancri A, which is part of a binary star system and one of just a handful of stars known to have at least five exoplanets—ranging from Neptune- to Jupiter-sized—in orbit around it. Systems have also been discovered in which one or more planets orbit both stars in a binary star system. Kepler-16b, for example, is an exoplanet comparable to Saturn in mass and size that follows a nearly circular 229-day orbit around two stars located some 196 light-years away.

Kepler-16b

B

A

orbit of star B

orbit of star A

size of Earth's orbit

size of Mercury's orbit

## PLANET ORBITING A BINARY STAR

The exoplanet Kepler-16b, discovered in 2011, orbits the binary star system Kepler-16. Here, the orbits of the two components of Kepler-16 (labeled stars A and B) are shown together with the orbit of Kepler-16b and, for comparison, the size of Earth's and Mercury's orbits around the Sun. Kepler-16b is thought to be made up of about half rock and half gas.

Upsilon Andromedae c

inclined, highly elliptical orbit

Upsilon Andromedae d

Upsilon (υ) Andromedae A

Upsilon Andromedae b

## INCLINED ORBITS

Upsilon (υ) Andromedae A is the primary member of a binary star system. Three of its four known planets, called Upsilon Andromedae b, c, and d, are shown here (the fourth planet orbits beyond planet d). The planets' orbits are inclined to each other, and planets c and d have orbits that are highly elliptical. Planet d resides in the system's habitable zone (see opposite). The innermost planet, Upsilon Andromedae b, orbits Upsilon (υ) Andromedae A every 4 days at a distance of 4.7 million miles (7.5 million km)—much closer than Mercury orbits the Sun.

## THE KEPLER MISSION

NASA's Kepler spacecraft was designed to carry out the first extensive search for transiting exoplanets, leading not just to the discovery of individual planets, but also to the first statistical evidence for how many planets might exist in the Milky Way. Kepler's primary mission was to "stare" at a single patch of sky in the constellation Cygnus, directing light collected through its 3.1-ft (0.95-m) telescope onto an advanced photometer (light meter) that continuously recorded the brightness of 150,000 main-sequence stars. When the pointing system that kept Kepler precisely aligned to its target area began to fail in 2013, the mission was redesigned to switch to target a new field of view every few months.

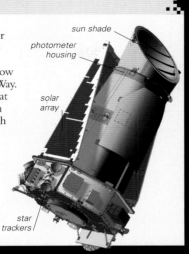

sun shade
photometer housing
solar array
star trackers

# ROCKY EXOPLANETS

From around 2005 onward, advances in the transit method (particularly using dedicated satellites such as Kepler) have seen the number of smaller planets catch up with and eventually overtake the known gas giants. The masses of these smaller, rocky worlds can be estimated from their diameter (the strength of the transit) and assumptions about their density and composition—they rarely have enough mass to affect the radial velocity of their parent star. Known rocky exoplanets range from Moon-sized worlds to super-Earths with several times our planet's mass, and even a couple of "mega-Earths" whose masses must approach those of Neptune. Missions such as NASA's Transiting Exoplanet Survey Satellite (TESS) and ESA's CHEOPS continue to expand our understanding of these intriguing worlds.

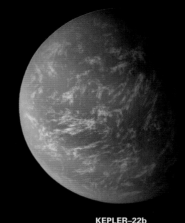

**EARTH AMONG THE EXOPLANETS**
Earth is shown here to scale with some significant rocky exoplanets. Proxima Centauri b is the closest exoplanet to Earth, while Kepler-37b is one of smallest worlds so far discovered. Kepler-10b is a searing-hot world that was the first to be discovered by Kepler, while Kepler-22b was the first known transiting exoplanet discovered within its star's habitable zone, potentially supporting a watery surface.

PROXIMA CENTAURI b | KEPLER-37b | EARTH | KEPLER-10b | KEPLER-22b

**PROXIMA CENTAURI b**
In 2016, astronomers announced the discovery of a roughly Earth-sized planet orbiting Proxima Centauri, our nearest stellar neighbor. Although technically in the habitable zone of its red dwarf star, Proxima b orbits at a distance of just 4.7 million miles (7.5 million km), putting it in the firing line of frequent violent flares and lessening the chances of this being a truly Earth-like world on our doorstep.

# LOOKING FOR EARTHS

If there is life elsewhere in the universe, it seems reasonable to expect to find it on a world similar to Earth: a rocky planet orbiting a main-sequence star. Evidence from the Kepler mission and other search programs suggests that about 20 percent of Sun-like stars have at least one giant planet. Current techniques can be used to estimate the orbital parameters of gas giants around these stars and thereby identify those with stable, circular orbits at a fair distance from their respective stars. Within a significant proportion of such systems, there is likely to be an inner zone where rocky terrestrial planets may have formed, some of them within the habitable zones associated with those stars. Although to date no exoplanet perfectly resembling Earth has been found, the signs are encouraging that one or more may be found in the reasonably near future. Once detected, analysis of light reflecting from the planets' atmospheres for telltale signs of life, such as oxygen and methane, should be possible—a capacity for this is already present in some existing telescopes involved in the exoplanet search.

**HABITABLE ZONE**
For life to develop on a planet, the planet must lie within the host star's "habitable zone," where liquid water can permanently exist on the surface. This zone's extent depends on the star's mass and luminosity.

**TRAPPIST-1 SYSTEM**
In 2015, three roughly Earth-sized planets were discovered orbiting the red-dwarf star TRAPPIST-1, 39.6 light-years away. A further four have joined them since, making the system a prime target in the hunt for truly Earth-like worlds. Three of the known planets orbit within the habitable zone, with planets d and e thought most likely to support an Earth-like climate.

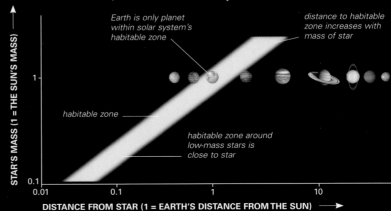

Earth is only planet within solar system's habitable zone
distance to habitable zone increases with mass of star
habitable zone
habitable zone around low-mass stars is close to star

STAR'S MASS (1 = THE SUN'S MASS)
DISTANCE FROM STAR (1 = EARTH'S DISTANCE FROM THE SUN)

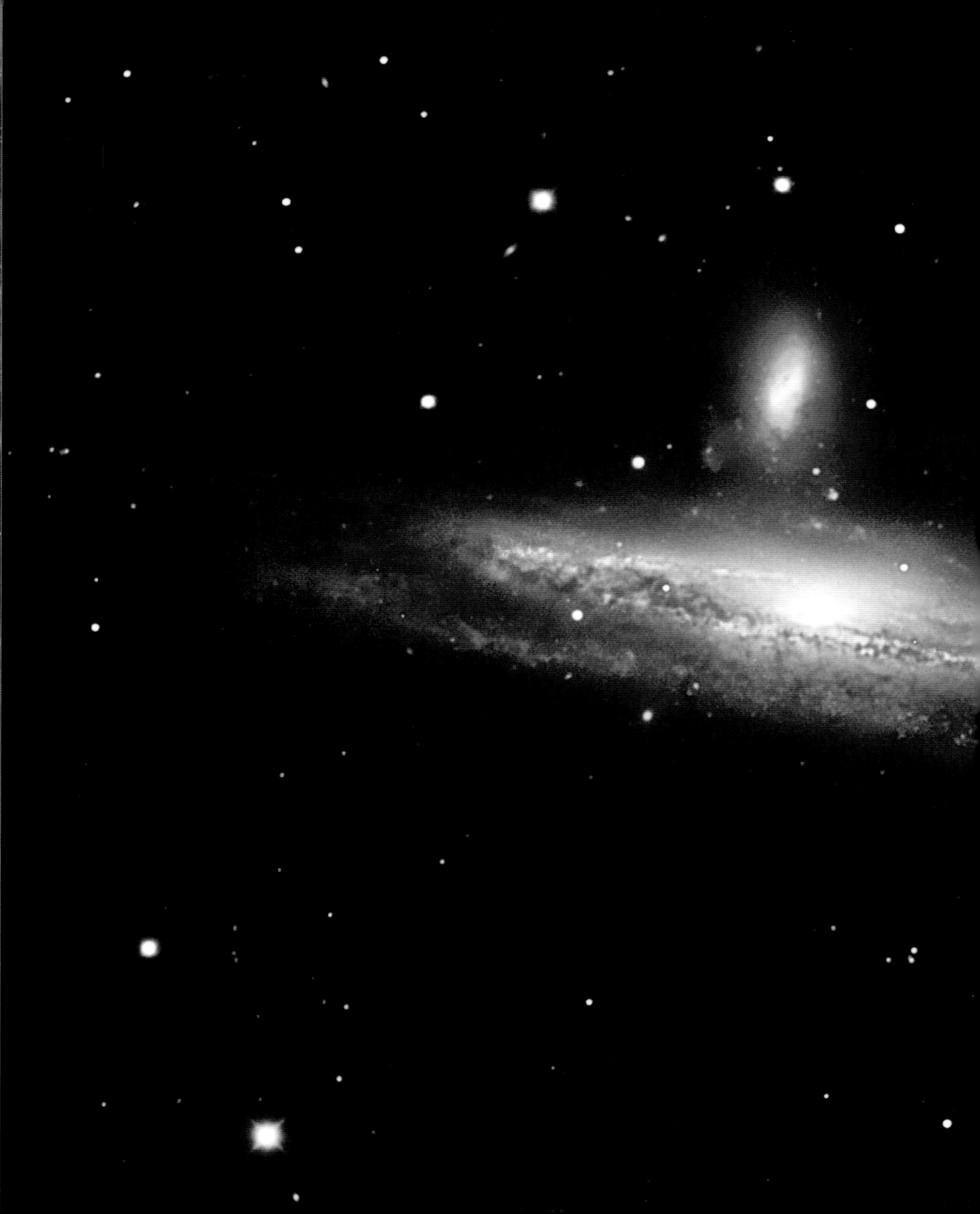

*"The history of astronomy is a history
of receding horizons."*

Edwin Hubble

OUTSIDE THE BOUNDS of the Milky Way stretch vast gulfs of space, the realm of the galaxies. The closest are on our own galactic doorstep—there is even a small galaxy currently in collision with the Milky Way. The farthest lie billions of light-years away, at the edge of the visible universe—their light has been traveling toward Earth for most of time. Galaxies range from great wheeling disks of matter to giant, diffuse globes of billions of stars and from starless clouds of gas to brilliant furnaces lit up by star formation. They are also violent—despite their stately motion over millions and billions of years, collisions are frequent and spectacular. Collisions disrupt galaxies, sending material spiraling into the supermassive black holes at their centers, fueling activity that may outshine ordinary galaxies many times over. Galaxies influence their surroundings and form constantly evolving clusters and superclusters. At the largest scale, it is these galaxy superclusters that define the structure of the universe itself.

**CLOSE ENCOUNTER**
Seventy million light-years from Earth, the spiral galaxy NGC 1531 is disrupted by a close encounter with the smaller NGC 1532. The smaller galaxy's gravity disrupts the larger one's spiral arms and distorts NGC 1531's shape, but it also triggers a huge wave of star birth seen in bright star purple-hued star clusters.

# BEYOND THE MILKY WAY

# TYPES OF GALAXY

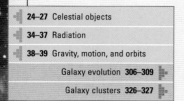
THROUGHOUT THE UNIVERSE, galaxies exist in enormous diversity. These vast wheels, globes, and clouds of material vary hugely in size and mass—the smallest contain just a few million stars, and the largest around a million million. Some are just a few thousand light-years across, and others can be a hundred times that size. Some contain only old red and yellow stars, while others are blazing star factories, full of young blue and white stars, gas, and dust. The features of galaxies are clues to their history and evolution, but astronomers have only recently begun to put the entire story together—and there are still many gaps in their knowledge.

**EDGE-ON SPIRALS**
NGC 1055 is a spiral galaxy that happens to lie edge-on to Earth. This image reveals that its dark dust lanes, silhouetted against the brightness of its stars, have been warped out of shape, probably in a close encounter with another galaxy.

## THE VARIETY OF GALAXIES

Galaxies can be classified by their shape, size, and color. At the most basic level, they are divided by shape into spiral, elliptical, and irregular galaxies. Edwin Hubble (see p.45) devised a more precise classification, still used today, that subdivides these galaxy shapes. Hubble classed spiral galaxies as types Sa to Sd—an Sa galaxy has tightly wound spiral arms, while an Sd very loose arms. Spirals with a bar across their center are classed as SBa to SBd. Hubble classed elliptical galaxies as E0 to E7 according to their shape in the sky—circular galaxies are E0, and elongated ellipses E7. Elliptical galaxies appear as two-dimensional ellipses, but in reality, they are three-dimensional ellipsoids ranging from roughly ball-shaped star clouds to cigar shapes. So Hubble's classification does not reflect their true geometry, since an E0 galaxy could be a cigar shape viewed end-on from Earth. Hubble also recognized an intermediate type of galaxy—the lenticular (type S0), with a spiral-like disk, a hub of old yellow stars, but no spiral arms. Finally, irregular galaxies (type Irr) are usually small and rich in gas, dust, and young stars, but have few signs of structure.

**NUMBERING ELLIPTICALS**
The class of an elliptical galaxy is found by dividing the difference between its long and short axes by the long-axis length and then multiplying by 10, making this galaxy, M110, an E6.

long axis = 8.7 arc-minutes

short axis = 3.4 arc-minutes

**IRREGULAR GALAXY**
Clouds of stars that lack clear disk- or ellipselike structure are called irregular galaxies. The Small Magellanic Cloud is one such irregular galaxy.

**ELLIPTICAL GALAXY**
Balls of stars, from perfect spheres, through egg shapes (such as M59, pictured here), to cigar-shaped ellipsoids, are called elliptical galaxies.

**SPIRAL GALAXY**
Vast, rotating disks of stars, dust, and gas are classed as spiral galaxies. Spirals have a ball-shaped nucleus inside a disk with spiral arms. M33 is a nearby spiral galaxy.

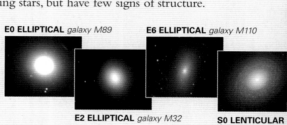

**E0 ELLIPTICAL** galaxy M89    **E6 ELLIPTICAL** galaxy M110

**E2 ELLIPTICAL** galaxy M32    **S0 LENTICULAR** galaxy NGC 2755

**HUBBLE'S CLASSIFICATION**
Hubble arranged his galaxy types in a fork shape, with ellipticals along the handle and spirals and barred spirals as prongs. This excludes irregular galaxies. He thought his scheme indicated the evolution of galaxies—today, astronomers know it is not so simple.

**Sb SPIRAL** galaxy NGC 4622

**Sa SPIRAL** galaxy NGC 7217    **Sc SPIRAL** the whirlpool galaxy (M51)

**SBa BARRED SPIRAL** galaxy NGC 660    **SBb BARRED SPIRAL** galaxy NGC 7479    **SBc BARRED SPIRAL** galaxy NGC 1300

## SPIRAL GALAXIES

Some 25–30 percent of galaxies in the nearby universe are spirals. In each one, a flattened disk of gas- and dust-rich material orbits a spherical nucleus, or hub, of old red and yellow stars, which is often distorted into a bar. Stars occur throughout the disk, but the brightest clusters of young blue and white stars are found only in the spiral arms. The space between the arms often looks empty viewed from Earth, but it is also full of stars. Above and below the disk is a spherical "halo" region, where globular clusters (see p.289) and stray stars orbit. Spiral galaxies rotate slowly—typically once every few hundred million years—but they do not behave like a solid object. Stars orbiting farther away take longer to complete an orbit than those close to the core. The resulting "differential rotation" is the key to understanding the spiral arms.

**BARRED SPIRAL**
Similar to our own galaxy, M83 (right) is a typical barred spiral, having a straight bar on either side of the galactic nucleus.

**ORBITS IN SPIRALS**
Stars in the disk of a spiral galaxy follow elliptical, nearly circular orbits in a single plane. Those in the hub have wildly irregular orbits at a multitude of angles.

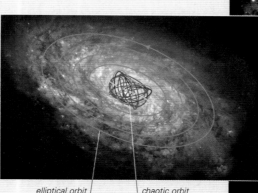

elliptical orbit    chaotic orbit

**FLOCCULENT SPIRAL**
The spiral galaxy NGC 2841 is flocculent, with bright stars clumped throughout the disk. Its star formation seems to be caused by local collapses of material rather than a large-scale density wave.

# SPIRAL ARMS

The continued presence of spiral arms in most disk-shaped galaxies was once a mystery. If the arms orbit more quickly near the nucleus, then during a galaxy's multi-billion-year lifetime, they would become tightly wrapped around the core. It now seems that the arms are in fact rotating regions of star formation, not rotating chains of stars themselves. The arms arise from a "density wave"— a zone that rotates far more slowly than the galaxy itself. The density wave is like a traffic jam—stars and other material slow down as they move into it and accelerate as they move out, but the jam itself advances only slowly. The increased density helps trigger the collapse of gas clouds and the start of star formation. The strength of the density wave varies between spirals. If the wave is strong, the result is a neat, "grand design" spiral with two clearly defined arms. If it is weak or nonexistent, disk stars will tend to form in localized regions, creating the more clumpy "flocculent" spirals.

**PERFECT GALAXY**
In this diagram of an ideal galaxy, objects follow neatly aligned elliptical orbits around the nucleus. They travel fastest when close to the nucleus and slowest when farthest away.

**SPIRAL REALITY**
In a real galaxy, the orbits do not line up neatly. The variety of alignments, coupled with the slower movement when farther from the nucleus, creates spiral zones in which objects are moving more slowly and so become bunched together.

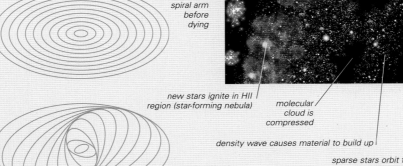

open clusters of longer-lived stars move out from spiral arm

young "OB" star clusters never move far from spiral arm before dying

new stars ignite in HII region (star-forming nebula)

molecular cloud is compressed

density wave causes material to build up

sparse stars orbit faster than the spiral arm and move into arm from behind

**DETAIL OF A SPIRAL ARM**
As material orbiting in a galaxy's disk approaches the denser region marked by the spiral arm, it packs together. Dark molecular clouds form, some of which turn into star-forming nebulae (see pp.238–239). New stars of all kinds ignite here, but the brightest ones soon die, so they always mark the spiral arms.

# ELLIPTICAL GALAXIES

Elliptical galaxies show little structure other than a simple ball shape. They span the range from the largest to the smallest galaxies. At one end, dwarf ellipticals are relatively tiny clusters of a few million stars, often very loosely distributed, appearing faint and diffuse. Such galaxies are scattered in the space between larger galaxies and must contain significant amounts of invisible material simply to hold them together. Some of this could be in a central black hole, but much of it seems to be mysterious "dark matter" (see p.27) scattered through the whole of the galaxy. At the other extreme lie the giant ellipticals—galaxies only found near the centers of large galaxy clusters and often containing many hundreds of billions of stars. Some giant ellipticals, called cD galaxies, have large outer envelopes of stars and even multiple concentrations of stars at their centers, suggesting they may have formed from the merging of smaller ellipticals. Almost all the stars in elliptical galaxies are yellow and red, and there is rarely any sign of star-forming gas and dust. The dominance of old, long-lived stars implies that any star formation in these galaxies has long since ended. Each star orbits the galaxy's dense core in its own path. The chances of collision are very remote, because stars are so small relative to the distances between them. With no gas and dust clouds to interact with, there is nothing else to flatten the stars into a single plane of rotation. Ellipticals are described according to their degree of elongation—how much they deviate from a perfect sphere (see p.302)—but the largest galaxies are always very close to perfect spheres.

**ORBITS IN ELLIPTICAL GALAXIES**
The orbits of stars in an elliptical galaxy vary wildly, from circles to very long ellipses, and are not confined to any specific direction.

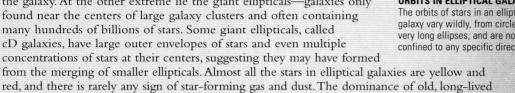

**GIANT ELLIPTICAL**
ESO 325-G004 is a giant elliptical galaxy at the heart of Abell S740, a cluster 450 million light-years away in Centaurus. Thousands of globular star clusters form pinpoints of light around the central galaxy.

**INTERMEDIATE GALAXY**
M49 in the Virgo Galaxy Cluster is a large elliptical of type E4. With a diameter of about 160,000 light-years, it is classed by some astronomers as a giant elliptical, although its mass is much less than that of the true giants.

**DWARF ELLIPTICAL**
The Leo I Galaxy is a nearby dwarf elliptical and one of the few we can study closely. With so few stars, there must be a large amount of dark matter holding the galaxy together with its gravity.

# LENTICULAR GALAXIES

At first glance, lenticular galaxies appear to be relatives of ellipticals—they are dominated by a roughly spherical nucleus of old red and yellow stars. However, around this nucleus, these galaxies also have a disk of stars and gas. This links them to spiral galaxies, and they are similar in overall size and general shape, although the nucleus is often considerably bigger than it would be in a spiral of similar size. The overall shape is often described as that of a lens, which is the root of the name "lenticular." The key difference between lenticulars and spirals is that lenticulars have no spiral arms and little sign of star-forming activity in their disks. Without the bright blue star clusters that illuminate the disks of spirals, lenticulars are sometimes hard to tell apart from ellipticals. Those that are face-on may be indistinguishable from ellipticals and misclassified. An edge-on spiral galaxy with a large nucleus can equally be misclassified as lenticular, because at oblique angles spiral structure is often invisible. Astronomers are uncertain how lenticular galaxies form, but they could be spiral galaxies that have lost most of their dust and gas.

**DUSTY LENTICULAR**
Lying 25 million light-years away, galaxy NGC 2787 is one of the closest lenticular galaxies. Dust lanes can be seen silhouetted against the nucleus, marking the plane of its disk.

elliptical orbits in the disk

chaotic orbits in the hub

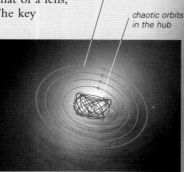

**ORBITS IN LENTICULAR GALAXIES**
Stars in the nucleus of a lenticular galaxy follow orbits with no specific plane, similar to those in an elliptical galaxy or a spiral nucleus. Gas and dust in the disk orbits in a more orderly plane.

**IRREGULAR DWARF**
The irregular dwarf galaxy NGC 4449 contains clusters of young bluish stars interspersed with dustier reddish regions of current star formation.

## GALAXIES AT DIFFERENT WAVELENGTHS

**COMBINED IMAGE OF NGC 1512**

Radiation of different wavelengths can reveal hidden structures within galaxies. The hottest stars appear brightest in ultraviolet, while cool, diffuse gas may be visible only in infrared. By overlaying images from different spectral regions, astronomers build up a full picture of a galaxy.

**FROM ULTRAVIOLET TO INFRARED**
These images of galaxy NGC 1512 increase in wavelength from left to right. Each wavelength is represented by a false color.

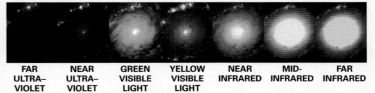

| FAR ULTRA-VIOLET | NEAR ULTRA-VIOLET | GREEN VISIBLE LIGHT | YELLOW VISIBLE LIGHT | NEAR INFRARED | MID-INFRARED | FAR INFRARED |

# IRREGULAR GALAXIES

Not all galaxies fit into the scheme of spirals, ellipticals, and lenticulars. Some of these misfit galaxies are colliding with companions or being pulled out of shape by a neighbor's gravity. These are usually classed under the catch-all term "peculiar" or "Pec." Many more are true irregulars (type Irr). These galaxies typically contain a lot of gas, dust, and hot blue stars. In fact, many irregulars are "starburst" galaxies, with great waves of star formation sweeping through them. Irregulars frequently have vast pink hydrogen-emission nebulae where star formation is taking place. Some irregulars show signs of structure—central bars and sometimes the beginnings of spiral arms. The Milky Way's brightest companion galaxies, the Large and Small Magellanic Clouds (see pp.310–311), are typical irregular galaxies.

**IRREGULAR STARBURST**
M82 is an irregular starburst galaxy crossed with dark dust lanes. It is undergoing an intense period of star birth.

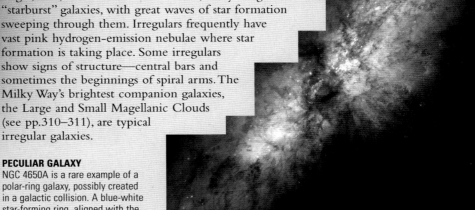

**PECULIAR GALAXY**
NGC 4650A is a rare example of a polar-ring galaxy, possibly created in a galactic collision. A blue-white star-forming ring, aligned with the poles, extends from the nucleus.

## ASTRONOMY FROM THE SOUTH POLE

Some of the best Earth-based observations of galaxies come from an automated observatory at the South Pole. The AASTO project takes advantage of the dryness on the Antarctic Plateau—the driest place on Earth. With no water vapor in the atmosphere, near-infrared light is not absorbed, so it reaches the ground unhindered.

# CENTRAL BLACK HOLES

Many, if not all, galaxies have a dark region within their nucleus that seems strange by contrast with the outer parts. The fast orbits of stars near galactic nuclei suggest an enormous concentration of mass in a tiny volume at the center of most spiral and elliptical galaxies—often billions of Suns' worth of material in a space little larger than the solar system. The only object that can reach such a density is a black hole (see p.26). Despite the tremendous gravity of this "supermassive" black hole, in most nearby galaxies the material has long since settled into steady orbits around it. With no material to absorb, the black hole remains dormant. When a gas cloud or other object comes too close, however, the black hole may awake, pulling in the stray material and heating it, producing radiation. The black hole may generate any type of radiation, from low-energy radio waves to high-energy X-rays. In extreme cases called "active galaxies" (see pp.320–321), the radiation from the nucleus is the galaxy's dominant feature.

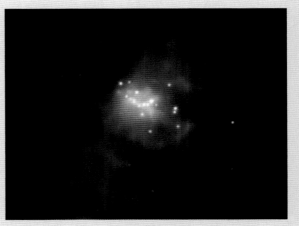

**HIDDEN SUPERMASSIVE BLACK HOLE**
An X-ray image of galaxy M82 shows glowing hot gas and intense point sources of X-rays. These are probably stellar-mass black holes surrounding a central supermassive black hole.

# GALAXY EVOLUTION

THE PROCESSES BY WHICH DIFFERENT types of galaxies form have puzzled astronomers for almost a century, but today, a new generation of telescopes capable of studying galaxies billions of light-years away is finally resolving some key questions. Light from these remote galaxies left on its long journey to Earth when the universe was very young, so it can reveal the secrets of the early stages of galactic evolution.

## THE DISTRIBUTION OF GALAXIES

Astronomers can only ever see a "snapshot" of a brief moment in a galaxy's long life story, so they have to build up a picture of galactic evolution by studying many individual galaxies. Such studies have revealed certain patterns, such as the fact that large elliptical galaxies are found only in substantial galaxy clusters. Changes in the type of galaxies seen at different distances—and therefore at different stages in cosmic history—can also reveal patterns in the way galaxies have developed. However, capturing the light of the most distant early galaxies is an enormous challenge, requiring techniques such as long-exposure deep-field photography and the use of gravitational lensing.

### GRAVITATIONALLY LENSED GALAXIES
This image of a small area in the constellation Hydra was taken by the Herschel Space Observatory and reveals more than 6,000 galaxies. The white squares indicate distant galaxies that have been gravitationally lensed by foreground galaxies. These distant galaxies are brighter at the submillimeter wavelengths detected by Herschel than at visible-light wavelengths.

### DUSTY LENTICULAR GALAXY
This image of the lenticular galaxy NGC 1316 in the constellation of Fornax was taken by the Hubble Space Telescope. It reveals a complex series of dust lanes and dust patches in the galaxy, indicating that it was formed from the merger of two galaxies rich in dust and gas.

#### EXPLORING SPACE

## HERSCHEL

Operational from 2009 to 2013, the European Space Agency's Herschel Space Observatory was designed to observe the longest infrared wavelengths (the far infrared and submillimeter wavelengths on the boundary with radio waves). Its primary mirror was 11.5 ft (3.5 m) in diameter and its instruments were cooled to –456°F (–271°C), enabling it to map some of the coolest and most distant objects in the universe.

secondary mirror

sun shield

primary mirror

**THE HERSCHEL SPACE OBSERVATORY**

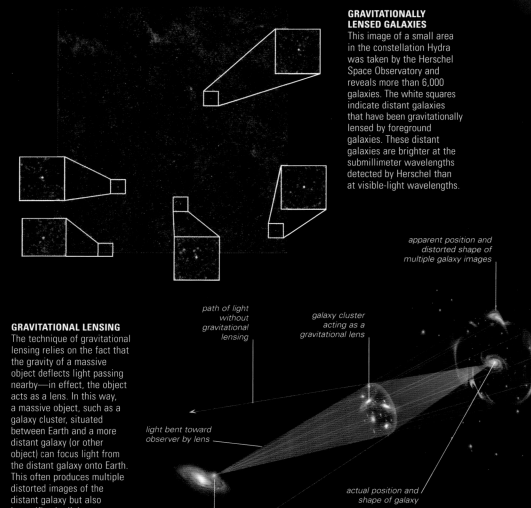

apparent position and distorted shape of multiple galaxy images

path of light without gravitational lensing

galaxy cluster acting as a gravitational lens

### GRAVITATIONAL LENSING
The technique of gravitational lensing relies on the fact that the gravity of a massive object deflects light passing nearby—in effect, the object acts as a lens. In this way, a massive object, such as a galaxy cluster, situated between Earth and a more distant galaxy (or other object) can focus light from the distant galaxy onto Earth. This often produces multiple distorted images of the distant galaxy but also intensifies its light.

light bent toward observer by lens

observer on Earth

actual position and shape of galaxy

# GALAXY FORMATION

Until recently, there were two main theories of galaxy formation. The first was a "top-down" scenario, in which galaxies coalesced out of huge clouds of matter, eventually becoming dense enough to form stars within them. The second was a "bottom-up" scenario in which small-scale structures formed first and gradually merged to create larger structures—galaxies. These two scenarios arose as a result of different ideas about the properties of dark matter (see p.27), specifically whether it is "hot" and fast-moving or "cold" and slow-moving. It now seems that the bottom-up model, with galaxy formation driven by the presence of relatively slow-moving cold dark matter (CDM), is correct. Computer simulations (such as the one shown below) suggest that in the early life of the universe, cold dark matter started to clump together in localized regions. These clumps acted as seeds, attracting yet more matter and eventually developing into protogalaxies and then mature galaxies. This happened in numerous localized regions throughout the universe, leading to the distribution of galaxies seen today.

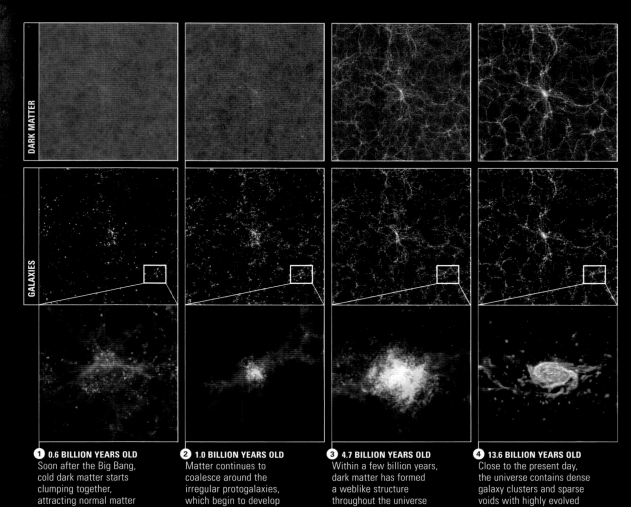

DARK MATTER

GALAXIES

**1** **0.6 BILLION YEARS OLD**
Soon after the Big Bang, cold dark matter starts clumping together, attracting normal matter and forming irregular protogalaxies (bottom).

**2** **1.0 BILLION YEARS OLD**
Matter continues to coalesce around the irregular protogalaxies, which begin to develop into larger galaxies (bottom).

**3** **4.7 BILLION YEARS OLD**
Within a few billion years, dark matter has formed a weblike structure throughout the universe and galaxies have become larger and more complex.

**4** **13.6 BILLION YEARS OLD**
Close to the present day, the universe contains dense galaxy clusters and sparse voids with highly evolved galaxies, such as ellipticals and spirals (bottom).

**BLACK HOLES IN GIANT ELLIPTICAL GALAXIES**
A comparison of the giant elliptical galaxies NGC 4621 (top) and NGC 4472 (bottom) reveals few stars in the core of NGC 4472 compared to the bright center of NGC 4621. It is thought that this "star deficit" is due to many of NGC 4472's stars being ejected during the violent collision of NGC 4472 with another galaxy and the merger of their central supermassive black holes.

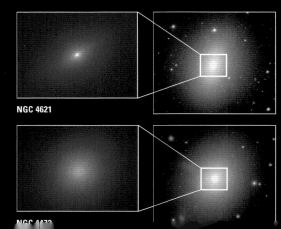

NGC 4621

NGC 4472

# THE ROLE OF BLACK HOLES

Since the 1990s, evidence has mounted to suggest that most if not all galaxies contain supermassive black holes at their centers, similar to the one at the center of our own Milky Way (see p.229). The masses of these black holes seem to be closely related to the overall sizes of the galaxies in which they lie, and a few galaxies even seem to contain two black holes in their cores. This suggests that black holes helped seed the formation of galaxies, and it also supports the "bottom-up" theory that larger galaxies are formed from mergers of smaller ones, with the central black holes ultimately joining together. However, the origin of these supermassive black holes is still unclear. Some theories suggest that the first black holes could have formed in the Big Bang or that they were created by the slow collapse of gas clouds around dark matter cores. Another possibility, which seems the most likely, is that they were

# GALAXY COLLISIONS

Relative to their size, galaxies are quite closely packed together—
although they are separated by distances of hundreds of thousands
of light-years, galaxies themselves are typically tens of thousands of
light-years across. Furthermore, the enormous gravity exerted by large
galaxies and their tendency to form within large-scale clusters allow
them to influence and attract one another. As a result, collisions and
close encounters between galaxies are comparatively common. In 1966,
US astronomer Halton Arp compiled the first catalog of galaxies that
did not fit neatly into the common categories of spiral, irregular, and
elliptical. With the benefit of more recent observations, it now seems
that most of Arp's unusual galaxies were the result of past collisions and
interactions between galaxies. Even some apparently normal galaxies are
now thought to have interacted with other galaxies in the past, and it
is also clear that many large galaxies are "cannibals," tearing apart and
ultimately absorbing smaller galaxies that stray too close. However,
during intergalactic collisions, individual stars rarely collide, and it
may take several billion years before the mutual gravity of colliding
galaxies finally pulls together most of their material into a single
combined cloud of stars.

**COLLIDING GALAXIES**
Situated about 450 million light-years from
Earth in the constellation Hercules, two
spiral galaxies (NGC 6050 and IC 1179,
also collectively known as ARP 272) are
colliding. Tidal forces in both galaxies
are triggering enormous waves of star
formation, manifested in the bright
clusters around their spiral arms.

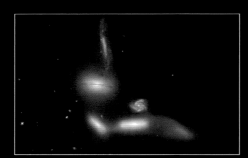

**SEYFERT'S SEXTET**
Despite its name, this group
contains only five galaxies—the
bright patch on the right is an
unwinding spiral arm. Only four
of the galaxies are at the same
distance from Earth, about 190
million light-years away; the face-
on spiral galaxy is about five times
that distance. The four nearest
galaxies are being distorted by
gravitational forces between them.

**THE SPLINTER GALAXY**
Also known as the Knife Edge Galaxy or
NGC 5907, this galaxy is an edge-on spiral
that lies about 40 million light-years from
Earth in the constellation Draco. It is
surrounded by extraordinary looping trails
of faint stars, nicknamed the "Ghost Stream,"
which are thought to be the remnants of
a smaller galaxy that has now been
consumed by NGC 5907.

# COLLISIONS AND EVOLUTION

The process of collision is now thought to play a key role in
transforming galaxies from one type to another. In the early
stages of a collision, stars that may have had relatively orderly
orbits are pushed into highly elongated and tilted paths, and
powerful shock waves passing through interstellar gas and dust
generate tremendous bursts of new star formation. In the longer
term, the remaining gas may become energized to such a degree
that it can escape the galaxy's gravity altogether, depriving it of
the means to continue star formation. In this way, spiral and
irregular galaxies can be transformed into ellipticals surrounded
by clouds of hot gas, as seen in the central regions of many
galaxy clusters. However, it has also been theorized that this
process can be reversed, at least in the relatively short term.
According to this theory, cold intergalactic gas is constantly
drawn in by the galaxy's gravity and can ultimately form a

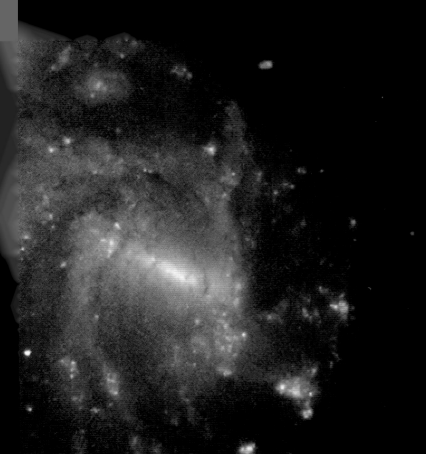

# TIDAL FORCES

As two galaxies approach each other, their gravitational fields interact and can affect their shapes. For example, because the galaxies' gravitational fields pull more strongly on the near side of each of the galaxies than on their more distant sides, their near sides become stretched out toward each other. Such gravitational distortion is greater on the less massive of the two galaxies because of the stronger gravity of the other, more massive galaxy. However, the disks of even large spiral galaxies can be warped by the gravity of relatively small neighbors. When spirals collide with one another, one or more of the spiral arms may unwind, transforming into a long trail of stars that stretches out on the opposite side from the collision. Among the best-known examples of these "tidal tails" are the ones associated with the Tadpole Galaxy and the Antennae Galaxies.

**THE TADPOLE GALAXY**
This galaxy, in the constellation Draco, has a tail that stretches for some 280,000 light-years and is thought to have formed when one of the spiral galaxy's arms unwound in a close encounter with a smaller galaxy.

# STARBURSTS

Intergalactic collisions can send immense shock waves through the galaxies involved, compressing large areas of interstellar gas and triggering enormous waves of star formation known as starbursts. During these events, star birth occurs much faster than normal, giving rise to huge "super star clusters" that may (if they survive) evolve into globular clusters. Starbursts are commonly seen in direct collisions, such as that of the Antennae Galaxies, but can also occur in close encounters between galaxies, as seen in the Cigar Galaxy (see p.314) due to its close encounter with Bode's Galaxy. Radiation from the numerous massive stars being formed, coupled with shock waves from supernovae as the heaviest stars rapidly age and explode, may blow gas and dust out of the galaxy, and it is this dispersion that may ultimately bring the starbursts to an end.

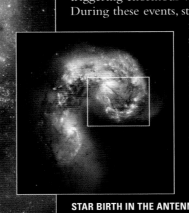

**STAR BIRTH IN THE ANTENNAE GALAXIES**
This image of the Antennae Galaxies (NGC 4038 and 4039), which lie about 45 million light-years from Earth, reveals stars being born in huge starburst regions. The newborn stars are a brilliant white-blue and are surrounded by glowing pink emission nebulae.

# GALAXIES

Astronomers are drawn naturally to the brightest, the most beautiful, and the most intriguing galaxies. However, of the 100 billion galaxies in the observable universe, only a minority are spectacular spirals and giant ellipticals. Astronomers are beginning to understand that most galaxies are relatively small and faint—diffuse balls and irregular clouds of stars. The faintest and most common galaxies are dwarf ellipticals, which are like oversized globular star clusters of only a few million stars. These feeble galaxies are visible only if they lie nearby in intergalactic terms. The most brilliant are the giant ellipticals, which can be 20 times as luminous as the Milky Way.

**BIG AND BRIGHT**
Spirals such as Bode's Galaxy, M81, may be the most attractive type of galaxy, but they are far from the most common. Making up less than 30 percent of all galaxies, they are outnumbered by smaller, fainter galaxies.

## DWARF ELLIPTICAL GALAXY

### SagDEG

| CATALOG NUMBER | None |
|---|---|
| DISTANCE | 88,000 light-years |
| DIAMETER | 10,000 light-years |
| MAGNITUDE 7.6 for M54 | star cluster in SagDEG |

**SAGITTARIUS**

The Sagittarius Dwarf Elliptical Galaxy, often called SagDEG, was until recently our closest known galactic neighbor. It was not found until 1994, and was supplanted only by the discovery of the even closer Canis Major Dwarf in 2003. SagDEG remained hidden for so long because, like all dwarf ellipticals, it is a very faint scattering of stars. It is also well disguised by its position behind the great Sagittarius star clouds that mark our galaxy's center. SagDEG is small and obscure, but it has at least four orbiting globular clusters, which are brighter and more obvious. One of these, M54, was discovered by Charles Messier more than 200 years before the parent galaxy was found.

**STAR DENSITY**
SagDEG's existence came to light only when a survey of Sagittarius found regions of increased star density—the bright patches in this image.

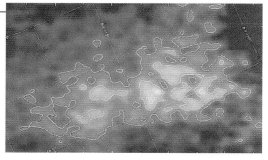

SagDEG's existence so close to our galaxy is a puzzle. It orbits the Milky Way in less than a billion years and so must have gone through several close encounters that should have ripped it apart and scattered its stars through the galactic halo. It has survived only due to a large amount of dark matter, producing more gravity than SagDEG's visible stars.

## IRREGULAR GALAXY

### Large Magellanic Cloud

| CATALOG NUMBER | None |
|---|---|
| DISTANCE | 179,000 light-years |
| DIAMETER | 20,000 light-years |
| MAGNITUDE 0.1 | |

**DORADO**

The Large Magellanic Cloud (LMC) bears the name of 16th-century explorer Ferdinand Magellan (see panel, opposite). However, cultures native to the Southern Hemisphere have recognized its existence since prehistoric times. Like its smaller counterpart, the Small Magellanic Cloud, the LMC appears from Earth to be a distinctive, isolated region of the Milky Way, some 10 degrees across, with its own areas of nebulosity and star clusters.

The LMC is in fact an irregular galaxy, orbiting the Milky Way roughly once every 1.5 billion years on a path that brought it to within 120,000 light-years of our galaxy at its closest approach around 250 million years ago. Although the LMC is irregular and is being distorted by the gravity of the Milky Way, it shows some signs of basic structure. Many of its stars are concentrated in a central barlike nucleus, curved at one end. Some astronomers have likened the LMC to a barred spiral with just one arm.

Like all irregular galaxies, the LMC is rich in gas, dust, and young stars, including some of the largest known regions of star birth. One such region is the magnificent Tarantula Nebula, also known as 30 Doradus. It is so brilliant that, if transported to the location of the Orion Nebula (see p.241)—only 1,500 light-years away in the Milky Way—it would be bright enough to cast shadows on Earth at night.

In recent times, the LMC was host to the only bright supernova since the invention of the telescope. Supernova 1987A (see p.266) was observed by astronomers around the world both during and after its explosion, and it has taught astronomers a lot about the final stages of the stellar life cycle.

**RADIO MAP**
This false-color radio image of the LMC is centered on the Tarantula Nebula. It shows intense radiation as red and black, indicating ionized hydrogen and star formation.

**SUPERNOVA BUBBLE**
This image shows a bubble of gas around the site of a supernova that exploded about 400 years ago in the LMC. The image is a composite from the Hubble Space Telescope and Chandra X-ray Observatory. Green and blue indicate hot, X-ray-emitting material, and pink shows the visible gas shell shocked by the blast wave from the supernova explosion. The bubble is about 23 light-years across and is expanding at over 11 million mph (18 million kph).

**TARANTULA NEBULA**
The Tarantula Nebula is home to not one but two huge star clusters. In the center of this image is R136, home to some of the most massive stars known, while at upper right is the slightly older Hodge 301, whose oldest stars have already gone supernova.

## IRREGULAR GALAXY

# Small Magellanic Cloud

**CATALOG NUMBER** NGC 292

**DISTANCE**
200,000 light-years

**DIAMETER**
10,000 light-years

**MAGNITUDE** 2.3

**TUCANA**

Like the Large Magellanic Cloud, the Small Magellanic Cloud (SMC) is an irregular galaxy in orbit around the Milky Way. It was in the SMC that Henrietta Leavitt discovered the Cepheid variable stars that were to unlock the secrets of the galactic distance scale (see pp.282, 356). Thanks to her discovery, astronomers know that the SMC is both more distant and genuinely smaller than the LMC, with around one-tenth of the larger cloud's mass. Like the LMC, the small cloud is also undergoing intense star formation. Some astronomers argue that the SMC also shows signs of a central barlike structure, but the case

is far from proven. It has one known globular cluster in orbit, but the SMC lies deceptively close in the sky to one of the Milky Way's largest globulars—47 Tucanae.

Both of the Magellanic Clouds are ultimately doomed to be torn to shreds and absorbed into our own galaxy. They have survived several close passes of the Milky Way, but now share their orbit with a trail of gas, dust, and stars torn away during

**CLOUD OF STARS**
The SMC forms a distinctive wedge-shaped cloud in southern skies. The pinkish areas in this optical photograph show the galaxy's major star-forming regions.

previous encounters. This "Magellanic Stream" has allowed astronomers to trace and refine their models for the orbits of the clouds.

## MAGELLAN'S DISCOVERY

The southernmost sky was not visible to Europeans until they visited the Southern Hemisphere. The Portuguese explorer Ferdinand Magellan was among the first to do so during his around-the-world voyage of 1519–1521. He was the first European to record two isolated patches of the Milky Way, which were later named after him.

**FERDINAND MAGELLAN**

---

## SC SPIRAL GALAXY

# Triangulum Galaxy

**CATALOG NUMBERS**
M33, NGC 598

**DISTANCE**
3 million light-years

**DIAMETER**
50,000 light-years

**MAGNITUDE** 5.7

**TRIANGULUM**

After the Andromeda Galaxy and the Milky Way, the Triangulum Galaxy (M33) is the third major member of the Local Group of galaxies. It is slightly more distant than the larger and brighter

Andromeda Galaxy (M31), and the two lie close to each other in the sky. M33 is affected by its larger neighbor's gravity, and it may even be in a long, slow orbit around the giant Andromeda spiral.

Seen from Earth, M33 is fainter and more diffuse than M31—partly because it is closer to face-on than edge-on, and partly because it really is less spectacular. However, the Triangulum Galaxy is more typical of spiral galaxies than its unusually bright companions. As with several Local Group galaxies, M33 is large and bright enough in the sky for its features to be cataloged, and several of them have NGC numbers. Most prominent is the star-forming region NGC 604, the largest emission nebula known. At 1,500 light-years across, it dwarfs anything in our own galaxy.

**FLOCCULENT SPIRAL**
M33 is an example of a flocculent spiral—a galaxy with arms that divide like split ends and separate into patches. The clumpy star clouds are thought to form due to localized changes in density.

**NEBULA NGC 604**
The most massive stars of NGC 604 are so hot that their fierce stellar winds are helping shape the nebula in the Triangulum Galaxy. A visible-light image (purple) captures NGC 604's filaments of gas and dust, while X-ray data (blue) reveals expanding bubbles of superhot gas between them.

**CLOUD DETAILS**
A near-infrared image of the LMC taken by the ESO's VISTA telescope reveals previously hidden stars tracing faint patterns that may indicate multiple spiral arms.

## Sb SPIRAL GALAXY

# Andromeda Galaxy

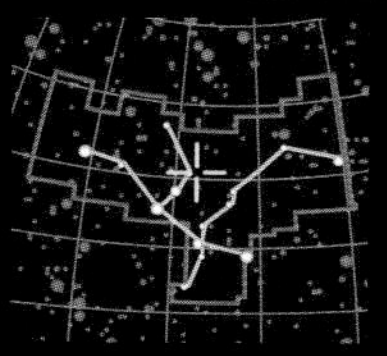

**CATALOG NUMBERS**
M31, NGC 224

**DISTANCE**
2.5 million light-years

**DIAMETER**
220,000 light-years

**MAGNITUDE** 3.4

**ANDROMEDA**

The Andromeda Galaxy (M31) is the closest major galaxy to the Milky Way and the largest member of the Local Group of galaxies. Its disk is twice as wide as our galaxy's.

M31's brightness and size mean it has been studied for longer than any other galaxy. First identified as a "little cloud" by Persian astronomer Al–Sufi (see p.421) in around 964 CE, it was for centuries assumed to be a nebula, at a similar distance to other objects in the sky. Improved telescopes revealed that this "nebula," like many others, had a spiral structure. Some

astronomers thought that M31 and other "spiral nebulae" might be solar systems in the process of formation, while others guessed correctly that they were independent systems of many stars. It was in the early 20th century that Edwin Hubble (see p.45) revealed the true nature of M31, at a stroke hugely increasing estimates of the size of the universe (see panel, opposite). Astronomers now know that M31, like the Milky Way, is a huge galaxy attended by a cluster of smaller orbiting galaxies, which occasionally fall inward under M31's gravity and are torn apart.

Despite being intensively studied, the Andromeda Galaxy still holds many mysteries, and it may not be

as typical a spiral galaxy as it appears. For example, despite its huge size, it seems to be only half as massive as the Milky Way, with a relatively sparse halo of dark matter. Despite this, M31 contains about a trillion stars, and its supermassive black hole has a mass of 150 million Suns, 40 times the mass of the Milky Way's central black hole. This big difference is surprising, because the black hole is thought generally to reflect the mass of its parent galaxy. Furthermore, studies at different wavelengths have revealed disruption in the galaxy's disk, possibly caused by an encounter with one of its satellite galaxies in the past few million years.

M31 and the Milky Way are moving toward each other, and they should collide and begin to coalesce in around 5 billion years.

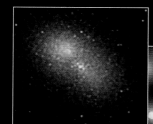

**DOUBLE CORE**
This Hubble close-up shows two distinct concentrations of stars at the very center of Andromeda. The central black hole lies within the fainter area.

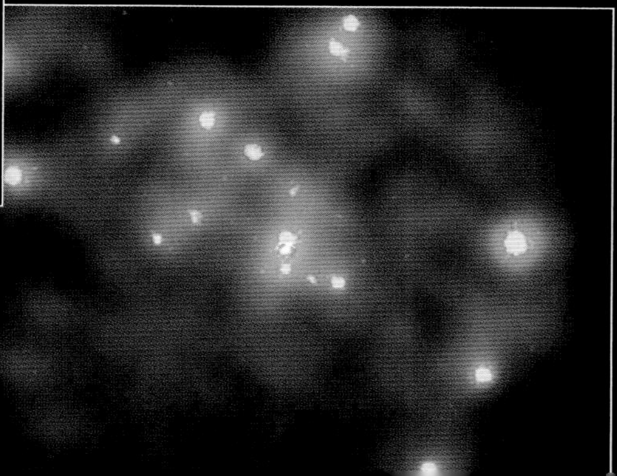

**CENTRAL BLACK HOLE**
This X-ray image of a small area of M31's core shows its central black hole as a blue dot—it is cool and inactive compared to the galaxy's other X-ray sources (yellow dots).

**GALACTIC NEIGHBORS**
Dark dust lanes are silhouetted against glowing gas and stars in this view of the Andromeda Galaxy and its two close companions, the dwarf elliptical galaxies M32 (upper left) and M110 (bottom).

## EXPLORING SPACE

## INTERGALACTIC DISTANCE

The study of M31 played a key role in the discovery that galaxies exist beyond our own. Although the spectra of galaxies suggested they shone with the light of countless stars, no one could measure their immense distance. In 1923, Edwin Hubble (see p.45) proved that M31 lay outside our galaxy. He found the true distance of M31 by calculating the luminosities of its Cepheid variable stars (see pp.282–283) and relating their true brightness to their apparent magnitude.

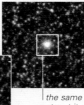

the same star at its brightest

Cepheid variable V1 at its faintest

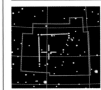

core

## Sb SPIRAL GALAXY

# Bode's Galaxy

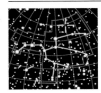

| CATALOG NUMBERS |
|---|
| M81, NGC 3031 |

**DISTANCE**
10.5 million light-years

**DIAMETER**
95,000 light-years

**MAGNITUDE** 6.9

**URSA MAJOR**

Bode's Galaxy, also known as M81, is one of the brightest spiral galaxies visible from the Northern Hemisphere. It is the dominant member of a galaxy group lying near the Local Group. The galaxy is named after Johann Elert Bode, a German astronomer who found it in 1774.

Bode's Galaxy has had a close encounter with M82, the Cigar Galaxy (see below),

### CLUSTERS REVEALED
This combined visible and ultraviolet image shows the hottest and brightest star clusters (blue and white blobs), lying in the core and spiral arms.

### X-RAY SOURCES
A Chandra X-ray image shows a strong X-ray source at the galaxy's core surrounded by smaller sources, probably X-ray binary stars.

in the past few tens of millions of years. The near miss created tidal forces that enhanced the density waves (see p.303) in M81. The rate of star birth around the density waves increased, highlighting the spiral arms. A long, straight dust lane along one side of the core could also have been created in the encounter.

By measuring the Doppler shifts of light from either side of the core, astronomers have found that the outer regions rotate more slowly than in most galaxies. This suggests that M81 has little of the dark matter that creates higher rotation rates in other galaxies.

### PERFECT SPIRAL
M81 is a beautifully symmetrical spiral galaxy tilted at an angle to our line of sight. This Hubble view shows star clusters, dust, and gas clouds in its spiral arms.

---

### GAS STREAMERS
Many of M82's most spectacular features are best seen at nonvisible wavelengths. This composite image shows visible light in green, hot X-rays in blue, infrared wavelengths in red, and hydrogen emissions in orange. Hot streamers of gas are clearly visible blowing away from the central starburst.

**OPTICAL IMAGE**

## IRREGULAR DISK GALAXY

# Cigar Galaxy

| CATALOG NUMBERS | M82, NGC 3034 |
|---|---|

**DISTANCE**
12 million light-years

**DIAMETER**
40,000 light-years

**MAGNITUDE** 8.9

**URSA MAJOR**

The brightest and most spectacular example of a "starburst galaxy," the Cigar Galaxy (M82) is an irregularly shaped cloud of stars that looks like a cigar from Earth. It is undergoing a period of intense star birth as a result of a close encounter with Bode's Galaxy (M81). The near miss has disrupted the galaxy's center, creating the dark dust lanes that obscure much of the core and triggering the creation of many massive, brilliant star clusters in an area a few thousand light-years across.

At infrared wavelengths, M82 is the brightest galaxy in the sky. In 2014, astronomers observed one of the closest recent supernovae to Earth here.

**X-RAY VIEW**

cluster of active black holes

### STARBURST GALAXY
The intense activity in M82's core is luminous at optical and X-ray wavelengths. The young stars illuminate the nebulae with visible light, while those that have rapidly completed their life cycle form active black holes, emitting X-rays.

## Sb SPIRAL GALAXY

# Black Eye Galaxy

| CATALOG NUMBERS |
|---|
| M64, NGC 4826 |

**DISTANCE**
19 million light-years

**DIAMETER**
51,000 light-years

**MAGNITUDE** 8.5

**COMA BERENICES**

This distinctive galaxy has a dark dust lane, running in front of its core, from which it gets its name. The dust lane is unusual because it arcs above the galaxy's core in an orbit of its own. Because it has not yet settled into the plane of the galaxy's rotation, it must have a recent origin and probably dates from the galaxy's absorption of a smaller galaxy that strayed too close. Another bizarre feature of the Black Eye Galaxy is that its outer regions are rotating in the opposite direction of the inner regions. This could be another effect of the collision.

**M64'S CENTRAL REGION AND DUST LANE**

## Sc SPIRAL AND IRREGULAR GALAXIES

# Whirlpool Galaxy

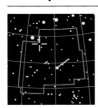

**CATALOG NUMBERS** M51, NGC 5194, NGC 5195

**DISTANCE**
31 million light-years

**DIAMETER**
100,000 light-years

**MAGNITUDE** 8.4

**CANES VENATICI**

Discovered by Charles Messier (see p.73) in 1773, the Whirlpool Galaxy is now known to be a pair of galaxies that is interacting—the brightest and clearest example of such a pair visible from Earth. The individual components are a spiral galaxy viewed face-on (NGC 5194) and a smaller irregular galaxy (NGC 5195). In visible light, the connection between them cannot be seen, but images at other wavelengths reveal an envelope of gas connecting the two. One effect of the interaction is to enhance the density wave in the larger galaxy, triggering increased star formation and making its spiral arms stand out very clearly. The Whirlpool was in fact the first "nebula" in which spiral structure was recognized, by William Parsons (see panel, right).

The interaction has also triggered increased activity in the cores of both of the galaxies—NGC 5195 is

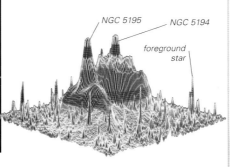

**CONTRASTING PAIR**
This infrared image, taken by the Spitzer Space Telescope, shows the Whirlpool Galaxy and its companion. The Whirlpool itself is rich in dust, which is colored red, while the companion is largely dust-free and appears blue.

undergoing a burst of star formation, which explains its unusual brightness, while NGC 5194's core is also much brighter than expected. It is even classified by some astronomers as an active Seyfert galaxy (see p.320).

The Whirlpool Galaxy is very bright despite its distance, indicating that it is

**LIGHT INTENSITY**
Plotting the intensity of light from different regions of M51 reveals the brightness of the two galactic cores (the twin peaks on the graph).

NGC 5195    NGC 5194

foreground star

large and luminous—it is similar in size to the Milky Way but brighter overall because of the large young star clusters in its spiral arms. It is thought to be the dominant member of a small group of galaxies, called simply the M51 group, which also includes the galaxy M63.

## WILLIAM PARSONS

William Parsons (1800–1867) was an Irish nobleman who used his great wealth to build the largest telescope of his time and made the first detailed studies of nebulae. In 1845, he made detailed drawings and noticed the spiral structure of some "nebulae," as galaxies were thought to be at the time. This was an important step to discovering that galaxies were not nebulae but separate star systems.

**PARSONS'S SKETCH OF M51**

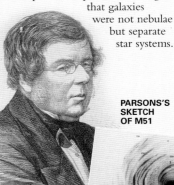

**LUMINOUS WHIRLPOOL**
This Hubble image combines data from different filters to reveal detail in M51, such as dark dust behind each spiral arm and bright pink regions of star birth.

BEYOND THE MILKY WAY

## SC SPIRAL GALAXY

# Pinwheel Galaxy

| CATALOG NUMBERS |
|---|
| M101, NGC 5457 |
| **DISTANCE** |
| 27 million light-years |
| **DIAMETER** |
| 170,000 light-years |
| **MAGNITUDE** 7.9 |

**URSA MAJOR**

Cataloged by Charles Messier (see p.73) as M101, the Pinwheel Galaxy is a bright, nearby spiral galaxy, but one that reveals its nature only when studied with powerful telescopes or seen on long-exposure photographs. Because it lies face-on to Earth, most of the Pinwheel's light is spread out across its disk, and a casual glance reveals only the bright central core. Detailed photographs show that M101 has an extensive, though rather lopsided, spiral-arm system, giving the appearance that the core is offset from the galaxy's true center. M101 is one of the largest spirals known—its visible diameter is more than twice that of our own galaxy. Its large, angular size in the sky (larger than a full moon) makes it one of the few galaxies whose individual regions can be isolated for study.

**RELATIVE RED SHIFT**
This computer image shows the red shift and blue shift of objects within M101, revealing its rotation. Yellow and red regions are moving away; green and blue parts are approaching.

**ASYMMETRICAL DISK**
M101's lopsided shape is thought to be caused by uneven distribution of mass in the disk affecting the orbit of its stars.

**DUST LANE**
The thick dust lane around the Sombrero Galaxy is silhouetted against its bright disk in this Hubble Space Telescope image.

## SA SPIRAL GALAXY

# Sombrero Galaxy

| CATALOG NUMBERS |
|---|
| M104, NGC 4594 |
| **DISTANCE** |
| 50 million light-years |
| **DIAMETER** |
| 50,000 light-years |
| **MAGNITUDE** 8.0 |

**VIRGO**

The dark dust lane and bulbous core of the Sombrero Galaxy (M104) give it a likeness to the traditional Mexican hat after which it is named. From Earth, we see the Sombrero Galaxy from 6 degrees above its equatorial plane—an ideal angle to provide a clear view of the core while also revealing the spiral arms. It is usually classified as an Sa or Sb spiral, although its core is unusually large and bright. Another odd feature is the dense swarm of globular star clusters orbiting the galaxy. More than 2,000 have been counted—10 times more than orbit the Milky Way.

In the galaxy's core is a disk of bright material tilted relative to the galaxy's plane. It is probably the accretion disk of a central supermassive black hole. X-ray emission from the region suggests some material is still being absorbed by the hole.

M104 was a late addition to Messier's catalog of celestial objects. He added it by hand to his copy of the catalog after discovering it in 1781. Several other astronomers also found it independently. One

**COMBINED VIEW**
This composite image shows the Sombrero at X-ray (blue), optical (green), and infrared (orange) wavelengths.

of these was William Herschel, who was the first to note the dark dust lanes that are M104's most distinctive feature. More recently, the Sombrero provided some of the first evidence for objects lying far beyond our own galaxy (see panel, below).

## VESTO SLIPHER

US astronomer Vesto Slipher (1875–1969) was one of the first to suggest that the universe is bigger than our galaxy. In 1912, at Lowell Observatory in Flagstaff, Arizona, he identified red-shifted lines in M104's spectrum. The lines told him the galaxy was receding at 2.25 million mph (3.6 million kph)—too fast for it to reside within the Milky Way.

# Spindle Galaxy

**CATALOG NUMBERS**
M102 (not confirmed),
NGC 5866

**DISTANCE**
40 million light-years

**DIAMETER**
60,000 light-years

**DRACO**          **MAGNITUDE** 9.9

The Spindle (NGC 5866) is an attractive galaxy oriented edge-on to observers on Earth. It is usually classified as a lenticular galaxy—a disk of stars, gas, and dust with a typical bulging core but with no sign of true spiral arms. However, spiral structure is hard to detect in an edge-on galaxy.

The Spindle Galaxy is the major member of the NGC 5866 Group, a small cluster of galaxies. Astronomers have measured the way these galaxies move and have found that the Spindle must contain an enormous mass of material—up to 1 billion solar masses, or 30 to 50 percent more than the Milky Way.

The Spindle Galaxy could be the mysterious entry number 102 in Charles Messier's catalog of astronomical features. Messier included the object at first without a location, then later gave coordinates that did not match any feasible object. Some believe that Messier had listed the Pinwheel Galaxy, M101, twice. More likely, however, is that M102 was the Spindle, and he added 5 degrees to his measurements in error.

### MASSIVE SPINDLE
From Earth, we see the Spindle Galaxy edge-on, giving it a cigar-shaped appearance with a fine silhouetted dust lane.

# M60

**CATALOG NUMBERS**
M60, NGC 4649

**DISTANCE**
58 million light-years

**DIAMETER**
120,000 light-years

**MAGNITUDE** 8.8

**VIRGO**

M60 is one of several giant elliptical galaxies in the Virgo galaxy cluster (see p.329), the central cluster in our own Local Supercluster of galaxies. The galaxy and its neighbor, M59, were discovered in 1779 by German astronomer Johann Köhler, who was observing a comet that passed close by. Charles Messier (see p.73) found them a few nights later and added them to his catalog of objects that might confuse comet hunters.

M60 is similar in diameter to many spiral galaxies but, as an E2 elliptical, it is very nearly spherical, containing a much larger volume. It probably has a mass of several trillion suns and is orbited by thousands of globular clusters. Using the Hubble Space Telescope to measure the motions of M60's stars, astronomers have discovered that a monster black hole of 4.5 billion solar masses lies at the galaxy's heart.

### CLOSE NEIGHBORS
M60 lies very close to the spiral M59 (upper right), and the two galaxies are thought to be interacting. In a billion years, M60 may even swallow M59 entirely.

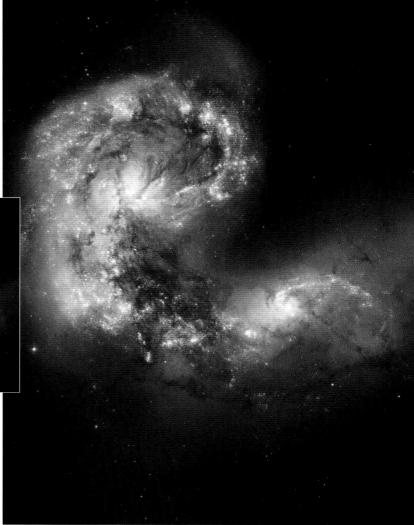

# Antennae Galaxies

**CATALOG NUMBERS**
NGC 4038, NGC 4039

**DISTANCE**
63 million light-years

**DIAMETER** 360,000
light-years (total)

**MAGNITUDE** 10.5

**CORVUS**

The Antennae Galaxies, NGC 4038 and 4039, are among the sky's most spectacular interacting galaxies. Seen from Earth, they appear as a central bright double-knot of material, with two long streamers of stars stretching in opposite directions, resembling an insect's antennae. However, powerful telescopes reveal that each streamer is in fact a spiral arm, uncurled from its parent galaxy by the tremendous gravitational forces of an intergalactic collision that began around 700 million years ago and continues today.

The Antennae have been studied for what they can tell us about galaxy collisions. Detailed images of the central region show that it is lit by hundreds of bright, intense star clusters. These are thought to be forming as gas clouds in the galaxies become compressed by the collision, triggering starbursts (see the Cigar Galaxy, p.314). Astronomers can use the clusters' redness to estimate their age—older clusters emit redder light because the brighter blue stars are the most massive and therefore the first to die.

### THE BIGGER PICTURE
A wide-field view of the Antennae taken from Earth reveals both the bright, distorted cores and the long, faint streamers formed by the disrupted spiral arms.

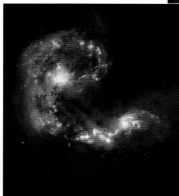

### CLOUDS AND CLUSTERS
Turbulent dust clouds and brilliant star clusters appear in a Hubble view of the colliding Antennae Galaxies at right. The image above—a composite of a Hubble visible-light view with microwave observations from the Atacama Large Millimeter Array in Chile—reveals clouds of dense, cold gas (pink, red, and yellow areas) from which new stars are forming.

## DISRUPTED SPIRAL GALAXY

# ESO 510-G13

| CATALOG NUMBER | ESO 510-G13 |
| --- | --- |
| DISTANCE | 150 million light-years |

| DIAMETER |
| --- |
| 105,000 light-years |

| MAGNITUDE |
| --- |
| 13.3 |

**HYDRA**

Despite being referred to only by a number rather than a name (its long designation comes from the European Southern Observatory's catalog), ESO 510-G13 is one of the most intriguing galaxies in the sky. It is an edge-on spiral with a clear dust lane marking its central plane. The dust lane has an obvious twist.

The most obvious explanation for the kink is that ESO 510-G13 has had a close encounter or collision with another galaxy in its recent past. Some astronomers have suggested that the collision is still going on, and the dust lane is the "ghost" of a galaxy that ESO 510-G13 has swallowed—as seen in the active galaxy Centaurus A (see p.322). Alternatively, the disk might have been warped by the gravity of a nearby galaxy. The galaxy responsible might be a small neighbor or a more distant but larger member of the same group. As their techniques and instruments improve, astronomers are finding this kind of distortion is common in spirals, although it often shows up more in the distribution of gas than in the stars, so it is usually most obvious at radio wavelengths. Our near neighbor M31 (see pp.312–313) has such a distortion, and the Milky Way seems to have one, too, perhaps caused by interaction with its own family of smaller neighbors.

### WARPED DISK
The bright core of ESO 510-G13 silhouettes the galaxy's warped dust lane in this image. The blue glow on the right is a huge area of bright young stars—evidence, perhaps, of a collision in the galaxy's recent history.

---

## SB0 BARRED SPIRAL GALAXY

# NGC 6782

| CATALOG NUMBER |
| --- |
| NGC 6782 |

| DISTANCE |
| --- |
| 183 million light-years |

| DIAMETER |
| --- |
| 82,000 light-years |

| MAGNITUDE | 12.7 |
| --- | --- |

**PAVO**

The Hubble Space Telescope imaged the apparently normal barred spiral galaxy NGC 6782 in 2001. Using ultraviolet detectors, it studied the pattern of the galaxy's hottest material. The image (see below) showed, in pale blue, two rings of stars so brilliant and hot that they emit most of their light as ultraviolet. The inner ring lies in the galaxy's bar and could have been ignited by tidal forces between the bar and the rest of the galaxy. The outer star ring is at the galaxy's edge.

**ULTRAVIOLET STAR RINGS**

---

## DISRUPTED SPIRAL GALAXIES

# The Mice

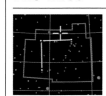

| CATALOG NUMBER |
| --- |
| NGC 4676 |

| DISTANCE |
| --- |
| 300 million light-years |

| DIAMETER |
| --- |
| 300,000 light-years |

| MAGNITUDE | 14.7 |
| --- | --- |

**COMA BERENICES**

The object classified as NGC 4676 is in fact a pair of colliding galaxies—called the Mice because they appear to have white bodies and long, narrow tails. As with the Antennae Galaxies (see p.317), the long streamers are the result of the spiral arms "unwinding" during the collision—though in this case, one of the arms lies edge-on to us and so appears to be long and straight despite being strongly curved away from us. Knots of bright blue stars in the streamers and the main bodies of the galaxies show where bursts of star formation are taking place. Computer simulations of the collision (see panel, right) suggest that the galaxies are now separating after a closest approach 160 million years ago.

### HIDDEN EXTENT
Image processing allows astronomers to amplify faint light from the outlying parts of the Mice, revealing their true shape and extent.

---

EXPLORING SPACE

## SIMULATING GALAXY COLLISIONS

The great challenge for astronomers studying colliding galaxies is that they can only ever see one stage in a story that unfolds over millions of years. Fortunately, today's supercomputers can help speed things up. By building "model" galaxies with simplified star clouds, gas, dust, and dark matter and then smashing them into each other in a computer, astronomers can measure how gravity affects the fate of the galaxies.

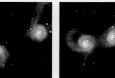  

**0 MY**    **400 MY**    **650 MY**

**1,000 MY**

### SPIRAL COLLISION SIMULATION
This computer simulation shows two spiral galaxies interacting and merging to form a large, irregular galaxy. Time is measured in millions of years (My).

### DESTINED TO UNITE
Although currently moving apart from a close encounter, the Mice are gravitationally locked together and doomed eventually to merge, perhaps resulting in the formation of a new giant elliptical galaxy.

# Cartwheel Galaxy

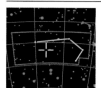

**CATALOG NUMBER**
ESO 350-G40

**DISTANCE**
500 million light-years

**DIAMETER**
150,000 light-years

**MAGNITUDE** 19.3

**SCULPTOR**

If the Cartwheel Galaxy looks unusual, it's because it is the victim of an intergalactic "hit-and-run." The Cartwheel was once a normal spiral galaxy. As we see the galaxy, it is recovering from a head-on collision with a smaller runaway galaxy many millions of years earlier in its history. Such events are rare in the cosmos—galactic collisions usually involve grazing encounters or a slow dance toward an eventual merger. The Cartwheel shows what happens when two galaxies pass

**CLOUDS IN THE CORE**
So-called "comet clouds," each a thousand light-years long, are found in the Cartwheel's core. They are thought to arise as hot, fast-moving gas set in motion by the collision plows through denser, slower-moving matter.

through each other at high speed while oriented at right angles to each other. The rotating density wave that is normally responsible for the spiral arms was disrupted in this case, resulting in the disappearance of the spiral structure. Meanwhile, a shock wave spread to the outer edge of the galaxy, creating a ring of vigorous star

formation. An inward-traveling shock wave is probably responsible for the core's unusual "bull's-eye" appearance.

For years, most astronomers suspected that one of the Cartwheel's two immediate neighbors was responsible for the collision. Both showed signs of being the culprit—a nearby small, blue galaxy has a

disrupted shape and vigorous star formation, while a yellow galaxy could have been stripped of its star-forming gas in the encounter. However, recent radio observations have shown a telltale stream of gas leading from the Cartwheel toward another small galaxy a quarter of a million light-years away.

**SPIRAL REGENERATION**
The "spokes" of the Cartwheel Galaxy (on the right) are the ghostly outlines of returning spiral arms.

---

# Hoag's Object

**CATALOG NUMBER**
PGC 54559

**DISTANCE**
500 million light-years

**DIAMETER**
120,000 light-years

**MAGNITUDE** 15.0

**SERPENS**

Hoag's Object is one of the most bizarre galaxies in the sky. Although its ring structure suggests parallels to the Cartwheel Galaxy (a spiral disrupted by a head-on collision, see above), there are no nearby galaxies that could have caused an impact. One of two theories might account for the shape of Hoag's Object and that of similar ring galaxies. The galaxies may be members of an unusual class of spiral in which the two arms develop into a circle. Alternatively, they may be former elliptical galaxies that have each swallowed another galaxy, creating a surrounding ring of star-forming material.

**SEE-THROUGH GALAXY**
The gap between Hoag's Object's core and its ring is truly transparent—a background galaxy can be seen through it near the top of this image. However, the gap could still contain large numbers of faint stars.

---

# Malin 1

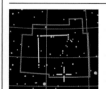

**CATALOG NUMBER**
None

**DISTANCE**
1 billion light-years

**DIAMETER**
600,000 light-years

**MAGNITUDE** 25.7

**COMA BERENICES**

Despite its dull appearance, Malin 1 is an extremely important galaxy. Discovered by accident in 1987, it is an enormous but faint spiral that is for some reason poor at forming stars. It seems that such low-surface-brightness galaxies could account for up to half the galaxies in the universe, though Malin 1 is one of the largest of the type.

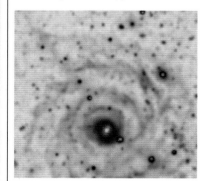

**MALIN 1 IN A NEGATIVE IMAGE**

# ACTIVE GALAXIES

material blasted from the nucleus expands into a lobe as it is slowed by the intergalactic medium

MANY GALAXIES ACROSS THE UNIVERSE show surprising features that mark them as out of the ordinary. Although there are several types of these strange galaxies, their unusual behavior can always be traced back to powerful activity in their nucleus—it seems that there is an underlying similarity between them, and for this reason, they are often studied together under the term "active galaxies."

## WHAT ARE ACTIVE GALAXIES?

Astronomers think that the features of active galaxies are linked to their central giant black holes. Most, if not all, galaxies have black holes with the mass of many millions of suns, known as supermassive black holes, at their nuclei (see p.305), but most such black holes are dormant—all material in these galaxies is in a stable orbit around the black hole. In active galaxies, matter is still falling inward, and as it falls, it is heated by intense gravity, generating a brilliant blast of radiation. As the black hole "engine" pulls matter in, the superheated material forms a spiraling accretion disk. The hot disk emits X-rays and other fierce, high-energy radiation. Around the outer edge of the disk, a dense torus (doughnut shape) of dust and gas forms. The intense magnetic field surrounding the black hole also catches some of the infalling material, firing it out as two narrow beams at the poles, at right angles to the plane of the accretion disk. These jets shine with radio-wavelength radiation due to the synchrotron mechanism (right).

jet of particles shooting from black hole's magnetic pole

star being ripped apart by intense gravity

location of black hole

spinning accretion disk of heated gas

torus of dust, typically 10 light-years across

jet expands into lobe thousands of light-years long

electron

magnetic field line

photon of radio-wavelength radiation

**BLACK-HOLE ENGINE**
The black hole of an active galactic nucleus is surrounded by a bright accretion disk and an outer dust cloud. Jets of material flow outward from the black hole's poles.

**SYNCHROTRON RADIATION**
As electrons from the black-hole jets move through the black hole's magnetic field, they are forced into spiral paths, releasing synchrotron radiation—a type of EM radiation that is most intense at long radio wavelengths.

## ACTIVE TYPES

Astronomers distinguish between four major types of active galaxies. Each displays its own set of active features, and in each case, these features are evidence of the violent activity at the nucleus. Radio galaxies are the most intense natural sources of radio waves in the sky. The emissions typically come from two huge lobes on either side of an apparently innocuous parent galaxy (and often linked to it by narrow jets). Seyfert galaxies are relatively normal spirals with a compact, luminous nucleus that may vary in brightness over just a few days. Quasars appear as starlike points of light that show similar but more extreme variability. Red-shifted lines in their spectra reveal that they are extremely distant galaxies—powerful modern telescopes can resolve them as galaxies with incredibly brilliant cores. They are more powerful and more distant cousins of the Seyfert galaxies. Finally, blazars (also known as BL Lacertae objects) are starlike variable points similar to quasars but with no significant lines in their spectra. The standard model of the black-hole engine (above) can explain the major features of each type—how the galaxy appears depends on the intensity of its activity and the angle at which we see it.

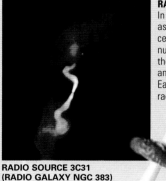

**RADIO GALAXY**
In a radio galaxy such as NGC 383, the central region of the nucleus is hidden by the edge-on dust ring, and observers on Earth see only the radio jets and lobes.

radio jet

dust ring

**RADIO SOURCE 3C31 (RADIO GALAXY NGC 383)**

**QUASAR**
In quasars, Earth-bound observers can see over the dust ring, and brilliant light from the nucleus and disk drowns out the light of the surrounding galaxy.

**QUASAR PG 0052+251**

**BLAZAR**
Blazars are active galaxies aligned so that observers on Earth look straight down the black-hole jet onto the nucleus. The galaxy is usually hidden by the brilliant light except in rare cases, such as Markarian 421.

**BLAZAR MARKARIAN 421**

**SEYFERT GALAXY**
In Seyfert galaxies such as M106, the nucleus and accretion disk are exposed to our view, as in a quasar, but the activity is weak.

**SEYFERT GALAXY M106**

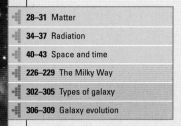

# THE HISTORY OF ACTIVE GALAXIES

The distribution of different types of active galaxies in the universe provides clues about how they evolve. Quasars and blazars are never seen close to Earth. They are always faint and distant, with red shifts indicating that they lie billions of light-years from Earth—we are seeing them as they were in much earlier times. Radio and Seyfert galaxies, in contrast, are scattered throughout the nearby universe, and radio jets are linked to both spiral and elliptical galaxies. So what happened to the quasars and blazars? It seems likely that they represent a brief phase in a galaxy's evolution, soon after its birth. At this time, material in the central regions would have had chaotic orbits, and the central black hole engine would have been fueled by a continuous supply of infalling stars, gas, and dust. As the black hole swept up the available matter, objects with stable orbits at a safe distance remained. Starved of fuel, the engine would have petered out, and the quasar became dormant—a normal galaxy such as the Milky Way. Today, such galaxies can become active again if they are involved in collisions that send new material falling in toward the black hole. Many nearby radio and Seyfert galaxies show evidence of recent collisions or close encounters, and some of these galaxies are close enough for infrared telescopes to image the dust rings around their cores directly (see p.323). However, levels of recent activity are restrained—even the most spectacular radio galaxies generate little energy compared to quasars, while Seyferts are the feeblest type of active galaxy.

disk of spiral galaxy

jet of particles emitting radio waves

active nucleus of galaxy containing an active black hole surrounded by a bright accretion disk and a dust ring

**ACTIVE GALAXY**
This idealized active galaxy is a spiral with a bright nucleus that hides an active black hole. From the black hole's poles blast two jets of particles, leaving at close to light speed, only slowing and billowing out into lobes many thousands of light-years away as the particles hit the intergalactic medium.

**NUDGED BACK INTO LIFE**
Optical images of Centaurus A clearly show the dark dust lane of a spiral colliding with this elliptical galaxy. The overlaid radio map shows the burst of activity—the jets and plumes—triggered by this event.

false-color radio image of jet of particles

dust lane (optical image)

optical view of galaxy's elliptical arrangement of stars

false-color radio image of galaxy's lobe

---

EXPLORING SPACE

## SUPERLUMINAL JETS

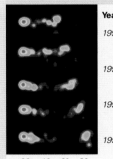

| Year |
| --- |
| 1992 |
| 1994 |
| 1996 |
| 1998 |

20  40  60  80
**Distance (light-years)**

Some quasars and blazars appear to defy the laws of physics. Image sequences, taken years apart, show jets of material blasting away from the nucleus, apparently traveling faster than the speed of light. This apparent motion is called "superluminal." In reality, it is an illusion created when jets traveling at very high speeds, of up to 99 percent of the speed of light, happen to be pointing almost directly toward us.

**TIME-LAPSE SEQUENCE**
These images show jet emissions from blazar 3C 279, taken at intervals of almost 2 years, and showing motion apparently five times the speed of light.

---

## IS THE MILKY WAY ACTIVE?

The Milky Way Galaxy, like any galaxy with a central black hole, has the potential to be active, and there is intriguing evidence that it might have burst into activity in the recent past. In 1997, scientists discovered a huge cloud of gamma-ray emission above the galactic center. The radiation has a distinctive frequency, suggesting it is the result of electrons encountering positrons—their antimatter equivalent (see p.31)—and annihilating in a burst of energy. The positrons might have been generated by activity at the core— perhaps an infall of matter into the black hole— and are now meeting scattered electrons in the outer galaxy and mutually annihilating to produce the distinctive glow. Because the clouds lie just 3,000 light-years from the galactic center, the activity must have occurred recently.

**ANTIMATTER FOUNTAIN**
This gamma-ray image traces positrons (antielectrons) around the Milky Way. The horizontal feature is the plane of the galaxy, with the fountain above it.

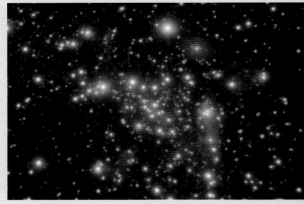

**GALACTIC CENTER**
This near-infrared image, taken using the Very Large Telescope in Chile, shows the center of the Milky Way. By following the motions of its central stars over more than 16 years, astronomers were able to determine that the supermassive black hole at the core is about 4 million times as massive as the Sun.

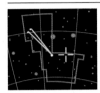

# ACTIVE GALAXIES

There are no simple rules governing the appearance of active galaxies. Some have a disrupted structure seen either in visible light or at other wavelengths, while others appear normal at first but radiate unusually large amounts of energy at certain wavelengths. In fact, the majority of galaxies show activity of one kind or another. However, a smaller proportion of galaxies have particularly active nuclei, powered by matter spiraling into their central black hole. These include Seyfert galaxies, radio galaxies, quasars, and blazars. The vast majority of known active galaxies are distant quasars. Objects lying nearer to the Milky Way, although less spectacularly violent, are at least close enough for astronomers to study in detail.

**JET FROM AN ACTIVE GALAXY**
Pictured in radio waves and false colors, this jet of particles blasted from the core of the galaxy M87 is a typical feature of active galaxies with black-hole engines.

---

## TYPE-II SEYFERT GALAXY

### Circinus Galaxy

| | |
|---|---|
| **CATALOG NUMBER** | ESO 97-G13 |
| **SHAPE** | Sb spiral |
| **DISTANCE** | 13 million light-years |
| **DIAMETER** | 37,000 light-years |
| **CIRCINUS** | **MAGNITUDE** 11.0 |

Although it is one of the nearest active galaxies to Earth, the spiral galaxy in Circinus went undiscovered until just a few decades ago. It remained hidden for so long partly because it lies just 4 degrees below the plane of the Milky Way and is obscured by star clouds. The full extent of the Circinus Galaxy's extraordinary nature was revealed only when it was observed by the Hubble Space Telescope in 1999. The galaxy is a Seyfert (see p.320)—a spiral with an unusually bright, compact region at its core, thought to result from material slowly drifting onto a massive central black hole. Hubble's infrared camera revealed how the galaxy's gas is concentrated in a central ring, just 250 light-years in diameter, around the black hole. Also apparent is a loose outer ring in the plane of the galaxy, around 1,300 light-years across, where great bursts of star formation are occurring. Finally, Hubble showed a cone-shaped cloud billowing above the plane of the galaxy. This is matter ejected by the magnetic fields of the black hole and glows as it is heated by the ultraviolet radiation from the nucleus.

**CONE OF MATTER**
The pinkish-white region near the core of the Circinus Galaxy shows where matter is being flung out, in a cone shape, from the central black hole into the gas cloud above the galaxy.

---

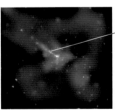

**RADIO CONTINUUM**     **RADIO (21-CM WAVELENGTH)**

**COMPOSITE VIEW**
Centaurus A has been imaged at various wavelengths (left and below). The image at far left is a composite at optical, microwave, and X-ray wavelengths.

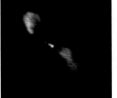

*jet*

**OPTICAL WAVELENGTHS**     **X-RAY WAVELENGTHS**

**DUSTY DISK**
This Hubble Space Telescope close-up of Centaurus A (right) reveals dark interstellar dust, glowing orange gas clouds, and brilliant blue star clusters formed in the collision between two galaxies.

---

## RADIO GALAXY

### Centaurus A

| | |
|---|---|
| **CATALOG NUMBER** | NGC 5128 |
| **SHAPE** | Peculiar elliptical |
| **DISTANCE** | 15 million light-years |
| **DIAMETER** | 80,000 light-years |
| **CENTAURUS** | **MAGNITUDE** 7.0 |

A ball of old yellow stars, NGC 5128 shows some features typical of an elliptical galaxy, but its most striking aspect is the dark dust lane that cuts across it, bisecting the uniform glow of stars with a ragged silhouette. What is more, the galaxy is at the center of a pair of vast radio lobes, 1 million light-years across. The name of this radio source, Centaurus A, is now the most widely used name for the galaxy itself. Astronomers have studied Centaurus A in detail at a range of wavelengths. The Hubble Space Telescope looked through the dust lanes with its infrared camera and found a huge accretion disk at the center—a sure sign of an active black hole pulling in matter at Centaurus A's core.

It is now generally agreed that NGC 5128 is an elliptical galaxy absorbing a spiral. The ghost of the spiral is shown by the dust lane and by the bright star clusters that stud it—perhaps generated by shock waves as the two galaxies merge.

## RADIO GALAXY

# M87

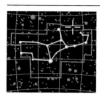

| | |
|---|---|
| **CATALOG NUMBERS** | M87, NGC 4486 |
| **SHAPE** | E1 giant elliptical |
| **DISTANCE** | 60 million light-years |
| **DIAMETER** | 120,000 light-years |
| **VIRGO** | |
| **MAGNITUDE** | 8.6 |

Lying at the heart of the Virgo Galaxy Cluster (see p.329), M87 is the closest example of a giant elliptical galaxy—a class of galaxy often found at the cores of old galaxy clusters. This huge ball of stars seems to have roughly the same diameter as the Milky Way but contains many more stars distributed across its spherical structure—probably several trillion. Long-exposure photographs reveal an extensive halo of more loosely scattered stars in a more elongated shape. The galaxy also has an unrivaled collection of globular star clusters in orbit—some astronomers estimate as many as 15,000 such groups.

What is more, M87 is an active galaxy—its location coincides with the Virgo A radio source and with a strong source of X-rays. There is even a sign of this activity that is visible at optical wavelengths in the form of a long, narrow jet of material being blasted from its interior. In 2019, the Event Horizon Telescope successfully imaged M87's enormous central black hole (see p.26).

### GALACTIC ERUPTION
Infrared images from the Spitzer Space Telescope show shock waves around Messier 87's high-energy jets.

*shock wave*

*supermassive black hole*

## TYPE-II SEYFERT GALAXY

# Fried Egg Galaxy

| | |
|---|---|
| **CATALOG NUMBER** | NGC 7742 |
| **SHAPE** | Sb spiral |
| **DISTANCE** | 72 million light-years |
| **DIAMETER** | 36,000 light-years |
| **PEGASUS** | |
| **MAGNITUDE** | 11.6 |

The small spiral galaxy NGC 7742 resembles a fried egg because of the intense yellow glow from its core. The core is much brighter than is usual for a galaxy of this size, because this is a Seyfert galaxy with a moderately active core. Seyferts emit radiation across a broad band of wavelengths—NGC 7742 is a type-II—a galaxy that is brightest in infrared light.

**CELESTIAL EGG**

# CARL SEYFERT

US astronomer Carl Seyfert (1911–1960) was the son of a pharmacist from Cleveland, Ohio. He studied at Harvard and went on to work at McDonald Observatory, then at Mount Wilson in California. It was here that he first identified the class of galaxies with unusually bright nuclei that bear his name (see p.320). In 1951, he also discovered Seyfert's Sextet, an interesting, compact cluster of galaxies (see p.329).

**SEYFERT'S OBSERVATORY**
At Nashville, Seyfert found time to give public lectures, as well as raising support and supervising the construction of the Arthur J. Dyer Observatory (above).

## RADIO GALAXY

# NGC 4261

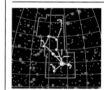

| | |
|---|---|
| **CATALOG NUMBER** | NGC 4261 |
| **SHAPE** | E1 elliptical |
| **DISTANCE** | 100 million light-years |
| **DIAMETER** | 60,000 light-years |
| **VIRGO** | |
| **MAGNITUDE** | 10.3 |

The elliptical galaxy NGC 4261 lies at the center of two great lobes of radio emission measuring 150,000 light-years from tip to tip. In many ways a typical radio galaxy, it is also one of the few active elliptical galaxies to have revealed its internal structure to astronomers. Infrared images from the Hubble Space Telescope pierced the obscuring clouds of stars to reveal an unexpectedly dense disk of dusty material, apparently spiraling onto the galaxy's central black hole.

Most elliptical galaxies are thought to be relatively dust-free, so where did the material in NGC 4261 come from? The most likely answer is that the elliptical galaxy has merged with a spiral in its relatively recent history. The spiral's individual stars have now become indistinguishable from the stars that were originally part of the elliptical galaxy, but the ghostly outline of the galaxy's gas and dust remains.

### RADIATING PLUMES
Combining optical and radio images of NGC 4261 reveals its full extent. The visible part of the galaxy is the white blob in the center, while the orange plumes mark the radio-emitting regions.

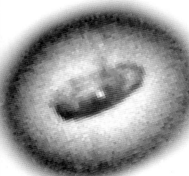

### ANCIENT REMAINS
A Chandra X-ray image of the galaxy's core reveals dozens of black holes and neutron stars around the supermassive central black hole, perhaps formed during its recent collision.

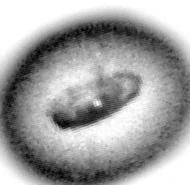

### DUST WHIRLPOOL
The Hubble Space Telescope's close-up image of the core reveals a dusty spiral of matter within a ring of glowing outer clouds. A distinct cone shows where matter is being flung off from the active galactic nucleus into the radio lobes.

## TYPE-I SEYFERT GALAXY

# NGC 5548

| | |
|---|---|
| **CATALOG NUMBER** | NGC 5548 |
| **SHAPE** | Sb spiral |
| **DISTANCE** | 220 million light-years |
| **DIAMETER** | 100,000 light-years |
| **BOÖTES** | |
| **MAGNITUDE** | 10.5 |

NGC 5548 is a type-I Seyfert galaxy—that is, a Seyfert that emits more ultraviolet and X-ray radiation than visible light. Like all Seyferts, it has a bright, compact core but, unlike the Fried Egg Galaxy (see above), its core is an intense blue-white. Using the Chandra X-ray Telescope, astronomers have detected an envelope of warm gas expanding around the core. The gas eventually forms two lobes of weak radio emission around the galaxy.

**HUBBLE IMAGE OF NGC 5548**

# NGC 1275

**CATALOG NUMBER**
NGC 1275

**SHAPE** Elliptical and
distorted spiral

**DISTANCE**
235 million light-years

**DIAMETER**
70,000 light-years

**PERSEUS**

**MAGNITUDE** 11.6

Despite being cataloged as a Seyfert
galaxy by Carl Seyfert himself (see
p.323), NGC 1275 has remained a
mystery. Recent observations have
shown that there are two objects—
one in front of the other.
A ghostly spiral galaxy,
revealed by its bright blue
star clusters, is responsible
for the dust lanes that
cross the bright central
region, but this brighter
region is in fact a separate
galaxy. Despite its Seyfert-
like core, it is an elliptical,
not a spiral. This galactic
giant lies at the heart of
the Perseus galaxy cluster,
and the foreground spiral
is racing toward it at 6.7
million mph (10.8 million
kph), its structure already
disrupted by the elliptical's
gravity. Adding to the
complexity, the elliptical
galaxy is also a radio source, and some
astronomers have argued that it shows
blazarlike activity (see BL Lacertae,
opposite). Whatever the details, NGC
1275 displays many of the typical
features of an active galactic nucleus.

**ATYPICAL ELLIPTICAL GALAXY**
NGC 1275 is unusual for an elliptical
galaxy in having a Seyfert-like core.
The dark dust lanes are the remains
of a now-disrupted separate spiral
galaxy in front of NGC 1275.

**CLUSTERS IN THE NUCLEUS**
The core of NGC 1275 offers
clues to the origin of globular
clusters—numerous globularlike
clusters are found here, but they
are composed of young blue
rather than old yellow stars.

# Cygnus A

**CATALOG NUMBER**
3C 405

**SHAPE** Pec (peculiar)

**DISTANCE**
600 million light-years

**DIAMETER**
120,000 light-years
(excluding radio lobes)

**CYGNUS**

**MAGNITUDE** 15.0

The most spectacular and powerful
radio galaxy in the nearby universe,
Cygnus A was discovered as soon as
radio telescopes began operating in
the 1950s. It features two huge lobes
of material emitting radio waves. The
lobes are visibly linked to their origin
at the heart of a faint, central, elliptical
galaxy by two long, narrow jets. From
lobe to lobe, the entire structure
extends over half a million light-years.

Despite its prominence in the
radio sky, mysteries still surround
Cygnus A, largely because of its great
remoteness. Early observations led
astronomers to believe the central
galaxy was in fact a pair of colliding
galaxies. Hubble Space Telescope
images suggested a resemblance to
NGC 5128, the Centaurus A Galaxy

(see p.322), which is thought to be
an elliptical galaxy that has recently
swallowed a spiral. Recent detection
of a large cloud of red-shifted gas
moving through the Cygnus A Galaxy
suggests that a collision may indeed
be the root cause of the activity.

Astronomers have also argued
about the origin of the "hot spots,"
where the radio lobes glow brightest

at either end. Studies by the Chandra
X-ray Telescope have shown that
Cygnus A lies at the center of a cloud
of hot but sparse gas. The jets have
blown out a football-shaped cavity
in the gas so vast that it dwarfs the
central galaxy. Tendrils of gas, which
are emitting X-rays and radio waves,
are also falling back down through the
cavity onto the poles of the galaxy,

**COMPLEX SOURCE**
This composite image shows Cygnus A's
radio jets in red, X-ray emission in blue, and
visible light in yellow. The central galaxy
is surrounded by a huge bubble of hot gas
through which the radio jets burst out.

drawn by its gravitational pull. The
hot spots are apparently created where
the outward blast of the jets collides
with the hot gas falling inward.

## BLAZAR (BL LAC OBJECT)

# BL Lacertae

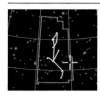

| CATALOG NUMBER | |
|---|---|
| BL Lac | |
| **SHAPE** Elliptical | |
| **DISTANCE** | |
| 1 billion light-years | |
| **DIAMETER** | |
| Unknown | |
| **LACERTA** | **MAGNITUDE** 12.4–17.2 |

BL Lacertae (BL Lac for short) was first cataloged as an irregular variable star by German astronomer Cuno Hoffmeister in the 1920s. Since then, astronomers' understanding of the object has

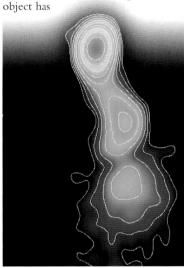

changed. For a variable star, it was very mysterious, showing rapid but completely unpredictable variations. At the same time, it displayed a totally featureless spectrum—it had neither the dark absorption lines seen in stars nor the bright emission lines found in galaxies (see p.35). It was not until 1969, when BL Lac was found to be a strong radio source, that astronomers realized it might be a new type of active galaxy. Today, it is seen as the founding member of a class of active galaxies called blazars or BL Lac objects. Blazars show many similarities to quasars but also some differences, most notably their featureless spectra.

The mystery of BL Lac was solved in the 1970s, when two astronomers blocked out or "occulted" BL Lac's bright core to study its surroundings. This revealed that it was embedded in a faint elliptical galaxy whose light was normally drowned out. Red-shifted lines in the spectrum of this galaxy confirmed BL Lac's great distance (see p.44). Today, blazars are accepted as rare cases in which Earth's position happens to align directly with the jet of material blasting out of an active galactic nucleus, with no obscuring material in the way.

### MAP OF A BLAZAR

This radio map of BL Lacertae shows the intensity of radiation (contour lines) and also its polarization (color)—an indication of magnetic field strength. The red object at the top is the galaxy's nucleus, while the lower regions are parts of a radio jet.

## QUASAR

# PKS 2349

| CATALOG NUMBER | |
|---|---|
| PKS 2349 | |
| **SHAPE** Disrupted | |
| **DISTANCE** | |
| 1.5 billion light-years | |
| **DIAMETER** | |
| Unknown | |
| **PISCES** | **MAGNITUDE** 15.3 |

The Hubble Space Telescope offered astronomers an unprecedented chance to study quasars in detail during the 1990s. One of their most intriguing subjects was the otherwise undistinguished quasar PKS 2349 (referred to by its designation in the catalog of the Australian Parkes radio telescope). For the first time, astronomers were able to see the faint host galaxies surrounding

### QUASAR CLOSE-UP

In Hubble's image of PKS 2349, the quasar is the bright central object, the companion galaxy is the smaller bright region above it, and the supposed host galaxy is the fainter ring extending from the quasar.

quasars, as well as other galaxies close to the quasars. The images showed that, in many cases, quasars do not just sit at the centers of their host galaxies, but are involved in violent interactions with neighboring galaxies and other quasars. PKS 2349 was referred to as a "smoking gun" because it showed these interactions so clearly. The quasar is surrounded by a ring of faint material that may mark the outline of its host galaxy—though, if so, the quasar itself is remarkably "off-center." A small companion galaxy about the size of the Large Magellanic Cloud (see p.310) also lies nearby and seems doomed to collide with the quasar itself.

---

## QUASAR

# 3C 273

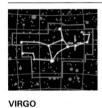

| CATALOG NUMBERS | |
|---|---|
| 3C 273, PKS 1226+02 | |
| **SHAPE** E4 elliptical | |
| **DISTANCE** | |
| 2.1 billion light-years | |
| **DIAMETER** 160,000 light-years (excluding jet) | |
| **VIRGO** | **MAGNITUDE** 12.8 |

The brightest quasar in the sky, 3C 273 was the second to be discovered. The existence of this radio source was already known when, in 1963, Australian astronomer Cyril Hazard used an occultation by the Moon (see p.69) to precisely establish its position, linking the radio source to what appeared to be an irregular variable star. The star's spectrum had a forest of unidentifiable dark emission lines (see p.35). Astronomers finally realized that

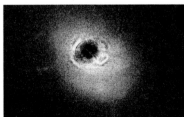
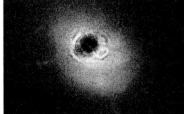

### HOST GALAXY

By blocking the light from 3C 273's nucleus, the Hubble Space Telescope was able to photograph detail (above) in the fainter surrounding galaxy, including traces of a spiral structure and a dust lane.

the lines could have been formed by hydrogen, oxygen, and magnesium if the light was heavily red-shifted and its source was racing away from us at 16 percent of the speed of light, or 107 million mph (173 million kph). We now know that the object is not a star, but a distant active galaxy.

### RADIO JET

An enormous jet of particles 100,000 light-years long streams out from the center of 3C 273. As the particles move away from the core (the white square), their energy diminishes, as shown in this image by the transition from blue (indicating X-rays) to red (infrared radiation).

## QUASAR

# 3C 48

| CATALOG NUMBERS | |
|---|---|
| 3C 48, PKS 0134+029 | |
| **SHAPE** SB interacting | |
| **DISTANCE** | |
| 2.8 billion light-years | |
| **DIAMETER** | |
| 100,000 light-years | |
| **TRIANGULUM** | **MAGNITUDE** 16.2 |

The radio source 3C 48 has a unique place in the history of the study of active galaxies. It was detected in the 1950s, and in 1960, Allan Sandage (see panel, below) confirmed that it coincided with a faint, blue, starlike object. The object's spectrum revealed strange emission lines (see p.35) that

### FIRST QUASAR

At first, 3C 48 is indistinguishable from foreground stars. It was only its unpredictable variability and radio emission that marked it out as something special.

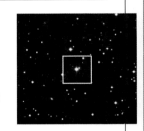

could not have been emitted by any known element. Studies of similar lines in the optical counterpart of 3C 273 (left) suggested that the lines of 3C 48 were hydrogen lines with a huge red shift, suggesting the object was extremely distant and receding at great speed. 3C 48 was therefore the first quasi-stellar object, or quasar, to be discovered.

## ALLAN SANDAGE

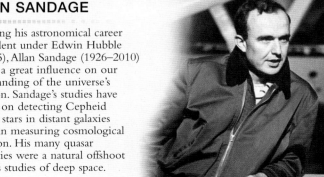

Beginning his astronomical career as a student under Edwin Hubble (see p.45), Allan Sandage (1926–2010) has had a great influence on our understanding of the universe's evolution. Sandage's studies have focused on detecting Cepheid variable stars in distant galaxies for use in measuring cosmological expansion. His many quasar discoveries were a natural offshoot from his studies of deep space.

# GALAXY CLUSTERS

GALAXIES ARE NATURALLY GATHERED.
Pulled together by their enormous gravity,
they cluster tightly, sometimes orbiting one
another, often colliding. As galaxies slowly
move within a cluster, the cluster's
structure changes. The evolution of
clusters can tell astronomers about
dark matter, and clusters can even be used as cosmic
"lenses" to peer back into the early universe.

## TYPES OF CLUSTERS

Some galaxy clusters are sparse, loose collections of galaxies.
The smallest clusters are usually termed "groups." The Local
Group (see p.328), of which the Milky Way is a member, is
one such cluster. Other clusters, such as the nearby Virgo
Cluster (see p.329), are denser, containing many hundreds of
galaxies in a chaotic distribution. Yet other clusters, such as
the Coma Cluster (see p.332), are even more dense, with
galaxies settled into a neat, spherical pattern around a
center dominated by giant elliptical galaxies. Although
clusters differ in density, the volume of space they occupy
is generally the same—a few million light-years across.
Not all galaxies exist in clusters—there are more isolated
"field galaxies" than there are cluster galaxies. Some
galaxy types do not exist outside clusters, however. Giant
ellipticals (see p.304) always lie near the center of large
clusters, as do vast, diffuse cD galaxies (below right). The
most numerous cluster components may be invisible,
including faint, diffuse dwarf elliptical galaxies and proposed
"dark galaxies." A dark galaxy would consist of hydrogen
gas and material too thin to condense and ignite stars. The
first such galaxy
may have been
found in the Virgo
Cluster in early 2005.

**DENSE CLUSTER**
This Hubble image shows a mix of yellow
elliptical and blue-white spiral galaxies in
cluster RXC J0031.1+1808, some 4 billion
light-years from Earth. Fainter arcs behind
this foreground cluster are created by
gravitational lensing of light from
far more distant galaxies.

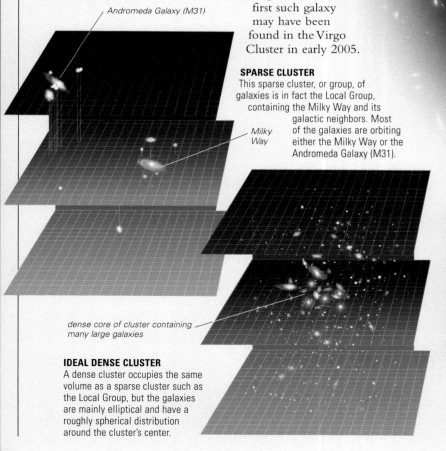

*Andromeda Galaxy (M31)*

*Milky Way*

**SPARSE CLUSTER**
This sparse cluster, or group, of
galaxies is in fact the Local Group,
containing the Milky Way and its
galactic neighbors. Most
of the galaxies are orbiting
either the Milky Way or the
Andromeda Galaxy (M31).

*dense core of cluster containing
many large galaxies*

**IDEAL DENSE CLUSTER**
A dense cluster occupies the same
volume as a sparse cluster such as
the Local Group, but the galaxies
are mainly elliptical and have a
roughly spherical distribution
around the cluster's center.

**DWARF ELLIPTICAL**
Most galaxies in the Local
Group, including the Sculptor
Dwarf, are dwarf ellipticals.
They are invisible in distant
clusters but must be present.

**cD GALAXY**
cD galaxies are similar
to giant ellipticals but
have extensive, sparse
outer haloes of stars.
They sometimes have
hints of multiple cores,
suggesting the merger
of several smaller
ellipticals. NGC 4889
(left) is a cD galaxy at
the heart of the dense
Coma Cluster.

**ABELL 2029**
This visible-light image of Abell 2029 shows that it is an old, regular, spherical cluster full of elliptical galaxies.

# THE INTERGALACTIC MEDIUM

Astronomers can estimate the overall mass of a galaxy cluster from the way in which its galaxies are moving, but also through the phenomenon of gravitational lensing—an effect of general relativity (see pp.42–43). When a compact cluster lies in front of more distant galaxies, its mass bends the light passing close to it and deflects distorted images of the distant galaxies toward Earth. By measuring the strength of this effect, it is possible to measure the mass of the cluster and model how it is distributed. Galaxy clusters contain far more mass than the visible galaxies can account for, and most of it is in the matter that permeates the space between galaxies. This intergalactic medium is distributed around the cluster's center rather than around the galaxies.

**INTERGALACTIC GAS**
An X-ray image of cluster Abell 2029 shows the hot gas cloud around its center. If not for the gravity of the cluster's dark matter, this gas would escape.

X-ray satellites such as Chandra have revealed the nature of part of this material—large galaxy clusters often contain huge clouds of sparse, hot gas glowing at X-ray wavelengths. Most is hydrogen, but heavier elements are present. It is thought to originate in the cluster galaxies and to be stripped away during encounters and collisions. Most of a cluster's mass is not gas, however, but dark matter.

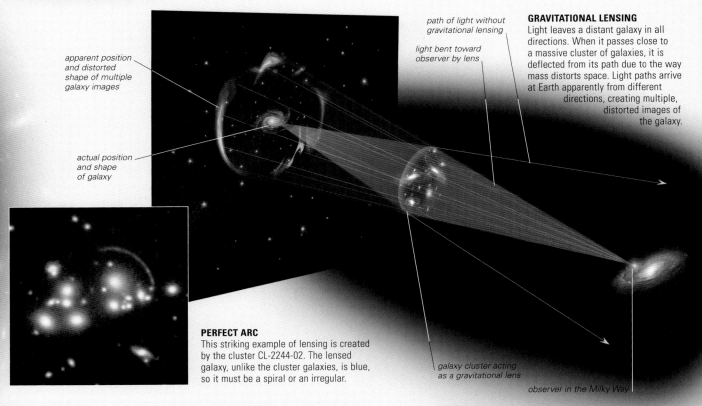

apparent position and distorted shape of multiple galaxy images

actual position and shape of galaxy

path of light without gravitational lensing

light bent toward observer by lens

**GRAVITATIONAL LENSING**
Light leaves a distant galaxy in all directions. When it passes close to a massive cluster of galaxies, it is deflected from its path due to the way mass distorts space. Light paths arrive at Earth apparently from different directions, creating multiple, distorted images of the galaxy.

**PERFECT ARC**
This striking example of lensing is created by the cluster CL-2244-02. The lensed galaxy, unlike the cluster galaxies, is blue, so it must be a spiral or an irregular.

galaxy cluster acting as a gravitational lens

observer in the Milky Way

# CLUSTER EVOLUTION

Astronomers have built a picture of cluster development that complements their models of galaxy evolution (see pp.306–309). According to their thinking, galaxy clusters start as loose collections of gas-rich spirals, irregulars, and small ellipticals. Because of their proximity and huge gravity, the spirals tend to merge, regenerating as spirals or forming ellipticals. Each interaction drives off more of the galaxies' free gas into the intergalactic medium. The high temperature and speed of atoms in this medium prevents their recapture by the cluster's galaxies. At this stage, the cluster is irregular, or "unrelaxed," and the pattern of galaxies and intergalactic gas is irregular and chaotic. However, as galaxies swing around each other, their random motions are eliminated, and they settle into a stable, spherical, "relaxed" distribution around the cluster's center. Eventually, even the largest elliptical galaxies begin to merge, forming giant ellipticals and cD galaxies. The hot gas, freed from ties to individual galaxies, sinks into the center of the cluster, where it lies evenly around the cluster's major elliptical galaxies. What remains is an old, spherical, relaxed cluster full of ellipticals.

**IRREGULAR AND RELAXED CLUSTERS**
The central regions of the Virgo Cluster (above) and the Coma Cluster (below) show the difference between an irregular and a more spherical (relaxed) pattern of galaxies.

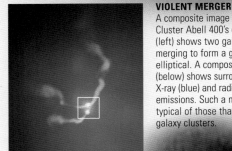

**VIOLENT MERGER**
A composite image of Cluster Abell 400's core (left) shows two galaxies merging to form a giant elliptical. A composite image (below) shows surrounding X-ray (blue) and radio (red) emissions. Such a merger is typical of those that shape galaxy clusters.

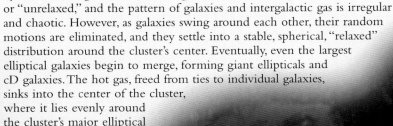

# GALAXY CLUSTERS

The shape and size of galaxy clusters are thought to be linked to their evolution. Clusters range from small groups comprising young, gas-rich irregular and spiral galaxies to highly evolved clusters dominated by giant ellipticals, with a central cloud of gas so hot that it emits X-rays. Astronomers can study details in nearby clusters that are too faint to see in

**DIVERSE CLUSTER**
Galaxy cluster Abell S0740 lies more than 450 million light-years away in Centaurus. The center of the cluster is dominated by a giant elliptical galaxy, while spirals and smaller ellipticals orbit farther out.

distant clusters. Earth's neighboring clusters do not offer a spectacle to stargazers, however, because clusters are so vast that their members are widely scattered across the sky. To appreciate clusters in a single picture, it is necessary to peer tens of millions of light-years into deep space.

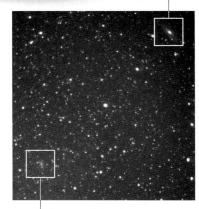

Andromeda Galaxy, M31

Triangulum Galaxy, M33

**LOCAL GROUP MEMBERS**
Because Earth is in the midst of the Local Group, the galaxies are scattered around the sky. However, two large members, M33 and M31, are near enough in the sky to appear in the same frame.

**IRREGULAR CLUSTER**

## Local Group

**DISTANCE**
0–5 million light-years

**NUMBER OF GALAXIES** 46

**BRIGHTEST MEMBERS**
Milky Way;
M31 (magnitude 3.5)

**ANDROMEDA AND TRIANGULUM**

The Local Group is the small galaxy cluster of which the Milky Way is a member. From Earth, its members appear dispersed throughout the sky, but some of its galaxies are grouped in the constellations of Andromeda and Triangulum. In space, the core of the group comprises about 30 members in a region just over 3 million light-years across. It is dominated by the Andromeda Galaxy (M31; see pp.312–313) and the Milky Way. Most of the smaller galaxies orbit close to one or

the other of these large spirals. The third large spiral in the group, M33 (see p.311), may also be trapped in a long orbit around M31.

Outnumbering these spirals is a host of dwarf elliptical and irregular galaxies. Examples include SagDEG and the two Magellanic Clouds (see pp.310–311), as well as M110 and M32, both ellipticals orbiting the M31

**BARNARD'S GALAXY**
This small, irregular galaxy (right), cataloged as NGC 6822, lies 1.7 million light-years away within the Local Group. It is rich in gas and dust, with many pinkish star-birth regions.

spiral. The Local Group appears to be relatively young. Its major galaxies are all spirals, and there is little matter in the space between galaxies—most of the cluster's gas is still trapped in the spirals. It is in an early state of cluster evolution. The Milky Way is currently colliding with the Magellanic Clouds and is heading inexorably toward an ultimate merger with M31.

**FORNAX DWARF GALAXY**
This dwarf spheroidal galaxy (left) has no obvious nucleus. Such faint and diffuse galaxies are easily missed in more distant galaxy clusters, but they are probably the most numerous.

**THE MILKY WAY GALAXY**
A major member of the Local Group is the Milky Way Galaxy. Earth is within the galaxy's disk, so our view is edge-on and stretched across the sky.

## IRREGULAR CLUSTER

# Sculptor Group

**SCULPTOR**

| | |
|---|---|
| **ALTERNATIVE NAME** | South Polar Group |
| **DISTANCE** | 9 million light-years to center |
| **NUMBER OF GALAXIES** | 19 (6 major) |
| **BRIGHTEST MEMBER** | NGC 253 (8.2) |

Lying just beyond the gravitational boundaries of the Local Group, the Sculptor Group is similar in size to the Local Group. It is also a young cluster of irregular and spiral galaxies, with no major ellipticals. It is possible that this group, the Local Group, and another group called Maffei 1 were once part of the same larger cluster.

The closest member to Earth is NGC 55, an irregular galaxy that, like the Large Magellanic Cloud (see p.310), shows enough structure for some astronomers to consider it a single-armed spiral. The dominant galaxy, however, is NGC 253. This large spiral is the same size as the Milky Way and more than twice the size of any other galaxy in the group.

**GALAXY NGC 253**
This large spiral dominates the Sculptor Group in this wide-field image. Most of the other galaxies are too faint to be seen without powerful telescopes.

**STARBURST GALAXY**
NGC 253 is a spiral starburst galaxy—a galaxy undergoing a surge of star formation. The surge may have been triggered by a series of supernovae.

## IRREGULAR CLUSTER

# Virgo Cluster

**VIRGO**

| | |
|---|---|
| **ALTERNATIVE NAME** | Virgo I Cluster |
| **DISTANCE** | 52 million light-years to center |
| **NUMBER OF GALAXIES** | 2,000 (160 major) |
| **BRIGHTEST MEMBER** | M49 (9.3) |

The Virgo Cluster is the nearest galaxy cluster worthy of the name; it is a dense collection of galaxies at the heart of the larger supercluster to which the Local Group also belongs. The contrast with smaller galaxy "groups" is striking—the Virgo Cluster contains around 160 major spiral and elliptical galaxies crammed into a volume little larger than that of the Local Group, along with more than 2,000 smaller galaxies. At its heart lie the giant ellipticals M87 (see p.323), M84, and M86, which are thought to have formed from the collisions of spirals over billions of years. Each giant elliptical seems to be at the center of its own subgroup of galaxies—the cluster has not yet settled to become uniform. The cluster's gravity influences a huge region, extending as far as the Local Group and beyond—the Milky Way and its neighbors are falling toward the Virgo Cluster at 900,000 mph (1.4 million kph).

**CENTER OF THE CLUSTER**
The Virgo Cluster's core has a high density of large galaxies. The two bright galaxies on the right are the ellipticals M84 and M86.

## EXPLORING SPACE

# X-RAY IMAGING AND CLUSTER GAS

Many galaxy clusters are strong sources of X-rays, and orbiting X-ray telescopes can reveal features that remain hidden in visible-light images. While some X-ray sources are located at the centers of the cluster galaxies, the majority of radiation often comes from diffuse gas clouds, independent of the individual galaxies. The process that strips gas out of the cluster galaxies (see p.327) also heats it to generate the X-rays. The distribution of gas offers clues to a cluster's age and history.

**FORNAX IN X-RAYS**
This image of the Fornax cluster shows X-ray-emitting gas in blue. Both central galaxies have trailing plumes of gas, suggesting that the entire cluster is moving through sparser clouds.

## REGULAR CLUSTER

# Fornax Cluster

**FORNAX**

| | |
|---|---|
| **CATALOG NUMBER** | Abell S 373 |
| **DISTANCE** | 65 million light-years to center |
| **NUMBER OF GALAXIES** | 54 major galaxies |
| **BRIGHTEST MEMBER** | NGC 1316 (9.8) |

Fornax is home to a relatively nearby galaxy cluster, centered at around the same distance as the Virgo Cluster. However, the Fornax Cluster is at a later stage of evolution than the younger Virgo group. Here, spiral galaxies are rare—the cluster's major galaxies are mostly ellipticals, distributed evenly around the giant elliptical NGC 1399. Dwarf galaxies lying between the major ones are also mostly small ellipticals, suggesting that the cluster formed long ago and that interactions between its galaxies have had time to strip away most of their star-forming gas (see p.327). This account of the cluster's evolution has recently been confirmed by the orbiting Chandra X-ray Observatory (see panel, left).

**GALAXY NGC 1365**
One of the Fornax Cluster's few spirals, NGC 1365 has a dust bar through its core.

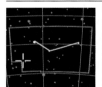

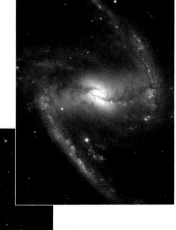

**CLUSTER CORE**
In the Fornax Cluster's central region lie NGC 1399 (center left) and NGC 1365 (bottom right). As a rule, elliptical galaxies predominate.

## COMPACT GROUP

# Seyfert's Sextet

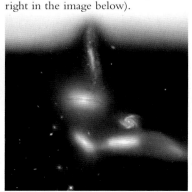

**SERPENS**

| | |
|---|---|
| **CATALOG NUMBERS** | NGC 6027 and NGC 6027A–C |
| **DISTANCE** | 190 million light-years |
| **NUMBER OF GALAXIES** | 4 |
| **BRIGHTEST MEMBER** | NGC 6027 (14.7) |

Seyfert's Sextet actually contains just four members—each a misshapen spiral galaxy locked to the others in a gravitational waltz within a region of space no larger than the Milky Way. The sextet, as seen from Earth, is completed by a small face-on spiral that happens to lie in the background and by a distorted star cloud (at lower right in the image below).

**QUARTET PLUS TWO**

**THE VIRGO CLUSTER**
Over 2,000 galaxies reside in the Virgo Cluster (see p.329), the nearest large cluster to us, some 50 million light-years away. The brightest of them are visible through amateur telescopes. Just below center is the elliptical galaxy M87 (see p.320), also known as the radio source Virgo A. M87 has an estimated mass of 2.4 trillion Suns, making it the largest galaxy in our region of the universe.

## REGULAR CLUSTER

# Hydra Cluster

**CATALOG NUMBER** Abell 1060

**DISTANCE** 160 million light-years

**NUMBER OF GALAXIES** 1,000+

**BRIGHTEST MEMBER** NGC 3311 (11.6)

**HYDRA**

The Hydra Cluster is similar in size to the huge Virgo Cluster (see p.329). It is the closest example of a "relaxed" cluster (see p.327) of mainly elliptical galaxies in a spherical distribution. Its hot X-ray gas also forms a spherical cloud around the core. The cluster is centered on two giant elliptical galaxies and an edge-on spiral, each 150,000 light-years across. These galaxies are interacting—the ellipticals' gravity has warped the spiral, while both ellipticals have distorted outer haloes. The cluster is the major member of the Hydra Supercluster, which adjoins the Local Supercluster (see pp.336–339).

### HEART OF THE HYDRA CLUSTER
In this image, the central giant ellipticals NGC 3309 and 3311 lie below the large, blue spiral NGC 3312. The two bright objects on either side are foreground stars.

### SPIRAL SILHOUETTE
NGC 3314, an unusual case of one spiral galaxy silhouetted against another, is one of Hydra's most beautiful objects.

## COMPACT GROUP

# Stephan's Quintet

**CATALOG NUMBER** Hickson 92

**DISTANCE** 340 million light-years (NGC 7320: 41 million light-years)

**NUMBER OF GALAXIES** 4/5

**BRIGHTEST MEMBER** NGC 7320 (13.6)

**PEGASUS**

First observed by French astronomer E. M. Stephan at the University of Marseilles in 1877, Stephan's Quintet appears to be a remarkably compact cluster of five galaxies. The galaxies are a mixture of spirals, barred spirals, and ellipticals and show clear signs of disruption from interactions. The largest galaxy as seen from Earth, NGC 7320, is probably a foreground object lying in front of a quartet of interacting galaxies. The spectral red shift (see p.35) of NGC 7320 is much smaller than those of the other four galaxies and instead matches that of several other galaxies close to it in the sky. Because it also appears physically different from the quartet, it seems likely that NGC 7320 is much closer and the unusual red shift is a normal result of the expansion of space (see p.44). However, a few astronomers claim that trails of material link NGC 7320 to other Quintet galaxies. If this is the case, then the red shift suggests that the galaxy is moving very fast relative to its neighbors and toward Earth, therefore reducing its overall speed of recession and its red shift. Or perhaps the red shift does not originate from its motion at all. These competing theories have turned Stephan's Quintet into a battleground for the small minority of astronomers who think that red shifts are not all caused by the expansion of space and that Hubble's Law (see p.44) does not always apply.

### FOUR OR FIVE?
The quintet consists of a quartet of yellow galaxies beside the white spiral NGC 7320. The contrasting appearance of NGC 7320 suggests it lies in front of the other galaxies.

### QUINTET CLOSE-UP
This detailed Hubble Space Telescope view of Stephan's Quintet shows chains of stars linking several of its interacting galaxies.

## REGULAR CLUSTER

# Coma Cluster

**CATALOG NUMBER** Abell 1656

**DISTANCE** 300 million light-years

**NUMBER OF GALAXIES** 3,000+

**BRIGHTEST MEMBER** NGC 4889 (13.2)

**COMA BERENICES**

Although it lies near the Virgo Cluster in the sky (see p.329), the Coma Cluster is much farther away. First recognized by William Herschel as a concentration of "fine nebulae" in 1785, this is one of the nearest highly evolved or "relaxed" galaxy clusters (see p.327). It is very dense, with over 3,000 galaxies, and is dominated by elliptical and lenticular galaxies. Because it is near the north galactic pole (and therefore free of the dense star fields of the Milky Way), it is well studied. Swiss-American astronomer Fritz Zwicky used Coma when he made the first measurements of galaxy movements within a cluster in the 1930s. He found the cluster contained many times more mass than its visible galaxies suggested—an idea that was not accepted until the 1970s. Overall, the cluster is moving away at 16 million mph (25 million kph). At the cluster's center lie the giant elliptical NGC 4889 and the lenticular galaxy NGC 4874. Most of the spirals and irregulars are in the outer regions. X-ray images show two distinct patches of cluster gas, suggesting that the cluster is absorbing a smaller cluster of galaxies. Like the Virgo and Hydra Clusters, Coma forms the core of its own galaxy supercluster.

### HIDDEN DWARFS
This combined visible and infrared image reveals dozens of faint green smudges, marking the location of dwarf galaxies that are too faint to see in visible light.

## IRREGULAR CLUSTER

# Hercules Cluster

**HERCULES**

| | |
|---|---|
| CATALOG NUMBER | Abell 2151 |
| DISTANCE | 500 million light-years |
| NUMBER OF GALAXIES | 100+ |
| BRIGHTEST MEMBER | NGC 6041A (14.4) |

The small Hercules Cluster is dominated by spiral and irregular galaxies, suggesting that it is in an early stage of development. In keeping with the best models of such clusters' formation (see p.327), it shows little sign of structure. Within the cluster, several pairs or groups of galaxies seem to be merging or interacting—encounters that will transform them into different kinds of galaxies and reduce their random movements until they become more evenly distributed. The most prominent of these mergers is NGC 6050, a pair of interlocking spiral galaxies near the cluster's center that may eventually form the core of a giant elliptical, such as those found in more evolved clusters.

**HERCULES FIELD**
This wide-field view captures most of the bright galaxies in Hercules and shows their irregular, "unrelaxed" distribution.

## REGULAR CLUSTER

# Abell 1689

**VIRGO**

| | |
|---|---|
| CATALOG NUMBER | Abell 1689 |
| DISTANCE | 2.2 billion light-years |
| NUMBER OF GALAXIES | 3,000+ |
| BRIGHTEST MEMBER | 2MASX J13112952-0120280 (16.5) |

Abell 1689 is one of the densest galaxy clusters known, with thousands of galaxies packed into a volume of space only 2 million light-years across. Its ball shape makes it a fine gravitational lens. By measuring the lensing power throughout the cluster, astronomers have worked out the distribution of the cluster's dark matter.

**DARK MATTER MAPPED IN ABELL 1689**

# GEORGE ABELL

George Abell (1927–1983) was a career astronomer and popularizer of science who carried out the first and most influential survey of galaxy clusters. After working on the Palomar Sky Survey during the 1940s and 1950s, using the Palomar Schmidt telescope, he turned his attention to analyzing the results, developing methods for distinguishing galaxy clusters from isolated field galaxies and classifying clusters into types.

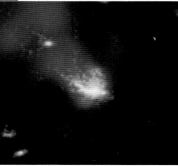

## REGULAR CLUSTER

# Abell 2065

**CORONA BOREALIS**

| | |
|---|---|
| CATALOG NUMBER | Abell 2065 |
| DISTANCE | 1 billion light-years |
| NUMBER OF GALAXIES | 1,000+ |
| BRIGHTEST MEMBER | PGC 54876 (16.0) |

Abell 2065, also known as the Corona Borealis Cluster, contains 400 or more large galaxies. A highly evolved cluster like the Coma Cluster (opposite), it emits X-rays from a diffuse cloud of hot gas. However, X-ray observations have found two distinct X-ray cores, suggesting that Abell 2065 may be two already ancient clusters merging together. The cluster lies at the center of the Corona Borealis Supercluster.

**THE CORONA BOREALIS CLUSTER**

## IRREGULAR CLUSTER

# Abell 2125

**URSA MINOR**

| | |
|---|---|
| CATALOG NUMBER | Abell 2125 |
| DISTANCE | 3 billion light-years |
| NUMBER OF GALAXIES | 1,000+ |
| BRIGHTEST MEMBER | Magnitude 17.0 |

Abell 2125 has been the subject of intense scrutiny from the orbiting Chandra X-Ray Observatory. The cluster lies close enough to Earth to see detail but so far away that images reaching Earth show an early and still active phase of its evolution, 3 billion years ago. Abell 2125 is therefore ideal for testing ideas on cluster formation.

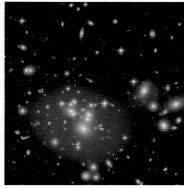

X-ray images reveal what optical ones cannot—that the cluster is forming from the merger of several smaller clusters. The most intense cloud of X-ray-emitting gas shows "clumpiness," which indicates it has recently come together. Spectra reveal that the cloud is enriched with heavy elements such as iron, and close-up images show gas actively being stripped away from galaxies such as C153. With it, the gas carries atoms of heavy metals created in supernova explosions, distributing them through the intergalactic medium. A fainter cloud of almost equal size, enveloping hundreds more galaxies, has remarkably few heavy elements, suggesting that the gas-stripping process becomes more powerful and thorough over time and that the cloud is much younger than its fainter neighbor.

Because X-ray evidence shows so much activity within the cluster, astronomers have also imaged it at other wavelengths. Infrared telescopes, for example, have revealed enormous bursts of star formation going on in galaxies far from the cluster center. One possible explanation is that, even at distances of up to 1 million light-years, the tidal forces from the center of a large cluster are enough to disrupt nearby galaxies and trigger starbursts.

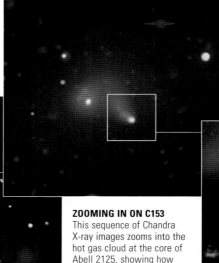

**ZOOMING IN ON C153**
This sequence of Chandra X-ray images zooms into the hot gas cloud at the core of Abell 2125, showing how gas is being stripped from galaxy C153 (right).

## REGULAR CLUSTER

# Abell 2218

**DRACO**

**CATALOG NUMBER**
Abell 2218

**DISTANCE**
2 billion light-years

**NUMBER OF GALAXIES**
250 or more

**BRIGHTEST MEMBER**
Unnamed galaxy (17.0)

Abell 2218 is a spectacular example of a highly evolved and extremely dense galaxy cluster. It contains more than 250 mostly elliptical galaxies in a volume of space roughly 1 million light-years across.

The cluster has taught astronomers much about galaxy clusters and about galaxies themselves. The cluster's density is so great that it affects the shape of the surrounding space, as predicted by Einstein's theory of general relativity (see p.42). Many more distant galaxies lie directly behind the cluster, and as light rays from these objects pass close to Abell 2218, their paths are deflected and focused toward Earth, in the same way that a magnifying lens focuses sunlight. This gravitational lensing (see p.327) brightens the images of galaxies that would otherwise be too far away to detect. It results in a series of distorted images of distant galaxies ringing the center of Abell 2218.

The galaxies beyond Abell 2218 lie much farther away, and therefore their images come from a much earlier time. Most of the lensed galaxies are blue-white, suggesting they are young irregulars and spirals very different from Abell 2218's own aged ellipticals. Some of the lensed galaxies align with X-ray sources, suggesting they are active galaxies. Recent studies yielded images of a galaxy so far beyond Abell 2218 that all its light has been red-

### HOLE IN THE COSMIC BACKGROUND
In this composite image of Abell 2218, yellow and red depict the X-ray-emitting gas around its core. The gas scatters the cosmic microwave background radiation, creating a hole, outlined here by contours.

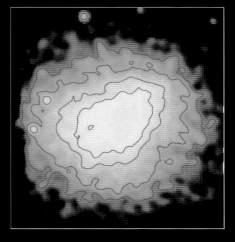

shifted into the infrared part of the spectrum. At the time, it was the most distant galaxy known, at 13 billion light-years from Earth. It must have formed shortly after the first stars, in the aftermath of the Big Bang.

Gravitational lensing can also reveal hidden properties of Abell 2218 itself. Because the strength of lensing depends on the cluster's density, it offers a measure of the distribution of all matter in the cluster—including the dark matter. Abell 2218 is one of the few galaxy clusters in which the pattern of visible matter (galaxies and X-ray-emitting gas) and the calculated distribution of dark matter do not match, suggesting the cluster is not as uniform as it appears in visible light.

Astronomers have now begun to use Abell 2218 to probe the origins of the universe. A phenomenon called the Sunyaev–Zel'dovich effect (see caption, opposite) creates holes and ripples in the cosmic microwave background radiation shining through the cluster. This happens because gas around Abell 2218's core scatters photons of microwave radiation, just as Earth's atmosphere scatters light. The strength of these ripples can be used to estimate the true diameter of the cluster's core, and therefore its distance from Earth, independently of its red shift. The red shift and distance can then be used together to find the expansion rate of the universe (see p.44).

EXPLORING SPACE

## MAPPING THE MISSING MASS

spikes coincide with galaxies

The total mass of a cluster can be up to five times that of its visible galaxies, but the distribution of the other, dark matter was a mystery until recently. Gravitational lensing now allows astronomers to measure the missing mass in clusters. By analyzing images of lensed galaxies, astronomers can pinpoint concentrations of mass distorting the light as it passes through the cluster.

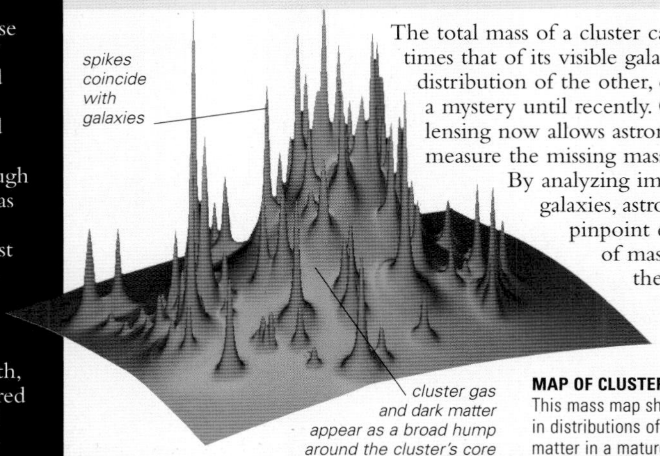

cluster gas and dark matter appear as a broad hump around the cluster's core

MAP OF CLUSTER CL0024+1654
This mass map shows the difference in distributions of visible and dark matter in a mature galaxy cluster.

BEYOND THE MILKY WAY

# GALAXY SUPERCLUSTERS

THE LARGEST-SCALE STRUCTURES in the universe are galaxy superclusters—collections of neighboring galaxy clusters that bunch together in chains and sheets stretching across the cosmos. These structures are echoes of those that formed in the Big Bang, and by studying the universe at these enormous scales, astronomers can learn about the way it formed and our place within it.

## GALAXY SUPERCLUSTERS

Just as galaxies are bound together by gravity into clusters, galaxy clusters themselves blur together at their edges to form even larger structures called superclusters. While individual clusters are typically about 10 million light-years in diameter (see p.326), superclusters are typically up to 200 million light-years across and merge with others at their edges. Where superclusters overlap, it is the gravitational behavior of individual clusters that determines to which supercluster they belong. The enormous size of superclusters and the great mass of galaxies in them allows them to modify the cosmological expansion of space (see pp.44–45), resulting in large-scale variations in the movement of galaxies. The best known example of this is a generalized flow of galaxies in our part of the universe, possibly toward a region known as the Great Attractor but more likely toward a more massive supercluster lying directly behind it.

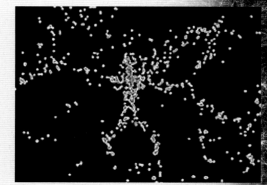

**PLOT OF GALAXIES**
This plot of a section of sky out to a distance of 1 billion light-years shows how galaxies cluster on the largest scale.

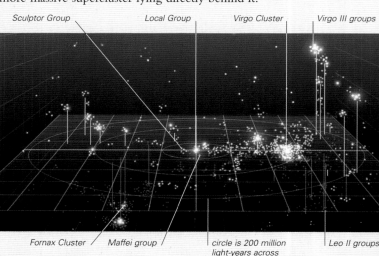

Sculptor Group    Local Group    Virgo Cluster    Virgo III groups

Fornax Cluster    Maffei group    circle is 200 million light-years across    Leo II groups

**NEARBY CLUSTERS**
This map shows the distribution of galaxy clusters within 100 million light-years of the Milky Way. Each point denotes a major galaxy—thousands of smaller ones are not included in the map.

**THE GREAT ATTRACTOR**
The sky in the direction of the Great Attractor is dominated by galaxies in the Norma Cluster, about 220 million light-years away. In 2016, astronomers found evidence for a Vela Supercluster of galaxies lying beyond it.

**MAPPING LANIAKEA**
In this artist's impression of the Laniakea Supercluster, each dot represents one of 100,000 galaxies spread across 520 million light-years of space, with bright areas indicating dense concentrations, such as the local supercluster region. The Milky Way's location is marked by a red dot near the top. Lines indicate the direction in which galaxies are flowing through space, while the yellow region indicates the extent of Laniakea's gravitational dominance.

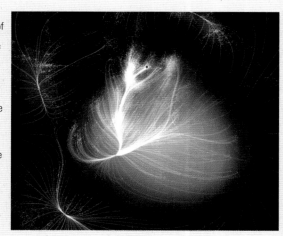

## THE LANIAKEA SUPERCLUSTER

A chain of galaxy clusters links our own small Local Group of galaxies to the Virgo Cluster, some 52 million light-years from Earth. Until recently, this much larger cluster, containing up to 2,000 galaxies, was thought to mark the gravitational heart of a distinct "Virgo Supercluster," encompassing the Local Group and at least 100 other bright galaxy clusters across 110 million light-years of space. However, the Virgo Supercluster always seemed relatively small and underweight compared to other known superclusters, and a 2014 reassessment of galaxy motions in the local universe saw it subsumed into a much larger supercluster called Laniakea. The Laniakea Supercluster encompasses the former Virgo, Hydra-Centaurus, Pavo-Indus, and Southern superclusters in a volume of space 520 million light-years across and containing 100,000 bright galaxies, with the Great Attractor at its center. The gravitational attraction binding Laniakea's distant extremes is too weak to hold it together against cosmic expansion driven by dark energy, so unlike its component parts, it will ultimately disintegrate.

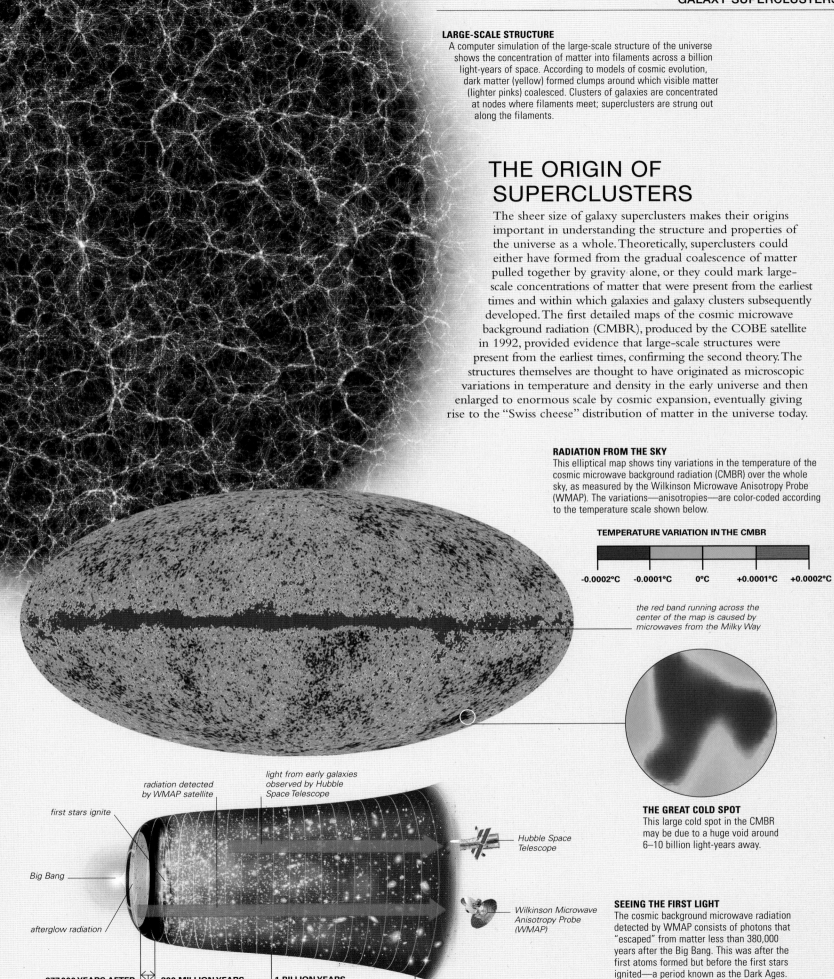

**LARGE-SCALE STRUCTURE**
A computer simulation of the large-scale structure of the universe shows the concentration of matter into filaments across a billion light-years of space. According to models of cosmic evolution, dark matter (yellow) formed clumps around which visible matter (lighter pinks) coalesced. Clusters of galaxies are concentrated at nodes where filaments meet; superclusters are strung out along the filaments.

# THE ORIGIN OF SUPERCLUSTERS

The sheer size of galaxy superclusters makes their origins important in understanding the structure and properties of the universe as a whole. Theoretically, superclusters could either have formed from the gradual coalescence of matter pulled together by gravity alone, or they could mark large-scale concentrations of matter that were present from the earliest times and within which galaxies and galaxy clusters subsequently developed. The first detailed maps of the cosmic microwave background radiation (CMBR), produced by the COBE satellite in 1992, provided evidence that large-scale structures were present from the earliest times, confirming the second theory. The structures themselves are thought to have originated as microscopic variations in temperature and density in the early universe and then enlarged to enormous scale by cosmic expansion, eventually giving rise to the "Swiss cheese" distribution of matter in the universe today.

**RADIATION FROM THE SKY**
This elliptical map shows tiny variations in the temperature of the cosmic microwave background radiation (CMBR) over the whole sky, as measured by the Wilkinson Microwave Anisotropy Probe (WMAP). The variations—anisotropies—are color-coded according to the temperature scale shown below.

**TEMPERATURE VARIATION IN THE CMBR**

| -0.0002°C | -0.0001°C | 0°C | +0.0001°C | +0.0002°C |

*the red band running across the center of the map is caused by microwaves from the Milky Way*

**THE GREAT COLD SPOT**
This large cold spot in the CMBR may be due to a huge void around 6–10 billion light-years away.

*radiation detected by WMAP satellite*

*light from early galaxies observed by Hubble Space Telescope*

*first stars ignite*

*Big Bang*

*afterglow radiation*

Hubble Space Telescope

Wilkinson Microwave Anisotropy Probe (WMAP)

**SEEING THE FIRST LIGHT**
The cosmic background microwave radiation detected by WMAP consists of photons that "escaped" from matter less than 380,000 years after the Big Bang. This was after the first atoms formed but before the first stars ignited—a period known as the Dark Ages. In contrast, Hubble can observe back only to about 400–800 million years after the Big Bang, when early galaxies had formed.

**377,000 YEARS AFTER BIG BANG**

**300 MILLION YEARS AFTER BIG BANG**

**1 BILLION YEARS AFTER BIG BANG**

**DARK AGES**

**PRESENT: 13.8 BILLION YEARS AFTER BIG BANG**

BEYOND THE MILKY WAY

# FILAMENTS AND VOIDS

At the largest scales measured, the universe reveals a clear overall structure. Galaxy superclusters join to form stringlike "filaments" or flat "sheets" around the edges of enormous and apparently empty regions known as "voids." While galaxy structure up to the level of clusters can be explained by the action of gravity since the Big Bang, the present age of the universe (13.7 billion years) is not nearly long enough for gravity alone to have organized the universe on the scale of filaments and voids. This indicates that large-scale cosmic structures are, in fact, expanded "echoes" of features from the earliest times. The first filaments to be discovered were galactic "walls" identified in the 1980s, and since then it has become clear that filaments contain not only luminous galaxies but also enormous clouds of hydrogen known as Lyman Alpha blobs. The first void, meanwhile, was discovered during a galaxy survey in 1978. Typically, voids are empty of both normal and dark matter, although some voids have been found to contain a few galaxies.

**5 BILLION LIGHT-YEARS ACROSS**  **150 MILLION LIGHT-YEARS ACROSS**  **4 MILLION LIGHT-YEARS ACROSS**

### THE COSMOLOGICAL PRINCIPLE
This principle is the assumption that, at the largest scales, the universe is essentially uniform in all its properties and in all directions, even though it is clearly not so at smaller scales. The principle seems to be borne out in practice—for example, when comparing the distribution of galaxies, as shown above.

# BETWEEN THE SUPERCLUSTERS

Studying luminous galaxies alone can give a deceptive view of the universe—not all normal matter produces detectable radiation, and dark matter neither produces radiation nor interacts with it. However, by analyzing light from distant quasars (see p.320), astronomers can measure the effects of intervening, but otherwise invisible, hydrogen clouds. As light from the quasars passes through such clouds, the hydrogen "imprints" it with absorption lines that form a pattern called a Lyman Alpha forest. The wavelengths of these absorption lines reveal the red shifts of the clouds and therefore their distance from Earth, allowing their distribution to be mapped. In addition, analyzing localized movements among galaxy clusters allows astronomers to map the distribution of dark matter. Both of these methods seem to confirm that the voids between superclusters are empty and that most normal and dark matter is concentrated around the visible galaxy filaments.

### GALAXY MAP FROM THE SLOAN SURVEY
Begun in 2000, the Sloan Digital Sky Survey is a major red-shift survey that has so far mapped more than a million objects, out to 2 billion light-years from Earth. On this map, each dot represents a galaxy, and the galaxies are plotted at distances from the center proportional to their distance from Earth.

*each dot is a separate galaxy, color-coded according to the average age of its stars: red dots contain older stars; blue and green are younger stars*

*unmapped sections are areas blocked from the telescope's view by the Milky Way*

*the edge of the survey map is about 2 billion light-years away from the Milky Way*

*dark regions in the mapped area are huge voids in space*

*filamentary structures are strings of galaxy clusters that are only partially mapped*

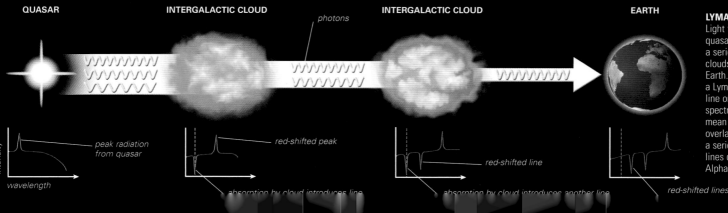

QUASAR    INTERGALACTIC CLOUD    *photons*    INTERGALACTIC CLOUD    EARTH

*peak radiation from quasar*    *red-shifted peak*    *red-shifted line*    *red-shifted lines*

*intensity*

*wavelength*

*absorption by cloud introduces line*    *absorption by cloud introduces another line*

### LYMAN-ALPHA FOREST
Light from a distant quasar passes through a series of hydrogen clouds on its way to Earth. Each superimposes a Lyman Alpha absorption line onto the quasar's spectrum, but red shifts mean the lines do not overlap. The result is a series of red-shifted lines called a Lyman-Alpha forest.

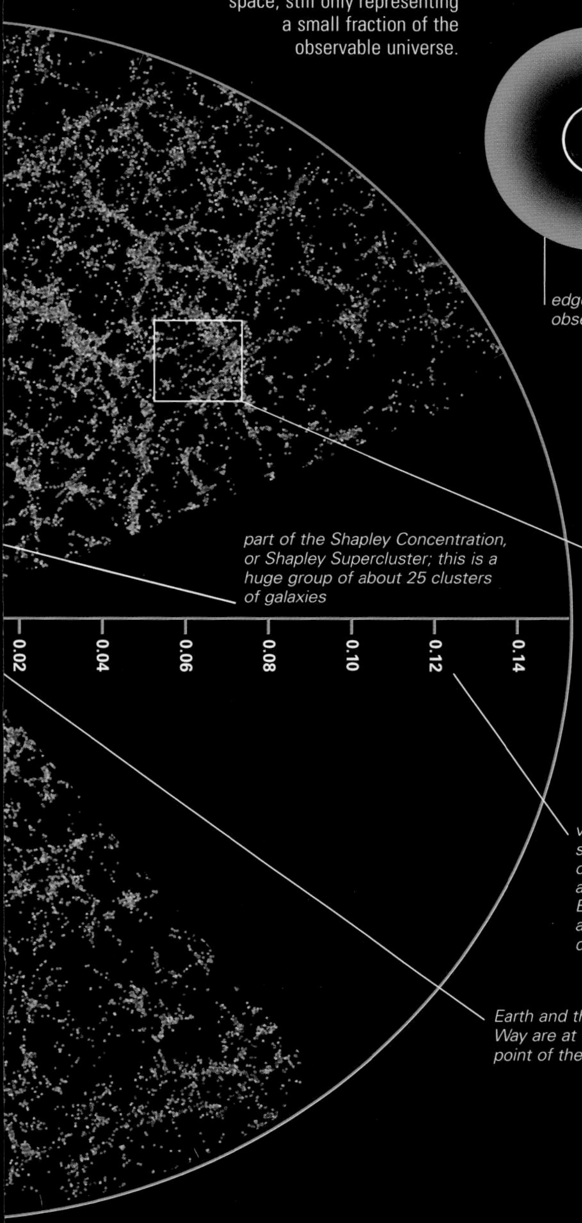

## REGION DETAILED BY GALAXY MAP
The galaxy map below covers two thin, wedge-shaped regions of space, still only representing a small fraction of the observable universe.

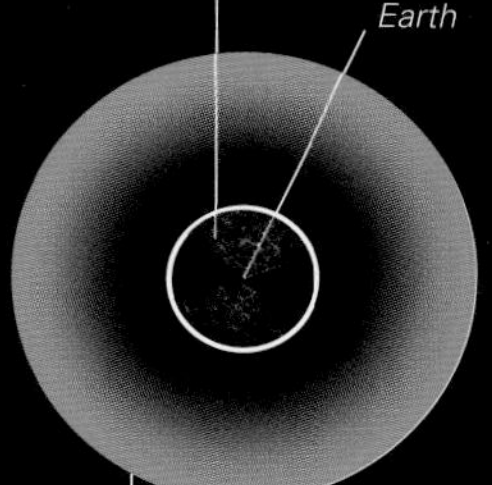

area depicted in survey galaxy map

Earth

edge of observable universe

part of the Shapley Concentration, or Shapley Supercluster; this is a huge group of about 25 clusters of galaxies

The Sloan Great Wall, a giant filament, is the largest known structure in the universe, at 1 billion light-years across

0.02   0.04   0.06   0.08   0.10   0.12   0.14

values on the red-shift scale are a measure of how fast galaxies are receding from Earth; they are also an indication of distance from Earth

Earth and the Milky Way are at the central point of the map

# MAPPING DEEP SPACE

While galaxy motions on a local scale are affected by gravitational influences such as the presence of superclusters, on the scale of the universe as a whole, these effects should become negligible in comparison to the overall cosmic expansion resulting from the Big Bang (see pp.48–51). According to Hubble's Law, the speed at which a far-off galaxy is moving away from us is, on average, proportional to its distance, and as a result the red shift in a distant galaxy's light can be used as a measure of its distance. The first large-scale survey of galaxy red shifts, carried out by the Harvard-Smithsonian Center for Astrophysics (CfA), started in 1977 and took 5 years to measure 13,000 galaxies. Since then, other surveys, such as the Sloan Digital Sky Survey and Two-degree-Field Galaxy Redshift Survey (2dFGRS), have mapped many more galaxies. These surveys have confirmed that the large-scale pattern of galaxy distribution remains essentially identical out to distances of billions of light-years.

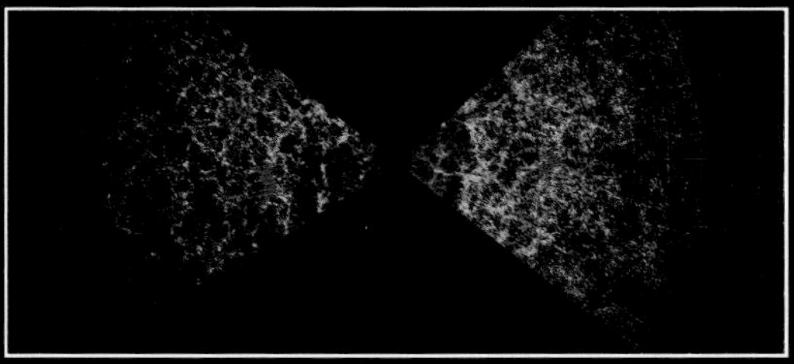

## OBSERVING A MILLION GALAXIES

Major galaxy red-shift surveys typically use multi-object spectrographs—devices that can simultaneously record the spectra of hundreds of objects. Instruments such as the Gemini multi-object spectrographs, mounted on two large telescopes in Hawaii and Chile, use special masks to separate light from the different objects before splitting it through a diffraction grating to obtain the spectra.

**GEMINI OBSERVATORY**
The Gemini observatory has two 26.9-ft (8.1-m) reflectors—one in Chile (shown above), the other in Hawaii—each fitted with a spectrograph for multi-object spectroscopy.

**GALAXY DISTRIBUTION PLOT FROM THE 2DFGRS**
Centered on Earth, this plot shows the positions of over 230,000 galaxies. The dots are galaxies and the colors indicate density, with dense regions redder and less dense ones bluer.

# ACCELERATING EXPANSION

One of the most remarkable recent astronomical discoveries has been the fact that cosmic expansion is accelerating. Studies of type 1a supernovae (see p.283) have revealed that they are unexpectedly faint in the most distant galaxies, which implies that they are farther away than they should be if the rate of expansion of the universe was constant or slowing down. Many cosmologists had expected the expansion of the universe to slow down as the initial impetus from the Big Bang began to fade, so the discovery that its expansion is getting faster implied that an important factor was missing from cosmological theories. Furthermore, the acceleration seems to have begun only around 5 billion years ago, with the universe slowing as predicted until then. Since its discovery in 1998, the accelerating expansion has been corroborated from other measurements, and it is now generally believed to be due to dark energy (see p.58). According to recent measurements, dark energy may be the most abundant form of mass-energy in the universe, accounting for almost 73 percent of the total.

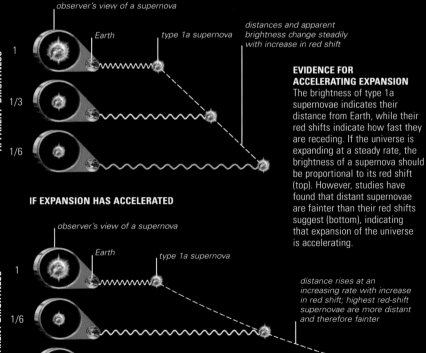

**IF EXPANSION HAS OCCURRED AT A STEADY RATE**

observer's view of a supernova

Earth          type 1a supernova

APPARENT BRIGHTNESS

1

1/3

1/6

distances and apparent brightness change steadily with increase in red shift

**IF EXPANSION HAS ACCELERATED**

observer's view of a supernova

Earth          type 1a supernova

APPARENT BRIGHTNESS

1

1/6

1/25

distance rises at an increasing rate with increase in red shift; highest red-shift supernovae are more distant and therefore fainter

**EVIDENCE FOR ACCELERATING EXPANSION**
The brightness of type 1a supernovae indicates their distance from Earth, while their red shifts indicate how fast they are receding. If the universe is expanding at a steady rate, the brightness of a supernova should be proportional to its red shift (top). However, studies have found that distant supernovae are fainter than their red shifts suggest (bottom), indicating that expansion of the universe is accelerating.

**TWO-MICRON ALL-SKY SURVEY (2MASS)**
This panoramic view of the entire sky at near-infrared wavelengths illustrates the distribution of galaxies beyond the Milky Way. The plane of the Milky Way runs across the center of this projection. Galaxies are color-coded by their red shift, from blue (the nearest), to green (intermediate distances), to red (the farthest). The purple area at top center right is the Virgo Cluster of galaxies.

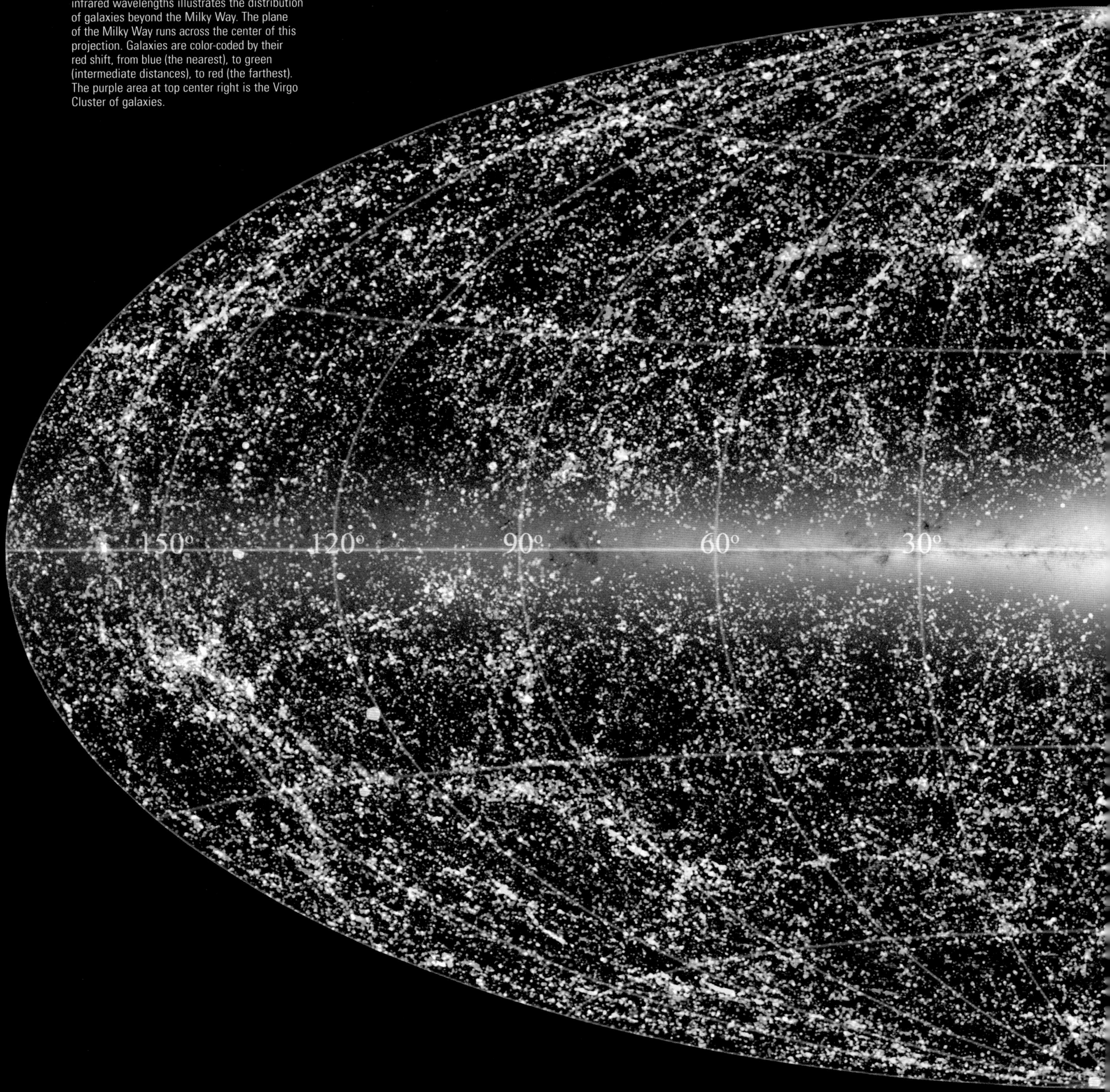

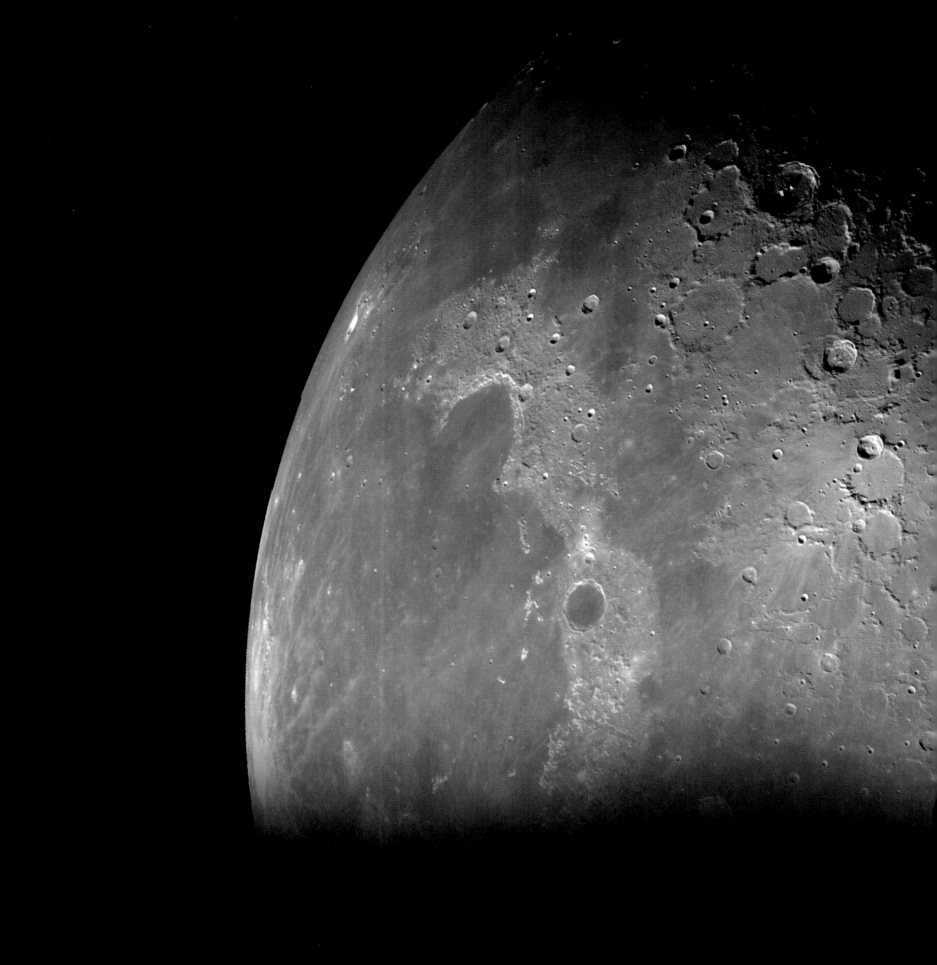

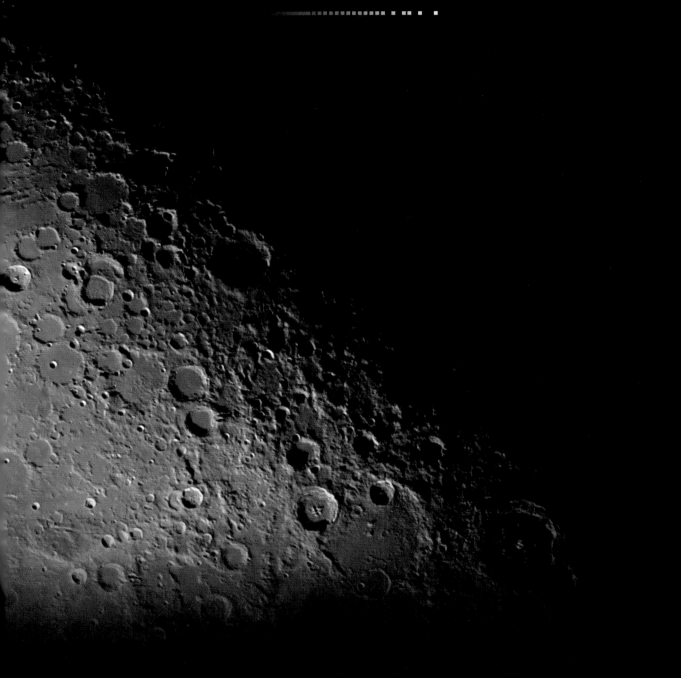

# THE NIGHT SKY

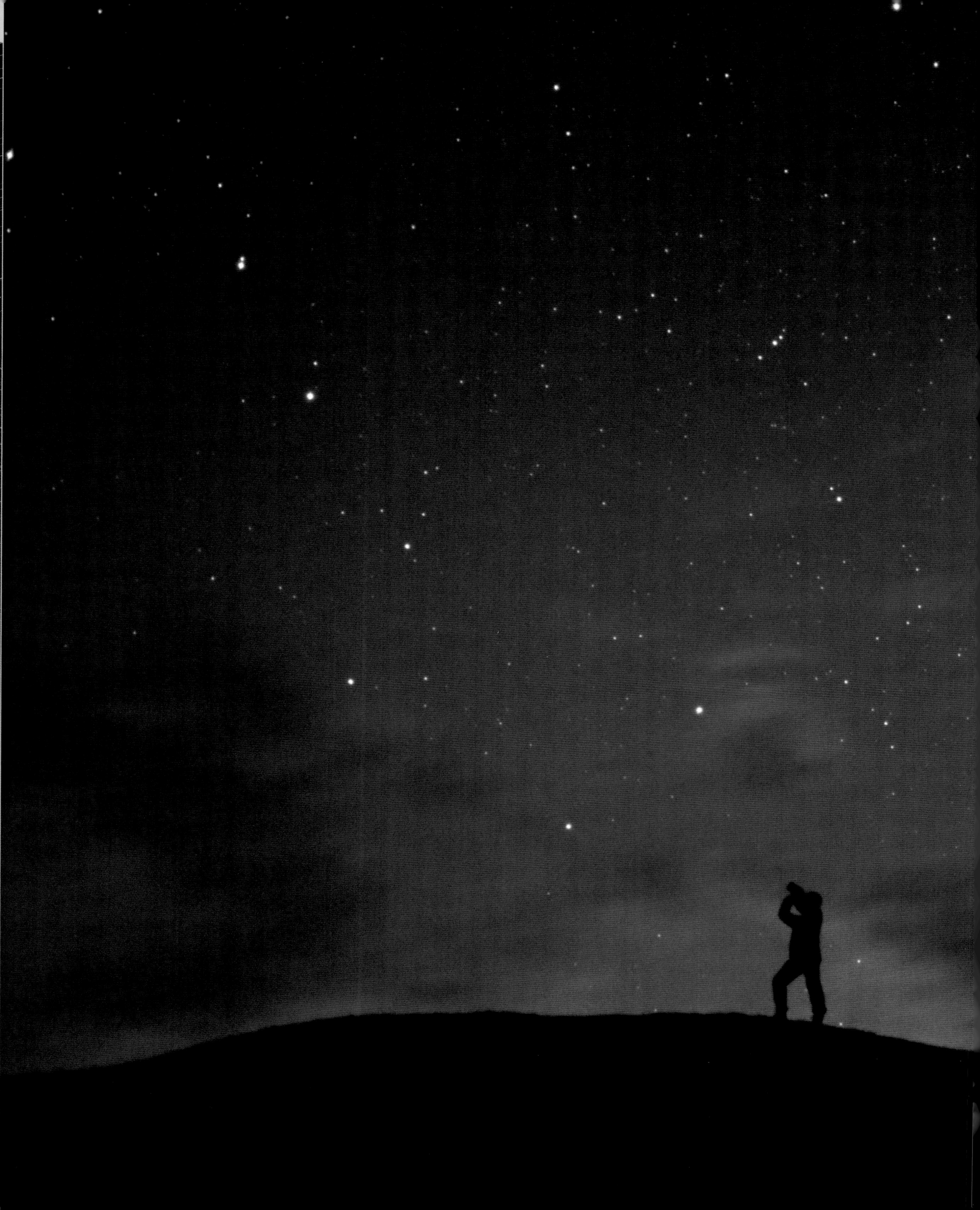

most cases, stars within a constellation lie in the same region of sky merely by chance, however, and are not related. Despite the apparent permanence of the skies, these patterns are not fixed, because all the stars are moving relative to Earth. Over time, the shape of all the constellations will change, and hundreds of thousands of years from now, they will be unrecognizable. Future generations will need to invent constellations of their own. But for now, 88 constellations fill our sky, interlocking like pieces of an immense jigsaw puzzle. Some are large, others small; some richly stocked with objects of note, others faint and seemingly barren. All are featured in the following pages.

**PATTERNS IN THE SKY**
As darkness falls, a stargazer scans the sky with binoculars. The familiar shape of the Big Dipper looms overhead, part of the constellation Ursa Major, the Great Bear. The north Pole Star, Polaris, can be seen high up on the right.

# THE CONSTELLATIONS

# THE HISTORY OF CONSTELLATIONS

THE FIRST CONSTELLATIONS were patterns of stars that ancient peoples employed for navigation, timekeeping, and storytelling. Recently, the pictorial aspect of constellations has become less significant, and they have become simply delineated regions of the sky, although the attraction of the myths and legends remains.

## EARLY CONSTELLATION LORE

The constellation system used today stems from patterns recognized by ancient Greek and Roman civilization. The earliest surviving account of ancient Greek constellations comes from the poet Aratus of Soli (c. 315–c. 245 BCE). His poem, the *Phaenomena*, written around 275 BCE, describes the sky in storybook fashion and identifies 47 constellations. It is based on a lost book of the same name by the Greek astronomer Eudoxus (c. 390–c. 340 BCE). Eudoxus reputedly introduced the constellations to the Greeks after learning them from priests in Egypt. These constellations had been adopted from Babylonian culture; they were originally created by the Sumerians around 2,000 BCE. However, the Greeks attached their own myths to the constellations detailed by Eudoxus, and Aratus's storybook of the stars proved immensely popular. Sometime in the 2nd century CE, it was joined by a more elaborate work of constellation lore called *Poetic Astronomy*, written by the Roman author Hyginus. Many editions of both these works were produced and translated over the centuries.

**ANTICANIS**
This page from a 9th-century edition of the star myths of Hyginus shows the constellation Canis Minor, here termed Anticanis. Hyginus's words, in Latin, form the shape of the dog's body.

## FILLING THE HEAVENLY SPHERE

The oldest surviving star catalog dates from the 2nd century CE and is contained in a book called the *Almagest*, written by the Greek astronomer and geographer Ptolemy (see panel, opposite). It records the positions and brightnesses of one thousand stars, arranged into 48 constellations, based on an earlier catalog by Hipparchus of Nicaea (c. 190–c. 120 BCE). In the 10th century CE, an Arab astronomer, al-Sufi (see p.421), updated the *Almagest* in his *Book of Fixed Stars*, which included Arabic names for many stars. These Arabic names are still used today, although often in corrupted form. No more constellations were introduced until the end of the 16th century, when Dutch explorers sailed to the East Indies. From there, they could observe the southern sky that was below the European horizon. Two navigators, Pieter Dirkszoon Keyser and Frederick de Houtman (see p.416), cataloged nearly 200 new southern stars, from which they and their mentor, Petrus Plancius (see p.358), a leading Dutch cartographer, created 12 new constellations. Plancius also created other northern constellations, forming them between those listed by Ptolemy. Nearly a century later, Johannes Hevelius (see p.384), a Polish astronomer, filled the remaining gaps in the northern sky, and in the mid-18th century, the French astronomer Nicolas Louis de Lacaille (see p.422) introduced another 14 constellations in the southern sky.

**POCKET GLOBE**
This pocket globe from the Science Museum, England, positions the Earth within a shell that depicts the surrounding celestial sphere. On the inside of the open shell are the constellations, drawn as mirror images.

**GLOBAL COVERAGE**
This beautiful celestial globe was made in 1692 by Vincenzo Coronelli, a celebrated Italian globe maker. As in all celestial globes, the constellation figures are shown reversed by comparison with their appearance in the sky.

# STAR CHARTS AND ATLASES

The first printed star chart was produced in 1515 by the great German artist Albrecht Dürer. Like a celestial globe, Dürer's chart depicted the constellations in reverse, showing the sky as it would be seen from an imaginary position outside the celestial sphere, but before long, charts were being made that could be compared directly with the sky. The finest early star atlas was *Uranometria*, produced in 1603 by the German astronomer Johann Bayer. This atlas remains one of the most beautiful examples of the celestial cartographer's art. Shortly after its publication, astronomy was revolutionized by the invention of the telescope. The first major star catalog and atlas of this new era was produced by England's first Astronomer Royal, John Flamsteed (1646-1719). *Atlas Coelestis* shows the Ptolemaic constellations visible from Greenwich, England, based on Flamsteed's own painstaking observations. The pinnacle of celestial mapping came in 1801, when Johann Bode, a German astronomer, published an atlas called *Uranographia*. Covering the entire sky, this atlas depicted over 100 constellations, some invented by Bode himself. Finally, in 1922, a list of 88 constellations was agreed upon by the International Astronomical Union, astronomy's governing body, which also defined the boundaries of each constellation. On modern star charts, the only sign of the traditional pictorial charts are the few lines that link the main stars, suggesting the overall shape of each constellation.

**SKETCHY FIGURES**
Leo, the Lion, an easily recognizable constellation of the zodiac, is here depicted on the *Atlas Coelestis*, by English astronomer John Flamsteed, published in 1729.

**HEAVENLY PICTURE BOOK**
Ancient people imagined gods, heroes, and beasts among the stars, and these figures were depicted on star charts until the 19th century. These charts, from John Flamsteed's *Atlas Coelestis* (1729), show those 48 constellations known to the ancient Greeks depicted on the northern and southern halves of the sky.

## PTOLEMY

Ptolemy (*c.* 100–170 CE) lived and worked in the great metropolis of Alexandria, Egypt, which was then part of the Greek empire. He was one of the last—and the greatest— of the ancient Greek astronomers. His Earth-centered model of the universe, outlined in the treatise *Almagest*, dominated astronomical theory for 1,400 years. Ptolemy also made a catalog of 1,022 stars in 48 constellations, based on earlier work by Hipparchus.

THE NIGHT SKY

# MAPPING THE SKY

The following pages divide the celestial sphere into six parts—
two polar regions and four equatorial regions—which show the
location of the 88 constellations. Each constellation is then profiled
in the following section. Each entry places the constellation and its
main features into the context of the rest of the sky.

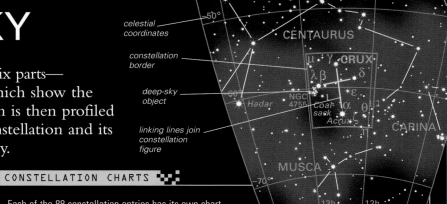

celestial coordinates

constellation border

deep-sky object

linking lines join constellation figure

## VISIBILITY MAPS

Not visible

Partially visible

Visible

80°N
60°N
40°N
20°N
0°
20°S
40°S
60°S

The entry for each constellation contains a map showing
the parts of the world from which it can be seen. The entire
constellation can be seen from the area shaded black, part is
visible from the area shaded gray, and it cannot be seen from the
area shaded white. Exact latitudes for full visibility are given
in the accompanying data set.

## CONSTELLATION CHARTS

Each of the 88 constellation entries has its own chart,
centered around the constellation area. These charts
show all stars brighter than magnitude 6.5. Within the
constellation borders, every star brighter than magnitude 5
is labeled. Deep-sky objects are represented by an icon.

### KEY TO STAR MAGNITUDES

-1.5–0    0–0.9    1.0–1.9    2.0–2.9    3.0–3.9    4.0–4.9    5.0–5.9    6.0–6.9

### DEEP-SKY OBJECTS

- Galaxy
- Globular cluster
- Open cluster
- Diffuse nebula
- Planetary nebula or supernova remnant
- Black hole or X-ray binary

## THE NORTH POLAR SKY

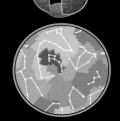

Almost in the
center of this chart
is the star Polaris, in
Ursa Minor, which
lies less than 1°
from the north
celestial pole. For
observers in the
Northern Hemisphere,
the stars around the pole
never set—they are
circumpolar. The viewer's
latitude will determine how
much of the sky is circumpolar:
the farther north, the larger the
circumpolar area. This chart shows
the sky from declinations 90° to 50°.

## STAR MAGNITUDES

⬤ -1  ⬤ 0  ⬤ 1  ⬤ 2  · 3  · 4  · 5    ⊛ Variable star

Star magnitudes shown here are for the equatorial and polar sky charts

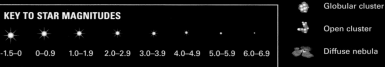

## THE GREEK ALPHABET

On most star charts, bright stars are identified by Greek letters according to a system invented by Johann Bayer (see p.72).

| | | | | | | | |
|---|---|---|---|---|---|---|---|
| Alpha | α | Eta | η | Nu | ν | Tau | τ |
| Beta | β | Theta | θ | Xi | ξ | Upsilon | υ |
| Gamma | γ | Iota | ι | Omicron | ο | Phi | φ |
| Delta | δ | Kappa | κ | Pi | π | Chi | χ |
| Epsilon | ε | Lambda | λ | Rho | ρ | Psi | ψ |
| Zeta | ζ | Mu | μ | Sigma | σ | Omega | ω |

## VISIBILITY ICONS

Beside every photograph is an icon indicating the kind of view it illustrates. Some photographs show the star or deep-sky object as it can be seen by the naked eye, through binoculars, or through amateur telescopes. Others are the result of CCD photography or show the view through professional observing equipment.

- 👁 Naked eye
- 🔭 Binoculars
- ✈ Telescope (amateur)
- 🖳 CCD
- ⚖ Professional equipment

## ALPHABETICAL INDEX OF THE 88 CONSTELLATIONS

The constellation entries are ordered by their position on the celestial sphere, beginning with Ursa Minor in the north and spiraling south in a clockwise direction before finishing with Octans. This alphabetical list provides an alternative way of locating constellation entries.

# THE SOUTH POLAR SKY

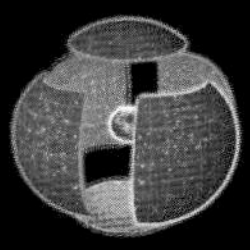

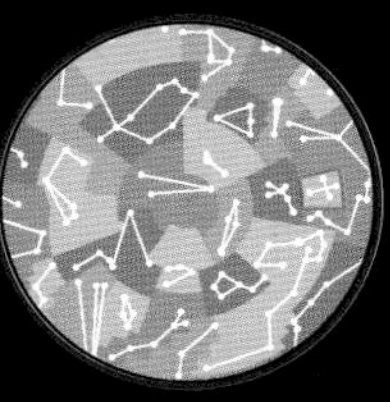

There is no southern equivalent of Polaris, the north Pole Star—in fact, the area around the south celestial pole is remarkably barren. This chart shows the sky from declinations -50° to -90°. Many of the stars on this chart are circumpolar for southern observers—that is, the stars never set and are always visible in the night sky. The farther south the viewer, the greater the amount of sky that is circumpolar.

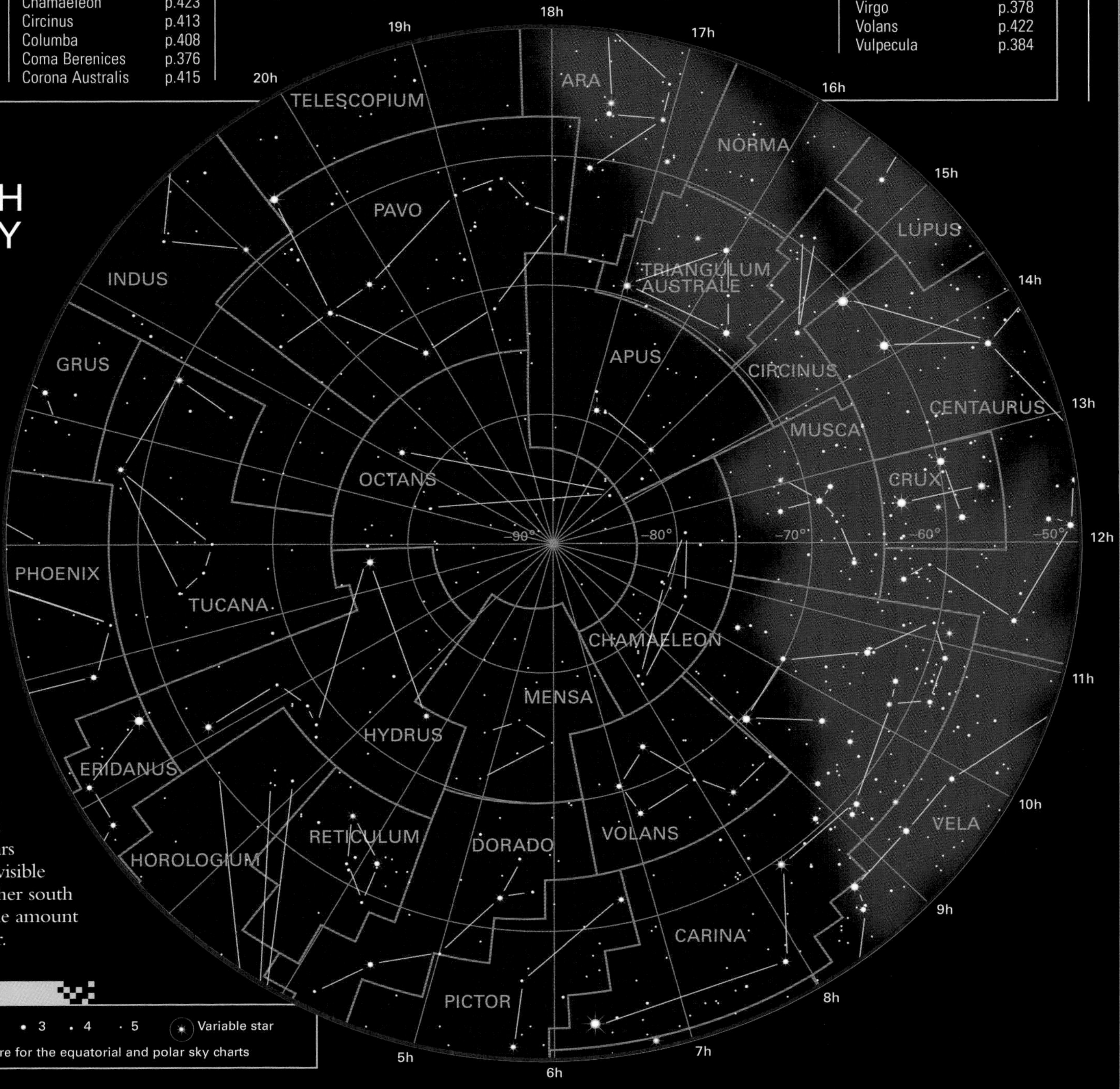

## STAR MAGNITUDES

✹ -1   ✷ 0   • 1   · 2   · 3   · 4   · 5   ⊛ Variable star

Star magnitudes shown here are for the equatorial and polar sky charts

# EQUATORIAL SKY CHART 1

This part of the sky is best placed for observation on evenings in September, October, and November. It contains the vernal equinox, in Pisces, which is the point at which the Sun's path, the ecliptic, crosses the celestial equator into the northern half of the sky. The Sun reaches this point in late March each year. The 0h line of right ascension also passes through this point; this is the celestial equivalent of 0° longitude (the prime meridian) on Earth. The most distinctive feature in this region of the night sky is the great Square of Pegasus— although one star in the square actually belongs to neighboring Andromeda.

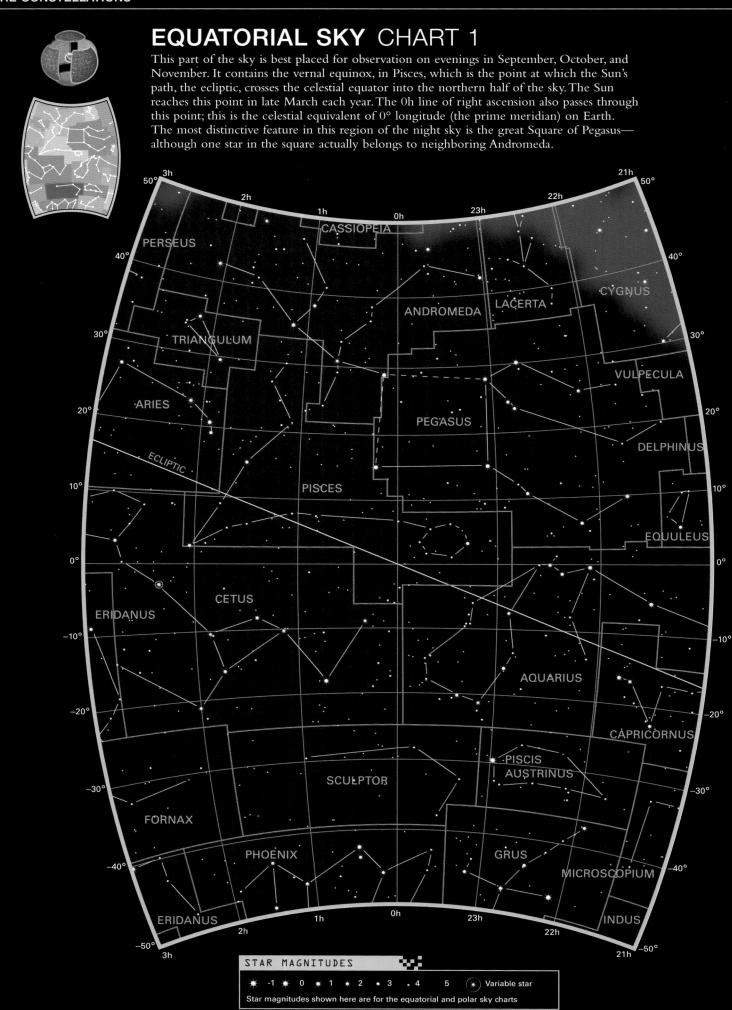

STAR MAGNITUDES

-1  0  1  2  3  4  5  Variable star

Star magnitudes shown here are for the equatorial and polar sky charts

# EQUATORIAL SKY CHART 2

This area of sky is best placed for observation on evenings in June, July, and August. It contains the point where the Sun reaches its most southerly declination each year, in Sagittarius. This happens around December 21, which is the longest day in the Southern Hemisphere and the shortest day in the Northern. Rich Milky Way star fields cross this region of sky, from Cygnus in the north to Sagittarius and Scorpius in the south. Hercules and Ophiuchus, both representing mythical giants, stand head to head in the north. Notable star patterns in the south are the Teapot asterism in Sagittarius and the curving tail of Scorpius, the Scorpion.

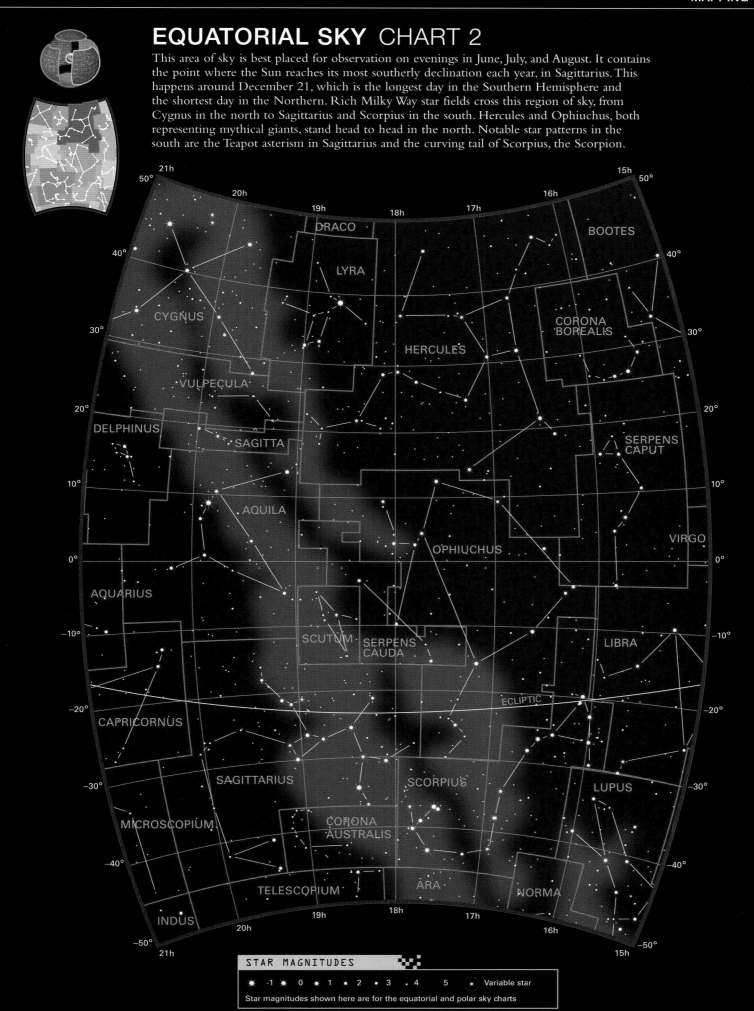

STAR MAGNITUDES

-1    0    1    2    •3    .4    5    ⊛ Variable star

Star magnitudes shown here are for the equatorial and polar sky charts

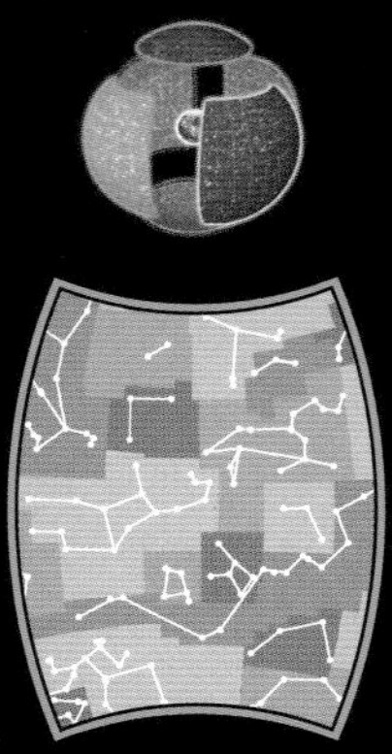

# EQUATORIAL SKY CHART 3

This region is best placed for observation on evenings in March, April, and May. It contains the point at which the Sun moves across the celestial equator into the Southern Hemisphere each year. This point lies in Virgo, and the Sun reaches it around September 21. In the northern constellation Boötes lies Arcturus, a notably orange-colored star whose visibility marks the arrival of northern spring. South of it is the zodiacal constellation of Virgo, whose brightest star is the blue-white Spica. Adjoining Virgo is Leo, one of the few constellations that genuinely resembles the animal it is said to represent—in this case, a crouching lion.

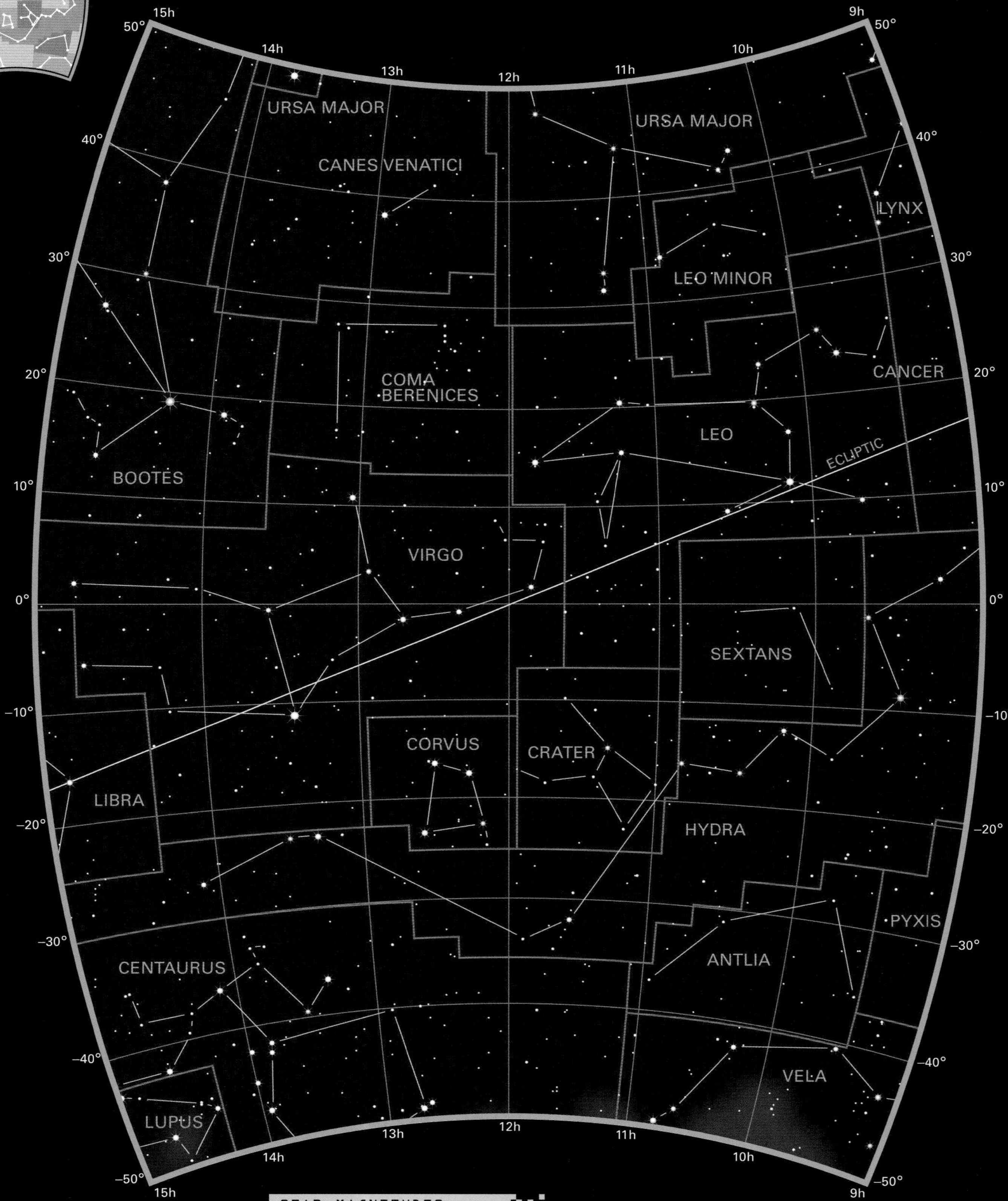

URSA MAJOR

CANES VENATICI

URSA MAJOR

LYNX

LEO MINOR

COMA BERENICES

LEO

CANCER

ECLIPTIC

BOÖTES

VIRGO

SEXTANS

CORVUS

CRATER

LIBRA

HYDRA

PYXIS

CENTAURUS

ANTLIA

VELA

LUPUS

STAR MAGNITUDES

-1   0   1   2   3   4   5   Variable star

Star magnitudes shown here are for the equatorial and polar sky charts

# EQUATORIAL SKY CHART 4

This region is best placed for observation on December, January, and February evenings.
It contains the point at which the Sun is farthest north of the celestial equator, on the border
of Taurus with Gemini. This occurs around June 21, when days are longest in the Northern
Hemisphere and shortest in the Southern. Glittering stars and magnificent constellations
abound in this region of sky, including the brightest star of all, Sirius in Canis Major.
A distinctive line of three stars marks the belt of Orion, while in Taurus the bright star
Aldebaran glints like the eye of the bull, along with the Hyades and Pleiades star clusters.

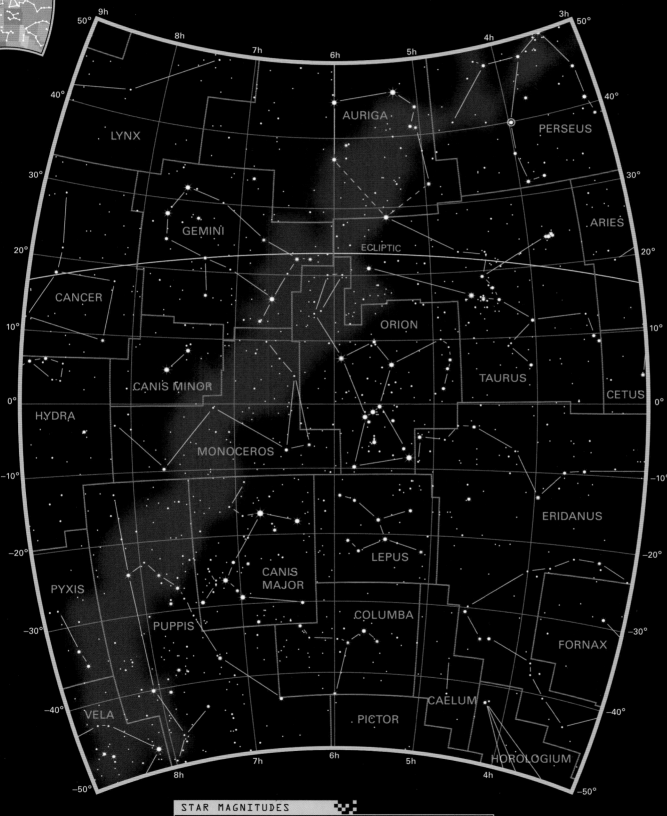

STAR MAGNITUDES

✦ -1  ✦ 0  ✦ 1  • 2  • 3  • 4  · 5  ⊙ Variable star

Star magnitudes shown here are for the equatorial and polar sky charts

LONG-TAILED BEAR ◉
The tail of the Little Bear curves away from the north Pole Star, Polaris (upper left). Unlike real bears, the celestial bears Ursa Minor and Ursa Major both have long tails.

## THE LITTLE BEAR

# Ursa Minor

**SIZE RANKING** 56

**BRIGHTEST STAR**
Polaris (α) 2.0

**GENITIVE**
Ursae Minoris

**ABBREVIATION** UMi

**HIGHEST IN SKY AT 10 PM**
May–July

**FULLY VISIBLE**
90°N–0°

Ursa Minor is an ancient Greek constellation, which is said to represent Ida, one of the nymphs who nursed the god Zeus when he was an infant (see panel, right). Ursa Minor contains the north celestial pole and also its nearest naked-eye star, Polaris or Alpha (α) Ursae Minoris (see pp.278–279), which is currently less than 1 degree from the north celestial pole. The distance between them is steadily decreasing due to precession (see p.64). They will come closest around 2100, when the separation will be about 0.5°.

The main stars of Ursa Minor form a shape known as the Little Dipper, reminiscent of the larger and brighter Big Dipper in Ursa Major, although its handle curves in the opposite direction. The two brightest stars in the bowl of the Little Dipper, Beta (β) and Gamma (γ) Ursae Minoris, are popularly known as the Guardians of the Pole.

**THE NORTH POLE STAR ✦ ◉**
This image of Polaris was taken through a telescope, but the star is clearly visible with the naked eye. Images taken by the Hubble Space Telescope have revealed that Polaris has a companion not visible from Earth.

## SPECIFIC FEATURES

Polaris, the north Pole Star, is a creamy white supergiant and a Cepheid variable (see p.282), but its brightness changes are too slight to be noticeable to the naked eye. With a telescope, an unrelated 8th-magnitude star can be seen nearby.

Two stars in the bowl of the Little Dipper—Gamma and Eta Ursae Minoris—are both wide doubles. Gamma is the brighter of the two, at magnitude 3.0, and its 5th-magnitude companion, 11 Ursae Minoris, can be seen with the naked eye or binoculars. Eta, at magnitude 5.0, can also be seen with the naked eye. It has a partner of magnitude 5.5, 19 Ursae Minoris; both stars are easily visible with binoculars. Each of the component stars in both Gamma and Eta lie at different distances from the Earth and are thus unrelated.

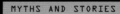

THE
LITTLE BEAR

URSA
MINOR

DRACO

*Polaris*

.80°

*Kochab*

*Pherkad*

4

•5

17h     16h     15h     14h

MYTHS AND STORIES

## NURSING NYMPHS

According to Greek mythology, at his birth, the infant Zeus was hidden from his murderous father, Cronus, and taken to a cave on the island of Crete, where he was nursed by two nymphs, usually named as Adrastea and Ida. In gratitude, Zeus later placed the nymphs in the sky as the Great Bear and the Little Bear, respectively.

**THE PROTECTED CHILD**
The infant Zeus is cared for by nymphs and shepherds, in the *Feeding of Jupiter* by the French artist Nicolas Poussin.

## THE DRAGON
# Draco

| | |
|---|---|
| **SIZE RANKING** | 8 |
| **BRIGHTEST STAR** | Eltanin (γ) 2.2 |
| **GENITIVE** | Draconis |
| **ABBREVIATION** | Dra |
| **HIGHEST IN SKY AT 10 PM** | April–August |
| **FULLY VISIBLE** | 90°N–4°S |

One of the ancient Greek constellations, Draco represents the dragon of Greek myth that was slain by Hercules (see panel, below). This large constellation winds for nearly 180° around the north celestial pole. Despite its size, Draco is not particularly easy to identify, apart from a lozenge shape marking the head. This is formed by four stars, including the constellation's brightest member, Gamma (γ) Draconis, popularly known as Eltanin, meaning "the serpent."

### SPECIFIC FEATURES
Double and multiple stars are a particular feature of Draco. Nu (ν) Draconis, the faintest of the four stars in the dragon's head, is a readily identifiable pair. It consists of identical white components of 5th magnitude and is considered to be among the finest doubles visible with binoculars. Psi (ψ) Draconis is a somewhat closer pair, with components of 5th and 6th magnitudes, and requires a small telescope to be divided. More challenging to discern is Mu (μ) Draconis, with its two 6th-magnitude stars, which requires a telescope with high magnification to be seen as double.

The wide pair of stars 16 and 17 Draconis is easily spotted with binoculars, and the brighter of the two—17 Draconis—can be further divided with a small telescope with high magnification, turning this into a triple star. A similar triple is 39 Draconis; when viewed with a small telescope with low magnification, it appears a double, but at higher magnification, the brighter star divides into a closer pair with components of magnitudes 5.0 and 8.0. Two more doubles that can readily be seen with a small telescope are Omicron (o) Draconis, with stars of 5th and 8th magnitudes, and 40 and 41 Draconis, which are both 6th-magnitude orange dwarfs.

In central Draco lies a planetary nebula made famous by a striking Hubble Space Telescope image: NGC 6543, or the Cat's Eye Nebula (see p.258). Processed in false color, the Hubble picture shows the nebula as red, but when seen through a small telescope it appears blue-green, as do all planetary nebulae.

**BEAR AND DRAGON** 👁
The long body of Draco curls around the stars of Ursa Minor, the Little Bear. The head of the dragon is easily identifiable.

**THE CAT'S EYE NEBULA** 🔭 💻
This amateur CCD image of NGC 6543 shows some of the color and structure captured by the Hubble Space Telescope, but visually the nebula appears as a blue-green ellipse.

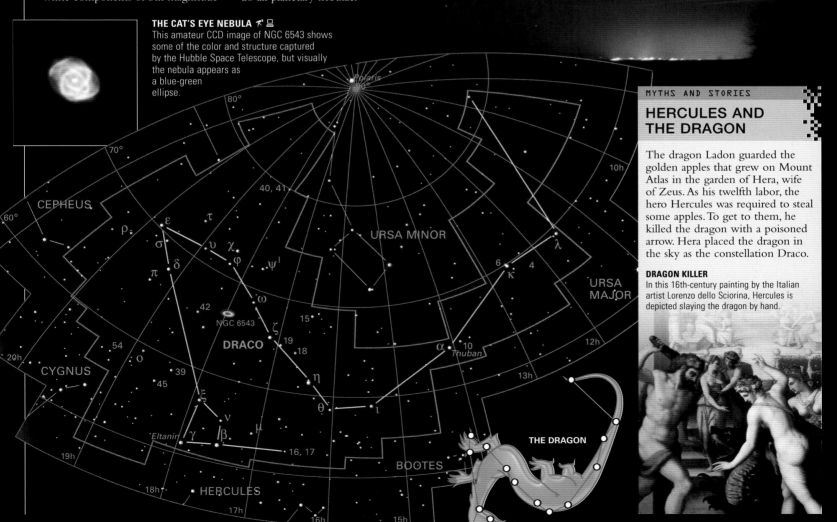

THE DRAGON

### MYTHS AND STORIES
## HERCULES AND THE DRAGON

The dragon Ladon guarded the golden apples that grew on Mount Atlas in the garden of Hera, wife of Zeus. As his twelfth labor, the hero Hercules was required to steal some apples. To get to them, he killed the dragon with a poisoned arrow. Hera placed the dragon in the sky as the constellation Draco.

**DRAGON KILLER**
In this 16th-century painting by the Italian artist Lorenzo dello Sciorina, Hercules is depicted slaying the dragon by hand.

## CEPHEUS

# Cepheus

**SIZE RANKING** 27

**BRIGHTEST STAR**
Alpha (α) 2.5

**GENITIVE** Cephei

**ABBREVIATION** Cep

**HIGHEST IN SKY AT 10 PM**
September–October

**FULLY VISIBLE**
90°N–1°S

Cepheus lies in the far northern sky between Cassiopeia and Draco. Its main stars form a distorted tower or steeple shape, yet this ancient Greek constellation in fact represents the mythical King Cepheus of Ethiopia, who was the husband of Queen Cassiopeia and the father of Andromeda. Cepheus is not a particularly prominent constellation.

## SPECIFIC FEATURES
The constellation's most celebrated star is Delta (δ) Cephei (see p.286), from which all Cepheid variables take their name. Just under 1,000 light-years away, this yellow-colored supergiant varies between magnitudes 3.5 and 4.4 every 5 days 9 hours.

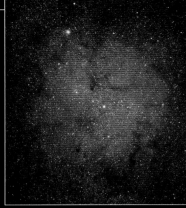

**IC 1396**
The Garnet Star or Mu Cephei (top left) lies on the edge of the large but faint nebula IC 1396. The nebula is centered on the 6th-magnitude multiple star Struve 2816.

These changes can be followed with the naked eye. Delta (δ) Cephei is also a double star; its 6th-magnitude, blue-white companion is visible through a small telescope.

A significant variable star of a different kind is Mu (μ) Cephei, which is a red supergiant that ranges anywhere between magnitudes 3.4 and 5.1 every 2 years or so. This supergiant is also known as the Garnet Star on account of its strong red coloration.

Nonvariable stars near Delta (δ) and Mu (μ) Cephei can be used to gauge the magnitude of these two variable stars at any given time. For example, they can be compared to Zeta (ζ) at magnitude 3.4, Epsilon (ε) at magnitude 4.2, or Lambda (λ) Cephei at magnitude 5.1 (see chart, below).

**THE KING** 👁
Shaped like a bishop's miter, Cepheus is not easy to pick out in the sky. He is flanked by his prominent wife, Cassiopeia, and Draco, the dragon.

CEPHEUS

## HENRIETTA LEAVITT

Henrietta Swan Leavitt (1868–1921) worked at Harvard College Observatory in the early 20th century. Her study of variable stars in the Small Magellanic Cloud led to the period-luminosity law. This law links the variation period of a Cepheid variable to its intrinsic brightness, which in turn can indicate distance. Her law remains fundamental to our knowledge of the scale of the universe.

**DETERMINATION**
By painstakingly measuring photographic plates, Henrietta Leavitt discovered 2,400 variable stars of all types.

**DELTA (Δ) AND MU (M) CEPHEI**

**MAGNITUDE KEY**

| | |
|---|---|
| ⭐ | 0.0–0.9 |
| ✦ | 1.0–1.9 |
| ✦ | 2.0–2.9 |
| ✴ | 3.0–3.9 |
| • | 4.0–4.9 |
| • | 5.0–5.9 |
| · | 6.0–6.9 |

# Cassiopeia

| | |
|---|---|
| **SIZE RANKING** | 25 |
| **BRIGHTEST STARS** | Schedar (α) 2.2, Gamma (γ) 2.2 |
| **GENITIVE** | Cassiopeiae |
| **ABBREVIATION** | Cas |
| **HIGHEST IN SKY AT 10 PM** | October–December |
| **FULLY VISIBLE** | 90°N–12°S |

This distinctive constellation of the northern sky is found within the Milky Way between Perseus and Cepheus and north of Andromeda. The large W shape formed by its five main stars is easily recognizable. It is an ancient Greek constellation, representing the mythical Queen Cassiopeia of Ethiopia.

## SPECIFIC FEATURES

Gamma (γ) Cassiopeiae (see p.285) is a hot, rapidly rotating star that occasionally throws off rings of gas from its equator, which causes unpredictable changes in its brightness. It has ranged between magnitudes 3.0 and 1.6, but it currently lies at magnitude 2.2, which makes it the joint brightest star in the constellation.

A variable with a more predictable cycle is Rho (ρ) Cassiopeiae, an intensely luminous, yellow-white supergiant that fluctuates between 4th and 6th magnitudes every 10 or 11 months. It is estimated that it lies more than 10,000 light-years away, which is exceptionally distant for a naked-eye star.

Eta (η) Cassiopeiae is an attractive stellar pair consisting of a yellow and a red star. Its components are of magnitudes 3.5 and 7.5 and can be seen through a small telescope. This pair forms a true binary; the fainter companion orbits the brighter star every 480 years.

Cassiopeia contains a number of open clusters within range of small instruments. Chief among them is M52 (see p.290), near the border with Cepheus. It is visible through binoculars as a somewhat elongated patch of light, and its individual stars—including a bright orange giant at one edge—can be seen through a small telescope. M103 is a small, elongated group best viewed through a small telescope. Nearby is a larger cluster, NGC 663, which is more suitable for binocular observation. NGC 457 is a looser star cluster containing the 5th-magnitude star Phi (φ) Cassiopeiae. This cluster's appearance has been likened to an owl—its two brightest stars mark the owl's eyes.

**M103** 📷 ⋌
M103's main feature is a chain of three stars like a mini Orion's belt. The northernmost member of the line (top) is not a true member of the cluster but lies closer to Earth.

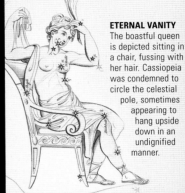

**M52** 📷 ⋌
Through binoculars, this cluster appears as a misty patch about one-third the diameter of a full moon. A telescope is needed to resolve its individual stars.

## THE VAIN QUEEN

Wife of Cepheus and mother of Andromeda, Queen Cassiopeia was notoriously vain. She enraged the Nereids, daughters of Poseidon, by boasting she was more beautiful. In punishment, Poseidon sent a sea monster to ravage her kingdom, which eventually led to the rescue of Andromeda by Perseus (see p.368).

**ETERNAL VANITY**
The boastful queen is depicted sitting in a chair, fussing with her hair. Cassiopeia was condemned to circle the celestial pole, sometimes appearing to hang upside down in an undignified manner.

**CASSIOPEIA**

**POLAR POINTER** 👁
The distinctive W shape formed by the main stars of Cassiopeia is easy to locate in the sky. The center of the W points toward the north celestial pole.

THE NIGHT SKY

# Camelopardalis

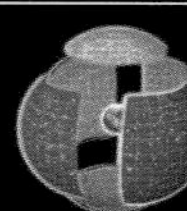

**SIZE RANKING** 18

**BRIGHTEST STAR**
Beta (β) 4.0

**GENITIVE**
Camelopardalis

**ABBREVIATION** Cam

**HIGHEST IN SKY AT 10 P.M.**
December–May

**FULLY VISIBLE**
90°N–3°S

This dim constellation of the far northern sky, representing a giraffe, was introduced in the early 17th century on a celestial globe created by the Dutch astronomer Petrus Plancius (see panel, below). The giraffe's long neck can be visualized as stretching around the north celestial pole toward Ursa Minor and Draco.

## SPECIFIC FEATURES

The brightest star in the constellation, Beta (β) Camelopardalis, is a double star whose fainter companion can be seen with a small telescope or even powerful binoculars. South of Beta (β) is 11 and 12 Camelopardalis, a wide double star with components of 5th and 6th magnitudes.

Within the giraffe's hindquarters is NGC 1502, a small open star cluster visible through binoculars or a small telescope. Binoculars also show a long chain of faint stars called Kemble's Cascade, which lead away from NGC 1502 toward Cassiopeia. This star feature is named after Lucian Kemble, a Canadian amateur astronomer who first drew attention to it in the late 1970s. None of the stars, however, are actually related.

NGC 2403 is a 9th-magnitude spiral galaxy that looks like a comet when seen through a small telescope. It is one of the brightest and closest galaxies to the Earth, outside the Local Group.

## KEMBLE'S CASCADE

In an area five times the diameter of a full moon, the stars of Kemble's Cascade seem to tumble down the sky. The small star cluster NGC 1502 can be seen in the lower left of the picture.

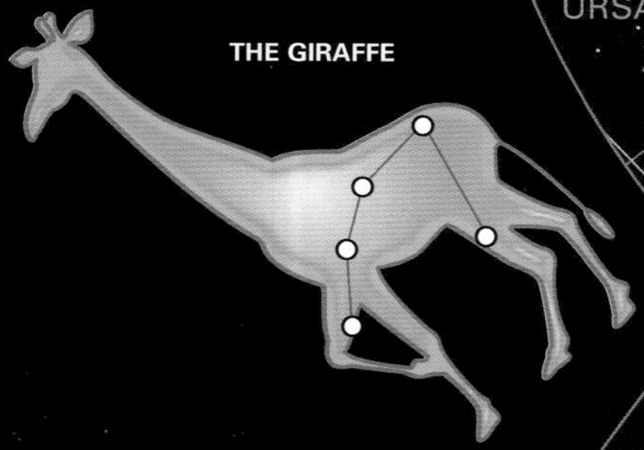

**THE GIRAFFE**

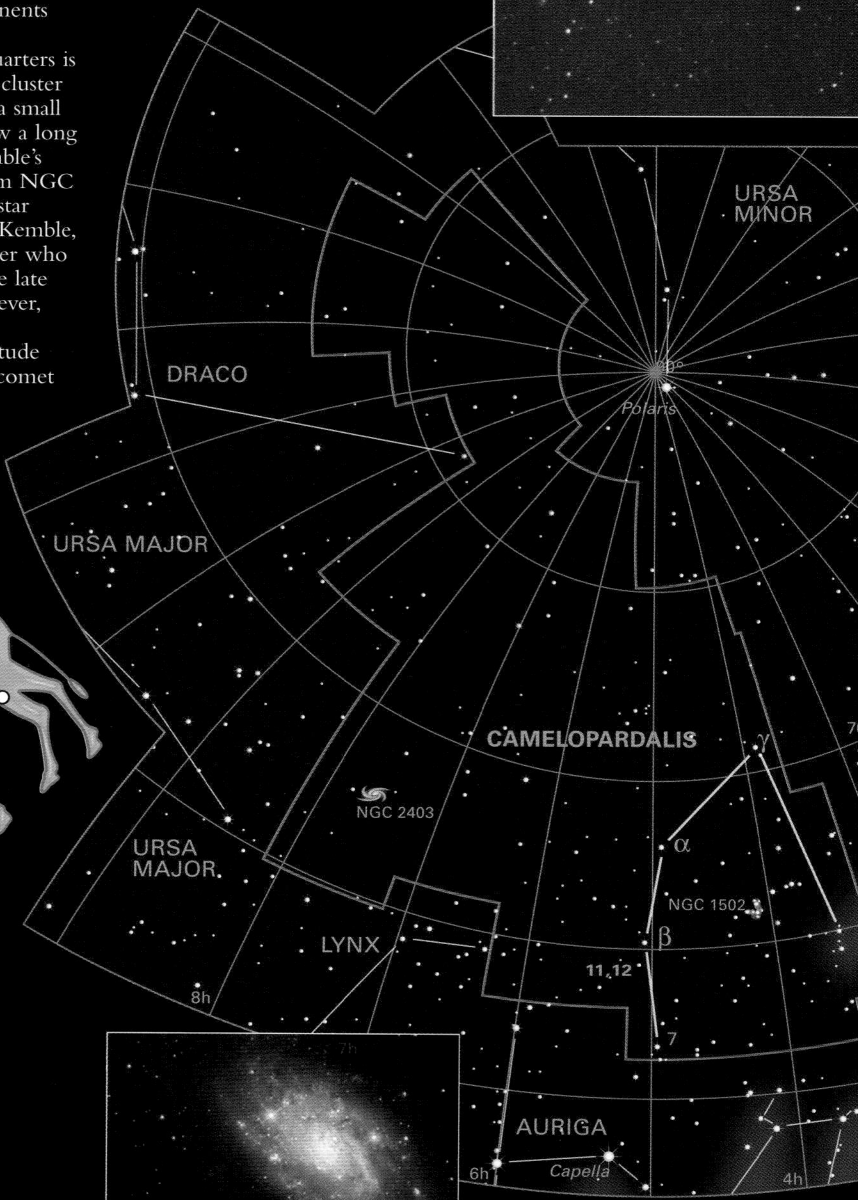

**NGC 2403**
Color images of this galaxy reveal the pink glow of large emission nebulae in its spiral arms. It is about 11 million light-years away.

## PARTIAL VIEW

It can be difficult to relate the figure of a giraffe to the stars of Camelopardalis. Here, the stars of the giraffe's legs are shown. The animal's long neck would stretch off the top of the picture.

## PETRUS PLANCIUS

This Dutch church minister was also an expert geographer and astronomer. Petrus Plancius (1552–1622) taught the navigators on the first Dutch sea voyages to the East Indies how to measure star positions. In turn, they produced for him a catalog of the southern stars divided into 12 new constellations, which Plancius depicted on his celestial globes. He also invented several constellations, such as Columba, Camelopardalis, and Monoceros, using some of the fainter stars visible from Europe.

# THE CHARIOTEER

## Auriga

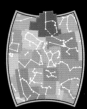

| | |
|---|---|
| **SIZE RANKING** | 21 |
| **BRIGHTEST STAR** | Capella (α) 0.1 |
| **GENITIVE** | Aurigae |
| **ABBREVIATION** | Aur |
| **HIGHEST IN SKY AT 10 P.M.** | December–February |
| **FULLY VISIBLE** | 90°N–34°S |

Auriga is easily identified in the northern sky by the presence of Capella (α), the most northerly first-magnitude star. Auriga lies in the Milky Way between Gemini and Perseus, to the north of Orion. The constellation represents a charioteer.

### SPECIFIC FEATURES

Auriga's outstanding feature is a chain of three large and bright open star clusters. All three will just fit within the same field of view in wide-angle binoculars. Of the trio, M38's stars are the most scattered and, when viewed with a small telescope, seem to form chains. The middle cluster is M36, the smallest cluster but also the easiest to spot, while M37 is the largest and contains the most stars, but these are faint. All three clusters lie about 4,000 light-years away.

The star-forming nebula IC 405 is located nearby. Bright light from 6th-magnitude AE Aurigae near its center lights up the surrounding gases.

Auriga also contains two extraordinary eclipsing binaries of long period. One is Zeta (ζ) Aurigae, which is an orange giant orbited by a smaller blue star that eclipses it every 2.7 years. This causes a 30 percent decrease in brightness for 6 weeks, from magnitude 3.7 to 4.0. More remarkable, however, is Epsilon (ε) Aurigae (see p.281). This intensely luminous giant star is orbited by a mysterious dark partner that eclipses it every 27 years—the longest interval of any eclipsing binary. During the eclipse, Epsilon's brightness is halved, from magnitude 3.0 to 3.8, and it remains dimmed for more than a year. Astronomers think that its companion star is enveloped in a disk of dust seen almost edge-on. The last eclipse occurred between 2009 and 2011. The next is due to start in 2036.

**SHARED STAR** 👁
Neighboring Beta (β) Tauri completes the charioteer figure. Auriga is usually identified as a king of Athens, Erichthonius.

**THE FLAMING STAR NEBULA**
AE Aurigae is a hot, massive star of magnitude 6 that lights up the surrounding cloud of gas and dust that is the Flaming Star Nebula, IC 405.

# THE LYNX

## Lynx

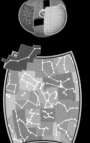

| | |
|---|---|
| **SIZE RANKING** | 28 |
| **BRIGHTEST STAR** | Alpha (α) 3.1 |
| **GENITIVE** | Lyncis |
| **ABBREVIATION** | Lyn |
| **HIGHEST IN SKY AT 10 P.M.** | February–March |
| **FULLY VISIBLE** | 90°N–28°S |

Lynx is a fair-sized but faint constellation in the northern sky. It was introduced in the late 17th century by Johannes Hevelius (see p.384), who wanted to fill the gap between Ursa Major and Auriga. Hevelius is reputed to have named it Lynx because only the lynx-eyed would be able to see it—Hevelius himself had very sharp eyesight. The animal he drew on his star chart, however, looked little like a real lynx.

### SPECIFIC FEATURES

Lynx contains many interesting double and multiple stars. For example, 12 Lyncis appears double with a small telescope, but with a telescope of 3 in (75 mm) or larger aperture the brighter star divides into two components of 5th and 6th magnitudes, which have an orbital period of about 900 years.

An easier triple to identify is 19 Lyncis. This consists of two stars of 6th and 7th magnitudes and a wider 8th-magnitude companion, all visible through a small telescope. A more challenging double star is 38 Lyncis, with components of 4th and 6th magnitudes. A telescope of 3-in (75-mm) aperture is required to separate the individual stars.

**THE LYNX**

**ELUSIVE FELINE** 👁
Lynx consists of nothing more than a few faint stars zigzagging between Ursa Major and Auriga. To spot it, sharp eyesight or binoculars are required.

THE GREAT BEAR

# Ursa Major

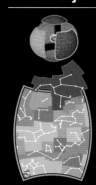

| | |
|---|---|
| **SIZE RANKING** 3 | |
| **BRIGHTEST STARS** Alpha (α) 1.8, Epsilon (ε) 1.8. | |
| **GENITIVE** Ursae Majoris | |
| **ABBREVIATION** UMa | |
| **HIGHEST IN SKY AT 10 P.M.** February–May | |
| **FULLY VISIBLE** 90°N–16°S | |

Ursa Major is one of the best-known constellations and a prominent feature of the northern sky. Seven of its stars form the familiar shape of the Big Dipper. But as a whole, Ursa Major is much larger than this; it is the third-largest constellation in the sky. The two stars in the Big Dipper's bowl farthest from the handle, Dubhe (α) and Merak (β), point toward the north Pole Star, Polaris, while the curved handle of the Big Dipper points toward the bright star Arcturus in the adjoining constellation of Boötes.

## SPECIFIC FEATURES

The Big Dipper is one of the most famous patterns in the sky. Its shape is formed by the stars Dubhe (α), Merak (β), Phecda (γ), Megrez (δ) Ursae Majoris, Alioth (ε), Mizar (ζ) (see p.276), and Alkaid (η). With the exception of Dubhe and Alkaid, these stars travel through space in the same direction, and they form what is known as a moving cluster.

Mizar (ζ), the second star in the Big Dipper's handle, is next to Alcor (see p.276), an eighth, fainter star in the Big Dipper, which can be seen with good eyesight. A small telescope reveals that Mizar also has a closer 4th-magnitude companion.

In southern Ursa Major lies another attractive double star, Xi (ξ) Ursae Majoris, which can easily be divided through only a small telescope. This pair, with components of 4th and 5th magnitudes, forms a true binary relationship, orbiting every 60 years, which is quick by the standards of visual binary stars.

One of the easiest galaxies to identify with binoculars is M81, which is in northern Ursa Major, and is also known as Bode's Galaxy (see p.314). This spiral galaxy is at an angle and can be seen on clear, dark nights as a slightly elongated patch of light. A telescope is needed to spot the rather more elongated shape of the smaller and fainter Cigar Galaxy (see p.314), or M82, which is found one diameter of a full moon away from Bode's Galaxy. This unusual-looking object is now thought to be a spiral galaxy, seen edge-on, mottled with dust clouds and undergoing a burst of star formation following an encounter with M81.

Another major spiral galaxy in this constellation is the Pinwheel Galaxy, M101 (p.316), which lies near the end of the Big Dipper's handle. Though larger than Bode's Galaxy, it is fainter and thus more difficult to see. An even greater challenge to find and identify, however, is the Owl Nebula, or M97, located under the bowl of the Big Dipper. This planetary nebula is one of the faintest objects in Charles Messier's catalog, and a telescope of around 3 in (75 mm) aperture is needed to make out its gray-green disk, which is three times larger than that of Jupiter. A telescope with an even larger aperture reveals the two dark patches, like an owl's eyes, that give rise to its popular name.

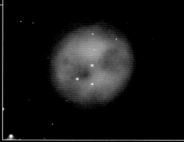

**THE OWL NEBULA** ✶ 🖥
The dark, owl-like eyes of the faint planetary nebula M97 are visible only through large telescopes or on photographs and CCD images such as this one.

**THE CIGAR GALAXY** ✶ 🖥
M82 is a peculiar-looking spiral galaxy, edge-on to us, that is undergoing a burst of star formation triggered by a close encounter with the larger and brighter spiral galaxy M81 about 300 million years ago.

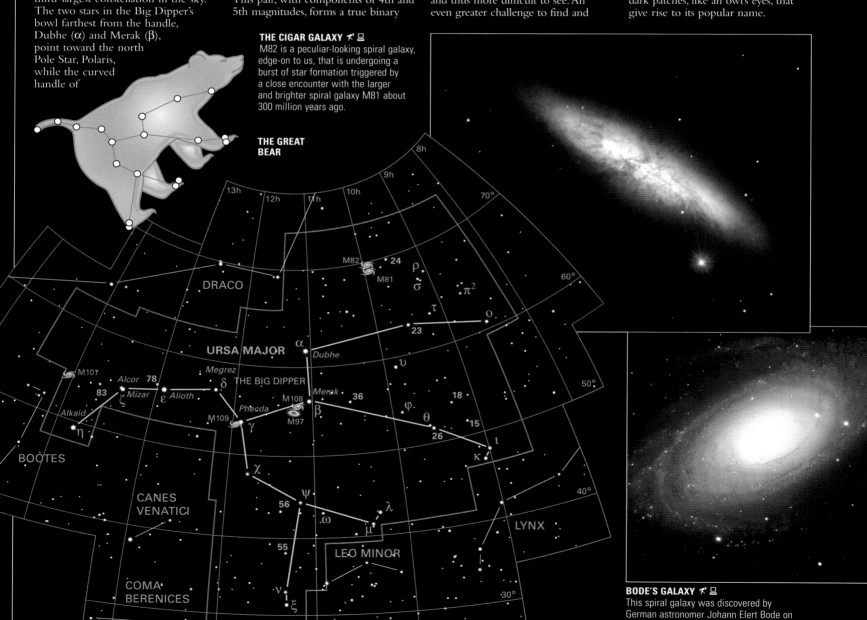

THE GREAT BEAR

**BODE'S GALAXY** ✶ 🖥
This spiral galaxy was discovered by German astronomer Johann Elert Bode on December 31, 1774. Located approximately 11 million light-years away, M81 is nevertheless one of the brightest and most visible galaxies in the sky.

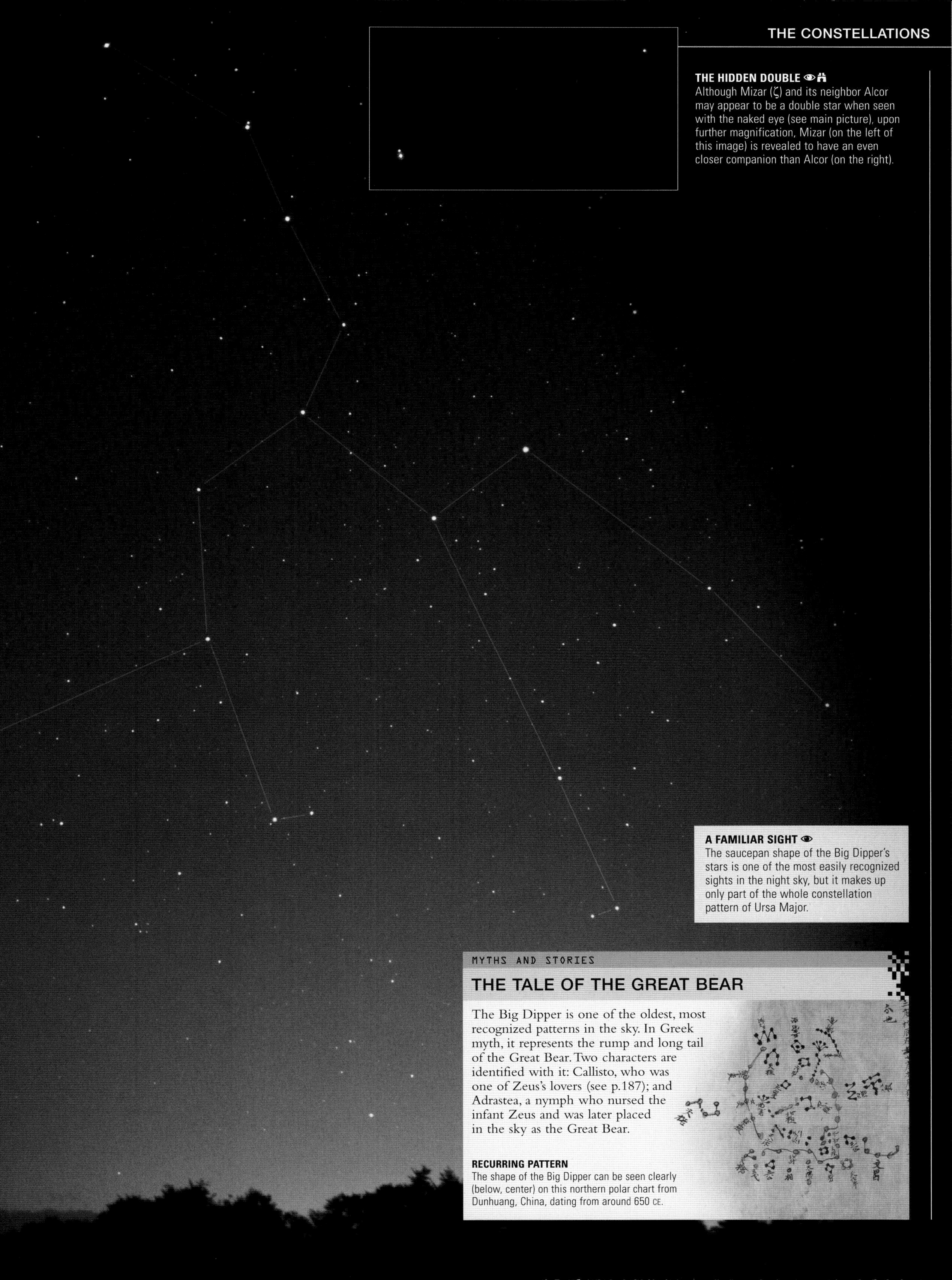

**THE HIDDEN DOUBLE** 👁 📷
Although Mizar (ζ) and its neighbor Alcor may appear to be a double star when seen with the naked eye (see main picture), upon further magnification, Mizar (on the left of this image) is revealed to have an even closer companion than Alcor (on the right).

**A FAMILIAR SIGHT** 👁
The saucepan shape of the Big Dipper's stars is one of the most easily recognized sights in the night sky, but it makes up only part of the whole constellation pattern of Ursa Major.

MYTHS AND STORIES

## THE TALE OF THE GREAT BEAR

The Big Dipper is one of the oldest, most recognized patterns in the sky. In Greek myth, it represents the rump and long tail of the Great Bear. Two characters are identified with it: Callisto, who was one of Zeus's lovers (see p.187); and Adrastea, a nymph who nursed the infant Zeus and was later placed in the sky as the Great Bear.

**RECURRING PATTERN**
The shape of the Big Dipper can be seen clearly (below, center) on this northern polar chart from Dunhuang, China, dating from around 650 CE.

# Canes Venatici

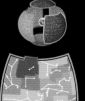

**SIZE RANKING** 38

**BRIGHTEST STAR**
Cor Caroli (α) 2.9

**GENITIVE** Canum
Venaticorum

**ABBREVIATION** CVn

**HIGHEST IN SKY AT 10 PM**
April–May

**FULLY VISIBLE**
90°N–37°S

Canes Venatici lies in the northern sky between Boötes and Ursa Major. This constellation represents two dogs held on a leash by the herdsman Boötes. It was formed by Johannes Hevelius (see p.384) at the end of the 17th century from stars that had previously been part of Ursa Major.

## SPECIFIC FEATURES

The constellation's brightest star, Alpha (α) Canum Venaticorum, is known as Cor Caroli, meaning Charles's Heart, in commemoration of King Charles I of England. This wide double star, with components of magnitudes 2.9 and 5.6, is easily separated with a small telescope. The brighter star is slightly variable, by about one-tenth of a magnitude, which is too small to be noticeable to the naked eye. Larger variation is found in Y Canum Venaticorum, a deep red supergiant popularly known as La Superba. It fluctuates between magnitudes 5.0 and 6.5 every 160 days or so.

### THE SUNFLOWER GALAXY
M63 is a spiral galaxy with patchy outer arms that is seen at an angle from Earth. The arms give rise to comparisons with the appearance of a sunflower. The star to its right in this photograph is of 9th magnitude.

### THE WHIRLPOOL GALAXY
The core of this beautiful spiral galaxy (also known as M51) appears as a point of light in a small telescope, as does its companion galaxy NGC 5195 (top) at the end of one arm.

Canes Venatici also contains some fine galaxies, such as the Whirlpool Galaxy (see p.315), or M51, which is found seven diameters of a full moon from the star at the end of the handle of the Big Dipper (in Ursa Major). The Whirlpool Galaxy was the first galaxy in which spiral form was detected—the observation being made in 1845 by William Parsons (see p.315) in Ireland. The galaxy appears as a round patch of light through binoculars, but a moderate-sized telescope is needed to make out the spiral arms. At the end of one of the arms lies a smaller galaxy, NGC 5195, which is passing close to M51.

Two spiral galaxies worth looking for through a small telescope are the Sunflower Galaxy (M63) and M94.

**THE HUNTING DOGS**

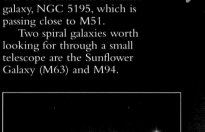

### GLOBULAR CLUSTER M3
This cluster is one of the biggest and brightest globular clusters in the northern sky. A telescope with a 4-in (100-mm) aperture is needed to resolve its individual stars.

**TWO BRIGHT STARS**
Canes Venatici represents a pair of hounds, but the unaided eye can see little more than the constellation's brightest stars, Cor Caroli and Beta Canum Venaticorum.

THE NIGHT SKY

# Boötes

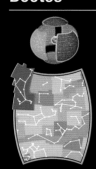

| | |
|---|---|
| **SIZE RANKING** | 13 |
| **BRIGHTEST STAR** | Arcturus (α) -0.1 |
| **GENITIVE** | Boötis |
| **ABBREVIATION** | Boo |
| **HIGHEST IN SKY AT 10 PM** | May–June |
| **FULLY VISIBLE** | 90°N–35°S |

The Greek constellation Boötes contains the brightest star north of the celestial equator, Arcturus—Alpha (α) Boötis—which is also the fourth-brightest star in the entire sky. This large and conspicuous constellation extends from Draco and the handle of the Big Dipper (in Ursa Major) to Virgo. Faint stars in the northern part of Boötes once formed the now-defunct constellation of Quadrans Muralis, which gave its name to the Quadrantid meteor shower that radiates from this area every January.

## SPECIFIC FEATURES

Arcturus is classified as a red giant, but as with most supposedly "red" stars, it actually looks orange to the unaided eye. Its coloring becomes stronger when viewed through binoculars. In billions of years, our Sun will swell into a red giant similar to this star.

Boötes is noted for its double stars, the most celebrated of which is Izar (see p.277), or Epsilon (ε) Boötis, at the heart of the constellation. To the naked eye, it appears of magnitude 2.4, but high magnification on a telescope of at least 3 in (75 mm) aperture reveals a close, 5th-magnitude companion that is blue-green in color, providing one of the most beautiful contrasts of all double stars.

Much easier to divide with any small telescope are Kappa (κ) and Xi (ξ) Boötis. Kappa's stars, with components of 5th and 7th magnitudes, are unrelated, but Xi, with stars also of 5th and 7th magnitudes, is a true binary with an orbital period of 150 years and has warm yellow-orange hues.

Easiest of all are the doubles Mu (μ) Boötis, with components of 4th and 6th magnitudes, and Nu (ν) Boötis, with two 5th-magnitude components—both are widely spaced enough to divide with binoculars.

**THE HERDSMAN**

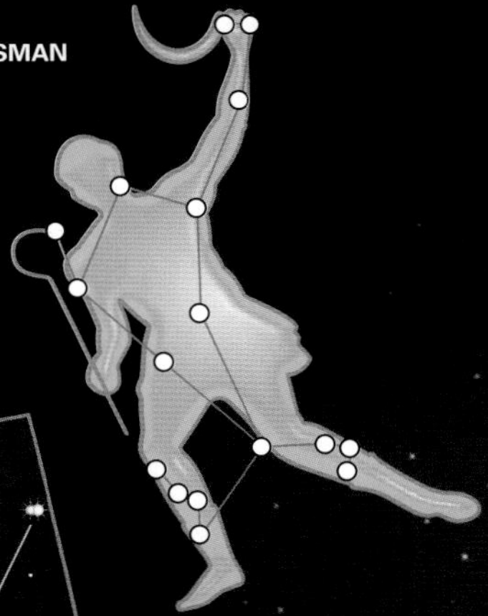

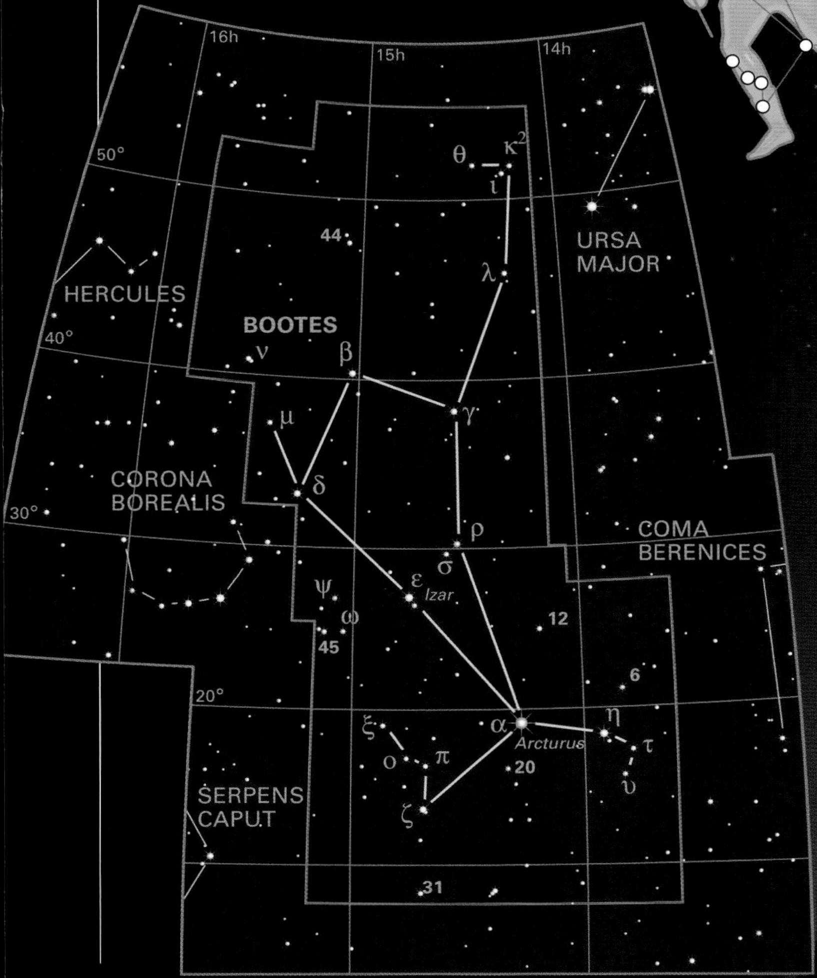

## THE BEAR-KEEPER

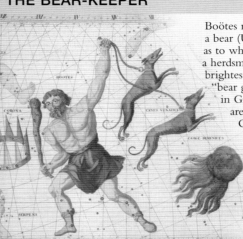

Boötes represents a man herding a bear (Ursa Major). Myths differ as to whether he is a hunter or a herdsman, as the constellation's brightest star, Arcturus, means "bear guard" or "bear-keeper" in Greek. The man's two dogs are represented by adjoining Canes Venatici. In Greek myth, Boötes was identified with Arcas, son of Zeus and Callisto.

### ADJACENT STARS

Boötes is depicted here, leading the two hunting dogs, in an 18th-century star chart by Sir James Thornhill.

**DOUBLE STAR IZAR** ⚲
Epsilon (ε) Boötis, which is also known as Izar or Pulcherrima, is a challenging double star consisting of a bright orange star with a fainter blue-green companion star.

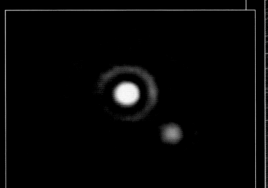

**KITE-SHAPED CONSTELLATION** 👁
Boötes, containing the bright star Arcturus, stands aloft in spring skies in the Northern Hemisphere. The crown of Corona Borealis can be seen to its left.

# Hercules

| | |
|---|---|
| **SIZE RANKING** | 5 |
| **BRIGHTEST STAR** | Kornephoros (β) 2.8 |
| **GENITIVE** | Herculis |
| **ABBREVIATION** | Her |
| **HIGHEST IN SKY AT 10 PM** | June–July |
| **FULLY VISIBLE** | 90°N–38°S |

This large but not particularly prominent constellation of the northern sky represents Hercules, the strong man of Greek myth. In the sky, Hercules is depicted clothed in a lion's pelt, brandishing a club and the severed head of the watchdog Cerberus and kneeling with one foot on the head of the celestial dragon, Draco—the tools and conquests of some of his 12 labors.

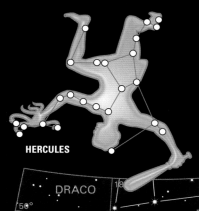

**HERCULES**

The most distinctive feature of this constellation is a quadrilateral of stars called the Keystone, which is composed of Epsilon (ε), Zeta (ζ), Eta (η), and Pi (π) Herculis.

### SPECIFIC FEATURES

Alpha (α) Herculis, (see p.285), or Rasalgethi, is actually the second-brightest star in Hercules. It fluctuates between 3rd and 4th magnitudes. As with most such erratic variables, Rasalgethi is a bloated red giant that pulsates in size, causing the brightness changes. A small telescope brings a 5th-magnitude blue-green companion star into view.

On one side of the Keystone lies M13, which is regarded as the finest globular cluster of northern skies. Under ideal conditions, M13 can be glimpsed with the naked eye, and through binoculars it appears like a hazy star half the width of a full moon. Slightly farther away from the Keystone is a second globular cluster— M92. This often overlooked cluster is smaller and fainter than M13, and when seen through binoculars, it can easily be mistaken for an ordinary star.

Several readily seen double stars are to be found in Hercules, including Kappa (κ) Herculis, with components of 5th and 6th magnitudes, and 100 Herculis, with its two 6th-magnitude stars. Positioned closer together, and hence requiring higher magnification, are 95 Herculis, with two 5th-magnitude components, and Rho (ρ) Herculis, with components of 5th and 6th magnitudes.

**UPSIDE DOWN** 👁
In the night sky, Hercules is positioned with his feet pointing toward the pole (top left in this picture) and his head pointing south.

**GLOBULAR CLUSTER M13** 🔭 ⚹
Through binoculars, this cluster appears as a rounded patch of light. It breaks up into countless starry points when viewed through a small telescope.

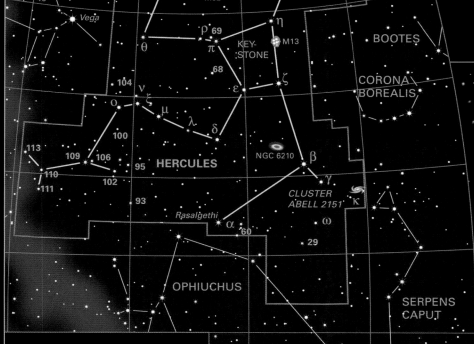

**THE HERCULES GALAXY CLUSTER** ⚌
Every fuzzy object in this picture is a faint galaxy in the cluster Abell 2151, some 500 million light-years away.

# yra

| | |
|---|---|
| **SIZE RANKING** | 52 |
| **BRIGHTEST STAR** | Vega (α) 0.0 |
| **GENITIVE** | Lyrae |
| **ABBREVIATION** | Lyr |
| **HIGHEST IN SKY AT 10 PM** | July–August |
| **FULLY VISIBLE** | 90°N–42°S |

yra lies on the edge of the Milky Way next to Cygnus and is a compact onstellation of the northern sky. It cludes Vega, or Alpha (α) Lyrae (see 253), which is the fifth-brightest star the sky and one of the so-called ummer Triangle of stars—the other vo being Deneb (in Cygnus) and ltair (in Aquila). The Lyrid meteors diate from a point near Vega round April 21–22 every year. Lyra presents the stringed instrument ayed by Orpheus (see panel, below).

## PECIFIC FEATURES

ega dazzles at magnitude 0.0, ppearing somewhat blue-white color to the unaided eye. It is

the standard star against which astronomers compare the color and brightness of all other stars.

The finest quadruple star in the sky—Epsilon (ε) Lyrae (see p.276)—is found three diameters of a full moon from Vega. Binoculars easily show it as a neat pair of 5th-magnitude white stars, but each of these has a closer companion that is brought into view with a telescope of 2.5–3 in (60–75 mm) aperture and high magnification. All four stars are linked by gravity and are in long-term orbit around each other.

Two other double stars near Vega that are easy to identify with binoculars are Zeta (ζ) and Delta (δ) Lyrae, each with components of 4th and 6th magnitudes. Beta (β) Lyrae is another double star, easily resolved by a small telescope into its cream and blue components. The brighter star (the cream one) is an eclipsing binary that fluctuates between magnitudes 3.3 and 4.4 every 12.9 days. Many years of study have established that Beta's two stars are so close that gas from the larger of the pair falls toward the smaller companion and some of it spirals off into space. Almost midway between Beta

and Gamma (γ) Lyrae lies the most photographed of Lyra's celestial treasures, the Ring Nebula (see p.257), or M57. This planetary nebula is shaped like a smoke ring and appears through a small telescope as a disk larger than that of Jupiter. Larger apertures are needed to make out the central hole. Studies with the Hubble Space Telescope have revealed that the "ring" is in fact a cylinder of gas thrown off from the central star, oriented almost end-on to the Earth.

**THE LYRE**

## HE RING NEBULA

ne of the most famous planetary nebulae the whole sky, the Ring Nebula, or M57, onsists of hot gas shed from a central star. s beautiful colors are revealed only in hotographs such as this one.

## ORPHEUS

Heartbroken Orpheus descended into the Underworld to retrieve his wife, Eurydice, who had been killed by a snake. His songs charmed Hades, god of the Underworld, who agreed to release Eurydice, provided Orpheus did not look back as he led her to the surface. At the last minute, Orpheus glanced behind him, and Eurydice faded away. Orpheus then roamed the Earth, disconsolately playing his lyre.

### ENTRANCED

Orpheus was said to have charmed even the rocks and streams with his music. In this 19th-century painting, he tames the wild animals with his songs.

**STRINGED INSTRUMENT**
Lyra, dominated by dazzling Vega, represents the harp played by Orpheus, the musician of Greek myth. Arab astronomers visualized the constellation as an eagle or vulture.

THE SWAN

# Cygnus

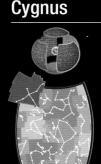

| | |
|---|---|
| **SIZE RANKING** | 16 |
| **BRIGHTEST STAR** | Deneb (α) 1.2 |
| **GENITIVE** | Cygni |
| **ABBREVIATION** | Cyg |
| **HIGHEST IN SKY AT 10 PM** | August–September |
| **FULLY VISIBLE** | 90°N–28°S |

Situated in a rich area of the Milky Way, Cygnus is one of the most prominent constellations of the northern sky and contains numerous objects of interest. The relatively large constellation depicts a swan in flight, but its main stars are arranged in the shape of a giant cross, hence its alternative popular name of the Northern Cross.

## SPECIFIC FEATURES
Cygnus's brightest star, Deneb—Alpha (α) Cygni—lies in the tail of the swan or at the top of the cross, depending on how the constellation is visualized. Deneb is an immensely luminous supergiant star located about 1,400 light-years away, making it the most distant 1st-magnitude star. It forms one corner of the northern Summer Triangle—a familiar sight in the skies of northern summers and southern winters—which is completed by Vega (in Lyra) and Altair (in Aquila).

The beak of the swan (or the foot of the cross) is marked by a double star, Beta (β) Cygni, known as Albireo. Its two stars are sufficiently far apart that they can be seen separately with ordinary binoculars, if steadily mounted, and they are easy targets for a small telescope. The brighter star, of magnitude 3.1, is orange, and the fainter star, magnitude 5.1, is blue-green.

A similar color difference is

### ALBIREO 🔭
Beta (β) Cygni, also known as Albireo, marks the beak of the swan. This double star, with its strikingly contrasting colors, is easily separated with a small telescope.

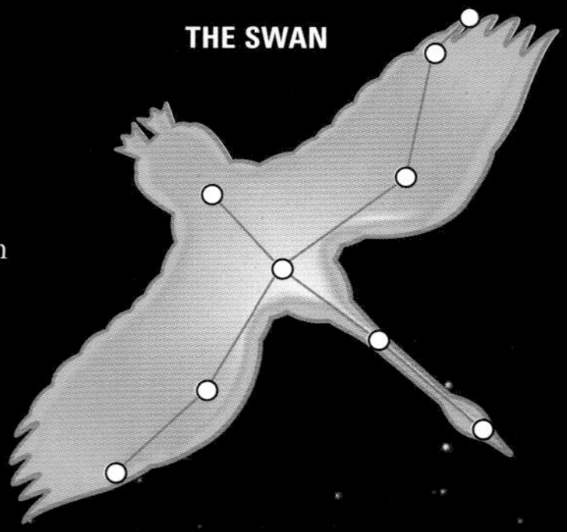

evident between Omicron-1 (o¹) Cygni, a 4th-magnitude orange star, and its wide 5th-magnitude companion, 30 Cygni, which has a noticeable bluish color when seen through binoculars. A 7th-magnitude star, again bluish, and even closer to Omicron-1, can also be seen with binoculars or a small telescope.

THE SWAN

### POISED IN FLIGHT 👁
Among the stars of Cygnus, it is comparatively easy to visualize a swan, with its wings outstretched, as it flies along the Milky Way.

Another pair of stars that is easy to spot with a small telescope is 61 Cygni (see p.252), which consists of two orange dwarfs of 5th and 6th magnitudes that orbit each other every 650 years. A large open star cluster, M39, covers an area of sky of similar size to a full moon near the constellation's border with Lacerta.

On clear nights, the Milky Way appears as a hazy band of light running through Cygnus, divided in two by an intervening cloud of dust known as the Cygnus Rift or the Northern Coalsack. The rift continues, via Aquila, into Ophiuchus.

Two large and remarkable nebulae are found in Cygnus, although neither is easy to identify. The glowing gas cloud of the North America Nebula

(NGC 7000), near Deneb, can be glimpsed through binoculars on clear, dark nights, but its full majesty becomes apparent only on long-exposure photographs or CCD images. The Veil Nebula is a diffuse nebula found in the wing of the swan. Again, it is best seen on photographs, although the brightest part—NGC 6992—can just be made out with binoculars or a small telescope and becomes more prominent with the addition of filters to the telescope. Considerably smaller but much easier to spot is the Blinking Planetary

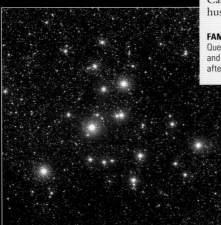

### OPEN CLUSTER M39 📷 🔭
M39 is the larger and brighter of the two Messier clusters in Cygnus and contains around 30 members arranged in a triangular shape, with a double star near the center. It lies 900 light-years away and is easily spotted with binoculars. Under good conditions, M39 is visible to the naked eye.

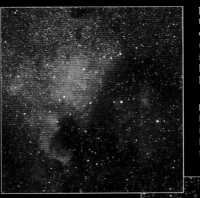

## LEDA AND THE SWAN

The swan represents the disguise adopted by Zeus for an illicit love tryst. The object of his desire is sometimes said to have been a nymph called Nemesis or, in a more popular version, Queen Leda of Sparta. After her union with Zeus, Leda is said to have given birth to either one or two eggs, according to different versions of the story, from which hatched Castor, Pollux, and their sister Helen of Troy. Pollux and Helen were reputedly the offspring of Zeus, but Castor was the son of Leda's husband, King Tyndareus.

**FAMILY GROUPING**
Queen Leda, the twins Castor and Pollux, and the swan are captured in this painting after the original by Leonardo da Vinci.

(NGC 6826) in the other wing of the swan, with a blue-green disk similar in size to that of Jupiter. It is popularly known as the Blinking Planetary because of an odd optical effect in which, as the observer looks alternately directly at it and off to one side, it appears to blink on and off.

Two objects of considerable astrophysical interest in Cygnus are beyond the reach of amateur observers. Cygnus A (see p.324) is a powerful radio source, the result of two galaxies in collision millions of light-years away. Cygnus X-1 (see p.272), near Eta (η) Cygni, is an intense X-ray source, thought to be a black hole orbiting a 9th-magnitude blue supergiant in our galaxy.

### NORTH AMERICA NEBULA 📷 🔭 🖥
In the tail of the swan lies NGC 7000, which is popularly known as the North America Nebula, on account of its similarity in shape to that continent.

### VEIL NEBULA 🔭 🖥
Splashed across an area wider than six full moons is the Veil Nebula, a loop of gas that is the remains of a star that exploded as a supernova thousands of years ago.

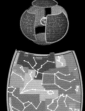

ANDROMEDA

# Andromeda

**SIZE RANKING** 19

**BRIGHTEST STARS**
Alpheratz (α) 2.1,
Mirach (β) 2.1

**GENITIVE**
Andromedae

**ABBREVIATION** And

**HIGHEST IN SKY AT 10 P.M.**
October–November

**FULLY VISIBLE**
90°N–37°S

This celebrated constellation of the northern skies depicts the daughter of the mythical Queen Cassiopeia, who is represented by a neighboring constellation. The head of the princess is marked by Alpheratz—Alpha (α) Andromedae—which is the star at the nearest corner of the Square of Pegasus, in another adjacent constellation. Long ago, Alpheratz was regarded as being shared with the constellation Pegasus, where it marked the navel of the horse. The star's official name—Alpheratz—is derived from an Arabic term that means "the horse's navel."

## SPECIFIC FEATURES
On a clear night, the farthest it is possible to see with the naked eye is about 2.5 million light-years, which is the distance to the Andromeda Galaxy (see pp.312–313), a huge spiral of stars similar to our own galaxy. Also known as M31, this galaxy spans several diameters of a full moon and lies high in the mid-northern sky on fall evenings. The naked eye sees it as a faint patch; it looks elongated, rather than spiral, because it is tilted at a steep angle toward the Earth. When looking at M31 through a telescope, low magnification must be used to give the widest field of view

and to concentrate the light. The small companion galaxies, M32 and M110, are difficult to see through a small telescope.

Gamma (γ) Andromedae, known also as Almach (see p.277), is a double star of contrasting colors. It consists of an orange giant star of magnitude 2.3 and a fainter blue companion, and it is easily seen as double through a small telescope.

The open star cluster NGC 752 spreads over an area larger than a full moon and can be identified with binoculars, but a small telescope is needed to resolve its individual stars of 9th magnitude and fainter.

NGC 7662, which is popularly known as the Blue Snowball, is one of the easiest planetary nebulae to identify, and it can be found through a small telescope.

**THE ANDROMEDA GALAXY**
Only the inner parts of M31 are bright enough to be seen with small instruments. CCD images such as this bring out the full extent of the spiral arms. Below M31 on this image lies M110, while M32 is on its upper rim.

**THE BLUE SNOWBALL**
When seen through a small telescope, NGC 7662 appears as a bluish disk. Its structure is brought out only on CCD images such as this one.

**MYTHS AND STORIES**

## HEROIC RESCUE

According to Greek mythology, Andromeda was chained to a rock on the seashore and offered as a sacrifice to a sea monster in atonement for the boastfulness of her mother, Queen Cassiopeia. The Greek hero Perseus, flying home after slaying Medusa, the Gorgon, noticed the maiden's plight. He responded by swooping down in his winged sandals and killing the sea monster. He then whisked Andromeda to safety and married her.

**DAMSEL IN DISTRESS**
The Flemish artist Rubens added the flying horse Pegasus to his 17th-century depiction of Andromeda's dramatic rescue by Perseus from captivity on the rock.

**HEAD TO TOE**
Andromeda is one of the original Greek constellations. Its brightest stars represent the princess's head (α), her pelvis (β), and her left foot (γ).

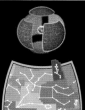

THE LIZARD

# Lacerta

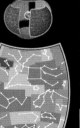

**SIZE RANKING** 68

**BRIGHTEST STAR**
Alpha (α) 3.8

**GENITIVE**
Lacertae

**ABBREVIATION** Lac

**HIGHEST IN SKY AT 10 P.M.**
September–October

**FULLY VISIBLE**
90°N–33°S

Lacerta consists of a zigzag of faint stars in the northern sky, squeezed between Andromeda and Cygnus like a lizard between rocks. It is one of the seven constellations invented by Johannes Hevelius (see p.384) during the late 17th century.

This constellation contains no objects of note for amateur astronomers, although BL Lacertae (see p.325), which was once thought

to be a peculiar 14th-magnitude variable star, has given its name to a class of galaxies with active nuclei called BL Lac objects or "blazars."

A BL Lac object is a type of quasar that shoots jets of gas from its center directly toward the Earth. Because we see these jets of gas head-on, these BL Lac objects tend to look starlike.

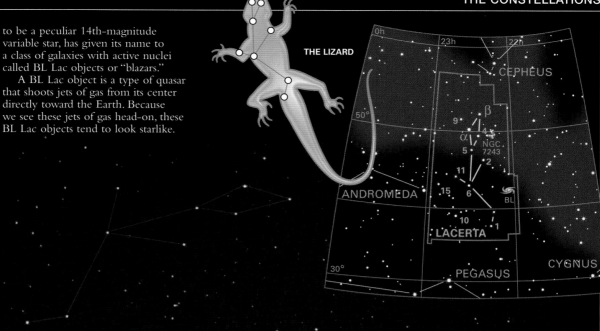

THE LIZARD

---

THE TRIANGLE

# Triangulum

**SIZE RANKING** 78

**BRIGHTEST STAR**
Beta (β) 3.0

**GENITIVE** Trianguli

**ABBREVIATION** Tri

**HIGHEST IN SKY AT 10 P.M.**
November–December

**FULLY VISIBLE**
90°N–52°S

This small northern constellation is to be found lying between Andromeda and Aries. It consists of little more than a triangle of three stars. Triangulum is one of the constellations known to the ancient Greeks, who visualized it as the Nile delta or the island of Sicily.

## SPECIFIC FEATURES
Triangulum contains the third-largest member of our Local Group of galaxies, M33 or the Triangulum Galaxy (see p.311). In physical terms, M33 is about one-third the size of the Andromeda Galaxy, or M31 (see pp.312–313), and is much fainter.

The spiral galaxy M33 appears as a large pale patch of sky. It is similar in size to a full moon, when viewed through binoculars or a small telescope on a dark, clear night. To see the spiral arms, a large telescope is needed. M33 looks like a starfish on long-exposure photographs.

There is little else of note in the constellation apart from 6 Trianguli. This yellow star has a magnitude of 5.2 and has a 7th-magnitude companion that can be detected through a small telescope.

THE TRIANGLE

**TRIANGULUM AND MARS** 👁
This image of the three stars that make up the shape of Triangulum also includes the planet Mars, passing through neighboring Pisces.

**M33** ✸ ⌨
The clouds of pinkish gas in the arms of M33 show up in CCD images of this spiral galaxy in the Local Group. It is presented almost face-on to the Earth.

THE NIGHT SKY

## THE VICTORIOUS HERO

# Perseus

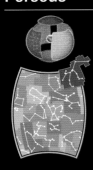

**SIZE RANKING** 24

**BRIGHTEST STAR**
Mirfak (α) 1.8

**GENITIVE** Persei

**ABBREVIATION** Per

**HIGHEST IN SKY AT 10 P.M.**
November–December

**FULLY VISIBLE**
90°N–31°S

Perseus is a prominent northern constellation lying in the Milky Way between Cassiopeia and Auriga. It is an original Greek constellation and represents Perseus, who was sent to slay Medusa, the Gorgon. In the sky, Perseus is depicted with his left hand holding the Gorgon's head, which is marked by Algol—Beta (β) Persei—a famous variable star (see p.276). His right hand brandishes his sword, marked by the twin clusters NGC 869 and NGC 884.

## SPECIFIC FEATURES

The constellation's brightest member—Mirfak, or Alpha (α) Persei—is of magnitude 1.8. It lies at the center of a group of stars known as the Alpha Persei Cluster or Melotte 20. Scattered over an area of sky that is several times the diameter of a full moon, the cluster is an excellent sight through binoculars.

Algol is an eclipsing binary consisting of two stars in close orbit, one much hotter and brighter than the other. Together they shine at magnitude 2.1, but every 69 hours the fainter star eclipses its companion. Over a period of five hours, the combined light of the pair drops to just one-third its normal value, a change that is readily noticeable to the naked eye. Algol's brightness returns to normal after another 5 hours. Predictions of Algol's eclipses can be found in astronomical annuals and magazines.

Rho (ρ) Persei is a variable of a different kind: it is a red giant that fluctuates by about 50 percent in brightness every 7 weeks or so.

Popularly termed the Double Cluster, the twin open clusters NGC 869 and NGC 884 are one of the showpieces of the northern sky. Each cluster contains hundreds of stars of 7th magnitude and fainter, and covers an area of sky similar to that of a full moon. They lie more than 7,000 light-years away in the Perseus spiral arm of our galaxy. Both clusters are noticeable to the naked eye as a brighter patch in the Milky Way near the border with Cassiopeia and can be seen well through binoculars or a small telescope.

M34 is a scattered open cluster of several dozen stars near the border with Andromeda. It covers an apparent area similar to that of a full moon and is easy to spot through binoculars.

**ALPHA PERSEI CLUSTER** 👁
Mirfak and its surrounding cluster lie above center. The Pleiades Cluster is lower right, and Capella, in Auriga, is lower left.

**THE VICTORIOUS HERO**

MYTHS AND STORIES

## MEDUSA

Perseus, the son of Zeus and Danaë, was sent to bring back the head of Medusa, the Gorgon, whose evil gaze turned everything to stone. He was given a bronze shield by the goddess Athena, a sword of diamond by Hephaestus, and winged sandals by Hermes. Looking only at Medusa's reflection in his shield, Perseus managed to decapitate the Gorgon.

**SUCCESSFUL MISSION**
Perseus proudly displays the severed head of Medusa, the Gorgon, in this neoclassical sculpture by Antonio Canova.

**DOUBLE CLUSTER** 👥 ✦
Of these two star clusters, NGC 869 (left) appears to be more densely packed. NGC 884 (right) contains some red giant stars, which its neighbor lacks.

# Aries

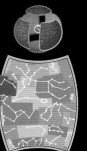

**SIZE RANKING** 39

**BRIGHTEST STAR**
Hamal (α) 2.0

**GENITIVE** Arietis

**ABBREVIATION** Ari

**HIGHEST IN SKY AT 10 P.M.**
November–December

**FULLY VISIBLE**
90°N–58°S

This not particularly conspicuous constellation of the zodiac is found between Pisces and Taurus. Its most recognizable features are three stars near the border with Pisces: Alpha (α), Beta (β), and Gamma (γ) Arietis, of 2nd, 3rd, and 4th magnitudes.

Aries depicts the golden-fleeced ram of Greek legend (see panel, below). Over 2,000 years ago, the vernal equinox—the point at which the ecliptic crosses the celestial equator—lay near the border of Aries and Pisces. The effect of precession (see p.64) has now moved the vernal equinox almost into Aquarius, but it is still called the first point of Aries.

## SPECIFIC FEATURES

Gamma was one of the first stars discovered to be double, and it was found by the English scientist Robert Hooke in 1664, when telescopes were still quite crude and it was not realized that double stars are numerous. To the naked eye, it appears of 4th magnitude, but when viewed through a small telescope it consists of nearly identical white stars of magnitudes 4.6 and 4.7.

Lambda (λ) Arietis, of 5th magnitude, has a companion of 7th magnitude that can be seen through large binoculars. Pi (π) Arietis, also of 5th magnitude, has a very close companion of 8th magnitude.

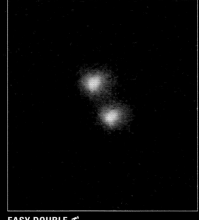

**EASY DOUBLE** ✶
Gamma (γ) Arietis is readily separable by a small telescope to reveal a pair of white stars, each of 5th magnitude.

THE RAM

PERSEUS
TRIANGULUM
ARIES
TAURUS
CETUS
PISCES
Hamal
Sheratan
Mesarthim
ECLIPTIC

## MYTHS AND STORIES

# THE GOLDEN FLEECE

Aries represents the ram whose golden fleece hung on a tree in Colchis on the Black Sea. Jason and the Argonauts undertook an epic voyage to bring this fleece back to Greece. Jason was aided in his task by Medea, who had fallen in love with him. She was the daughter of King Aeetes, who owned the fleece. Medea bewitched the serpent guarding the fleece so that Jason could steal it. Taking Medea and the fleece with him, Jason then sailed away in the *Argo*.

**GOLDEN MOMENT**
Watched by an admiring Medea, Jason removes the glittering fleece from the oak tree on which it hung at Colchis, in this illustration by L. du Bois-Reymond.

**LEGENDARY RAM** ⊙
From a crooked line formed by three faint stars, ancient astronomers visualized the figure of a crouching ram, with its head turned back over its shoulder.

# Taurus

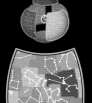

**SIZE RANKING** 17

**BRIGHTEST STAR**
Aldebaran (α) 0.85

**GENITIVE** Tauri

**ABBREVIATION** Tau

**HIGHEST IN SKY AT 10 P.M.**
December–January

**FULLY VISIBLE**
88°N–58°S

Taurus is a large and prominent northern constellation of the zodiac, and it contains a wealth of objects including the Pleiades and Hyades star clusters (see p.291 and p.290) and M1, the Crab Nebula (see pp.270–271). Its stars represent the head and fore-quarters of a mythical Greek bull. The Hyades cluster is centered on the bull's face, while the constellation's brightest star, Aldebaran—Alpha (α) Tauri (see p.256)—is its glinting eye. Elnath (or Elnath)—Beta (β) Tauri—and Zeta (ζ) Tauri mark the tips of the bull's long horns. Each November, the Taurid meteors appear to radiate from a point south of the Pleiades.

## SPECIFIC FEATURES

Aldebaran is a red giant whose color is clearly apparent to the naked eye. As with many red giants, it is slightly variable in brightness, but the amount is only about

### THE CRAB NEBULA
This supernova reveals the beauty of a massive star's violent death throes. Convoluted filaments of gas expand away from the site of the supernova explosion, which was seen from Earth in 1054 CE.

one-tenth of a magnitude either side of its average value of 0.85 and is barely noticeable. Although Aldebaran appears to be part of the Hyades cluster, it is 67 light-years away—less than half the cluster's distance—and is superimposed only by chance.

The main stars of the Hyades are arranged in a V-shape that is the width of over 10 diameters of a full moon. More than a dozen stars are visible with the unaided eye, and dozens more come into view through binoculars. At 150 light-years away, the Hyades is the closest major star cluster to Earth. On one arm of the Hyades' V-shape is a wide double star, Theta (θ) Tauri. At magnitude 3.4, the brighter of the pair, Theta-1 (θ¹), is also the brightest member of the Hyades. Another double star that is easy to spot is Sigma (σ) Tauri, which has two 5th-magnitude components, near Aldebaran. The apex of the Hyades cluster points toward Lambda (λ) Tauri, an eclipsing binary of the same type as Algol (in Perseus). It varies between magnitudes 3.4 and 3.9 in a cycle lasting just under 4 days.

An even brighter star cluster is the Pleiades, which hovers over the bull's shoulders. Although popularly known as the Seven Sisters, after a group of mythical Greek nymphs (see panel, opposite), the Pleiades in fact contains nine named stars: the seven sisters and their parents, Atlas and Pleione. The brightest member is Alcyone (see p.277), which is of magnitude 2.9 and lies near the center of the cluster. The Pleiades covers an area of sky three times the width of a full moon. On long-exposure photographs of the Pleiades, a surrounding haze is visible. This was once thought to be leftover gas and dust from the

The Hyades (lower left) is the larger of these two dazzling star clusters; the Pleiades (upper right) is a tighter bunch that appears hazy at first glance—good viewing conditions are needed to see all nine named stars with the naked eye.

stars' formation, but it is now recognized as an unrelated cloud into which the cluster has drifted.

The first object on Charles Messier's list of cometlike objects (see p.73), M1 is the remains of a star that exploded as a supernova in 1054 CE. It was given its popular name, the Crab Nebula, by the Irish astronomer William Parsons (see p.315) in 1844, because he

thought the filaments of gas that protruded from the supernova remnant resembled the legs of a crab. The Crab Nebula is found two diameters of a full moon away from Zeta Tauri. Through a small telescope, it appears as a faint elliptical glow several times larger than the disk of Jupiter. Large apertures are needed to make out the level of detail seen by Parsons.

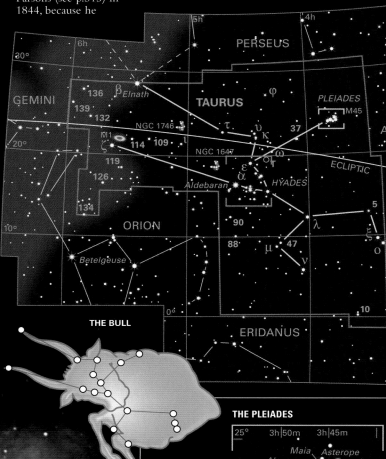

**THE BULL**

**THE PLEIADES**

Maia  Asterope
Alcyone      Taygeta
Celaeno
Pleione         Electra
Atlas
η      Merope

**THE HYADES**

ε

δ³
δ²  δ¹
Aldebaran
α    75
σ²           63
89    85    θ¹ 70
γ
81 80 θ² 71
ρ              58
76  π  60  57

**MAGNITUDE KEY**

| | |
|---|---|
| ✦ | 0.0–0.9 |
| ✦ | 1.0–1.9 |
| ✦ | 2.0–2.9 |
| ✦ | 3.0–3.9 |
| ✦ | 4.0–4.9 |
| • | 5.0–5.9 |
| • | 6.0–6.9 |

**RAGING BULL** 👁
Taurus, the celestial bull, thrusts his star-tipped horns into the night air. The bull is said to represent a disguise adopted by Zeus in a Greek myth. The bright reddish "star" seen here on the bull's back, below the Pleiades, is actually the planet Mars.

MYTHS AND STORIES

## THE LOST PLEIAD

The popular name for the Pleiades is the Seven Sisters, although only six stars are easily visible to the naked eye. Two myths have arisen to explain the "missing" Pleiad. One myth says that the star that shines least brightly is Merope, the only one of the seven sisters to marry a mortal. Another story says that it is Electra, who could not bear to stay and watch the fall of Troy, the city founded by her brother.
The names of the stars in the cluster do not follow either of these legends, however, as the faintest named member is actually Asterope.

**WANDERING STAR**
This 19th-century painting, *The Lost Pleiad*, depicts the separation of one of the Pleiades from her sisters.

## THE TWINS

# Gemini

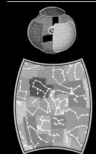

| | |
|---|---|
| **SIZE RANKING** 30 | |
| **BRIGHTEST STAR** Pollux (β) 1.2 | |
| **GENITIVE** Geminorum | |
| **ABBREVIATION** Gem | |
| **HIGHEST IN SKY AT 10 P.M.** January–February | |
| **FULLY VISIBLE** 90°N–55°S | |

This prominent zodiacal constellation represents the mythical twins Castor and Pollux, who were the sons of Queen Leda of Sparta and the brothers of Helen of Troy (see Leda and the Swan, p.367). The constellation is easily identifiable within the northern sky because of its two brightest stars, which are named after the twins. Even though it is labeled Beta (β) Geminorum, Pollux is brighter than Castor, or Alpha (α) Geminorum (see p.276). The two stars mark the heads of the twins, while their feet lie bathed in the Milky Way. In mid-December each year, the Geminid meteors radiate from a point in Gemini near Castor.

## SPECIFIC FEATURES

Castor is a remarkable multiple star. To the naked eye, it appears as a single entity of magnitude 1.6, but through a small telescope with suitably high magnification, it divides into a sparkling blue-white duo of 2nd and 3rd magnitudes. The two stars form a genuine binary, with an orbital period of 450 years, which also has a 9th-magnitude red dwarf companion. Although these three stars cannot be divided further visually, each is a spectroscopic binary, bringing the total number of stars in the Castor system to six.

Although Castor and Pollux are named after twins, the stars themselves are far from identical. Being an orange giant, Pollux is noticeably warmer-toned than Castor. It is also closer to Earth, lying only 34 light-years away, compared to Castor's 51 light-years.

The open star cluster M35 lies at the feet of the twins. Under clear skies, this cluster can be glimpsed with the naked eye, but it is more easily found with binoculars, through which it appears as an elongated, elliptical patch of starlight spanning the same apparent width as a full moon. When viewed through a small telescope, its individual stars seem to form chains or curved lines.

Two variable stars of note in Gemini are Zeta (ζ) Geminorum (see p.286), which is a Cepheid variable that ranges between magnitudes 3.6 and 4.2 every 10.2 days, and Eta (η) Geminorum (see p.284), which is a red giant whose brightness can vary anywhere between magnitudes 3.1 and 3.9. This constellation also contains the

Eskimo Nebula, or NGC 2392 (see p.259), a planetary nebula with a bluish disk similar in size to that of the globe of Saturn and visible through a small telescope. Larger telescope apertures are needed to reveal the nebula's surrounding fringe of gas, reminiscent of an Inuit parka, that gives NGC 2392 its popular name. An alternative name for this nebula is the Clown-Face Nebula.

### THE TWINS

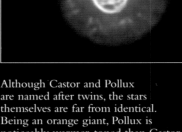

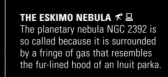

**THE ESKIMO NEBULA** ✦ ⌨
The planetary nebula NGC 2392 is so called because it is surrounded by a fringe of gas that resembles the fur-lined hood of an Inuit parka.

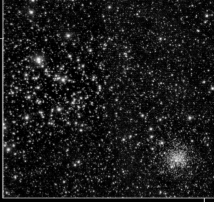

**LARGE AND SMALL CLUSTER** 🔭 ✦
The large star cluster M35 is visible through binoculars; larger telescopes reveal a fainter and more distant cluster, NGC 2158 (bottom right), in the same field of view.

**CELESTIAL TWINS** 👁
Castor and Pollux, the twins of the Greek myth, stand side by side in the sky between Taurus and Cancer. The bright "star" in the middle of Gemini in this picture is actually the planet Saturn.

# THE CRAB

# Cancer

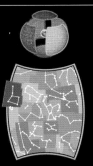

| | |
|---|---|
| **SIZE RANKING** | 31 |
| **BRIGHTEST STAR** | Beta (β) 3.5 |
| **GENITIVE** | Cancri |
| **ABBREVIATION** | Cnc |
| **HIGHEST IN SKY AT 10 P.M.** | February–March |
| **FULLY VISIBLE** | 90°N–57°S |

Cancer is the faintest of the 12 zodiacal constellations, lying in the northern sky between Gemini and Leo, and it represents the crab of Greek mythology (see panel, right). Cancer includes the major open star cluster M44 (see p.290), which is alternatively known as the Beehive Cluster, the Manger Cluster, or Praesepe—which is the Latin for both "hive" and "manger." It also includes the stars Gamma (γ) and Delta (δ) Cancri, which represent two donkeys feeding at the manger. These two stars are sometimes known as Asellus Borealis and Asellus Australis, the northern and southern asses.

## SPECIFIC FEATURES

Iota (ι) Cancri is a 4th-magnitude yellow giant with a nicely contrasting 7th-magnitude blue-white companion. The companion is just detectable through 10 × 50 binoculars, and it is easy to identify through a small telescope. Another double star that can be seen through a small telescope is Zeta (ζ) Cancri. Its components, of 5th and 6th magnitude, form a binary star with an orbital period of more than 1,000 years.

The Beehive Cluster (M44) is a large open cluster at the heart of Cancer, located between Gamma (γ) and Delta (δ) Cancri. The ancient Greeks could see the cluster as a misty spot with the unaided eye, but in modern urban skies it is unlikely to be visible without binoculars. This cluster consists of a scattering of stars of 6th magnitude and fainter. It appears to cover an area more than three times wider than the diameter of a full moon, and although it can be seen through binoculars, it is too wide to fit in the field of view of most telescopes.

The Beehive Cluster's glory overshadows another open cluster, M67, which is smaller and denser yet still the width of a full moon in the sky. It lies about 2,600 light-years away—more distant than the Beehive Cluster, which is just under 600 light-years away. M67 can be found with binoculars, but a telescope is needed to resolve individual stars. At an estimated age of around 5 billion years, it is one of the oldest open clusters known—it is also of an age similar to Earth's.

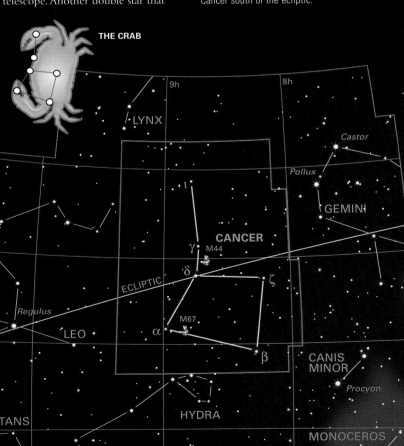

**M67** 🔭
Inferior to M44, but still worthy of note, M67 can be found with binoculars in the region of Cancer south of the ecliptic.

### THE BEEHIVE CLUSTER 🔭
Also known as the Manger, M44 is an open cluster located between the two asses feeding from the manger, Gamma (γ) (center, top) and Delta (δ) Cancri (center, bottom).

## MYTHS AND STORIES

# A SMALL VICTORY

According to the Greek story, a crab attacked Hercules during his fight with the many-headed Hydra but was crushed underfoot during the struggle. Such a minor role befits this faint constellation.

### SCUTTLING AWAY
A small crab can be seen in the foreground of this 18th-century engraving, *Hercules Fights the Lernean Hydra*.

**THE CRAB**

Castor
Pollux
GEMINI
LYNX
ι
CANCER
γ  M44
δ
ζ
ECLIPTIC
Regulus
M67
α
LEO
β
CANIS MINOR
HYDRA
Procyon
TANS
MONOCEROS
9h
8h

### HIDDEN CRAB 👁
Cancer is the faintest constellation in the zodiac, but it contains a major star cluster, M44, which is just visible in this photograph as a hazy patch near the center of the constellation.

THE NIGHT SKY

# Leo Minor

| | |
|---|---|
| **SIZE RANKING** | 64 |
| **BRIGHTEST STAR** | 46 Leonis Minoris 3.8 |
| **GENITIVE** | Leonis Minoris |
| **ABBREVIATION** | LMi |
| **HIGHEST IN SKY AT 10 P.M.** | March–April |
| **FULLY VISIBLE** | 90°N–48°S |

This small, insignificant constellation, adjacent to Leo in the northern sky, represents a lion cub, although this is not suggested by the pattern of its stars. It was introduced in the 17th century by the Polish astronomer Johannes Hevelius (see p.384).

## SPECIFIC FEATURES

Unusually, this constellation has no star labeled Alpha. This is due to an error by the 19th-century English astronomer Francis Baily, who assigned the Greek letters to the constellation's stars. When doing so, he overlooked assigning a Bayer letter to the brightest star, 46 Leonis Minoris, which should have been recorded as Alpha (α), although he did label the second-brightest star as Beta (β) Leonis Minoris.

Although Leo Minor contains no objects of interest for users of binoculars or a small telescope, Beta (β) is a close double star that can be separated by a telescope with very large aperture. It has a magnitude of 4.2, and its component stars orbit each other every 37 years.

### THE LION CUB 👁

Having located the distinctive shape of the Sickle in Leo (top, right), look north of it to find the faint stars of Leo Minor.

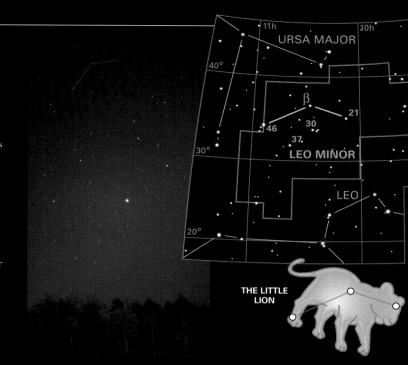

**THE LITTLE LION**

---

# Coma Berenices

| | |
|---|---|
| **SIZE RANKING** | 42 |
| **BRIGHTEST STAR** | Beta (β) 4.2 |
| **GENITIVE** | Comae Berenices |
| **ABBREVIATION** | Com |
| **HIGHEST IN SKY AT 10 P.M.** | April–May |
| **FULLY VISIBLE** | 90°N–56°S |

Coma Berenices represents the flowing locks of Queen Berenice of Egypt, which she cut off as a tribute to the gods after the safe return of her husband, Ptolemy III, from battle during the 3rd century BCE. It is a faint but interesting northern constellation, lying between Leo and Boötes. In the mid-16th century, it was named as a constellation by the German cartographer Caspar Vopel. Before then, its stars were regarded as forming the tail of Leo.

## SPECIFIC FEATURES

The Coma Star Cluster, also known as Melotte 111, is the constellation's main feature. It comprises several dozen faint stars, which fan out distinctively for several diameters of a full moon southward from Gamma (γ) Comae Berenices. This open cluster, which is seen to best advantage through binoculars, has been imagined as both the bushy tip of a lion's tail and a lock of Berenice's hair.

Coma Berenices contains numerous galaxies in its southern half. Most of these are members of the Virgo Cluster, such as M85, M88, M99, and M100, but two notable exceptions, M64 (see p.314) and NGC 4565, are closer to the Earth.

Popularly known as the Black Eye Galaxy, M64 is a spiral galaxy tilted at an angle to the Earth, which is seen as an elliptical patch of light through a small telescope; it is best seen with a telescope with an aperture of 6 in (150 mm) or more. A dust cloud near the galaxy's nucleus creates the "black eye" effect.

NGC 4565, another spiral galaxy, lies edge-on to the Earth and is more difficult to spot. It appears long and thin when viewed through a telescope with a 4-in (100-mm) aperture, and a lane of dark dust is revealed in long-exposure photographs.

### THE BLACK EYE GALAXY 🔭 💻

The spiral galaxy M64 sports a large, dark dust cloud near its core, giving it the appearance of a blackened eye.

**NGC 4565** 🔭 💻

Seen edge-on, this spiral galaxy displays a lane of dark dust along its spiral arms when viewed through larger apertures.

**BERENICE'S HAIR**

### MANE OF HAIR 👁

The distinctive splay of the Coma Star Cluster marks out Coma Berenices in the night sky. Leo's hindquarters can be seen closer to the horizon.

## THE LION

# Leo

**SIZE RANKING** 12

**BRIGHTEST STAR**
Regulus (α) 1.4

**GENITIVE** Leonis

**ABBREVIATION** Leo

**HIGHEST IN SKY AT 10 P.M.**
March–April

**FULLY VISIBLE**
82°N–57°S

The outline stars of Leo really do bear a marked resemblance to a crouching lion, in this large constellation of the zodiac, located just north of the celestial equator. It is one of the easiest constellations to recognize. The pattern of six stars that marks the lion's head and chest is known as the Sickle and is shaped like a reversed question mark or a hook. The Leonid meteors radiate from the region of the Sickle every November (see pp.220–221).

SPECIFIC FEATURES
Regulus—Alpha (α) Leonis (see p.253)—lies at the foot of the Sickle. It is the faintest of the first-magnitude stars, at magnitude 1.4, and its wide companion is of 8th magnitude.

The double star Algieba, or Gamma (γ) Leonis, consists of components of magnitudes 2.2 and 3.5. Both stars are orange giants, and they orbit each other every 600 years or so. A nearby star—40 Leonis—is unrelated.

Zeta (ζ) Leonis is a wide triple star, consisting of a 3rd-magnitude star with a 6th-magnitude companion to both the north and south, which can be seen with binoculars. All three stars are at different distances from Earth and, hence, they are unrelated.

A pair of spiral galaxies, M65 and M66, can be glimpsed with a small telescope beneath the hindquarters of Leo. A fainter pair of spirals, M95 and M96, lie under the lion's body, as does an elliptical galaxy, M105, 1 one degree away.

**THE LION**

**ALGIEBA** ✈
This beautiful pair of golden-colored orange giants is clearly visible through small telescopes.

**LEO TRIPLET** ✈ ⌨
A trio of galaxies lies near Theta (θ) Leonis: M65 (lower right); M66 (lower left); and the edge-on spiral NGC 3628 (top). Although NGC 3628 appears the largest on photographs, it is less bright than the others and is difficult to see through small telescopes.

**THE BIG CAT** 👁
The crouching lion is a distinctive sight in the night sky. The pattern of its stars is disturbed here by the presence of Jupiter under the lion's body.

### MYTHS AND STORIES

## HERCULES AND THE LION

Leo represents the mythical lion that lived in a cave near the Greek town of Nemea, terrorizing the area and emerging to attack and devour local inhabitants. As the first of the 12 labors in his quest for immortality, Hercules was sent by his cousin Eurystheus to kill the lion. Finding that the creature's hide was impervious to his arrows, Hercules instead wrestled and strangled the beast. He then used the lion's own razor-sharp claws to cut off its pelt, which he wore victoriously as a cloak.

**THE HERO AND THE BEAST**
Hercules grapples with the Nemean Lion in a sculpture by the 16th-century Flemish artist Jean de Boulogne, or Giambologna.

# Virgo

**SIZE RANKING** 2

**BRIGHTEST STAR**
Spica (α) 1.0

**GENITIVE** Virginis

**ABBREVIATION** Vir

**HIGHEST IN SKY AT 10 P.M.**
April–June

**FULLY VISIBLE**
67°N–75°S

Virgo straddles the celestial equator, between Leo and Libra. It is the largest constellation of the zodiac, and the second-largest overall. The constellation depicts a Greek virgin goddess (see panel, right). Virgo contains the Virgo Cluster (see p.329), the closest large cluster of galaxies to Earth, which is some 50 million light-years away and which extends over the border of Virgo into Coma Berenices. The Sun is in Virgo during the September equinox each year.

## SPECIFIC FEATURES
Gamma (γ) Virginis, or Porrima (see p.253), is a binary star with the relatively short period of 169 years. As a result of this short period, the effects of the two stars' orbital motions can easily be followed through amateur telescopes. As seen from Earth, the two stars were closest together in 2005, when a telescope with an aperture of 10 in (250 mm) was needed to separate them. By 2012, the stars had moved far enough apart that they could be divided by

a telescope of only 2.4 in (60 mm). For the rest of the 21st century, it will be possible to split the components of Gamma Virginis with a small-aperture telescope. Both of the stars are of magnitude 3.5.

In the upper part of Virgo's body lie the numerous galaxies of the Virgo Cluster. None is easy to see with a small instrument. The brightest members are giant ellipticals, notably M49, M60 (see p.317), M84, M86, and M87 (see p.323). M87 is a strong radio and X-ray source also known as Virgo A. Long-exposure photographs show it is ejecting a jet of gas, like certain quasars.

The Sombrero Galaxy (see p.316), or M104, is Virgo's best-known galaxy. This spiral is about two-thirds as far away as the Virgo Cluster. It is oriented almost edge-on to the Earth, so that a dark lane of dust in the galaxy's plane crosses its central bulge. The bulge may be all that can be seen through a small telescope; the dust lane is only revealed when seen through a large-aperture telescope or on long-exposure photographs.

The brightest quasar in the sky, 3C 273 (see p.325), also lies in the bowl of Virgo. However, it is much more distant than the Virgo Cluster. Through most telescopes, it appears as nothing more than a 13th-magnitude star. Only professional equipment will reveal it as the center of an active galaxy, which is some 2,000 million light-years away from Earth.

**MYTHS AND STORIES**

## THE VIRGIN GODDESS

Virgo is usually identified as Dike, the Greek goddess of justice, who abandoned the Earth and flew up to heaven when human behavior deteriorated. Neighboring Libra represents her scales of justice. Virgo is also visualized as Demeter, the harvest goddess, who holds an ear of wheat, which is represented by the constellation's brightest star, Spica.

**BOUNTIFUL OFFERINGS**
Demeter presented Triptolemus, a prince of Eleusis, with a chariot drawn by winged dragons and grains of wheat to sow crops wherever he traveled.

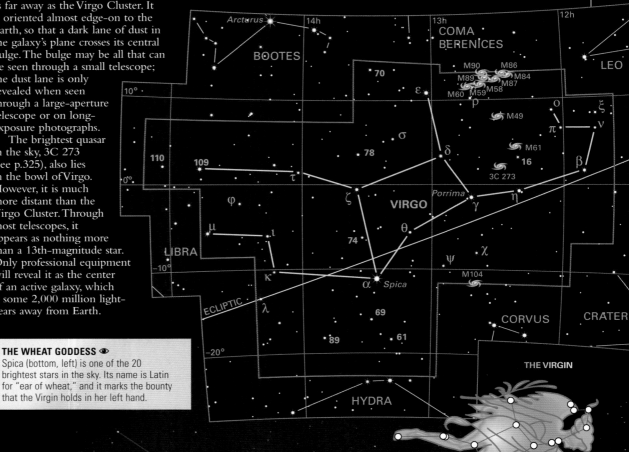

**THE WHEAT GODDESS** 👁
Spica (bottom, left) is one of the 20 brightest stars in the sky. Its name is Latin for "ear of wheat," and it marks the bounty that the Virgin holds in her left hand.

**THE SOMBRERO GALAXY** ✦
The Sombrero Galaxy (M104) is a spiral galaxy with a large central bulge, seen almost edge-on, and resembling a Mexican hat. It lies about 30 million light-years away.

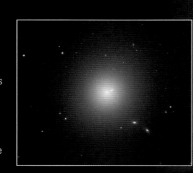

**M87** ✦ 🖥
Through a small telescope, the giant elliptical galaxy M87 appears as a rounded glow, but photographs and CCD images reveal the jet of gas that is being expelled from its highly active nucleus. Here, the jet is just visible near the top right of the core.

# Libra

**SIZE RANKING** 29

**BRIGHTEST STAR**
Beta (β) 2.6

**GENITIVE** Librae

**ABBREVIATION** Lib

**HIGHEST IN SKY AT 10 P.M.**
May–June

**FULLY VISIBLE**
60°N–90°S

This constellation of the zodiac lies just south of the celestial equator between Virgo and Scorpius. Originally, the ancient Greeks visualized the constellation as the claws of the neighboring Scorpius, which is why Libra's brightest stars have names that mean "northern claw" and "souwthern claw." Libra's present-day identification as Virgo's scales of justice became more common in Roman times.

## SPECIFIC FEATURES

Zubenelgenubi (Arabic for "the southern claw") or Alpha (α) Librae is a wide double star of 3rd and 5th magnitudes and is easily divisible with binoculars or even sharp unaided eyesight. To the north of this pair is the constellation's brightest star,

Zubeneschamali ("the northern claw") or Beta (β) Librae, which shows a greenish tinge when viewed through binoculars or a telescope. This highly unusual coloring is due, presumably, to the chemical composition of Zubeneschamali's outer layers.

In the heart of the constellation lies Iota (ι) Librae, a double with stars of 5th and 6th magnitudes which can be viewed through binoculars. A small telescope will reveal the closer 9th-magnitude companion

of the brighter star. Mu (μ) Librae, with components of 6th and 7th magnitude, is a more difficult pair to separate; a telescope with 3 in (75 mm) aperture is needed.

Delta (δ) Librae is an eclipsing variable. Every 2 days 8 hours, it rises and falls between 5th and 6th magnitudes. This change can be easily followed with binoculars.

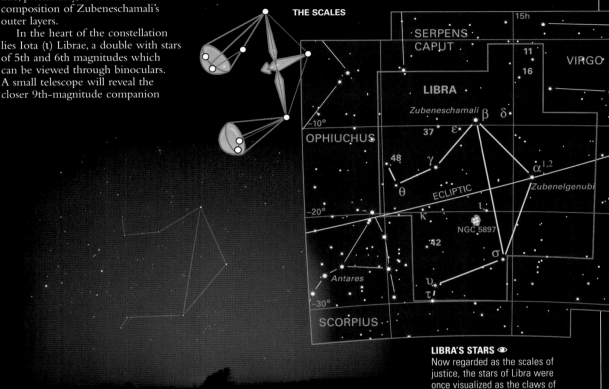

THE SCALES

**LIBRA'S STARS** 👁
Now regarded as the scales of justice, the stars of Libra were once visualized as the claws of the adjacent scorpion, Scorpius.

---

# Corona Borealis

**SIZE RANKING** 73

**BRIGHTEST STAR**
Alphecca (α) 2.2

**GENITIVE**
Coronae Borealis

**ABBREVIATION** CrB

**HIGHEST IN SKY AT 10 P.M.**
June

**FULLY VISIBLE**
90°N–50°S

Corona Borealis is a small but distinctive constellation in the northern sky, between Boötes and Hercules, consisting of a horseshoe shape of seven stars. It is one of the original Greek constellations and represents the crown worn by Princess Ariadne (see panel, right).

## SPECIFIC FEATURES

The arc of the northern crown contains the remarkable variable star R Coronae Borealis (see p.287), a yellow supergiant normally of 6th magnitude, which shows sudden dips in brightness. These fades, which are due to a build-up of sooty particles in its atmosphere, occur every few years and can last for months.

Corona Borealis has three double stars of note for small-instrument users, although none is particularly bright. Nu (ν) Coronae Borealis is a pair of 5th-magnitude red giants divisible with binoculars. Zeta (ζ) Coronae Borealis is a blue-white pair, with

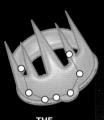

THE NORTHERN CROWN

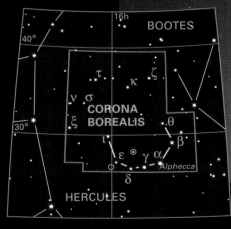

**CROWN OF STARS** 👁
Like a celestial tiara, the seven main stars of Corona Borealis form a distinctive arc between Boötes and Hercules.

components of 5th and 6th magnitudes—an attractive sight when seen through a small telescope—while Sigma (σ) Coronae Borealis is a yellow pair with components of 6th and 7th magnitudes, which can also be split with a small telescope.

## PRINCESS ARIADNE

Ariadne, daughter of King Minos of Crete, helped Theseus slay the Minotaur, a gruesome creature that was half bull, half human. Theseus sailed off with Ariadne to the island of Naxos, where he then abandoned her. The god Dionysus looked down on the princess and was overcome. At their wedding, Ariadne wore a jewel-studded crown, which Dionysus threw into the sky, where the crown's jewels were changed into stars.

**CROWNING GLORY**
Dionysus, known as Bacchus by the Romans, holds Ariadne's jeweled crown in this painting by the 17th-century French artist Eustache Le Sueur.

## THE SERPENT

# Serpens

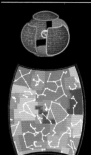

**SIZE RANKING** 23

**BRIGHTEST STAR**
Unukalhai (α) 2.6

**GENITIVE**
Serpentis

**ABBREVIATION** Ser

**HIGHEST IN SKY AT 10 PM**
June–August

**FULLY VISIBLE**
74°N–64°S

Although counted as a single constellation, Serpens is in fact split into two separate areas, and is thus unique. It is one of the original 48 Greek constellations and straddles the celestial equator. Serpens represents a huge snake coiled around Ophiuchus, who holds the head (Serpens Caput) in his left hand and the tail (Serpens Cauda) in his right. In Greek mythology, snakes were a symbol of rebirth, because of the fact that they shed their skins. Ophiuchus represents the great healer Asclepius, who was reputedly able to revive the dead (see panel, opposite).

## SPECIFIC FEATURES

The Eagle Nebula (see pp.244–245) in Serpens Cauda was made world-famous by a spectacular Hubble Space Telescope picture of dark columns of dust embedded within its glowing gas. Unfortunately, the dust columns show up only through a telescope of large aperture and on long-exposure photographs such as those from the Hubble Space Telescope.

The Eagle Nebula contains a star cluster, M16, which can be spotted readily through binoculars or a small telescope. It appears as a hazy patch covering an area of sky that is similar in size to a full moon. Another open cluster that is visible through binoculars is IC 4756, which appears

about twice the size of M16. It is situated in Serpens Cauda near the tip of the serpent's tail.

Close to the border with Virgo lies M5, which is about 25,000 light-years away. Its condensed center appears as a faint area about half the size of a full moon, when viewed with binoculars, while the curving chains of stars in its outskirts are revealed only through a telescope with an aperture of 4 in (100 mm) or more.

Delta (δ) Serpentis, near the serpent's head, is a binary with components of 4th and 5th magnitudes. It is divisible using high powers of magnification on a small telescope.

Theta (θ) Serpentis, near the serpent's tail, is a pair of white stars that are easily split through a small telescope. This wide double star has components of magnitude 4.6 and 5.0.

**M5** 👥 🔭
This is one of the finest globular clusters in northern skies. M5 is noticeably elliptical in shape when viewed through a telescope.

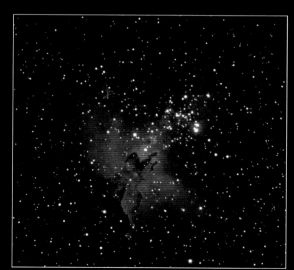

**THE EAGLE NEBULA** 🔭 💻 📷
This image was captured by a professional 13-ft 1-in (4-m) telescope. It can only be seen well with telescopes of large aperture.

**SERPENTINE STARS** 👁
The upper part of the snake (above, right) contains Unukalhai (α), which derives its name from the Arabic for "the serpent's neck."

# Ophiuchus

**SIZE RANKING** 11

**BRIGHTEST STAR**
Rasalhague (α) 2.1

**GENITIVE**
Ophiuchi

**ABBREVIATION** Oph

**HIGHEST IN SKY AT 10 PM**
June–July

**FULLY VISIBLE**
59°N–75°S

This large constellation straddling the celestial equator depicts a man holding a snake. The head of Ophiuchus adjoins Hercules in the north, while his feet rest on Scorpius in the south. The Sun passes through Ophiuchus in the first half of December, but despite this, the constellation is not regarded as a true member of the zodiac.

Ophiuchus was the site of the last supernova explosion seen in our galaxy, which appeared in 1604. It far outshone all other stars and is known as Kepler's Star (see p.273) after Johannes Kepler, who wrote about it in *De stella nova* (see p.68).

## SPECIFIC FEATURES
Lying on the edge of the Milky Way, in the direction of the center of our galaxy, Ophiuchus contains numerous star clusters. Messier cataloged seven globular clusters, although none is particularly prominent. M10 and M12 (see p.295) are both near the center of the constellation and detectable through binoculars on a clear night. Better sights for binoculars are two large and scattered open clusters, NGC 6633 and IC 4665.

An amazing multiple star is Rho (ρ) Ophiuchi, lying near Antares (in neighboring Scorpius). This 5th-magnitude star has a 7th-magnitude companion on either side of it, which are best viewed through binoculars. Another 6th-magnitude companion that is much closer to the central star can be identified through a small telescope using high magnification. The complex nebulosity in this area, including around Antares, is revealed only in long-exposure photographs.

The beautiful double star 70 Ophiuchi consists of yellow and orange dwarfs, with components of 4th and 6th magnitudes, while the double star 36 Ophiuchi is a pair of orange dwarfs with components of 5th magnitude.

Barnard's Star is the most celebrated star in Ophiuchus and is the second-closest star to the Sun. Even though this red dwarf is a mere 5.9 light-years away, its light output is so feeble that it appears as only magnitude 9.5, and it is too faint to see without a telescope. Barnard's Star is moving so quickly relative to the background stars that its change in position is noticeable over a matter of only a few years (see chart, right).

**THE SERPENT HOLDER**

**INTRICATE NEBULOSITY** 💻
Complex nebulosity extends from the area around Rho (ρ) Ophiuchi (at the top of the image below), southward to Antares (bottom).

**BARNARD'S STAR MOVEMENT**

**MAGNITUDE KEY**

| | |
|---|---|
| ✦ | 0.0–0.9 |
| ✦ | 1.0–1.9 |
| ✦ | 2.0–2.9 |
| ✦ | 3.0–3.9 |
| • | 4.0–4.9 |
| • | 5.0–5.9 |
| · | 6.0–6.9 |

**M10** 🔭 ✶
The large globular cluster M10 is some 14,000 light-years away. Like its neighbor M12, it is detectable through binoculars on a clear night.

**SNAKE MAN** 👁
Ophiuchus represents a man wrapped in the coils of a huge snake, the constellation Serpens. The ecliptic runs through Ophiuchus, and planets can be seen within its borders.

## MYTHS AND STORIES

### ASCLEPIUS

Ophiuchus is identified with Asclepius, the Greek god of medicine who reputedly had the power to revive the dead. Hades, god of the Underworld, feared that this ability endangered his trade in dead souls and asked Zeus to strike Asclepius down. Zeus then placed the great healer among the stars.

**RESTORATIVE POWERS**
Asclepius is watched as he heals a female patient, in this 5th-century BCE marble relief from Piraeus, Greece.

## THE SHIELD

# Scutum

| | |
|---|---|
| **SIZE RANKING** | 84 |
| **BRIGHTEST STAR** | Alpha (α) 3.8 |
| **GENITIVE** | Scuti |
| **ABBREVIATION** | Sct |
| **HIGHEST IN SKY AT 10 PM** | July–August |
| **FULLY VISIBLE** | 74°N–90°S |

This minor constellation is situated in a rich area of the Milky Way, between Aquila and Sagittarius, south of the celestial equator. It was introduced by Johannes Hevelius (see p.384) in the late 17th century. He gave it the name *Scutum Sobiescianum*, meaning Sobieski's Shield, to honor his patron, King John Sobieski of Poland.

### SPECIFIC FEATURES
Delta (δ) Scuti is the prototype of a class of variable star that pulsates in size every few hours, changing brightness by only a few tenths of a magnitude. Delta itself varies between magnitude 4.6 and 4.8 in less than 5 hours, but the change is only detectable with sensitive instruments. Far more obvious is R Scuti, an orange supergiant that rises and falls between magnitudes 4.2 and 8.6 in a 20-week cycle.

Near R Scuti is the beautiful Wild Duck Cluster (M11), which appears as a smudgy glow half the apparent width of a full moon when viewed through binoculars. This open cluster gained its popular name because its

**SCUTUM STAR CLOUD** 👁🔭
One of the brightest parts of the Milky Way lies in Scutum and is known as the Scutum Star Cloud. The bright spot at center left is the Wild Duck Cluster.

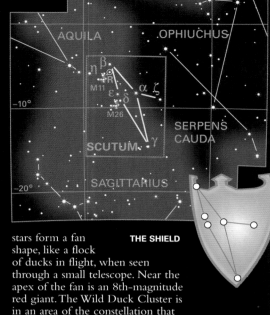

stars form a fan shape, like a flock of ducks in flight, when seen through a small telescope. Near the apex of the fan is an 8th-magnitude red giant. The Wild Duck Cluster is in an area of the constellation that is known as the Scutum Star Cloud. This rich star field is located just south of Beta (β) Scuti.

**THE SHIELD**

**WILD DUCKS** 🔭
Seen through a small telescope, M11 looks like the V-shaped flight pattern of wildfowl. This effect is less apparent on photographs.

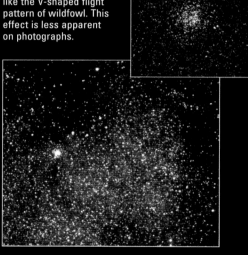

**SOBIESKI'S SHIELD** 👁
Scutum has no bright stars of its own, but it lies in an area of the Milky Way, between Aquila and Sagittarius, that is particularly rich with stars.

---

## THE ARROW

# Sagitta

| | |
|---|---|
| **SIZE RANKING** | 86 |
| **BRIGHTEST STAR** | Gamma (γ) 3.5 |
| **GENITIVE** | Sagittae |
| **ABBREVIATION** | Sge |
| **HIGHEST IN SKY AT 10 PM** | August |
| **FULLY VISIBLE** | 90°N–69°S |

Sagitta was known to the ancient Greeks, who believed it represented an arrow shot by either Apollo, Hercules, or Eros. It is the third-smallest constellation, lying in the Milky Way between Vulpecula and Aquila in the northern sky. It is faint and easily overlooked.

### SPECIFIC FEATURES
There is little of note in Sagitta for users of small instruments. Zeta (ζ) Sagittae is a 5th-magnitude star with a 9th-magnitude companion that is visible in a small telescope, but it is not a particularly impressive double. S Sagittae is a Cepheid variable that halves in brightness every 8.4 days before recovering again, as it swings between magnitudes 5.2 and 6.0.

Midway along the shaft of the arrow is M71, a modest globular cluster detectable with binoculars but better seen through a telescope. M71 lacks the central condensation typical of most globulars and instead looks more like a dense open cluster.

WZ Sagittae is a dwarf nova variable (see Novae, p.282). It is rarely in outburst.

**ARROW IN FLIGHT** 👁
The small arrow Sagitta flies over the stars of Aquila, the eagle, and toward Delphinus, the dolphin.

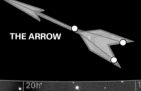

**THE ARROW**

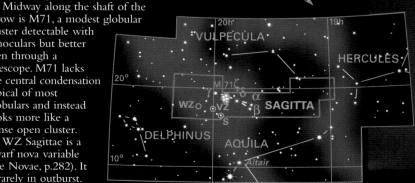

## THE EAGLE

# Aquila

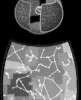

| | |
|---|---|
| **SIZE RANKING** | 22 |
| **BRIGHTEST STAR** | Altair (α) 0.8 |
| **GENITIVE** | Aquilae |
| **ABBREVIATION** | Aql |
| **HIGHEST IN SKY AT 10 PM** | July–August |
| **FULLY VISIBLE** | 78°N–71°S |

Aquila depicts an eagle in flight (see panel, right). It lies on the celestial equator in a rich area of the Milky Way near Cygnus, Scutum, and Sagittarius, yet there are no deep-sky objects of particular note within it. Aquila's brightest star, Altair or Alpha (α) Aquilae (see p.252), forms one corner of the northern Summer Triangle of stars, completed by Vega (in Lyra) and Deneb (in Cygnus).

Altair is flanked by 4th-magnitude Alshain, or Beta (β) Aquilae, and 3rd-magnitude Tarazed, or Gamma (γ) Aquilae, which form a distinctive trio.

### SPECIFIC FEATURES

Aquila's main feature of interest is Eta (η) Aquilae (see p.286), which is one of the brightest Cepheid variables. Eta ranges between magnitudes 3.5 and 4.4 on a cycle of 7.2 days. As with all members of this class, it is a brilliant supergiant. Its distance is estimated at 1,400 light-years.

The constellation also has some faint double stars that can readily be split with a small telescope: 15 Aquilae, with stars of 5th and 7th magnitudes, and 57 Aquilae, with two 6th-magnitude components.

**THE HOOK** ⊙🔭
This easily recognizable group of stars in southern Aquila includes Lambda (λ) Aquilae (center left) and branches into neighboring Scutum.

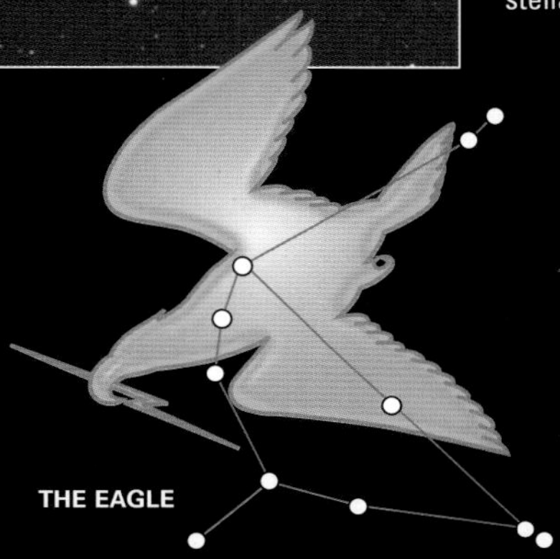

**THE EAGLE**

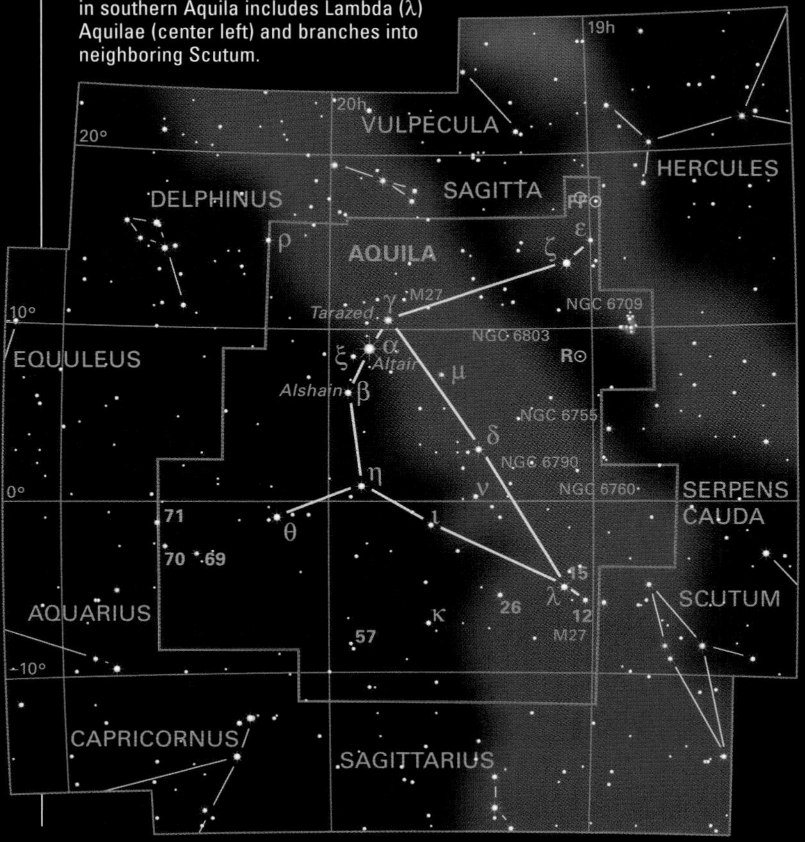

### MYTHS AND STORIES

## WINGED CARRIERS

The eagle has at least two identifications in Greek mythology. It was the bird that carried the thunderbolts for the god Zeus, and in one myth Zeus sent an eagle, or took the form of an eagle, to carry the shepherd boy Ganymede up to Mount Olympus, where he was made a servant of the gods. Zeus had spied the boy tending sheep in a field and had become infatuated with him. Ganymede is represented by neighboring Aquarius.

**ON EAGLE'S WINGS**
The beautiful youth Ganymede is carried aloft by an eagle in Peter Paul Rubens's 17th-century painting *The Abduction of Ganymede*.

**STELLAR TRIO** ⊙
Altair, the constellation's brightest star, is flanked by 3rd-magnitude Tarazed (top), which has a noticeably orange color, and 4th-magnitude Alshain (bottom), forming an attractive stellar trio.

**SWOOPING ACROSS THE SKIES** ⊙
The eagle swoops across the evening skies in the second half of the year. Its main star, Altair, is the most southerly of those that form the northern Summer Triangle. Aquila points toward the stars of Capricornus.

THE NIGHT SKY

THE FOX

# Vulpecula

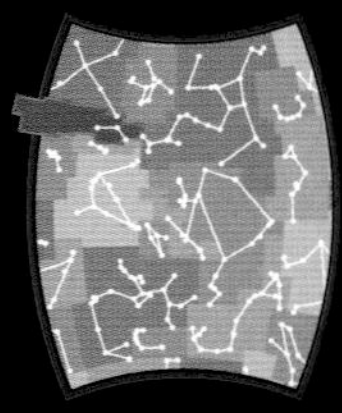

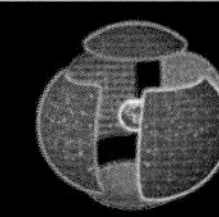

| | |
|---|---|
| **SIZE RANKING** | 55 |
| **BRIGHTEST STAR** | Alpha (α) 4.4 |
| **GENITIVE** | Vulpeculae |
| **ABBREVIATION** | Vul |
| **HIGHEST IN SKY AT 10 PM** | August–September |
| **FULLY VISIBLE** | 90°N–61°S |

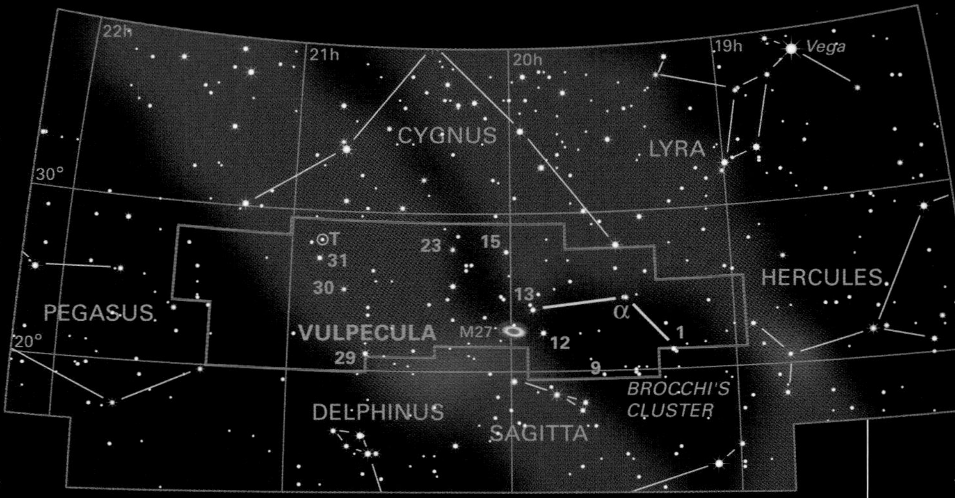

This small, faint northern constellation lies in the Milky Way, south of Cygnus. When it was first introduced in the late 17th century by the Polish astronomer Johannes Hevelius (see panel, below), it was

**THE FOX**

named *Vulpecula cum Anser* (the fox with the goose). Its name has since been simplified to Vulpecula. Despite its relative obscurity, it contains two unmissable objects for binocular users.

**SPECIFIC FEATURES**
The brightest star in the constellation, Alpha (α) Vulpeculae, is a 4th-magnitude red giant with a 6th-magnitude orange star nearby, which is visible with binoculars. The two lie at different distances and are unrelated.

Brocchi's Cluster is one of the binocular treasures of the sky. This grouping of 10 stars, with components ranging from 5th to 7th magnitude, is better known as the Coathanger because of its shape: a line of six stars forms the bar of the hanger, while the remaining four are

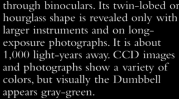

the hook. All the stars are unrelated, however, and so do not form a true cluster. The Coathanger's shape is therefore the delightful product of a chance alignment.

Popularly known as the Dumbbell Nebula, M27 is the easiest planetary nebula to spot in the sky. It appears as a rounded patch, about one-third the

size of a full moon, when viewed through binoculars. Its twin-lobed or hourglass shape is revealed only with larger instruments and on long-exposure photographs. It is about 1,000 light-years away. CCD images and photographs show a variety of colors, but visually the Dumbbell appears gray-green.

**THE DUMBBELL NEBULA** ♒ ✦ ⌨
Reputedly the easiest planetary nebula to spot, M27 can be found with binoculars on dark nights. A telescope is needed to make out the twin lobes that give rise to its popular name.

**THE COATHANGER** ♒
Perhaps the most charming of all star clusters is Brocchi's Cluster, also known as the Coathanger. This group of stars, easily visible through binoculars, appears to mark out the shape of a simple coathanger.

## JOHANNES HEVELIUS

Johannes Hevelius (1611–1687) was born and worked in the town of Danzig, Germany (now Gdansk, Poland), where he established an observatory equipped with the finest instruments of his time. Among his legacies was a star catalog and atlas, published posthumously by his assistant and second wife, Elizabeth, introducing seven new constellations and filling the gaps in the northern skies.

**JOINT EFFORT**
Johannes Hevelius and his wife Elizabeth measured star positions with a large sextant. This instrument is commemorated in one of the constellations Hevelius invented, Sextans.

**FOX IN THE MILKY WAY** 👁
Vulpecula is a shapeless constellation sandwiched between the more easily recognizable pattern of Sagitta, the arrow (at the left of this picture), and the head of the swan, Cygnus.

## THE DOLPHIN

# Delphinus

**SIZE RANKING** 69

**BRIGHTEST STAR**
Rotanev (β) 3.6

**GENITIVE** Delphini

**ABBREVIATION** Del

**HIGHEST IN SKY AT 10 PM**
August–September

**FULLY VISIBLE**
90°N–69°S

This small but distinctive constellation is situated between Aquila and Pegasus. According to Greek myth, Delphinus represents the dolphin that saved the poet and musician Arion from drowning after he leaped into the sea to escape robbers onboard a ship. Alternatively, the constellation is said to depict one of the dolphins sent by Poseidon to bring the sea nymph Amphitrite to him to marry. It is one of the constellations listed by the astronomer Ptolemy (see p.347).

The whole constellation was once popularly known as Job's Coffin, presumably because of the boxlike shape of its area, although sometimes this name is restricted to the diamond asterism formed by the four brightest

**GAMMA DELPHINI** ⚲
Gamma (γ) Delphini is an attractive double star. Although both the component stars are usually described as yellow, some observers see the fainter star as bluish.

stars: Sualocin (α), Rotanev (β), and Gamma (γ) and Delta (δ) Delphini. Who coined the name Job's Coffin and when is not known.

### SPECIFIC FEATURES
Gamma (γ) Delphini is normally described as an attractive orange-yellow double star. Its components are of 4th and 5th magnitudes, and they are easily separated by a small telescope.

The fainter and closer double star Struve 2725, which has components of 7th and 8th magnitudes, can also be seen through a small telescope and is visible in the same field of view as Gamma (γ) Delphini.

## NICCOLÒ CACCIATORE

Alpha (α) and Beta (β) Delphini bear the unusual names Sualocin and Rotanev. When reversed, these names spell Nicolaus Venator. This is the Latinized name of Niccolò Cacciatore (1780–1841), an Italian astronomer who was assistant to Giuseppe Piazzi, the director of the Palermo Observatory, Sicily. Cacciatore defied convention by surreptitiously naming two stars after himself in the Palermo star catalog of 1814. No one realized what he had done until much later, by which time the star names had become established.

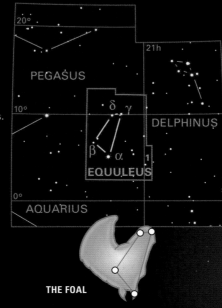

THE DOLPHIN

**THE PLAYFUL DOLPHIN** 👁
The kite-shaped Delphinus, on the edge of the Milky Way near Cygnus, brings to mind a dolphin jumping from ocean waters.

## THE FOAL

# Equuleus

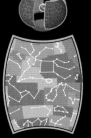

**SIZE RANKING** 87

**BRIGHTEST STAR**
Alpha (α) 3.9

**GENITIVE** Equulei

**ABBREVIATION** Equ

**HIGHEST IN SKY AT 10 PM**
September

**FULLY VISIBLE**
90°N–77°S

The second-smallest constellation in the sky represents the head of a young horse, or foal, and lies next to the larger celestial horse, Pegasus. No myths or legends are associated with Equuleus, which is thought to have

been added to the sky by the Greek astronomer Ptolemy in his 2nd-century-CE compendium of the original Greek constellations.

### SPECIFIC FEATURES
Gamma (γ) Equulei is a wide double star with components of 5th and 6th magnitudes and is easily separated with binoculars. Its two stars are unrelated. The 5th-magnitude double star 1 Equulei—labeled as Epsilon (ε) Equulei on some maps—has a 7th-magnitude companion, which can be seen through a small telescope, and a fainter true companion, which can be seen only through instruments with larger apertures. Other than these two double stars, there is nothing of note in Equuleus for users of binoculars or small telescopes.

THE FOAL

**THE HORSE'S HEAD** 👁
Equuleus consists of a small area of faint stars wedged between Pegasus and Delphinus and is easily overlooked.

## THE WINGED HORSE
# Pegasus

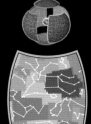

**SIZE RANKING** 7

**BRIGHTEST STARS**
Beta (β) 2.4,
Epsilon (ε) 2.4

**GENITIVE** Pegasi

**ABBREVIATION** Peg

**HIGHEST IN SKY AT 10 P.M.**
September–October

**FULLY VISIBLE**
90°N–53°S

Pegasus lies north of the zodiacal
constellations Aquarius and Pisces,
in low northern declinations, and it
adjoins Andromeda. It was one of
the original 48 Greek constellations.
Pegasus represents the flying horse
ridden by the hero Bellerophon,
although he is sometimes wrongly
identified as the steed of Perseus
(see panel, below). Although only
the forequarters of the horse are
indicated by stars, the constellation
is still the seventh-largest in the sky.

### SPECIFIC FEATURES
The Great Square of Pegasus is
formed by the stars Alpha (α), Beta
(β), and Gamma (γ) Pegasi, plus Alpha
(α) Andromedae. Long ago, the fourth
star of the Square was also known as
Delta (δ) Pegasi and was shared with

### THE GREAT SQUARE 👁
The most distinctive feature of
this constellation is the Great
Square of Pegasus, which forms
the horse's body.

Andromeda, but now it is
exclusively Andromeda's. A
line of more than 30 full moons
would fit into the Square,
yet for such a large area it is
surprisingly devoid of stars,
its brightest one being Upsilon
(υ) Pegasi, of magnitude 4.4.
Therefore, when the Great
Square is viewed through
polluted skies, it may seem
completely empty.

The constellation's two
brightest stars are Beta, a
red giant that varies between
magnitudes 2.3 and 2.7, and
Epsilon (ε) Pegasi, a yellow
star of magnitude 2.4 with
a wide 8th-magnitude
companion that can be seen
through a small telescope.

Not far from Epsilon
lies the globular cluster M15
(see p.295), which is one
of the finest such objects
in northern skies. It is just
at the limit of naked-eye
visibility when viewed
through clear skies.

Just outside the Great
Square of Pegasus is a 5th-magnitude
star called 51 Pegasi, which was the
first star beyond the Sun confirmed to
have a planet in orbit around it. This
planet was discovered in 1995, and
its mass is about half that of Jupiter.

**M15** 🔭 ✶
A telescope
with 6-in (150-mm)
aperture resolves this
cluster into individual
stars. It is over 30,000
light-years away.

**THE WINGED HORSE**

### MYTHS AND STORIES
## BELLEROPHON AND PEGASUS

Pegasus the winged horse was
born from the body of Medusa,
the Gorgon, when she was
decapitated by Perseus. He flew
to Mount Helicon, home of the
Muses, where he stamped on the
ground and brought forth a spring
called Hippocrene, the "horse's
fountain." With the aid of a golden
bridle from Athena, the hero
Bellerophon tamed Pegasus, and
rode the horse on his successful
mission to kill the fire-breathing
monster Chimera. Bellerophon
later attempted to ride Pegasus up
to Olympus to join the gods, but
he fell off. The horse arrived safely.

**TAKING FLIGHT**
Pegasus beats his wings, as though attempting
to ascend to the skies, in this statue in
Powerscourt Gardens, Dublin, Ireland.

## THE WATER CARRIER

# Aquarius

**SIZE RANKING** 10

**BRIGHTEST STARS**
Sadalmelik (α) 2.9,
Sadalsuud (β) 2.9

**GENITIVE**
Aquarii

**ABBREVIATION** Aqr

**HIGHEST IN SKY AT 10 P.M.**
June–July

**FULLY VISIBLE**
65°N–86°S

This large constellation of the zodiac is visualized as a youth (or, sometimes, an older man) pouring water from a jar. It is is found between Capricornus and Pisces, near the celestial equator. The stars Gamma (γ), Zeta (ζ), Eta (η), and Pi (π) Aquarii form a Y-shaped grouping that makes up the Water Jar, from which a stream of stars represents water flowing toward Piscis Austrinus. In early May each year, the Eta Aquariid meteor shower radiates from the area of the water jar.

In Greek myths and stories, Aquarius represents Ganymede—a beautiful shepherd boy to whom the god Zeus took a liking, too. Zeus dispatched his eagle (or, in some stories, turned himself into an eagle) to carry Ganymede up to Mount Olympus, where he became a cup bearer to the gods. The Eagle is represented by neighboring Aquila.

### THE SATURN NEBULA ✦ 💻
NGC 7009's resemblance to the ringed planet Saturn is most evident when it is viewed through a large telescope or on a CCD image.

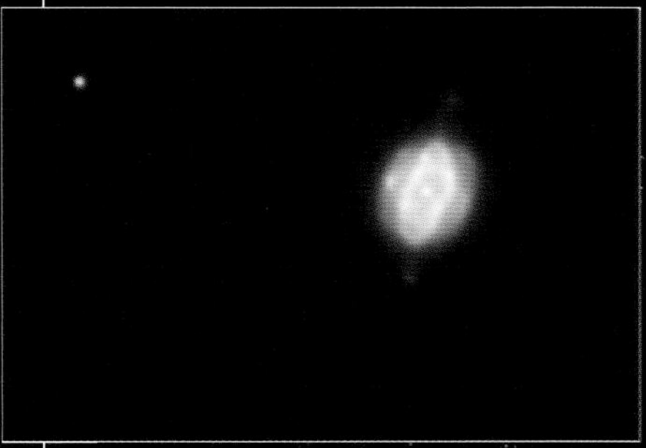

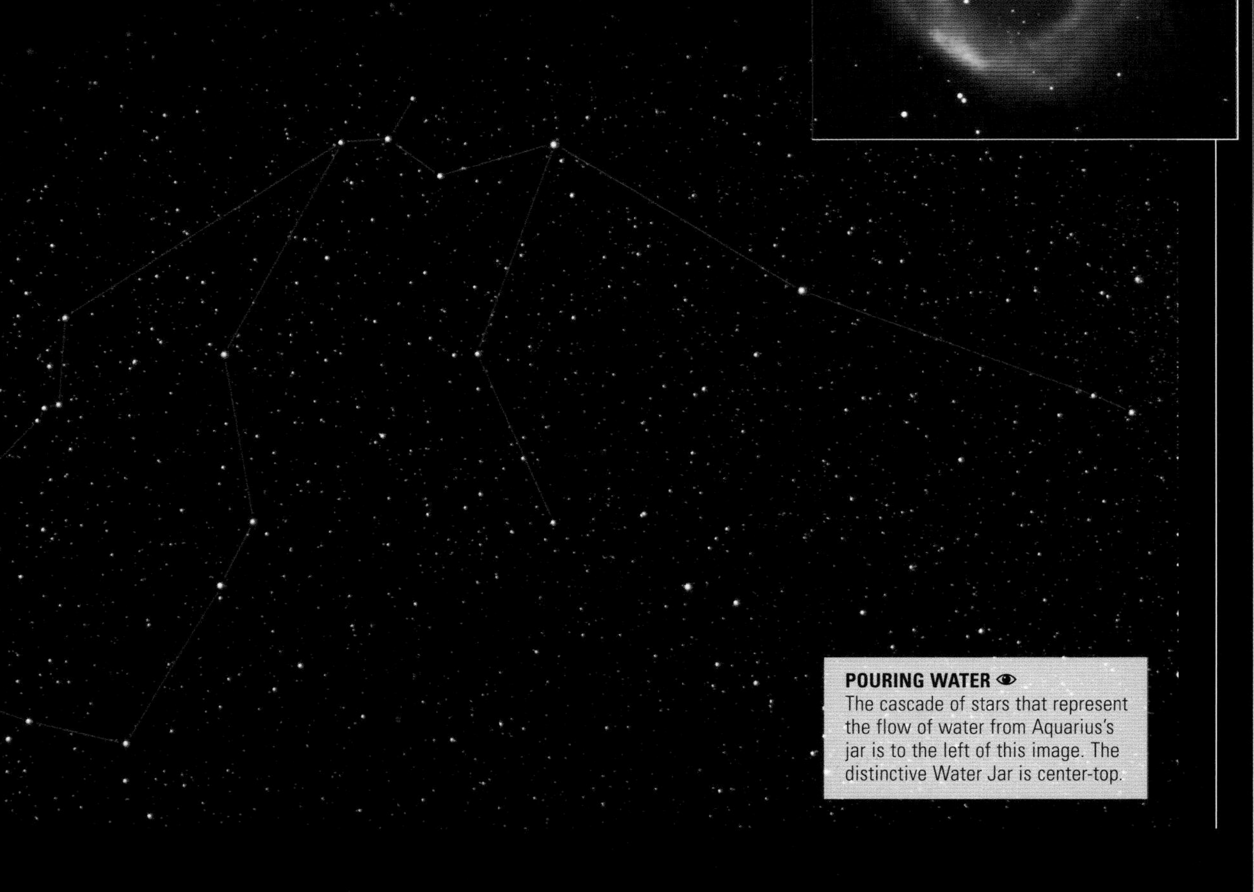

### SPECIFIC FEATURES
Zeta—the star at the center of the Water Jar group—is a close binary of 4th-magnitude stars just at the limit of resolution with a telescope of 2.4 in (60 mm) aperture. Located near the border with Equuleus, the globular cluster M2 appears as a fuzzy star when viewed through binoculars or a small telescope.

Aquarius contains two of the best-known planetary nebulae in the sky. The Helix Nebula (NGC 7293; see p.257) is thought to be the closest planetary nebula to Earth, being about 650 light-years away. It is therefore one of the largest of such

nebulae in apparent size, at almost half the diameter of a full moon. However, because its light is spread over such a large area, the Helix Nebula can be identified only when skies are clear and dark. Visually, this nebula appears as a pale gray patch, showing none of the beautiful colors captured on photographs.

The second planetary nebula—the Saturn Nebula (NGC 7009)—is easier to spot, appearing to be of a size similar to the disk of Saturn when viewed with a small telescope. Its faint extensions on either side, rather like the rings of Saturn, give rise to the object's popular name.

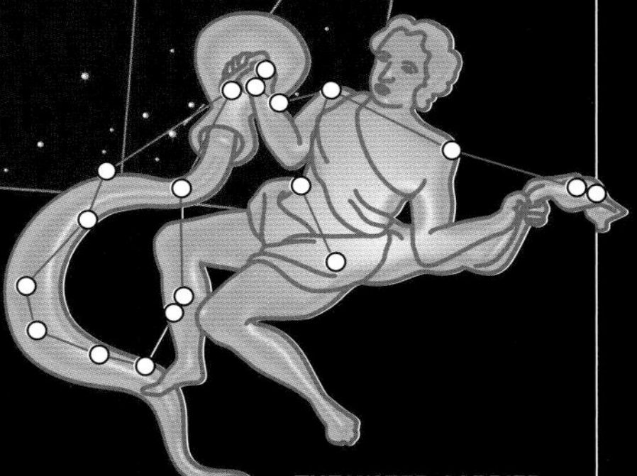

**THE WATER CARRIER**

### THE HELIX NEBULA 🔭 ✦ 💻
NGC 7293 is visible as a pale rounded patch through binoculars under dark skies, but its detailed structure and approximate colors are brought out in CCD images such as this.

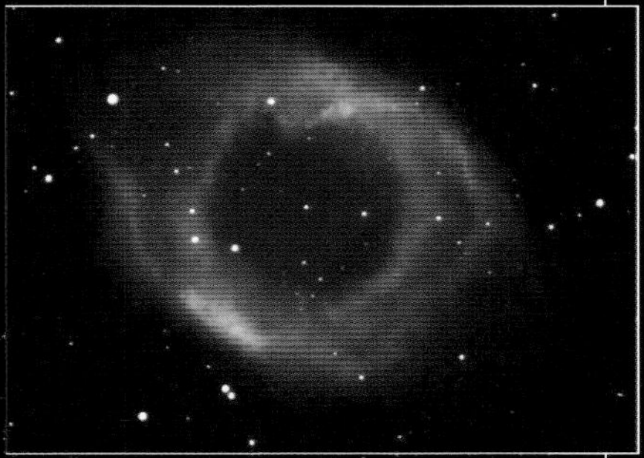

### POURING WATER 👁
The cascade of stars that represent the flow of water from Aquarius's jar is to the left of this image. The distinctive Water Jar is center-top.

## THE FISH

# Pisces

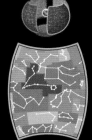

**SIZE RANKING** 14

**BRIGHTEST STAR**
Eta (η) 3.6

**GENITIVE**
Piscium

**ABBREVIATION** Psc

**HIGHEST IN SKY AT 10 P.M.**
October–November

**FULLY VISIBLE**
83°N–56°S

This zodiacal constellation represents two mythical fish (see panel, right). Its main claim to fame is that it contains the vernal equinox, which is the point where the Sun crosses the celestial equator into the Northern Hemisphere each year in March— on star maps, this is where 0h right ascension intersects 0° declination. Because of the slow wobble of the Earth, known as precession (see p.64), the point of the vernal equinox is gradually moving along the celestial equator and will enter Aquarius in about 2600 CE.

### SPECIFIC FEATURES
The most distinctive feature of Pisces is the ring of seven stars lying south of the Great Square of Pegasus. Known as the Circlet, this ring marks the body of one of the fish. It includes TX Piscium (also known as 19 Piscium), a deep-orange-colored red giant that fluctuates irregularly between magnitudes 4.8 and 5.2.

Alrescha (α) is a close pair of stars of 4th and 5th magnitudes that can be separated with a telescope with an aperture of 4 in (100 mm). These two stars form a true binary with an orbital period of more than 3,000 years. Zeta (ζ) and Psi-1 (ψ¹) Piscium are two more doubles that can be divided with a small telescope.

A beautiful face-on spiral galaxy, M74, lies just over two diameters of a full moon from the constellation's brightest star, Eta (η) Piscium. It appears as a round, bright glow through a small telescope; the spiral arms only show up well through a telescope with larger aperture and on long-exposure photographs.

**THE CIRCLET** 👁 🔭
The body of the southerly fish is marked by a ring of stars called the Circlet. One of the stars, TX Piscium, is a red giant of variable brightness, which appears noticeably orange through binoculars.

**DIVERGENCE** 👁
Pisces represents a pair of fish tied together by their tails with ribbon. The point where the two ribbons are knotted together is marked by the star Alpha (α) Piscium.

## MYTHS AND STORIES

### EROS AND APHRODITE

Ancient Greek myths concerning the origins of the constellation of Pisces are rather vague. In one myth, Aphrodite and her son Eros transformed themselves into fish and plunged into the Euphrates to escape the fearsome monster Typhon. In another version of the same story, two fish swam up and carried Aphrodite and Eros to safety on their backs.

**RESCUE AT SEA**
In this 17th-century painting by the Flemish artist Jacob Jordaens, Aphrodite and Eros are carried away on the back of a fish.

**THE FISH**

**M74** 🔭 📷
The spiral galaxy M74 is seen face-on and appears as a rounded glow when viewed through a small telescope. Larger apertures are needed to see its spiral arms.

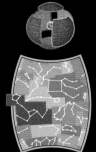

# Cetus

**SIZE RANKING** 4

**BRIGHTEST STAR**
Diphda (β) 2.0

**GENITIVE** Ceti

**ABBREVIATION** Cet

**HIGHEST IN SKY AT 10 P.M.**
October–December

**FULLY VISIBLE**
65°N–79°S

Cetus is represented on old star charts as an unlikely-looking, almost comical, hybrid sea monster, although the figure is also sometimes referred to as a whale. It is one of the original 48 Greek constellations listed by Ptolemy in his *Almagest*. It is a large but not very obvious constellation found in the equatorial region of the sky, and it lies south of the zodiacal constellations Pisces and Aries. Cetus is home to the celebrated variable star, Mira (ο) (see p.285), as well as a peculiar spiral galaxy, M77.

## SPECIFIC FEATURES
Menkar (α) is the second-brightest star in the constellation. It forms part of the loop of stars that mark the sea monster's head, and it has a wide and unrelated 6th-magnitude companion that is visible through binoculars. Positioned near the neck of the sea monster is Gamma (γ) Ceti, a close double star

**THE SEA MONSTER**

that is more challenging to divide than Menkar. High magnification on a telescope is required to see the two component stars, of 4th and 7th magnitudes.

Mira (ο) is the prototype of a common type of red giant that pulsates in size over months or years. Mira can reach magnitude 2 at its brightest—although magnitude 3 is more usual—while at its faintest it drops to magnitude 10. Hence, depending on how much it has swollen or contracted within its 11-month cycle, Mira can be either a naked-eye star or one that is visible only with a telescope.

Tau (τ) Ceti is 11.9 light-years away. Its temperature and brightness make it the most Sun-like of all Earth's nearby stars. Tau is, however, surrounded by a swarm of asteroids and comets, which would subject any local planets to

**M77** ✈ 🖳
Because it is a Seyfert galaxy, the spiral galaxy M77 looks like a fuzzy star through smaller telescopes—only its extremely bright core can be seen.

devastating bombardments. Thus the prospects for life in its vicinity seem rather slim.

M77 is found near Delta (δ) Ceti. This spiral galaxy is the brightest example of a Seyfert galaxy (see Types of Active Galaxies, p.320). Related to quasars, Seyfert galaxies are a class of galaxies that have extremely bright centers. M77 is oriented face-on toward the Earth, although only its core is visible through a small telescope, and it looks only like a small, round patch. M77 lies just under 50 million light-years away.

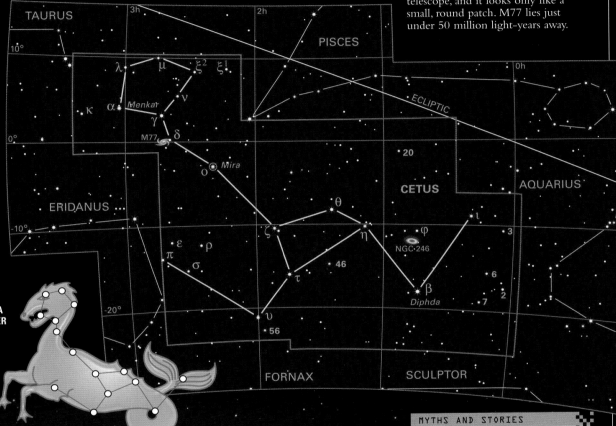

## MYTHS AND STORIES

### THE SEA MONSTER

Cetus was the sea monster sent to devour the princess Andromeda in the famous Greek myth (see p.368). On his return from killing Medusa the Gorgon, Perseus spied Andromeda's plight and swooped down on the sea monster as it attacked, stabbing it repeatedly with his sword in a fury of blood and foam, and leaving its water-logged corpse on the beach for the local people to pillage.

**MYTHICAL MONSTER**
Old star charts depict Cetus with enormous jaws and a coiled tail, its flippers dipped in the neighboring constellation, the river Eridanus.

**LURCHING MONSTER** 👁
Cetus is large but not particularly prominent. Its most celebrated star is the variable red giant Mira (ο), which for much of the time is too faint to be seen with the naked eye.

THE NIGHT SKY

## THE HUNTER

# Orion

| | |
|---|---|
| **SIZE RANKING** | 26 |
| **BRIGHTEST STARS** | Rigel (β) 0.2, Betelgeuse (α) 0.5 |
| **GENITIVE** | Orionis |
| **ABBREVIATION** | Ori |
| **HIGHEST IN SKY AT 10 PM** | December–January |
| **FULLY VISIBLE** | 79°N–67°S |

Orion is one of the most glorious constellations in the sky, representing a giant hunter or warrior followed by his dogs, Canis Major and Canis Minor (see panel, below). Its most distinctive feature is Orion's belt, formed by a line of three 2nd-magnitude stars almost exactly on the celestial equator. A complex of stars and nebulosity represents the sword that hangs from Orion's belt and contains the great star-forming region of M42, the Orion Nebula (p.241). In October each year, the Orionid meteors seem to radiate from a point near Orion's border with Gemini.

### SPECIFIC FEATURES

Marking one shoulder of Orion is Betelgeuse—Alpha (α) Orionis (see p.256)—a red supergiant hundreds of times larger than the Sun. Betelgeuse varies irregularly in brightness between magnitudes 0.0 and 1.3, but it averages around magnitude 0.5. It is about 500 light-years away, and it is closer to Earth than any of the other bright stars in Orion.

Betelgeuse contrasts noticeably in color with Rigel—Beta (β) Orionis—an even more luminous blue supergiant, which marks one of Orion's feet. Apart from the rare times when Betelgeuse is at its maximum magnitude, Rigel is the brightest star in the constellation. Rigel lies 860 light-years from Earth—almost twice as far away as Betelgeuse. Its 7th-magnitude companion can be picked out from its surrounding glare using a small telescope. Two other easily seen double stars are in Orion's belt. Delta (δ) Orionis has a 7th-magnitude companion, which is visible through a small telescope or binoculars. It is a greater challenge to reveal the close 4th-magnitude companion of Zeta (ζ) Orionis—this requires a telescope with an aperture of at least 3 in (75 mm).

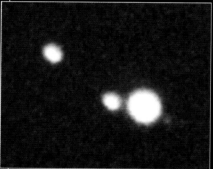

**MULTIPLE COMPANIONS** ✵
Sigma (σ) Orionis is a remarkable multiple star with three fainter companions—two on one side and an even fainter one on the opposite side—appearing similar to a planet orbited by moons.

## MYTHS AND STORIES

### THE GREAT HUNTER

In Greek mythology, Orion was a tall and handsome man and the son of Poseidon, god of the sea. The Greek poet Homer, in his *Odyssey*, described Orion as a great hunter who brandished a club of bronze. Despite his hunting prowess, Orion was killed by a mere scorpion, some say in retribution for his boastfulness. In the sky, Orion is placed opposite the constellation of Scorpius and, each night, the hunter flees below the horizon as the scorpion rises.

**HUNTER AND WARRIOR**
This depiction of Orion is from an ancient manuscript based on the *Book of Fixed Stars*, which was written by the Arabic astronomer al-Sufi around 964 CE.

**BRIGHT HUNTER** 👁
Orion, the hunter, is one of the most magnificent and easily recognizable constellations. A line of three stars makes up his belt, while an area of star clusters and nebulae forms his sword.

The real treasures of this constellation lie in the area around Orion's sword. NGC 1981, for example, appears as a large, scattered cluster of stars through binoculars; its brightest stars are of 6th magnitude. NGC 1977 is an elongated patch of nebulosity surrounding the stars 42 and 45 Orionis. Nearby is the Orion Nebula, an enormous star-forming cloud of gas, 1,500 light-years away, covering an area of sky wider than two diameters of a full moon. Its glowing gas appears multicolored on photographs and CCD images, yet visually it looks only gray-green because the eye is not sensitive to colors in

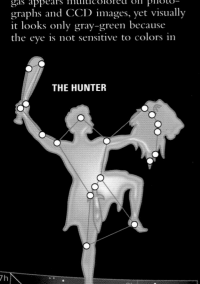

**THE HUNTER**

faint objects. On clear nights, it appears to the naked eye as a hazy patch of light and is obvious through any form of optical aid. An extension of the Orion Nebula bears a separate number, M43, but both are part of the same cloud. At the center of M42 lies a multiple star, Theta-1 ($\theta^1$) Orionis (see p.281), better known as the Trapezium because it appears as a group of four stars of 5th to 8th magnitude when seen through a small telescope. To one side of the nebula lies Theta-2 ($\theta^2$) Orionis, a double star with components of 5th and 6th magnitudes that can be separated with binoculars. At the tip of Orion's sword lies Iota ($\iota$) Orionis, a double, with components of 3rd and 7th magnitudes, divisible with a small telescope. Struve 747 is a wider double star nearby, with components of 5th and 6th magnitudes.

Even more impressive is the multiple star Sigma ($\sigma$) Orionis (p.281). A small telescope shows that the main 4th-magnitude star has two 7th-magnitude companions on one side and a closer 9th-magnitude component on the other.

Extending from the belt star Zeta ($\zeta$) Orionis is a strip of bright nebulosity, IC 434, against which is silhouetted the Horsehead Nebula (see p.240). This is probably the best-known dark nebula in the sky, and it shows well on photographs. To see it visually requires a large telescope and a dark viewing site.

**THE TRAPEZIUM** ✶ 🖥
At the heart of the Orion Nebula lies a multiple star called the Trapezium ($\theta^1$) (center, right). Its four stars are visible through small telescopes, but with larger apertures, two additional stars can also be seen.

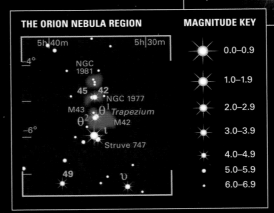

**THE ORION NEBULA REGION**

| **MAGNITUDE KEY** | |
|---|---|
| ✦ | 0.0–0.9 |
| ✦ | 1.0–1.9 |
| ✧ | 2.0–2.9 |
| ✧ | 3.0–3.9 |
| ⋆ | 4.0–4.9 |
| • | 5.0–5.9 |
| · | 6.0–6.9 |

5h 40m     5h 30m
-4°
NGC 1981
45  42
NGC 1977
M43  $\theta^1$ Trapezium
$\theta^2$  M42
-6°
Struve 747

49          $\upsilon$

**THE ORION NEBULA** ✶ 🖥
To the naked eye and through binoculars, the Orion Nebula (M42) appears only as a misty patch south of Orion's belt, its heart lit up by newborn stars. Its full beauty and its pinkish color become apparent only on photographs and CCD images such as this.

GEMINI
NGC 2175
69
$\xi$
$\nu$
$\mu$
$\alpha$
Betelgeuse
MONOCEROS
56
51
M78
NGC 2024
IC 434
42
TAURUS
Aldebaran
15  11
$o^1$
$o^2$
$\lambda$ $\phi^1$
$\phi^2$
32  $\gamma$
Bellatrix
$\omega$
$\psi^2$ 23
$\psi^1$  $\rho$
$\delta$
$\epsilon$  31
22
$\sigma$
$\eta$
M 42
29
$\kappa$
$\tau$  $\beta$
Rigel
**ORION**
$\pi^1$
$\pi^2$
$\pi^3$
$\pi^4$
$\pi^6$ $\pi^5$
ERIDANUS
CANIS MAJOR
Sirius
LEPUS

7h    6h    5h    4h
20°
0°
-10°
-20°

**THE HORSEHEAD NEBULA** 🖥 ⚖
Looking like a knight in a celestial chess game, the Horsehead Nebula is a curiously shaped dark dust cloud silhouetted against IC 434, a backdrop of glowing hydrogen. It lies to the south of Zeta ($\zeta$) Orionis (center left) in Orion's belt.

**THE NIGHT SKY**

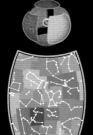

## THE GREATER DOG
# Canis Major

| | |
|---|---|
| **SIZE RANKING** | 43 |
| **BRIGHTEST STARS** | Sirius (α) -1.4, Adhara (ε) 1.5 |
| **GENITIVE** | Canis Majoris |
| **ABBREVIATION** | CMa |
| **HIGHEST IN SKY AT 10 PM** | January–February |
| **FULLY VISIBLE** | 56°N–90°S |

This southern constellation contains the brightest star in the entire sky: Sirius, or Alpha (α) Canis Majoris (see p.252). It forms a triangle with Procyon (in Canis Minor) and Betelgeuse (in Orion). Canis Major was known to the ancient Greeks as one of the two dogs following Orion, the hunter (see panel, below).

## SPECIFIC FEATURES

Sirius is a more powerful star than the Sun, giving out about 20 times as much light, and it is among the closest stars to Earth, being 8.6 light-years away. In combination, these factors give Sirius an apparent brightness twice that of the second-brightest star, Canopus (in Carina). Sirius is accompanied by a faint white dwarf, Sirius B (see p.268), which orbits it every 50 years.

M41 is a large open cluster that is visible as a hazy patch to the naked eye. Its stars, which are scattered over an area about the size of a full moon, are revealed with binoculars, while telescopes show chains of stars radiating from its center.

Around Tau (τ) Canis Majoris is NGC 2362, which is best viewed with a telescope. Also nearby is UW Canis Majoris, an eclipsing binary.

### ORION'S HUNTING DOG 👁
The great dog stands on its hind legs in the sky, holding brilliant Sirius in its jaws like a sparkling ball.

**NGC 2362** ✈
The brightest member of this neat cluster of stars is the 4th-magnitude blue supergiant Tau (τ) Canis Majoris, which is almost at its center.

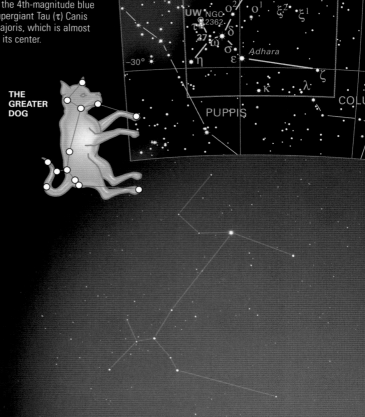

THE GREATER DOG

## THE LITTLE DOG
# Canis Minor

| | |
|---|---|
| **SIZE RANKING** | 71 |
| **BRIGHTEST STAR** | Procyon (α) 0.4 |
| **GENITIVE** | Canis Minoris |
| **ABBREVIATION** | CMi |
| **HIGHEST IN SKY AT 10 PM** | February |
| **FULLY VISIBLE** | 89°N–77°S |

Canis Minor is one of the original Greek constellations and lies virtually on the celestial equator. It is usually identified as the smaller of Orion's two hunting dogs.

The constellation is easily identified by its brightest star— Procyon, or Alpha (α) Canis Minoris (see p.284)—which forms a large sparkling triangle with two other 1st-magnitude stars: Betelgeuse (in Orion) and Sirius (in Canis Major). Other than that, the constellation contains little of particular note to small telescope users.

## SPECIFIC FEATURES
Procyon is the eighth-brightest star in the sky. It is somewhat cooler and fainter than the other dog star, Sirius, and also more distant, lying 11.4 light-years away. It has a white dwarf partner, Procyon B, which is visible only with a very large telescope.

THE LITTLE DOG

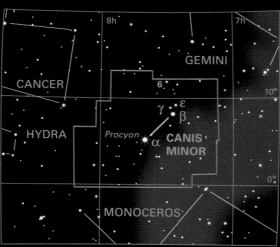

### LONE STAR 👁
Unlike the distinctive constellation of the Greater Dog, Canis Minor consists of little more than its brightest star, Procyon.

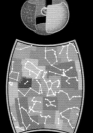

# Monoceros

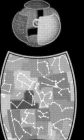

| | |
|---|---|
| **SIZE RANKING** | 35 |

**BRIGHTEST STAR**
Alpha (α) 3.9

**GENITIVE**
Monocerotis

**ABBREVIATION** Mon

**HIGHEST IN SKY AT 10 PM**
January–February

**FULLY VISIBLE**
78°N–78°S

Monoceros is often overlooked, because it is overshadowed by neighboring Orion, Gemini, and Canis Major. It is easy to locate, however, since it is situated on the celestial equator in the middle of the large triangle formed by the brilliant 1st-magnitude stars Betelgeuse (in Orion), Procyon (in Canis Minor), and Sirius (in Canis Major).

Although none of the stars of Monoceros is bright, the Milky Way passes through it and it contains many deep-sky objects of interest.

The constellation was introduced in the early 17th century by the Dutch astronomer and cartographer Petrus Plancius and depicts the unicorn, a mythical animal with religious symbolism.

## SPECIFIC FEATURES

Beta (β) Monocerotis (see p.281) is regarded as one of the finest triple stars in the sky for small instruments. It is readily separated to show an arc of three 5th-magnitude stars.

The double star Epsilon (ε) Monocerotis is labeled 8 Monocerotis on some charts. Its components, of 4th and 7th magnitudes, are easily spotted through a small telescope.

Prime among Monoceros's most celebrated clusters and nebulae is NGC 2244, an elongated group of stars of 6th magnitude and fainter. Surrounding the cluster is a glorious nebula known as the Rosette Nebula, although it is faint and seen well only on CCD images and photographs.

NGC 2264 is another combination of open cluster and nebula. This triangular group can be seen through binoculars or a small telescope. Its brightest member is 5th-magnitude S Monocerotis—an intensely hot and luminous star that is slightly variable. CCD images and photographs show a surrounding nebulosity into which encroaches a dark wedge called the Cone Nebula (see p.242).

M50 is an open cluster about half the apparent size of a full moon. It is visible through binoculars, but a telescope is needed to resolve its individual stars.

NGC 2232 is larger and more scattered, and its brightest stars are visible through binoculars.

**FRAMED BEAST** 👁
Monoceros occupies the space within the bright triangle of stars formed by Sirius (seen here at upper right), Betelgeuse (upper left), and Procyon (bottom center).

**THE CONE NEBULA** ⚲ ☐ ⚖
This tapering region of dark gas and dust intrudes into brighter nebulosity at the southern end of the star cluster NGC 2264. The Cone Nebula is visible only on images taken with a large telescope, as here.

**THE ROSETTE NEBULA** ⚲ ☐
The flowerlike form of the Rosette Nebula glows like a pink carnation in this CCD image. At its center is the star cluster NGC 2244, which can readily be identified through binoculars.

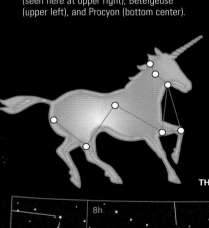

**THE UNICORN**

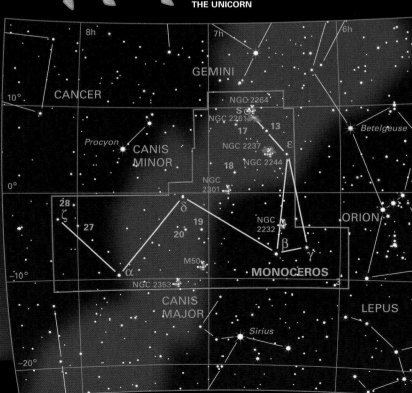

## THE WATER SNAKE

# Hydra

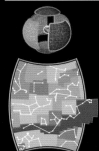

**SIZE RANKING** 1

**BRIGHTEST STAR**
Alphard (α) 2.0

**GENITIVE** Hydrae

**ABBREVIATION** Hya

**HIGHEST IN SKY AT 10 P.M.**
February–June

**FULLY VISIBLE**
54°N–83°S

## HERCULES AND THE HYDRA

The Hydra was a serpent with nine heads, one of them immortal, which lived in a swamp near the town of Lerna, emerging to ravage crops and cattle. As the second of his labors, Hercules was sent to kill the monster. He flushed it from its lair with flaming arrows and cut off each head in turn, ending with the immortal head, which he buried under a rock.

**DEADLY BLOWS**
Hercules battles with the Hydra in this sculpture by François-Joseph Bosio (1768–1845), which is exhibited in the Tuileries gardens, Paris.

Hydra depicts the multiple-headed monster that fought and was killed by Hercules in the second of his labors (see panel, right). During the struggle, a crab joined forces with the Hydra but was crushed underfoot by Hercules; it was later commemorated as the constellation Cancer. Although the Hydra had nine heads, it is represented in the sky with a single head—presumably its immortal one.

The constellation is the largest of all 88 and stretches for more than a quarter of the way around the sky from its head, south of Cancer and just north of the celestial equator, to its tail in the Southern Hemisphere between Libra and Centaurus. Despite its size, there is little to mark out this constellation other than a group of six stars of modest brightness, which forms the head of the water snake.

### SPECIFIC FEATURES

Hydra's brightest star is 2nd-magnitude Alphard, or Alpha (α) Hydrae. Alphard means "the solitary one," and this name reflects its position in an otherwise blank area of sky. This orange-colored giant is in fact the only star in the constellation brighter than magnitude 3.0. It is about 180 light-years away. Epsilon (ε) Hydrae is a close binary star with components of contrasting colors that can be divided with a telescope with an aperture of at least 3 in (75 mm) and high magnification. The yellow and blue component stars are of magnitude 3.4 and 6.7 and have an orbital period of nearly 600 years.

M48 is an open star cluster on the border with Monoceros. It lies nearly 2,000 light-years away. M48 is larger than a full moon and it is seen well through binoculars or a small telescope. It contrasts with the globular cluster M68 (see p.295), which resembles a fuzzy star when viewed through binoculars or a small telescope.

M83 is a spiral galaxy, toward the Hydra's tail, that lies about 15 million light-years away. Through a small telescope, it appears as an elongated glow, but a telescope of larger aperture will reveal its spiral structure and its noticeable central "bar," which may be similar to the bar that is thought to lie across the center of the Milky Way Galaxy.

The planetary nebula known as the Ghost of Jupiter, or NGC 3242, is to be found near the star Mu (μ) Hydrae, in the central part of Hydra's body.

**LONG SERPENT** 👁
The Hydra's head, at the right in this photograph, lies south of Cancer while the tip of its tail lies far to the left, south of the stars of Libra.

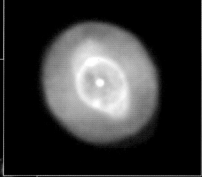

**THE GHOST OF JUPITER**
When viewed through a small telescope, the planetary nebula NGC 3242 appears as an ethereal, blue-green, elliptical glow about the size of the planet Jupiter, hence its popular name—the Ghost of Jupiter.

**M83**
This magnificent face-on spiral galaxy is to be found lying on the border of Hydra and Centaurus. M83 has a central "bar" of stars and gas, and it is sometimes known as the Southern Pinwheel.

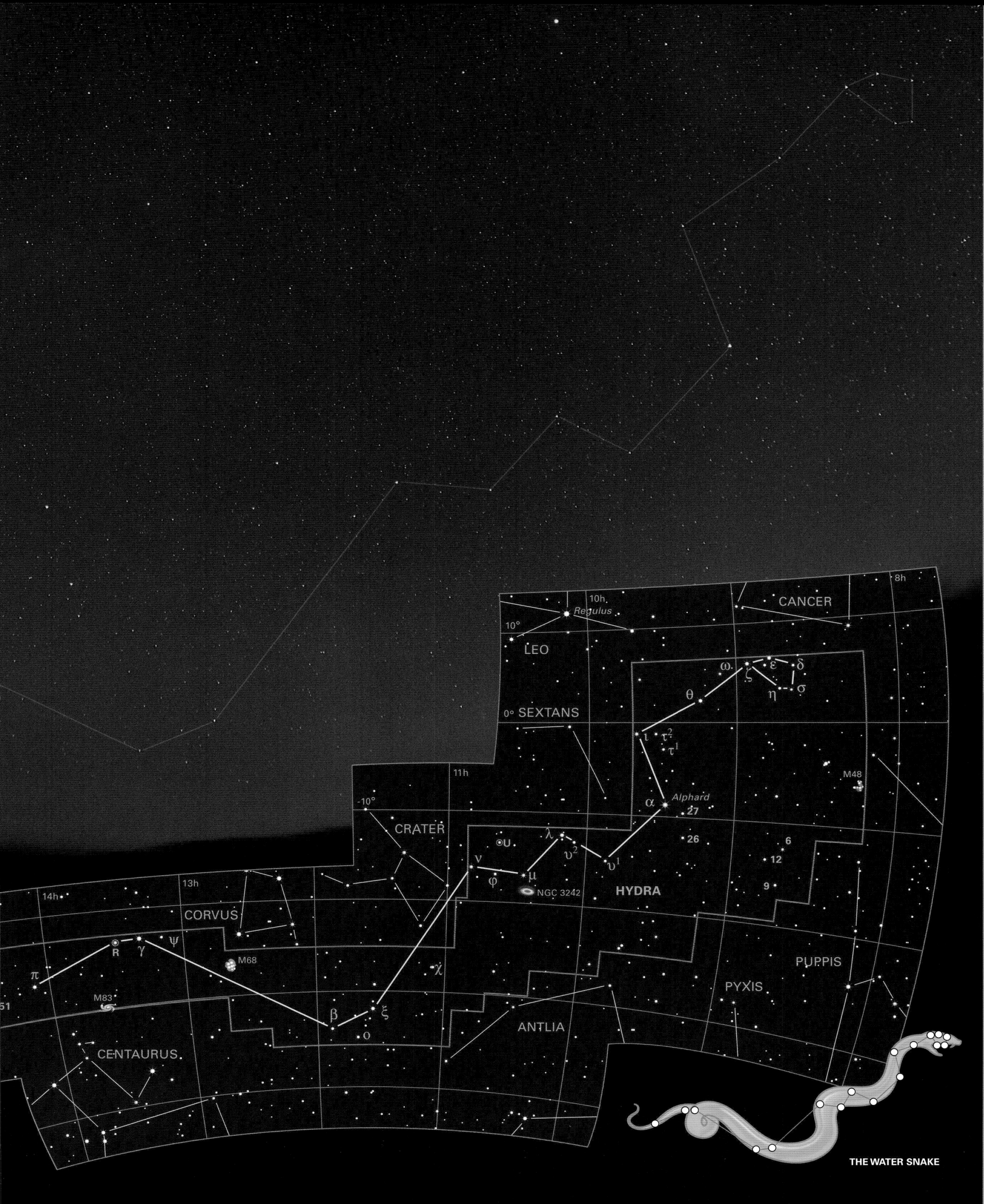

CANCER

10h,
*Regulus*

10°

LEO

0° SEXTANS

ω ζ ε δ
θ η σ

ι τ²
τ¹

M48

-10°

α *Alphard*
27

11h

CRATER

26

6

λ

⊙U

ν

υ²

12

φ μ

υ¹

9

13h

NGC 3242

**HYDRA**

14h

CORVUS

PUPPIS

χ

M68

PYXIS

R γ ψ

π

β ξ

ANTLIA

M83

o

51

CENTAURUS

**THE WATER SNAKE**

## THE AIR PUMP

# Antlia

| | |
|---|---|
| **SIZE RANKING** | 62 |
| **BRIGHTEST STAR** | Alpha (α) 4.3 |
| **GENITIVE** | Antliae |
| **ABBREVIATION** | Ant |
| **HIGHEST IN SKY AT 10 PM** | March–April |
| **FULLY VISIBLE** | 49°N–90°S |

This constellation was one of those introduced in the mid-18th century by the French astronomer Nicolas Louis de Lacaille (see p.422) to commemorate scientific and technical inventions—in this case, an air pump designed by the French physicist Denis Papin for his experiments on gases.

## SPECIFIC FEATURES

Zeta (ζ) Antliae appears as a wide pair of 6th-magnitude stars when viewed through binoculars. The brighter of the pair has a companion of 7th magnitude.

NGC 2997 is an elegant spiral galaxy inclined at about 45 degrees to our line of sight. Unfortunately, it is just too faint to be well seen through a small telescope, although it can be captured beautifully on photographs and CCD images. NGC 2997 is about 35 million light-years away.

### NGC 2997 ✦ 🖵
This classic spiral galaxy reveals pink clouds of hydrogen gas dotted along its spiral arms in CCD images.

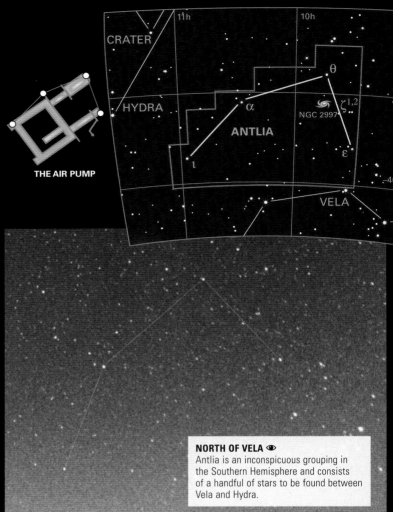

THE AIR PUMP

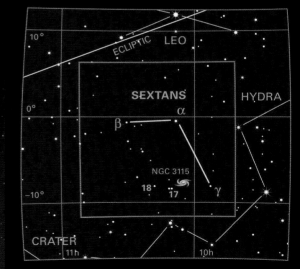

**NORTH OF VELA** 👁
Antlia is an inconspicuous grouping in the Southern Hemisphere and consists of a handful of stars to be found between Vela and Hydra.

## THE SEXTANT

# Sextans

| | |
|---|---|
| **SIZE RANKING** | 47 |
| **BRIGHTEST STAR** | Alpha (α) 4.5 |
| **GENITIVE** | Sextantis |
| **ABBREVIATION** | Sex |
| **HIGHEST IN SKY AT 10 PM** | March–April |
| **FULLY VISIBLE** | 78°N–83°S |

Representing a sextant used for taking star positions in the days before telescopes were invented, Sextans was introduced in the late 17th century by the Polish astronomer Johannes Hevelius (see p.384), who used such a device when cataloging the stars.

## SPECIFIC FEATURES

Two unrelated stars of 6th magnitude, 17 and 18 Sextantis, form a line-of-sight double star, which shows neatly through binoculars.

In the same part of the constellation lies NGC 3115, which is popularly named the Spindle Galaxy because of its highly elongated shape. This lenticular galaxy is detectable through a small telescope.

**HEVELIUS'S SEXTANT** 👁
Sextans is difficult to pick out with the naked eye because it is a faint and unremarkable constellation. It lies on the celestial equator south of Leo.

THE SEXTANT

**THE SPINDLE GALAXY** ✦ 🖵
NGC 3115 is a lenticular galaxy seen edge-on from Earth, so it appears highly elliptical in shape when viewed through a telescope. It is just over 30 million light-years from us.

THE CUP

# Crater

| | |
|---|---|
| **SIZE RANKING** | 53 |
| **BRIGHTEST STAR** | Delta (δ) 3.6 |
| **GENITIVE** | Crateris |
| **ABBREVIATION** | Crt |
| **HIGHEST IN SKY AT 10 PM** | April |
| **FULLY VISIBLE** | 65°N–90°S |

Crater is a faint constellation representing a cup. Although larger than Corvus, to which it is linked in Greek myth (see panel, right), Crater contains no objects that might be of interest to users of small telescopes.

This area once contained two other constellations that have since been dropped by astronomers. In the late 18th century, a French astronomer, J. J. Lalande, introduced Felis, the cat, between Hydra and Antlia, while others introduced Noctua, the night owl, on the tail of Hydra (see panel illustration, right).

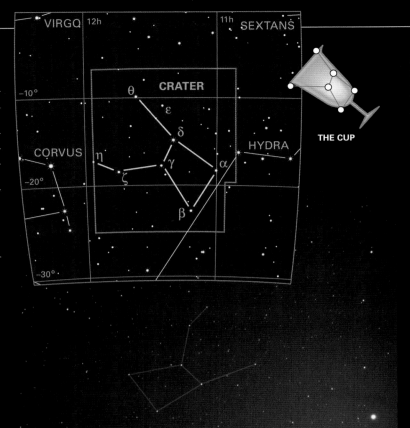

THE CUP

## CROW AND CUP

Crater and Corvus feature together in a Greek myth in which the god Apollo sent the crow (Corvus) to fetch water in a cup (Crater). On the way, the greedy crow stopped to eat figs. As an alibi, the crow snatched up a water snake (Hydra) and blamed it for delaying him, but Apollo saw through the deception and banished the trio to the skies.

### HISTORIC DEPICTION
The area around Crater as shown in *Urania's Mirror*, a set of 19th-century constellation cards.

### CELESTIAL VESSEL 👁
Crater is to be found lying next to Corvus on the back of Hydra, the water snake. This undistinguished constellation is also known as the Goblet or the Chalice.

---

THE CROW

# Corvus

| | |
|---|---|
| **SIZE RANKING** | 70 |
| **BRIGHTEST STAR** | Gamma (γ) 2.6 |
| **GENITIVE** | Corvi |
| **ABBREVIATION** | Crv |
| **HIGHEST IN SKY AT 10 PM** | April–May |
| **FULLY VISIBLE** | 65°N–90°S |

The four brightest stars of Corvus—Beta (β), Gamma (γ), Delta (δ), and Epsilon (ε) Corvi—form a distinctive keystone shape in this small constellation south of Virgo. Oddly, the star labeled Alpha (α) Corvi, at magnitude 4.0, is significantly fainter than all of these. Corvus is one of the original 48 Greek constellations and represents a crow, the sacred bird of the Greek god Apollo.

## SPECIFIC FEATURES
Delta is a double star with components of 3rd and 9th magnitudes. It is divisible through a small telescope.

Corvus also boasts a remarkable pair of interacting galaxies: NGC 4038 and 4039. At 10th magnitude, they are too faint to be seen through a small telescope, but photographs reveal this as a graphic example of a galactic collision. When the galaxies passed each other, gravity pulled out stars and gas to create a shape like an insect's feelers, hence their popular name, the Antennae (see p.317).

### THE ANTENNAE 🖥 ⚊
As NGC 4038 and 4039 sweep past each other, gravity draws out long streams of dust and gas from them. The streams extend off the top and bottom of this picture.

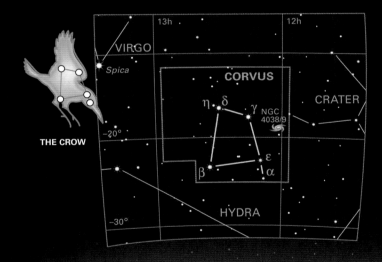

THE CROW

### PECKING BIRD 👁
Corvus, the crow, is linked in legend with neighboring Crater, the cup. The crow is visualized as pecking at Hydra, the water snake, on whose back it stands.

## THE CENTAUR

# Centaurs

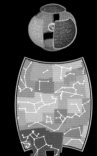

**SIZE RANKING** 9

**BRIGHTEST STARS**
Rigil Kentaurus (α)
-0.3, Hadar (β) 0.6

**GENITIVE** Centauri

**ABBREVIATION** Cen

**HIGHEST IN SKY AT 10 PM**
April–June

**FULLY VISIBLE**
25°N–90°S

This dominating constellation of the southern skies contains a variety of notable objects, including the closest star to the Sun and a most unusual galaxy. Centaurus represents the centaur Chiron (see panel, right), who had the torso of a man and the four legs of a horse.

## SPECIFIC FEATURES

Alpha (α) Centauri (see p.252), or Rigil Kentaurus, is a fabulous multiple star. To the naked eye, it is the third-brightest star in the sky. Its system includes two Sun-like stars, which appear so bright because they are only 4.3 light-years away, closer than any other stars bar one—Proxima Centauri (see p.252), which is about 0.1 light-years closer. However, Proxima is of only magnitude 11 and lies four diameters of a full moon away from its brighter companions. Its position means it is outside Alpha's telescopic field of view, so identification is difficult.

Although it bears a Greek letter, Omega (ω) Centauri is not a star but a globular cluster, the largest and brightest visible from Earth. To the naked eye, it is a large, hazy star, and a small telescope is required to resolve the brightest individual members of this globular cluster.

Almost due north of Omega is the peculiar galaxy NGC 5128, also known as the radio source Centaurus A (see p.322). This object is thought to result from the merger of a giant elliptical galaxy and a spiral galaxy. Photographs show a dark band of dust across the galaxy's center, the remains of the spiral galaxy, but larger apertures are needed to make out this feature visually. NGC 5128 is the brightest galaxy outside the Local Group and, at a distance of about 12 million light-years, is the closest peculiar galaxy to us.

The planetary nebula NGC 3918, or the Blue Planetary, is easily identified through a small telescope. It appears like a larger version of the disk of Uranus. Also in Centaurus are two interesting open star clusters NGC 3766 and NGC 5460.

**THE CENTAUR**

## CHIRON

The wise and scholarly centaur Chiron was the offspring of Cronus, king of the Titans, and the sea nymph Philyra. He lived in a cave, from where he taught hunting, medicine, and music to the offspring of the gods. His most successful pupil was Asclepius, son of Apollo, who became the greatest healer of the ancient world. Chiron was immortalized among the stars after Hercules accidentally shot him with a poisoned arrow.

**TEACHER OF THE GODS**
This Roman fresco shows Chiron teaching Achilles, his foster son, to play the lyre. They are in Chiron's cave on Mount Pelion.

**ALPHA CENTAURI** ⚲
The two yellow stars, of magnitudes 0.0 and 1.3, of this beautiful double star orbit each other every 80 years. They are easily separated through a small telescope.

**CELESTIAL CENTAUR** 👁
The brilliant stellar pairing of Alpha (α) and (β) Beta Centauri guides the eye to Centaurus, the celestial centaur. The familiar pattern of Crux, the Southern Cross, lies beneath the centaur's body.

# Lupus

**SIZE RANKING** 46

**BRIGHTEST STAR**
Alpha (α) 2.3

**GENITIVE** Lupi

**ABBREVIATION** Lup

**HIGHEST IN SKY AT 10 PM**
May–June

**FULLY VISIBLE**
34°N–90°S

Lupus is a southern constellation lying on the edge of the Milky Way between the better-known figures of Centaurus and Scorpius. It contains numerous double stars of interest to amateur observers.

It was one of the original 48 constellations familiar to the ancient Greeks, who visualized it as a wild animal speared by Centaurus (see panel, below).

## SPECIFIC FEATURES

Kappa (κ) Lupi, with components of magnitudes 3.9 and 5.7, and Xi (ξ) Lupi, with components of magnitudes 5.1 and 5.6, are two doubles that are easy to spot through a small telescope.

Pi (π) Lupi can be divided into matching 5th-magnitude components through a telescope with an aperture of 3 in (75 mm). Even more challenging is 4th-magnitude Mu (μ) Lupi, which has a wide

7th-magnitude companion visible through a small telescope. Its primary star, however, is a close double, needing an aperture of at least 4 in (100 mm) to separate.

The 3rd-magnitude Epsilon (ε) Lupi has a companion of 9th magnitude, and Eta (η) Lupi is a 3rd-magnitude star with an 8th-magnitude companion.

NGC 5822 is a rich open cluster within the Milky Way. Its brightest stars are of only 9th magnitude, so it is not particularly prominent. It lies 2,400 light-years away.

**NGC 5822** ✈
This large open cluster in southern Lupus contains more than 100 stars of 9th magnitude and fainter. It can be seen through binoculars or a small telescope.

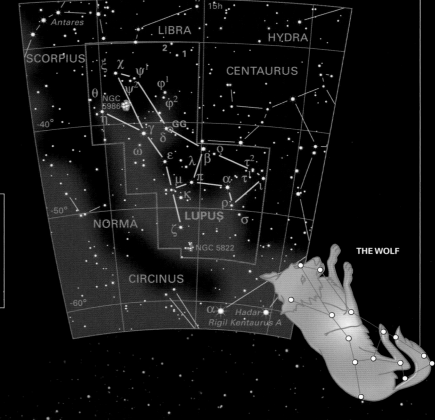

**THE WOLF**

## THE LANCED BEAST

To the ancient Greeks and Romans, Lupus represented a wild animal of unspecified nature that had been speared, by neighboring Centaurus, on a long pole called a thyrsus. In consequence, Centaurus and Lupus were often regarded as a combined figure. The identification of Lupus as a wolf seems to have become common during Renaissance times.

### IN MIRROR IMAGE
This medieval Arabic illustration shows Centaurus holding Lupus and the thyrsus, which has become a bunch of leaves or flowers.

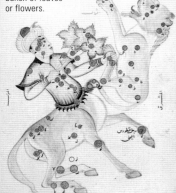

**BESTIAL OFFERING** 👁
Here, Lupus is partly surrounded by the stars of Centaurus. In Greek myth, the centaur killed the beast and carried it to the altar, Ara.

# Sagittarius

**SIZE RANKING** 15

**BRIGHTEST STAR**
Epsilon (ε) 1.8

**GENITIVE** Sagittarii

**ABBREVIATION** Sgr

**HIGHEST IN SKY AT 10 PM**
July–August

**FULLY VISIBLE**
44°N–90°S

This prominent zodiacal constellation is found between Scorpius and Capricornus in the southern celestial hemisphere. It includes a highly recognizable star pattern called the Teapot, with a pointed lid (λ) and large spout (γ, ε, and δ). The handle of the Teapot is sometimes also called the Milk Dipper.

The Milky Way is particularly broad and rich in Sagittarius, because the center of our galaxy (Sagittarius A) lies in this direction. The exact center of the galaxy is thought to coincide with a radio source known as Sagittarius A★, near where the borders of Sagittarius, Ophiuchus, and Scorpius meet. Sagittarius boasts more Messier objects than any other constellation—it has 15 in all.

Although old star charts depicted this constellation as a centaur, in Greek mythology, Sagittarius was identified as a different type of creature known as a satyr. He is usually said to be Crotus, son of Pan, who invented archery and went hunting on horseback. He is seen aiming his bow at neighboring Scorpius.

## SPECIFIC FEATURES

Beta (β) Sagittarii appears to the naked eye as a pair of 4th-magnitude stars. The more northerly (and slightly brighter) of the two stars has a 7th-magnitude companion. All three stars are at different distances, and thus are unrelated.

Probably the finest object for binoculars is M8, the Lagoon Nebula (see p.243), which extends for three times the width of a full moon. It contains the cluster NGC 6530, with stars of 7th magnitude and fainter, as well as the 6th-magnitude blue supergiant 9 Sagittarii.

The Trifid Nebula, M20 (see p.246), is so named because it is trisected by dark lanes of dust. Visually, it is far less impressive than its photographic representation, and little more than the faint double star at its center can be identified through a small instrument.

On the northern border of Sagittarius with Scutum lies another frequently photographed object—the

Omega Nebula, M17. The loose cluster of stars within it can be detected through binoculars.

M22 is one of the finest globular clusters in the entire sky. Under good conditions, it is visible to the naked eye. Through a small telescope, it is somewhat elliptical in outline, while one with an aperture of 3 in (75 mm) will resolve its brightest stars.

M23 is a large open cluster visible through binoculars near the border with Ophiuchus. M25 is another binocular cluster, while M24 is a bright Milky Way star field the length of four diameters of a full moon.

**M22** 🔭 ⚹
This prominent globular cluster lies near the lid of the Teapot. Through binoculars, it appears as a woolly ball about two-thirds the apparent diameter of a full moon.

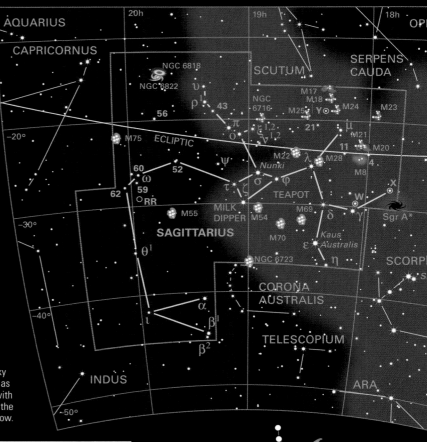

**THE LAGOON NEBULA** 🔭 ⚹ 🖥
One of the largest nebulae in the sky is M8, which appears in binoculars as an elongated, milky patch of light with embedded stars, including those in the cluster NGC 6530, which make it glow.

**THE ARCHER**

**THE TRIFID NEBULA** ⚹ 🖥
The pinkish emission of the Trifid Nebula contrasts with the blue reflection nebula to its north, as revealed on long-exposure photographs and CCD images. At its heart is a faint double star, which is overexposed on this image.

**THE OMEGA NEBULA** 🔭 ⚹ 🖥
M17 can be glimpsed through binoculars but shows up better through a telescope. It resembles the Greek capital letter omega (Ω). However, others see it as a swan, hence its alternative name, the Swan Nebula.

**MOUNTED BOWMAN** 👁
The stars that make up the outline of
Sagittarius, the Archer, lie in front of
dense Milky Way star fields toward the
center of our galaxy. North is to the left
in this photograph.

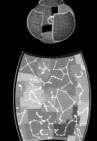

## THE SCORPION

# Scorpius

**SIZE RANKING** 33

**BRIGHTEST STAR**
Antares (α) 1.0
(variable)

**ABBREVIATION** Sco

**GENITIVE** Scorpii

**HIGHEST IN SKY AT 10 P.M.**
June–July

**FULLY VISIBLE**
44°N–90°S

This beautiful and easily recognizable zodiacal constellation is situated in the southern sky. It depicts a scorpion (see panel, below) whose raised tail is marked by a curve of stars extending into a rich area of the Milky Way toward the center of the Galaxy.

## SPECIFIC FEATURES

Antares, or Alpha (α) Scorpii (see p.256), is a red supergiant hundreds of times larger than the Sun. It fluctuates from about magnitude 0.9 to 1.2 every 4 to 5 years.

Normally, Delta (δ) Scorpii is of magnitude 2.3, but in the year 2000 it unexpectedly began to brighten by over 50 percent. Whether it will remain at its new magnitude or return to its previous value is unknown.

Beta (β) Scorpii is a line-of-sight pair with components of 3rd and 5th magnitudes, while Omega (ω) Scorpii is an even wider unrelated pair, with stars of 4th magnitude. A small telescope easily splits Nu (ν) Scorpii into a double with components of 4th and 6th magnitudes. Mu (μ) Scorpii is another naked-eye pair, with stars of 3rd and 4th magnitudes.

More complex is Xi (ξ) Scorpii, a white and orange pair of stars of 4th and 7th magnitudes. In the same field of view a fainter and wider pair

can also be seen. All four stars are gravitationally linked, making this a genuine quadruple.

The open cluster M7 is visible to the naked eye as a hazy patch. It has dozens of stars of 6th magnitude and fainter scattered over an area twice the apparent width of a full moon. About twice as distant is M6, which is known as the Butterfly Cluster (see p.290) because of its shape when viewed through binoculars. On one wing lies BM Scorpii, a variable orange giant. Near Antares, M4 (see p.294) is one of the closest globular clusters to us, at 7,000 light-years away.

Just too far south to have featured on Charles Messier's list (see p.73) is the open cluster NGC 6231. Its brightest member, 5th-magnitude Zeta (ζ) Scorpii, has a 4th-magnitude companion much closer to us.

The strongest X-ray source in the sky is Scorpius X-1. This consists of a 13th-magnitude blue star orbited by a neutron star.

**GLITTERING CLUSTERS** 👁 🔭
Two prominent star clusters, M6 and M7, adorn the tail of Scorpius in the Milky Way. M6 is at the center of this photograph; M7 is bottom left.

**THE SCORPION**

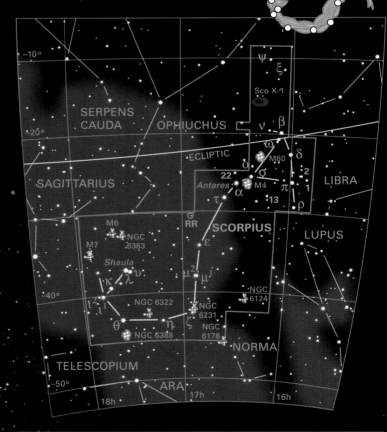

**STING IN THE TAIL** 👁
This view of Scorpius has south at the top and shows the scorpion raising its curving tail as though to strike. Its heart is marked by the red star Antares.

MYTHS AND STORIES

## THE DEATH OF ORION

In Greek mythology, Scorpius was the scorpion that stung Orion to death. According to one story, the scorpion was sent by Artemis, the goddess of hunting, after Orion had tried to attack her, while another account relates how Mother Earth dispatched the scorpion to humble Orion after he had boasted that he could kill any wild beast.

**MISPLACED FOOT**
Like other old star charts, Jean Fortin's *Atlas Céleste* shows the foot of Ophiuchus awkwardly overlapping Scorpius.

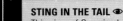

## THE SEA GOAT

# Capricornus

**SIZE RANKING** 40

**BRIGHTEST STAR** Deneb Algedi (δ) 2.9

**GENITIVE** Capricorni

**ABBREVIATION** Cap

**HIGHEST IN SKY AT 10 P.M.** August–September

**FULLY VISIBLE** 62°N–90°S

This is the smallest constellation of the zodiac and not at all prominent; it is situated in the southern sky between Sagittarius and Aquarius. In Greek myth, Capricornus represents the goatlike god Pan (see panel, right), who jumped into a river and became part fish to escape from the monster Typhon.

### SPECIFIC FEATURES

Alpha (α) Capricorni is a wide pairing of unrelated 4th-magnitude stars. They can be separated through binoculars or even with good eyesight. Alpha-1 ($\alpha^1$) Capricorni is a yellow supergiant nearly 900 light-years away, while Alpha-2 ($\alpha^2$) is a yellow giant less than one-eighth that distance from Earth.

Beta (β) Capricorni is a 3rd-magnitude yellow giant with a 6th-magnitude blue-white companion that can be seen through a small telescope or even good binoculars.

The modest globular cluster M30 is visible as a hazy patch through a small telescope.

**M30** 🔭 ✈

Chains of stars extending like fingers from the northern side of this cluster are visible through a large telescope.

**THE SEA GOAT**

**CAPRICORNUS AND MARS** 👁
Mars is seen here to the left of Capricornus, whose stars form a roughly triangular shape depicting Pan as half goat, half fish.

### PAN

This Greek god of the countryside had the hind legs and horns of a male goat but the body of a human. He created the pipes of Pan, also known as the syrinx, from reeds of different lengths.

**PAN-PIPER**
This stone statue of Pan playing his reed pipes is to be found in the garden of a castle in Schwetzingen, Germany.

---

## THE MICROSCOPE

# Microscopium

**SIZE RANKING** 66

**BRIGHTEST STARS** Gamma (γ) 4.7, Epsilon (ε) 4.7

**GENITIVE** Microscopii

**ABBREVIATION** Mic

**HIGHEST IN SKY AT 10 P.M.** August–September

**FULLY VISIBLE** 45°N–90°S

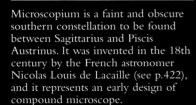

Microscopium is a faint and obscure southern constellation to be found between Sagittarius and Piscis Austrinus. It was invented in the 18th century by the French astronomer Nicolas Louis de Lacaille (see p.422), and it represents an early design of compound microscope.

### SPECIFIC FEATURES

The orange giant Alpha (α) Microscopii, of 5th magnitude, has a 10th-magnitude companion that is visible through an amateur telescope.

**THE MICROSCOPE**

**UNDER THE MICROSCOPE** 👁
Microscopium is a faint and almost featureless constellation. It lies near Capricornus and the much more conspicuous Sagittarius.

## THE SOUTHERN FISH

# Piscis Austrinus

| | |
|---|---|
| **SIZE RANKING** | 60 |
| **BRIGHTEST STAR** | Fomalhaut (α) 1.2 |
| **GENITIVE** | Piscis Austrini |
| **ABBREVIATION** | PsA |
| **HIGHEST IN SKY AT 10 PM** | September–October |
| **FULLY VISIBLE** | 53°N–90°S |

Piscis Austrinus was known to the ancient Greeks, including Ptolemy in the 2nd century CE. It depicts a fish, which was said to be the parent of the two fish represented by the zodiacal constellation Pisces.

This constellation has also been called Piscis Australis. It is made prominent in the Southern Hemisphere by the presence of 1st-magnitude Fomalhaut, or Alpha (α) Piscis Austrini (see p.253). This blue-white star lies 25 light-years away.

### NEVER-ENDING DRINK 👁
In the sky, water from the jar of the adjacent Aquarius, the Water Carrier, flows toward the mouth of the fish, marked by Fomalhaut. The star's name is an Arabic term meaning "fish's mouth."

## SPECIFIC FEATURES
Beta (β) Piscis Austrini is a wide double star with components of 4th and 8th magnitudes. It is divisible through a small telescope.

More difficult to separate with a small telescope is Gamma (γ) Piscis Austrini, a closer pair of stars of 5th and 8th magnitudes.

**THE SOUTHERN FISH**

---

## THE SCULPTOR

# Sculptor

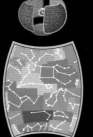

| | |
|---|---|
| **SIZE RANKING** | 36 |
| **BRIGHTEST STAR** | Alpha (α) 4.3 |
| **GENITIVE** | Sculptoris |
| **ABBREVIATION** | Scl |
| **HIGHEST IN SKY AT 10 PM** | October–November |
| **FULLY VISIBLE** | 50°N–90°S |

This unremarkable southern constellation was introduced in the 18th century by the French astronomer Nicolas Louis de Lacaille (see p.422). He originally described it as representing a sculptor's studio, although the name has since been shortened.

Sculptor contains the south pole of our galaxy—that is, the point 90 degrees south of the plane of the Milky Way. As a result, we can see numerous far-off galaxies in this direction, since they are unobscured by intervening stars or nebulae.

### SPECIFIC FEATURES
Epsilon (ε) Sculptoris is a binary star that can be separated with a small telescope. Its components, of 5th and 9th magnitudes, have an orbital period of more than 1,000 years.

The spiral galaxy NGC 253 is seen nearly edge-on, so that it appears highly elongated. Under good sky conditions, it can be picked up through binoculars or a small telescope. Nearby lies the fainter and smaller globular cluster NGC 288. Another spiral galaxy is NGC 55, which is similar in size and shape to NGC 253.

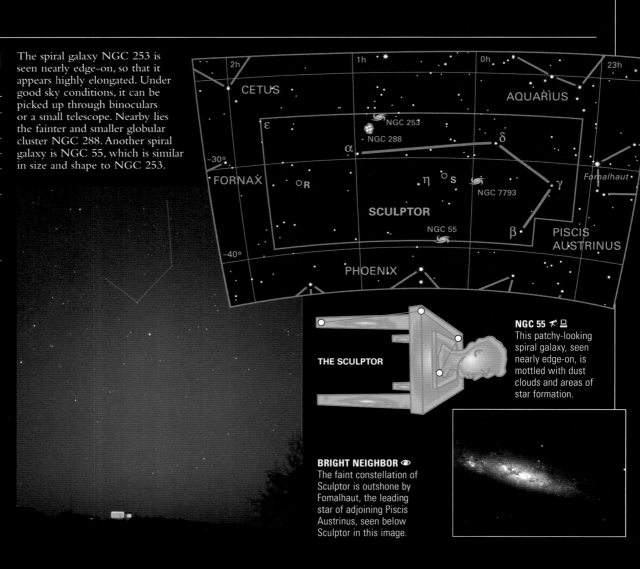

**THE SCULPTOR**

### NGC 55 🏹 💻
This patchy-looking spiral galaxy, seen nearly edge-on, is mottled with dust clouds and areas of star formation.

### BRIGHT NEIGHBOR 👁
The faint constellation of Sculptor is outshone by Fomalhaut, the leading star of adjoining Piscis Austrinus, seen below Sculptor in this image.

## THE FURNACE

# Fornax

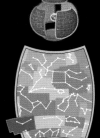

THE FURNACE

| | |
|---|---|
| SIZE RANKING | 41 |
| BRIGHTEST STAR | Alpha (α) 3.9 |
| GENITIVE | Fornacis |
| ABBREVIATION | For |
| HIGHEST IN SKY AT 10 PM | November–December |
| FULLY VISIBLE | 50°N–90°S |

A handful of faint stars makes up this undistinguished constellation of the southern sky. Fornax is situated on the edge of Eridanus and Cetus, and it represents a furnace used by chemists for distillation. It was originally known by the name Fornax Chemica, the chemical furnace, but this has since been shortened to Fornax.

### SPECIFIC FEATURES

The brightest star in the constellation, 4th-magnitude Alpha (α) Fornacis, has a yellow companion, which orbits it every 300 years. This 7th-magnitude star is visible through a small telescope.

On the border of Fornax with Eridanus lies a small cluster of galaxies known as the Fornax Cluster (see p.329). It is about 65 million light-years away, and its brightest member—the peculiar spiral NGC 1316—is a radio source known as Fornax A. Another prominent member of the Fornax Cluster is the beautiful barred spiral galaxy NGC 1365.

**THE FORNAX CLUSTER** ⚡ ⚘ ▱
Most of the galaxies in this cluster in southern Fornax are ellipticals, including the 10th-magnitude NGC 1399 (left of center in this photograph). Standing out among the elliptical galaxies is the large barred spiral NGC 1365 (bottom right).

**NGC 1365** ⚡ ⚘ ▱
This barred spiral galaxy is the largest in the Fornax Cluster and is about as massive as the Milky Way. It can be identified through a moderate-sized telescope.

**PROTECTED POSITION** 👁
Fornax is tucked into a bend in the celestial river, Eridanus. It was introduced by Nicolas Louis de Lacaille during the 18th century.

## THE CHISEL

# Caelum

THE CHISEL

| | |
|---|---|
| SIZE RANKING | 81 |
| BRIGHTEST STAR | Alpha (α) 4.4 |
| ABBREVIATION | Cae |
| GENITIVE | Caeli |
| HIGHEST IN SKY AT 10 PM | December–January |
| FULLY VISIBLE | 41°N–90°S |

Sandwiched between Eridanus and Columba is this small and faint southern constellation, which was introduced in the 18th century by the French astronomer Nicolas Louis de Lacaille (see p.422). It represents a stonemason's chisel.

### SPECIFIC FEATURES

Gamma (γ) Caeli is a double star, consisting of an orange giant of magnitude 4.6, with an 8th-magnitude companion. Because they are positioned close together, a modest-sized telescope is required in order to separate them.

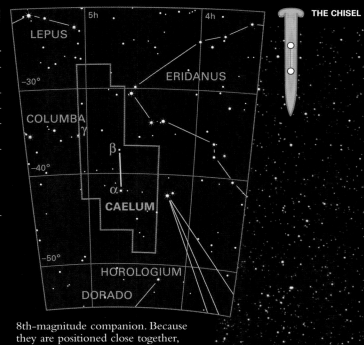

**ENGRAVED IN STONE** 👁
Beta (β) and Alpha (α) Caeli mark the shaft of the celestial chisel, which points toward the constellations Dorado and Reticulum in the south.

## THE RIVER
# Eridanus

**SIZE RANKING** 6

**BRIGHTEST STAR**
Achernar (α) 0.5

**GENITIVE** Eridani

**ABBREVIATION** Eri

**HIGHEST IN SKY AT 10 PM**
November–January

**FULLY VISIBLE**
32°N–89°S

This large constellation represents a river meandering from the foot of Taurus south to Hydrus. Its range in declination of 58 degrees is the greatest of any constellation.

The only star of any note in Eridanus is 1st-magnitude Achernar, or Alpha (α) Eridani, which lies at the southern tip of the constellation. The name Achernar is of Arabic origin and means "river's end."

Eridanus features in the story of Phaethon, son of the sun god Helios, who attempted to drive his father's chariot across the sky. He lost control and fell into the river below. This river has been identified with two real ones: the Nile in Egypt and the Po in Italy.

## SPECIFIC FEATURES

For all its size, Eridanus is short on objects of interest for a small telescope. The best is the multiple star Omicron-2 (o²) Eridani (see p.276), also known as 40 Eridani, which includes both a red dwarf and a white one. To the eye, it appears as a 4th-magnitude orange star, but a small telescope reveals a 10th-magnitude companion, the white dwarf. This is the easiest white dwarf to spot with a small telescope. It forms a binary with a fainter red dwarf, although this star may require a telescope with a slightly larger aperture to be detectable.

Two double stars of note are Theta (θ) Eridani, consisting of white stars of 3rd and 4th magnitudes divisible through a small telescope, and 32 Eridani, a contrasting pair of orange and blue stars of 5th and 6th magnitudes, also within range of a small telescope.

The galaxy NGC 1300 is estimated to lie around 75 million light-years away and is too faint for viewing through a small telescope. However, it shows up beautifully on photographs.

### MULTIPLE STAR

The primary star of Omicron-2 (o²) Eridani is in the center of this photograph, while its white-dwarf and red-dwarf companions overlap each other to the right.

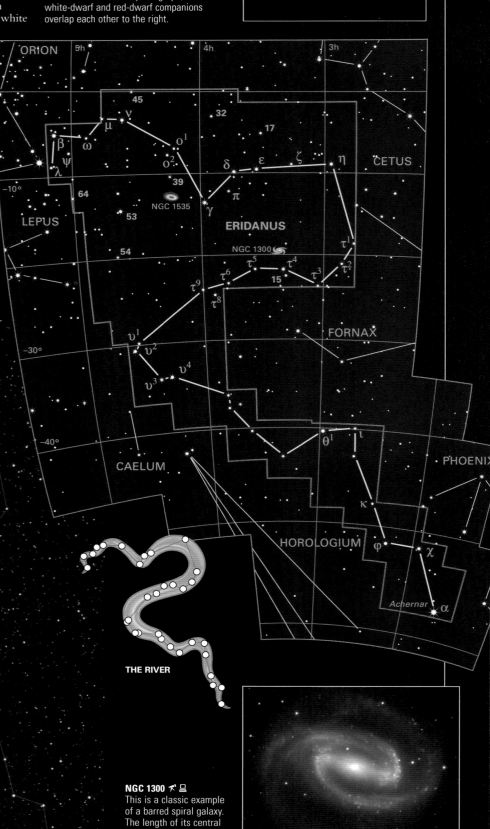

### CELESTIAL RIVER 👁
Eridanus has its source next to Rigel (in Orion) and flows south to Achernar. It is fully visible to almost all of the Southern Hemisphere and half of the Northern.

**THE RIVER**

### NGC 1300 ✺ 🖥
This is a classic example of a barred spiral galaxy. The length of its central bar is greater than the diameter of the Milky Way, being 150,000 light-years across.

## THE HARE

# Lepus

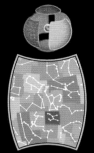

| | |
|---|---|
| SIZE RANKING | 51 |
| BRIGHTEST STAR | Arneb (α) 2.6 |
| GENITIVE | Leporis |
| ABBREVIATION | Lep |
| HIGHEST IN SKY AT 10 PM | January |
| FULLY VISIBLE | 62°N–90°S |

Lepus is often overlooked because it is surrounded by sparkling Orion and Canis Major, yet it is worthy of note. It is one of the constellations known to the ancient Greeks.

**M79**
This somewhat sparse 8th-magnitude globular cluster, 42,000 light-years away, has long, starry arms that give it the appearance of a starfish.

## SPECIFIC FEATURES

Gamma (γ) Leporis is a 4th-magnitude yellow star with a 6th-magnitude orange companion, which is visible through binoculars. Another double star is Kappa (κ) Leporis, a 4th-magnitude star with a close companion of 7th magnitude. It is difficult to separate through telescopes of small aperture.

NGC 2017 is a compact group of stars in what seems to be a chance alignment. Thus it is not a true star cluster at all.

Near the border with Eridanus lies R Leporis, an intensely red variable star of the same type as Mira (in Cetus). Its brightness ranges from 6th to 12th magnitude every 14 months or so.

The globular cluster M79 can be seen though a small telescope. In the same field of view lies Herschel 3752, a triple star with components of 5th, 7th, and 9th magnitudes.

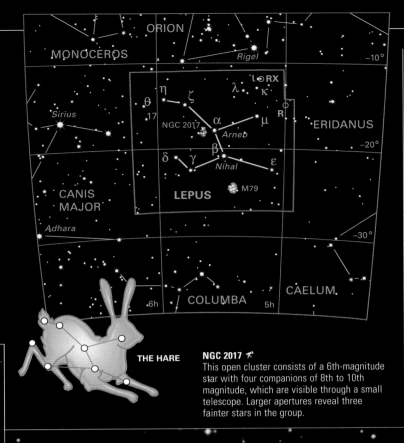

THE HARE

**NGC 2017**
This open cluster consists of a 6th-magnitude star with four companions of 8th to 10th magnitude, which are visible through a small telescope. Larger apertures reveal three fainter stars in the group.

### MYTHS AND STORIES

## A RUNNING HARE

According to Greek mythology, there were no hares on the island of Leros until one man introduced a pregnant female. Soon everyone was raising hares, but they became pests, destroying crops and reducing the population to starvation. The inhabitants eventually drove the hares out of the island and put the image of the hare among the stars as a reminder that one can have too much of a good thing.

### THE HUNTER AND HUNTED

One of Orion's dogs chases the hare in this 15th-century Flemish illustration, which was based on the *Liber Floridus* of Lambertus, compiled during the Middle Ages.

**SAFE HAVEN**
Lepus, the celestial hare, crouches under the feet of Orion, like an animal trying to hide from its hunter. Orion's dogs, Canis Major and Canis Minor, lie nearby.

THE NIGHT SKY

## THE DOVE
# Columba

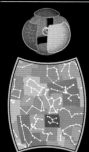

| | |
|---|---|
| **SIZE RANKING** | 54 |

**BRIGHTEST STAR**
Phact (α) 2.7

**GENITIVE** Columbae

**ABBREVIATION** Col

**HIGHEST IN SKY AT 10 P.M.**
January

**FULLY VISIBLE**
46°N–90°S

The Dutch theologian and astronomer Petrus Plancius (see p.358) formed this southern constellation in the late 16th century from stars near Lepus and Canis Major that had not previously been allocated to any constellation. It supposedly represents Noah's dove (see panel, right).

### SPECIFIC FEATURES
Fifth-magnitude Mu (μ) Columbae is a fast-moving star apparently thrown out from the area of the Orion Nebula about 2.5 million years ago. Astronomers think that it was once a member of a binary system that was disrupted by a close encounter with another star. The other member of the former binary, moving away from Orion in the opposite direction, is 6th-magnitude AE Aurigae.
NGC 1851 is a modest globular cluster that is visible as a faint patch through a small telescope.

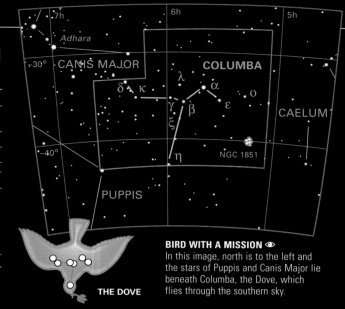

**BIRD WITH A MISSION** ◉
In this image, north is to the left and the stars of Puppis and Canis Major lie beneath Columba, the Dove, which flies through the southern sky.

**THE DOVE**

### MYTHS AND STORIES
## NOAH'S DOVE

In the Biblical story of the Flood, Noah loaded an ark with a male and female of every kind of animal on Earth. It then rained for 40 days and 40 nights, drowning everything except the animals aboard the ark. When the rain abated, Noah sent out a dove to find dry land. The dove came back with an olive stem in its beak—a sure sign that the waters were at last receding.

**WINGED MESSENGER**
The dove returns to Noah's Ark, carrying an olive branch, in this illustration by a 10th-century Catalan monk named Emeterio.

## THE COMPASS
# Pyxis

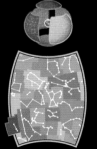

**SIZE RANKING** 65

**BRIGHTEST STAR**
Alpha (α) 3.7

**GENITIVE** Pyxidis

**ABBREVIATION** Pyx

**HIGHEST IN SKY AT 10 P.M.**
February–March

**FULLY VISIBLE**
52°N–90°S

Pyxis is a faint and unremarkable southern constellation lying next to Puppis on the edge of the Milky Way. It represents a ship's magnetic compass. The constellation was introduced in the 18th century by the French astronomer Nicolas Louis de Lacaille (see p.422).

### SPECIFIC FEATURES
T Pyxidis is a recurrent nova—that is, one that has undergone several recorded outbursts. Six eruptions have been seen since 1890, the last being in 2011. During these outbursts, it has brightened to 6th or 7th magnitude. It is likely to brighten again at any time and become visible through binoculars.

**THE COMPASS**

**COMPASS BEARINGS** ◉
In this image of the scattered stars of Pyxis, the Compass, north is on the left, and the stars of adjacent Puppis are to be found above Pyxis.

THE STERN

# Puppis

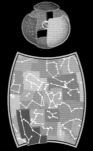

**SIZE RANKING** 20

**BRIGHTEST STAR**
Naos (ζ) 2.2

**GENITIVE** Puppis

**ABBREVIATION** Pup

**HIGHEST IN SKY AT 10 P.M.**
January–February

**FULLY VISIBLE**
39°N–90°S

This rich southern constellation straddling the Milky Way was originally part of the ancient Greek constellation of Argo Navis (the ship of Jason and the Argonauts, see p.410) until it was divided into three parts in the 18th century. Puppis, representing the ship's stern, is the largest part. The stars of each section retained their original Greek letters, and in the case of Puppis the lettering now starts at Zeta (ζ) Puppis, a star that is also known as Naos.

## SPECIFIC FEATURES

Third-magnitude Xi (ξ) Puppis has a wide and unrelated 5th-magnitude companion that is visible through binoculars, while k Puppis consists of a pair of nearly identical stars with components of 5th magnitude that can be split through a small telescope.

L Puppis is a wide naked-eye and binocular pair, of which L$^2$ Puppis is a variable red giant that ranges between 3rd and 6th magnitudes every 5 months or so.

M46 and M47 are a pair of open clusters that together create a brighter patch in the Milky Way. Both appear of similar size to a full moon. M46 is the richer of the two, while M47 is the closer, being about 1,500 light-years away—that is, less than one-third of the distance of its apparent neighbor. The cluster M93 lies about 3,500 light-years away.

NGC 2477 is an open cluster that looks almost like a globular cluster when seen through binoculars, while NGC 2451 is more scattered and has the 4th-magnitude orange giant c Puppis near its center.

### NGC 2477 🔭✦

This is one of the richest open clusters, containing an estimated 2,000 stars. It is about 4,000 light-years away. The star below NGC-2477 in this picture—b Puppis—is of magnitude 4.5.

THE STERN

### STERN OF THE ARGO 👁
The stars of Puppis, representing the stern of the Argo, are seen here rising behind thin clouds. Sirius (in Canis Major) lies near the left edge of this picture.

### SHARP CLUSTER 🔭✦
M93 is an attractive open cluster for viewing through binoculars or a small telescope. It is shaped like an arrowhead with two orange giants near its tip.

### M46 AND NEBULA 🔭✦
A small planetary nebula, seen here below center, seems to be part of M46 but in fact lies in the foreground.

THE NIGHT SKY

## THE SAILS

# Vela

| | |
|---|---|
| **SIZE RANKING** | 32 |
| **BRIGHTEST STAR** | Gamma (γ) 1.8 |
| **GENITIVE** | Velorum |
| **ABBREVIATION** | Vel |
| **HIGHEST IN SKY AT 10 PM** | February–April |
| **FULLY VISIBLE** | 32°N–90°S |

In the 18th century, the ancient Greek constellation Argo Navis (the ship of Jason and the Argonauts—see panel, below) was divided into three parts, one of which was Vela, which represents the ship's sails. Because the stars labeled Alpha (α) and Beta (β) in the former Argo Navis are now in Carina, to the south, the labeling of the stars in Vela starts with Gamma (γ) Velorum, or Regor (see p.253).

Between Gamma and Lambda (λ) Velorum are found the gaseous strands of the Vela supernova remnant (see p.269)—the supernova could have been seen from Earth around 11,000 years ago—while Delta (δ) and Kappa (κ) Velorum combine with two stars in Carina to form the False Cross (sometimes mistaken for the true Southern Cross).

### SPECIFIC FEATURES
Gamma Velorum is the brightest example of a Wolf–Rayet star, a rare type of star that has lost its outer layers, thereby exposing its ultra-hot interior. A 4th-magnitude companion is visible through a small telescope or good binoculars. In addition, two

wider companions, with components of 8th and 9th magnitudes, are visible through a telescope.

IC 2391 is the best star cluster in Vela for the naked eye or binoculars. It is a group of several dozen stars covering a greater area than a full moon. To the north of it is another binocular cluster, IC 2395.

NGC 2547 is an open cluster half the size of a full moon and can be identified through binoculars or a small telescope.

Popularly known as the Eight-Burst Nebula, NGC 3132 has complex loops that are revealed only through a large telescope or on long-exposure photographs. A small telescope will show the nebula's disk, of similar apparent size to Jupiter, and the 10th-magnitude star at its center.

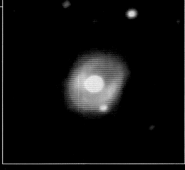

**THE EIGHT-BURST NEBULA** ☄ 💻
The planetary nebula NGC 3132 has loops of gas that interlock like figure-eights, hence the object's popular name.

**UNDER SAIL** 👁
Vela represents the mainsail of the *Argo*, the ship of Jason and the Argonauts, sailing through the southern sky in the quest for the golden fleece.

THE SAILS

ANTLIA

PYXIS

PUPPIS

VELA

NGC 3132

ψ

λ

NGC 3201

γ

μ

NGC 3228

IC 2395

NGC 2547

IC 2391

φ

κ

O

IC 2488

CENT-AURUS

CARINA

11h

10h

9h

8h

−40°

−50°

**IC 2391** 👁 🔭
Omicron (o) Velorum, at magnitude 3.6, is the brightest member of IC 2391, a scattered cluster that lies some 500 light-years from Earth in the southern reaches of Vela.

## THE KEEL

# Carina

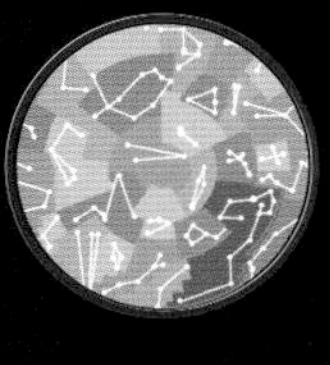

| | |
|---|---|
| **SIZE RANKING** | 34 |
| **BRIGHTEST STAR** | Canopus (α) -0.6 |
| **GENITIVE** | Carinae |
| **ABBREVIATION** | Car |
| **HIGHEST IN SKY AT 10 PM** | January–April |
| **FULLY VISIBLE** | 14°N–90°S |

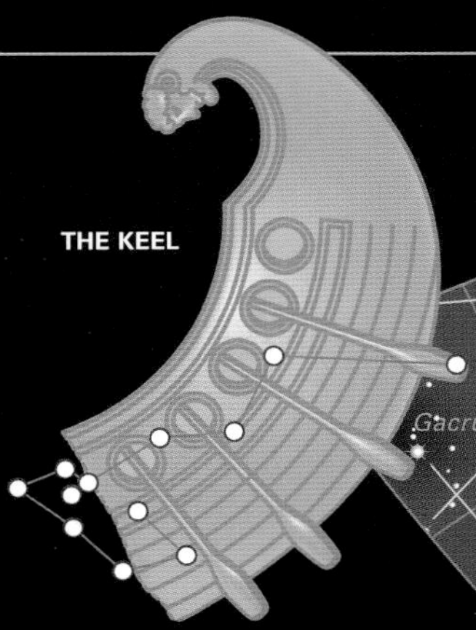

**THE KEEL**

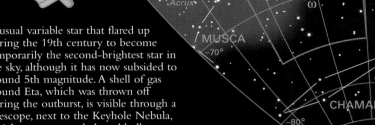

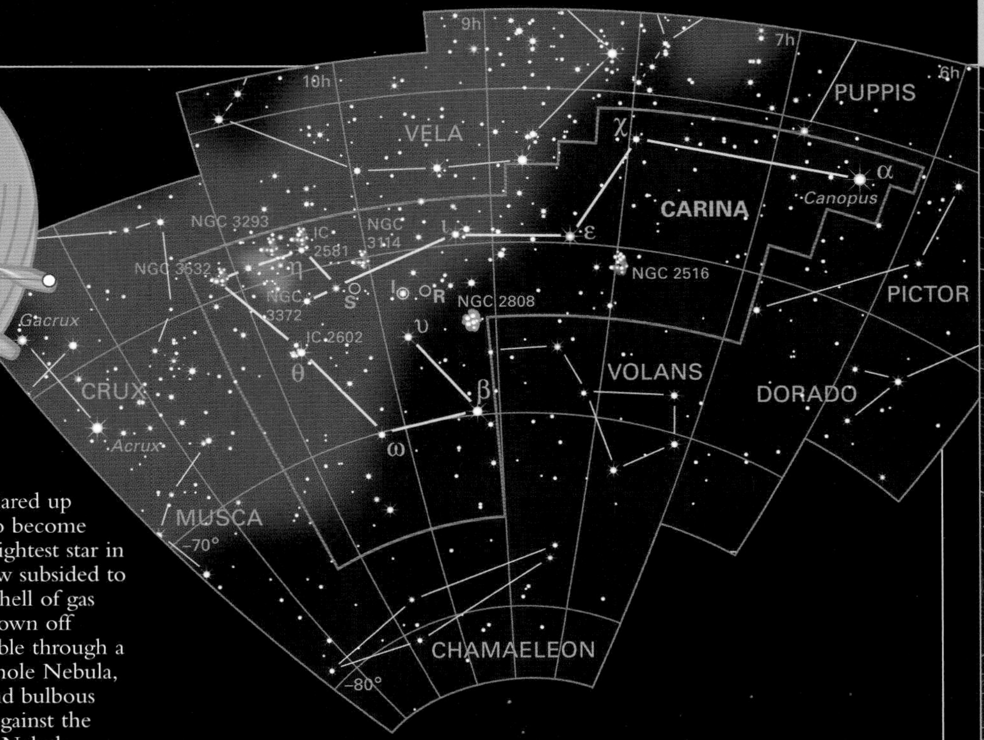

Carina is a major southern constellation that was originally part of the larger figure of Argo Navis, which depicted a ship, until that was split up in the 18th century. Carina represents the ship's keel.

Its most prominent star, Canopus, or Alpha (α) Carinae, is a white supergiant 310 light-years away and second in brightness only to Sirius in the entire sky. The stars Epsilon (ε) and Iota (ι) Carinae form a pseudo "southern cross," known as the False Cross, in conjunction with two stars in neighboring Vela.

## SPECIFIC FEATURES

Splashed across the Milky Way near the border with Centaurus and Vela is the Carina Nebula, NGC 3372 (see p.247), a patch of glowing gas four diameters of a full moon wide. It is visible to the eye and well seen through binoculars. The densest and brightest part of the nebula is around Eta (η) Carinae (see p.262), an

unusual variable star that flared up during the 19th century to become temporarily the second-brightest star in the sky, although it has now subsided to around 5th magnitude. A shell of gas around Eta, which was thrown off during the outburst, is visible through a telescope, next to the Keyhole Nebula, which appears as a dark and bulbous cloud of dust silhouetted against the glowing gas of the Carina Nebula.

A glorious sight through binoculars, another treasure is IC 2602, an open cluster known as the Southern Pleiades. Twice the apparent size of a full moon, it contains several stars visible to the naked eye—the brightest being 3rd-magnitude Theta (θ) Carinae.

Among Carina's naked-eye clusters is NGC 3532. At its widest point, this elongated group of stars is twice the width of a full moon. NGC 3114 is about the same apparent size as a full moon, its brightest individual members being visible through binoculars. NGC 2516 is sparser and appears cross-shaped through binoculars. Its brightest star is a 5th-magnitude red giant.

## ELONGATED CLUSTER
What appears to be the brightest member of NGC 3532, in the lower left of this photograph, is in fact an extremely luminous background star some four times farther off.

## THE CARINA NEBULA
The brightest part of this immense cloud of glowing gas is V-shaped (shown here), while the star Eta (η) Carinae itself (below left of center) is a peculiar variable that appears as a hazy orange ellipse.

**EVEN KEEL**
Carina represents the keel and hull of the Argonauts' ship, the *Argo*. The blade of the steering oar is marked by Canopus, Carina's brightest star.

THE NIGHT SKY

# Crux

| | |
|---|---|
| **SIZE RANKING** | 88 |
| **BRIGHTEST STARS** | Acrux (α) 0.8, Mimosa (β) 1.3 |
| **GENITIVE** | Crucis |
| **ABBREVIATION** | Cru |
| **HIGHEST IN SKY AT 10 P.M.** | April–May |
| **FULLY VISIBLE** | 25°N–90°S |

Crux lies in a rich area of the Milky Way. Although it is the smallest constellation, it is instantly recognizable and is squeezed between the legs of the centaur, Centaurus. The longer axis of the Southern Cross, as Crux is popularly known, points toward the south celestial pole. Its stars were known to the ancient Greeks (see panel, below), who regarded them as part of Centaurus. They were made into a separate constellation in the 16th century.

## SPECIFIC FEATURES

Alpha (α) Crucis or Acrux is the most southerly first-magnitude star. It is a glittering double that is readily divisible through a small telescope. The two components are of magnitudes 1.3 and 1.8. A wider 5th-magnitude star can be seen through binoculars; it is not related to Acrux.

The star at the top of the cross is the 2nd-magnitude red giant Gamma (γ) Crucis or Gacrux. It has an unrelated 6th-magnitude companion visible through binoculars. Nearby, Mu (μ) Crucis is a wide pair of 4th- and 5th-magnitude stars easily separated through a small telescope or even good binoculars.

One of the gems of the southern sky is the Jewel Box Cluster (see p.294), or NGC 4755, visible to the naked eye as a brighter patch within the Milky Way near Beta (β) Crucis or Mimosa. Its individual stars, the brightest being of 6th magnitude, cover about one-third the width of the full Moon. They can be viewed through binoculars or a small telescope. Near the center is a ruby-colored

supergiant that contrasts with the blue-white sparkle of the other stars, producing a resemblance to a casket of jewels, hence the popular name.

The Coalsack Nebula is to be found beside the Jewel Box. This dark cloud of dust blocks light from the stars of the Milky Way behind it. It spans the width of 12 full Moons and extends into neighboring Centaurus and Musca, so is prominent to the naked eye and through binoculars.

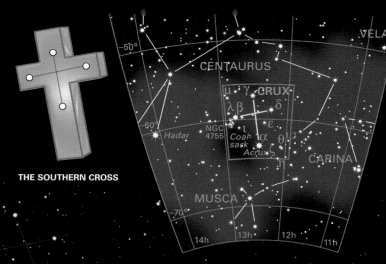

**THE SOUTHERN CROSS**

**THE COALSACK** 👁🔭
The Coalsack Nebula, which lies next to the stars of the Southern Cross, is a smudgy cloud of interstellar dust silhouetted against the bright background of the Milky Way.

**JEWELS OF THE SKIES** 🔭✦
The Jewel Box Cluster is a sparkling group of stars just north of the Coalsack Nebula, although the cluster is almost 10 times more distant from the Earth.

**SIGN OF THE CROSS** 👁
Four prominent stars make up the Southern Cross, one of the most famous of all celestial patterns, which appears on the flags of several nations.

EXPLORING SPACE

## REDISCOVERING STARS

When European seafarers returned from exploring the southern latitudes in the 15th and 16th centuries, they reported stars they had never seen before. Among these explorers was Amerigo Vespucci (1454–1512), an Italian who in 1501 charted Alpha (α) and Beta (β) Centauri and the stars of Crux. Astronomers later realized that these stars had been known to the ancient Greeks but that precession (see p.64) had subsequently carried them below the horizon in Europe.

**AMERIGO VESPUCCI**
This imaginative view of Amerigo Vespucci observing the Southern Cross with an astrolabe was painted by the 16th-century Flemish artist Joannes Stradanus (Hans van der Straet).

## THE FLY
# Musca

| | |
|---|---|
| **SIZE RANKING** | 77 |
| **BRIGHTEST STAR** | Alpha (α) 2.7 |
| **GENITIVE** | Muscae |
| **ABBREVIATION** | Mus |
| **HIGHEST IN SKY AT 10 P.M.** | April–May |
| **FULLY VISIBLE** | 14°N–90°S |

This modest constellation is to be found in the Milky Way south of Crux and Centaurus. In fact, the southern tip of the dark Coalsack Nebula extends into it from Crux.

Musca is one of the southern constellations introduced at the end of the 16th century by the Dutch navigator-astronomers Pieter Dirkszoon Keyser and Frederick de Houtman. It represents a fly.

### SPECIFIC FEATURES
Theta (θ) Muscae is a double star with components of 6th and 8th magnitude, divisible through a small telescope. The fainter component is an example of a Wolf–Rayet star—a hot star that has lost its outer layers. Musca also has a globular cluster, known as NGC 4833 (see p.295).

**NGC 4833** 🔭 ☄
This globular cluster is just visible through binoculars. Individual stars can be resolved with a telescope of 4 in (100 mm) aperture.

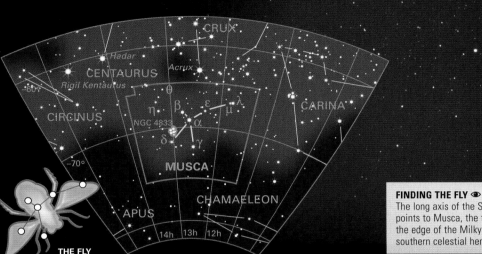

**THE FLY**

**FINDING THE FLY** 👁
The long axis of the Southern Cross points to Musca, the fly, which lies on the edge of the Milky Way within the southern celestial hemisphere.

## THE COMPASSES
# Circinus

| | |
|---|---|
| **SIZE RANKING** | 85 |
| **BRIGHTEST STAR** | Alpha (α) 3.2 |
| **GENITIVE** | Circini |
| **ABBREVIATION** | Cir |
| **HIGHEST IN SKY AT 10 P.M.** | May–June |
| **FULLY VISIBLE** | 19°N–90°S |

Circinus represents a pair of dividing compasses, as used by surveyors and navigators. It is one of the figures introduced in the 18th century by the French astronomer Nicolas Louis de Lacaille (see p.422).

This small southern constellation is squeezed awkwardly in between Centaurus and Triangulum Australe. It lies next to Alpha (α) Centauri, so it is not difficult to locate.

### SPECIFIC FEATURES
Circinus contains little of interest for amateur astronomers. Alpha (α) Circini, however, is its one star of note. It is situated against the background of the Milky Way and is easy to identify, being a double with components of 3rd and 9th magnitudes. These are divisible through a small telescope.

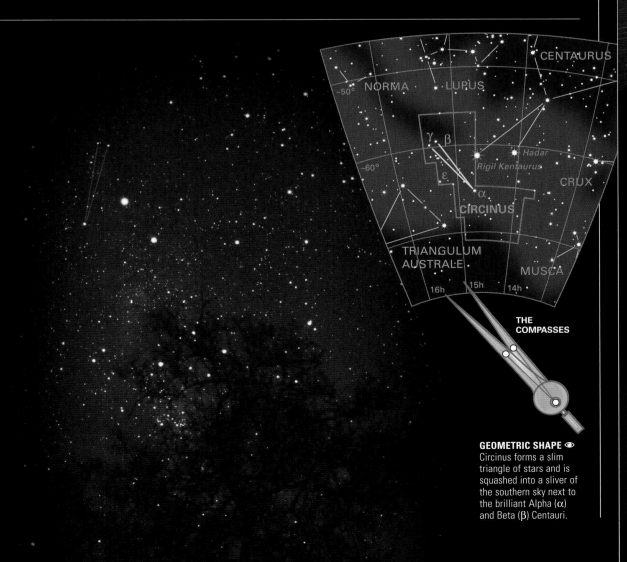

**THE COMPASSES**

**GEOMETRIC SHAPE** 👁
Circinus forms a slim triangle of stars and is squashed into a sliver of the southern sky next to the brilliant Alpha (α) and Beta (β) Centauri.

## THE SET SQUARE

# Norma

**SIZE RANKING** 74

**BRIGHTEST STAR**
Gamma-2 (γ²) 4.0

**GENITIVE** Normae

**ABBREVIATION** Nor

**HIGHEST IN SKY AT 10 PM**
June

**FULLY VISIBLE**
29°N–90°S

Norma was introduced in the 1750s by the Frenchman Nicolas Louis de Lacaille (see p.422) and was originally known as *Norma et Regula*, the square and ruler. It is an unremarkable southern constellation lying in the Milky Way between Lupus and the zodiacal constellation of Scorpius.

The stars that Lacaille designated Alpha (α) and Beta (β) have since been incorporated into Scorpius.

### SPECIFIC FEATURES
At magnitude 4.0, Gamma-2 (γ²) Normae is the constellation's brightest star, and it is one-half of a naked-eye double together with Gamma-1 (γ¹) Normae, of magnitude 5.0. The two stars lie at widely different distances and hence are unrelated.

Two other doubles in the constellation that are readily separated through a small telescope are Epsilon (ε) Normae, with components of 5th and 7th magnitudes, and Iota-1 (ι¹) Normae, with components of 5th and 8th magnitudes.

NGC 6087 is a large open cluster that has radiating chains of stars, which are visible through binoculars. Near its center is its brightest star, S-Normae—a Cepheid variable that ranges in brightness from magnitude 6.1 to 6.8 every 9.8 days.

**NGC 6067**
This rich cluster covers an area of sky about half the apparent diameter of a full moon. It is seen against the backdrop of the Milky Way.

**RIGHT ANGLE**
Norma's most distinctive feature is a right-angled trio of three faint stars, which is somewhat difficult to identify among the rich Milky Way star fields.

**THE SET SQUARE**

---

## THE SOUTHERN TRIANGLE

# Triangulum Australe

**SIZE RANKING** 83

**BRIGHTEST STAR**
Alpha (α) 1.9

**GENITIVE** Trianguli
Australis

**ABBREVIATION** TrA

**HIGHEST IN SKY AT 10 PM**
June–July

**FULLY VISIBLE**
19°N–90°S

Triangulum Australe is one of the constellations of the southern sky that was introduced at the end of the 16th century by the Dutch navigators Pieter Dirkszoon Keyser and Frederick de Houtman. It is the smallest of the 12 they identified.

Although smaller than its northern counterpart, Triangulum, this constellation contains brighter stars and so is more prominent.

### SPECIFIC FEATURES
NGC 6025 lies on Triangulum Australe's northern border with Norma. It is 2,700 light-years away from Earth. This open cluster is noticeably elongated in shape and is about one-third the apparent diameter of a full moon. It is easily seen through binoculars.

Alpha (α) Trianguli Australis is an orange giant whose color shows prominently through binoculars.

There is nothing else in the constellation to attract users of small telescopes.

**SOUTHERN TRIPLET**
Triangulum Australe is an easily recognized triangle of stars, lying in the Milky Way near brilliant Alpha (α) and Beta (β) Centauri, which here are visible on the right.

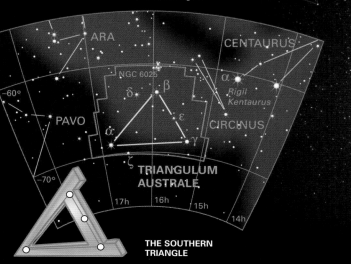

**THE SOUTHERN TRIANGLE**

# Ara

| | |
|---|---|
| **SIZE RANKING** | 63 |
| **BRIGHTEST STARS** | Alpha (α) 2.8, Beta (β) 2.8 |
| **GENITIVE** | Arae |
| **ABBREVIATION** | Ara |
| **HIGHEST IN SKY AT 10 PM** | June–July |
| **FULLY VISIBLE** | 22°N–90°S |

Ara was visualized by the ancient Greeks as the altar on which the gods of Olympus swore an oath of allegiance before their battle with the Titans for control of the universe (see

THE ALTAR

## MYTHS AND STORIES

### TITANOMACHIA

Titanomachia, or the Clash of the Titans, was the 10-year war for dominance of the universe between the gods on Mount Olympus, led by Zeus, and the Titans on Mount Othrys. In gratitude for their victory, Zeus placed the altar of the gods in the sky.

**VICTORY PANEL**
Part of the battle of the gods and Titans is here depicted in the Zeus Altar of Pergamon, which was sculpted in Greece *c.* 180 BCE.

panel, left). This southern constellation lies within the Milky Way and is situated south of Scorpius.

SPECIFIC FEATURES
The attractive open cluster NGC 6193 consists of about 30 stars of 6th magnitude and fainter. It can be viewed through binoculars.

NGC 6397 is among the closest globular clusters to us, being around 10,000 light-years away, and can be seen well through binoculars or a small telescope. Like NGC 6193, it appears relatively large—both being over half the apparent width of a full moon.

Ara contains no stars of particular interest to users of small telescopes.

**NGC 6397** 📷 ✦
The globular cluster NGC 6397 has a condensed center and scattered outer regions in which chains and sprays of stars can be traced.

**INCENSE BURNER** 👁
Ara, the celestial altar, is oriented with its top facing south. Incense burning on the altar might give off the "smoke" of the Milky Way above it.

---

## Corona Australis

| | |
|---|---|
| **SIZE RANKING** | 80 |
| **BRIGHTEST STARS** | Alpha (α) 4.1, Beta (β) 4.1 |
| **GENITIVE** | Coronae Australis |
| **ABBREVIATION** | CrA |
| **HIGHEST IN SKY AT 10 PM** | July–August |
| **FULLY VISIBLE** | 44°N–90°S |

The small southern constellation of Corona Australis lies under the feet of Sagittarius. It comprises stars of 4th magnitude and fainter, and it was one of the 48 constellations recognized by the ancient Greek astronomer Ptolemy (see p.347).

SPECIFIC FEATURES
Gamma (γ) Coronae Australis is a binary star with components of 5th magnitude. The pair orbit each other every 122 years, and they are slowly moving apart as seen from Earth. This means the components are becoming easier to view individually. Meanwhile, a 4-in (100-mm) aperture is needed to divide this challenging star.

Kappa (κ) Coronae Australis is an unrelated double with components of 6th magnitude, which are readily divided through a small telescope.

The modest globular cluster NGC 6541 covers about one-third the apparent diameter of a full moon. It is visible through a small telescope or binoculars.

**SOUTHERN ARC** 👁
Corona Australis is an attractive arc of stars that represents a crown or laurel wreath.

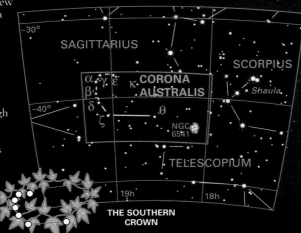

THE SOUTHERN CROWN

## THE TELESCOPE
# Telescopium

| | |
|---|---|
| SIZE RANKING | 57 |
| BRIGHTEST STAR | Alpha (α) 3.5 |
| GENITIVE | Telescopii |
| ABBREVIATION | Tel |
| HIGHEST IN SKY AT 10 PM | July–August |
| FULLY VISIBLE | 33°N–90°S |

Telescopium is an almost entirely undistinguished southern constellation near Sagittarius and Corona Australis. It was invented by the French astronomer Nicolas Louis de Lacaille (see p.422) to commemorate the telescope. Its pattern of stars represents one of the aerial telescopes used at the Paris observatory. These were refractors with extremely long focal lengths—to reduce chromatic aberration—suspended from tall poles by ropes and pulleys.

## SPECIFIC FEATURES
Delta (δ) Telescopii is an unrelated pair of stars with components of 5th magnitude. It can be divided with binoculars or even good eyesight.

**THE TELESCOPE**

**LONG VIEW** 👁
Telescopium depicts an early design of refracting telescope with a long tube supported by a flimsy mounting—a far cry from the massive reflectors of today.

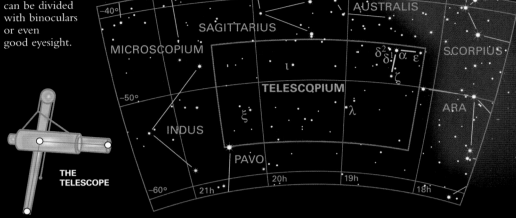

---

## THE INDIAN
# Indus

| | |
|---|---|
| SIZE RANKING | 49 |
| BRIGHTEST STAR | Alpha (α) 3.1 |
| GENITIVE | Indi |
| ABBREVIATION | Ind |
| HIGHEST IN SKY AT 10 PM | August–October |
| FULLY VISIBLE | 15°N–90°S |

This southern constellation was introduced in the late 16th century by Pieter Dirkszoon Keyser and Frederick de Houtman (see panel, right). It represents a human figure with a spear and arrows, although it remains unclear whether this is supposed to be a native of the East Indies (as discovered by the Dutch explorers during their expeditions) or a native of the Americas.

## SPECIFIC FEATURES
Fifth-magnitude Epsilon (ε) Indi is one of the closest stars to us, being 11.8 light-years away. Somewhat smaller and cooler than the Sun, it appears pale orange in color.
   Theta (θ) Indi is a 4th-magnitude star with a companion of 7th magnitude that can be identified through a small telescope.

**CONCEALED FIGURE** 👁
Only a vivid imagination could discern the figure of a human in the constellation of Indus, which comprises a few faint stars next to the distinctive figures of Grus and Tucana.

EXPLORING SPACE
## DUTCH VOYAGES OF DISCOVERY

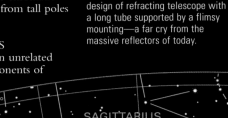

As well as exploring the southern oceans, Dutch traders and navigators charted the southern sky. On the first Dutch expedition to the East Indies in 1595 were two Dutch navigator–astronomers, Pieter Dirkszoon Keyser (c. 1540–1596) and Frederick de Houtman (1571–1627). Keyser died during the voyage, but his celestial observations, along with those of de Houtman, were returned to the Dutch cartographer Petrus Plancius (see p.358) and formed the basis for 12 new constellations, all of which are still recognized.

**FAMILY OF EXPLORERS**
The first Dutch expedition to the East Indies consisted of four ships and was led by Cornelis de Houtman, the brother of Frederick, who was on the trip as a navigator.

**THE INDIAN**

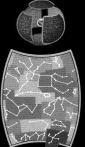

## THE CRANE

# Grus

| | |
|---|---|
| **SIZE RANKING** | 45 |
| **BRIGHTEST STAR** | Alnair (α) 1.7 |
| **GENITIVE** | Gruis |
| **ABBREVIATION** | Gru |
| **HIGHEST IN SKY AT 10 PM** | September–October |
| **FULLY VISIBLE** | 33°N–90°S |

Grus represents a long-necked wading bird—a crane—although it has also been depicted as a flamingo. It is a constellation of the southern sky and is situated between Piscis Austrinus and Tucana. Grus was introduced at the end of the 16th century by the Dutch navigator–astronomers Pieter Dirkszoon Keyser and Frederick de Houtman (see panel, opposite).

### SPECIFIC FEATURES
Delta (δ) Gruis is a pair of 4th-magnitude giants, with one yellow component and one red one, while Mu (μ) Gruis is a pair of 5th-magnitude yellow giants. Both pairs are divisible with the naked eye. They appear double due to chance alignments and are not true binaries.

Beta (β) Gruis is a red giant whose brightness ranges from magnitude 2.0 to 2.3, with no set period.

**THE CRANE**

### SHOWING THE WAY 👁
Two wide doubles—Delta (δ) and Mu (μ) Gruis—appear along the extended neck of Grus, the Crane, which points to the lower right in this image.

## THE PHOENIX

# Phoenix

| | |
|---|---|
| **SIZE RANKING** | 37 |
| **BRIGHTEST STAR** | Ankaa (α) 2.4 |
| **GENITIVE** | Phoenicis |
| **ABBREVIATION** | Phe |
| **HIGHEST IN SKY AT 10 PM** | October–November |
| **FULLY VISIBLE** | 32°N–90°S |

Phoenix lies at the southern end of Eridanus, next to that constellation's brightest star, Achernar. It is the largest of the 12 southern constellations introduced during the late 16th century by the Dutch navigator–astronomers Pieter Dirkszoon Keyser and Frederick de Houtman (see panel, opposite). It represents the mythical bird that was supposedly born from the ashes of its predecessor (see panel, right).

### SPECIFIC FEATURES
Zeta (ζ) Phoenicis is a variable double consisting of a 4th-magnitude star with an 8th-magnitude companion. The brighter star is an eclipsing binary and varies between magnitudes 3.9 and 4.4 every 1.7 days.

## MYTHICAL BIRD

According to legend, the phoenix was said to live for 500 years. At the end of its life span, it built a nest of cinnamon bark and incense, on which it died, some say in fire. A baby phoenix was born from its ancestor's remains. The death and rebirth of the phoenix has been seen as symbolic of the daily rising and setting of the Sun.

**FUNERAL PYRE**
The phoenix is consumed by fire in this 18th-century German copper engraving from *Bilderbuch für Kinder.*

### PHOENIX FALLING 👁
The stars of Phoenix sink toward the western horizon in the morning sky, with Grus below it. North is to the right in this photograph.

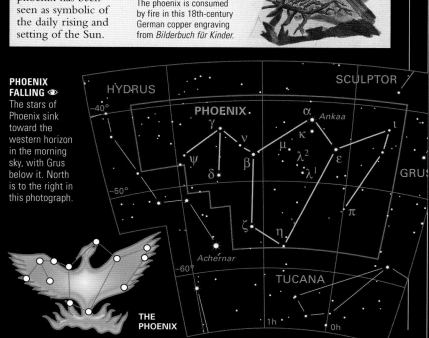

**THE PHOENIX**

# Tucana

| | |
|---|---|
| **SIZE RANKING** | 48 |
| **BRIGHTEST STAR** | Alpha (α) 2.9 |
| **GENITIVE** | Tucanae |
| **ABBREVIATION** | Tuc |
| **HIGHEST IN SKY AT 10 P.M.** | September–November |
| **FULLY VISIBLE** | 14°N–90°S |

This far-southern constellation is to be found at the end of the celestial river, Eridanus. It represents the large-beaked tropical bird that is native to South and Central America.

Tucana was introduced in the late 16th century by the Dutch navigator–astronomers Pieter Dirkszoon Keyser and Frederick de Houtman (see p.416).

### SPECIFIC FEATURES
Tucana contains the Small Magellanic Cloud (see p.311), the lesser of the two satellite galaxies that accompany our own galaxy. To the naked eye, it appears like a detached patch of the Milky Way and is seven times wider than the apparent diameter of a full moon. Star fields and clusters within the Small Magellanic Cloud can be detected through binoculars or a small telescope. It is about 200,000 light-years away.

Two globular clusters lie near the Small Magellanic Cloud, although both are actually foreground objects in our galaxy and so are not associated with the Cloud. The more prominent of the two is 47 Tucanae (see p.294), which looks like a hazy 4th-magnitude star to the naked eye. It apparently covers the same area of sky as a full moon when viewed through binoculars or a small telescope. In the entire sky, only Omega (ω) Centauri is a more impressive globular cluster than 47 Tucanae. NGC 362, the other globular cluster in Tucana, is smaller and fainter and requires binoculars or a small telescope to be seen.

Beta (β) Tucanae is a naked-eye or binocular double with stars of 4th and 5th magnitudes. The brighter component can be further separated through a telescope. Kappa (κ) Tucanae, near NGC 362, is a double star of 5th and 7th magnitudes divisible through a small telescope.

**THE SMC** 👁 🔭 ✦
This neighboring mini-galaxy, the Small Magellanic Cloud, appears noticeably elongated. To its right in this image lies 47 Tucanae, or NGC 104, a globular cluster in our galaxy.

**47 TUCANAE** 👁 🔭 ✦
This bright globular cluster looks like a fuzzy star on wide-angle photographs like the one above right, but telescopes reveal it to be an immense swarm of stars.

**THE TOUCAN**

**GRUS**

**PHOENIX**

**INDUS**

**ERIDANUS**

*Achernar*

β

ζ η
ε

γ

α

ν

**TUCANA**

δ

**PAVO**

κ

NGC 362

**47**
NGC 104

**HYDRUS**

SMC

–60°

–70°

**OCTANS**

2h 1h 0h

**BIRD OF THE SOUTHERN SKIES** 👁
The Toucan's huge beak points downward as the constellation sets toward the western horizon. North is to the right in this picture.

## THE LITTLE WATER SNAKE

# Hydrus

| | |
|---|---|
| SIZE RANKING | 61 |
| BRIGHTEST STAR | Beta (β) 2.8 |
| GENITIVE | Hydri |
| ABBREVIATION | Hyi |
| HIGHEST IN SKY AT 10 P.M. | October–December |
| FULLY VISIBLE | 8°N–90°S |

Hydrus was introduced in the late 16th century by the Dutch navigator–astronomers Pieter Dirkszoon Keyser and Frederick de Houtman (see p.416). It is a constellation of the far-southern sky and is situated between the Large Magellanic Cloud (see p.310) and the Small Magellanic Cloud (see p.311).

This constellation represents a small water snake. It should not be confused with the larger constellation Hydra, also identified as a water snake, which has been recognized since the time of the ancient Greeks.

## SPECIFIC FEATURES

Pi (π) Hydri is a wide double of 6th-magnitude red giants, although they lie at different distances from us and hence are unrelated. It can be split readily through binoculars.

Pi-1 (π¹) is of magnitude 5.6 and is to be found just over 700 light-years away. Pi-2 (π²) lies much closer to us, being about 490 light-years away; it is of magnitude 5.7.

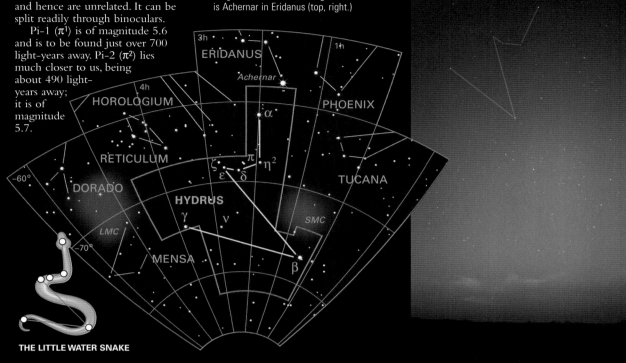

**HYDRUS AND ACHERNAR** 👁
The sinuous little water snake winds its way across southern skies between the two Magellanic Clouds. The brightest star near it is Achernar in Eridanus (top, right.)

**THE LITTLE WATER SNAKE**

---

## THE PENDULUM CLOCK

# Horologium

| | |
|---|---|
| SIZE RANKING | 58 |
| BRIGHTEST STAR | Alpha (α) 3.9 |
| GENITIVE | Horologii |
| ABBREVIATION | Hor |
| HIGHEST IN SKY AT 10 P.M. | November–December |
| FULLY VISIBLE | 23°N–90°S |

Horologium represents a pendulum clock, as used in observatories. Some depictions show its brightest star, Alpha (α) Horologii, marking the clock's pendulum (as in the illustration here), while others include it as one of the clock weights.

This faint and unremarkable constellation of the southern sky lies near the foot of Eridanus and was introduced by the French astronomer Nicolas Louis de Lacaille (see p.422).

## SPECIFIC FEATURES

R Horologii is a red-giant variable star of the same type as Mira (in Cetus). It ranges between 5th and 14th magnitudes every 13 months or so.

### NGC 1261 ✦ ⬭
The best deep-sky object in Horologium for amateur instruments is NGC 1261, a compact globular cluster of 8th magnitude more than 50,000 light-years from us.

NGC 1261 is a modest globular cluster dimly detectable through a small telescope.

Arp–Madore 1 (AM1) is another globular cluster of note within the constellation Horologium. It is the most distant known globular cluster from the Sun, being nearly 400,000 light-years away. Because it is of 16th magnitude, a large telescope is needed to detect it.

**THE PENDULUM CLOCK**

**STELLAR CLOCK** 👁
The shape of Horologium is reminiscent of a clock with a long pendulum—unlike many of the shapeless constellations invented by de Lacaille.

THE NIGHT SKY

# Reticulum

| | |
|---|---|
| **SIZE RANKING** | 82 |
| **BRIGHTEST STAR** | Alpha (α) 3.3 |
| **GENITIVE** | Reticuli |
| **ABBREVIATION** | Ret |
| **HIGHEST IN SKY AT 10 P.M.** | December |
| **FULLY VISIBLE** | 23°N–90°S |

Reticulum is a small constellation in the southern sky, near the Large Magellanic Cloud (see p.310). It was introduced by the French astronomer Nicolas Louis de Lacaille (see p.422) and represents the reticule, or grid, in his eyepiece, which he used for measuring star positions.

## SPECIFIC FEATURES
Zeta (ζ) Reticuli is a yellow double star. Its 5th-magnitude components can be split through binoculars.

THE NET

### CASTING THE NET 👁
This rhomboidal group of stars lies near the Large Magellanic Cloud, which is too faint to be seen here in the morning sky. The star at upper right is Achernar (in Eridanus).

# Pictor

| | |
|---|---|
| **SIZE RANKING** | 59 |
| **BRIGHTEST STAR** | Alpha (α) 3.2 |
| **GENITIVE** | Pictoris |
| **ABBREVIATION** | Pic |
| **HIGHEST IN SKY AT 10 P.M.** | December–February |
| **FULLY VISIBLE** | 26°N–90°S |

Pictor was invented by the French astronomer Nicolas Louis de Lacaille (see p.422), who imagined it as an artist's easel, complete with palette. He originally called it *Equuleus Pictoris*, although that name has since been shortened. It is a faint constellation of the southern sky, and it is situated beside the constellations Puppis and Columba.

## SPECIFIC FEATURES
Beta (β) Pictoris is 63 light-years away. It is of special interest because, in 1984, astronomers discovered a disk of dust and gas orbiting this blue-white star of magnitude 3.9. The circumstellar disk is thought to be a planetary system in the process of formation. The planets of our solar system are believed to have developed from a similar disk that existed around the Sun shortly after its formation.
    Iota (ι) Pictoris is a double star with components of 6th-magnitude. These are readily separated through a small telescope.

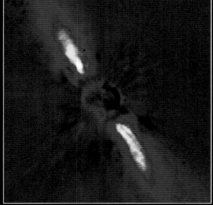

### BETA PICTORIS 🖥 ♿
The bright areas on this professional false-color image indicate the circumstellar disk. Distortions in the shape may be due to a planetary system forming around the star.

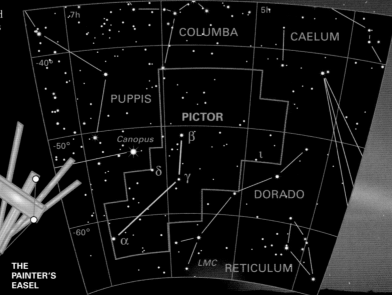

THE PAINTER'S EASEL

### DIVIDING LINE 👁
Pictor consists of little more than a crooked line of stars between brilliant Canopus (in Carina), seen here on the left, and the Large Magellanic Cloud.

# THE GOLDFISH
# Dorado

| | |
|---|---|
| **SIZE RANKING** | 72 |
| **BRIGHTEST STAR** | Alpha (α) 3.3 |
| **GENITIVE** | Doradus |
| **ABBREVIATION** | Dor |
| **HIGHEST IN SKY AT 10 P.M.** | December–January |
| **FULLY VISIBLE** | 20°N–90°S |

Dorado is one of the southern constellations introduced in the late 16th century by the Dutch navigator–astronomers Pieter Dirkszoon Keyser and Frederick de Houtman (p.416). Although known as the goldfish, Dorado in fact represents the dolphinfish found in tropical waters, not common aquarium and pond fish. The constellation has also been depicted as a swordfish.

Most of the Large Magellanic Cloud (see p.310) is contained within Dorado, although this mini-galaxy also extends into Mensa. The first recorded mention of the Large Magellanic Cloud is credited to al-Sufi (see panel, below).

## SPECIFIC FEATURES
The Large Magellanic Cloud is a satellite galaxy of the Milky Way. It is situated some 179,000 light-years away from the Earth and, at first sight, looks like a detached part of the Milky Way. Its numerous star clusters and

nebulous patches are brought into view through binoculars or a small telescope.

A remarkable object in the Large Magellanic Cloud is the Tarantula Nebula, or NGC 2070. It is bright enough to be visible with the naked eye and can be seen well through binoculars. A cluster of newborn stars at the heart of the Tarantula Nebula can be detected through binoculars or a small telescope, while photographs show its looping extremities, like a spider's legs, from which this large nebula of glowing gas gets its popular name.

In February 1987, a supernova flared up in the Large Magellanic Cloud. Supernova 1987A, as it was called, reached 3rd magnitude in May of that year, and this made it the brightest supernova visible from Earth since 1604. It remained visible to the naked eye for 10 months.

Beta (β) Doradus is one of the brightest Cepheid variables, ranging between magnitudes 3.5 and 4.1 every 9.8 days, while R Doradus is an erratic red giant that varies from 5th to 6th magnitude every 11 months or so.

## SUPERNOVA 1987A
This supernova has faded since its dramatic flare-up in 1987. To its upper left in this image is the spider-like Tarantula Nebula.

## THE LMC
The brighter of the two mini-galaxies that accompany our own, the Large Magellanic Cloud appears elongated in shape. It includes the Tarantula Nebula (here on its upper-left edge).

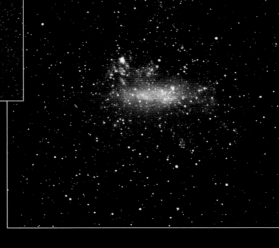

## HEADING SOUTH
Dorado, the Goldfish, swims through the southern skies, apparently on its way to the south celestial pole.

## AL-SUFI

Abd al-Rahman al-Sufi (903–986 CE), known also by his Latinized name, Azophi, was an Arabic astronomer. Around 964 CE, he produced the *Book of the Fixed Stars*—an updated version of Ptolemy's *Almagest*—which introduced many star names still in use today. Later editions of the book contained Arabic illustrations of the constellations (like the one below).

### CONSTELLATION PORTRAIT
A version of al-Sufi's *Book of the Fixed Stars* was produced in the 16th century by a Persian artist. It included this image of Boötes.

**THE GOLDFISH**

# Volans

| | |
|---|---|
| **SIZE RANKING** | 76 |
| **BRIGHTEST STARS** | Beta (β) 3.8, Gamma (γ) 3.8 |
| **GENITIVE** | Volantis |
| **ABBREVIATION** | Vol |
| **HIGHEST IN SKY AT 10 PM** | January–March |
| **FULLY VISIBLE** | 14°N–90°S |

This small and faint constellation of the southern sky between Carina and the Large Magellanic Cloud (see p.310) was introduced in the late 16th century by the Dutch navigator–astronomers Pieter Dirkszoon Keyser and Frederick de Houtman (see p.416). It represents the tropical fish that uses its outstretched fins as wings to glide through the air.

## SPECIFIC FEATURES
Although it lies on the edge of the Milky Way, Volans is surprisingly bereft of deep-sky objects. It does, however, contain two good double stars, one of them being

4th-magnitude Gamma (γ) Volantis, which is jointly the brightest star in the constellation. This orange star has a yellow companion of 6th magnitude. They form a beautiful double when viewed through a small telescope.

Epsilon (ε) Volantis is another interesting double, although it is not as colorful as Gamma. Its components, which are of 4th and 7th magnitudes, can be detected readily through a small telescope.

THE FLYING FISH

**FISH IN FLIGHT** 👁
The Flying Fish leaps into the evening sky above the eastern horizon. Beneath it here are the Milky Way and the stars of Carina and Vela, with the False Cross at left.

# Mensa

| | |
|---|---|
| **SIZE RANKING** | 75 |
| **BRIGHTEST STAR** | Alpha (α) 5.1 |
| **GENITIVE** | Mensae |
| **ABBREVIATION** | Men |
| **HIGHEST IN SKY AT 10 PM** | December–February |
| **FULLY VISIBLE** | 5°N–90°S |

The French astronomer Nicolas Louis de Lacaille (see panel, right) introduced this constellation. It commemorates Table Mountain near the modern Cape Town, South Africa, which is close to where he set up his observatory. When viewing the wispy appearance of the Large Magellanic Cloud (see p.310) in Mensa, de Lacaille may have recalled the clouds sometimes seen over the real Table Mountain. It is the only constellation that de Lacaille did not name after a scientific or artistic tool.

Mensa is the faintest of all 88 constellations, and its brightest star, Alpha (α) Mensae, is of only 5th magnitude. Its main point of interest is that part of the Large Magellanic Cloud overlaps into it from neighboring Dorado. Other than this cloud, there is nothing to attract the casual observer to this small constellation of the south polar region of the sky.

**TABLE TOP** 👁
The far-southern constellation Mensa appears in this photograph above pink-tinged clouds in the dawn sky.

THE TABLE MOUNTAIN

## NICOLAS LOUIS DE LACAILLE

This French astronomer charted the southern skies in 1751–1752 from Cape Town, South Africa. Nicolas Louis de Lacaille (1713–1762) observed the positions of nearly 10,000 stars, producing a catalog and star chart on which he introduced 14 new constellations. Most of these represented instruments of the arts and sciences.

**SOUTHERN VIEWPOINT**
Lacaille observed the stars from near Table Mountain, which is covered by an attractive "tablecloth" of clouds in this photograph.

THE CHAMELEON

# Chamaeleon

| | |
|---|---|
| **SIZE RANKING** | 79 |
| **BRIGHTEST STARS** | Alpha (α) 4.1, Gamma (γ) 4.1 |
| **GENITIVE** | Chamaeleontis |
| **ABBREVIATION** | Cha |
| **HIGHEST IN SKY AT 10 PM** | February–May |
| **FULLY VISIBLE** | 7°N–90°S |

Chamaeleon was named after the lizard that can change its skin color to match its surroundings. It is a small, faint constellation of the south polar region of the sky.

The constellation was introduced at the end of the 16th century by the Dutch navigator–astronomers Pieter Dirkszoon Keyser and Frederick de Houtman (see p.416).

## SPECIFIC FEATURES

Delta (δ) Chamaeleontis is a wide pair of unrelated stars of 4th and 5th magnitudes. They are easily seen through binoculars.

NGC 3195 is a planetary nebula of similar apparent size to Jupiter, but it is relatively faint, so it requires a moderate-sized telescope to be seen.

**THE CHAMELEON**

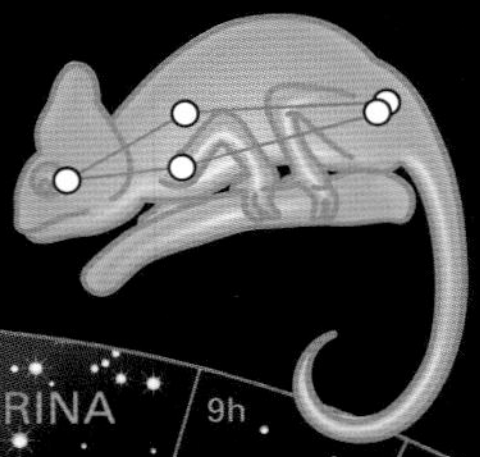

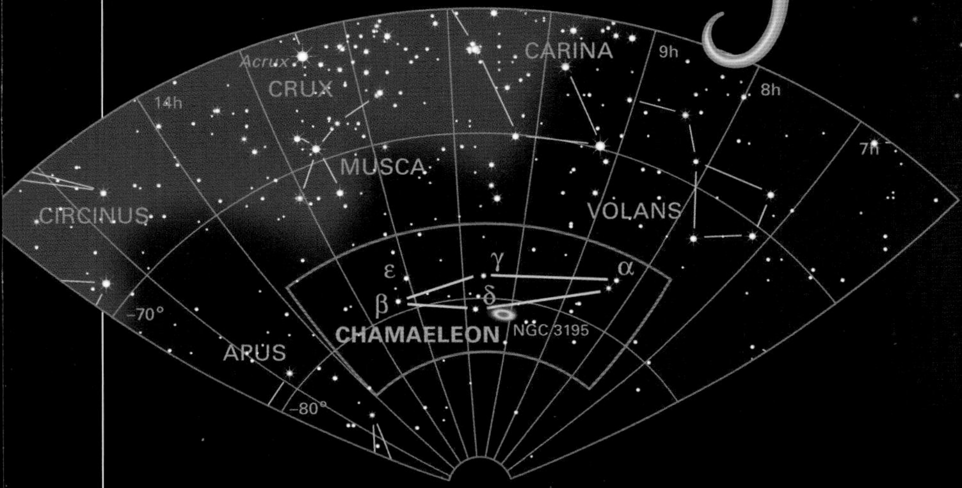

**CAMOUFLAGE ARTIST** 👁
Chamaeleon lies close to the south celestial pole, which is to the left of it in this picture. To the north of this constellation are found the rich Milky Way star fields of Carina.

---

THE BIRD OF PARADISE

# Apus

| | |
|---|---|
| **SIZE RANKING** | 67 |
| **BRIGHTEST STAR** | Alpha (α) 3.8 |
| **GENITIVE** | Apodis |
| **ABBREVIATION** | Aps |
| **HIGHEST IN SKY AT 10 PM** | May–July |
| **FULLY VISIBLE** | 7°N–90°S |

The constellation Apus is situated in the almost featureless area around the south celestial pole. It was invented in the late 16th century by the Dutch navigator–astronomers Pieter Dirkszoon Keyser and Frederick de Houtman (see p.416).

## SPECIFIC FEATURES

Delta (δ) Apodis is a wide pair of unrelated 5th-magnitude red giants, while Theta (θ) Apodis is a red giant that varies somewhat erratically between 5th and 7th magnitudes every 4 months or so.

**THE BIRD OF PARADISE**

**EXOTIC BIRD** 👁
Apus, which is south of the distinctive Triangulum Australe, represents a bird of paradise but is a disappointing tribute to such an exotic bird.

THE NIGHT SKY

## THE PEACOCK

# Pavo

| | |
|---|---|
| **SIZE RANKING** | 44 |
| **BRIGHTEST STAR** | Peacock (α) 1.9 |
| **GENITIVE** | Pavonis |
| **ABBREVIATION** | Pav |
| **HIGHEST IN SKY AT 10 PM** | July–September |
| **FULLY VISIBLE** | 15°–90°S |

Pavo is one of the far-southern constellations that were introduced at the end of the 16th century by the Dutch navigator–astronomers Pieter Dirkszoon Keyser and Frederick de Houtman (see p.416). It represents the peacock of southeast Asia, which the Dutch explorers encountered on their travels. In more recent times, its brightest star, 2nd-magnitude Alpha (α) Pavonis, was given the name Peacock.

In Greek mythology, the peacock was the sacred bird of Hera, wife of Zeus, who traveled through the air in a chariot drawn by these birds. It was Hera who placed the markings on the tail of the peacock after an episode involving Zeus and one of his illicit loves, Io. Although Zeus had disguised Io as a white cow, Hera suspected something was amiss and set the 100-eyed Argus to keep watch on the heifer. Her husband retaliated by sending his son Hermes to release Io. In order to overcome Argus, Hermes told him tales and played music on his reed pipe until the watchman's eyes closed one by one. When Argus was finally asleep, Hermes chopped off his head and set Io free. In his memory, Hera then placed the eyes of Argus on the peacock's tail.

The constellation Pavo is to be found on the edge of the Milky Way south of Sagittarius and next to another exotic bird, the toucan (the constellation Tucana).

## SPECIFIC FEATURES

Kappa (κ) Pavonis is one of the brighter Cepheid variables. Its fluctuations, between magnitudes 3.9 and 4.8 every 9.1 days, can be followed with the naked eye.

Xi (ξ) Pavonis is a double star with components of unequal brightness—4th and 8th magnitudes. The fainter star is difficult to identify with the smallest-aperture telescopes because its brighter neighbor overwhelms it.

NGC 6752 is one of the largest and brightest globular clusters in the sky. It is just at the limit of naked-eye visibility but readily located through binoculars. It covers half the apparent width of a full moon. A telescope with an aperture of 3 in (75 mm) or more will resolve its brightest individual stars.

The large spiral galaxy NGC 6744 is presented virtually face-on to the Earth. It is visible as an elliptical haze in a telescope of small to moderate aperture. NGC 6744 lies about 30 million light-years away.

**NGC 6744** 📷 💻
This beautiful barred spiral galaxy in Pavo is detectable through a small telescope. The Milky Way might appear like this when viewed from the outside.

**NGC 6752** 📷 💻
The fine globular cluster NGC 6752 remains little known because of its far-southern declination. The bright star seen above right of it in this image is a foreground object in our galaxy.

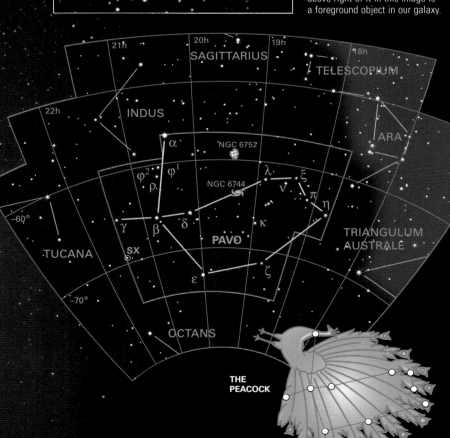

**CELESTIAL DISPLAY** 👁
The constellation Pavo, the Peacock, is depicted fanning its tail across the southern skies, in imitation of a real-life peacock when attracting a mate.

THE PEACOCK

## THE OCTANT

# Octans

| | |
|---|---|
| **SIZE RANKING** | 50 |
| **BRIGHTEST STAR** | Nu (ν) 3.8 |
| **GENITIVE** | Octantis |
| **ABBREVIATION** | Oct |
| **HIGHEST IN SKY AT 10 PM** | October |
| **FULLY VISIBLE** | 0°–90°S |

This constellation, which originally was also known as *Octans Nautica* or *Octans Hadleianus*, contains the south celestial pole. It was introduced in the 18th century by the French astronomer Nicolas Louis de Lacaille (see p.422).

The area of sky in which Octans lies is quite barren. Within naked-eye range, the nearest star to the south celestial pole is Sigma (σ) Octantis. It is of only magnitude 5.4 and thus far from prominent.

Because of the effect of precession (see p.64), the positions of the celestial poles are constantly changing. As a result, the south celestial pole is moving farther away from Sigma and toward the constellation of Chamaeleon. There are no bright stars in this area either, so the region of the south celestial pole will remain blank for another 1,500 years, when the pole will pass just over a degree away from 4th-magnitude Delta (δ) Chamaeleontis.

Octans represents an instrument known as an octant, which was used by navigators to help them find their position (see panel, below). It was invented by the English instrument maker John Hadley (1682–1744).

## SPECIFIC FEATURES

Lambda (λ) Octantis is a double star that is divisible with a small telescope. The components are of 5th and 7th magnitudes.

**SOUTHERN STAR TRAILS** 👁
Curving star trails, drawn out by the Earth's rotation on this long-exposure photograph, emphasize the barren nature of the area around the south celestial pole.

**AT THE POLE** 👁
Octans comprises only a scattering of faint stars. There is no bright star to mark the southern pole, which lies near the center of this image.

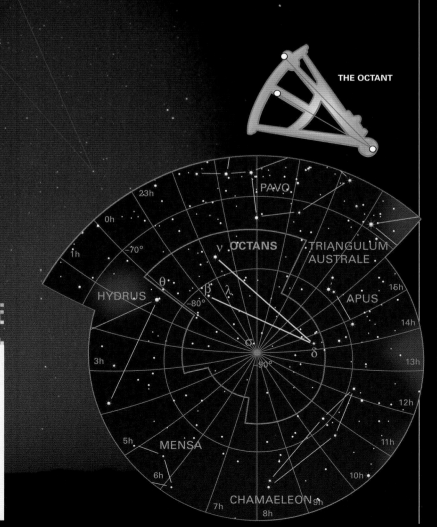

**THE OCTANT**

EXPLORING SPACE

## NAVIGATION

In 1731, the British mathematician John Hadley built a device called a doubly reflecting octant. The navigator sighted the horizon through a telescope and adjusted a movable arm until the reflected image of the Sun or a star overlay the direct view of the horizon. The altitude of the Sun or star could be read off a scale, from which the navigator could deduce his latitude.

**OCTANT**
This wood and brass octant is by Browning of Boston. In later designs, the arc was extended from one-eighth of a circle to one-sixth, and the octant became the modern sextant.

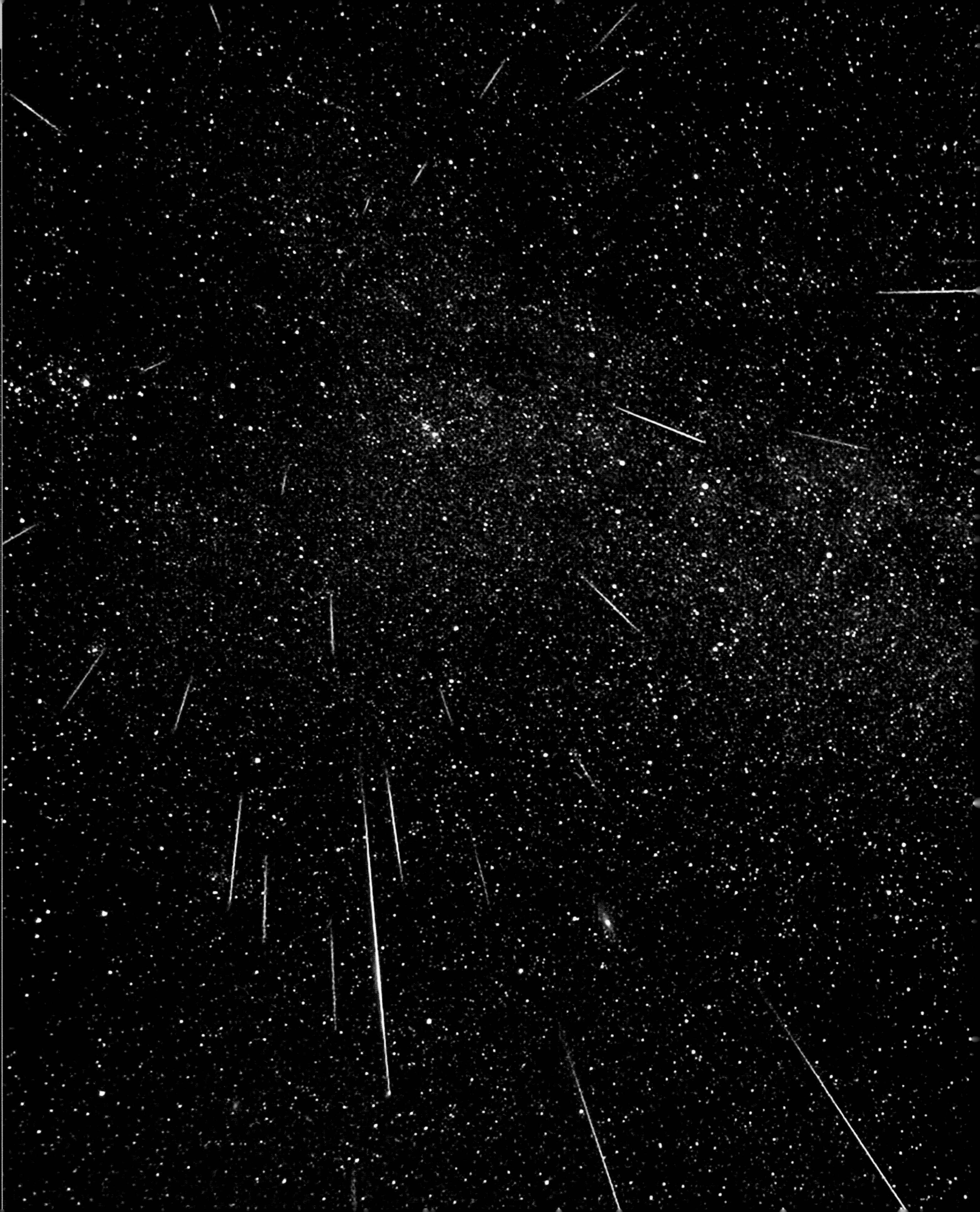

*"If the stars should appear one night in a thousand years, how would men believe and adore; and preserve for many generations the remembrance of the city of God which had been shown! But every night come out these envoys of beauty, and light the universe with their admonishing smile."*

Ralph Waldo Emerson

AS THE EARTH MAKES its year-long journey around the Sun, the night sky changes its appearance and the stars seem to move from east to west. Depending on the observer's location, some stars are circumpolar and always visible, but others are seen only at certain times of the year. For example, some stars are seen well in the evening sky in January, but are invisible 6 months later, when the Earth has moved around its orbit to the opposite side of the Sun. The following section tracks seasonal changes in the night sky for observers in both the Northern and Southern Hemispheres. As well as covering the regular annual cycles of the stars and constellations, it charts the positions of the planets and provides an observer's guide to celestial events, such as meteor showers and eclipses of the Sun and Moon.

**THE PERSEID METEORS**
This composite image shows the Perseid meteor shower that occurs in August every year. Also visible is the star-studded band of the Milky Way.

# MONTHLY SKY GUIDE

# USING THE SKY GUIDES

THIS MONTH-BY-MONTH GUIDE to the night sky features charts that show the whole sky as it appears from most places on Earth's surface. It complements the CONSTELLATIONS section, in which detailed maps show smaller areas of sky. For each month, text, tables, and supporting charts identify good objects for observation and show the positions of the planets.

## SPECIAL EVENTS

### PHASES OF THE MOON

| | FULL MOON | NEW MOON |
|------|-----------|----------|
| 2020 | January 10 | January 24 |
| 2021 | January 28 | January 13 |
| 2022 | January 17 | January 2 |
| 2023 | January 6 | January 21 |
| 2024 | January 25 | January 11 |
| 2025 | January 13 | January 29 |
| 2026 | January 3 | January 18 |

## MONTHLY HIGHLIGHTS AND PLANET LOCATORS

For each month of the year, a double-page introduction highlights different phenomena in the sky. Dates of special events, such as phases of the Moon and eclipses, are listed year-by-year in a table. The main text describes those stars, deep-sky objects, and meteor showers that feature prominently in that particular month—this text complements the whole-sky charts that follow. The introductory pages also feature a planet locator chart. This map shows the band of sky that lies on either side of the ecliptic, the plane close to which the planets always appear. These charts should be used in conjunction with the extra information supplied in the Special Events table, as well as the whole-sky charts and the individual constellation entries (see pp.354–425).

**SPECIAL EVENTS CALENDAR** △
The introduction to each month contains a Special Events table, which lists the dates of full and new moons and events such as lunar and solar eclipses and planetary conjunctions and transits (see p.69). This table also lists the dates when Mercury is at greatest elongation.

*each month of the year has its own introductory pages*

*the text highlights the most prominent stars, deep-sky objects, and meteor showers*

*observation from northern and southern latitudes is covered separately in the text*

*photographs illustrate some of the most interesting features to be observed*

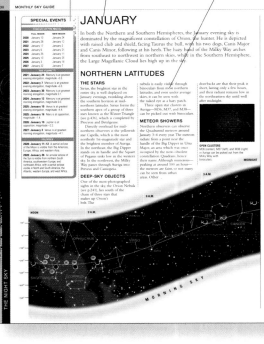

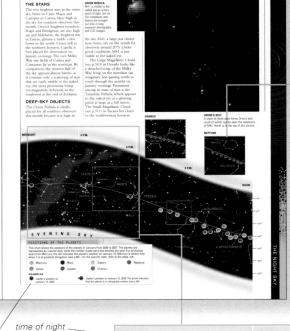

△ **THE OUTER PLANETS**
The two outermost planets, Uranus and Neptune, are shown on magnified insets of the main chart because they move relatively slowly through our sky.

*ecliptic*

*Earth's axis of rotation*

*celestial sphere*

*celestial equator*

*position of planet shown by colored dot*

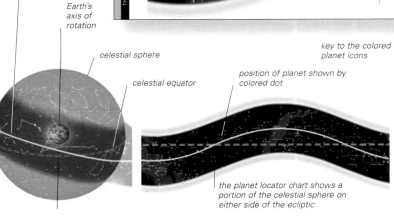

*key to the colored planet icons*

*the planet locator chart shows a portion of the celestial sphere on either side of the ecliptic*

*time of night (in local time) when this area of sky lies on the meridian (an imaginary line running north–south)*

*arrow indicates that the planet is in retrograde motion (see p.68)*

*declination coordinates*

*celestial equator*

*time when an area of sky is visible: evening sky (from sunset to midnight) or morning sky (from midnight to sunrise)*

*ecliptic*

**PLANET LOCATOR CHARTS** △
These charts show the positions of the planets at 10:00 p.m. local standard time on the 15th day of the month. Each planet is represented by a different-colored dot, and the number inside the dot refers to a particular year. Each chart shows the planets' positions in relation to the 13 constellations along the ecliptic (see p.65), the area in which the planets are always found.

**THE INNER PLANETS** ▷
The six planets closest to the Sun are represented on the main body of the chart. Bands along the top and bottom indicate in local time when that area of sky is highest in the sky. However, local sunset and sunrise times will affect the darkness of the sky and thus the visibility of the planets.

# THE WHOLE-SKY CHARTS

The introduction to each month is followed by two whole-sky charts. These show the position of the stars at 10:00 p.m. local time on the 15th day of the month for both the Northern and Southern Hemispheres. They project the half of the celestial sphere (see pp. 62–63) that would be visible to a viewer under perfect conditions—that is, without any obstruction to the horizon. Any given star rises 4 minutes earlier each night compared to the previous night. Thus, the night sky changes subtly from one night to the next and even more dramatically from one month to another. To use the whole-sky charts, determine the color-coded horizon and zenith for your location (below), turn to the appropriate month, and position yourself and the whole-sky chart (right).

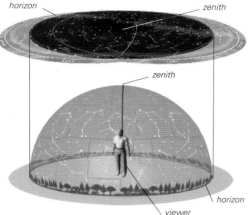

horizon        zenith

zenith

horizon

viewer

### △ CELESTIAL SPHERE
Each whole-sky chart shows an area that equals more than half a celestial sphere because it combines three different projections of the night sky, as seen from three different latitudes on Earth. Each month, the sky charts show the night sky as it appears from 60°–20°N on the Northern Hemisphere chart and from 0°–40°S on the Southern Hemisphere chart.

### HORIZONS AND ZENITHS ▷
The stars located near the center of each chart can be seen on the zenith (the point directly overhead), while the stars near the chart's edge appear close to the horizon. Color-coded lines and crosses are used to identify the horizon and zenith on each of the three latitude projections on each monthly chart.

M52

CEPHEUS

60°N
40°N
20°N
0°
20°S
40°S

### △ LINES OF LATITUDES
Determine the latitude line that is closest to your geographical location and use the color-coding on the sky charts to find the view from your location. Note that a 10° difference in latitude has little effect on the stars that can be seen.

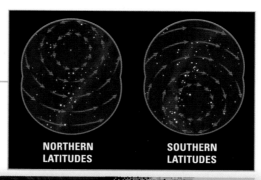

**NORTHERN LATITUDES**    **SOUTHERN LATITUDES**

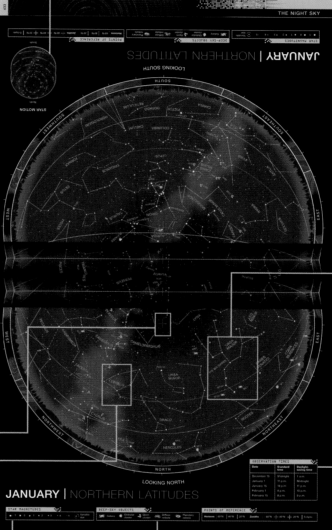

JANUARY | NORTHERN LATITUDES

### ▽ STAR MAGNITUDE
Stars that appear brighter than magnitude 6 are illustrated on the whole-sky charts. This key can be used to gauge their magnitude. About 25 prominent stars are also labeled with their popular names.

### ▽ DEEP-SKY OBJECTS
Icons are used to represent a selection of deep-sky objects of interest to the amateur astronomer.

### ◁ STAR-MOTION DIAGRAMS
These diagrams show the direction in which the stars appear to move as the night progresses. Stars near the equator appear to move from east to west, while circumpolar stars circle around the celestial poles without setting.

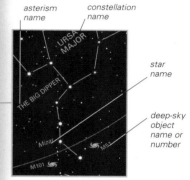

### △ ORIENTATION
To view the sky to the north, face north and hold the map flat, with the NORTH label closest to your body. One of the color-coded lines around the near edge of the map will relate to the horizon in front of you. To view the south, turn around and reposition the map.

asterism name        constellation name

URSA MAJOR

THE BIG DIPPER

star name

Mizar

deep-sky object name or number

M101

### △ MAIN FEATURES
All 88 constellations are featured on the whole-sky charts, as well as any notable deep-sky objects within their boundaries. Well-known and easily recognizable stars, star clusters, and asterism patterns (see p.72) are also labeled.

### OBSERVATION TIMES

| Date | Standard time | Daylight saving time |
|---|---|---|
| December 15 | Midnight | 1 a.m. |
| January 1 | 11 p.m. | Midnight |
| January 15 | 10 p.m. | 11 p.m. |
| February 1 | 9 p.m. | 10 p.m. |
| February 15 | 8 p.m. | 9 p.m. |

### △ OBSERVING TIMES
Each chart shows the sky as it appears at 10:00 p.m. local standard time, mid-month. However, this view can also be seen at other times of the month, as well as 1 hour later, when local daylight saving time is in use. To view the sky at a time before or after 10:00 p.m., you may need to consult a different monthly chart.

## DEEP-SKY OBJECTS

Galaxy | Globular cluster | Open cluster | Diffuse nebula | Planetary nebula

## STAR MAGNITUDES

-1   0   1   2   3   4   5   Variable star

THE NIGHT SKY

# JANUARY

In both the Northern and Southern Hemispheres, the January evening sky is dominated by the magnificent constellation of Orion, the hunter. He is depicted with raised club and shield, facing Taurus the bull, with his two dogs, Canis Major and Canis Minor, following at his heels. The hazy band of the Milky Way arches from southeast to northwest in northern skies, while in the Southern Hemisphere, the Large Magellanic Cloud lies high up in the sky.

## NORTHERN LATITUDES

### THE STARS

Sirius, the brightest star in the entire sky, is well displayed on January evenings, twinkling above the southern horizon at mid-northern latitudes. Sirius forms the southern apex of a group of three stars known as the Winter Triangle (see p.436), which is completed by Procyon and Betelgeuse.

Directly overhead for mid-northern observers is the yellowish star Capella, which is the most northerly 1st-magnitude star and the brightest member of Auriga. In the northeast, the Big Dipper stands on its handle and the Square of Pegasus sinks low in the western sky. In the northwest, the Milky Way passes through Auriga into Perseus and Cassiopeia.

### DEEP-SKY OBJECTS

One of the most-photographed sights in the sky, the Orion Nebula (see p.241), lies south of the chain of three stars that makes up Orion's belt. The nebula is easily visible through binoculars from most northern latitudes, and even under average skies, it can be seen with the naked eye as a hazy patch.

Three open star clusters in Auriga—M36, M37, and M38—can be picked out with binoculars.

### METEOR SHOWERS

Northern observers can observe the Quadrantid meteors around January 3–4 every year. The meteors radiate from a point near the handle of the Big Dipper in Ursa Major, an area which was once occupied by the now-obsolete constellation Quadrans, hence their name. Although numerous—peaking at around 100 an hour—the meteors are faint, so not many can be seen from urban areas. Other drawbacks are that their peak is short, lasting only a few hours, and their radiant remains low in the northeastern sky until well after midnight.

**OPEN CLUSTERS**
M36 (center), M37 (left), and M38 (right) in Auriga can be picked out from the Milky Way with binoculars.

## SPECIAL EVENTS

### PHASES OF THE MOON

| | FULL MOON | NEW MOON |
|---|---|---|
| 2020 | January 10 | January 24 |
| 2021 | January 28 | January 13 |
| 2022 | January 17 | January 2 |
| 2023 | January 6 | January 21 |
| 2024 | January 25 | January 11 |
| 2025 | January 13 | January 29 |
| 2026 | January 3 | January 18 |
| 2027 | January 22 | January 7 |

### THE PLANETS

**2021: January 24** Mercury is at greatest evening elongation, magnitude –0.3.

**2022: January 7** Mercury is at greatest evening elongation, magnitude –0.3.

**2023: January 30** Mercury is at greatest morning elongation, magnitude 0.1.

**2024: January 12** Mercury is at greatest morning elongation, magnitude 0.0.

**2025: January 10** Venus is at greatest evening elongation, magnitude –4.4.

**2025: January 16** Mars is at opposition, magnitude –1.4.

**2026: January 10** Jupiter is at opposition, magnitude –2.2.

**2027: January 3** Venus is at greatest morning elongation, magnitude –4.1.

### ECLIPSES

**2028: January 11–12** A partial eclipse of the Moon is visible from the Americas, Europe, Africa, and western Asia.

**2028: January 26** An annular eclipse of the Sun is visible from northern South America, southwestern Europe, and northwest Africa, with a partial eclipse visible in North and South America, the Atlantic, western Europe, and west Africa.

MIDNIGHT

3 A.M.

6 A.M.

NOON

9 A.M.

Arcturus

LEO

27

Regulus

CANCER

10°

0°

OPHIUCHUS

VIRGO

–10°

Spica

–20°

22

20

23

24

24

27

Antares

LIBRA

–30°

21

SAGITTARIUS

SCORPIUS

Shaula

–40°

MORNING SKY

–50°

# SOUTHERN LATITUDES

## THE STARS

The two brightest stars in the entire sky, Sirius in Canis Major and Canopus in Carina, blaze high in the sky for southern observers this month. Orion's brightest members, Rigel and Betelgeuse, are also high up, and Aldebaran, the brightest star in Taurus, glistens a ruddy color lower in the north. Closer still to the northern horizon, Capella is best placed for observation on January evenings. The rich Milky Way star fields of Carina and Centaurus lie in the southeast. By comparison, the western half of the sky appears almost barren, as it contains only a scattering of stars that are easily visible to the naked eye, the most prominent being 1st-magnitude Achernar, in the southwest at the end of Eridanus.

## DEEP-SKY OBJECTS

The Orion Nebula is ideally placed for all southern observers this month because it is high in

**ORION NEBULA**
M41 is visible to the naked eye as a hazy patch of light, but its full complexity and beauty are brought out only in long-exposure photographs and CCD images.

the sky. M41, a large star cluster near Sirius, sits on the zenith for observers around 20°S. Under good conditions, M41 is just visible to the naked eye.

The Large Magellanic Cloud (see p.310) in Dorado looks like a detached scrap of the Milky Way lying on the meridian (an imaginary line passing north to south through the zenith) on January evenings. Prominent among its mass of stars is the Tarantula Nebula, which appears to the naked eye as a glowing patch as large as a full moon. The Small Magellanic Cloud (see p.311) in Tucana lies closer to the southwestern horizon.

**ORION'S BELT**
A chain of three stars forms Orion's belt, south of which can be seen the nebulosity of M42. North is to the top of this picture.

**URANUS**

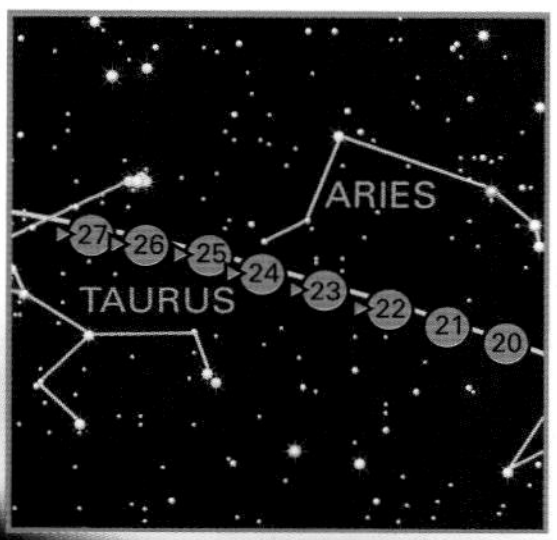

**NEPTUNE**

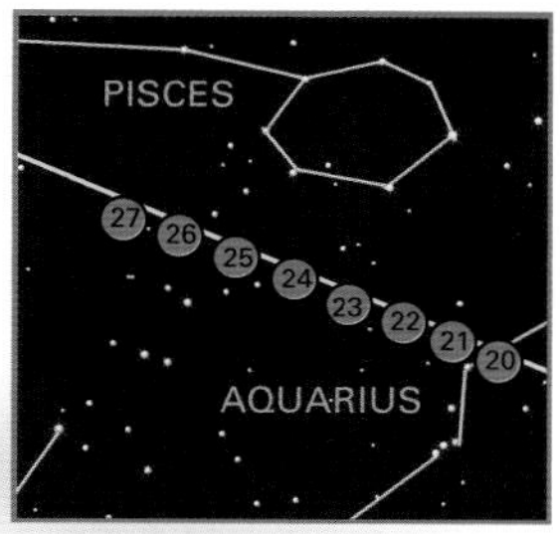

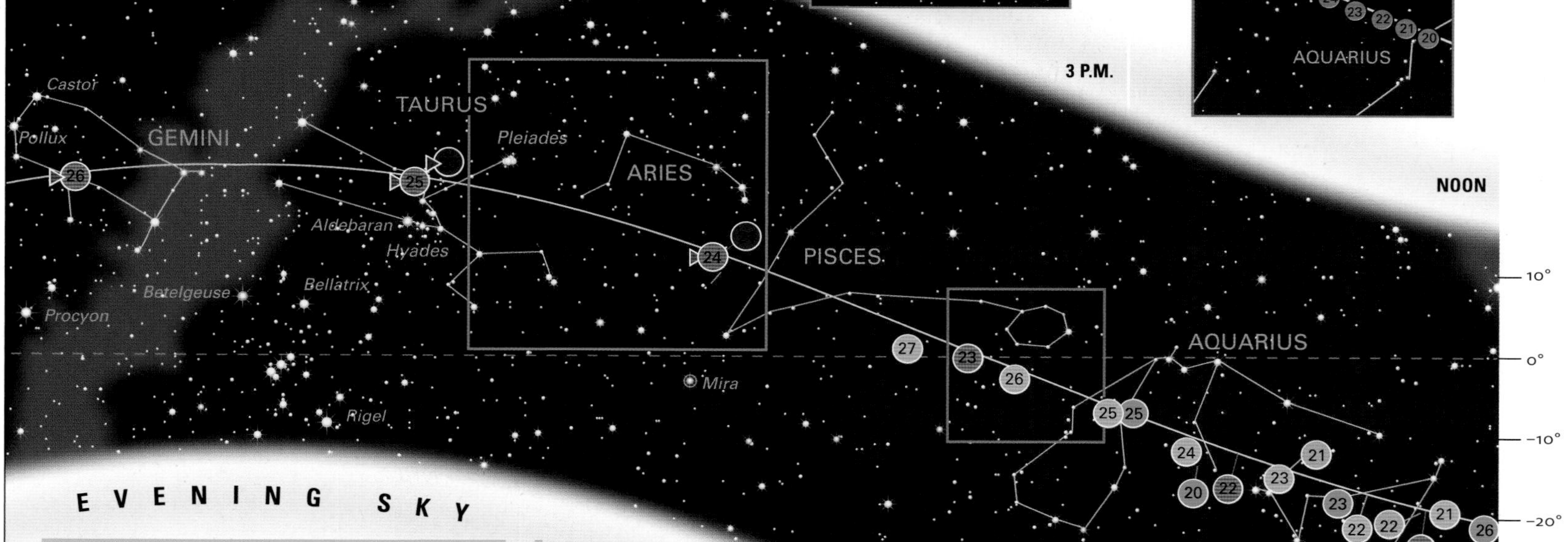

MIDNIGHT  9 P.M.  6 P.M.  3 P.M.  NOON

**EVENING SKY**

## POSITIONS OF THE PLANETS

This chart shows the positions of the planets in January from 2020 to 2027. The planets are represented by colored dots, while the number inside each dot denotes the year. For all planets apart from Mercury, the dot indicates the planet's position on January 15. Mercury is shown only when it is at greatest elongation (see p.68)—for the specific date, refer to the table, left.

- ⬤ Mercury
- ⬤ Mars
- ⬤ Saturn
- ⬤ Neptune
- ⬤ Venus
- ⬤ Jupiter
- ⬤ Uranus

**EXAMPLES**

(20) Jupiter's position on January 15, 2020

▶(24) Jupiter's position on January 15, 2024. The arrow indicates that the planet is in retrograde motion (see p.68).

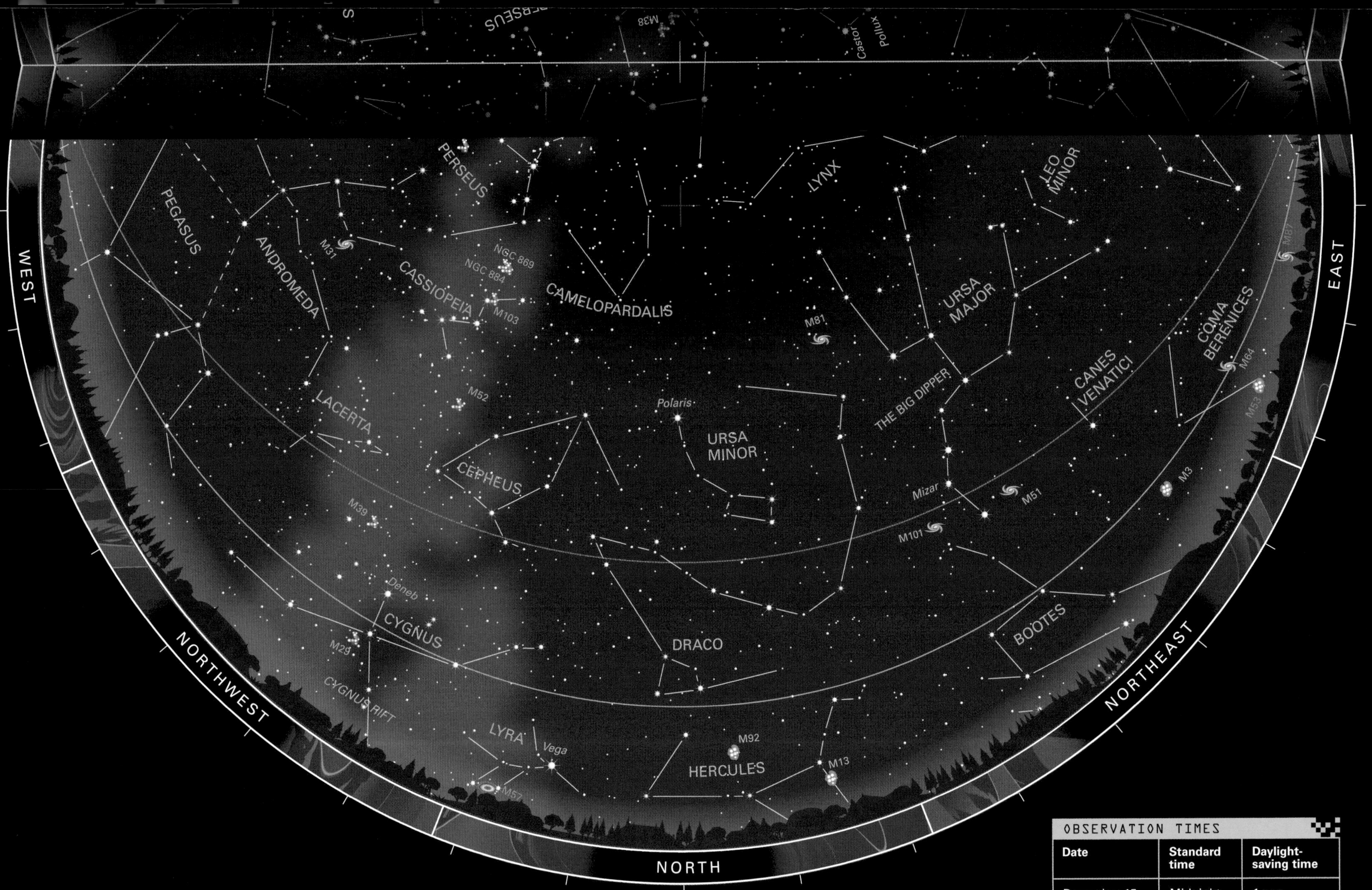

# JANUARY | NORTHERN LATITUDES

## LOOKING NORTH

### STAR MAGNITUDES

- -1
- 0
- 1
- 2
- 3
- 4
- 5
- Variable star

### DEEP-SKY OBJECTS

- Galaxy
- Globular cluster
- Open cluster
- Diffuse nebula
- Planetary nebula

### POINTS OF REFERENCE

| Horizons | | | Zeniths | | |
|---|---|---|---|---|---|
| 60°N | 40°N | 20°N | 60°N | 40°N | 20°N |
| | | | | | Ecliptic |

### OBSERVATION TIMES

| Date | Standard time | Daylight-saving time |
|---|---|---|
| December 15 | Midnight | 1 a.m. |
| January 1 | 11 p.m. | Midnight |
| January 15 | 10 p.m. | 11 p.m. |
| February 1 | 9 p.m. | 10 p.m. |
| February 15 | 8 p.m. | 9 p.m. |

WEST

NORTHWEST

NORTH

NORTHEAST

EAST

WEST

PEGASUS

ANDROMEDA

LACERTA

CYGNUS

Cygnus Rift

Deneb

LYRA

Vega

M57

HERCULES

M92

M13

DRACO

CEPHEUS

M39

M52

CASSIOPEIA

M103

NGC 884

NGC 869

PERSEUS

CAMELOPARDALIS

Polaris

URSA MINOR

LYNX

Castor

Pollux

M38

PERSEUS

M81

THE BIG DIPPER

Mizar

M101

URSA MAJOR

LEO MINOR

BOOTES

CANES VENATICI

M51

M3

COMA BERENICES

M64

M53

M87

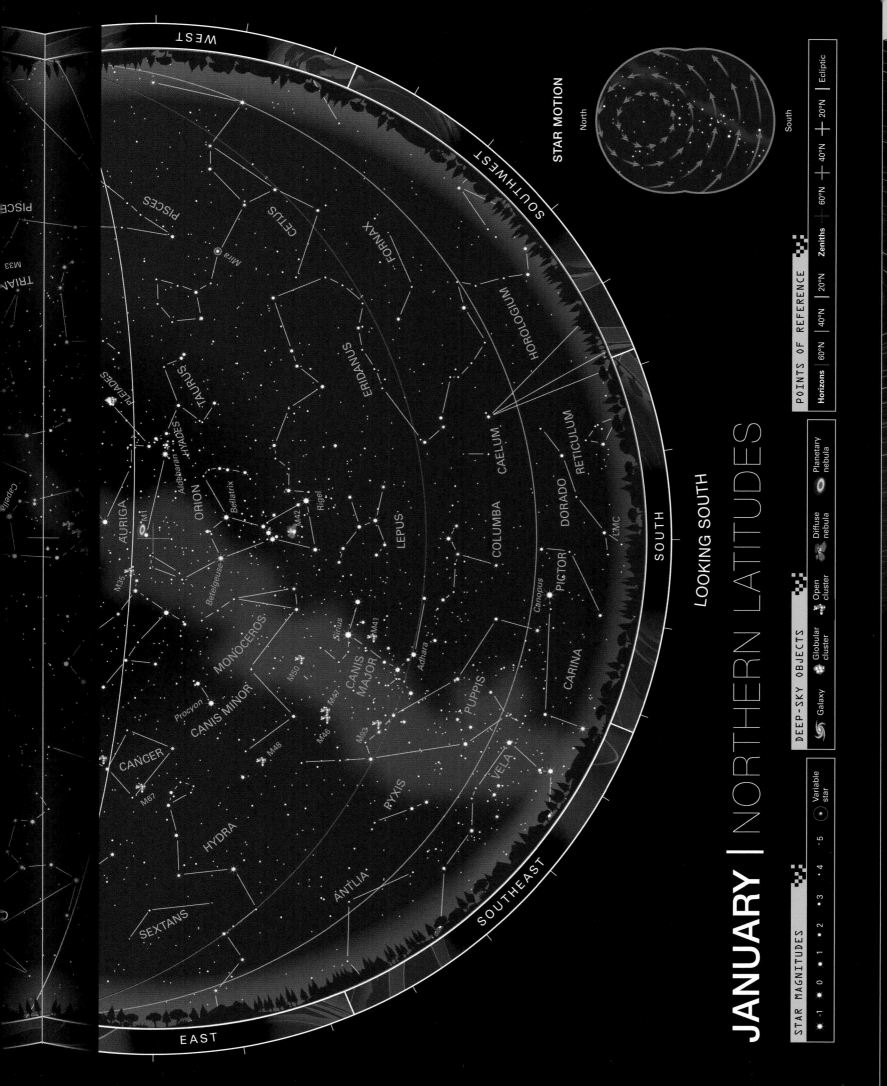

# JANUARY | NORTHERN LATITUDES

LOOKING SOUTH

THE NIGHT SKY

**STAR MOTION**

North

South

**POINTS OF REFERENCE**

| Horizons | 60°N | 40°N | 20°N | Zeniths | 60°N | 40°N | 20°N | Ecliptic |

**DEEP-SKY OBJECTS**

Galaxy | Globular cluster | Open cluster | Diffuse nebula | Planetary nebula

**STAR MAGNITUDES**

-1 | 0 | 1 | 2 | 3 | 4 | 5 | Variable star

WEST

SOUTHWEST

SOUTH

SOUTHEAST

EAST

WEST

PISCES

PISCES

M33

TRIAN

Capella

PLEIADES

HYADES

TAURUS

Aldebaran

AURIGA

M1

M35

ORION

Bellatrix

Betelgeuse

M42

Rigel

CETUS

Mira

ERIDANUS

FORNAX

HOROLOGIUM

CAELUM

COLUMBA

LEPUS

RETICULUM

DORADO

PICTOR

LMC

Canopus

CARINA

VELA

PYXIS

PUPPIS

M93

M46

M47

CANIS MAJOR

Sirius

Adhara

M41

M50

MONOCEROS

Procyon

CANIS MINOR

CANCER

M67

M48

HYDRA

ANTLIA

SEXTANS

# JANUARY | SOUTHERN LATITUDES

## LOOKING NORTH

### STAR MAGNITUDES

- ·-1
- ·0
- ·1
- ·2
- ·3
- ·4
- ·5
- ⊙ Variable star

### DEEP-SKY OBJECTS

- 🌀 Galaxy
- ✦ Globular cluster
- ✦ Open cluster
- ✧ Diffuse nebula
- ⊙ Planetary nebula

### POINTS OF REFERENCE

| Horizons | | | | Zeniths | | | |
|---|---|---|---|---|---|---|---|
| 0° | 20°S | 40°S | | 0° | 20°S | 40°S | Ecliptic |

### OBSERVATION TIMES

| Date | Standard time | Daylight-saving time |
|---|---|---|
| December 15 | Midnight | 1 a.m. |
| January 1 | 11 p.m. | Midnight |
| January 15 | 10 p.m. | 11 p.m. |
| February 1 | 9 p.m. | 10 p.m. |
| February 15 | 8 p.m. | 9 p.m. |

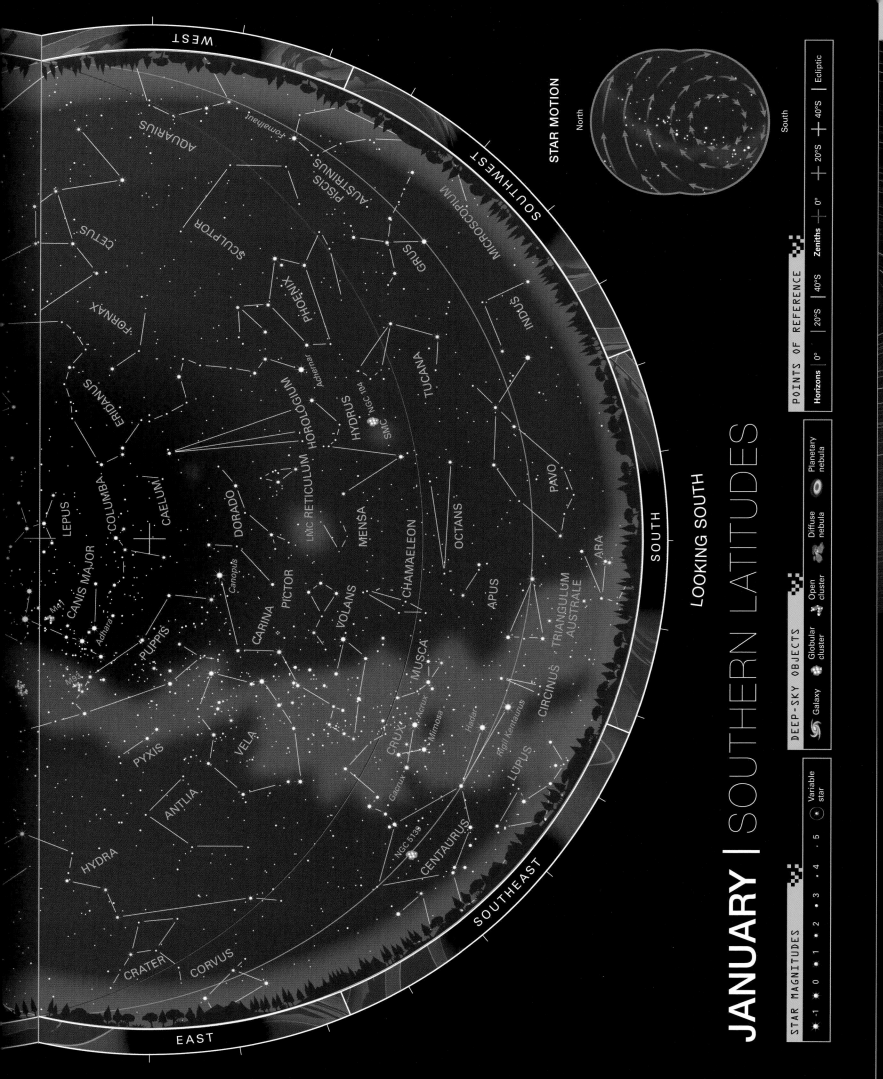

# JANUARY | SOUTHERN LATITUDES

LOOKING SOUTH

WEST

SOUTHWEST

SOUTH

SOUTHEAST

EAST

**STAR MOTION**

North

South

**POINTS OF REFERENCE**

| Horizons | 0° | 20°S | 40°S | Zeniths | 0° | 20°S | 40°S | Ecliptic |

**DEEP-SKY OBJECTS**

Galaxy | Globular cluster | Open cluster | Diffuse nebula | Planetary nebula

**STAR MAGNITUDES**

-1 | 0 | 1 | 2 | 3 | 4 | 5 | Variable star

AQUARIUS

CETUS

FORNAX

ERIDANUS

LEPUS

COLUMBA

CANIS MAJOR

CAELUM

PICTOR

DORADO

CARINA

PUPPIS

PYXIS

VELA

ANTLIA

HYDRA

CRATER

CORVUS

CENTAURUS

LUPUS

CIRCINUS

TRIANGULUM AUSTRALE

CRUX

MUSCA

CHAMAELEON

APUS

ARA

PAVO

INDUS

OCTANS

MENSA

VOLANS

RETICULUM

HOROLOGIUM

HYDRUS

TUCANA

LMC

SMC

PHOENIX

SCULPTOR

PISCIS AUSTRINUS

GRUS

MICROSCOPIUM

Fomalhaut

Achernar

NGC 104

Canopus

Adhara

M41

M93

Gacrux

Acrux

Mimosa

Hadar

Rigil Kentaurus

NGC 5139

THE NIGHT SKY

# FEBRUARY

Castor and Pollux, the brightest stars in the northern zodiacal constellation of Gemini, lie close to the celestial meridian (the imaginary north–south line in the sky) on February evenings, as does Procyon in Canis Minor, which adjoins Gemini to the south. In the Southern Hemisphere, Carina, Puppis, and Vela—the three constellations that once formed the large ancient Greek constellation Argo Navis, ship of the Argonauts—are high in the sky.

## NORTHERN LATITUDES

### THE STARS

Gemini is almost overhead as seen from mid-northern latitudes in February, with the faintest of the zodiacal constellations, Cancer, close by but slightly lower in the sky. South of Gemini, the sparkling Winter Triangle formed by Sirius (in Canis Major), Betelgeuse (in Orion), and Procyon (in Canis Minor) remains prominent. Taurus, the Bull, backs away from Orion toward the western horizon, with Auriga and Perseus higher above it. Close to the northwest horizon is the W-shaped Cassiopeia. Leo, the Lion, is moving into the eastern sky, with the familiar figure of the Big Dipper above it in the northeast.

### DEEP-SKY OBJECTS

M35, a large open star cluster at the feet of Gemini, is easily seen through binoculars. The Beehive Cluster (see p.290)—also known as M44 or Praesepe—lies nearby in Cancer. Through binoculars, the Beehive is visible as a scattering of stars three times wider than a full moon; under ideal conditions, it can be glimpsed by the naked eye as a hazy patch—it was known to the ancient Greeks.

The Milky Way runs through Monoceros, an often-overlooked constellation framed by the Winter Triangle, which contains several open star clusters. One of the most notable of these clusters, NGC 2244, is visible through binoculars. It is located at the heart of the elusive Rosette Nebula, which is seen well only in photographs.

**THE WINTER TRIANGLE**
Brilliant Sirius (bottom) forms a prominent triangle in the northern winter sky with Procyon (top left) and Betelgeuse (top right).

**NEPTUNE**

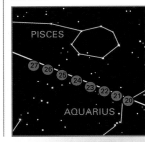

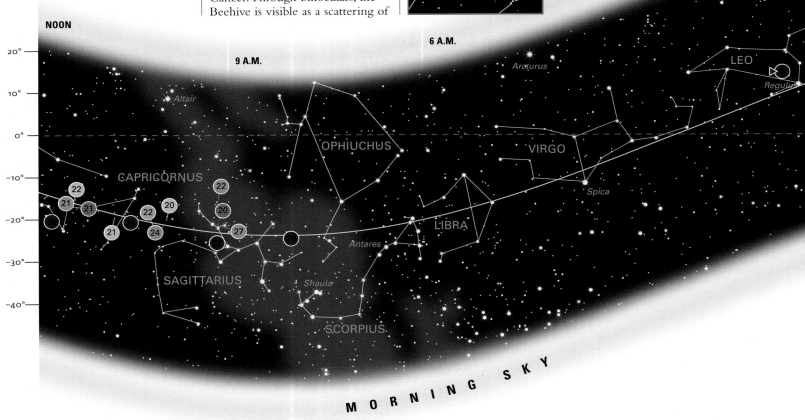

# SOUTHERN LATITUDES

## THE STARS

Sirius (see p.252) and Canopus, the two brightest stars in the entire sky, remain high for southern observers throughout February, while Achernar, the 1st-magnitude star at the end of the celestial river Eridanus, sinks toward the southwestern horizon. In the southeast, Crux, the Southern Cross, enters the scene, followed by the bright stars of Centaurus. Higher up is the False Cross, which is formed by four stars in Vela and Carina and is sometimes mistaken for the true Southern Cross.

Due north lie Castor (see p.276) and Pollux in Gemini. Orion is also high in the sky, with Taurus lower in the northwest. As seen from the most southerly latitudes, Perseus has already set and Auriga is following. Meanwhile, looking northeast, the distinctive shape of Leo, the Lion, has come into view.

## DEEP-SKY OBJECTS

The Milky Way, which meanders from southeast to northwest this month, contains numerous star clusters, of which M46 and M47, adjacent in Puppis, are prominent. Both clusters are at the edge of naked-eye visibility and look superb through binoculars. Two other open clusters that can be seen excellently through binoculars are NGC 2451 and NGC 2477, also in Puppis; farther south, in Vela, IC 2391 and IC 2395 are also good examples.

Outside the boundaries of the Milky Way, the open cluster M41 is found south of Sirius, while in the north, the Beehive Cluster (see p.290), or M44, is well positioned for observation in both February and March. In Carina, another open cluster, NGC 2516, is prominent. The Large Magellanic Cloud and the Tarantula Nebula are on view, south of Canopus, in the constellation Dorado.

**FINDING THE SOUTH CELESTIAL POLE**
The south celestial pole (left) is not marked by a bright star, but it can be located by intersecting two imaginary lines. One is the extension of the long axis of Crux. The other is at right angles to the line joining Alpha (α) and Beta (β) Centauri.

**URANUS**

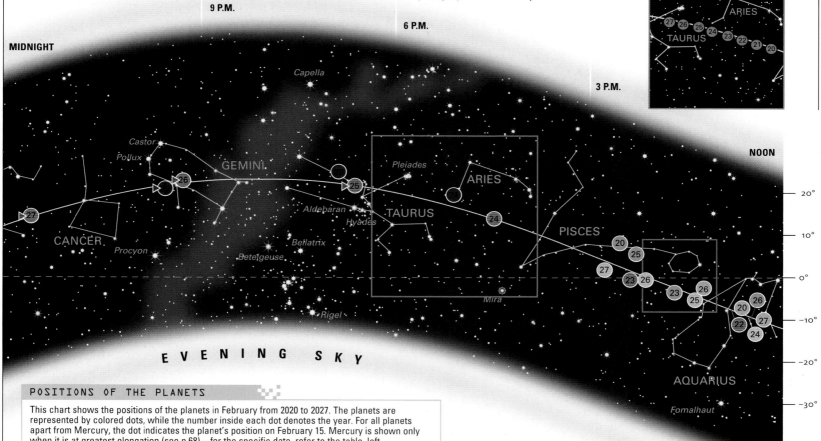

## POSITIONS OF THE PLANETS

This chart shows the positions of the planets in February from 2020 to 2027. The planets are represented by colored dots, while the number inside each dot denotes the year. For all planets apart from Mercury, the dot indicates the planet's position on February 15. Mercury is shown only when it is at greatest elongation (see p.68)—for the specific date, refer to the table, left.

- Mercury
- Venus
- Mars
- Jupiter
- Saturn
- Uranus
- Neptune

**EXAMPLES**

Jupiter's position on February 15, 2020

Jupiter's position on February 15, 2025. The arrow indicates that the planet is in retrograde motion (see p.68).

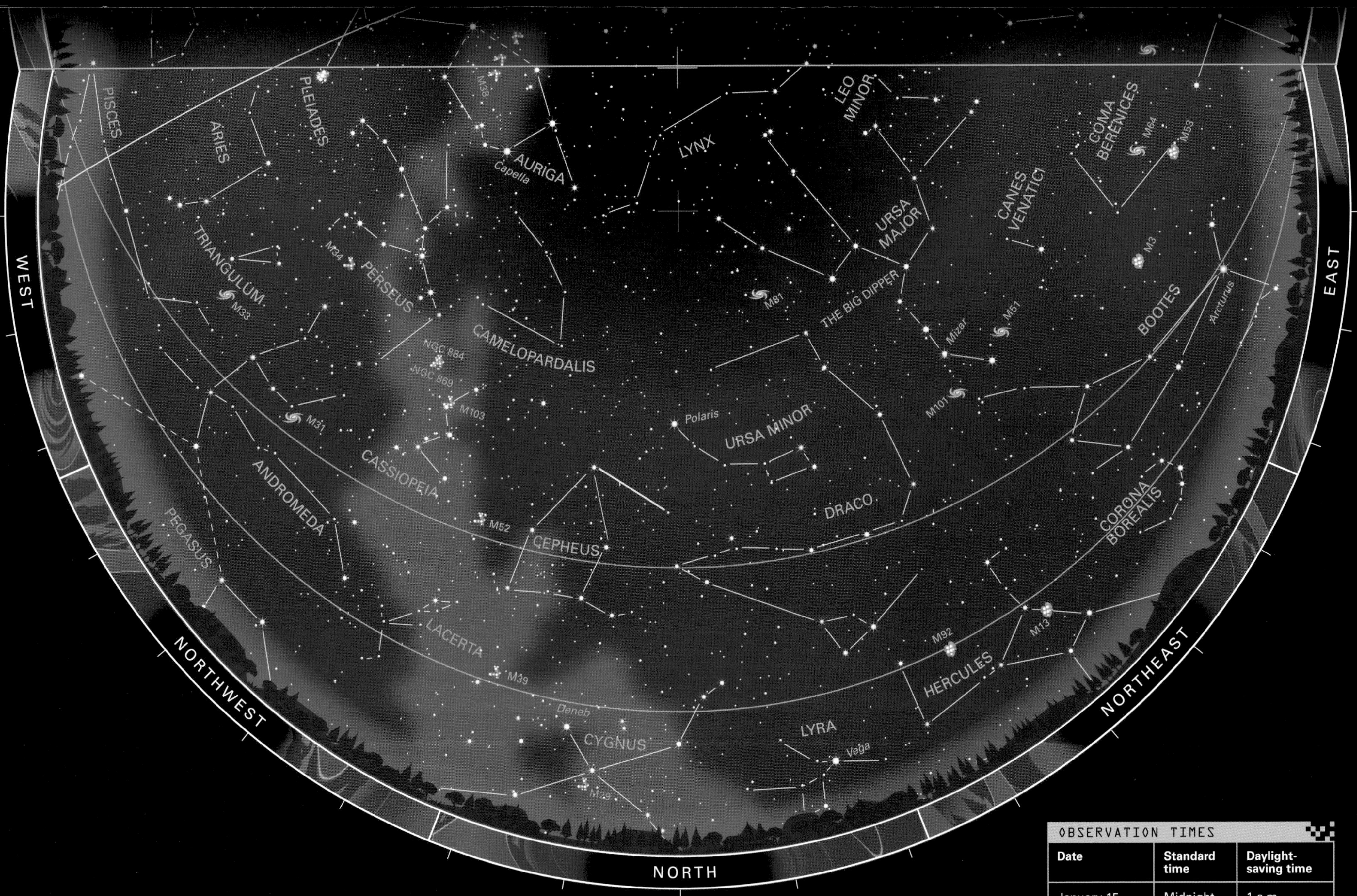

# FEBRUARY | NORTHERN LATITUDES

## LOOKING NORTH

### STAR MAGNITUDES

-1 · 0 · 1 · 2 · 3 · 4 · 5

○ Variable star

### DEEP-SKY OBJECTS

🌀 Galaxy

● Globular cluster

✦ Open cluster

☁ Diffuse nebula

◉ Planetary nebula

### POINTS OF REFERENCE

**Horizons** | 60°N | 40°N | 20°N

**Zeniths** | 60°N | 40°N | 20°N | Ecliptic

### OBSERVATION TIMES

| Date | Standard time | Daylight-saving time |
|---|---|---|
| January 15 | Midnight | 1 a.m. |
| February 1 | 11 p.m. | Midnight |
| February 15 | 10 p.m. | 11 p.m. |
| March 1 | 9 p.m. | 10 p.m. |
| March 15 | 8 p.m. | 9 p.m. |

WEST

NORTHWEST

NORTH

NORTHEAST

EAST

WEST

PISCES

PEGASUS

ARIES

TRIANGULUM

M33

ANDROMEDA

M31

PLEIADES

M34

PERSEUS

M38

CASSIOPEIA

NGC 884

NGC 869

M103

Capella

AURIGA

M52

CAMELOPARDALIS

M39

CEPHEUS

LACERTA

Deneb

CYGNUS

M29

Polaris

URSA MINOR

LYNX

M81

LEO MINOR

LYRA

Vega

DRACO

THE BIG DIPPER

URSA MAJOR

M101

Mizar

CANES VENATICI

HERCULES

M92

M51

M13

COMA BERENICES

M64

CORONA BOREALIS

BOOTES

M3

M53

Arcturus

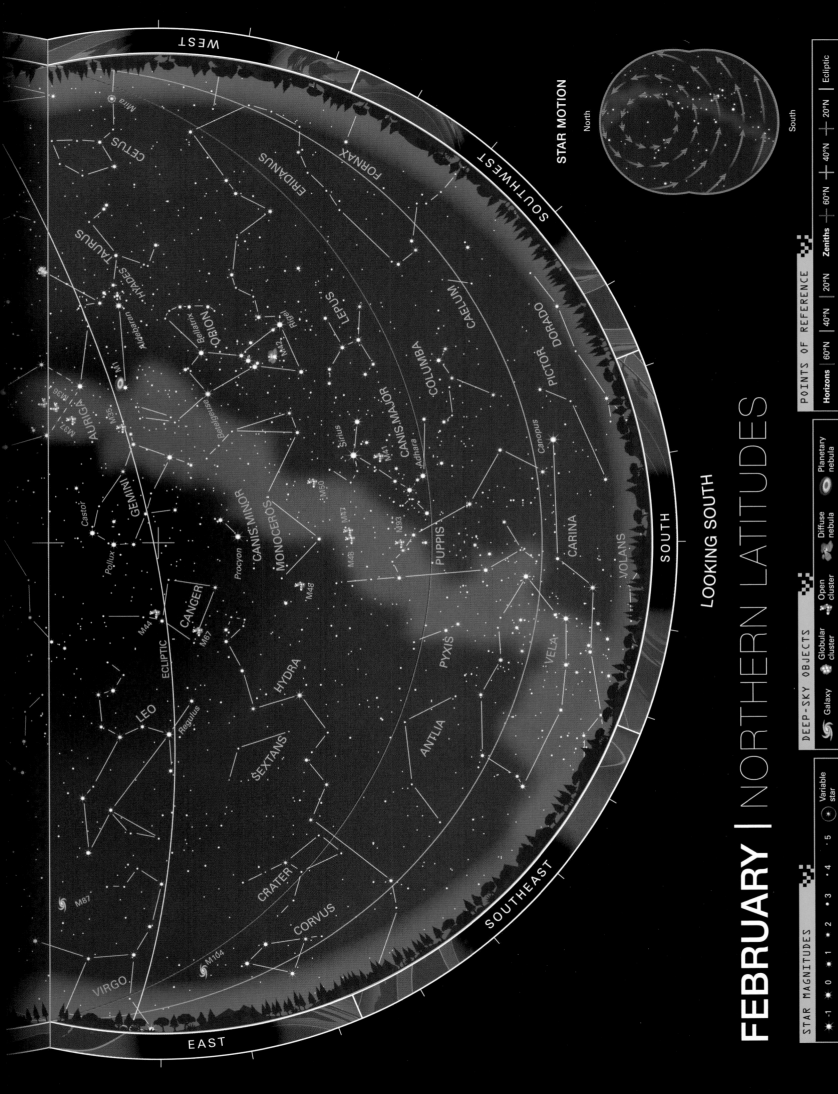

# FEBRUARY | NORTHERN LATITUDES

LOOKING SOUTH

THE NIGHT SKY

WEST

SOUTHWEST

SOUTH

SOUTHEAST

EAST

WEST

**STAR MOTION**

North

South

**STAR MAGNITUDES**

| -1 | 0 | 1 | 2 | 3 | 4 | 5 | Variable star |

**DEEP-SKY OBJECTS**

Galaxy · Globular cluster · Open cluster · Diffuse nebula · Planetary nebula

**POINTS OF REFERENCE**

| Horizons | 60°N | 40°N | 20°N | Zeniths | 60°N | 40°N | 20°N | Ecliptic |

## Constellations and objects

CETUS

TAURUS

HYADES

Aldebaran

M1

AURIGA

M37

M36 M38

M35

GEMINI

Castor

Pollux

CANCER

M44

M67

LEO

Regulus

ORION

Bellatrix

Betelgeuse

Rigel

M42

ERIDANUS

FORNAX

LEPUS

CANIS MINOR

Procyon

MONOCEROS

M50

M48

M46

M47

M93

CANIS MAJOR

Sirius

M41

Adhara

COLUMBA

CAELUM

PICTOR

DORADO

Canopus

CARINA

PUPPIS

PYXIS

VOLANS

VELA

ANTLIA

HYDRA

SEXTANS

CRATER

CORVUS

M104

VIRGO

M87

ECLIPTIC

Mira

THE NIGHT SKY

# FEBRUARY | SOUTHERN LATITUDES

## LOOKING NORTH

### OBSERVATION TIMES

| Date | Standard time | Daylight-saving time |
| --- | --- | --- |
| January 15 | Midnight | 1 a.m. |
| February 1 | 11 p.m. | Midnight |
| February 15 | 10 p.m. | 11 p.m. |
| March 1 | 9 p.m. | 10 p.m. |
| March 15 | 8 p.m. | 9 p.m. |

# FEBRUARY | SOUTHERN LATITUDES

LOOKING SOUTH

THE NIGHT SKY

WEST

EAST

SOUTHWEST

SOUTH

SOUTHEAST

STAR MOTION

North

South

**POINTS OF REFERENCE**

| Horizons | 0° | 20°S | 40°S | Zeniths | 0° | 20°S | 40°S | Ecliptic |
|---|---|---|---|---|---|---|---|---|

**DEEP-SKY OBJECTS**

Galaxy | Globular cluster | Open cluster | Diffuse nebula | Planetary nebula

**STAR MAGNITUDES**

-1 | 0 | 1 | 2 | 3 | 4 | 5 | Variable star

CETUS

FORNAX

ERIDANUS

SCULPTOR

PHOENIX

Achernar

HOROLOGIUM

CAELUM

DORADO

RETICULUM

PICTOR

GRUS

TUCANA

SMC

NGC 104

MENSA

HYDRUS

LMC

INDUS

VOLANS

CARINA

CHAMAELEON

OCTANS

PAVO

Canopus

PUPPIS

VELA

MUSCA

Acrux

CRUX

Gacrux

Mimosa

TRIANGULUM AUSTRALE

APUS

ARA

PYXIS

ANTLIA

Rigil Kentaurus

Hadar

CIRCINUS

NORMA

HYDRA

CENTAURUS

NGC 5139

LUPUS

CRATER

M83

CORVUS

M104

VIRGO

Spica

CANIS MAJOR

LEPUS

M41

Adhara

M93

COLUMBA

# MARCH

Nights grow shorter in the Northern Hemisphere but longer in the Southern Hemisphere as the Sun moves toward the equinox on March 20. On that date, the Sun lies exactly on the celestial equator, and all over the world, day and night are of equal length. For northern observers, Orion and the other brilliant constellations of winter are departing toward the western horizon, while for southern observers, the rich star fields of Carina and Centaurus are moving to center stage.

## NORTHERN LATITUDES

### THE STARS

The distinctive sickle-shaped group of stars that makes up the head of Leo, the Lion, takes pride of place in the northern evening sky this month, with the fainter stars of Cancer to its right. Below it, in the south, lies a blank-looking area of sky occupied by the faint constellations Sextans, Crater, and Hydra. The only notable star in this area is 2nd-magnitude Alphard (in Hydra)—which, appropriately, means "the solitary one"—lying on the north–south meridian.

The saucepan shape of the Big Dipper rides high in the northeast, its handle pointing down toward the bright star Arcturus, in Boötes, which is the harbinger of northern spring. Closer again to the horizon is Spica in Virgo. In the west, the stars of Gemini and Auriga remain high, with Taurus and Orion lower down. Sirius twinkles near the southwest horizon.

### DEEP-SKY OBJECTS

The beautiful spiral galaxy M81 (see p.314) in northern Ursa Major lies near the north–south meridian on March evenings and is detectable through binoculars in clear skies. Farther south, the Beehive cluster (see p.290), or M44, in Cancer remains well positioned for observation.

**THE SICKLE OF LEO**
The stars that represent the head and neck of Leo, the Lion, form a distinctive shape like a sickle or a reversed question mark.

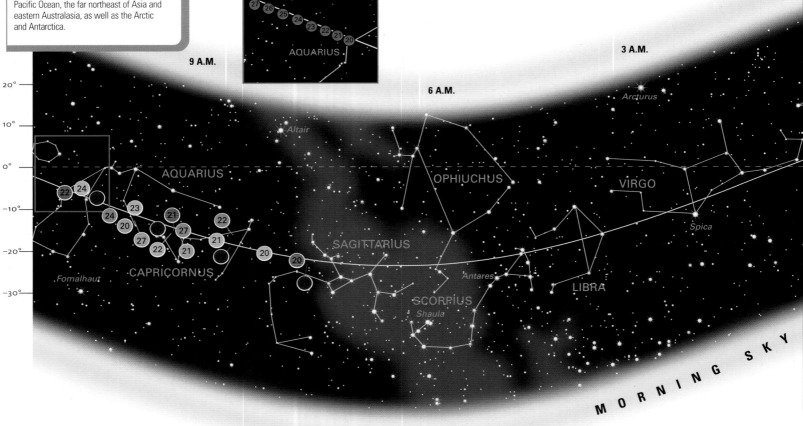

NEPTUNE

PISCES

AQUARIUS

9 A.M.

6 A.M.

3 A.M.

MIDNIGHT

Altair

Arcturus

AQUARIUS

OPHIUCHUS

VIRGO

Spica

SAGITTARIUS

Antares

LIBRA

Fomalhaut

CAPRICORNUS

SCORPIUS

Shaula

MORNING SKY

# SOUTHERN LATITUDES

## THE STARS

Leo, the Lion, and its brightest star Regulus (see p.253) are high in the northern half of the sky for all southern observers, with Castor (see p.276) and Pollux in Gemini lower in the northwest. Sirius (see p.252) still sparkles high in the western sky, but Orion sinks on its side toward the western horizon. Almost overhead for observers in midlatitudes is Alphard, the brightest star in the constellation Hydra, which sprawls across an otherwise barren region of sky toward the southeast horizon.

Spica, the brightest star in Virgo, is well-placed in the east, and Canopus, in Carina, is prominent in the southwest sky. However, the main focus of attention is in the southeast, where the Southern Cross, Crux, now rides high along with brilliant Alpha (α) and Beta (β) Centauri— Rigil Kentaurus (see p.252) and Hadar—which point toward it.

## DEEP-SKY OBJECTS

An open star cluster popularly known as the Southern Pleiades, IC 2602 lies close to the meridian on March evenings. Its brightest member, 3rd-magnitude Theta (θ) Carinae, is easily visible to the naked eye, and binoculars reveal at least two dozen more members.

Four degrees to the north of the Southern Pleiades lies a large glowing region visible to the naked eye, NGC 3372, also known as the Carina Nebula (see p.247), which contains the erratic variable star Eta (η) Carinae (see p.262). Farther north, between Antlia and Vela, telescopes will pick up the planetary nebula NGC 3132, also known as the Eight-Burst Nebula. On view in the southwest sky are the Large Magellanic Cloud and the Tarantula Nebula (in Dorado).

**THE FALSE CROSS**
Two stars in Vela (top left and center right) and two in Carina (center left and bottom right) form the False Cross in the southern sky.

**URANUS**

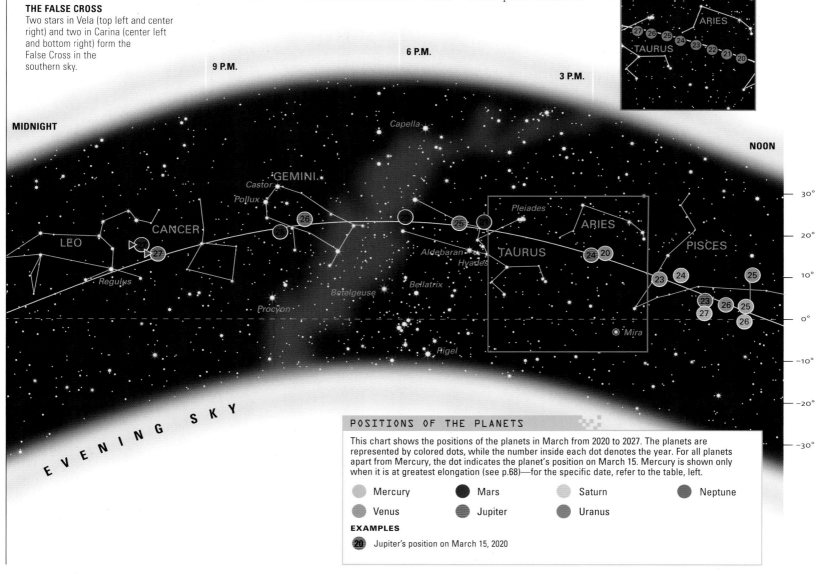

**POSITIONS OF THE PLANETS**

This chart shows the positions of the planets in March from 2020 to 2027. The planets are represented by colored dots, while the number inside each dot denotes the year. For all planets apart from Mercury, the dot indicates the planet's position on March 15. Mercury is shown only when it is at greatest elongation (see p.68)—for the specific date, refer to the table, left.

- Mercury
- Venus
- Mars
- Jupiter
- Saturn
- Uranus
- Neptune

**EXAMPLES**

20 Jupiter's position on March 15, 2020

# MARCH | NORTHERN LATITUDES

## LOOKING NORTH

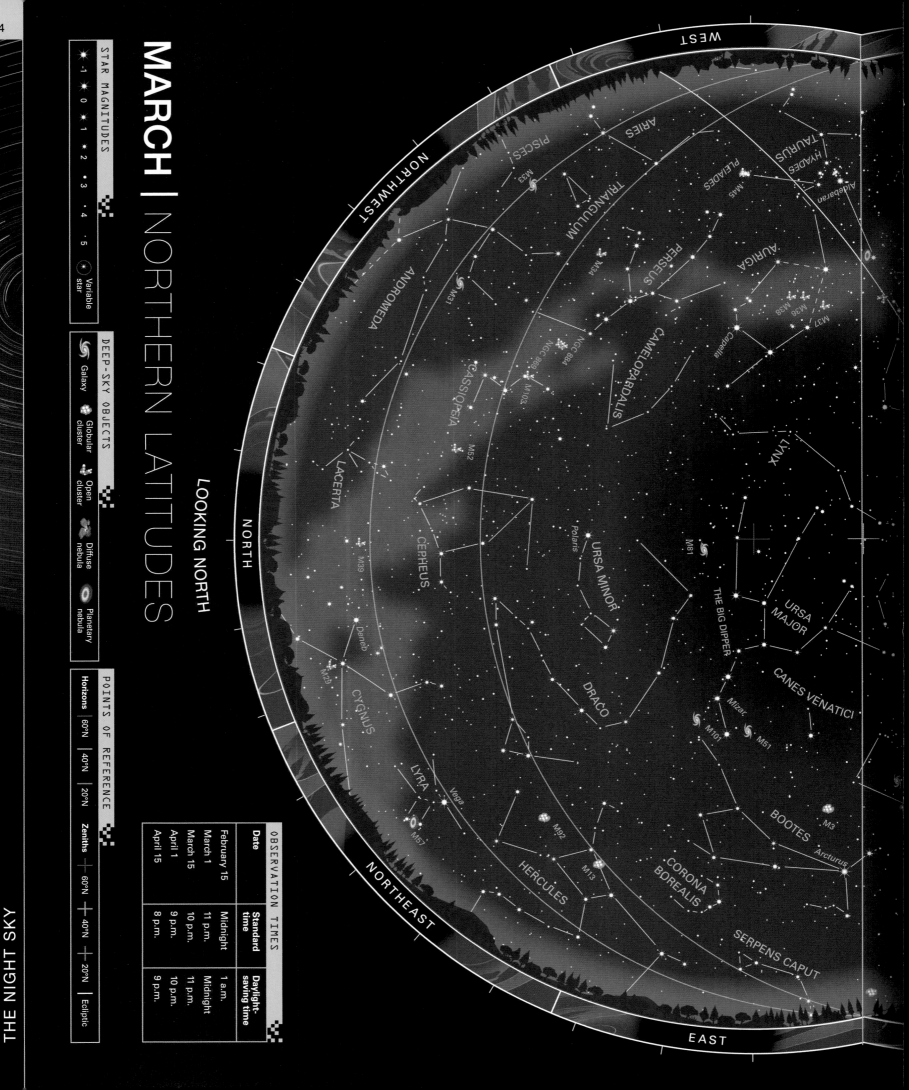

DEEP-SKY OBJECTS

- Galaxy
- Globular cluster
- Open cluster
- Diffuse nebula
- Planetary nebula

POINTS OF REFERENCE

| Horizons | | | |
|---|---|---|---|
| 60°N | 40°N | 20°N | |

| Zeniths | | | |
|---|---|---|---|
| 60°N | 40°N | 20°N | Ecliptic |

OBSERVATION TIMES

| Date | Standard time | Daylight-saving time |
|---|---|---|
| February 15 | Midnight | 1 a.m. |
| March 1 | 11 p.m. | Midnight |
| March 15 | 10 p.m. | 11 p.m. |
| April 1 | 9 p.m. | 10 p.m. |
| April 15 | 8 p.m. | 9 p.m. |

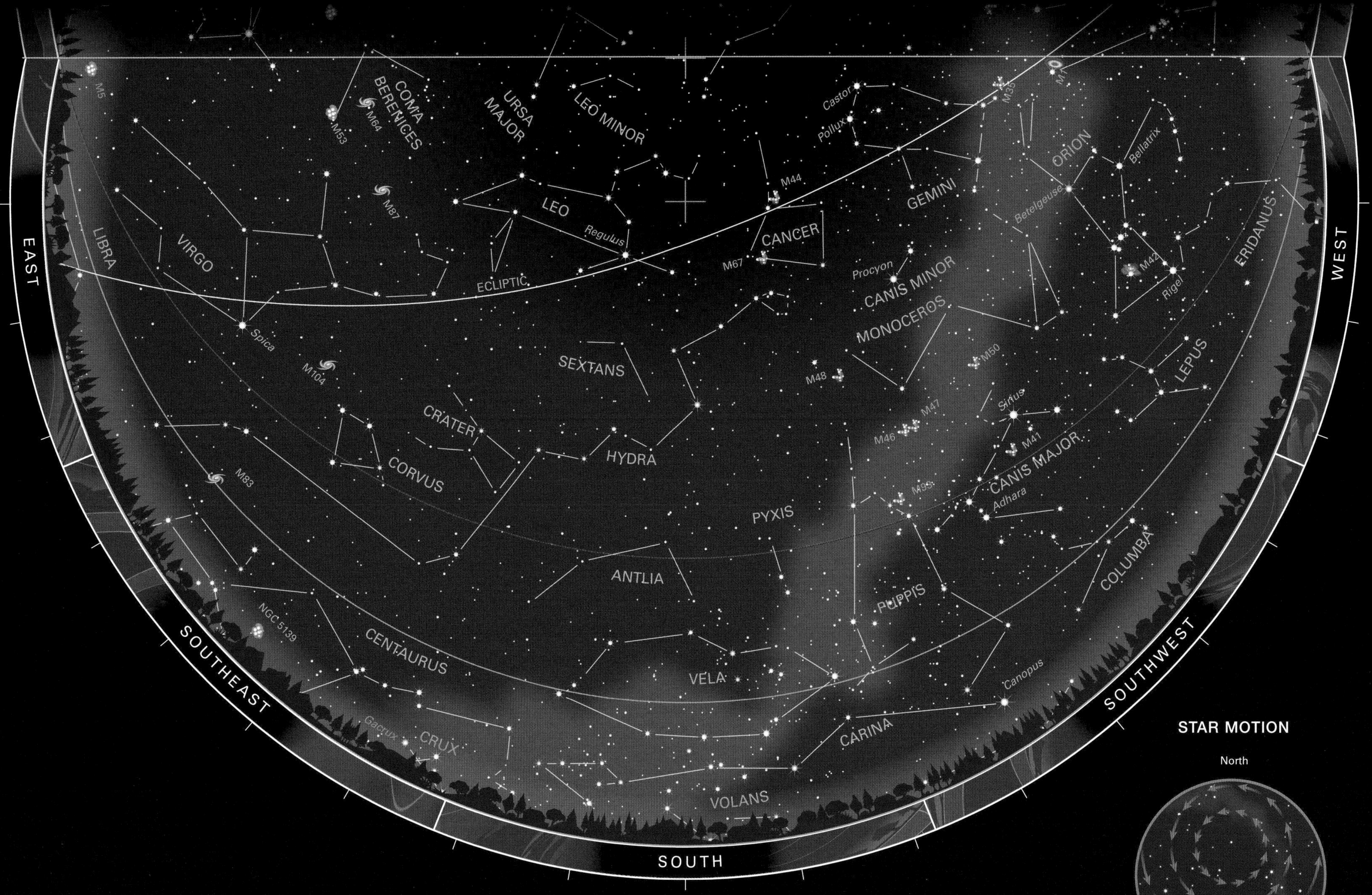

# MARCH | NORTHERN LATITUDES

**THE NIGHT SKY**

**LOOKING SOUTH**

**STAR MOTION**

North

South

**POINTS OF REFERENCE**

| Horizons | 60°N | 40°N | 20°N | Zeniths | 60°N | 40°N | 20°N | Ecliptic |

**DEEP-SKY OBJECTS**

Galaxy · Globular cluster · Open cluster · Diffuse nebula · Planetary nebula

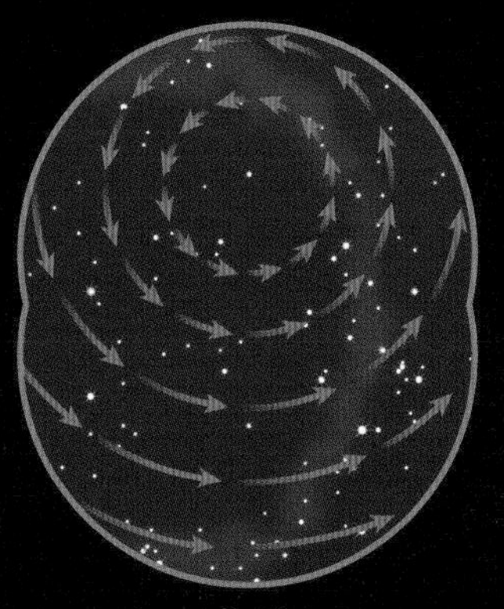

**STAR MAGNITUDES**

-1 · 0 · 1 · 2 · 3 · 4 · 5 · Variable star

WEST

SOUTHWEST

SOUTH

SOUTHEAST

EAST

WEST

EAST

ERIDANUS

ORION

Bellatrix

Rigel

Betelgeuse

M42

M78

M35

GEMINI

Castor

Pollux

CANCER

M44

M67

LEPUS

COLUMBA

CANIS MAJOR

Sirius

Adhara

M41

MONOCEROS

M50

CANIS MINOR

Procyon

M48

M46

M47

M93

PUPPIS

Canopus

CARINA

PYXIS

VELA

VOLANS

ANTLIA

SEXTANS

HYDRA

CRATER

CORVUS

CENTAURUS

CRUX

Gacrux

Acrux

NGC 5139

M83

M104

LEO MINOR

LEO

Regulus

URSA MAJOR

COMA BERENICES

M64

M53

M87

VIRGO

Spica

LIBRA

M5

ECLIPTIC

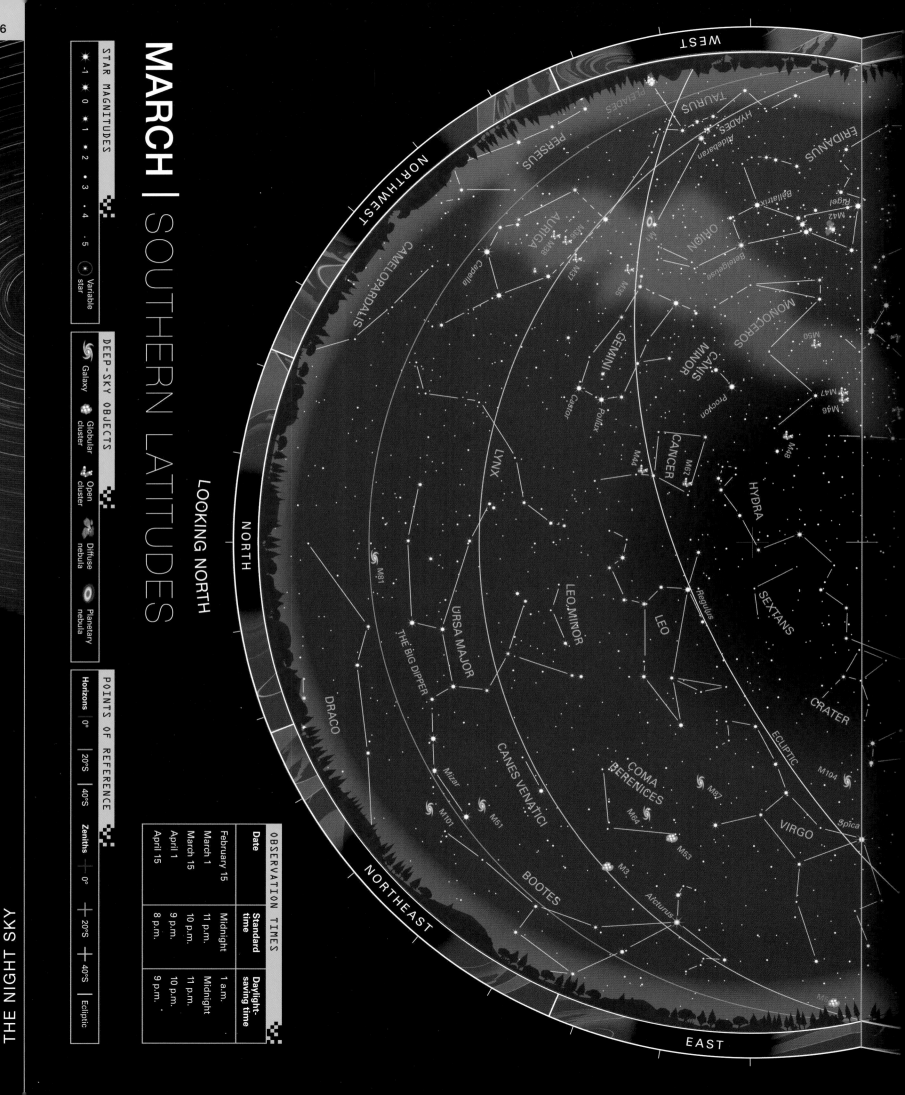

# MARCH | SOUTHERN LATITUDES

## LOOKING NORTH

### STAR MAGNITUDES

| | |
|---|---|
| ✦ | -1 |
| ✦ | 0 |
| • | 1 |
| • | 2 |
| • | 3 |
| · | 4 |
| · | 5 |
| ⊙ | Variable star |

### DEEP-SKY OBJECTS

| | |
|---|---|
| 🌀 | Galaxy |
| ⚬ | Globular cluster |
| ✾ | Open cluster |
| ✺ | Diffuse nebula |
| ⬭ | Planetary nebula |

### POINTS OF REFERENCE

| Horizons | 0° | 20°S | 40°S | Zeniths |
|---|---|---|---|---|
| | + | + | + | 0° |
| | + | + | + | 20°S |
| | | | + | 40°S |
| | | | | Ecliptic |

### OBSERVATION TIMES

| Date | Standard time | Daylight-saving time |
|---|---|---|
| February 15 | Midnight | 1 a.m. |
| March 1 | 11 p.m. | Midnight |
| March 15 | 10 p.m. | 11 p.m. |
| April 1 | 9 p.m. | 10 p.m. |
| April 15 | 8 p.m. | 9 p.m. |

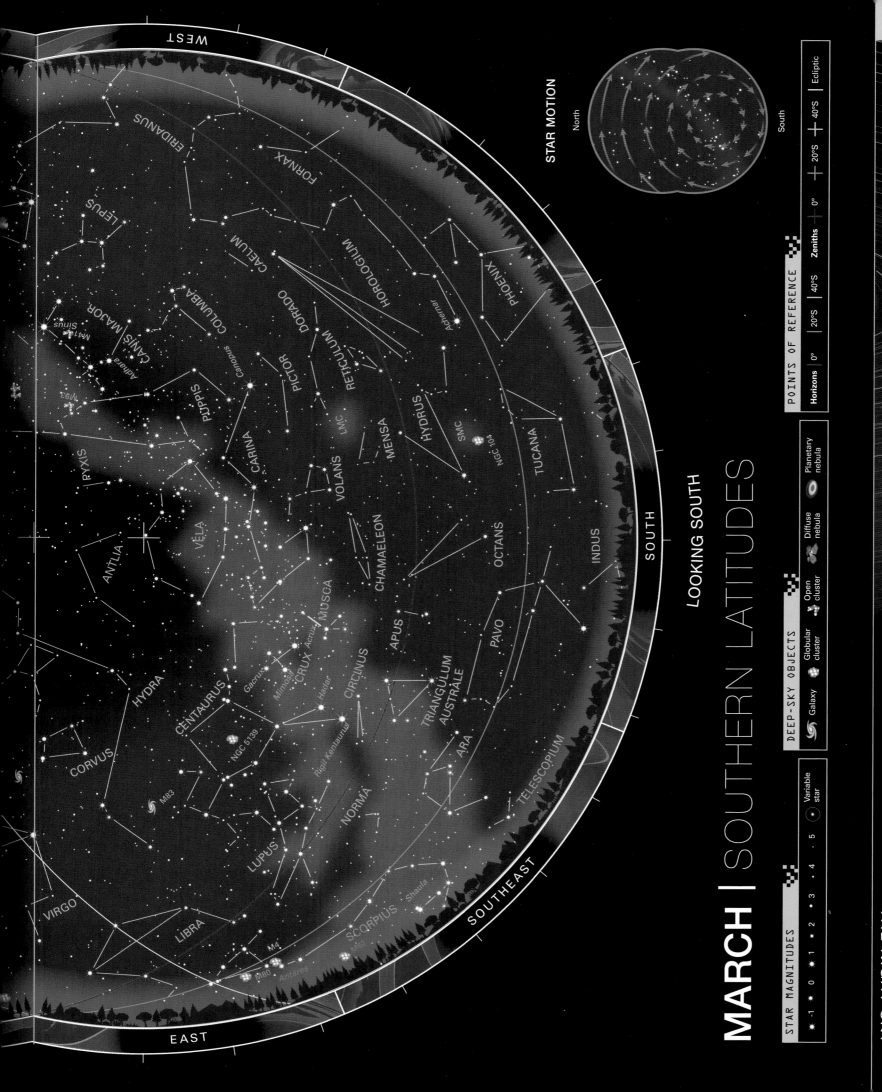

# MARCH | SOUTHERN LATITUDES

LOOKING SOUTH

**STAR MOTION**

North

South

## POINTS OF REFERENCE

| Horizons | 0° | 20°S | 40°S | Zeniths | 0° | 20°S | 40°S | Ecliptic |

## DEEP-SKY OBJECTS

Galaxy · Globular cluster · Open cluster · Diffuse nebula · Planetary nebula

## STAR MAGNITUDES

-1 · 0 · 1 · 2 · 3 · 4 · 5 · Variable star

THE NIGHT SKY

---

### Constellations and objects labelled on the chart

WEST · EAST · SOUTH · SOUTHEAST

ERIDANUS · FORNAX · LEPUS · CAELUM · DORADO · HOROLOGIUM · RETICULUM · PICTOR · COLUMBA · CANIS MAJOR · Sirius · M41 · Adhara · PUPPIS · Canopus · CARINA · PYXIS · VELA · ANTLIA · M93 · HYDRA · VOLANS · MENSA · HYDRUS · PHOENIX · Achernar · SMC · NGC 104 · TUCANA · LMC · CHAMAELEON · OCTANS · INDUS · APUS · PAVO · MUSCA · CRUX · Acrux · Gacrux · Mimosa · Hadar · CIRCINUS · TRIANGULUM AUSTRALE · ARA · TELESCOPIUM · CENTAURUS · NGC 5139 · Rigil Kentaurus · NORMA · M83 · CORVUS · LUPUS · VIRGO · LIBRA · SCORPIUS · Shaula · M4 · M62 · M80 · Antares

## SPECIAL EVENTS

**PHASES OF THE MOON**

| | FULL MOON | NEW MOON |
|---|---|---|
| **2020** | April 8 | April 23 |
| **2021** | April 27 | April 12 |
| **2022** | April 16 | April 1, 30 |
| **2023** | April 6 | April 20 |
| **2024** | April 23 | April 8 |
| **2025** | April 13 | April 27 |
| **2026** | April 2 | April 17 |
| **2027** | April 20 | April 7 |

**PLANETS**

**2022: April 5** Mars and Saturn appear about half a Moon-width apart in the low eastern predawn sky.

**2022: April 29** Mercury is at greatest evening elongation, magnitude 0.5.

**2022: April 30** Venus and Jupiter appear half a Moon-width apart in the low eastern predawn sky.

**2023: April 11** Mercury is at greatest evening elongation, magnitude 0.3.

**2025: April 21** Mercury is at greatest morning elongation, magnitude 0.6.

**2025: April 29** Venus and Saturn appear seven Moon-widths apart in the low eastern predawn sky.

**2026: April 3** Mercury is at greatest morning elongation, magnitude 0.5.

**ECLIPSES**

**2022: April 30** A partial eclipse of the Sun is visible from the southeast Pacific and south South America.

**2023: April 20** An annularm total eclipse of the Sun is visible from Indonesia, Australia, and Papua New Guinea. It begins annular, farther along its path becomes total, and ends annular.

**2024: April 8** A total eclipse of the Sun is visible from Mexico, central US, and east Canada. A partial eclipse is visible from North and Central America.

# APRIL

One of the most familiar patterns in the sky, the seven stars that make up the Big Dipper lie overhead from mid-northern latitudes, with the crouching figure of Leo, the Lion, reigning farther south. In the eastern sky, the daffodil-colored Arcturus, in Boötes, announces the arrival of spring in the north. In southern latitudes, the Southern Cross lies close to the north–south meridian, and Alpha ($\alpha$) and Beta ($\beta$) Centauri—Rigil Kentaurus and Hadar—are high in the southeast.

## NORTHERN LATITUDES

### THE STARS

On April evenings, the Big Dipper is high in the sky. The stars in the bowl point north to Polaris (see pp.278–279), the north Pole Star, while following the curve of its handle leads to Arcturus, in Boötes, which is the brightest star north of the celestial equator. Continuing this curve leads to Spica, the brightest star in Virgo, close to the southeastern horizon. South of Leo and Virgo, the sprawling figure of Hydra occupies a large but mostly blank area of sky. By April, most of the stars of winter have disappeared in the west, although Gemini remains on view and Capella, in Auriga, twinkles in the northwest.

### DEEP-SKY OBJECTS

M81 (see p.314), the beautiful spiral galaxy in northern Ursa Major, is well placed for observation this month. A large open star cluster worthy of attention can be found in Coma Berenices and consists of a scattering of stars of 5th magnitude and fainter fanned out over an area of sky several times wider than a full moon. Known as the Coma Star Cluster, this is best viewed through wide-angle binoculars. To its south is the Virgo Cluster (see p.329); a telescope is needed to see its numerous but faint member galaxies.

NEPTUNE

PISCES
27
26
25
24
23
22
21
20
AQUARIUS

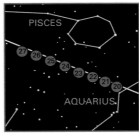

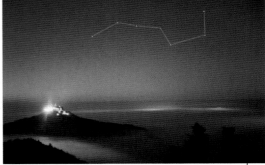

**THE BIG DIPPER**
The familiar shape of the Big Dipper can be seen high in the sky on northern spring evenings.

### METEOR SHOWER

One of the weaker annual meteor showers, the Lyrids reaches its peak around April 21–22, when a dozen or so meteors per hour can be seen radiating from a point near Vega (see p.253) in Lyra. Although not numerous, Lyrids are bright and fast. Rates are highest toward dawn, when Vega is highest in the sky, and they are much lower for a day or so on either side of the peak.

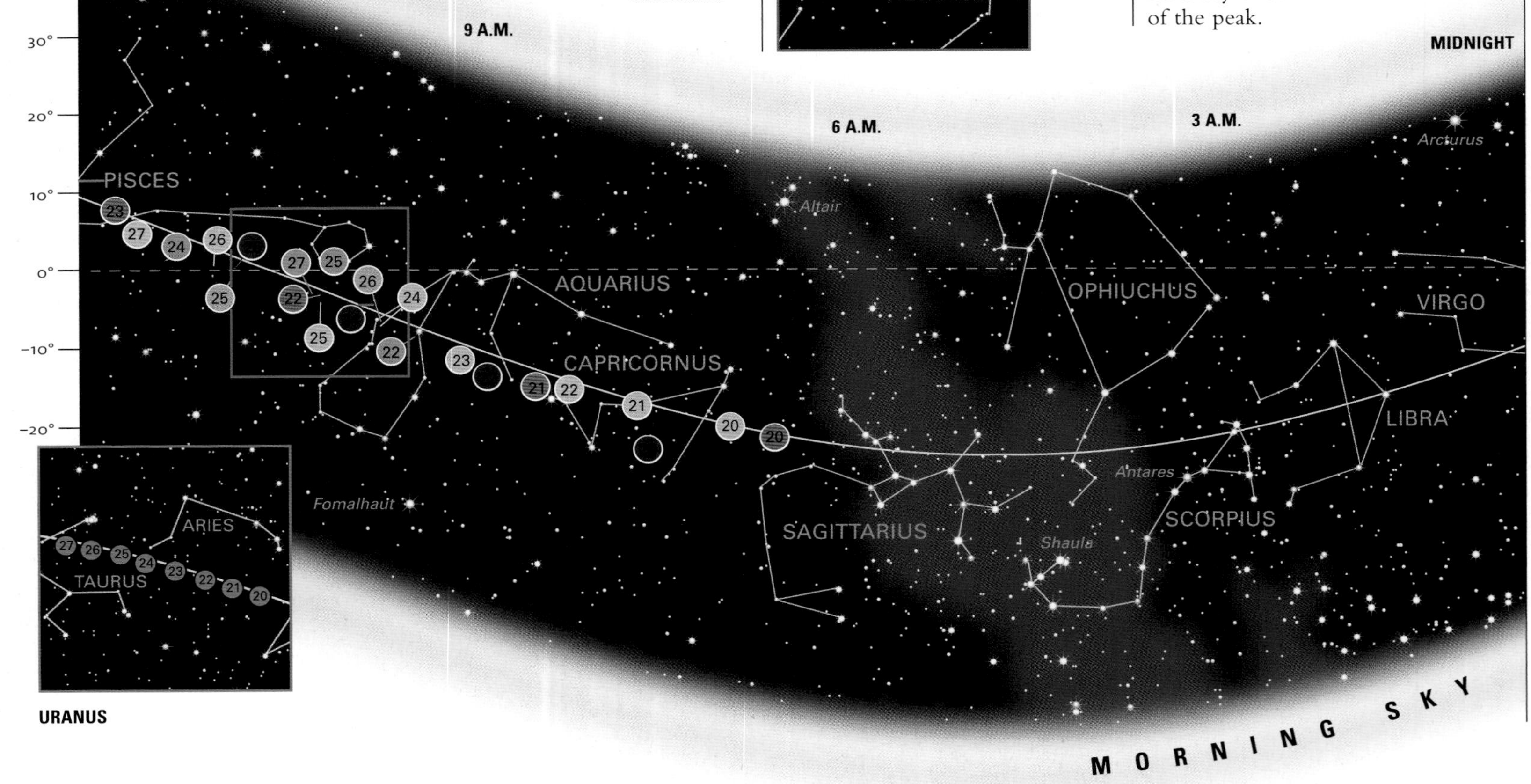

30°
20°
10°
0°
-10°
-20°
9 A.M.
6 A.M.
3 A.M.
MIDNIGHT
PISCES
AQUARIUS
CAPRICORNUS
Fomalhaut
Altair
OPHIUCHUS
SAGITTARIUS
Shaula
Antares
SCORPIUS
LIBRA
VIRGO
Arcturus
MORNING SKY
ARIES
TAURUS
27
26
25
24
23
22
21
20

URANUS

# SOUTHERN LATITUDES

## THE STARS

In the Southern Hemisphere, Crux lies almost on the north–south meridian line, with Rigil Kentaurus (see p.252) and Hadar—Alpha (α) and Beta (β) Centauri—slightly to its lower left. Antares, in Scorpius, is rising in the southeast, while Canopus, in Carina, sinks low in the southwest. Hydra's long body meanders overhead, its head adjoining Cancer in the northwest and its tail ending between Libra and Centaurus in the southeast. Spica, the brightest star in Virgo, is high in the east. Leo lies in the north, with Arcturus, in Boötes, in the northeast. Observers north of latitude 40°S can see the Big Dipper low on the northern horizon.

## DEEP-SKY OBJECTS

Next to the Southern Cross, an apparent gap in the rich stream of the Milky Way is visible to the naked eye. This is, in fact, a dark nebula, known as the Coalsack, which obscures the light of the background stars. On its edge is the Jewel Box cluster (see.294), or NGC 4755, which looks like a hazy star to the naked eye.

On display in Carina is the cluster IC 2602 and the Carina Nebula (see p.247), or NGC 3372. To the east, among the rich star fields of Centaurus, is the globular cluster NGC 5139 or Omega (ω) Centauri, which looks like a hazy 4th-magnitude star. In the north of the sky, members of the Virgo Cluster are well placed for telescopic observation this month.

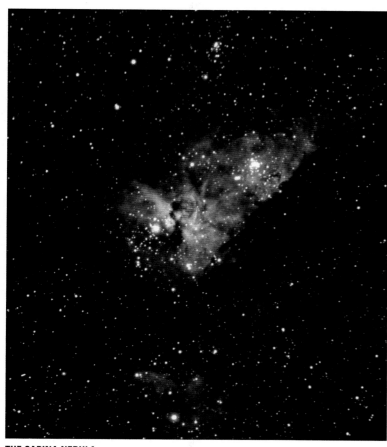

**THE CARINA NEBULA**
This huge nebula in the southern Milky Way is visible to the naked eye. Eta (η) Carinae (center left) is a peculiar variable star that is surrounded by a glowing shell of gas.

**THE COALSACK**
This dark cloud of dust (center left), next to the Southern Cross, is silhouetted against the bright background of the Milky Way.

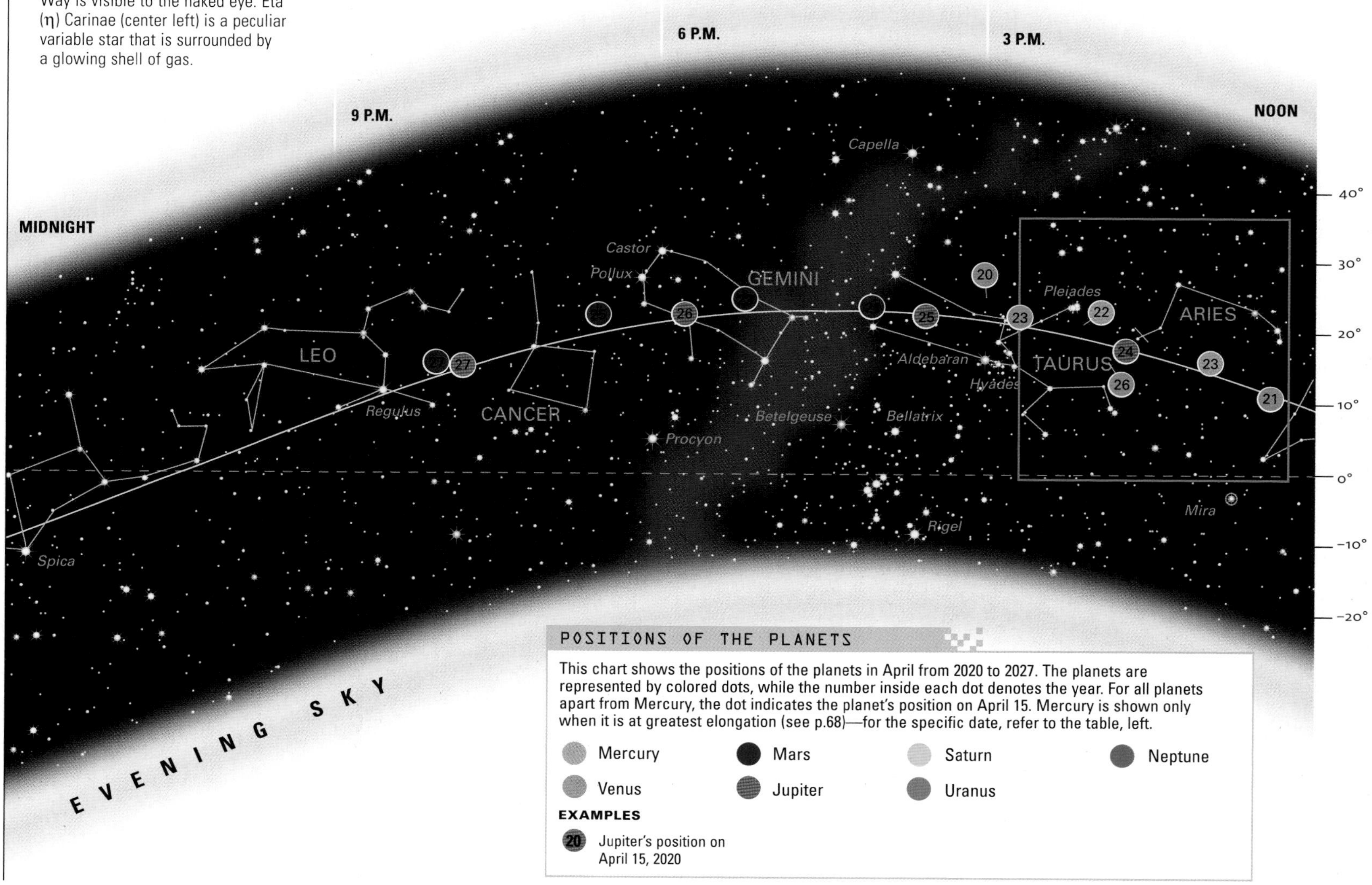

POSITIONS OF THE PLANETS

This chart shows the positions of the planets in April from 2020 to 2027. The planets are represented by colored dots, while the number inside each dot denotes the year. For all planets apart from Mercury, the dot indicates the planet's position on April 15. Mercury is shown only when it is at greatest elongation (see p.68)—for the specific date, refer to the table, left.

- Mercury
- Venus
- Mars
- Jupiter
- Saturn
- Uranus
- Neptune

**EXAMPLES**

20 Jupiter's position on April 15, 2020

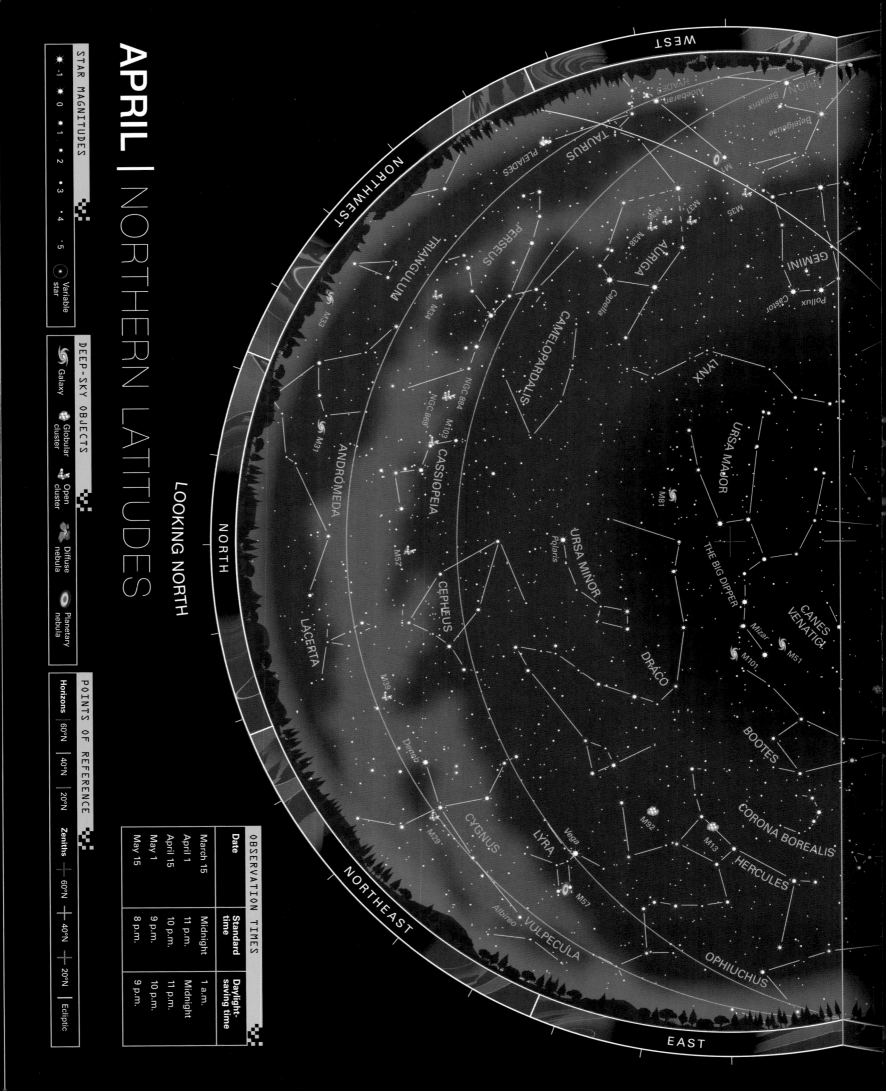

# APRIL | NORTHERN LATITUDES

## LOOKING NORTH

| OBSERVATION TIMES | | |
|---|---|---|
| Date | Standard time | Daylight-saving time |
| March 15 | Midnight | 1 a.m. |
| April 1 | 11 p.m. | Midnight |
| April 15 | 10 p.m. | 11 p.m. |
| May 1 | 9 p.m. | 10 p.m. |
| May 15 | 8 p.m. | 9 p.m. |

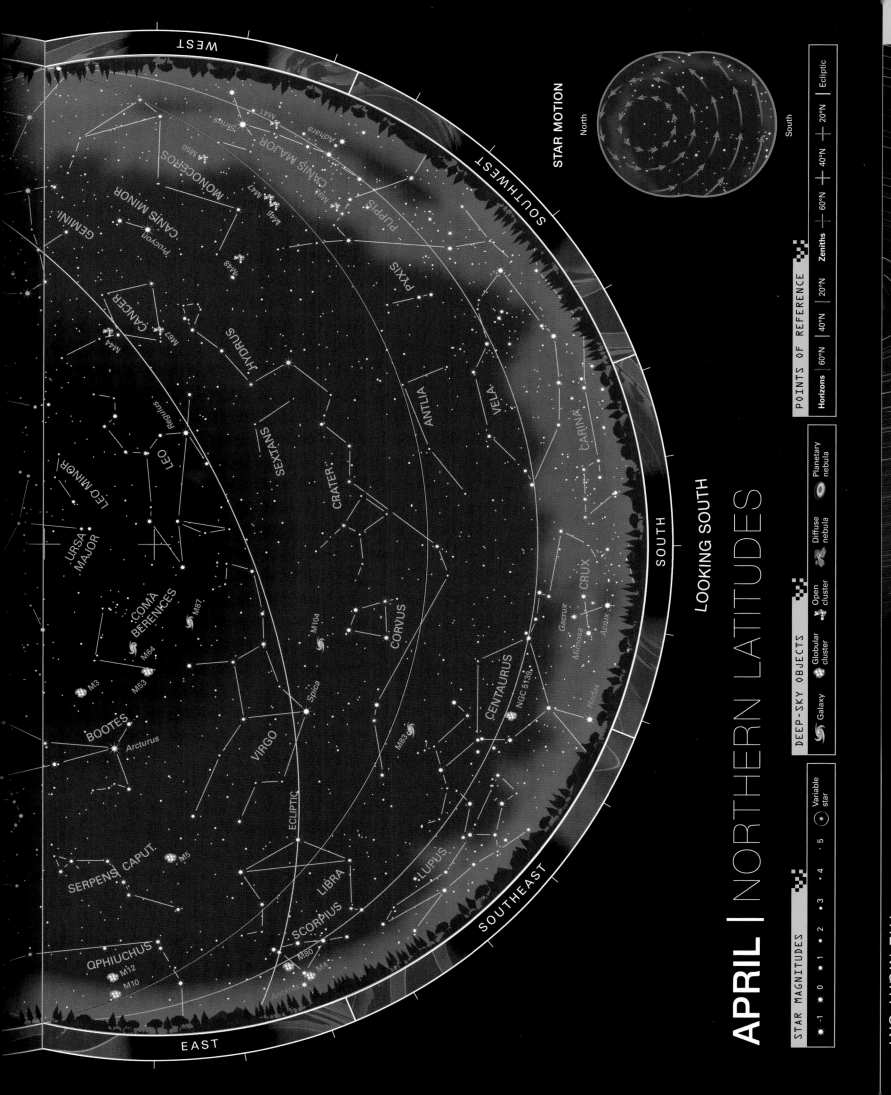

# APRIL | NORTHERN LATITUDES

LOOKING SOUTH

THE NIGHT SKY

STAR MOTION

North

South

**POINTS OF REFERENCE**

| Horizons | 60°N | 40°N | 20°N | Zeniths | 40°N | 20°N | Ecliptic |

**DEEP-SKY OBJECTS**

Galaxy | Globular cluster | Open cluster | Diffuse nebula | Planetary nebula

**STAR MAGNITUDES**

-1 | 0 | 1 | 2 | 3 | 4 | 5 | Variable star

WEST

SOUTHWEST

SOUTH

SOUTHEAST

EAST

GEMINI
CANIS MINOR
MONOCEROS
CANIS MAJOR
Sirius
Adhara
M41
M50
M46
M47
M93
PUPPIS
PYXIS
Procyon
CANCER
M44
M67
HYDRA
ANTLIA
VELA
CARINA
SEXTANS
LEO
Regulus
URSA MAJOR
LEO MINOR
CRATER
CORVUS
M104
M87
COMA BERENICES
M64
M53
M3
Spica
VIRGO
CRUX
Gacrux
Mimosa
Acrux
Hadar
CENTAURUS
NGC 5139
M83
BOOTES
Arcturus
ECLIPTIC
LIBRA
LUPUS
SERPENS CAPUT
M5
SCORPIUS
M80
M4
Antares
OPHIUCHUS
M12
M10

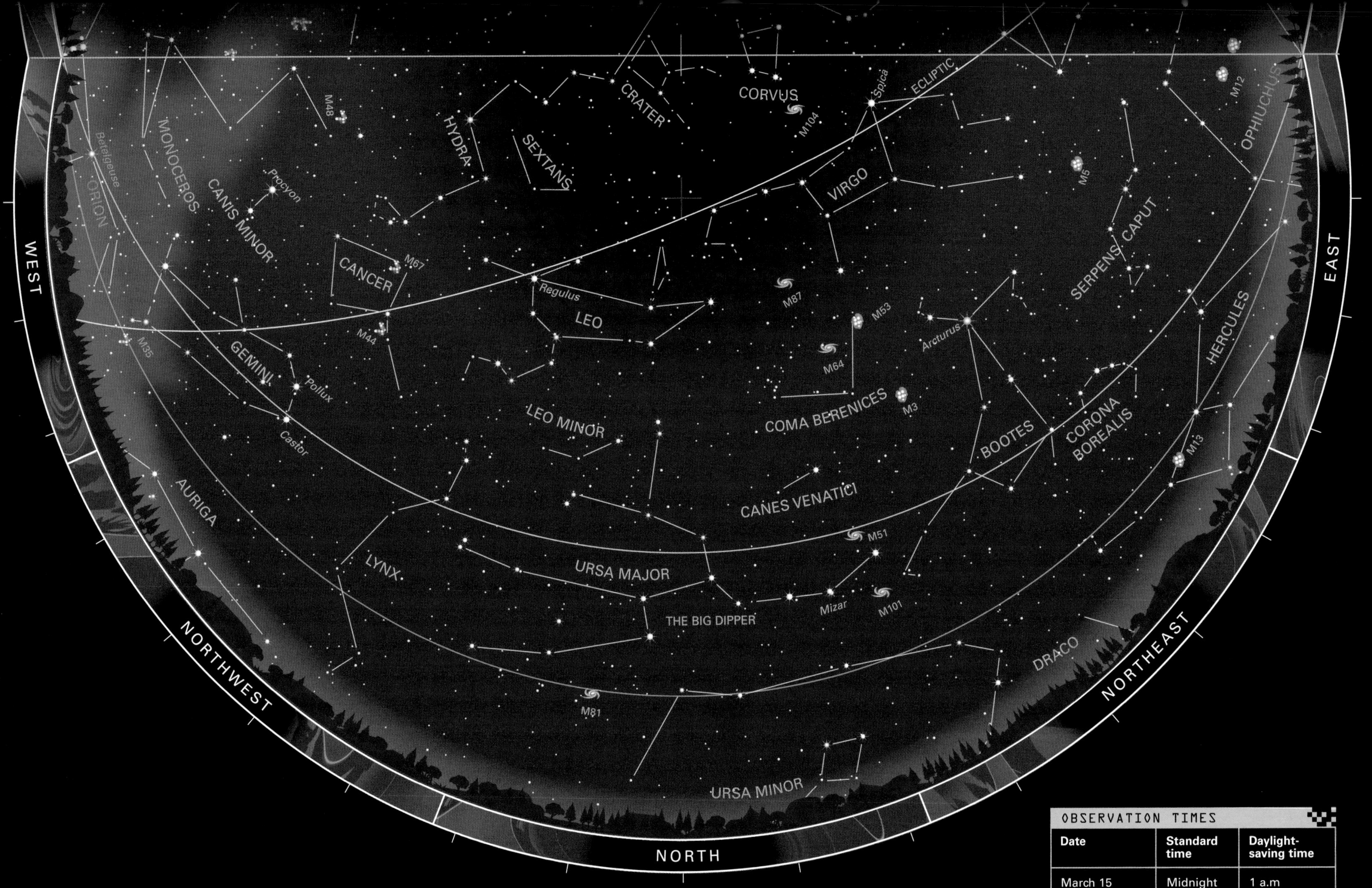

# APRIL | SOUTHERN LATITUDES

## LOOKING NORTH

### STAR MAGNITUDES

- -1
- 0
- 1
- 2
- 3
- 4
- 5
- Variable star

### DEEP-SKY OBJECTS

- Galaxy
- Globular cluster
- Open cluster
- Diffuse nebula
- Planetary nebula

### POINTS OF REFERENCE

| Horizons | 0° | 20°S | 40°S |
| --- | --- | --- | --- |
| Zeniths | 0° | 20°S | 40°S | Ecliptic |

### OBSERVATION TIMES

| Date | Standard time | Daylight-saving time |
| --- | --- | --- |
| March 15 | Midnight | 1 a.m |
| April 1 | 11 p.m. | Midnight |
| April 15 | 10 p.m. | 11 p.m |
| May 1 | 9 p.m. | 10 p.m |
| May 15 | 8 p.m. | 9 p.m |

WEST

NORTHWEST

NORTH

URSA MINOR

DRACO

NORTHEAST

EAST

ORION

Betelgeuse

MONOCEROS

AURIGA

GEMINI

Castor

Pollux

CANIS MINOR

Procyon

M35

M48

CANCER

M44

M67

Regulus

HYDRA

LYNX

LEO MINOR

LEO

SEXTANS

URSA MAJOR

THE BIG DIPPER

CRATER

Mizar

CANES VENATICI

COMA BERENICES

M64

M53

M87

CORVUS

M104

M81

M101

M51

M3

VIRGO

Spica

ECLIPTIC

Arcturus

BOOTES

CORONA BOREALIS

SERPENS CAPUT

M5

M13

HERCULES

M12

OPHIUCHUS

# APRIL | SOUTHERN LATITUDES

LOOKING SOUTH

THE NIGHT SKY

STAR MOTION

North

South

## POINTS OF REFERENCE

| | | | |
|---|---|---|---|
| Horizons | 0° | 20°S | 40°S |
| Zeniths | 0° | 20°S | 40°S | Ecliptic |

## STAR MAGNITUDES

-1 • 0 • 1 • 2 · 3 · 4 · 5    ⊕ Variable star

## DEEP-SKY OBJECTS

🌀 Galaxy   ⬤ Globular cluster   ✴ Open cluster   Diffuse nebula   Planetary nebula

WEST

SOUTHWEST

SOUTH

SOUTHEAST

EAST

ORION

LEPUS

CANIS MAJOR

Sirius

M50

M41

Adhara

M93 M46 M47

COLUMBA

CAELUM

ERIDANUS

HOROLOGIUM

RETICULUM

Achernar

PHOENIX

TUCANA

NGC 104

SMC

HYDRUS

OCTANS

MENSA

LMC

DORADO

PICTOR

VOLANS

CARINA

PUPPIS

VELA

PYXIS

ANTLIA

HYDRA

CORVUS

CENTAURUS

NGC 5139

M83

Gacrux

CRUX

Mimosa

Acrux

Hadar

MUSCA

CHAMAELEON

APUS

PAVO

INDUS

TELESCOPIUM

ARA

Rigil Kentaurus

CIRCINUS

TRIANGULUM AUSTRALE

NORMA

LUPUS

LIBRA

SCORPIUS

Antares

M4

M80

M19

M62

Shaula

M6

M7

CORONA AUSTRALIS

SAGITTARIUS

M69

M54

M28

M8

M21 M24

M23

M9

M10

OPHIUCHUS

# MAY

As summer approaches, the days get longer in the Northern Hemisphere, restricting early evening observation, while in the Southern Hemisphere, the opposite is true, as the days become shorter and the nights longer. For northern observers, the Big Dipper is high up in the sky and Virgo is due south. Observers south of the equator are treated to the sight of the brilliant stars of Centaurus (the Centaur) and Crux (the Southern Cross) at their highest.

## NORTHERN LATITUDES

### THE STARS

The tip of the handle of the Big Dipper lies on the north–south meridian this month. The second star in the handle, Mizar, has a fainter companion, Alcor, which is visible to the naked eye (see p.276). The curved handle of the Big Dipper points toward orange Arcturus in Boötes, also high up. Almost due south is Spica, the brightest star in Virgo.

Gemini, the last of the winter constellations, begins to set in the northwest. As it departs, the stars of summer rise in the east, led by the brilliant blue-white star Vega (see p.253) in Lyra. For those observers at lower northerly latitudes, Antares and the stars of Scorpius begin to appear over the southeastern horizon.

### DEEP-SKY OBJECTS

Two large and relatively bright galaxies are well positioned for observation in May. South of the Big Dipper's handle is the Whirlpool Galaxy (see p.315), or M51, while to the north of the handle is M101, which is larger but less prominent. On clear nights, each appears as a faint patch of light through binoculars; a telescope is needed to see their spiral structures. The fan-shaped Coma Star Cluster is well positioned, as is the Virgo Cluster of galaxies (see p.329).

### METEOR SHOWER

The Eta Aquariid meteor shower is visible this month, but because the radiant lies virtually on the celestial equator, the shower is not seen well in far northerly latitudes.

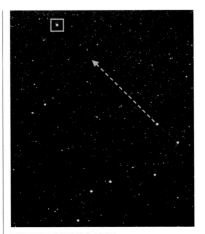

**FINDING THE POLE STAR**
Alpha (α) and Beta (β) Ursae Majoris, in the bowl of the Big Dipper, point toward the north Pole Star, Polaris (in green box).

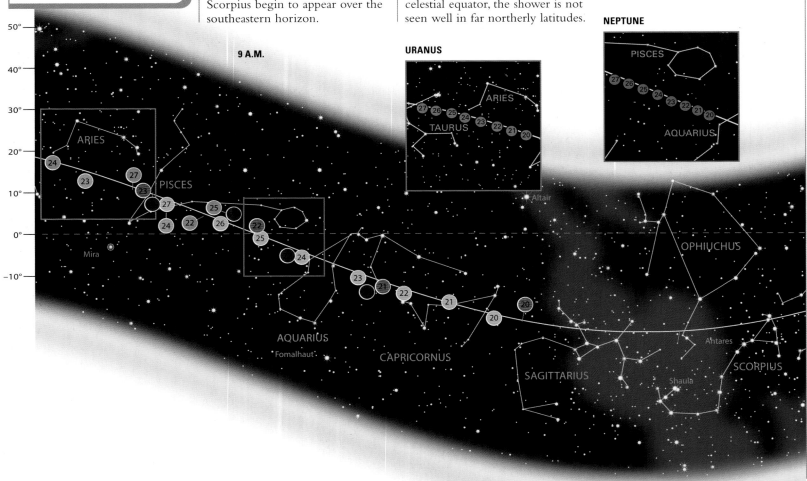

9 A.M.

URANUS

NEPTUNE

MORNING SKY

# SOUTHERN LATITUDES

## THE STARS

The constellation Crux and the two bright stars in Centaurus that act as a pointer to it, Alpha (α) Centauri—or Rigil Kentaurus (see p.252)—and Beta (β) Centauri—Hadar, are high in the southern sky in May. Crux is to the west of the north–south meridian, and Rigil Kentaurus and Hadar are on the eastern side. Although Rigil Kentaurus is usually described as the closest naked-eye star to the Sun, it actually consists of two yellowish stars that form a double star that is easily divided through a small telescope. The brightest member of the Southern Cross, Acrux—Alpha (α) Crucis—is also a double star that is divisible with a small telescope, but its component stars are blue-white.

Spica, in Virgo, lies high overhead with orange Arcturus, in Boötes, in the north. Leo sinks toward the northwestern horizon, while in the southeast, Scorpius and Sagittarius are coming into view—a sign that the southern winter is approaching.

## DEEP-SKY OBJECTS

The largest and brightest globular cluster in the sky, NGC 5139, or Omega (ω) Centauri, appears to the naked eye as a hazy 4th-magnitude star lying virtually on the north–south meridian this month. To the north of it lies NGC 5128, a peculiar radio-emitting galaxy also known as Centaurus A (see p.322), which is one of the easiest galaxies to find with binoculars. Another bright galaxy located near the meridian is M83, a spiral galaxy that is positioned face-on to Earth.

In Crux, the dark Coalsack Nebula and the sparkling Jewel Box (see p.294) remain prominent.

## METEOR SHOWER

The Eta Aquariid meteor shower reaches its peak around May 5–6, when 30 or so fast-moving meteors can be seen radiating each hour from near the star Eta (η) Aquarii, located almost exactly on the celestial equator. However, this part of the sky does not rise very high until around 3 A.M., and the meteor shower is best seen from equatorial and southerly locations, where May nights are longer. The Eta Aquariids are caused by dust from Halley's Comet (see p.217).

### RICH STAR FIELDS

Alpha (α) and Beta (β) Centauri (left) point toward the constellation Crux (right). The Coalsack Nebula (bottom right), most of which lies within Crux, obscures a large area of stars in the Milky Way.

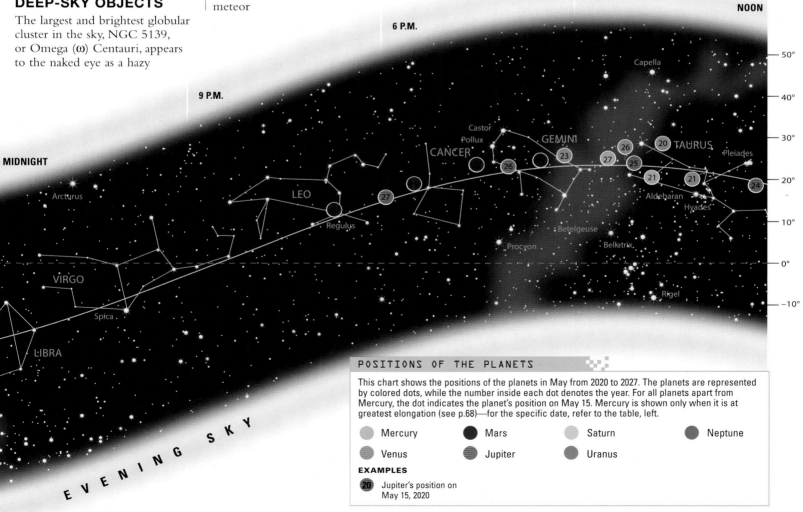

## POSITIONS OF THE PLANETS

This chart shows the positions of the planets in May from 2020 to 2027. The planets are represented by colored dots, while the number inside each dot denotes the year. For all planets apart from Mercury, the dot indicates the planet's position on May 15. Mercury is shown only when it is at greatest elongation (see p.68)—for the specific date, refer to the table, left.

- Mercury
- Venus
- Mars
- Jupiter
- Saturn
- Uranus
- Neptune

**EXAMPLES**

20 Jupiter's position on May 15, 2020

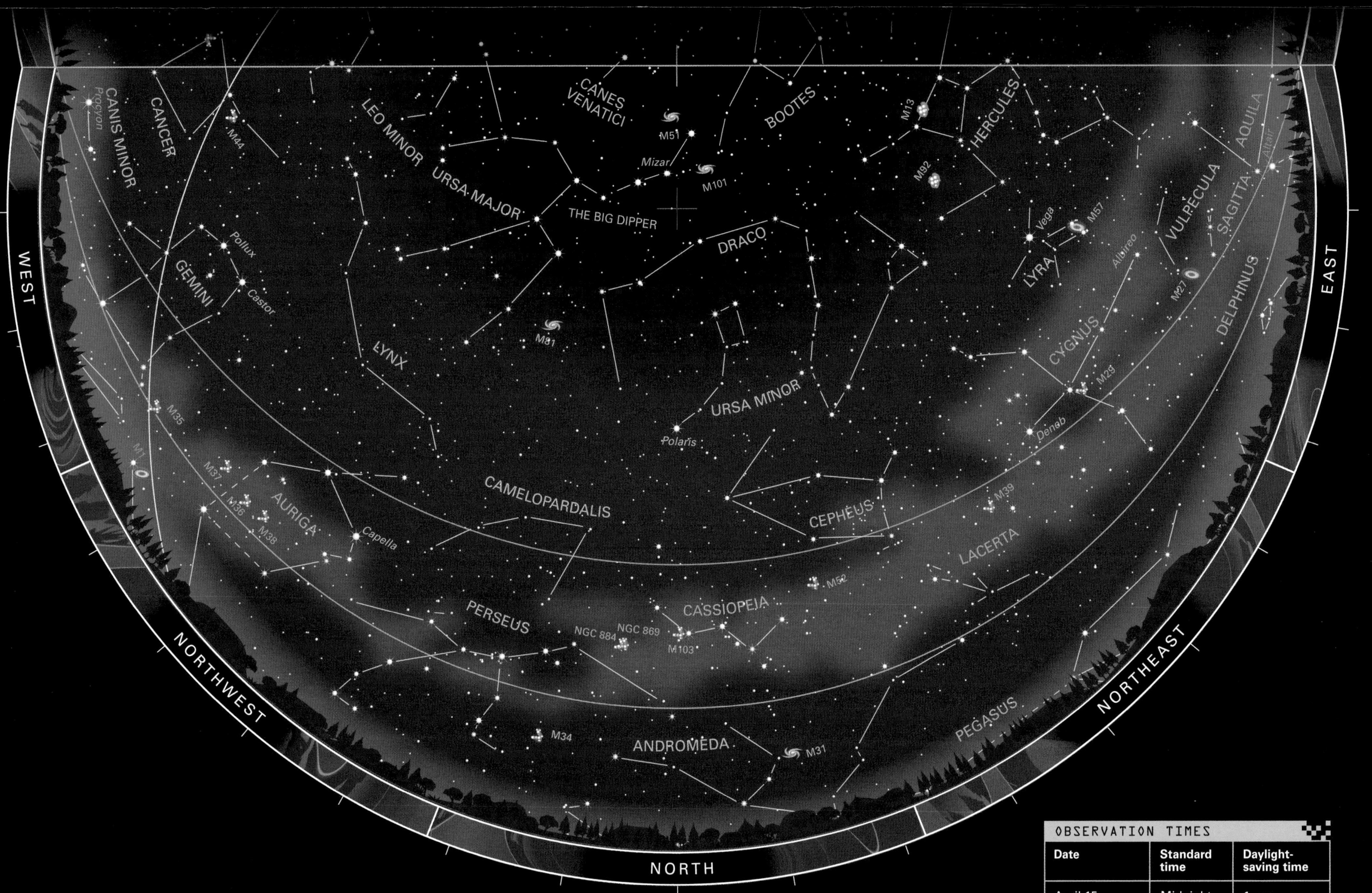

# MAY | NORTHERN LATITUDES

## LOOKING NORTH

### STAR MAGNITUDES

| | |
|---|---|
| ★ | -1 |
| ★ | 0 |
| ⋆ | 1 |
| • | 2 |
| • | 3 |
| • | 4 |
| · | 5 |
| ⊙ | Variable star |

### DEEP-SKY OBJECTS

| | |
|---|---|
| 🌀 | Galaxy |
| ✸ | Globular cluster |
| ⁂ | Open cluster |
| ❀ | Diffuse nebula |
| ◎ | Planetary nebula |

### POINTS OF REFERENCE

| Horizons | 60°N | 40°N | 20°N | |
|---|---|---|---|---|
| Zeniths | 60°N | 40°N | 20°N | Ecliptic |

### OBSERVATION TIMES

| Date | Standard time | Daylight-saving time |
|---|---|---|
| April 15 | Midnight | 1 a.m. |
| May 1 | 11 p.m. | Midnight |
| May 15 | 10 p.m. | 11 p.m. |
| June 1 | 9 p.m. | 10 p.m. |
| June 15 | 8 p.m. | 9 p.m. |

WEST

NORTHWEST

NORTH

NORTHEAST

EAST

WEST

CANIS MINOR
Procyon
CANCER
GEMINI
M44
M35
Pollux
Castor
M37
M36
M38
AURIGA
Capella
LYNX
CAMELOPARDALIS
PERSEUS
M34
NGC 884
NGC 869
M103
CASSIOPEIA
ANDROMEDA
M31
M52
CEPHEUS
LACERTA
M39
PEGASUS
DELPHINUS
M29
Deneb
CYGNUS
SAGITTA
Albireo
VULPECULA
M27
AQUILA
Altair
M57
LYRA
Vega
HERCULES
M13
M92
BOÖTES
URSA MINOR
Polaris
DRACO
URSA MAJOR
THE BIG DIPPER
Mizar
M51
M101
CANES VENATICI
LEO MINOR
M81

# MAY | NORTHERN LATITUDES

LOOKING SOUTH

STAR MOTION

North

South

## POINTS OF REFERENCE

| Horizons | 60°N | 40°N | 20°N | Zeniths | 60°N | 40°N | 20°N | Ecliptic |

## DEEP-SKY OBJECTS

Galaxy | Globular cluster | Open cluster | Diffuse nebula | Planetary nebula

## STAR MAGNITUDES

-1 | 0 | 1 | 2 | 3 | 4 | 5 | Variable star

WEST

SOUTHWEST

SOUTH

SOUTHEAST

EAST

PYXIS

HYDRA

M67

SEXTANS

ANTLIA

VELA

LEO

Regulus

CRATER

CORVUS

M104

Spica

VIRGO

M87

COMA BERENICES

M64

M53

M3

Arcturus

CENTAURUS

Gacrux

Mimosa

CRUX Acrux

M83

BOOTES

CORONA BOREALIS

SERPENS CAPUT

M5

LIBRA

ECLIPTIC

LUPUS

Hadar

Rigil Kentaurus

CIRCINUS

HERCULES

M12

M10

OPHIUCHUS

M9

M80

M4

Antares

M19

M62

Shaula

SCORPIUS

NORMA

ARA

M14

SERPENS CAUDA

M16

M18

M24

M21

M23

M8

M6

M17

M25

M22

M26

M11

M28

M20

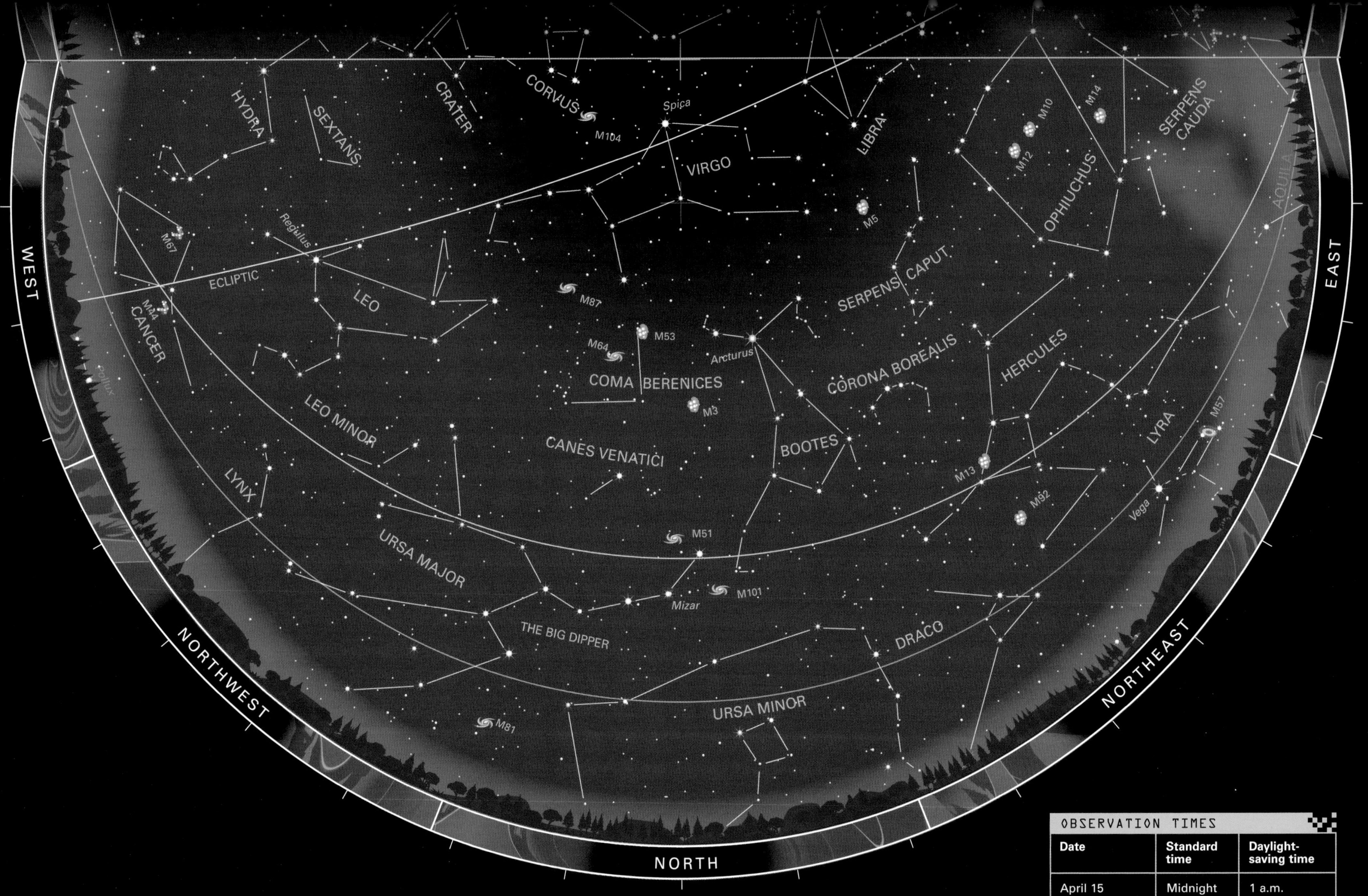

# MAY | SOUTHERN LATITUDES

## LOOKING NORTH

| OBSERVATION TIMES | | |
|------|------|------|
| Date | Standard time | Daylight-saving time |
| April 15 | Midnight | 1 a.m. |
| May 1 | 11 p.m. | Midnight |
| May 15 | 10 p.m. | 11 p.m. |
| June 1 | 9 p.m. | 10 p.m. |
| June 15 | 8 p.m. | 9 p.m. |

# MAY | SOUTHERN LATITUDES

LOOKING SOUTH

STAR MOTION

STAR MAGNITUDES

| | | | | | |
|---|---|---|---|---|---|
| -1 | 0 | 1 | 2 | 3 | 4 | 5 |

Variable star

Ecliptic

DEEP-SKY OBJECTS

Galaxy | Globular cluster | Open cluster | Diffuse nebula | Planetary nebula

POINTS OF REFERENCE

| Horizons | 0° | 20°S | 40°S |
|---|---|---|---|
| Zeniths | 0° | 20°S | 40°S |

North

South

THE NIGHT SKY

## SPECIAL EVENTS

### PHASES OF THE MOON

| | FULL MOON | NEW MOON |
|---|---|---|
| 2020 | June 5 | June 21 |
| 2021 | June 24 | June 10 |
| 2022 | June 14 | June 29 |
| 2023 | June 4 | June 18 |
| 2024 | June 22 | June 6 |
| 2025 | June 11 | June 25 |
| 2026 | June 30 | June 15 |
| 2027 | June 19 | June 4 |

### PLANETS

**2020: June 4** Mercury is at greatest evening elongation, magnitude 0.7.

**2022: June 16** Mercury is at greatest morning elongation, magnitude 0.7.

**2023: June 4** Venus is at greatest evening elongation, magnitude –4.3

**2023: June 4** Venus is at greatest morning elongation, magnitude –4.3.

**2026: June 15** Mercury is at greatest evening elongation, magnitude 0.7.

**2028: June 26** Mercury is at greatest morning elongation, magnitude 0.7.

### ECLIPSES AND TRANSITS

**2020: June 21** An annular eclipse of the Sun is visible from central Africa, south Asia, China, and the Pacific.

**2021: June 10** An annular eclipse of the Sun is visible from north Canada, Greenland, and Russia.

# JUNE

Northern nights are at their shortest, and southern ones at their longest, around the solstice on June 21, the date on which the Sun reaches its farthest point north of the celestial equator. In the northern sky, Arcturus and the other stars of Boötes stand high, and the giant Summer Triangle of Vega (in Lyra), Deneb (in Cygnus), and Altair (in Aquila) lies in the eastern half of the sky. Southern observers enjoy a rich band of constellations in the Milky Way during their long winter nights.

## NORTHERN LATITUDES

### THE STARS

The bowl of the Little Dipper, in Ursa Minor, stands high above the northern horizon with the sinuous body of Draco, the Dragon, winding around it. The horseshoe shape of Corona Borealis, the Northern Crown, lies on the north–south meridian with the head of Serpens, the Serpent, below it, while Arcturus, in Boötes, is high in the western half of the sky. In this area of sky, Arcturus is the base of a large Y-shaped pattern of bright stars completed by Epsilon (ε) and Gamma (γ) Boötis plus Alpha (α) Corona Borealis (also known as Alphecca). Leo is setting in the west, and Spica, in Virgo, is low in the southwest. In the eastern sky, the bright stars Vega (see p.253), Deneb, and Altair (see p.252) mark the corners of the Summer Triangle, best seen in late summer and fall. Ruddy Antares and the stars of Scorpius twinkle low on the southern horizon—June and July are the best months of the year for far-northern observers to see Scorpius in the evening sky.

### DEEP-SKY OBJECTS

The brightest globular cluster in northern skies, M13, is high up on summer evenings. It can be found along one side of the Keystone of Hercules, a quadrangle of stars that forms the torso of the constellation Hercules. M13 appears as a fuzzy

**NOCTILUCENT CLOUDS**
These high-altitude clouds can be seen on summer nights, illuminated by the Sun's rays that come over the horizon around midnight.

6th-magnitude star through binoculars, and it can be glimpsed by the naked eye under good conditions. It can be compared to M5, another 6th-magnitude globular cluster visible through binoculars. M5 lies in the head of Serpens and is usually regarded as the second-best northern globular cluster. Near the handle of the Big Dipper, the spiral galaxies M51 and M101 remain well positioned for observation.

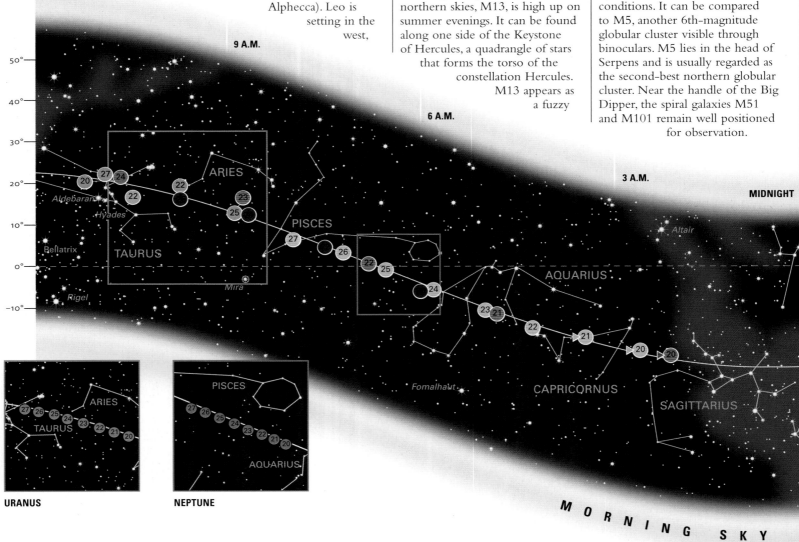

URANUS

NEPTUNE

# SOUTHERN LATITUDES

## THE STARS

A rich band of constellations can be seen across the sky, from southwest to northeast, along the path of the Milky Way. Crux (the Southern Cross) and Centaurus (the Centaur) are in the southwest, to the right of the celestial meridian. The lesser-known constellations Lupus, Norma, and Triangulum Australe are on the meridian. Ruddy Antares (see p.256) is overhead, with the curving tail of Scorpius, the Scorpion, extending to the southeast. Next to its tail are the dense star fields of Sagittarius in the Milky Way. Along the Milky Way to the east is Altair (see p.252) in the constellation Aquila, while Vega (see p.253) is low in the northeast. Arcturus and Spica are high in the northwest.

## DEEP-SKY OBJECTS

Heading away from Scorpius and toward the Milky Way and the center of the galaxy, two magnificent open star clusters, M6 and M7, are positioned near the end of the Scorpion's tail. Both clusters are visible to the naked eye, and they appear

### THE SCORPION'S LAIR
Orange-red Antares, the star at the heart of Scorpius, and the curved line of stars marking the Scorpion's tail are distinctive sights in June skies. Hovering over the "sting" in the tail are two prominent star clusters, M6 and M7 (bottom left).

magnificent through binoculars. M7 is the larger and brighter of the two; it appears twice the width of a full moon. Another prominent open cluster in Scorpius is NGC 6231, positioned next to Zeta (ζ) Scorpii.

The globular cluster Omega (ω) Centauri, or NGC 5139, and the peculiar galaxy NGC 5128, or Centaurus A, remain well placed for observation this month, as do the Coalsack Nebula and the Jewel Box Cluster (see p.264), in Crux, and the spiral galaxy M83 (in the constellation Hydra).

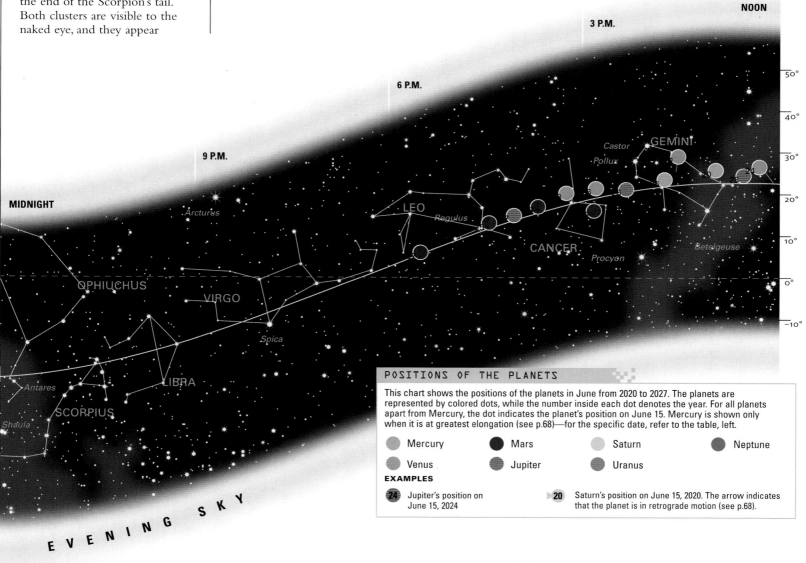

NOON

3 P.M.

6 P.M.

9 P.M.

MIDNIGHT

50°
40°
30°
20°
10°
0°
−10°

GEMINI
Castor
Pollux
LEO
Regulus
Arcturus
CANCER
Procyon
Betelgeuse
OPHIUCHUS
VIRGO
Spica
LIBRA
Antares
SCORPIUS
Shaula

E V E N I N G   S K Y

## POSITIONS OF THE PLANETS

This chart shows the positions of the planets in June from 2020 to 2027. The planets are represented by colored dots, while the number inside each dot denotes the year. For all planets apart from Mercury, the dot indicates the planet's position on June 15. Mercury is shown only when it is at greatest elongation (see p.68)—for the specific date, refer to the table, left.

- Mercury
- Venus
- Mars
- Jupiter
- Saturn
- Uranus
- Neptune

**EXAMPLES**

24 — Jupiter's position on June 15, 2024

20 — Saturn's position on June 15, 2020. The arrow indicates that the planet is in retrograde motion (see p.68).

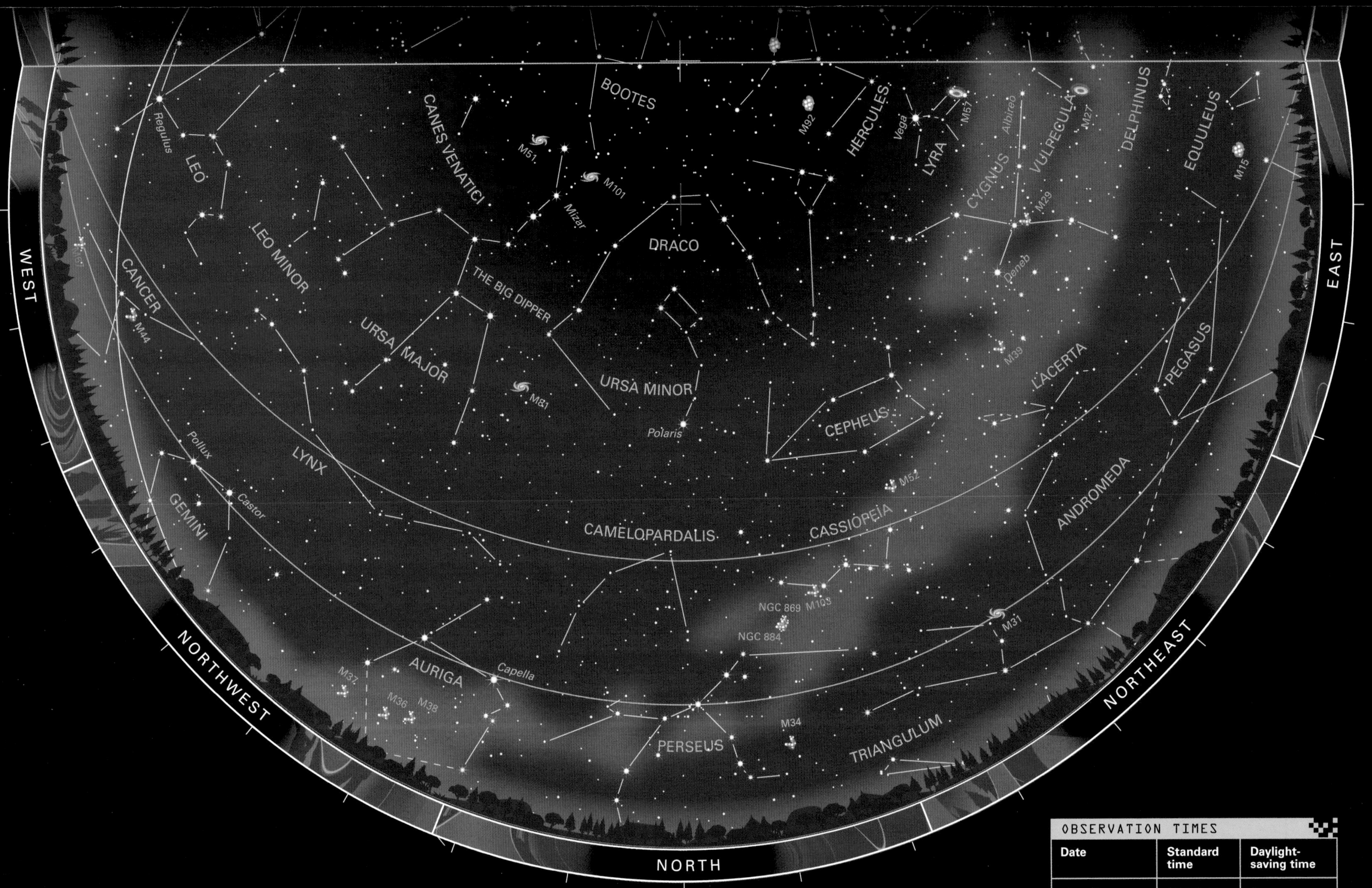

THE NIGHT SKY

# JUNE | NORTHERN LATITUDES

LOOKING NORTH

STAR MAGNITUDES

- -1
- 0
- 1
- 2
- 3
- 4
- 5
- Variable star

DEEP-SKY OBJECTS

- Galaxy
- Globular cluster
- Open cluster
- Diffuse nebula
- Planetary nebula

POINTS OF REFERENCE

Horizons | 60°N | 40°N | 20°N
Zeniths | 60°N | 40°N | 20°N | Ecliptic

OBSERVATION TIMES

| Date | Standard time | Daylight-saving time |
|---|---|---|
| May 15 | Midnight | 1 a.m. |
| June 1 | 11 p.m. | Midnight |
| June 15 | 10 p.m. | 11 p.m. |
| July 1 | 9 p.m. | 10 p.m. |
| July 15 | 8 p.m. | 9 p.m. |

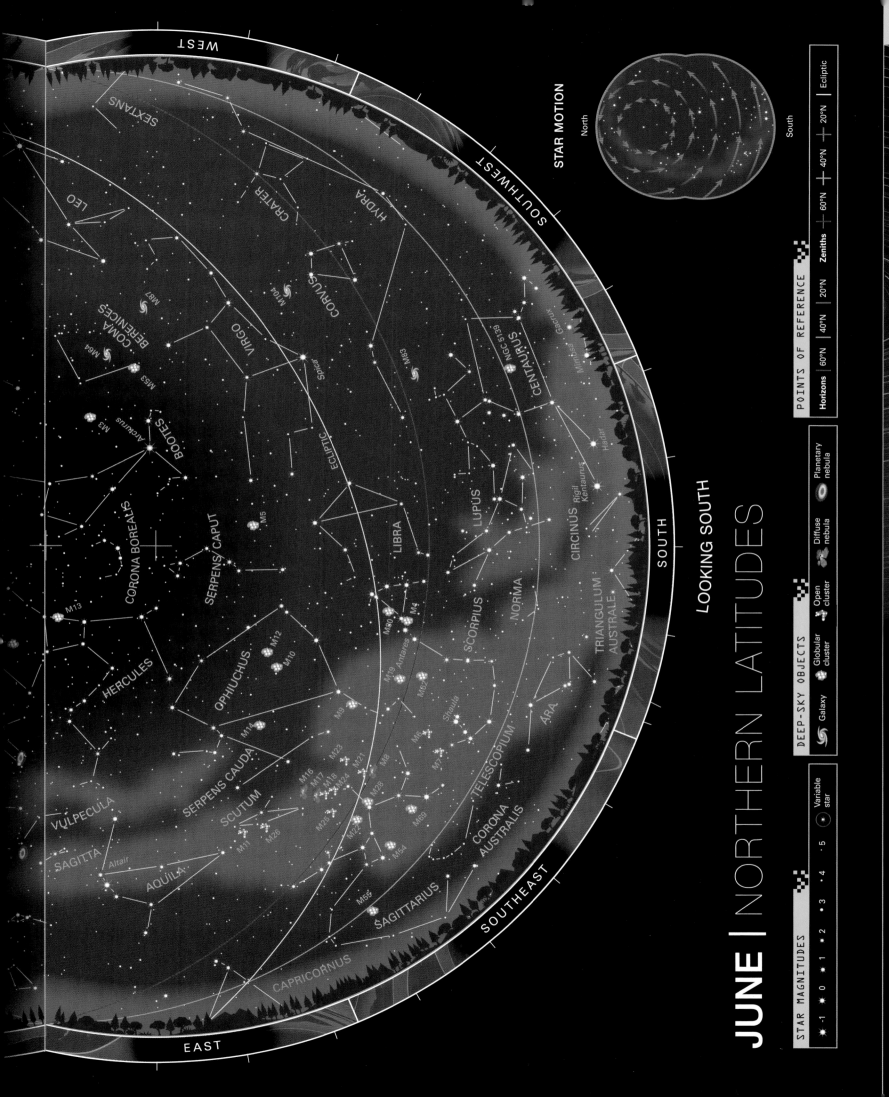

# JUNE | NORTHERN LATITUDES

**LOOKING SOUTH**

**STAR MOTION**

North

South

**POINTS OF REFERENCE**

| Horizons | 60°N | 40°N | 20°N | Zeniths | 60°N | 40°N | 20°N | Ecliptic |

**DEEP-SKY OBJECTS**

| Galaxy | Globular cluster | Open cluster | Diffuse nebula | Planetary nebula |

| Variable star |

**STAR MAGNITUDES**

| -1 | 0 | 1 | 2 | 3 | 4 | 5 |

**THE NIGHT SKY**

WEST

SOUTHWEST

SOUTH

SOUTHEAST

EAST

WEST

SEXTANS

LEO

CRATER

HYDRA

CORVUS

VIRGO

COMA BERENICES

M87

M64

M53

M3

BOÖTES

Arcturus

CORONA BOREALIS

SERPENS CAPUT

M5

M104

ECLIPTIC

Spica

CENTAURUS

NGC 5139

Gacrux

Mimosa

Rigil Kentaurus

Hadar

LIBRA

LUPUS

CIRCINUS

NORMA

M83

HERCULES

M13

OPHIUCHUS

M12

M10

M14

SERPENS CAUDA

SCUTUM

M9

M19  Antaros

M30

M4

M62

Shaula

SCORPIUS

TRIANGULUM AUSTRALE

ARA

VULPECULA

SAGITTA

Altair

AQUILA

M11

M26

M23

M16

M17

M18

M24

M25

M22

M21

M28

M8

M20

M6

M7

M69

TELESCOPIUM

CORONA AUSTRALIS

M54

M55

SAGITTARIUS

CAPRICORNUS

# JUNE | SOUTHERN LATITUDES

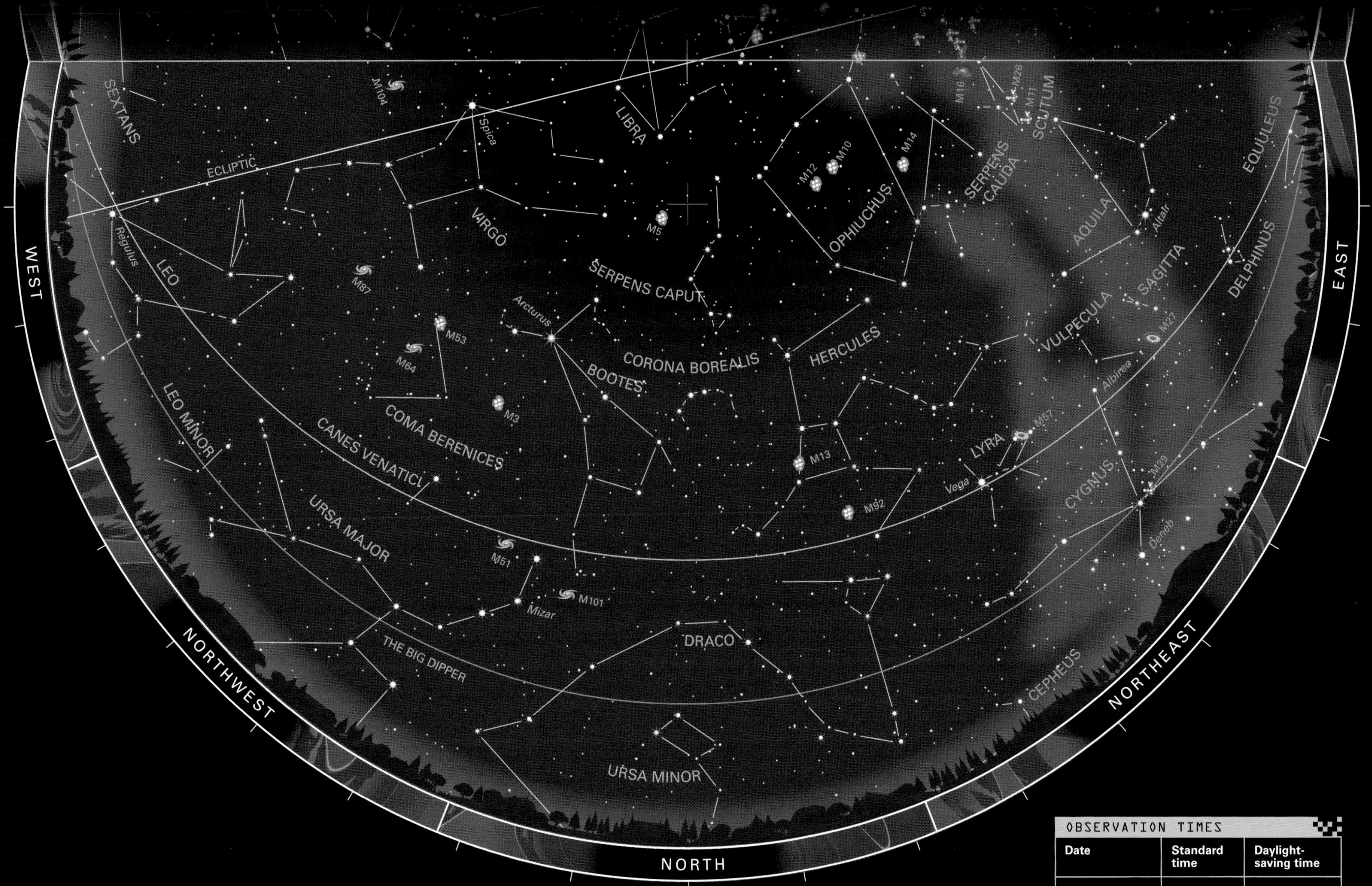

## LOOKING NORTH

### OBSERVATION TIMES

| Date | Standard time | Daylight-saving time |
|------|---------------|----------------------|
| May 15 | Midnight | 1 a.m. |
| June 1 | 11 p.m. | Midnight |
| June 15 | 10 p.m. | 11 p.m. |
| July 1 | 9 p.m. | 10 p.m. |
| July 15 | 8 p.m. | 9 p.m. |

# JUNE | SOUTHERN LATITUDES

## LOOKING SOUTH

THE NIGHT SKY

**STAR MOTION**

North

South

**POINTS OF REFERENCE**

| Horizons | 0° | 20°S | 40°S | Zeniths | 0° | 20°S | 40°S | Ecliptic |

**DEEP-SKY OBJECTS**

Galaxy | Globular cluster | Open cluster | Diffuse nebula | Planetary nebula

**STAR MAGNITUDES**

-1 | 0 | 1 | 2 | 3 | 4 | 5 | Variable star | Ecliptic

# JULY

The strong man of Greek mythology, Hercules, lies overhead as seen from mid-northern latitudes, between the bright stars Vega (in Lyra) and Arcturus (in Boötes). South of Hercules is another large constellation, Ophiuchus, which represents a man encircled by a serpent, Serpens. In southern skies, the Milky Way passes overhead from the southwest to the northeast. The zodiacal constellations Scorpius and Sagittarius stand high in the Milky Way's richest part.

## NORTHERN LATITUDES

### THE STARS

Overhead lies Hercules, which is a large but not particularly striking constellation. Its most distinctive feature is a quadrangle formed by four stars, called the Keystone. North of Hercules lies the lozenge-shaped head of Draco, the Dragon. Between Draco and the north celestial pole is the bowl of the Little Dipper, in Ursa Minor.

Arcturus, in Boötes, remains prominent in the western sky. Spica, in Virgo, is lower in the southwest, and the Big Dipper dips low in the northwest. In the eastern half

of the sky, the stars of the Summer Triangle climb ever higher, while the Square of Pegasus appears closer to the eastern horizon.

Low in the south are the rich constellations Scorpius and Sagittarius. This is the best month for northern observers to see the two most southerly zodiacal figures in the evening sky.

### DEEP-SKY OBJECTS

Ophiuchus, the large constellation between Hercules and Scorpius, contains numerous globular clusters, although only two of them, M10 and M12, are of any note. The most impressive deep-sky objects in Ophiuchus are the open clusters IC 4665 and NGC 6633, both

good binocular sights. The globular clusters M13, in Hercules, and M5, in the head of Serpens, remain well positioned this month.

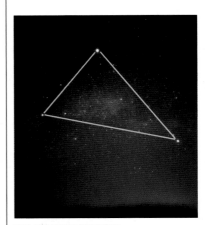

**THE SUMMER TRIANGLE**
Deneb (left), Vega (top), and Altair (right) form a prominent triangle that remains visible well into fall in northern skies.

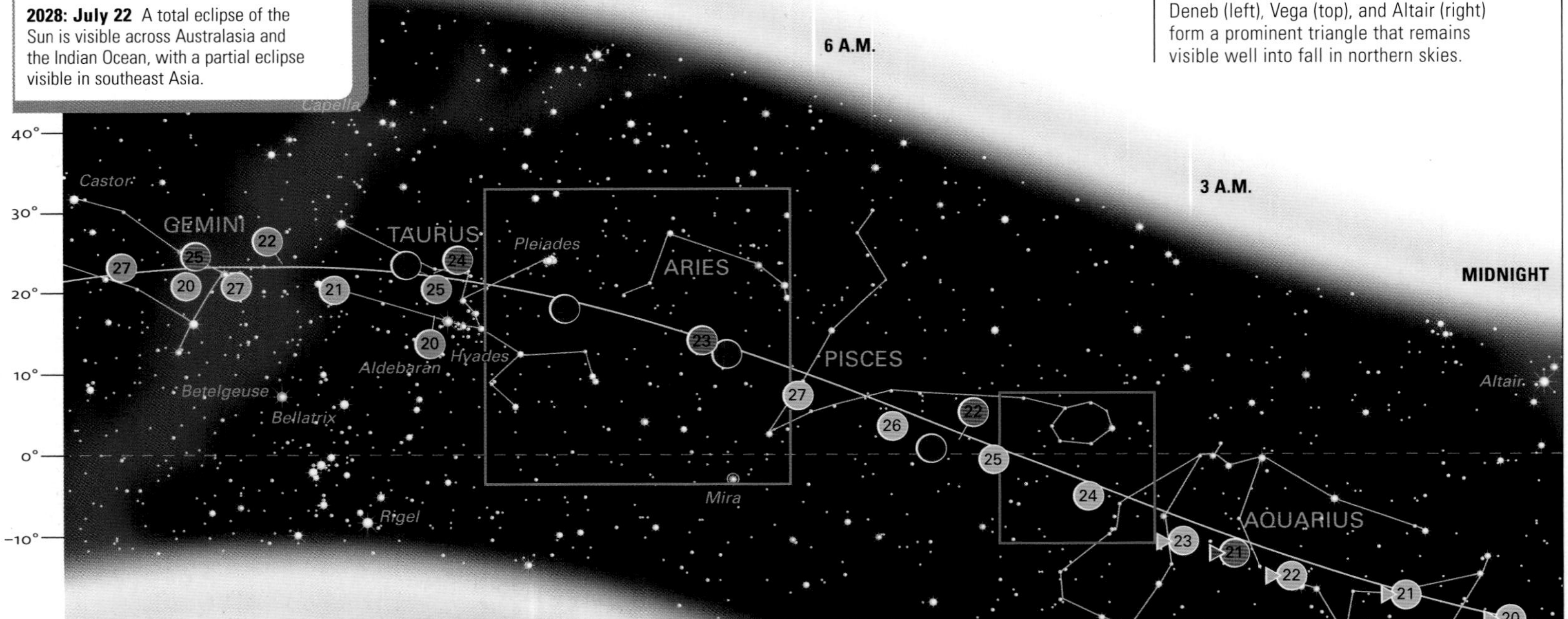

9 A.M.

6 A.M.

3 A.M.

MIDNIGHT

40°

30°

20°

10°

0°

–10°

Castor

GEMINI

27 25 22

20 27

Capella

TAURUS

24

25

21

20

Pleiades

Aldebaran

Hyades

ARIES

23

Betelgeuse

Bellatrix

Rigel

Mira

PISCES

27

26

22

25

24

AQUARIUS

23

21

22

21

20

Altair

Fomalhaut

CAPRICORNUS

SAGITTARIUS

MORNING SKY

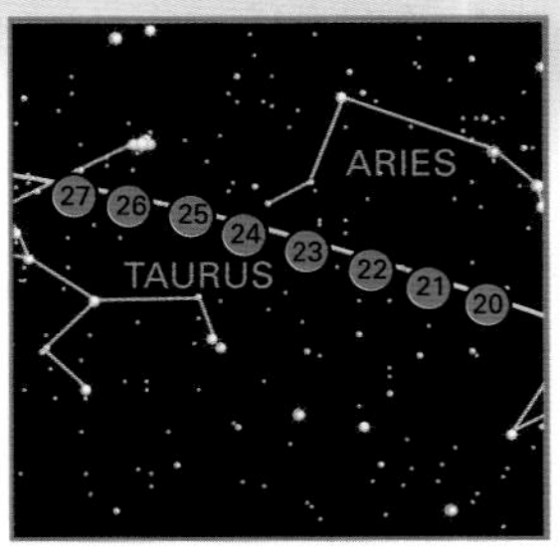

URANUS

27 26 25 24 23 22 21 20
ARIES
TAURUS

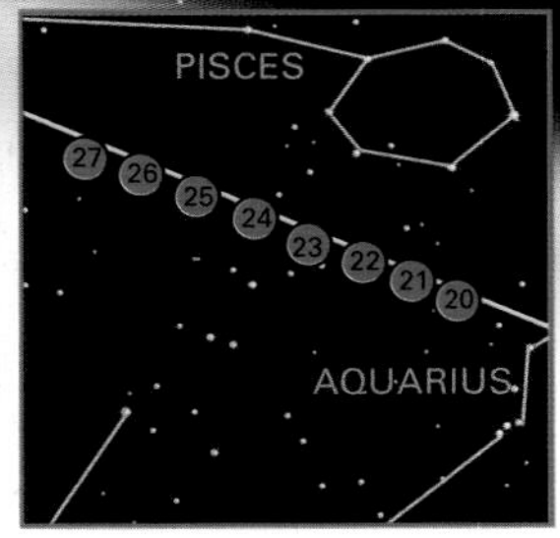

NEPTUNE

PISCES

27 26 25 24 23 22 21 20

AQUARIUS

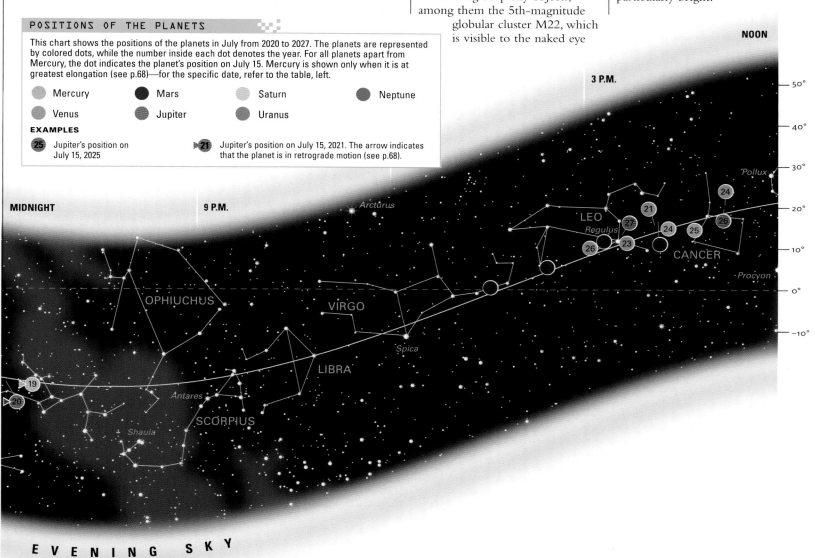

**TOWARD THE CENTER OF THE GALAXY**
The center of the galaxy cannot be seen directly, because it is obscured behind the dense Milky Way star fields of Sagittarius and Scorpius. The exact center is thought to be marked by an intense radio source called Sagittarius A* (boxed).

# SOUTHERN LATITUDES

## THE STARS

The curved tail of Scorpius and the asterism known as the Teapot, formed from the main stars of Sagittarius, are virtually overhead for southern observers. The Milky Way is particularly dense and bright toward Sagittarius and Scorpius because this is the view toward the center of the galaxy.

Alpha (α) and Beta (β) Centauri—Rigil Kentaurus (see p.252) and Hadar—are in the southwest, pointing down to Crux, the Southern Cross. Spica (in Virgo) is in the eastern sky; Arcturus (in Boötes) is in the northwest; and Vega (see p.253), in Lyra, is in the north. Altair (see p.252), in Aquila, is high in the northeast, and observers about 30°S or closer to the equator can see Deneb, in Cygnus, low in the northeast. In the southeast, the 1st-magnitude Fomalhaut, in Piscis Austrinus, enters the scene.

## DEEP-SKY OBJECTS

Sagittarius is well stocked with outstanding deep-sky objects, among them the 5th-magnitude globular cluster M22, which is visible to the naked eye under good conditions. The Lagoon Nebula (see p.243), or M8, an elongated gas cloud containing the star cluster NGC 6530, can be seen well through binoculars. To the north, in Serpens Cauda, the tail of the Serpent, lies the cluster M16—visible through binoculars—embedded in the much fainter Eagle Nebula (see pp.244–245).

Other famous deep-sky objects in Sagittarius, such as the Trifid Nebula, M20 (see p.246), need to be seen through a telescope. However, one particularly bright patch of the Milky Way, M24, is prominent to the naked eye. In adjoining Scorpius, the bright open clusters M6 and M7 remain high in the sky.

## METEOR SHOWER

The Delta Aquariids, the best southern meteor shower, is active in July and August, reaching a peak around July 29. At best, perhaps 20 meteors an hour can be seen radiating from the southern half of Aquarius, but they are not particularly bright.

## POSITIONS OF THE PLANETS

This chart shows the positions of the planets in July from 2020 to 2027. The planets are represented by colored dots, while the number inside each dot denotes the year. For all planets apart from Mercury, the dot indicates the planet's position on July 15. Mercury is shown only when it is at greatest elongation (see p.68)—for the specific date, refer to the table, left.

- Mercury
- Venus
- Mars
- Jupiter
- Saturn
- Uranus
- Neptune

**EXAMPLES**

25 Jupiter's position on July 15, 2025

21 Jupiter's position on July 15, 2021. The arrow indicates that the planet is in retrograde motion (see p.68).

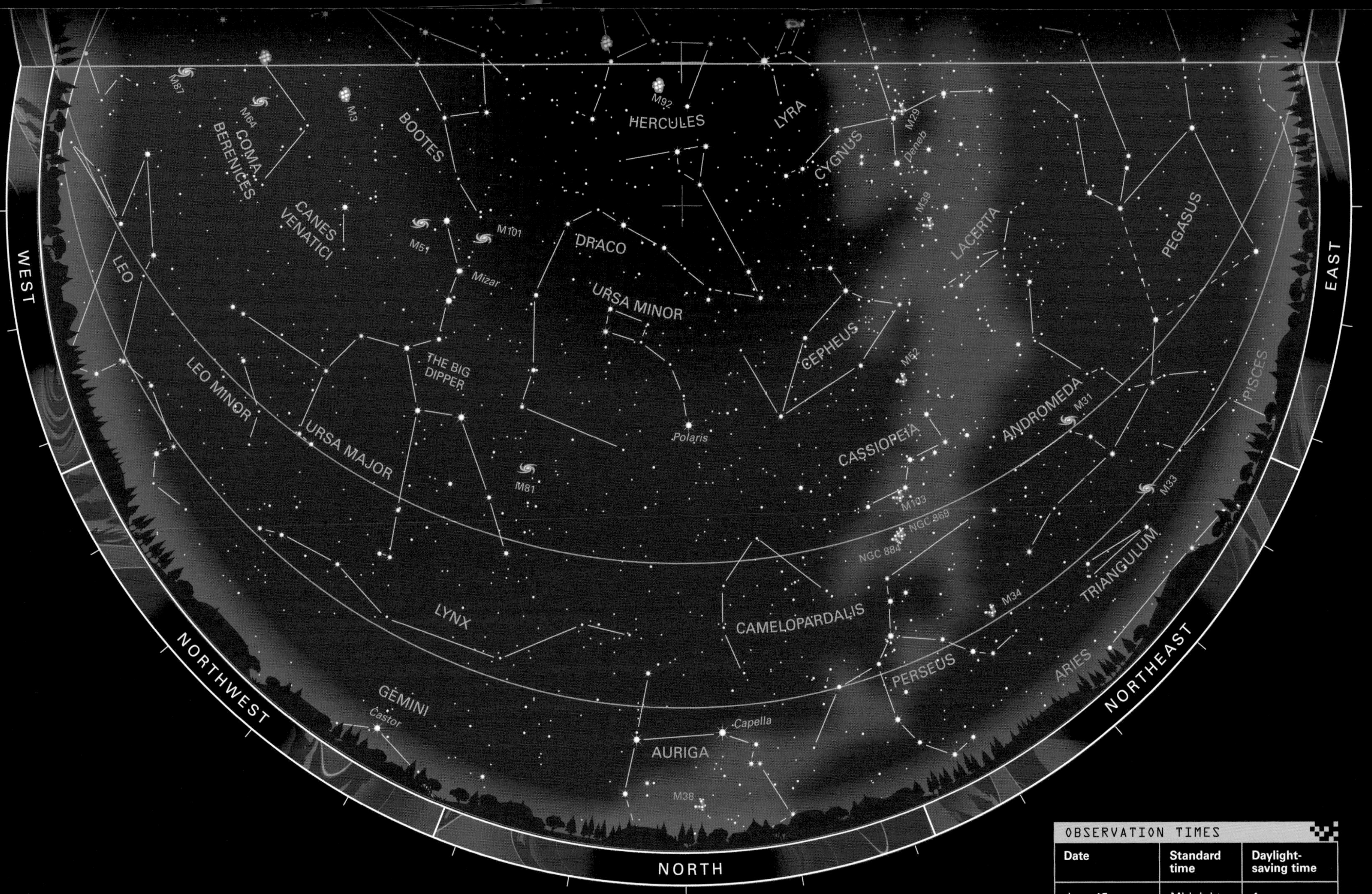

# JULY | NORTHERN LATITUDES

## LOOKING NORTH

### STAR MAGNITUDES

| | |
|---|---|
| ★ | -1 |
| ★ | 0 |
| ⋆ | 1 |
| · | 2 |
| · | 3 |
| · | 4 |
| · | 5 |
| ⊙ | Variable star |

### DEEP-SKY OBJECTS

| | |
|---|---|
| 🌀 | Galaxy |
| ✦ | Globular cluster |
| ✧ | Open cluster |
| ✿ | Diffuse nebula |
| ◉ | Planetary nebula |

### POINTS OF REFERENCE

**Horizons**

| 60°N | 40°N | 20°N |

**Zeniths**

| + 60°N | + 40°N | + 20°N | Ecliptic |

### OBSERVATION TIMES

| Date | Standard time | Daylight-saving time |
|---|---|---|
| June 15 | Midnight | 1 a.m. |
| July 1 | 11 p.m. | Midnight |
| July 15 | 10 p.m. | 11 p.m. |
| August 1 | 9 p.m. | 10 p.m. |
| August 15 | 8 p.m. | 9 p.m. |

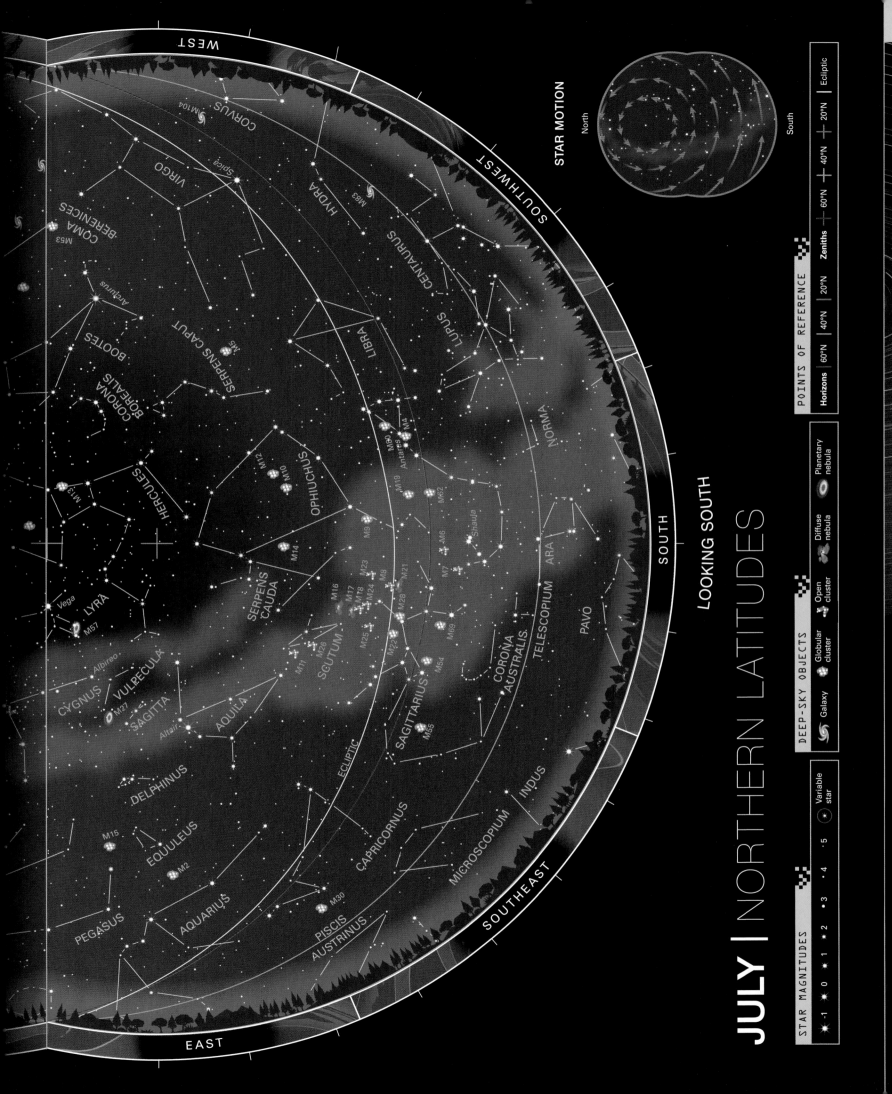

# JULY | NORTHERN LATITUDES

## STAR MOTION

North

South

## POINTS OF REFERENCE

Horizons | 60°N | 40°N | 20°N

Zeniths | 60°N | 40°N | 20°N

## DEEP-SKY OBJECTS

Galaxy | Globular cluster | Open cluster | Diffuse nebula | Planetary nebula

## STAR MAGNITUDES

-1 | 0 | 1 | 2 | 3 | 4 | 5

Variable star

Ecliptic

LOOKING SOUTH

THE NIGHT SKY

WEST

EAST

SOUTHWEST

SOUTH

SOUTHEAST

WEST

COMA BERENICES

M63

VIRGO

Spica

M104

CORVUS

HYDRA

M83

CENTAURUS

LUPUS

NORMA

ARA

BOOTES

Arcturus

CORONA BOREALIS

SERPENS CAPUT

M5

LIBRA

HERCULES

M13

OPHIUCHUS

M12

M10

M14

M90

M4

Antares

M19

M62

Shaula

M6

M7

TELESCOPIUM

PAVO

INDUS

LYRA

Vega

M57

SERPENS CAUDA

M9

M23

M16

M17

M18

M24

M8

M21

M28

M54

SAGITTARIUS

CORONA AUSTRALIS

MICROSCOPIUM

CYGNUS

Albireo

VULPECULA

M27

SAGITTA

Altair

AQUILA

SCUTUM

M11

M26

M25

M22

M69

M55

ECLIPTIC

DELPHINUS

CAPRICORNUS

EQUULEUS

M15

M2

AQUARIUS

M30

PEGASUS

PISCIS AUSTRINUS

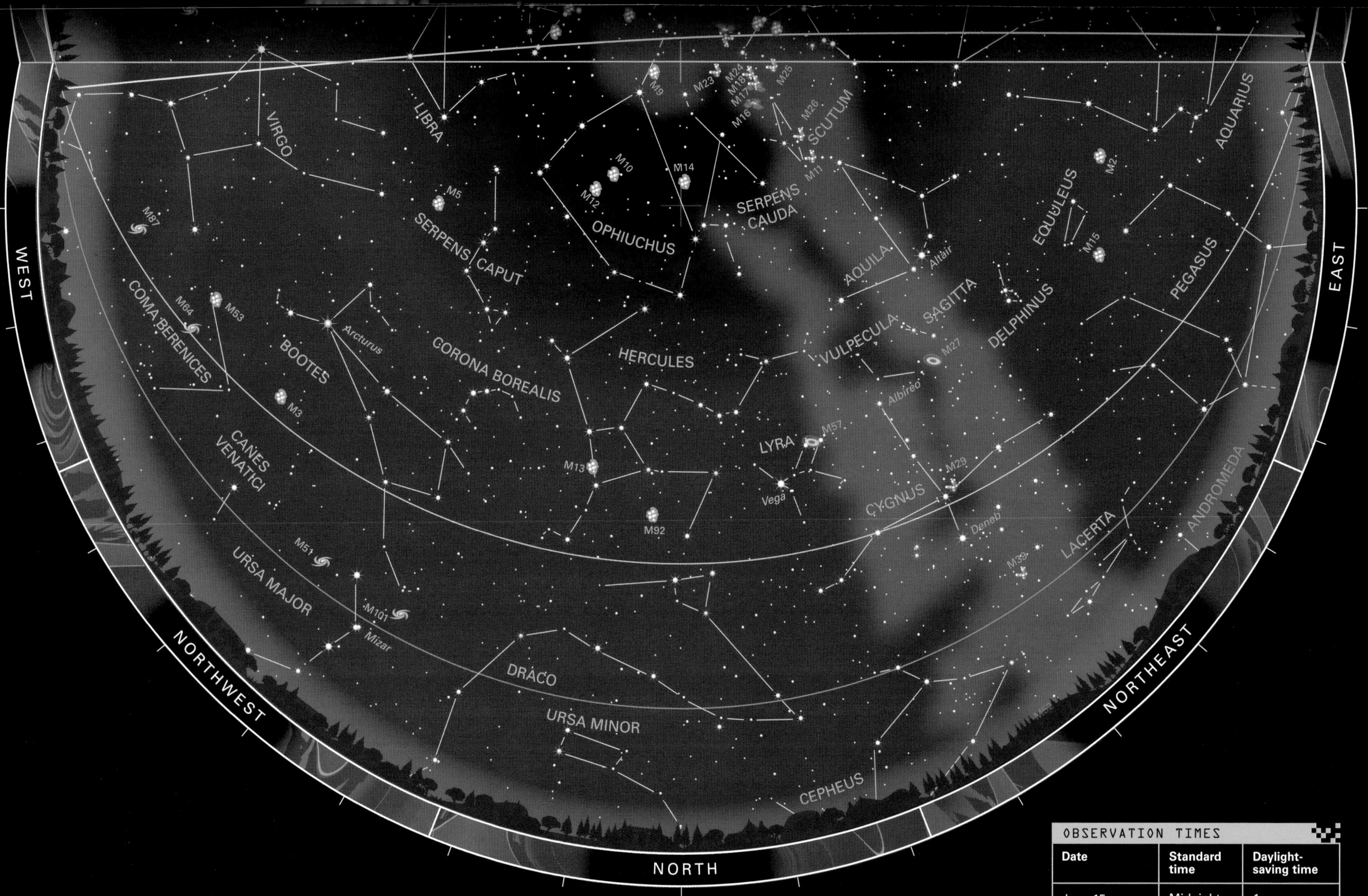

# JULY | SOUTHERN LATITUDES

## LOOKING NORTH

**STAR MAGNITUDES**

- ·-1
- ·0
- ·1
- ·2
- ·3
- ·4
- ·5
- Variable star

**DEEP-SKY OBJECTS**

- Galaxy
- Globular cluster
- Open cluster
- Diffuse nebula
- Planetary nebula

**POINTS OF REFERENCE**

| Horizons | 0° | 20°S | 40°S | Zeniths |
|---|---|---|---|---|
| | | | | 0° |
| | | | | 20°S |
| | | | | 40°S |
| | | | | Ecliptic |

**OBSERVATION TIMES**

| Date | Standard time | Daylight-saving time |
|---|---|---|
| June 15 | Midnight | 1 a.m. |
| July 1 | 11 p.m. | Midnight |
| July 15 | 10 p.m. | 11 p.m. |
| August 1 | 9 p.m. | 10 p.m. |
| August 15 | 8 p.m. | 9 p.m. |

WEST

NORTHWEST

NORTH

NORTHEAST

EAST

VIRGO
LIBRA
COMA BERENICES
M87
M64
M53
CANES VENATICI
M3
BOOTES
Arcturus
SERPENS CAPUT
M5
URSA MAJOR
M51
M101
Mizar
DRACO
CORONA BOREALIS
OPHIUCHUS
M12
M10
M9
M14
SERPENS CAUDA
M23
M24
M18
M17
M16
M25
M26
HERCULES
M13
M92
URSA MINOR
SCUTUM
M11
CEPHEUS
LYRA
Vega
M57
VULPECULA
AQUILA
Altair
CYGNUS
Albireo
M29
SAGITTA
M27
Deneb
DELPHINUS
M39
LACERTA
EQUULEUS
PEGASUS
M15
M2
ANDROMEDA
AQUARIUS

# JULY | SOUTHERN LATITUDES

LOOKING SOUTH

STAR MOTION

North

South

**POINTS OF REFERENCE**

Horizons | 0° | 20°S | 40°S | Zeniths | 0° | 20°S | 40°S | Ecliptic

**DEEP-SKY OBJECTS**

Galaxy | Globular cluster | Open cluster | Diffuse nebula | Planetary nebula

**STAR MAGNITUDES**

-1 | 0 | 1 | 2 | 3 | 4 | 5 | Variable star

THE NIGHT SKY

## Constellations and objects

WEST
SOUTHWEST
SOUTH
SOUTHEAST
EAST

VIRGO
Spica
M104
CORVUS
CRATER
HYDRA
ANTLIA
M83
LIBRA
CENTAURUS
VELA
NGC 5139
Mimosa
Gacrux
CRUX
Acrux
Hadar
Rigil Kentaurus
CIRCINUS
MUSCA
CARINA
LUPUS
NORMA
TRIANGULUM AUSTRALE
APUS
CHAMAELEON
VOLANS
PICTOR
Canopus
ARA
SCORPIUS
M80
M4
Antares
M62
M19
Shaula
M6
M7
CORONA AUSTRALIS
TELESCOPIUM
PAVO
OCTANS
MENSA
DORADO
LMC
RETICULUM
M21
M8
M28
M69
M54
M22
INDUS
TUCANA
HYDRUS
SMC
NGC 104
Achernar
HOROLOGIUM
ERIDANUS
M55
SAGITTARIUS
MICROSCOPIUM
GRUS
PHOENIX
SCULPTOR
CAPRICORNUS
M30
PISCIS AUSTRINUS
Fomalhaut
AQUARIUS
ECLIPTIC

# AUGUST

The Summer Triangle formed by the bright stars Vega (in Lyra), Deneb (in Cygnus), and Altair (in Aquila) lies on the north–south celestial meridian in the northern sky this month. The cross-shaped figure of Cygnus, the swan, stands out against the background of the Milky Way, which passes overhead in mid-northern latitudes. In the southern sky, the rich Milky Way star fields in Sagittarius and Scorpius, toward the center of the galaxy, remain well placed for observation.

## SPECIAL EVENTS

### PHASES OF THE MOON

| | FULL MOON | NEW MOON |
|---|---|---|
| **2020** | August 3 | August 19 |
| **2021** | August 22 | August 8 |
| **2022** | August 12 | August 27 |
| **2023** | August 1, 31 | August 16 |
| **2024** | August 19 | August 4 |
| **2025** | August 9 | August 23 |
| **2026** | August 28 | August 12 |
| **2027** | August 17 | August 2, 31 |

### PLANETS

**2020: August 13** Venus is at greatest morning elongation, magnitude –4.3.

**2021: August 2** Saturn is at opposition, magnitude 0.2.

**2021: August 9** Jupiter is at opposition, magnitude –2.9.

**2022: August 14** Saturn is at opposition, magnitude 0.3.

**2022: August 10** Mercury is at greatest evening elongation, magnitude 0.5.

**2023: August 26** Mercury is at greatest evening elongation, magnitude 0.6.

**2023: August 27** Saturn is at opposition, magnitude 0.4.

**2025: August 19** Mercury is at greatest morning elongation, magnitude 0.2.

**2026: August 2** Mercury is at greatest morning elongation, magnitude 0.5.

**2026: August 15** Venus is at greatest evening elongation, magnitude –4.0.

### ECLIPSES

**2026: August 12** A total eclipse of the Sun is visible from northern Spain, the North Atlantic, Greenland, Iceland, and the Arctic, with a partial eclipse visible in northwest Africa, Europe, and northern North America.

**2026: August 27–28** A partial eclipse of the Moon is visible across the Americas, Europe, Africa, the Atlantic and Pacific Oceans, and Antarctica.

## NORTHERN LATITUDES

### THE STARS

Blue-white Vega (see p.253), in the constellation Lyra, is the first bright star to appear overhead as the sky darkens on August evenings. Next to Lyra is Cygnus, popularly known as the Northern Cross. The star at the head of Cygnus, Albireo, is a beautifully colored double star, easily divided by the smallest of telescopes. South of Cygnus is Aquila, the Eagle, from where the Milky Way continues, via Scutum, toward Sagittarius and Scorpius in the southwest. Hercules and Ophiuchus remain well placed in the southwest, and Arcturus, in Boötes, is lower in the west. In the east, the Square of Pegasus leads the stars of fall into view.

### DEEP-SKY OBJECTS

The August skies are stocked with deep-sky objects for northern observers. The Milky Way is divided by a dark dust cloud known as the Cygnus Rift, which extends southwestward from Cygnus into Ophiuchus. South of Cygnus, in the obscure constellation Vulpecula, is the planetary nebula M27, popularly known as the Dumbbell, the easiest such object to see through binoculars. Another celebrated planetary nebula, the Ring Nebula (see p.257) or M27, in Lyra, can be found with a telescope. The Wild Duck Cluster, or M11, in Scutum is a 6th-magnitude open cluster visible through binoculars.

### METEOR SHOWER

The year's top meteor shower, the Perseids, reaches a peak around August 12, although some activity can be seen for a week or so on either side of this date. Perseid meteors are bright: at best, an average of one a minute can be seen streaking away from northern Perseus. Most Perseids are seen after midnight, because Perseus does not rise high before then.

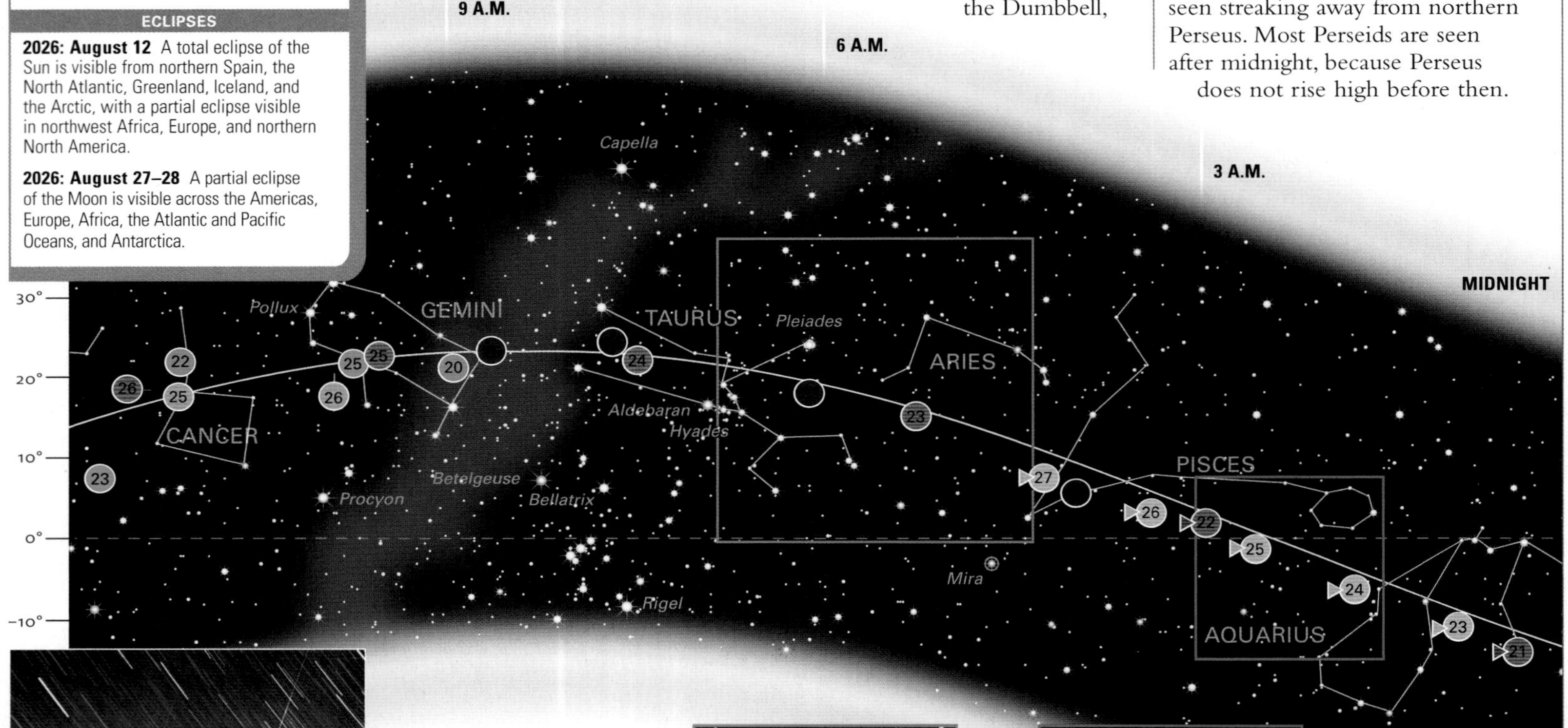

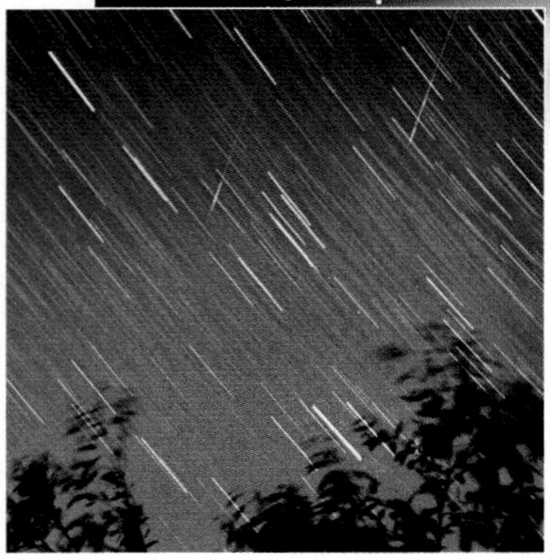

**PERSEID METEORS**
Mild nights in mid-August are ideal for lying outside and watching members of the Perseid meteor shower flash across the northern sky.

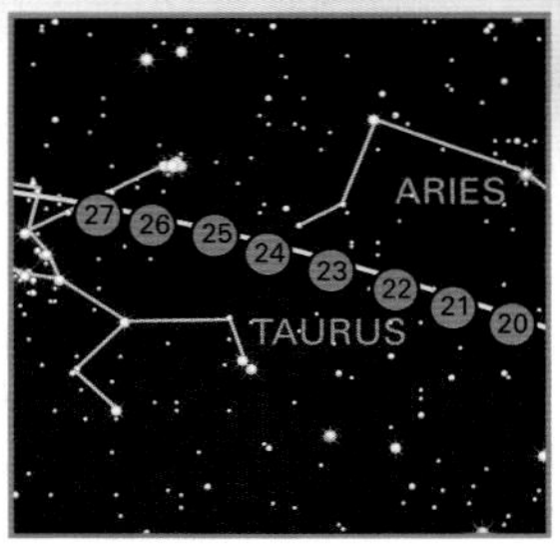

**URANUS**

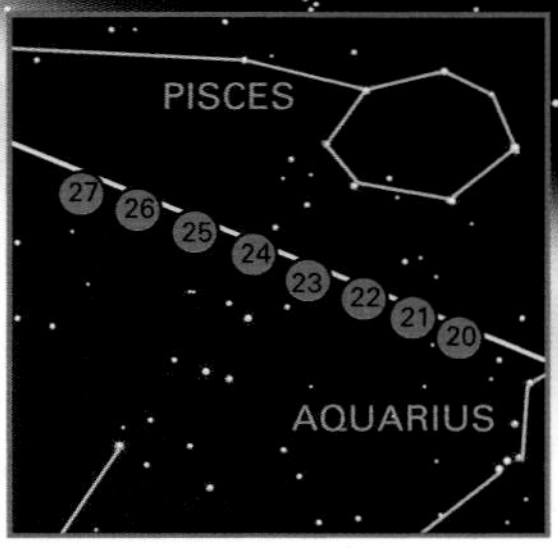

**NEPTUNE**

# SOUTHERN LATITUDES

## THE STARS

Sagittarius and its Milky Way star fields remain high overhead, with Scorpius to the southwest of it. Alpha (α) and Beta (β) Centauri—Rigil Kentaurus (see p.252) and Hadar—are low on the southwestern horizon. To the north are Altair (in Aquila), Vega (in Lyra), and Deneb (in Cygnus), the stars that form the northern Summer Triangle—this is the best time of year to see them in the evening sky from southern latitudes. The Square of Pegasus is rising in the northwest. Fomalhaut, in the constellation Piscis Austrinus, is high in the east, with Achernar, in Eridanus, lower in the southeast. The Small Magellanic Cloud (see p.311) is visible midway between Achernar and the south celestial pole.

celestial meridian earlier in the year, such as the Lagoon Nebula (see p.243), M22 in Sagittarius, M16 in Serpens Cauda, and M6 and M7 in Scorpius. In addition, this month southern observers can see the Wild Duck Cluster (M11) in Scutum and, looking north of the equator, the Dumbbell Nebula (M27) in Vulpecula and the Ring Nebula (M57) in Lyra (see p.257).

## DEEP-SKY OBJECTS

The best deep-sky objects to view in the southern sky on August evenings are those that passed the

**THE LAGOON NEBULA IN SAGITTARIUS**
Among the dense star fields of the Milky Way lies the Lagoon Nebula (bottom right), also known as M8, in Sagittarius (right).

**SAGITTARIUS**
The Teapot asterism (bottom), formed by eight stars in Sagittarius, is a familiar pattern in summer skies.

### POSITIONS OF THE PLANETS

This chart shows the positions of the planets in August from 2020 to 2027. The planets are represented by colored dots, while the number inside each dot denotes the year. For all planets apart from Mercury, the dot indicates the planet's position on August 15. Mercury is shown only when it is at greatest elongation (see p.68)—for the specific date, refer to the table, left.

- ⬤ Mercury
- ⬤ Venus
- ⬤ Mars
- ⬤ Jupiter
- ⬤ Saturn
- ⬤ Uranus
- ⬤ Neptune

**EXAMPLES**

㉓ Jupiter's position on August 15, 2023

▶⑳ Jupiter's position on August 15, 2020. The arrow indicates that the planet is in retrograde motion (see p.68).

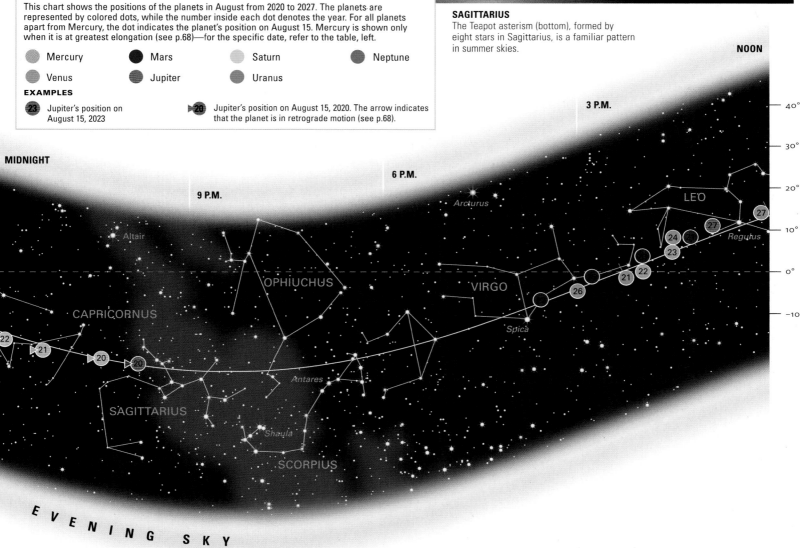

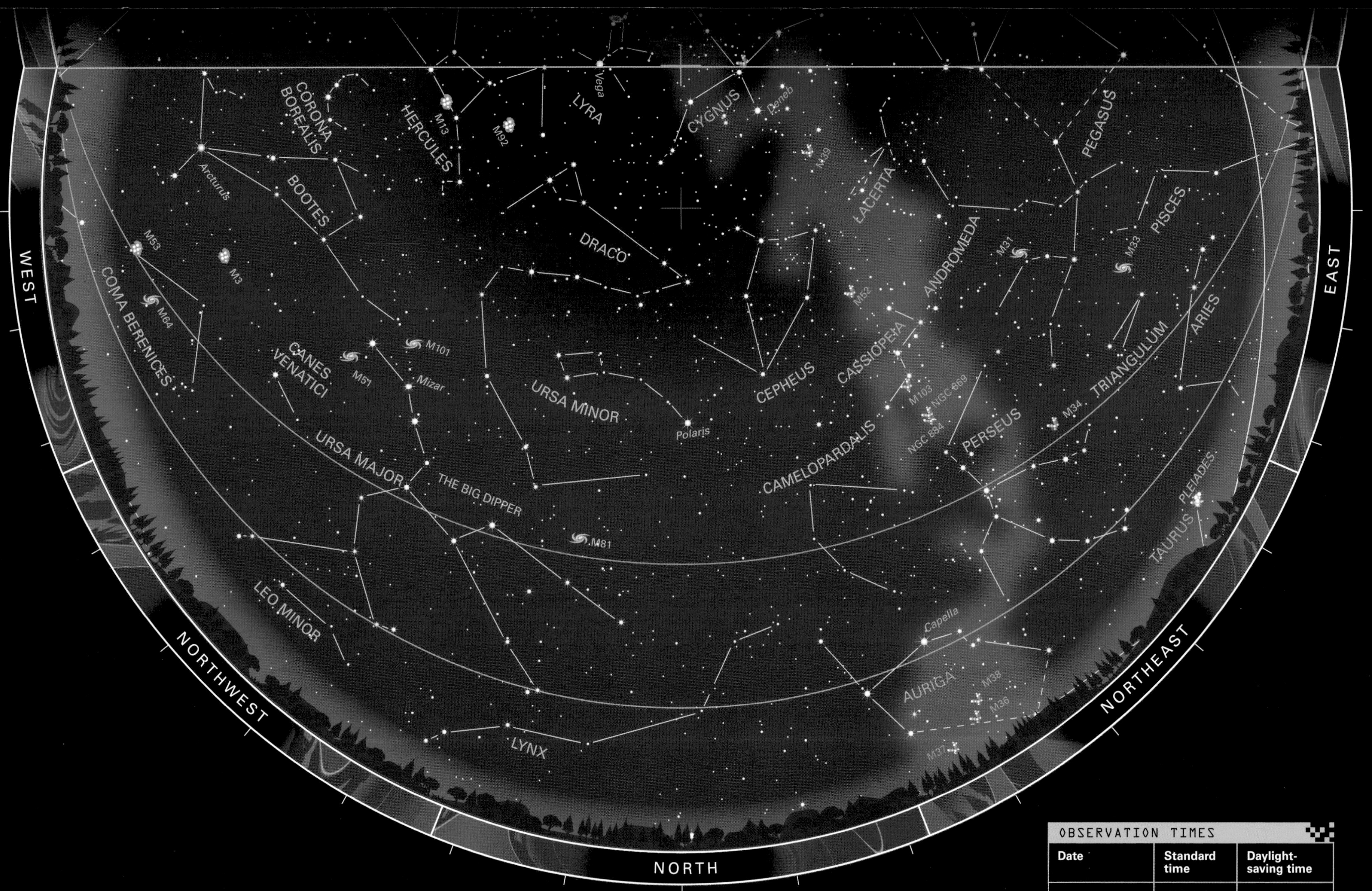

# AUGUST | NORTHERN LATITUDES

## LOOKING NORTH

**STAR MAGNITUDES**

| | | | | | | | |
|---|---|---|---|---|---|---|---|
| -1 | 0 | 1 | 2 | 3 | 4 | 5 | Variable star |

**DEEP-SKY OBJECTS**

- Galaxy
- Globular cluster
- Open cluster
- Diffuse nebula
- Planetary nebula

**POINTS OF REFERENCE**

| **Horizons** | 60°N | 40°N | 20°N |
| --- | --- | --- | --- |
| **Zeniths** | 60°N | 40°N | 20°N |
| | | | Ecliptic |

**OBSERVATION TIMES**

| Date | Standard time | Daylight-saving time |
| --- | --- | --- |
| July 15 | Midnight | 1 a.m. |
| August 1 | 11 p.m. | Midnight |
| August 15 | 10 p.m. | 11 p.m. |
| September 1 | 9 p.m. | 10 p.m. |
| September 15 | 8 p.m. | 9 p.m. |

WEST

NORTHWEST

COMA BERENICES

M64

M53

M63

Arcturus

M3

CORONA BOREALIS

BOÖTES

CANES VENATICI

M51

URSA MAJOR

Mizar

M101

THE BIG DIPPER

LEO MINOR

LYNX

HERCULES

M13

M92

URSA MINOR

M81

DRACO

Polaris

LYRA

Vega

CEPHEUS

CYGNUS

Deneb

CAMELOPARDALIS

CASSIOPEIA

M39

M52

LACERTA

NGC 884

M103

NGC 869

ANDROMEDA

AURIGA

Capella

PERSEUS

M31

M38

M36

M34

TRIANGULUM

M33

PEGASUS

PISCES

ARIES

TAURUS

PLEIADES

NORTHEAST

NORTH

NORTHWEST

EAST

WEST

# AUGUST | NORTHERN LATITUDES

**LOOKING SOUTH**

**STAR MOTION**

North

South

## POINTS OF REFERENCE

| Horizons | 60°N | 40°N | 20°N | Zeniths | 60°N | 40°N | 20°N | Ecliptic |
|----------|------|------|------|---------|------|------|------|----------|

## DEEP-SKY OBJECTS

Galaxy | Globular cluster | Open cluster | Diffuse nebula | Planetary nebula

## STAR MAGNITUDES

-1 | 0 | 1 | 2 | 3 | 4 | 5 | Variable star

**THE NIGHT SKY**

WEST

SOUTHWEST

SOUTH

SOUTHEAST

EAST

VIRGO
M5
SERPENS CAPUT
LIBRA
OPHIUCHUS
M12
M10
HERCULES
M14
SERPENS CAUDA
M23
M16
M17
M18
M24
M19
M62
M80
M4
Antares
Shaula
SCORPIUS
LUPUS
NORMA
ARA
M6
M7
M8
M21
M25
M20
M28
M22
M9
M54
M69
SAGITTARIUS
CORONA AUSTRALIS
TELESCOPIUM
SCUTUM
M11
M26
AQUILA
Altair
Ecliptic
M55
PAVO
INDUS
MICROSCOPIUM
GRUS
LYRA
M57
VULPECULA
M71
Albireo
SAGITTA
M27
DELPHINUS
CYGNUS
M29
EQUULEUS
M2
M15
CAPRICORNUS
PISCIS AUSTRINUS
M30
Fomalhaut
SCULPTOR
PHOENIX
PEGASUS
AQUARIUS
PISCES
CETUS

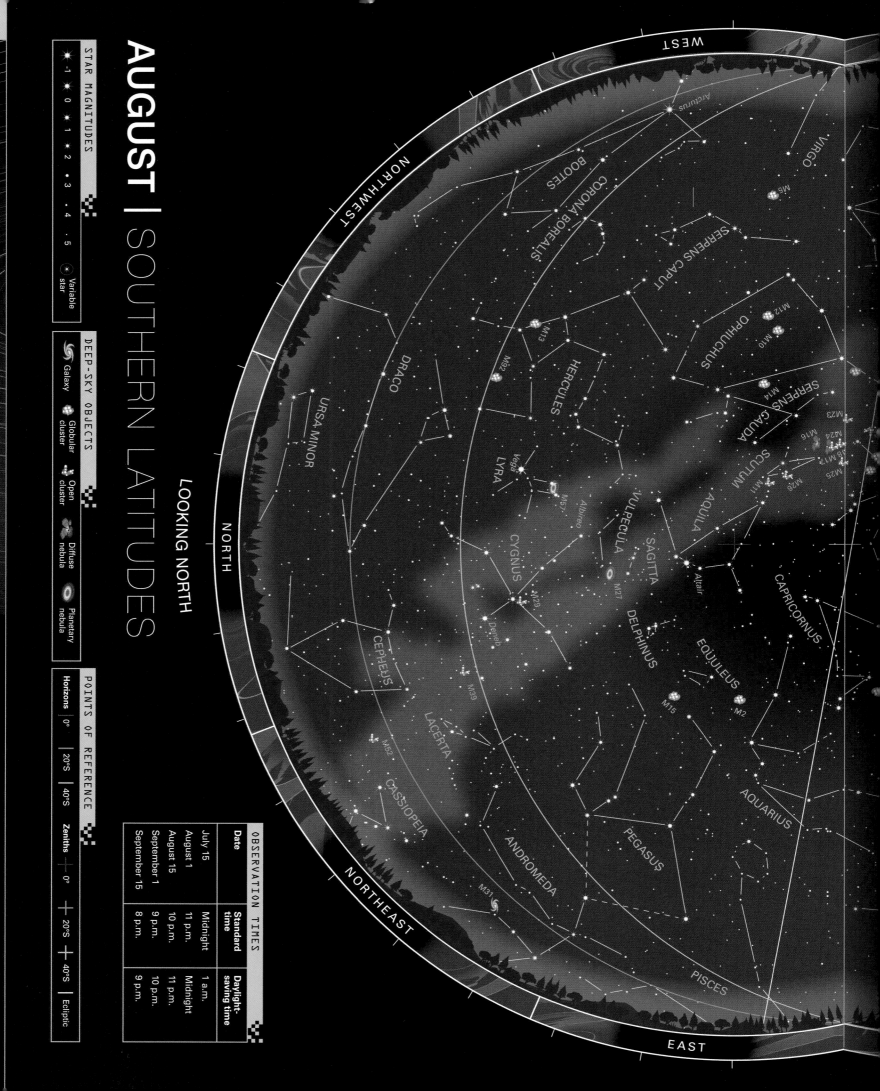

# AUGUST | SOUTHERN LATITUDES

LOOKING NORTH

OBSERVATION TIMES

| Date | Standard time | Daylight-saving time |
|---|---|---|
| July 15 | Midnight | 1 a.m. |
| August 1 | 11 p.m. | Midnight |
| August 15 | 10 p.m. | 11 p.m. |
| September 1 | 9 p.m. | 10 p.m. |
| September 15 | 8 p.m. | 9 p.m. |

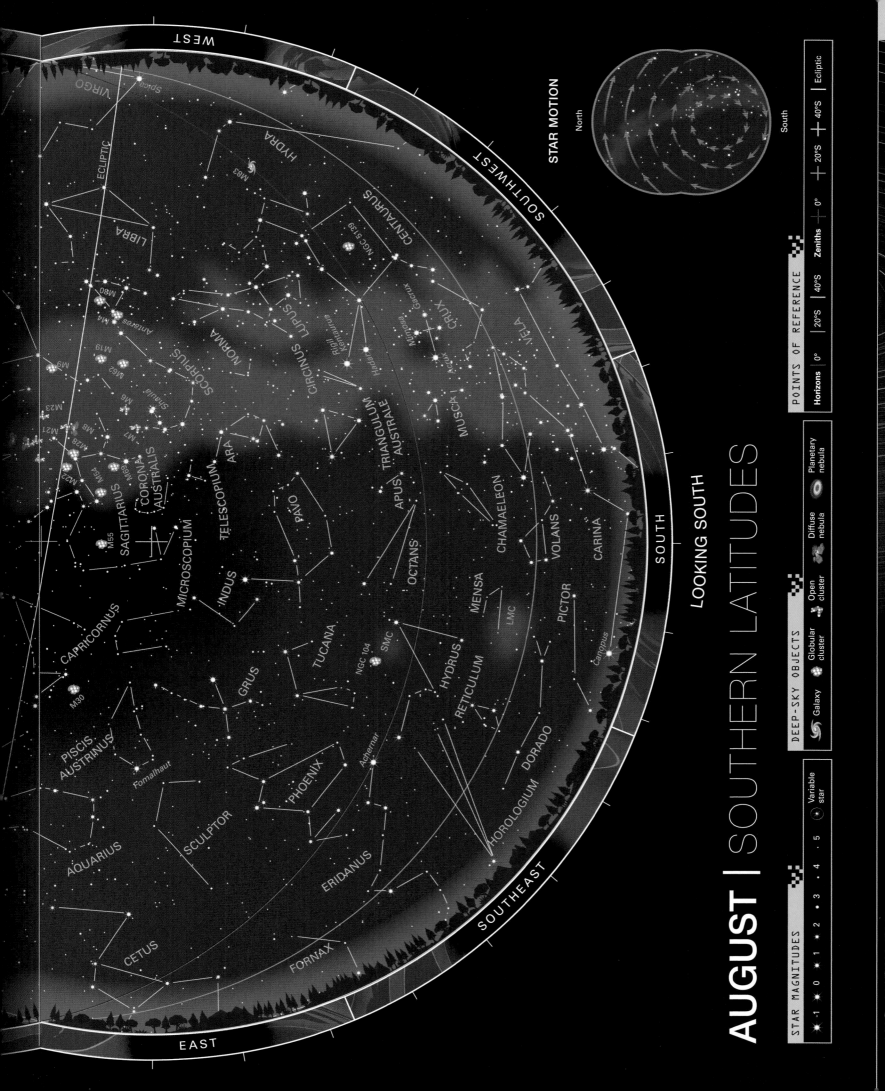

# AUGUST | SOUTHERN LATITUDES

LOOKING SOUTH

THE NIGHT SKY

STAR MOTION

North

South

**POINTS OF REFERENCE**

| Horizons | 0° | 20°S | 40°S | Zeniths | 0° | 20°S | 40°S |

**DEEP-SKY OBJECTS**

Galaxy | Globular cluster | Open cluster | Diffuse nebula | Planetary nebula

**STAR MAGNITUDES**

-1 | 0 | 1 | 2 | 3 | 4 | 5 | Variable star | Ecliptic

WEST

SOUTHWEST

SOUTH

SOUTHEAST

EAST

WEST

VIRGO
Spica
LIBRA
ECLIPTIC
HYDRA
M83
CENTAURUS
NGC 5139
LUPUS
NORMA
CIRCINUS
Rigil Kentaurus
Hadar
CRUX
Mimosa
Acrux
Gacrux
MUSCA
VELA
SCORPIUS
Antares
M4
M80
M19
M62
Shaula
M6
M7
M23
M21
M28
M8
M22
M54
M69
ARA
TELESCOPIUM
TRIANGULUM AUSTRALE
APUS
CHAMAELEON
CARINA
Canopus
SAGITTARIUS
CORONA AUSTRALIS
M55
MICROSCOPIUM
INDUS
PAVO
OCTANS
MENSA
LMC
VOLANS
PICTOR
CAPRICORNUS
GRUS
TUCANA
NGC 104
SMC
HYDRUS
RETICULUM
DORADO
HOROLOGIUM
M30
PISCIS AUSTRINUS
Fomalhaut
SCULPTOR
PHOENIX
Achernar
ERIDANUS
AQUARIUS
CETUS
FORNAX

## SPECIAL EVENTS

### PHASES OF THE MOON

| | FULL MOON | NEW MOON |
|---|---|---|
| 2020 | September 2 | September 17 |
| 2021 | September 20 | September 7 |
| 2022 | September 10 | September 25 |
| 2023 | September 29 | September 15 |
| 2024 | September 18 | September 3 |
| 2025 | September 7 | September 21 |
| 2026 | September 26 | September 11 |
| 2027 | September 16 | September 30 |

### PLANETS

**2021: September 14** Mercury is at greatest evening elongation, magnitude 0.4.

**2022: September 26** Jupiter is at opposition, magnitude −2.9.

**2023: September 2** Mercury is at greatest morning elongation, magnitude −0.2.

**2024: September 5** Mercury is at greatest morning elongation, magnitude 0.0.

**2024: September 8** Saturn is at opposition, magnitude 0.6.

**2025: September 21** Saturn is at opposition, magnitude 0.6.

**2027: September 24** Mercury is at greatest evening elongation, magnitude 0.3.

### ECLIPSES

**2024: September 18** A partial eclipse of the Moon is visible from North and South America, Europe, and Africa.

**2025: September 7** A total eclipse of the Moon is visible from Europe, Africa, Asia, and Australia.

**2025: September 21** A partial eclipse of the Sun is visible from the south Pacific, New Zealand, and Antarctica.

# SEPTEMBER

Northern nights grow longer as the Sun approaches the celestial equator, but in the Southern Hemisphere, the nights shorten. On September 22–23, the Sun lies on the celestial equator, and day and night are of equal length worldwide. The rich band of constellations along the Milky Way, from Cygnus in the north to Sagittarius and Scorpius in the south, begins to give way this month to fainter constellations, many of them with watery associations, such as Capricornus, Aquarius, and Pisces.

## NORTHERN LATITUDES

### THE STARS

Cepheus, high up in the north, is best placed for evening observation this month and next. Its most celebrated star is Delta (δ) Cephei, the prototype of a class of pulsating variables. Deneb in Cygnus, Vega (see p.253) in Lyra, and Altair (see p.252) in Aquila—the stars of the Summer Triangle—remain high in the western half of the sky, while the Square of Pegasus is high in the east with Cassiopeia between it and the north celestial pole. The bright star Fomalhaut (see p.253) in

Piscis Austrinus is low in the south with Aquarius above it. A cascade of faint stars suggests the flow of water from the water carrier's urn toward the southern fish, Piscis Austrinus. For observers at high northern latitudes, this is the best time of year to see the zodiacal constellation Capricornus in the evening sky, lying low in the south to the right of Fomalhaut.

### DEEP-SKY OBJECTS

Near Deneb in Cygnus lies one of the most distinctive nebulae in the sky, NGC 7000, popularly called

the North America Nebula, on account of its shape. Under clear, dark skies, it can be detected with binoculars, but it is best seen on long-exposure photographs. Another object of note in Cygnus is the open star cluster M39, which is visible through binoculars. The 6th-magnitude globular cluster M15, also visible through binoculars, is not far from the star Enif—Epsilon (ε) Pegasi—which marks the horse's nose in Pegasus.

**URANUS**

9 A.M.

6 A.M.

3 A.M.

MORNING SKY

**THE HARVEST MOON**
The full moon that occurs closest to the northern fall equinox is termed the Harvest Moon because its light was said to assist farmers working late in the fields.

# SOUTHERN LATITUDES

## THE STARS

Scorpius is low in the west, with Sagittarius and the densest regions of the Milky Way above it. The large northern Summer Triangle of Altair, Vega, and Deneb is visible in the northwest, while in the southwest, Alpha (α) and Beta (β) Centauri—Rigil Kentaurus (see p.252) and Hadar—are visible from latitude 20°S and farther south. The Square of Pegasus dominates the northeastern sky.

First-magnitude Fomalhaut (see p.253) in Piscis Austrinus is almost overhead, along with Capricornus and Aquarius. Achernar, the bright star at the end of the celestial river Eridanus, is high in the southeast, as is the Small Magellanic Cloud (see p.311). A group of constellations with exotic names, such as Phoenix, Tucana, Grus, and Pavo, is spread across the southern half of the sky.

## DEEP-SKY OBJECTS

Aquarius contains two famous planetary nebulae, although neither is particularly easy to find through small instruments. The Helix Nebula (see p.257), or NGC 7293, is the nearest planetary nebula to us. Its size means that its light is spread out over such a large area that clear skies are essential to glimpse it through binoculars or a low-power telescope. The Saturn Nebula, NGC 7009, is so named because, when seen through a large telescope, it appears to have rings like the planet Saturn. A small telescope shows the Saturn Nebula simply as a greenish disk.

Also in Aquarius is the globular cluster M2, which resembles a fuzzy star when seen through binoculars. To the north of this is another globular cluster that can be viewed through binoculars, M15 in Pegasus.

**THE SMALL MAGELLANIC CLOUD**
This small satellite galaxy (left) appears beside the globular cluster 47 Tucanae (right), which is in the foreground in our own galaxy.

**NEPTUNE**

### POSITIONS OF THE PLANETS

This chart shows the positions of the planets in September from 2020 to 2027. The planets are represented by colored dots, while the number inside each dot denotes the year. For all planets apart from Mercury, the dot indicates the planet's position on September 15. Mercury is shown only when it is at greatest elongation (see p.68)—for the specific date, refer to the table, left.

- Mercury
- Venus
- Mars
- Jupiter
- Saturn
- Uranus
- Neptune

**EXAMPLES**

25 Jupiter's position on September 15, 2025

20 Saturn's position on September 15, 2020. The arrow indicates the planet is in retrograde motion (see p.68).

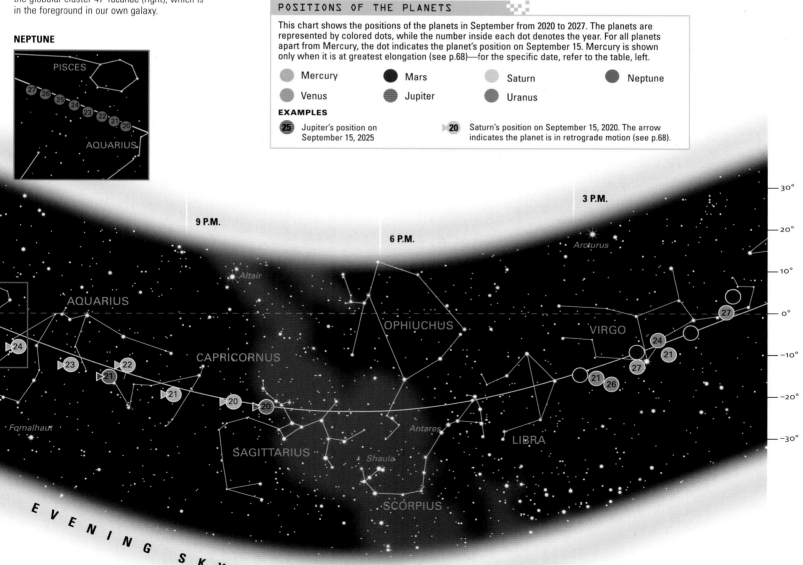

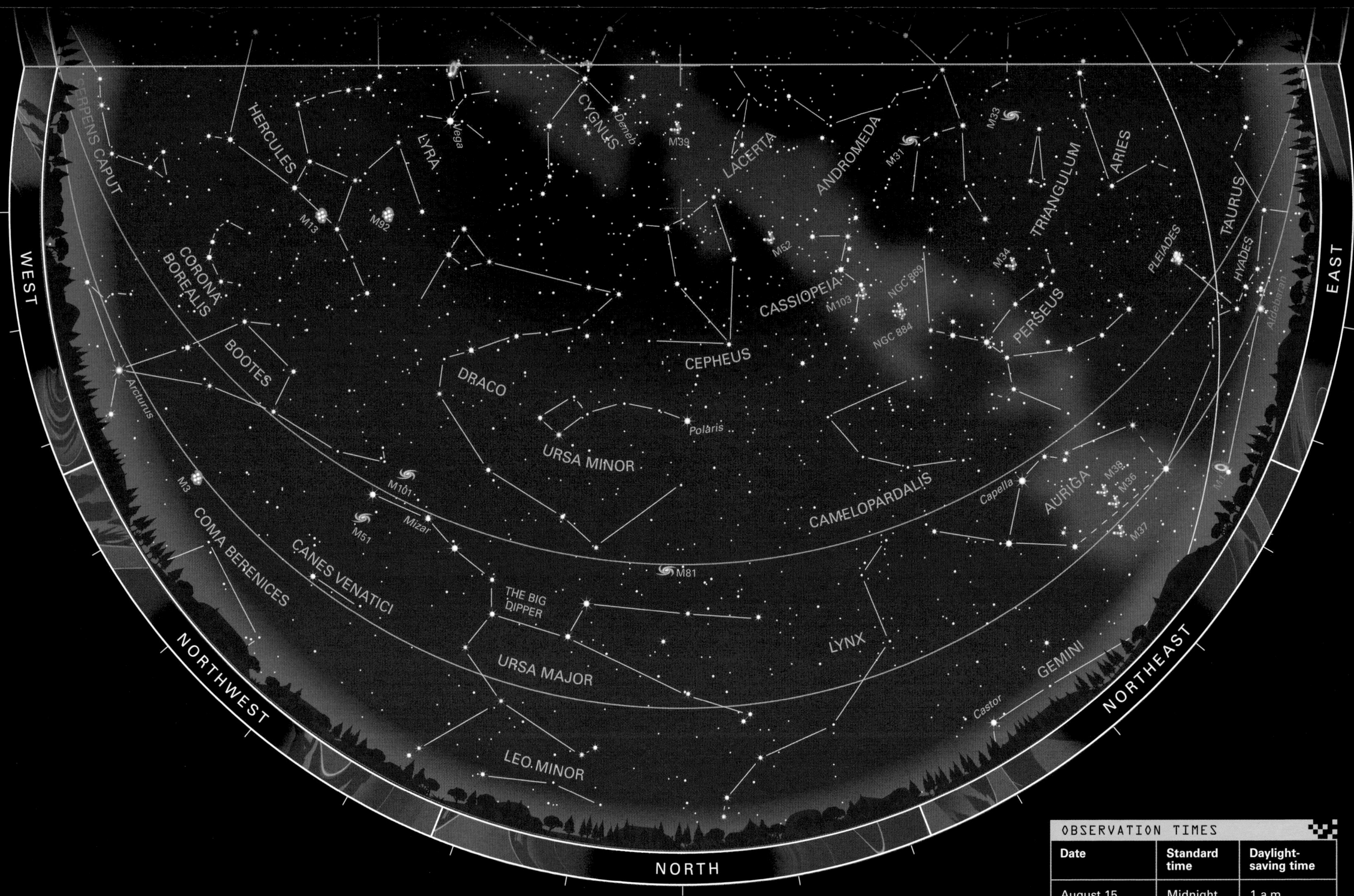

THE NIGHT SKY

# SEPTEMBER | NORTHERN LATITUDES

## LOOKING NORTH

### STAR MAGNITUDES

| | |
|---|---|
| ✦ | -1 |
| ✦ | 0 |
| ✦ | 1 |
| ⋆ | 2 |
| · | 3 |
| · | 4 |
| · | 5 |
| ⊙ | Variable star |

### DEEP-SKY OBJECTS

| | |
|---|---|
| 🌀 | Galaxy |
| ✳ | Globular cluster |
| ⁂ | Open cluster |
| ☁ | Diffuse nebula |
| ◉ | Planetary nebula |

### POINTS OF REFERENCE

| Horizons | 60°N | 40°N | 20°N |
|---|---|---|---|
| **Zeniths** | 60°N | 40°N | 20°N | Ecliptic |

### OBSERVATION TIMES

| Date | Standard time | Daylight-saving time |
|---|---|---|
| August 15 | Midnight | 1 a.m. |
| September 1 | 11 p.m. | Midnight |
| September 15 | 10 p.m. | 11 p.m. |
| October 1 | 9 p.m. | 10 p.m. |
| October 15 | 8 p.m. | 9 p.m. |

WEST

NORTHWEST

SERPENS CAPUT

Arcturus

COMA BERENICES

CORONA BOREALIS

BOOTES

HERCULES

CANES VENATICI

M3

M51

M13

M81

DRACO

M92

LYRA

Vega

LEO MINOR

URSA MAJOR

THE BIG DIPPER

URSA MINOR

CEPHEUS

Mizar

M101

Polaris

CYGNUS

Deneb

M81

NORTH

LYNX

CAMELOPARDALIS

CASSIOPEIA

M103

M52

LACERTA

M39

ANDROMEDA

NGC 884

NGC 869

M31

Castor

GEMINI

Capella

AURIGA

PERSEUS

M34

TRIANGULUM

M33

NORTHEAST

M38

M36

M37

ARIES

PLEIADES

M1

HYADES

TAURUS

Aldebaran

EAST

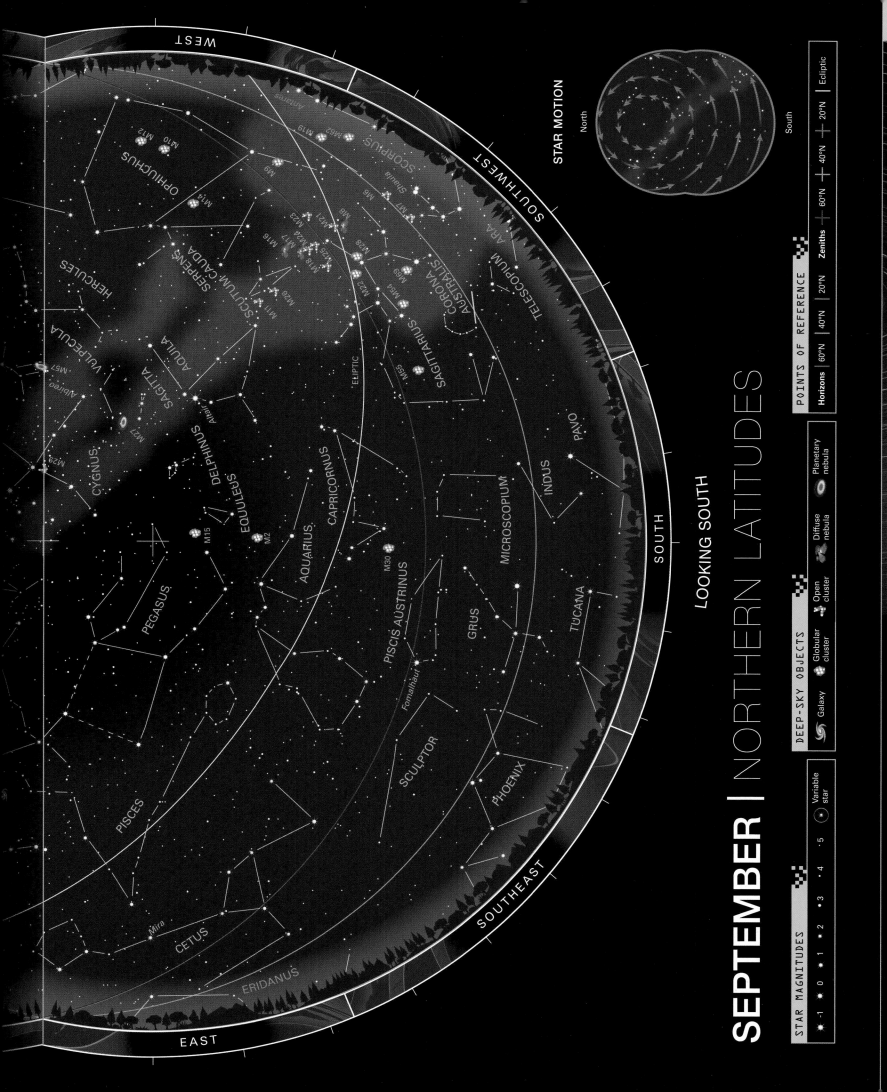

# SEPTEMBER | NORTHERN LATITUDES

LOOKING SOUTH

THE NIGHT SKY

WEST

SOUTHWEST

SOUTH

SOUTHEAST

EAST

STAR MOTION

North

South

**STAR MAGNITUDES**

-1   0   1   2   3   4   5

Variable star

**DEEP-SKY OBJECTS**

Galaxy   Globular cluster   Open cluster   Diffuse nebula   Planetary nebula

**POINTS OF REFERENCE**

Horizons   60°N   40°N   20°N   Zeniths   60°N   40°N   20°N   Ecliptic

### Constellations and objects labeled on the chart

OPHIUCHUS
HERCULES
SERPENS CAUDA
VULPECULA
CYGNUS
SAGITTA
AQUILA
DELPHINUS
EQUULEUS
PEGASUS
AQUARIUS
CAPRICORNUS
PISCES
CETUS
ERIDANUS
SCULPTOR
PISCIS AUSTRINUS
MICROSCOPIUM
GRUS
PHOENIX
TUCANA
INDUS
PAVO
TELESCOPIUM
CORONA AUSTRALIS
SAGITTARIUS
SCORPIUS
ARA

M12, M10, M14, M9, M19, M62, M23, M21, M20, M8, M17, M18, M24, M25, M16, M26, M11, M28, M22, M54, M69, M70, M55, M15, M2, M30, M57, M27, M29, M71

Albireo, Altair, Fomalhaut, Mira, Antares, Shaula

Ecliptic

# SEPTEMBER | SOUTHERN LATITUDES

## LOOKING NORTH

### STAR MAGNITUDES

* -1
* 0
* 1
* 2
* 3
* 4
* 5
* Variable star

### DEEP-SKY OBJECTS

* Galaxy
* Globular cluster
* Open cluster
* Diffuse nebula
* Planetary nebula

### POINTS OF REFERENCE

| Horizons | 0° | 20°S | 40°S |
| Zeniths | 0° | 20°S | 40°S | Ecliptic |

### OBSERVATION TIMES

| Date | Standard time | Daylight-saving time |
| --- | --- | --- |
| August 15 | Midnight | 1 a.m. |
| September 1 | 11 p.m. | Midnight |
| September 15 | 10 p.m. | 11 p.m. |
| October 1 | 9 p.m. | 10 p.m. |
| October 15 | 8 p.m. | 9 p.m. |

# SEPTEMBER | SOUTHERN LATITUDES

LOOKING SOUTH

STAR MOTION

North

South

STAR MAGNITUDES

-1 · 0 · 1 · 2 · 3 · 4 · 5 | ⊙ Variable star

DEEP-SKY OBJECTS

Galaxy | Globular cluster | Open cluster | Diffuse nebula | Planetary nebula

POINTS OF REFERENCE

Horizons | 0° | 20°S | 40°S | Zeniths | 0° | 20°S | 40°S | Ecliptic

**Constellations and objects labeled:**

OPHIUCHUS, LIBRA, LUPUS, CENTAURUS, SCORPIUS, NORMA, CIRCINUS, CRUX, ARA, TRIANGULUM AUSTRALE, CORONA AUSTRALIS, TELESCOPIUM, SAGITTARIUS, CAPRICORNUS, MICROSCOPIUM, INDUS, PAVO, APUS, MUSCA, CHAMAELEON, CARINA, VELA, OCTANS, GRUS, TUCANA, MENSA, VOLANS, PICTOR, AQUARIUS, PISCIS AUSTRINUS, SCULPTOR, PHOENIX, HYDRUS, RETICULUM, DORADO, PUPPIS, CETUS, FORNAX, ERIDANUS, HOROLOGIUM, CAELUM, COLUMBA

**Stars and objects:** M80, M4, Antares, M19, M62, M9, M16, M17, M18, M21, M23, M24, M25, M28, M22, M54, M69, M70, M55, M30, Shaula, Rigil Kentaurus, Hadar, Agena, Acrux, Mimosa, Gacrux, NGC 6397, NGC 5139, SMC, NGC 104, LMC, Fomalhaut, Achernar, Canopus

WEST

EAST

SOUTHWEST

SOUTH

SOUTHEAST

THE NIGHT SKY

## SPECIAL EVENTS

**PHASES OF THE MOON**

| | FULL MOON | NEW MOON |
|---|---|---|
| **2020** | October 1, 31 | October 16 |
| **2021** | October 20 | October 6 |
| **2022** | October 9 | October 25 |
| **2023** | October 28 | October 14 |
| **2024** | October 17 | October 2 |
| **2025** | October 7 | October 21 |
| **2026** | October 26 | October 10 |
| **2027** | October 15 | October 29 |

**PLANETS**

**2020: October 1** Mercury is at greatest evening elongation, magnitude 0.3.

**2020: October 13** Mars is at opposition, magnitude –2.6.

**2021: October 25** Mercury is at greatest morning elongation, magnitude –0.3.

**2021: October 29** Venus is at greatest evening elongation, magnitude –4.4.

**2022: October 8** Mercury is at greatest morning elongation, magnitude –0.3.

**2023: October 23** Venus is at greatest morning elongation, magnitude –4.4.

**2025: October 29** Mercury is at greatest evening elongation, magnitude 0.1.

**2026: October 4** Saturn is at opposition, magnitude 0.4.

**ECLIPSES**

**2022: October 25** A partial eclipse of the Sun is visible from Europe, northeast Africa, the Middle East, and west Asia.

**2023: October 14** An annular eclipse of the Sun is visible from west US, southern North America, Colombia, and Brazil.

**2023: October 28** A partial eclipse of the Moon is visible from the east of North and South America, Europe, Africa, Asia, and Australia.

**2024: October 2** An annual eclipse of the Sun is visible from south Chile and south Argentina.

# OCTOBER

The Square of Pegasus takes center stage in the northern skies in both hemispheres, a sign of the arrival of the northern fall and the southern spring. Northeast of it lies the Andromeda Galaxy, the closest large galaxy to the Earth. South of the Square, a band of faint zodiacal constellations crosses the sky, from Aries in the east to Capricornus in the southwest.

## NORTHERN LATITUDES

### THE STARS

The Square of Pegasus lies high in the sky from mid-northern latitudes. From one corner of the Square, the stars of Andromeda extend northeastward toward Perseus and Cassiopeia. Capella twinkles above the horizon in Auriga, lower in the northeast. In the north, the Big Dipper is at its lowest, and it is below the horizon for observers south of about latitude 30°N.

Directly beneath the Square of Pegasus is a loop of stars known as the Circlet, representing the body of one of the fish in the zodiacal constellation of Pisces. Fomalhaut (see p.253) in Piscis Austrinus is low on the southern horizon beneath the stars of Aquarius. In the western sky, the Summer Triangle lingers, while in the east, Taurus leads the stars of winter into view.

### DEEP-SKY OBJECTS

October evenings are a good time to view the Andromeda Galaxy, M31 (see pp.312–313). It can be seen as an elongated misty patch with the naked eye, if skies are not too polluted, and it is easily visible through binoculars, spanning a greater width than a full moon. High in the north, M52, an open cluster near Cassiopeia, is visible through binoculars. Between this and the Square of Pegasus lies an often-overlooked planetary nebula, NGC 7662, nicknamed the Blue Snowball. A small telescope is needed to see it.

### METEOR SHOWER

One of the year's lesser showers, the Orionids, reaches a peak of some 25 meteors an hour around October 21. They radiate from northern Orion, near the border with Gemini. This area rises late, so the meteors are best seen after midnight.

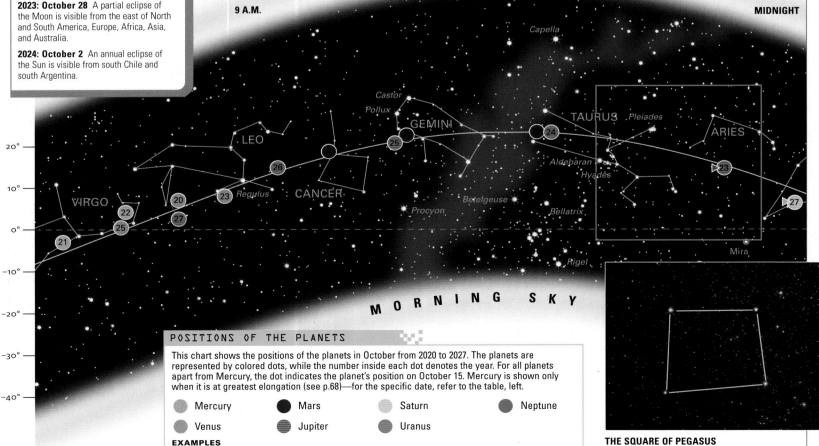

**POSITIONS OF THE PLANETS**

This chart shows the positions of the planets in October from 2020 to 2027. The planets are represented by colored dots, while the number inside each dot denotes the year. For all planets apart from Mercury, the dot indicates the planet's position on October 15. Mercury is shown only when it is at greatest elongation (see p.68)—for the specific date, refer to the table, left.

Mercury · Mars · Saturn · Neptune · Venus · Jupiter · Uranus

**EXAMPLES**

20 Jupiter's position on October 15, 2020

22 Jupiter's position on October 15, 2022. The arrow indicates the planet is in retrograde motion (see p.68).

**THE SQUARE OF PEGASUS**
This huge square in the northern fall sky is composed of three stars in Pegasus and one in Andromeda (top left).

# SOUTHERN LATITUDES

## THE STARS

In contrast to the sparkling skies of southern winter, the constellations of October evenings are mostly faint and unremarkable. One star that stands out is 1st-magnitude Fomalhaut (see p.253), almost overhead in the constellation Piscis Austrinus. In the northwest sky is Altair (see p.252) in Aquila and, high in the north, the Square of Pegasus. Between Pegasus and Fomalhaut lies Aquarius, the Water Carrier. More constellations with watery associations fill the eastern part of the sky—Pisces, the Fish; Cetus, the Sea Monster or the Whale; and Eridanus, the River. The constellation Eridanus ends at the bright star Achernar, high in the south. The Small Magellanic Cloud (see p.310) is lower in the south, with the Large Magellanic Cloud (see p.310) now in view in the southeast. Canopus in Carina is also visible in the southeast for those farther south of the equator than 20°S.

**FAMILIAR ASTERISMS**
The Circlet of Pisces (left) and the Y-shaped Water Jar of Aquarius (right) are two easily recognizable star patterns in the October evening sky.

## DEEP-SKY OBJECTS

Tucana contains the second-best globular cluster in the sky, 47 Tucanae, or NGC 104, which is visible to the naked eye as a fuzzy star and appears impressive through binoculars. It covers the same area of sky as a full moon, near the Small Magellanic Cloud, but it lies much closer to us—about 15,000 light-years away—in our own galaxy. On the edge of the SMC, NGC 362 is another, fainter globular cluster, also in our galaxy.

October and November evenings are the best time for southern observers to view the Andromeda Galaxy, M31 (see pp.312–313), which lies low in the northern sky. Near it is another member of our Local Group of galaxies, M33, a smaller spiral galaxy that is less easy to see. In clear, dark skies, it can be glimpsed through binoculars or a low-power telescope as a large, rounded patch.

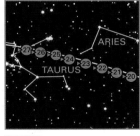

**URANUS**

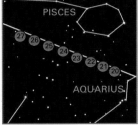

**NEPTUNE**

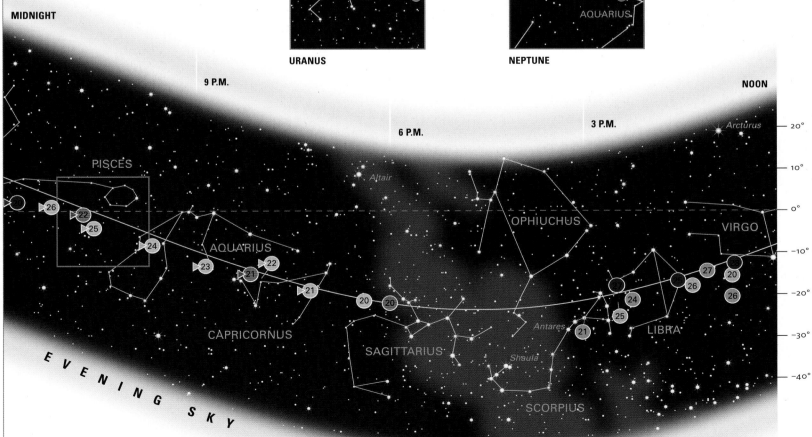

MIDNIGHT

9 P.M.

6 P.M.

3 P.M.

NOON

20°
10°
0°
−10°
−20°
−30°
−40°

EVENING SKY

PISCES

AQUARIUS

CAPRICORNUS

SAGITTARIUS

OPHIUCHUS

SCORPIUS

LIBRA

VIRGO

Altair

Antares

Shaula

Arcturus

THE NIGHT SKY

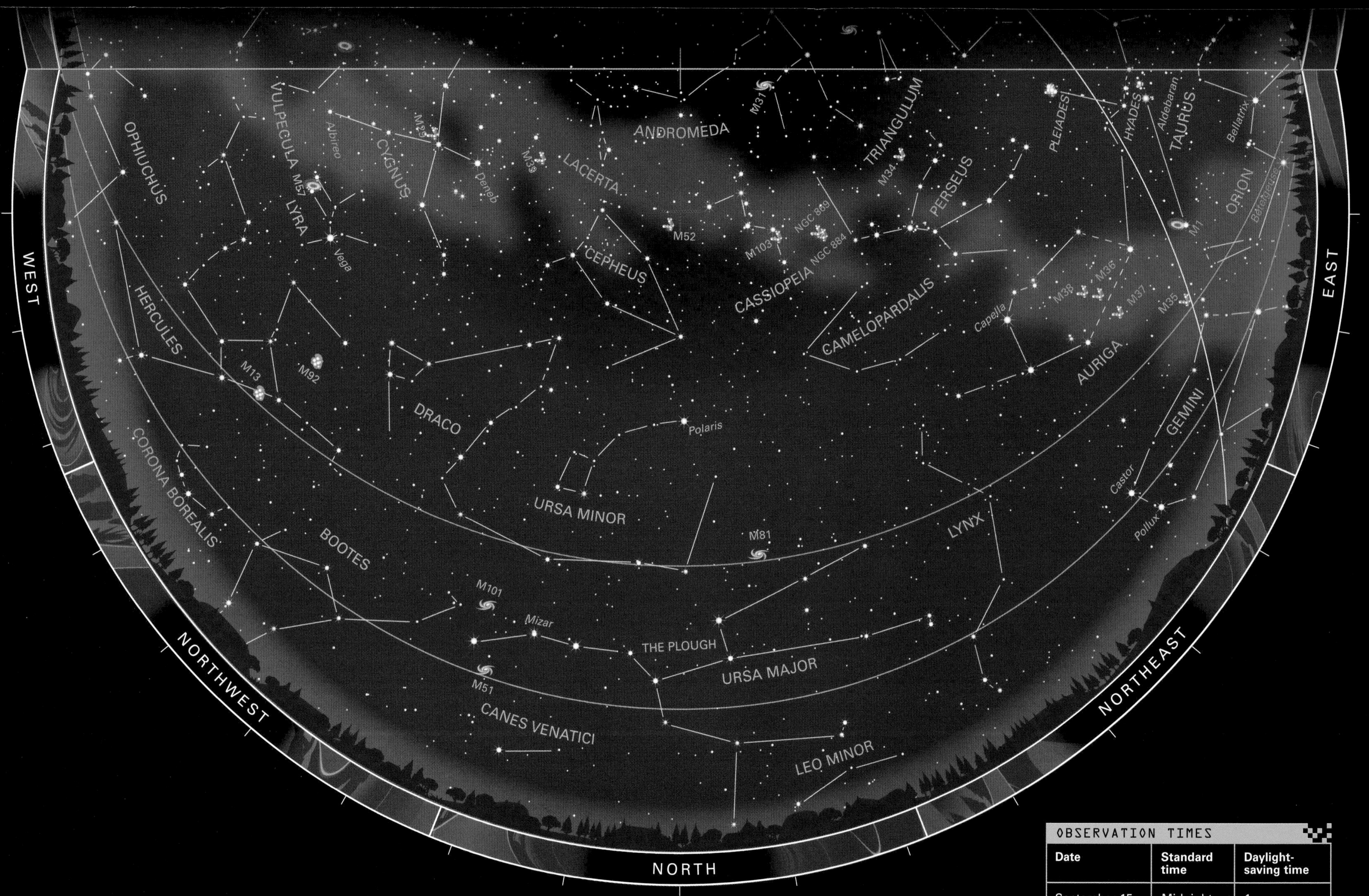

# OCTOBER | NORTHERN LATITUDES

## LOOKING NORTH

### STAR MAGNITUDES

- -1
- 0
- 1
- 2
- 3
- 4
- 5
- Variable star

### DEEP-SKY OBJECTS

- Galaxy
- Globular cluster
- Open cluster
- Diffuse nebula
- Planetary nebula

### POINTS OF REFERENCE

| Horizons | 60°N | 40°N | 20°N |
| Zeniths | 60°N | 40°N | 20°N | Ecliptic |

### OBSERVATION TIMES

| Date | Standard time | Daylight-saving time |
| --- | --- | --- |
| September 15 | Midnight | 1 a.m. |
| October 1 | 11 p.m. | Midnight |
| October 15 | 10 p.m. | 11 p.m. |
| November 1 | 9 p.m. | 10 p.m. |
| November 15 | 8 p.m. | 9 p.m. |

WEST

NORTHWEST

NORTH

NORTHEAST

EAST

WEST

OPHIUCHUS
HERCULES
CORONA BOREALIS
M13
M92
LYRA
Vega
Albireo
VULPECULA M57
BOOTES
CYGNUS
M39
Deneb
DRACO
CANES VENATICI
M51
Mizar
M101
URSA MINOR
CEPHEUS
LACERTA
M52
ANDROMEDA
THE PLOUGH
Polaris
M31
CASSIOPEIA
M103
NGC 869
NGC 884
URSA MAJOR
M81
CAMELOPARDALIS
TRIANGULUM
M34
LEO MINOR
PERSEUS
LYNX
Capella
PLEIADES
AURIGA
M38
M36
M37
HYADES
Castor
Pollux
GEMINI
M35
Aldebaran
TAURUS
M1
ORION
Bellatrix
Betelgeuse

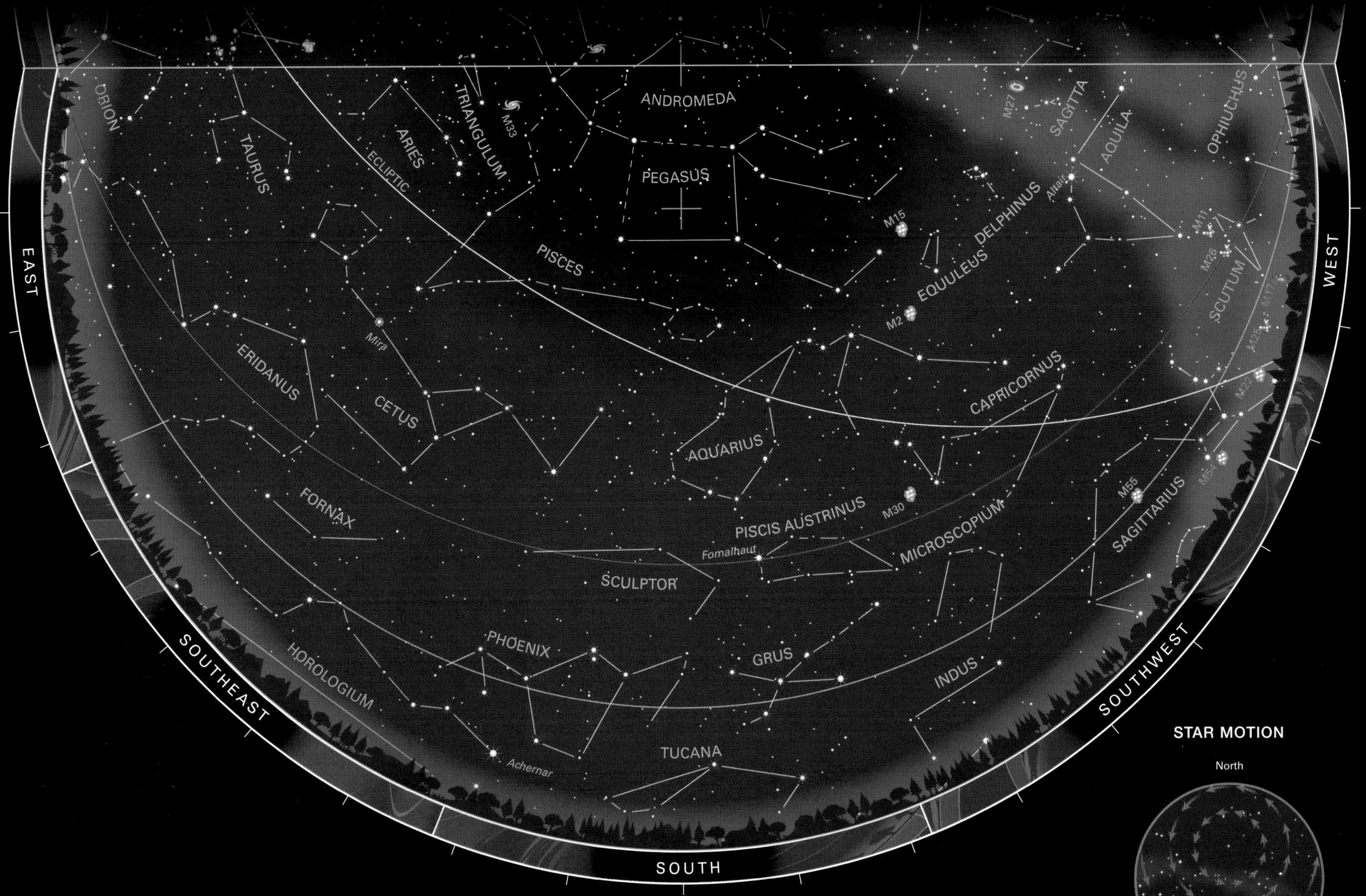

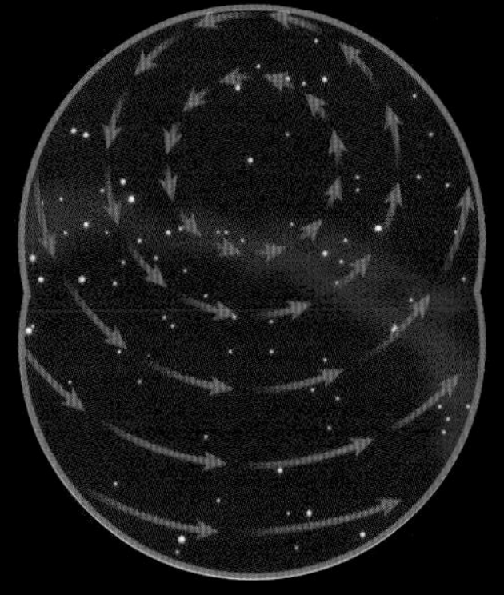

# OCTOBER | NORTHERN LATITUDES

LOOKING SOUTH

THE NIGHT SKY

**STAR MOTION**

North

South

**POINTS OF REFERENCE**

| Horizons | 60°N | 40°N | 20°N | Zeniths | 60°N | 40°N | 20°N | Ecliptic |
|---|---|---|---|---|---|---|---|---|

**DEEP-SKY OBJECTS**

Galaxy | Globular cluster | Open cluster | Diffuse nebula | Planetary nebula

**STAR MAGNITUDES**

-1 · 0 · 1 · 2 · 3 · 4 · 5 | Variable star

WEST

WEST

SOUTHWEST

SOUTH

SOUTHEAST

EAST

WEST

OPHIUCHUS

AQUILA

SAGITTA

SCUTUM

M11

M26

M16

M17

M55

SAGITTARIUS

M54

DELPHINUS

EQUULEUS

Altair

M27

CAPRICORNUS

MICROSCOPIUM

INDUS

M15

M2

M30

AQUARIUS

PISCIS AUSTRINUS

Fomalhaut

GRUS

TUCANA

PEGASUS

ANDROMEDA

PISCES

SCULPTOR

PHOENIX

Achernar

M33

TRIANGULUM

ARIES

ECLIPTIC

Mira

CETUS

FORNAX

HOROLOGIUM

ERIDANUS

TAURUS

ORION

# OCTOBER | SOUTHERN LATITUDES

## LOOKING NORTH

### STAR MAGNITUDES

- -1
- 0
- 1
- 2
- 3
- 4
- 5
- Variable star

### DEEP-SKY OBJECTS

- Galaxy
- Globular cluster
- Open cluster
- Diffuse nebula
- Planetary nebula

### POINTS OF REFERENCE

| Horizons | 0° | 20°S | 40°S |
| --- | --- | --- | --- |
| Zeniths | 0° | 20°S | 40°S | Ecliptic |

### OBSERVATION TIMES

| Date | Standard time | Daylight-saving time |
| --- | --- | --- |
| September 15 | Midnight | 1 a.m. |
| October 1 | 11 p.m. | Midnight |
| October 15 | 10 p.m. | 11 p.m. |
| November 1 | 9 p.m. | 10 p.m. |
| November 15 | 8 p.m. | 9 p.m. |

# OCTOBER | SOUTHERN LATITUDES

LOOKING SOUTH

WEST

SOUTHWEST

SOUTH

SOUTHEAST

EAST

STAR MOTION

North

South

**STAR MAGNITUDES**

-1   0   1   2   3   4   5   Variable star

**DEEP-SKY OBJECTS**

Galaxy   Globular cluster   Open cluster   Diffuse nebula   Planetary nebula

**POINTS OF REFERENCE**

Horizons   0°   20°S   40°S   Zeniths   0°   20°S   40°S   Ecliptic

SCUTUM
M26
M11
M25
M24 M18 M16
M23
M28 M21
M22
M9
M19
M4
M54
M8 M6 M7
M20
Shaula
Antares
SERPENS CAUDA
SAGITTARIUS
M55
SCORPIUS
M62
CORONA AUSTRALIS
TELESCOPIUM
LUPUS
CAPRICORNUS
NORMA
MICROSCOPIUM
ARA
TRIANGULUM AUSTRALE
M30
INDUS
PAVO
CIRCINUS
Rigil Kentaurus
PISCIS AUSTRINUS
Fomalhaut
GRUS
TUCANA
APUS
Hadar
CENTAURUS
Mimosa
CRUX
Gacrux
Acrux
MUSCA
SCULPTOR
PHOENIX
NGC 104
SMC
OCTANS
CHAMAELEON
PHOENIX
HYDRUS
MENSA
LMC
CETUS
FORNAX
Achernar
HOROLOGIUM
RETICULUM
DORADO
VOLANS
CARINA
VELA
ERIDANUS
CAELUM
PICTOR
PUPPIS
Canopus
COLUMBA
LEPUS
Rigel
M42
ORION
CANIS MAJOR
Adhara

THE NIGHT SKY

## SPECIAL EVENTS

### PHASES OF THE MOON

| | FULL MOON | NEW MOON |
|---|---|---|
| 2020 | November 30 | November 15 |
| 2021 | November 19 | November 4 |
| 2022 | November 8 | November 29 |
| 2023 | November 27 | November 13 |
| 2024 | November 15 | November 1 |
| 2025 | November 5 | November 20 |
| 2026 | November 24 | November 9 |
| 2027 | November 14 | November 28 |

### PLANETS

**2020: November 10** Mercury is at greatest morning elongation, magnitude −0.3.

**2023: November 3** Jupiter is at opposition, magnitude −2.9. At midnight, it is visible to the south from northern latitudes and to the north from southern latitudes.

**2024: November 16** Mercury is at greatest evening elongation, magnitude −0.1.

**2026: November 20** Mercury is at greatest morning elongation, magnitude −0.3.

**2027: November 4** Mercury is at greatest morning elongation, magnitude −0.3.

### ECLIPSES AND TRANSITS

**2021: November 19** A partial eclipse of the Moon is visible from North and South America, northern Europe, eastern Asia, Australia, and the Pacific.

**2022: November 8** A total eclipse of the Moon is visible from Asia, Australia, the Pacific, and North and South America.

# NOVEMBER

Cassiopeia lies overhead for northern observers, as the Milky Way runs from Cygnus in the west to Gemini in the east. The large figures of Pisces, the Fish, and Cetus, the Sea Monster or Whale, are spread across the equatorial region of the sky, while in the southern sky, the Large and Small Magellanic Clouds are high up.

## NORTHERN LATITUDES

### THE STARS

All the main characters in the Perseus and Andromeda myth (see p.368) are on display in the November evening sky. Cetus contains a remarkable variable star, Mira (see p.285). It is easily visible to the naked eye when at maximum brightness, every 11 months or so, but the rest of the time it fades out of sight. High in the west is the Square of Pegasus, with the stars of the Summer Triangle lower in the northwest.

### DEEP-SKY OBJECTS

Two open star clusters, NGC 457 and NGC 663, are easy to see with binoculars in Cassiopeia. Even better are NGC 869 and 884, a pair known as the Double Cluster, embedded in the Milky Way between Perseus and Cassiopeia. The Andromeda Galaxy, M31 (see pp.312–313), remains high up this month.

### METEOR SHOWERS

The Taurids have a broad peak in the first week of the month, when around 10 meteors an hour may be seen coming from the region south of the Pleiades Cluster. Although not numerous, the meteors are long-lasting and often bright. A second meteor shower in November, the Leonids, radiates from the head of Leo, reaching a peak around November 17. Usually, no more than 10 meteors per hour are seen, but surges of activity occur every 33 years or so. High activity is not expected again until around 2032.

**THE PRINCESS, HERO, KING, AND QUEEN**
Joined in Greek myth, Andromeda (right), Perseus (bottom), Cepheus (top), and Cassiopeia (center) appear together in northern skies in November.

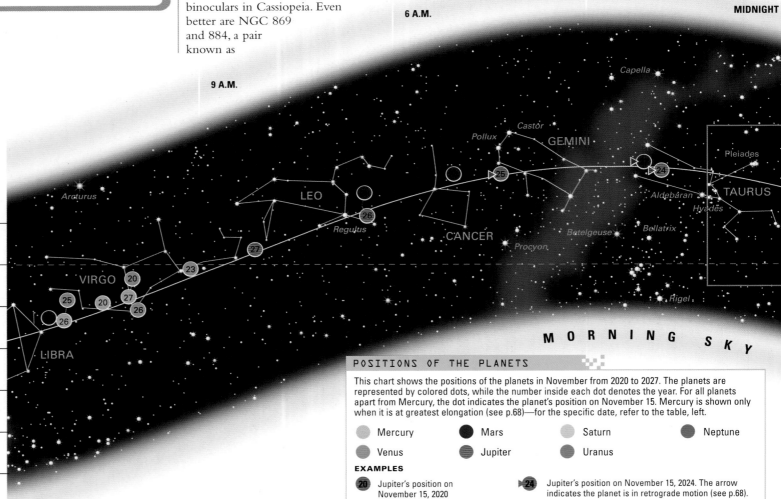

9 A.M.

6 A.M.

3 A.M.

MIDNIGHT

Capella

Castor

Pollux

GEMINI

Pleiades

Arcturus

LEO

Aldebaran

TAURUS

Hyades

Regulus

CANCER

Betelgeuse

Bellatrix

Procyon

Rigel

23

VIRGO

20

27

20

25

20

26

27

26

LIBRA

10°

0°

−10°

−20°

−30°

−40°

−50°

M O R N I N G   S K Y

### POSITIONS OF THE PLANETS

This chart shows the positions of the planets in November from 2020 to 2027. The planets are represented by colored dots, while the number inside each dot denotes the year. For all planets apart from Mercury, the dot indicates the planet's position on November 15. Mercury is shown only when it is at greatest elongation (see p.68)—for the specific date, refer to the table, left.

- Mercury
- Mars
- Saturn
- Neptune
- Venus
- Jupiter
- Uranus

**EXAMPLES**

**20** Jupiter's position on November 15, 2020

**24** Jupiter's position on November 15, 2024. The arrow indicates the planet is in retrograde motion (see p.68).

# SOUTHERN LATITUDES

## THE STARS

Achernar, the bright star at the end of Eridanus, lies high in the south on November evenings. The other stars of Eridanus extend to Orion, which is rising in the east. Aldebaran and the stars of Taurus are in the northeast, and the Square of Pegasus is high in the northwest. Aquarius is in the west, with Fomalhaut (see p.253) in Piscis Austrinus in the southwest. The Large and Small Magellanic Clouds (see p.310 and p.311) are high in the south. Brilliant Canopus in Carina is in the southeast, with Sirius (see p.252) in Canis Major rising in the east. Overhead is Cetus, containing the long-period variable star Mira.

## DEEP-SKY OBJECTS

South of the head of Cetus is M77, the brightest of the Seyfert-type galaxies (see p.320). Seyferts are spiral galaxies with unusually bright centers caused by hot gas spiraling around a massive black hole. A telescope is required to see M77.

In the south, the globular cluster 47 Tucanae is still in view near the meridian. The Large Magellanic Cloud, with the Tarantula Nebula, NGC 2070, is in the southeast, but it is best seen in January. In the north, the galaxies M31 and M33 are visible, while the Pleiades (see p.291) and Hyades Clusters (see p.290) are moving higher in the east.

### URANUS

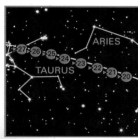

### NEPTUNE

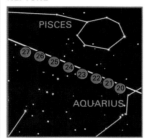

**MIDNIGHT**

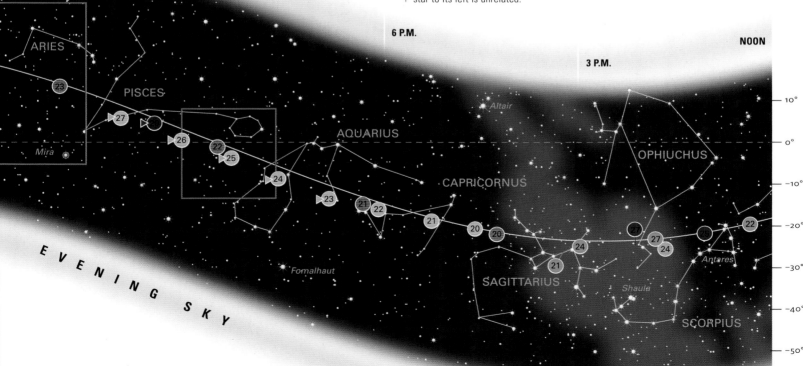

**CLASSIC VARIABLE**
The long-period variable star Mira (center) appears strongly red when near maximum brightness. The 9th-magnitude star to its left is unrelated.

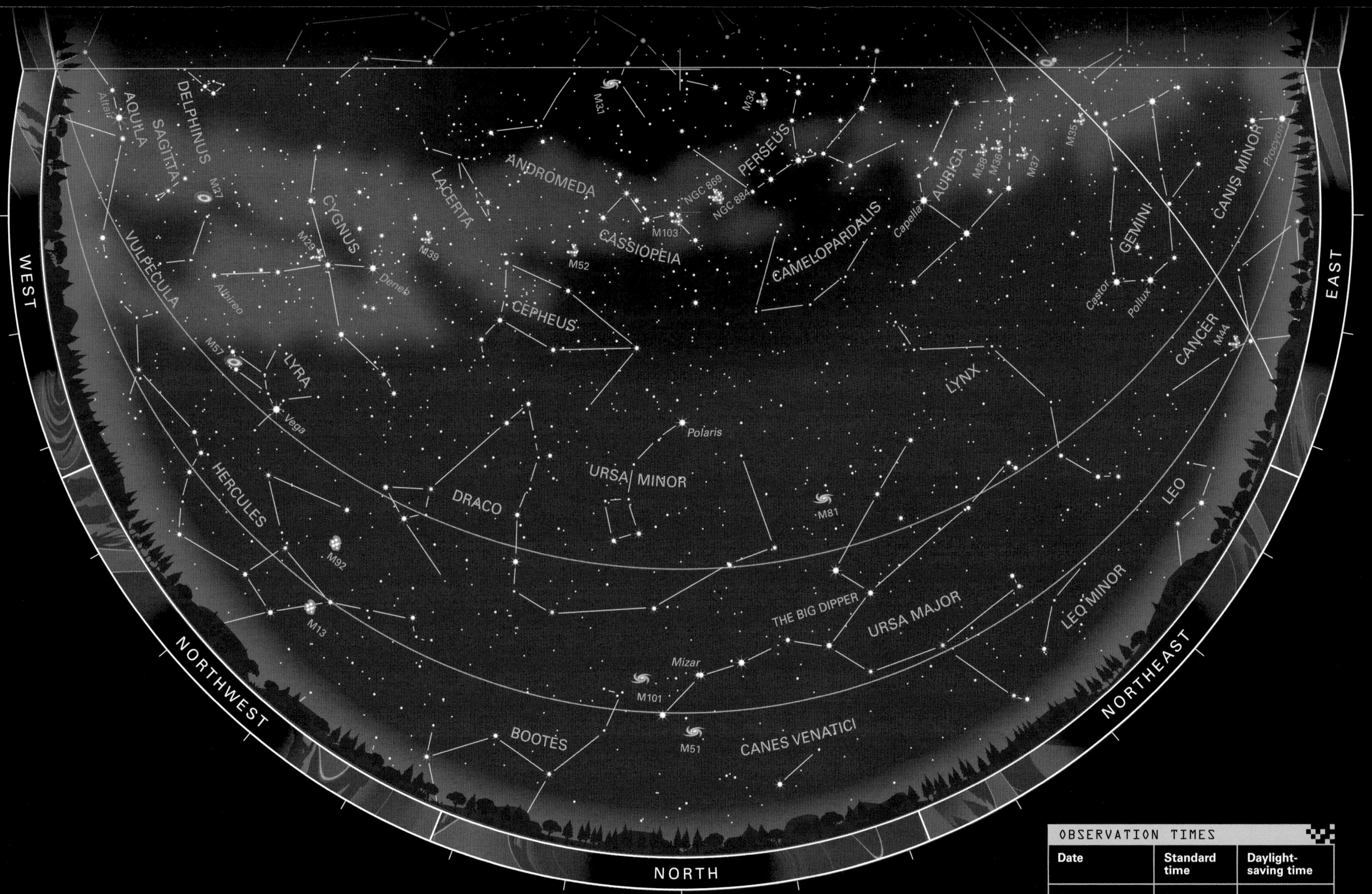

# NOVEMBER | NORTHERN LATITUDES

## LOOKING NORTH

### OBSERVATION TIMES

| Date | Standard time | Daylight-saving time |
|---|---|---|
| October 15 | Midnight | 1 a.m. |
| November 1 | 11 p.m. | Midnight |
| November 15 | 10 p.m. | 11 p.m. |
| December 1 | 9 p.m. | 10 p.m. |
| December 15 | 8 p.m. | 9 p.m. |

# NOVEMBER | NORTHERN LATITUDES

LOOKING SOUTH

WEST

EAST

SOUTHEAST

SOUTH

SOUTHWEST

WEST

**STAR MOTION**

North

South

## Constellations and objects

AQUILA

EQUULEUS

M2

M15

CAPRICORNUS

M30

MICROSCOPIUM

AQUARIUS

PISCIS AUSTRINUS

Fomalhaut

PISCES

GRUS

PEGASUS

ANDROMEDA

SCULPTOR

PHOENIX

TUCANA

ARIES

TRIANGULUM M33

Ecliptic

CETUS

Mira

FORNAX

ERIDANUS

HOROLOGIUM

Achernar

RETICULUM

PLEIADES M45

TAURUS

HYADES

Aldebaran

CAELUM

DORADO

M1

ORION

Bellatrix

Rigel

M42

LEPUS

COLUMBA

Betelgeuse

MONOCEROS

CANIS MAJOR

M50

Sirius

M41

## STAR MAGNITUDES

-1  0  1  2  3  4  5 · Variable star

## DEEP-SKY OBJECTS

Galaxy | Globular cluster | Open cluster | Diffuse nebula | Planetary nebula

## POINTS OF REFERENCE

| Horizons | 60°N | 40°N | 20°N | Zeniths | 20°N | 40°N | 60°N | Ecliptic |

THE NIGHT SKY

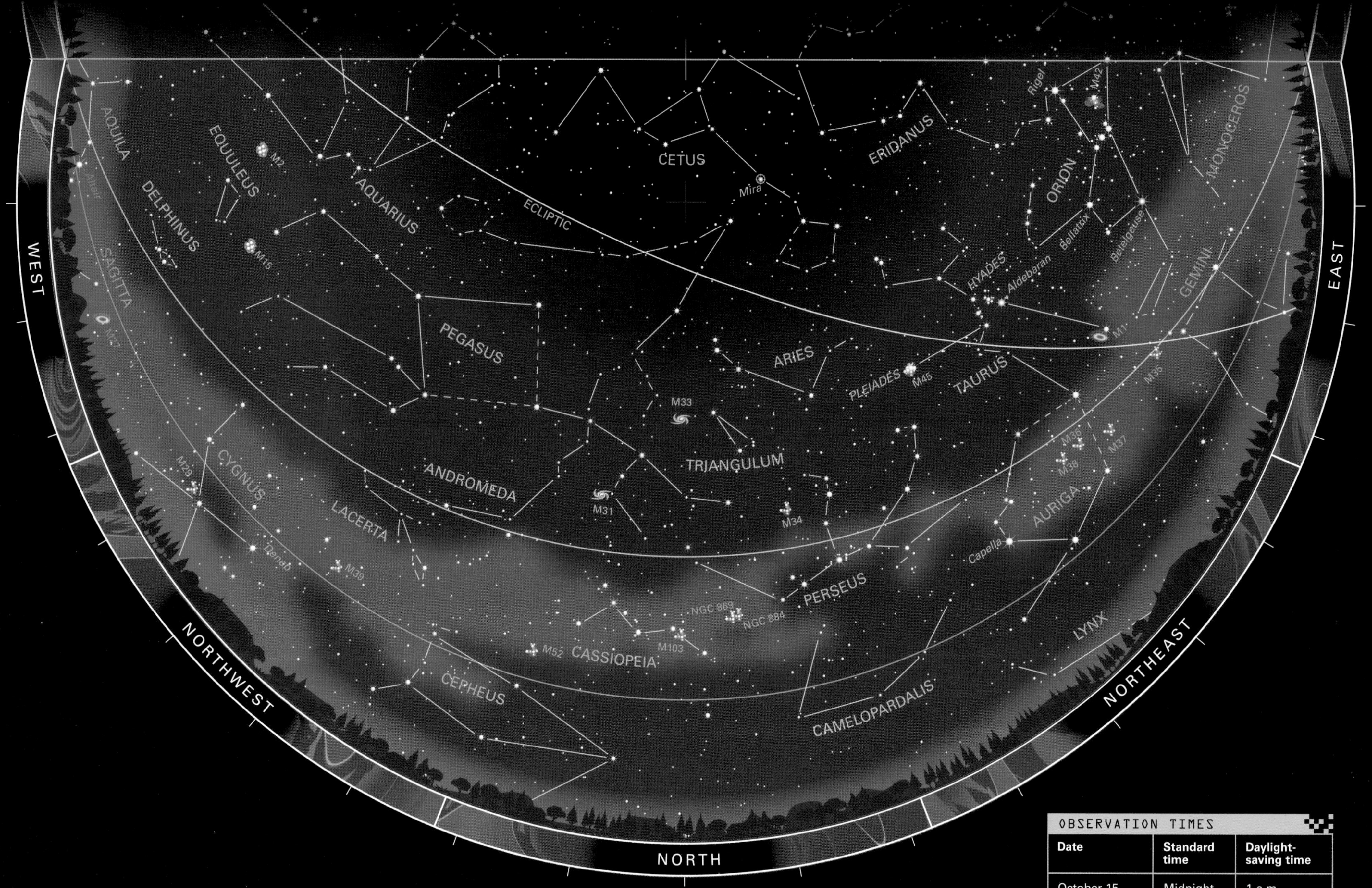

# NOVEMBER | SOUTHERN LATITUDES

## LOOKING NORTH

THE NIGHT SKY

### STAR MAGNITUDES

- ✦ -1
- ✦ 0
- ✦ 1
- ✦ 2
- • 3
- • 4
- · 5
- ⊙ Variable star

### DEEP-SKY OBJECTS

- 🌀 Galaxy
- ⬤ Globular cluster
- ✳ Open cluster
- ❀ Diffuse nebula
- ⬮ Planetary nebula

### POINTS OF REFERENCE

| Horizons | | | Zeniths | | |
|---|---|---|---|---|---|
| 0° | 20°S | 40°S | 0° | 20°S | 40°S | Ecliptic |

### OBSERVATION TIMES

| Date | Standard time | Daylight-saving time |
|---|---|---|
| October 15 | Midnight | 1 a.m. |
| November 1 | 11 p.m. | Midnight |
| November 15 | 10 p.m. | 11 p.m. |
| December 1 | 9 p.m. | 10 p.m. |
| December 15 | 8 p.m. | 9 p.m. |

Constellations and objects labeled on the chart: WEST, NORTHWEST, NORTH, NORTHEAST, EAST, SAGITTA, AQUILA, Altair, DELPHINUS, CYGNUS, Deneb, EQUULEUS, M15, M2, AQUARIUS, M39, LACERTA, ANDROMEDA, PEGASUS, CEPHEUS, M52, CASSIOPEIA, M103, M31, NGC 869, NGC 884, TRIANGULUM, M33, ARIES, CETUS, Mira, PERSEUS, M34, CAMELOPARDALIS, PLEIADES, M45, ERIDANUS, TAURUS, HYADES, Aldebaran, ORION, Rigel, Capella, AURIGA, Bellatrix, M38, M36, M37, M1, Betelgeuse, M42, LYNX, M35, GEMINI, MONOCEROS, ECLIPTIC

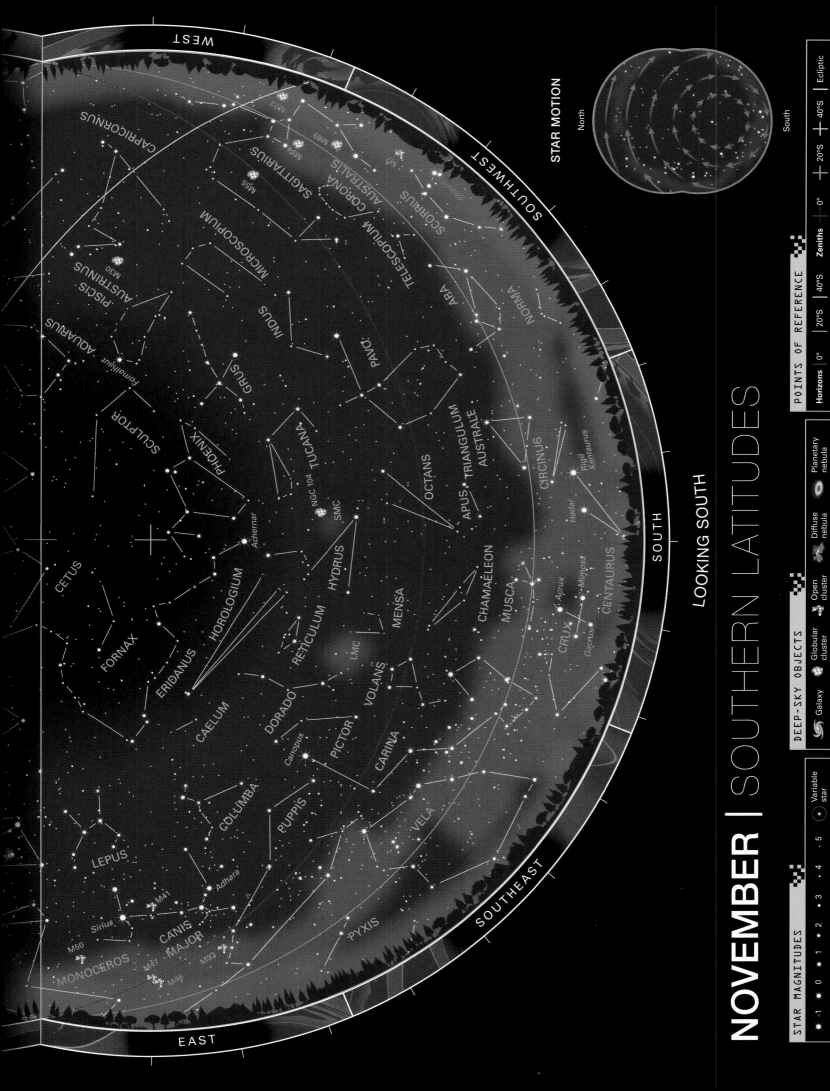

# NOVEMBER | SOUTHERN LATITUDES

## LOOKING SOUTH

## THE NIGHT SKY

**STAR MOTION**

North

South

**STAR MAGNITUDES**

-1    0    1    2    3    4    5    Variable star

**DEEP-SKY OBJECTS**

Galaxy    Globular cluster    Open cluster    Diffuse nebula    Planetary nebula

**POINTS OF REFERENCE**

Horizons    0°    20°S    40°S    Zeniths    0°    20°S    40°S    Ecliptic

WEST

WEST

SOUTHWEST

SOUTH

SOUTHEAST

EAST

### Constellation and star labels

CAPRICORNUS

M75

M54   M69

M70

SAGITTARIUS

M55

CORONA AUSTRALIS

M7

SCORPIUS

M6

Shaula

MICROSCOPIUM

TELESCOPIUM

ARA

NORMA

PISCIS AUSTRINUS

M30

AQUARIUS

Fomalhaut

GRUS

INDUS

PAVO

CIRCINUS

Rigil Kentaurus

Hadar

SCULPTOR

PHOENIX

TUCANA

NGC 104

SMC

OCTANS

APUS

TRIANGULUM AUSTRALE

CENTAURUS

Mimosa

CETUS

Achernar

HYDRUS

CHAMAELEON

MUSCA

Acrux

CRUX

Gacrux

FORNAX

HOROLOGIUM

RETICULUM

LMC

MENSA

ERIDANUS

CAELUM

DORADO

VOLANS

CARINA

VELA

Canopus

PICTOR

PUPPIS

COLUMBA

LEPUS

PYXIS

M41

Adhara

Sirius

CANIS MAJOR

M50

M47

M46

M93

MONOCEROS

# DECEMBER

The Sun reaches its farthest point south of the celestial equator this month, on December 21–22. As a result, Northern Hemisphere nights are the longest of the year, while in the Southern Hemisphere, they are the shortest. Earth has now completed another annual circuit of the Sun, and the evening stars end the year as they began, with the tableau of Orion and Taurus returning to center stage.

## NORTHERN LATITUDES

### THE STARS

Overhead lies Perseus, containing the famous variable star Algol (see p.276). From Perseus, the Milky Way leads northwestward to Cassiopeia and Cygnus, which is out of sight for those at around 20°N or closer to the equator. In the other direction, the Milky Way extends southeastward via Auriga and past Taurus to Gemini and the northern arm of Orion. The Square of Pegasus is in the west, while the Winter Triangle of Betelgeuse (see p.256) in Orion, Procyon (see p.284) in Canis Minor, and Sirius (see p.252) in Canis Major dominates the southeast. By comparison with the richness of this southeastern part of the sky, the southwest seems dull and empty, as it is occupied by the faint constellations Aries, Pisces, and Cetus. As the year ends, Sirius lies due south around midnight.

### DEEP-SKY OBJECTS

Large, bright clusters of stars abound in the December evening sky. In central Perseus, a few dozen stars cluster around the constellation's brightest member, Alpha (α) Persei or Mirfak. They form a group known as the Alpha Persei Cluster, which covers several diameters of a full moon and is a fine sight through binoculars.

In Taurus lies probably the finest open cluster in the entire sky, the Pleiades or M45 (see p.291). At least six members are visible to normal eyesight, but binoculars bring dozens more into view. Taurus contains an even larger cluster, the Hyades (see p.290), a V-shaped grouping which outlines the Bull's face. In addition to these groupings, the Double Cluster in Perseus, NGC 869 and NGC 884, already encountered in November, remains well placed.

### METEOR SHOWER

The year's second-best meteor shower, the Geminids, reaches a peak around December 13–14, when up to one meteor per minute can be seen radiating from a point near Castor in Gemini. Lesser activity is seen for a few days before the peak, but numbers fall off rapidly afterward.

#### PLANETS

**2020: December 21** Jupiter and Saturn appear less than a quarter of a Moon-width apart in the low western dusk sky.

**2021: December 29** Mercury and Venus appear about eight Moon-widths apart in the low western dusk sky.

**2022: December 8** Mars is at opposition, magnitude –1.8. At midnight, it is visible to the south from northern latitudes and to the north from southern latitudes.

**2022: December 21** Mercury is at greatest evening elongation, magnitude –0.3.

**2023: December 4** Mercury is at greatest evening elongation, magnitude –0.2.

**2024: December 7** Jupiter is at opposition, magnitude –2.8.

**2024: December 25** Mercury is at greatest morning elongation, magnitude –0.1.

**2025: December 7** Mercury is at greatest morning elongation, magnitude –0.2.

**2028: December 31** Mercury is at greatest evening elongation, magnitude –0.3.

#### ECLIPSES

**2020: December 14** A total eclipse of the Sun is visible from the south Pacific, Chile, Argentina, and the south Atlantic. A partial eclipse is visible from the Pacific, southern South America, and Antarctica.

**2021: December 4** A total eclipse of the Sun is visible from Antarctica. A partial eclipse is visible from Antarctica, South Africa, and the southern Atlantic.

**2028: December 31** A total eclipse of the Moon is visible across the Arctic, eastern Europe, Asia, and Australasia.

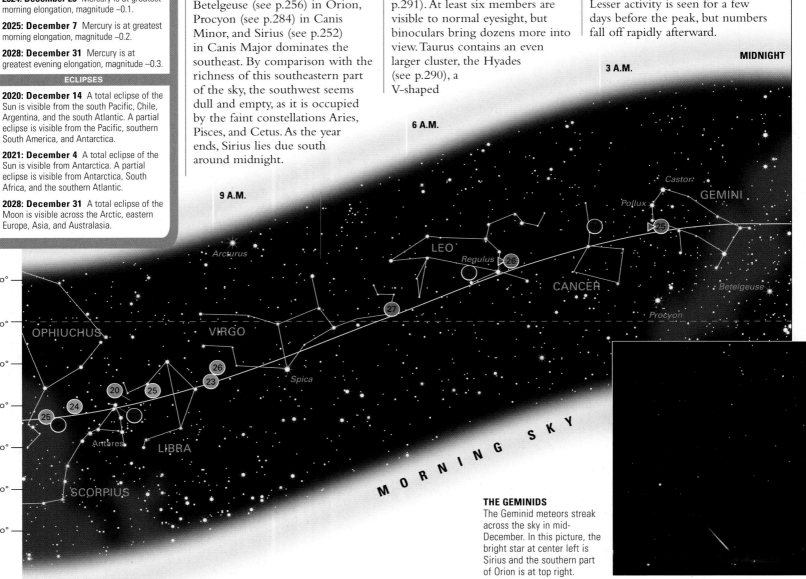

**THE GEMINIDS**
The Geminid meteors streak across the sky in mid-December. In this picture, the bright star at center left is Sirius and the southern part of Orion is at top right.

# SOUTHERN LATITUDES

## THE STARS

The distinctive figures of Orion and Taurus are high in the northeast, with Gemini and Auriga closer to the horizon. Perseus lies low in the north, while the Square of Pegasus sets in the northwest, followed by Pisces. Fomalhaut (see p.253) in Piscis Austrinus is in the southwest.

Eridanus, the River, meanders southwestward from the foot of Orion, ending at the bright star Achernar. Brighter Canopus is high in the southeast in Carina. The Large and Small Magellanic Clouds (see p.310 and p.311) lie high in the south, on either side of the celestial meridian. In the east, Betelgeuse in Orion, Procyon in Canis Minor, and Sirius in Canis Major form a large triangle, which is a sign of the approaching southern summer.

## DEEP-SKY OBJECTS

December and January evenings are the best time for southern observers to see the Pleiades (see p.291) and Hyades (see p.290), two large and prominent open star clusters north of the equator in Taurus. The Large Magellanic Cloud, containing the Tarantula Nebula, NGC 2070, is high in the southeast, but it is better seen in January. Overall, the southern evening sky is bereft of prominent deep-sky objects near the celestial meridian this month.

**THE LARGE MAGELLANIC CLOUD**
The LMC (bottom) lies deep in the southern sky between the bright stars Canopus (left) and Achernar (top right). The small pink patch on the LMC is the Tarantula Nebula.

### POSITIONS OF THE PLANETS

This chart shows the positions of the planets in December from 2020 to 2027. The planets are represented by colored dots, while the number inside each dot denotes the year. For all planets apart from Mercury, the dot indicates the planet's position on December 15. Mercury is shown only when it is at greatest elongation (see p.68)—for the specific date, refer to the table, left.

- Mercury
- Venus
- Mars
- Jupiter
- Saturn
- Uranus
- Neptune

**EXAMPLES**

20 Jupiter's position on December 15, 2020

23 Jupiter's position on December 15, 2023. The arrow indicates the planet is in retrograde motion (see p.68).

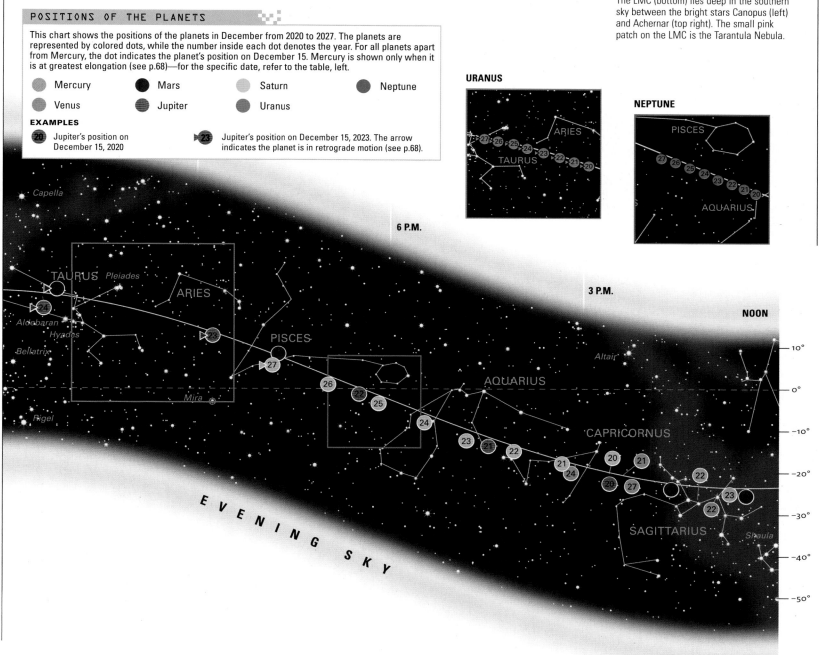

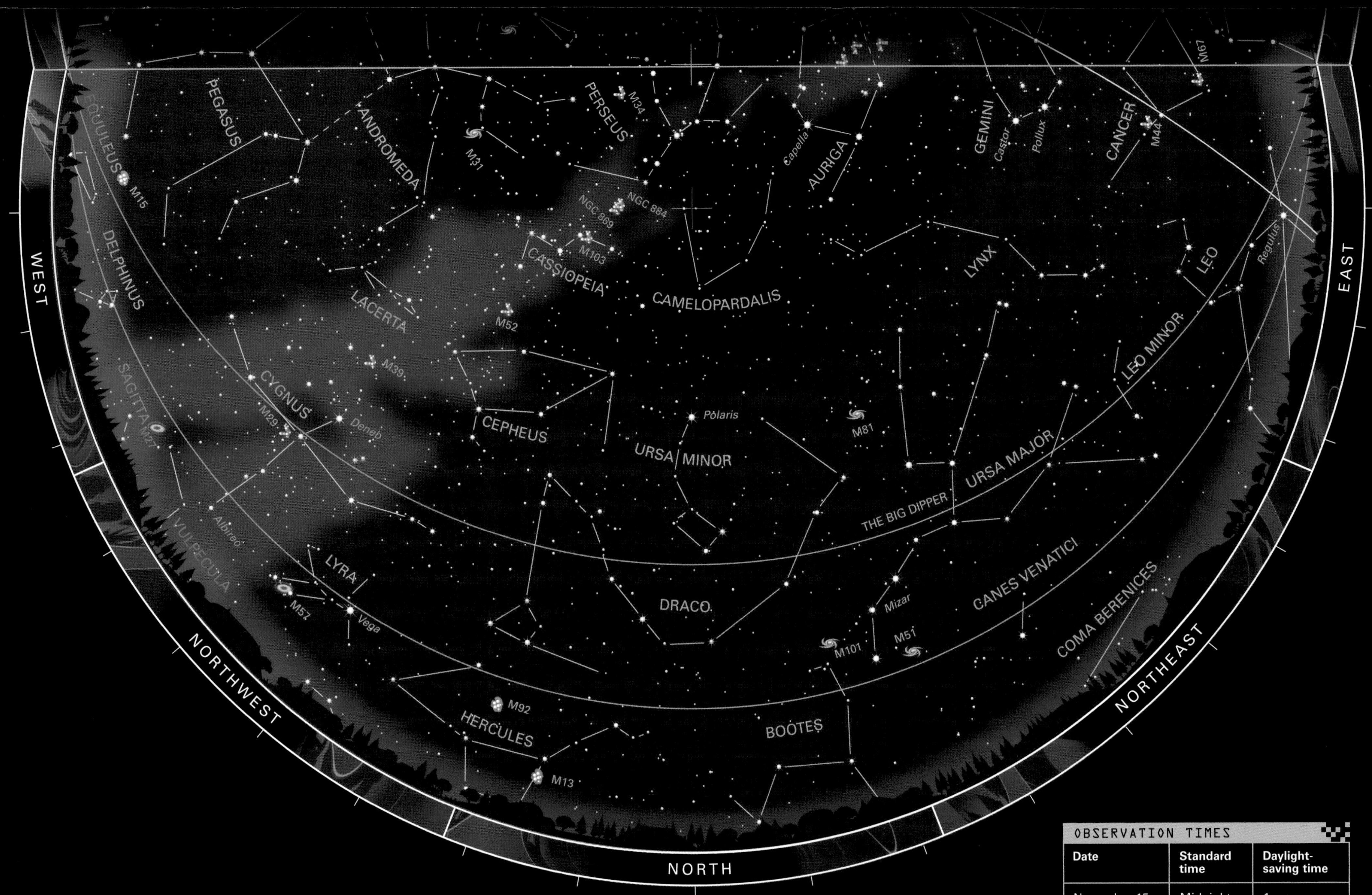

# DECEMBER | NORTHERN LATITUDES

## LOOKING NORTH

### STAR MAGNITUDES

- -1
- 0
- 1
- 2
- 3
- 4
- 5
- ⊙ Variable star

### DEEP-SKY OBJECTS

- Galaxy
- Globular cluster
- Open cluster
- Diffuse nebula
- Planetary nebula

### POINTS OF REFERENCE

**Horizons**

| 60°N | 40°N | 20°N |
|------|------|------|

**Zeniths**

| 60°N | 40°N | 20°N | Ecliptic |
|------|------|------|----------|

### OBSERVATION TIMES

| Date | Standard time | Daylight-saving time |
|------|---------------|----------------------|
| November 15 | Midnight | 1 a.m. |
| December 1 | 11 p.m. | Midnight |
| December 15 | 10 p.m. | 11 p.m. |
| January 1 | 9 p.m. | 10 p.m. |
| January 15 | 8 p.m. | 9 p.m. |

WEST

NORTHWEST

NORTH

NORTHEAST

EAST

WEST

DELPHINUS

SAGITTA M27

VULPECULA

Albireo

CYGNUS

PEGASUS

M15

M29

Deneb

LACERTA

ANDROMEDA

M39

M31

LYRA

M57

Vega

CEPHEUS

M52

CASSIOPEIA

M103

NGC 869

NGC 884

PERSEUS

M34

HERCULES

M92

M13

URSA MINOR

Polaris

DRACO

CAMELOPARDALIS

BOÖTES

THE BIG DIPPER

M81

AURIGA

Capella

M101

Mizar

M51

CANES VENATICI

URSA MAJOR

LYNX

GEMINI

Castor

Pollux

COMA BERENICES

LEO MINOR

CANCER

M44

LEO

M67

Regulus

# DECEMBER | NORTHERN LATITUDES

LOOKING SOUTH

STAR MOTION

North

South

## POINTS OF REFERENCE

| Horizons | 60°N | 40°N | 20°N | Zeniths | 20°N | 40°N | 60°N | Ecliptic |

## DEEP-SKY OBJECTS

Galaxy | Globular cluster | Open cluster | Diffuse nebula | Planetary nebula

## STAR MAGNITUDES

-1 | 0 | 1 | 2 | 3 | 4 | 5 | Variable star

THE NIGHT SKY

WEST

SOUTHWEST

SOUTH

SOUTHEAST

EAST

WEST

PEGASUS

ANDROMEDA

M33

TRIANGULUM

ARIES

PERSEUS

PLEIADES

AURIGA

M38

M36

M37

M1

TAURUS

Aldebaran

HYADES

ORION

Bellatrix

M42

Betelgeuse

GEMINI

M35

CANIS MINOR

Procyon

CANCER

M67

HYDRA

MONOCEROS

M50

M48

M47

M46

M93

PUPPIS

Sirius

Adhara

M41

CANIS MAJOR

LEPUS

COLUMBA

PICTOR

Canopus

DORADO

RETICULUM

CAELUM

HOROLOGIUM

ERIDANUS

FORNAX

CETUS

Mira

PISCES

ECLIPTIC

AQUARIUS

PISCIS AUSTRINUS

Fomalhaut

SCULPTOR

PHOENIX

HYDRUS

Achernar

Rigel

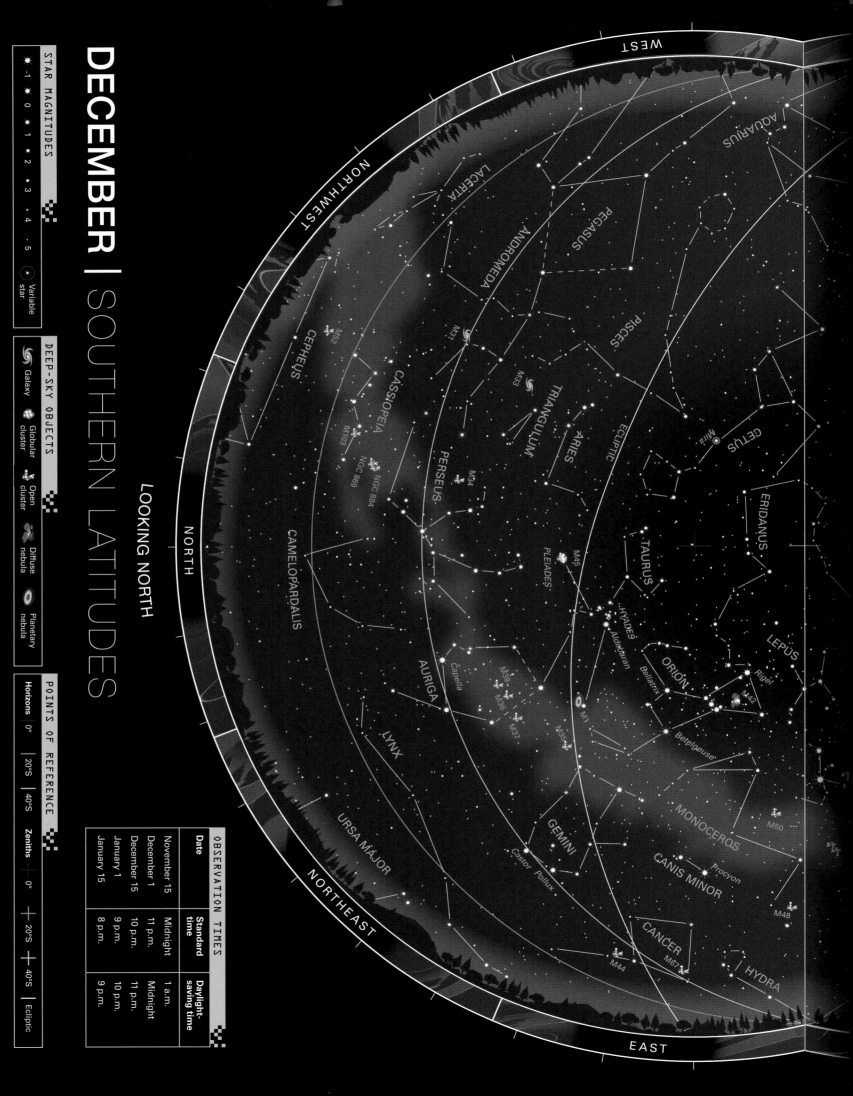

THE NIGHT SKY

# DECEMBER | SOUTHERN LATITUDES

LOOKING NORTH

## OBSERVATION TIMES

| Date | Standard time | Daylight-saving time |
|---|---|---|
| November 15 | Midnight | 1 a.m. |
| December 1 | 11 p.m. | Midnight |
| December 15 | 10 p.m. | 11 p.m. |
| January 1 | 9 p.m. | 10 p.m. |
| January 15 | 8 p.m. | 9 p.m. |

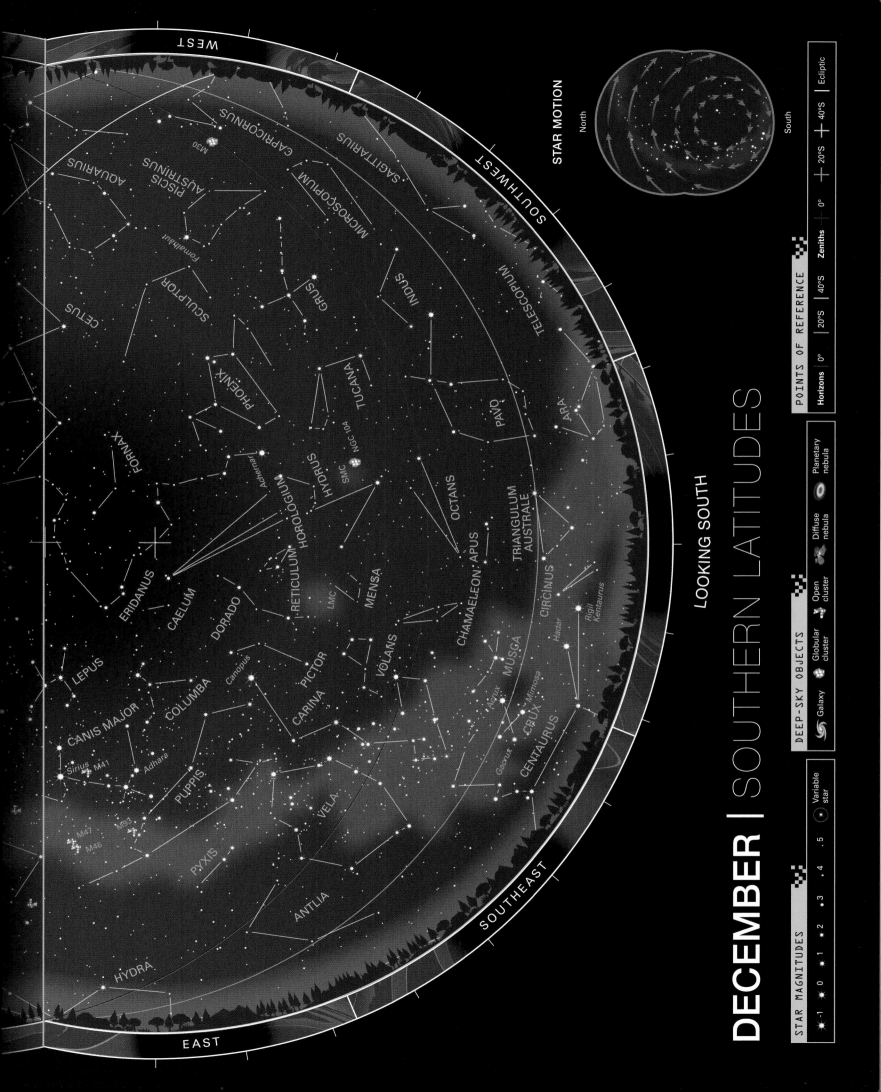

# DECEMBER | SOUTHERN LATITUDES

LOOKING SOUTH

STAR MOTION

North

South

## POINTS OF REFERENCE

| Horizons | 0° | 20°S | 40°S | Zeniths | 0° | 20°S | 40°S | Ecliptic |

## DEEP-SKY OBJECTS

Galaxy · Globular cluster · Open cluster · Diffuse nebula · Planetary nebula

## STAR MAGNITUDES

-1 · 0 · 1 · 2 · 3 · 4 · 5 · Variable star

WEST

SOUTHWEST

SOUTHEAST

EAST

CETUS

AQUARIUS

PISCIS AUSTRINUS

Fomalhaut

M30

CAPRICORNUS

SAGITTARIUS

MICROSCOPIUM

GRUS

INDUS

TELESCOPIUM

SCULPTOR

PHOENIX

TUCANA

NGC 104

SMC

PAVO

ARA

FORNAX

Achernar

HYDRUS

OCTANS

Rigil Kentaurus

RETICULUM

HOROLOGIUM

TRIANGULUM AUSTRALE

ERIDANUS

CAELUM

DORADO

LMC

MENSA

CHAMAELEON

APUS

CIRCINUS

Hadar

LEPUS

COLUMBA

PICTOR

Canopus

VOLANS

MUSCA

Acrux

Mimosa

CRUX

Gacrux

CENTAURUS

CANIS MAJOR

CARINA

Sirius

M41

Adhara

VELA

PUPPIS

M47

M46

M93

PYXIS

ANTLIA

HYDRA

THE NIGHT SKY

# GLOSSARY

## A

**absolute magnitude** see *magnitude*.

**absorption line** see *spectral line*.

**absorption nebula** see *nebula*.

**accelerating universe** A universe that expands at an accelerating rate. Current evidence indicates that the expansion of our universe had been slowing down under the action of gravity until about 6 billion years ago, but that since then it has been accelerating. The acceleration is believed to be driven by the repulsive influence of dark energy. See also *dark energy*.

**accretion (1)** The colliding and sticking together of small, solid particles and bodies to make progressively larger ones. **(2)** The process by which a body grows in mass by accumulating matter from its surroundings. An **accretion disk** is a disk of gas that revolves around a star or a compact object such as a white dwarf, neutron star, or black hole and has been drawn in from a companion star or from neighboring gas clouds.

**active galaxy** A galaxy that emits an exceptional amount of energy over a wide range of wavelengths, from radio waves to X-rays. An **active galactic nucleus (AGN)** is the compact, highly luminous core of an active galaxy that, in many cases, varies markedly in brightness over time, and is thought to be powered by the accretion of gas onto a supermassive black hole. See also *black hole, galaxy*.

**active prominence** see *prominence*.

**albedo** The ratio of the amount of light reflected by a body, such as a planet or a part of a planet's surface, to the amount of light that it receives from the Sun. Albedo values range from 0, for a perfectly dark object that reflects nothing, to 1, for a perfect reflector.

**altazimuth mounting** A mounting that enables a telescope to be rotated in altitude (around a horizontal axis) and in azimuth (around a vertical axis). Many large modern telescopes are mounted in this way, using computer-controlled motors to drive the telescope in altitude and azimuth so as to track the motion of an object across the sky. See also *altitude, azimuth, equatorial mounting*.

**altitude** The angular distance between the horizon and a celestial body. Altitude takes values from 0° (for an object on the horizon) to 90° (for an object that is directly overhead). See also *azimuth*.

**annual parallax** see *parallax*.

**annual proper motion** see *proper motion*.

**annular eclipse** see *eclipse*.

**antimatter** Material composed of antiparticles. See *antiparticle*.

**antiparticle** An elementary particle that has the same mass as a particle of ordinary matter but exactly opposite values of other quantities such as spin and electrical charge. For example, the antiparticle of the negatively charged electron is the positively charged **positron**. If a particle and its antiparticle collide, both are annihilated and converted into energy.

**aperture** The clear diameter of the objective lens or primary mirror of a telescope or other optical instrument.

**aphelion** The point on its elliptical orbit at which a body such as a planet, asteroid, or comet is at its greatest distance from the Sun.

**apogee** The point on its elliptical orbit around Earth at which a body such as the Moon or a spacecraft is at its greatest distance from Earth. See also *perigee*.

**apparent magnitude** see *magnitude*.

**arachnoid** A type of structure, found on the surface of Venus, that consists of concentric circular or oval fractures or ridges, together with a complex network of fractures or ridges that radiate outward. Its name derives from its superficial resemblance to a spiderweb. Typical diameters range from 30–110 miles (50 to 175 km).

**asterism** A conspicuous pattern of stars that is not itself a constellation. A well-known example is the Big Dipper, which forms part of the constellation Ursa Major (the Great Bear). See also *constellation*.

**asteroid** One of the vast number of small bodies that revolve independently around the Sun. Their diameters range from a few yards (meters) to around 600 miles (1,000 km). While the greatest concentration of asteroids is in the **Main Belt**, which lies between the orbits of Mars and Jupiter, asteroids are found throughout the solar system. A **near-Earth asteroid (NEA)** is a body whose orbit comes close to, or intersects, the orbit of Earth. Formally, a near-Earth asteroid is defined as one that has a perihelion distance of less than 1.3 times Earth's mean distance from the Sun. See also *Kuiper Belt*.

**astronomical unit (au)** A unit of distance measurement equal to the semimajor axis of Earth's elliptical orbit, equivalent to the average of the maximum and minimum distances between Earth and Sun. 1 au = 92,956,000 miles (149,598,000 km).

**atom** A basic building block of matter that is the smallest unit of a chemical element possessing the characteristics of that element. It consists of a nucleus of protons and neutrons, surrounded by a cloud of electrons. An atom has the same number of orbiting electrons as it has protons, so it is neutral (has no electrical charge). The chemical identity of an atom is determined by the number of protons in its nucleus (its **atomic number**). An atom of hydrogen (the simplest and lightest element) consists of a single proton and a single electron. See also *electron, neutron, proton*.

**aurora** A glowing, fluctuating display of light that is produced when charged particles entering a planet's upper atmosphere, usually in the vicinity of its north and south magnetic poles, collide with atoms and stimulate them to emit light.

**autumnal equinox** see *equinox*.

**azimuth** The angle between the north point on an observer's horizon and a celestial object, measured in a clockwise direction around the horizon. The azimuth of due north is 0°, due east 90°, due south 180°, and due west 270°. See also *altitude*.

## B

**background radiation** see *cosmic microwave background radiation*.

**barred spiral galaxy** A galaxy that has spiral arms emanating from the ends of an elongated, bar-shaped, nucleus. See also *galaxy, spiral galaxy*.

**baryon** A particle, composed of three quarks, that is acted on by the strong nuclear force. Examples include protons and neutrons, the building blocks of atomic nuclei.

**Big Bang** The event in which the universe was born. According to Big Bang theory, the universe originated a finite time ago in an extremely hot, dense initial state and ever since then has been expanding. The Big Bang was the origin of space, time, and matter.

**Big Crunch** The final state that will be reached by the universe if it eventually ceases to expand and then collapses in on itself.

**Big Rip** The tearing apart of all forms of structure in the universe—galaxy clusters, galaxies, stars, planets, atoms, and elementary particles—that is expected to occur should the repulsive effect of dark energy become infinitely strong in a finite time. See also *dark energy*.

**binary star** Two stars that revolve around each other under the influence of their mutual gravitational attraction. Each member star orbits the center of mass of the system, a point that lies closer to the more massive of the two stars. A **spectroscopic binary** is a system in which the two stars are too close to be resolved into separate points of light, but whose binary nature is revealed by its spectrum. The combined spectrum of the two stars contains two sets of spectral lines that shift in wavelength as the stars revolve around each other. An **eclipsing binary** is a system in which each star alternately passes in front of the other, cutting off all or part of its light and causing a periodic variation in the combined light of the two stars. See also *Doppler effect, spectral line*.

**BL Lacertae object** A type of active galaxy that has no detectable absorption or emission lines in its spectrum but which is believed to be similar to a quasar. The name derives from an object in the constellation Lacerta that was at first thought to be a variable star. See also *quasar*.

**black body** An idealized body that absorbs and reemits all the radiation that falls on its surface and which is a perfect radiator. A black body emits a continuous spectrum of radiation (**black-body radiation**) that peaks in brightness at a wavelength that depends on its surface temperature— the higher the temperature, the shorter the wavelength of peak brightness. See also *spectrum*.

**black dwarf star** A white dwarf star that has cooled to such a low temperature that it emits no detectable light. There has not been enough time since the origin of the universe for any star to cool down enough to become a black dwarf. See also *brown dwarf star, white dwarf star*.

**black hole** A compact region of space, surrounding a collapsed mass, within which gravity is so powerful that no material object, light, or any other kind of radiation can escape to the outside universe. The radius of a black hole is called the **Schwarzschild radius**, and its boundary is known as the **event horizon**. The greater its mass, the larger its radius. When a body collapses to form a black hole, all of its mass becomes compressed into a central point, a point of infinite density called a **singularity**. A **stellar-mass black hole** forms when the core of a high-mass star collapses; its mass is likely to be in the region of 3–100 times the mass of the Sun. A **supermassive black hole**, with a mass in the region of a few million to a few billion solar masses, is an object that forms when a very large mass collapses, or a number of black holes merge into one, in the core of a galaxy. See also *active galaxy, singularity*.

**blazar** The most variable type of active galaxy, which includes BL Lacertae objects and the most violently variable quasars. See also *active galaxy, BL Lacertae object, quasar.*

**blue shift** The displacement of spectral lines to shorter wavelengths that occurs when a light source is approaching an observer. See also *Doppler effect, red shift, spectral line.*

**Bok globule** A compact dark nebula, roughly spherical in shape, which contains 1 to 1,000 solar masses of gas and dust and has a diameter of between 0.1 and a few light-years. Globules of this kind are believed to be cool concentrations of gas and dust that eventually will collapse to form protostars. They are named after Dutch-born astronomer Bart Bok, who made a detailed study of these objects. See also *protostar.*

**brown dwarf star** A body that forms out of a contracting cloud of gas in the same way as a star, but, because it contains too little mass, never becomes hot enough to ignite the nuclear-fusion reactions that power a normal star. With less than 8 percent of the Sun's mass, a brown dwarf glows dimly at infrared wavelengths, fading gradually as it cools down.

# C

**caldera** A bowl-shaped depression caused by the collapse of a volcanic structure into an emptied magma chamber. A caldera is usually found at the summit of shield volcanoes such as those found on Venus, Earth, and Mars.

**captured rotation** See *synchronous rotation.*

**carbonaceous chondrite** see *chondrite.*

**cataclysmic variable** see *variable star.*

**catadioptric telescope** A type of telescope that combines mirror and lens components, rather than one or the other, to bring light to a focus. Schmidt–Cassegrain telescopes are a popular type of catadioptric telescope. See also *reflecting telescope, Schmidt–Cassegrain telescope.*

**celestial equator** A great circle on the celestial sphere that is a projection of Earth's own equator onto the celestial sphere. See also *celestial sphere, great circle.*

**celestial poles** The two points at which the line of Earth's axis, extended outward, meets the celestial sphere and around which the stars appear to revolve. The north celestial pole lies directly above Earth's North Pole and the south celestial pole directly above Earth's South Pole. See also *celestial sphere.*

**celestial sphere** An imaginary sphere that surrounds Earth. As Earth rotates from west to east, the sphere appears to rotate from east to west. In order to define the positions of stars and other celestial bodies, it is convenient to think of them as being attached to the inside surface of this sphere. See also *celestial equator, celestial poles.*

**center of mass** The point within an isolated system of bodies around which those bodies revolve. Where the system consists of two bodies (for example, a binary star), it is located at a point on a line joining their centers. If both bodies have the same mass, the center of mass lies midway between them, whereas if one body is more massive than the other, it lies closer to the more massive of the two.

**Cepheid variable** A type of variable star that increases and decreases in brightness in a regular, periodic way. Cepheids are pulsating variables, which vary in brightness as they expand and contract. The more luminous the Cepheid, the longer its period of variation. See also *variable star.*

**Chandrasekhar limit** The maximum possible mass for a white dwarf star. If the mass of a white dwarf exceeds this limit, which is about 1.4 solar masses, gravity will overwhelm its internal pressure and it will collapse. The limit was first calculated by Indian astrophysicist Subrahmanyan Chandrasekhar in 1931. See also *white dwarf star.*

**charge-coupled device (CCD)** An electronic imaging device that consists of a large array of tiny light-sensitive elements. The image of an object is constructed by reading off the electrical charges that accumulate in each element during an exposure.

**chondrite** A stony meteorite that contains a large number of small, spherical objects called **chondrules**. A **carbonaceous chondrite** is one that is rich in carbon, carbon compounds, and volatile materials. Carbonaceous chondrites are thought to be some of the least-altered primitive remnants of the protoplanetary disk from which the solar system formed. See also *meteorite, protoplanetary disk.*

**chromosphere** The thin layer in the Sun's atmosphere that lies between the photosphere (the visible surface) and the corona. Its faint, reddish-pink light can be seen directly during a total eclipse of the Sun when the Moon hides the dazzling photosphere. See also *photosphere.*

**circumpolar** A term used to describe a star, or other celestial body, that remains above the horizon at all times when viewed from a particular place on Earth's surface.

**circumstellar disk** A flattened, disk-shaped cloud of gas and dust that surrounds a star. A disk of this kind is usually associated with a young or newly forming star, in which case it is composed of material from the original dusty gas cloud that collapsed to form the central star. See also *protoplanetary disk.*

**closed universe** A universe that is curved in such a way that space is finite but has no discernible boundary (analogous to the surface of a sphere). A universe will be closed if its average density exceeds a particular value called the **critical density**. In the absence of a repulsive force, a closed universe will eventually cease to expand and will then collapse. See also *flat universe, open universe, oscillating universe.*

**coma** The cloud of gas and dust that surrounds the nucleus of a comet and which comprises its glowing "head." See also *comet.*

**comet** A small body composed mainly of dust-laden ice that revolves around the Sun, usually in a highly elongated orbit. Each time it approaches the Sun, gas and dust evaporate from its nucleus (the solid core of the comet) to form an extensive cloud, called the coma, and one or more tails. See also *coma, tail.*

**conjunction** A close alignment in the sky of two celestial bodies, which occurs when both bodies lie in the same direction as viewed from Earth. When a planet lies directly on the opposite side of the Sun from Earth, it is said to be at **superior conjunction**. If a planet passes between Earth and the Sun (Mercury and Venus are the only planets that can do this), it is said to be at **inferior conjunction**. See also *opposition.*

**constellation** One of 88 regions of the celestial sphere. Each constellation contains a grouping of stars joined by imaginary lines to represent a figure. The constellations are officially referred to by the Latin names of these figures. Many have been named after mythological characters or creatures (such as Orion, the Hunter) but some after more mundane objects (for example, Sextans, the Sextant). See also *asterism.*

**continuous spectrum** see *spectrum.*

**convection** The transport of heat by rising bubbles or plumes of hot liquid or gas. In a **convection cell**, rising streams of hot material cool, spread out, and then sink down to be reheated, so maintaining a continuous circulation.

**core (1)** The dense central region of a planet. **(2)** The central region of a star within which energy is generated by means of nuclear-fusion reactions. **(3)** A dense concentration of material within a gas cloud.

**Coriolis effect** The tendency of a wind or current to be deflected from its initial direction as a consequence of a planet's rotation. In the case of Earth, the deflection is to the right in the Northern Hemisphere and to the left in the Southern Hemisphere.

**corona** The outermost region of the atmosphere of the Sun or a star. The solar corona has an extremely low density and a very high temperature (about 2–9 million degrees Fahrenheit/ 1–5 million degrees Celsius). It cannot be observed except during a total eclipse of the Sun. See also *eclipse, solar wind.*

**coronal mass ejection** A huge, rapidly expanding bubble of plasma that is ejected from the Sun's corona. Containing billions of tons of material in the form of ions and electrons, together with associated magnetic fields, a typical coronal mass ejection propagates outward through interplanetary space at a speed of several hundred miles (kilometers) per second. See also *corona, ion, plasma.*

**cosmic microwave background radiation (CMBR)** Remnant radiation from the Big Bang, which is detectable as a faint distribution of microwave radiation across the whole sky. See also *Big Bang.*

**cosmic rays** Highly energetic subatomic particles, such as electrons, protons, and atomic nuclei, that hurtle through space at speeds close to the speed of light.

**cosmological constant** An extra term in Einstein's relativity equations which, if it has a positive value, corresponds to a repulsive force that could cause the universe to expand at an accelerating rate. Modern cosmologists associate the constant with a quantity called **vacuum energy** (residual energy that, according to quantum theory, exists even in a vacuum), one of the possible forms of the dark energy believed to permeate the universe. See also *dark energy.*

**cosmological red shift** see *red shift.*

**cosmology** The study of the nature, structure, origin, and evolution of the universe.

**crater** A bowl- or saucer-shaped depression in the surface of a planet or satellite, or at the summit of a volcano. Many have raised walls and some have a central peak. An **impact crater** is one excavated by a meteorite, asteroid, or comet impact, whereas a **volcanic crater** is the cavity from which a volcano discharges material. Raised walls are created by accumulation of ejected material.

**critical density** see *flat universe.*

**crust** The thin, rocky, outermost layer of a planet or major planetary satellite, which, like Earth, has separated into several layers, with the densest material toward its center and the least dense at its surface.

# D

**dark energy** A little-understood form of energy that appears to comprise about 70 percent of the total amount of mass and energy in the universe. It exerts a repulsive effect and is believed to be causing the expansion of the universe to accelerate. See also *accelerating universe.*

**dark matter** Matter that exerts a gravitational influence on its surroundings but does not emit detectable amounts of radiation. Dark matter appears to make up a large fraction of the total amount of mass contained in galaxies, galaxy clusters, and the universe as a whole.

**dark nebula** see *nebula*.

**declination** The angular distance of a celestial body north or south of the celestial equator. Declination is positive (+) if the object is north of the celestial equator and negative (-) if it is south of the celestial equator. A star on the celestial equator has a declination of 0°, whereas a star at one of the celestial poles has a declination of 90°. See also *celestial equator, celestial sphere, right ascension*.

**declination axis** see *equatorial mounting*.

**diffuse nebula** A luminous cloud of gas and dust. The term "diffuse" refers to the cloud's fuzzy appearance and to the fact that it cannot be resolved into individual stars. See also *nebula*.

**direct motion** see *retrograde motion*.

**direct rotation** see *retrograde rotation*.

**Doppler effect** The observed change in the wavelength or frequency of radiation that is caused by the motion of its source toward or away from an observer. See also *blue shift, red shift*.

**double star** Two stars that appear close together in the sky. If the two stars revolve around each other, the system is called a binary. An **optical double star** consists of two stars that appear to be close together only because they happen to lie in almost exactly the same direction when viewed from Earth; they lie at different distances and are not physically connected. See also *binary star*.

**dwarf planet** A celestial body that orbits the Sun and has sufficient mass and gravity to be spherical, but has not cleared the region around its orbit of other bodies, and is not a satellite.

**dwarf star** An alternative name for a main-sequence star that was originally devised to distinguish main-sequence stars, such as the Sun, from the much more luminous giant stars on the Hertzsprung–Russell diagram. See also *Hertzsprung–Russell diagram, main sequence*.

# E

**eccentricity (e)** A measure of how much an ellipse deviates from a perfect circle. Eccentricity takes a value between 0 and 1; a circle has eccentricity of 0, and the most elongated ellipses approach an eccentricity of 1. See also *ellipse*.

**eclipse** The passage of one celestial body into the shadow cast by another. A **lunar eclipse** occurs when the Moon passes into Earth's shadow and a **solar eclipse** when part of Earth's surface enters the shadow cast by the Moon. A **total lunar eclipse**

takes place when the whole of the Moon lies within the dark cone of Earth's shadow, and a **partial lunar eclipse** when only part of the Moon is in the shadow. During a **total solar eclipse**, the Sun is completely obscured by the dark disk of the Moon. A **partial solar eclipse** occurs when only part of the Sun's surface is hidden. If the Moon passes directly between the Sun and Earth when it is close to apogee, it will appear smaller than the Sun, and its dark disk will be surrounded by a ring, or annulus, of sunlight; an event of this kind is called an **annular eclipse**. See also *apogee*.

**eclipsing binary** see *binary star*.

**ecliptic** The track along which the Sun appears to travel around the celestial sphere, relative to the background stars, in the course of a year. It is equivalent to the plane of Earth's orbit.

**ejecta** Material thrown outward by the blast of an impact. Ejecta, which is produced when a meteorite strikes the surface of a planet or moon and excavates a crater, consists of freshly exposed material that may be markedly brighter than the adjacent surface. Sometimes the ejected material forms extensive streaks, or rays, which radiate from the point of impact. An **ejecta blanket** is a continuous sheet of deposited ejecta that surrounds a crater. See also *crater*.

**electromagnetic (EM) radiation** Oscillating electric and magnetic disturbances that propagate energy through space in the form of waves (electromagnetic waves). Examples include light and radio waves.

**electromagnetic spectrum** The complete range of electromagnetic radiation from the shortest wavelengths (gamma rays) to the longest wavelengths (radio waves).

**electron** A lightweight fundamental particle with negative electrical charge. A cloud of electrons surrounds the nucleus of an atom. The number of orbiting electrons in an atom is the same as the number of protons in its nucleus.

**ellipse** An oval curve drawn around two points called **foci** (singular: **focus**) such that the total distance from one focus to any point on the curve and then back to the other focus is constant. The maximum diameter of an ellipse is the **major axis**, and half of this diameter is the **semimajor axis**. The two foci lie on the major axis; the greater their separation, the more elongated the ellipse. See also *eccentricity, orbit*.

**elliptical galaxy** A galaxy that appears round or elliptical in shape and normally contains very little gas or dust. See also *galaxy*.

**elongation** The angle between the Sun and a planet, or other solar system body, when viewed from Earth. The elongation of a planet is 0° when it is in conjunction with the Sun and 180° when it is at opposition. **Greatest**

**elongation** is the maximum possible elongation of a body, such as Mercury or Venus, that lies inside the orbit of Earth. See also *conjunction, opposition*.

**emission line** see *spectral line*.

**emission nebula** see *nebula*.

**equatorial mounting** A mounting that allows a telescope to be turned around two axes, one of which (the **polar axis**) is parallel to, and the other (the **declination axis**) perpendicular to, Earth's axis of rotation. The telescope can follow the motion of a celestial object across the sky by being driven around the polar axis in the opposite direction of Earth's rotation at a rate of one revolution per sidereal day. See also *declination, right ascension, sidereal time*.

**equinox** An occasion when the Sun is vertically overhead at a planet's equator, and day and night have equal duration for the whole planet. In the case of Earth, the northern **vernal equinox** is the point at which the Sun crosses the celestial equator from south to north, on or around March 20 each year, and the northern **autumnal equinox** is the point at which the Sun crosses the celestial equator from north to south, on or around September 22. See also *right ascension*.

**eruptive prominence** see *prominence*.

**eruptive variable** see *variable star*.

**escape velocity** The minimum speed at which a projectile must be launched in order to recede forever from a massive body and not fall back. The escape velocity at Earth's surface is 25,200 mph (40,320 km/h).

**event horizon** see *black hole*.

**extrasolar planet (exoplanet)** A planet that revolves around a star other than the Sun.

# F

**facula (plural: faculae)** A patch of enhanced brightness on the solar photosphere that may be seen in a white-light image of the Sun, usually near the edge of the Sun's visible disk where the background brightness is lower. Faculae correspond to regions that are hotter than their immediate surroundings. They are associated with active solar regions but may appear before, and persist after, any sunspots that develop in those regions. See also *photosphere, sunspot*.

**Fraunhofer line** One of the 574 dark absorption lines in the spectrum of the Sun that were identified by the 19th-century German optician and instrument-maker Joseph von Fraunhofer. See also *spectral line*.

**flare star** A faint, cool, red dwarf star that displays sudden, short-lived increases in luminosity caused by extremely powerful flares that occur above its surface. See also *red dwarf star*.

**flat universe** A universe in which the overall net curvature of space is zero.

In such a universe, space is flat in the sense that, apart from localized distortions caused by massive bodies, its large-scale geometry is Euclidean and light rays travel in straight lines. A universe will be flat if its overall average density is equal to a particular value, called the **critical density**. See also *closed universe, open universe, oscillating universe*.

**focal length** The distance between the center of a lens, or the front surface of a concave mirror, and the point at which it forms a sharp image of a very distant object.

**frequency** The number of wave crests of a wave motion that pass a given point in one second. In the case of an electromagnetic wave (for example, light) the frequency is equal to the speed of light divided by the wavelength. See also *electromagnetic radiation*.

**fusion (nuclear fusion)** The process by which atomic nuclei are joined together during energetic collisions to form heavier atomic nuclei, with an associated release of large amounts of energy. Stars are powered by fusion reactions that take place in their central cores. In a main-sequence star such as the Sun, fusion reactions convert hydrogen into helium. See also *main sequence*.

# G

**galactic cluster** see *open cluster*.

**galaxy** A large aggregation of stars and clouds of gas and dust. Galaxies, which may be elliptical, spiral, or irregular in shape, contain from a few million to several trillion stars and have diameters ranging from a few thousand to over a hundred thousand light-years. The Sun is a member of the **Milky Way galaxy**, which is also sometimes known as **the Galaxy**. See also *Milky Way*.

**galaxy cluster** An aggregation of galaxies held together by gravity. Clusters that contain up to a few tens of member galaxies are called groups. Larger clusters are divided into **regular** and **irregular clusters**, depending on their degree of structure. The most richly populated regular clusters (**rich clusters**) contain up to several thousand galaxies.

**galaxy supercluster** A cluster of galaxy clusters, which is a loose aggregation of up to about ten thousand galaxies, spread through a volume of space with a diameter of up to about 200 million light-years. See also *galaxy cluster*.

**Galilean moon** One of the four largest natural satellites of the planet Jupiter, which were discovered in 1610 by the Italian astronomer Galileo Galilei. In order of distance from the planet, they are Io, Europa, Ganymede, and Callisto.

**gamma radiation** Electromagnetic radiation with extremely short wavelengths (shorter than X-rays) and very high frequencies. Gamma radiation occupies the shortest-wavelength region of the spectrum. See also *electromagnetic radiation, electromagnetic spectrum.*

**gamma-ray burst (GRB)** A sudden burst of gamma radiation from a source in a distant galaxy. Gamma-ray bursts are the most powerful explosive events in the present-day universe. They may be triggered by collisions between neutron stars or black holes or by an extreme version of a supernova called a **hypernova**.

**gas planet (gas giant)** A large planet that, like Jupiter or Saturn, consists predominantly of hydrogen and helium. Beneath its thick gaseous atmosphere, the pressure is so great that hydrogen and helium exist in liquid form. See also *rocky planet.*

**gegenschein** A very faint patch of light that sometimes may be seen on a clear, moonless night in the region of sky directly opposite the position of the Sun. It is caused by sunlight that has been reflected back toward Earth by interplanetary dust particles lying beyond the orbit of Earth. See also *zodiacal light.*

**general theory of relativity** see *relativity.*

**geocentric (1)** Treated as being viewed from the center of Earth. **(2)** Having Earth at the center (of a system). **Geocentric coordinates** are a system of positional measurements (such as right ascension and declination) that are treated as being measured from the center of Earth. A satellite that is traveling around Earth is in a **geocentric orbit**. **Geocentric cosmology** was the ancient theory that the Sun, Moon, planets, and stars revolved around a central Earth. See also *heliocentric.*

**giant star** A star that is larger and much more luminous than a main-sequence star of the same surface temperature. See also *Hertzsprung–Russell diagram, main sequence, red giant.*

**globular cluster** A near-spherical cluster of between 10,000 and more than 1 million stars. Globular clusters, which consist of very old stars, are located predominantly in the halos of galaxies. See also *open cluster.*

**gravitation** see *gravity.*

**gravitational lens** A massive body, or a distribution of mass (such as a galaxy cluster), whose gravitational field deflects light rays from a more distant background object, thereby acting as a lens to produce a magnified or distorted image, or images, of that background object.

**gravitational wave** A wavelike distortion of space that propagates at the speed of light. Although waves of this kind have not yet been detected directly, there is strong indirect evidence that they exist.

**gravity** The attractive force that acts between material bodies, particles, and photons. According to the theory of gravity developed in the 17th century by Isaac Newton (**Newtonian gravitation**), the force of gravity acting between two bodies is proportional to the product of their masses divided by the square of the distance between their centers. For example, if the distance between the bodies is doubled, the force of attraction is reduced to one quarter of its previous value. See also *relativity.*

**great circle** A circle on the surface of a sphere, the plane of which passes through the center of the sphere and which exactly divides the sphere into two equal hemispheres. Its name derives from the fact that it is the largest circle that can be drawn on the surface of a sphere. See also *celestial equator, meridian.*

**greenhouse effect** The process by which atmospheric gases make the surface of a planet hotter than would be the case if the planet had no atmosphere. Incoming sunlight is absorbed at the surface of a planet and reradiated as infrared radiation, which is then absorbed by **greenhouse gases** such as carbon dioxide, water vapor, and methane. Part of this trapped radiation is reradiated back down toward the ground, so raising its temperature.

# H

**HII region** A glowing region of ionized hydrogen surrounding one or more hot, highly luminous stars. An HII region is often just a part of a more extensive cloud of gas and dust, the remainder of which has not been ionized and is not shining. See also *ion, nebula.*

**halo** A spherical region surrounding a galaxy that contains a distribution of globular clusters, thinly scattered stars, and some gas. A **dark-matter halo** is a distribution of dark matter within which a galaxy is embedded.

**heliocentric (1)** Treated as being viewed from the center of the Sun. **(2)** Having the Sun at the center (of a system). **Heliocentric coordinates** specify the position of an object as seen from the center of the Sun. A body that is revolving round the Sun follows a **heliocentric orbit**. **Heliocentric cosmology** is a model of the universe, such as the one proposed in 1543 by Nicolaus Copernicus, in which the planets revolve around a central Sun.

**heliosphere** The region of space around the Sun within which the solar wind and interplanetary magnetic field are confined by the pressure of the interstellar medium. Its boundary is called the **heliopause**. See also *interstellar medium, solar wind.*

**helium burning** The generation of energy by means of fusion reactions that convert helium into carbon and oxygen. Helium burning takes place in the core of a star that has left the main sequence and become a red giant, and it may occur again, later in a star's evolution, in a shell surrounding the core. See also *fusion, main sequence, red-giant star.*

**Hertzsprung–Russell (HR) diagram** A diagram on which stars are plotted as points according to their luminosity and surface temperature. Luminosity (or absolute magnitude) is plotted on the vertical axis, and surface temperature (or spectral type or color) is plotted on the horizontal axis. Astrophysicists use the Hertzsprung–Russell diagram to classify stars. Depending on a star's position on the diagram, it may be classified as, for example, a main-sequence star, a giant, or a dwarf.

**Hubble constant** see *Hubble's law.*

**Hubble's law** The observed relationship between the red shifts in the spectra of remote galaxies and their distances, which implies that the speeds at which galaxies are receding are directly proportional to their distances. The **Hubble constant** (or **Hubble parameter**)—denoted by the symbol $H_0$—is the constant of proportionality that relates speed of recession to distance.

**hydrogen burning** The generation of energy by means of fusion reactions that convert hydrogen into helium. Hydrogen burning takes place in the core of a main-sequence star. When a star has consumed all the available hydrogen in its core, the core contracts and hydrogen burning then continues in a thin shell surrounding the core. See also *fusion, main sequence, proton–proton reaction.*

**hypernova** see *gamma-ray burst.*

# I

**impact crater** see *crater.*

**inclination** The angle at which one plane is tilted relative to another. The inclination of a planetary orbit is the angle between its plane and the plane of the ecliptic (the plane of Earth's orbit). The inclination of a planet's equator is the angle between the plane of its orbit and the plane of its equator. See also *ecliptic, orbit.*

**inferior conjunction** see *conjunction.*

**inferior planet** A planet that travels around the Sun on an orbit that is inside the orbit of Earth. The two inferior planets are Mercury and Venus. See also *superior planet.*

**inflation** A sudden, short-lived episode of accelerating expansion thought to have occurred at a very early stage in the history of the universe (about $10^{-35}$ seconds after the beginning of time). See also *Big Bang.*

**infrared radiation** Electromagnetic radiation with wavelengths longer than visible light but shorter than microwaves or radio waves. Infrared is the dominant form of radiation emitted from many cool astronomical objects, such as interstellar dust clouds. See also *electromagnetic radiation.*

**interstellar medium** The gas and dust that permeates the space between the stars within a galaxy.

**ion** A particle or system of particles with a net electrical charge. Positive ions are commonly formed when an atom loses one or more of its electrons, whereas negative ions result from an excess of electrons. Ions may form from complexes of former atoms. The process by which an atom or complex gains or loses an electron to become charged is called **ionization**. See also *electron, photon.*

**irregular cluster** see *galaxy cluster.*

**irregular galaxy** A galaxy that has no well-defined structure or symmetry.

**isotope** Any one of two or more forms of a particular chemical element, the atoms of which contain the same number of protons but different numbers of neutrons. For example, helium-3 and helium-4 are isotopes of helium; a nucleus of helium-4 (the heavier, and more common, isotope) contains two protons and two neutrons, whereas a nucleus of helium-3 contains two protons and one neutron. See also *atom, nucleus.*

# K

**Kepler's laws of planetary motion** Three laws, devised in the early 17th century by Johannes Kepler, that describe the orbital motion of planets around the Sun. In essence, the first law states that each planet's orbit is an ellipse, the second shows that a planet's speed varies as it travels around its orbit, and the third links its orbital period (the time taken to travel around the Sun) to its average distance from the Sun.

**Kuiper Belt (Edgeworth–Kuiper Belt)** A flattened distribution of icy planetesimals that orbit the Sun at distances in the region of 30–100 times Earth's distance from the Sun; it is the source of many of the shorter-period comets. See also *Oort Cloud, planetesimal.*

# L

**lenticular galaxy** A galaxy that is shaped like a convex lens. It has a central bulge that merges into a disk, but no spiral arms. See also *galaxy, spiral galaxy.*

**lepton** A fundamental particle, such as an electron or a neutrino, that is not acted on by the strong nuclear force.

**light-year (ly)** A unit of distance equal to the distance light travels in one year—5,878 billion miles (9,460 billion km).

**limb** The edge of the observed disk of the Sun, the Moon, or a planet.

**Local Group** The small cluster of more than 40 member galaxies to which the Milky Way galaxy belongs. The other major members are the spiral galaxies M31 (the Andromeda Galaxy) and M33. Most of the members are small (or dwarf) elliptical or irregular galaxies. See also *galaxy cluster*.

**local sidereal time** see *sidereal time*.

**luminosity** The total amount of energy emitted in one second by a source of radiation, such as the Sun or a star. The luminosity of a star can be expressed in watts or in units of solar luminosity (the luminosity of the Sun is 3.8 x 10²⁶ watts). Stars are divided into luminosity classes denoted by Roman numerals. See also *magnitude*.

**lunar eclipse** see *eclipse*.

# M

**MACHO** An acronym for MAssive Compact Halo Object, a very low-luminosity object—such as a planet, brown dwarf, exceedingly dim white dwarf, or black hole—that exists in the halo of a galaxy but is usually too faint to be seen directly. MACHOs are believed to account for a relatively small proportion of the unseen dark matter in a galaxy's halo. See also *dark matter, halo*.

**magnetic field** The region of space surrounding a magnetized body within which its magnetic influence affects the motion of an electrically charged particle.

**magnetosphere** The region of space around a planet within which the motion of charged particles is controlled by the planetary magnetic field rather than the solar wind and the associated interplanetary magnetic field. The shape of a planet's magnetosphere is influenced by the solar wind, which squeezes it inward on the Sun-facing side and drags it out to form an elongated "tail" (a **magnetotail**) on the opposite, or downstream, side. See also *solar wind*.

**magnification** The increase in the apparent angular size of an object when viewed through an optical instrument, such as a telescope. The magnification of a telescope is equal to the focal length of its objective lens or primary mirror divided by the focal length of its eyepiece.

**magnitude (absolute and apparent)** **Apparent magnitude** is a measure of the apparent brightness of an object as seen in the sky. The fainter the object, the higher the numerical value of its magnitude. The faintest stars visible to the naked eye are of magnitude 6, whereas the brightest objects in the sky have negative apparent magnitudes. A star said to be of 1st magnitude has a magnitude of 1.49 or less, a star of 2nd magnitude has a value of 1.50 to 2.49, and so on. **Absolute magnitude** is the apparent magnitude a star would have if it were located at a standard distance of 10 parsecs (32.6 light-years) from Earth. See also *luminosity, parsec*.

**Main Belt** see *asteroid*.

**main sequence** A band that slopes diagonally from the upper left (hot, high-luminosity region) to the lower right (cool, low-luminosity region) of the Hertzsprung–Russell diagram and contains about 90 percent of stars. Main-sequence stars, such as the Sun, shine by converting hydrogen in their cores to helium. See also *dwarf star, Hertzsprung–Russell diagram*.

**major axis** see *ellipse*.

**mantle** The rocky layer that lies between the core and the crust of a rocky (Earth-like) planet or a major planetary satellite. See also *core, crust*.

**mare (plural: maria)** A relatively smooth, dark, lava-filled basin on the surface of the Moon. The name derives from the Latin for "sea."

**massive compact halo object** see *MACHO*.

**meridian (1)** A great circle on the surface of Earth or another astronomical body that passes through the north and south poles and crosses the equator at right angles. **(2)** A great circle on the celestial sphere that passes through the north and south celestial poles and crosses the celestial equator at right angles. An observer's local meridian passes through the celestial pole, the zenith, and the north and south points of the horizon. See also *celestial sphere, great circle*.

**Messier catalog** A widely used catalog of nebulous objects (most of them nebulae, star clusters, and galaxies) that was published in 1781 by the French astronomer Charles Messier. Objects contained in this catalog are designated by the letter "M" followed by a number. For example, M31 is the Andromeda Galaxy and M42 is the Orion Nebula. See also *New General Catalog*.

**meteor** The short-lived streak of light seen when a meteoroid plunges into Earth's atmosphere and is heated to incandescence by friction. A **sporadic meteor** is one that appears at a random time from a random direction. A **meteor shower** is a substantial number of meteors that appear to radiate from a common point in the sky (the **radiant**) when Earth is passing through a stream of meteoroids. See also *meteorite, meteoroid*.

**meteorite** A rocky or metallic meteoroid that survives passage through the atmosphere and reaches Earth's surface in one piece or in fragments. See also *meteor, meteoroid*.

**meteoroid** A lump or small particle of rock, metal, or ice orbiting the Sun in interplanetary space. Sizes of meteoroids range from a small fraction of an inch (a fraction of a millimeter) to a few yards (meters). Some are debris from collisions between asteroids. Others are particles released by comets; these spread out along cometary orbits to form meteoroid streams. See also *asteroid, comet, meteor, meteorite*.

**Milky Way (1)** The spiral galaxy that contains the Sun, sometimes also referred to as the **Milky Way galaxy** or **the Galaxy**. **(2)** A faint, misty band of light that stretches across the night sky and consists of the combined light of vast numbers of stars and nebulae that lie in the disk and spiral arms of our galaxy. See also *galaxy*.

**Mira variable** A class of long-period variable stars named after the star Mira—Omicron (o) Ceti—in the constellation Cetus. Mira variables are cool, giant pulsating stars that vary in brightness over periods ranging from 100 days to more than 500 days. See also *variable star*.

**molecular cloud** A cool, dense cloud of gas and dust in which the temperature is sufficiently low to enable atoms to join together to form molecules such as molecular hydrogen ($H_2$) or carbon monoxide (CO), and within which conditions are favorable for star formation.

**moon** Also known as a **natural satellite**, a body that orbits a planet. **The Moon** is Earth's natural satellite. Orbiting Earth at a mean distance of 239,000 miles (384,000 km) in a period of 27.3 days, it has a diameter of 2,159 miles (3,476 km). See also *satellite*.

**moon dog** See *sun dog*.

**multiple star** A system consisting of two or more stars bound together by gravity and revolving around each other (a system of just two stars is also called a binary). See also *binary star*.

# N

**near-Earth asteroid** see *asteroid*.

**nebula (plural: nebulae)** A cloud of gas and dust in interstellar space. The name derives from the Latin for "cloud." There are several types of luminous nebula (nebulae that shine). An **emission nebula** is a cloud of gas and dust that contains one or more extremely hot, young, high-luminosity stars; ultraviolet light emitted by these stars causes the surrounding gas to glow. Nebulae of this kind are also called HII regions because they contain a large proportion of ionized hydrogen. A **reflection nebula** is observed when the dust particles within a cloud are lit up by light from a neighboring bright star. Other types of luminous nebulae include planetary nebulae (shells of gas puffed out by dying stars) and supernova remnants (the debris of exploded stars). A **dark nebula** (or **absorption nebula**) is a dust-laden cloud that blocks out light from background stars and appears as a dark patch in the sky. See also *diffuse nebula, HII region, planetary nebula, supernova*.

**neutrino** A fundamental particle of exceedingly low mass, which has zero electrical charge and which travels at very close to the speed of light.

**neutron** A particle, composed of three quarks, that has zero electrical charge and a mass fractionally greater than that of a proton. Neutrons are found in the nuclei of atoms. See also *atom*.

**neutron star** An exceedingly dense, compact star that is composed almost entirely of tightly packed neutrons. A typical neutron star has a diameter of around 6 miles (10 km) yet has about the same mass as the Sun. A neutron star forms when the core of a high-mass star collapses, triggering a supernova explosion. See also *pulsar, supernova*.

**New General Catalog (NGC)** A catalog of nebulae, clusters, and galaxies that was published in 1888 by the Danish astronomer John L. E. Dreyer. Objects in this catalog are denoted by "NGC" followed by a number. For example, the Andromeda Galaxy is NGC 224. See also *Messier catalog*.

**Newton's laws of motion** Three laws describing the behavior of moving bodies that were set out by Isaac Newton in 1687. Newton's first law states that a body continues to move in a straight line at a constant speed unless acted on by a force. The second law shows how a force causes a body to accelerate in the direction along which an applied force is acting. The third law states that for any force there is an equal and opposite reaction force.

**Newtonian gravity** see *gravity*.

**nova (plural: novae)** A star that suddenly brightens by a factor of thousands or more, then fades back to its original brightness over a period of weeks or months. The flareup occurs when a fusion reaction is triggered on the surface of a white dwarf by gas flowing from a companion star. The name derives from the Latin for "new," because the rapid brightening produces what appears to be a new star. See also *white dwarf, fusion*.

**nuclear bulge** see *spiral galaxy*.

**nuclear fusion** see *fusion*.

**nucleus (plural: nuclei) (1)** The compact central core of an atom, which consists of a number of positively charged protons and neutral neutrons. The nucleus of a hydrogen atom consists of a single proton. **(2)** The solid, ice-rich body of a comet. **(3)** The central core of a galaxy, within which stars are relatively densely packed together.

# O

**occultation** The passage of one body in front of another, which causes the more distant one to be wholly or partially hidden. The term is usually used to describe the passage of a body of larger apparent size in front of a body of smaller apparent size—

for example, when the Moon passes in front of a star or when a planet (such as Jupiter) passes in front of one of its moons.

**Oort Cloud (Oort–Öpik Cloud)** A spherical distribution of trillions of icy planetesimals and cometary nuclei that surrounds the solar system and extends out to a radius of about 1.6 light-years from the Sun. It provides the reservoir from which long-period and "new" comets originate. Its existence was proposed in 1950 by Dutch astronomer Jan H. Oort (a similar idea had also been suggested by Estonian astronomer Ernst J. Öpik). See also *comet, planetesimal*.

**open cluster** A loose cluster of up to a few thousand stars that lies in or close to the plane of the Milky Way galaxy. Member stars of each cluster formed from the same cloud of gas and dust, and have closely similar ages and chemical compositions. Clusters of this kind are also known as **galactic clusters**. See also *globular cluster*.

**open universe** A universe in which the average density is less than the critical density that is needed to halt its expansion and which, therefore, will expand forever. See also *closed universe, flat universe, oscillating universe*.

**opposition** The position of a planet when it is exactly on the opposite side of Earth from the Sun. Its elongation is then 180°, and it is highest in the sky at midnight. See also *conjunction, elongation*.

**optical double star** see *double star*.

**orbit** The path of a body that is moving within the gravitational field of another. The orbit of a planet around a star or a satellite around a planet will normally be an ellipse or, exceptionally, a circle (a circle is a special case of an ellipse).

**orbital period** The period of time during which a body travels once around its orbit. The **sidereal orbital period** is the time taken by one body to revolve around another (for example, the Moon around Earth) measured relative to the background stars.

**oscillating universe** A universe that expands and contracts in a cyclic fashion. The collapse of such a universe at the end of one cycle triggers a new Big Bang that initiates the next cycle. See *closed universe, flat universe, open universe*.

# P

**parallax** The apparent shift in position of an object when it is observed from different locations. **Stellar parallax** is the apparent shift in position of a relatively nearby star when viewed from different points on Earth's orbit. **Annual parallax** is the maximum angular displacement of a star from its mean position due to parallax. The greater the distance of a star, the smaller its parallax.

**parhelic circle** See *sun dog*.

**parsec (pc)** The distance at which a star would have an annual parallax of one second of arc (one second of angular measurement). One parsec is equivalent to 3.26 light-years, or 19,200 billion miles (30,900 billion km). See also *parallax*.

**parselene** See *sun dog*.

**penumbra (1)** The lighter, outer part of the shadow cast by an opaque body. An observer within the penumbra can see part of the illuminating source. See also *eclipse*. **(2)** The less dark and less cool outer region of a sunspot. See also *sunspot, umbra*.

**perigee** The point on its orbit at which a body that is revolving around Earth is at its closest to Earth. See also *apogee*.

**perihelion** The point on its orbit at which a planet, or other solar system body, is at its closest to the Sun.

**phase** The proportion of the visible hemisphere of the Moon or a planet that is illuminated by the Sun at any particular instant.

**photon** An individual package, or quantum, of electromagnetic energy, which may be envisaged as a "particle" of light. The shorter the wavelength of the radiation and higher the frequency, the greater the energy of the photon. See also *electromagnetic radiation*.

**photosphere** The thin, gaseous layer at the base of the solar atmosphere, from which the Sun's visible light is emitted and which corresponds to the visible surface of the Sun.

**planet** A body that is much less massive than a star, revolves around a star, and shines by reflecting that star's light. As a general guide, an orbiting body is considered to be a planet (rather than a brown dwarf) if its mass is less than about 13 times the mass of Jupiter. See also *brown dwarf star*.

**planetary nebula** A glowing shell of gas ejected by a star at a late stage in its evolution.

**planetesimal** One of the large number of small bodies, composed of rock or ice, that formed within the solar nebula and from which the planets were eventually assembled through the process of accretion.

**plasma** A completely ionized gas state of matter that consists of equal numbers of positively charged ions and negatively charged electrons. Plasmas usually have very high temperatures. Examples include the solar corona and solar wind, both of which consist predominantly of protons and electrons. See also *corona, solar wind*.

**polar axis** see *equatorial mounting*.

**positron** see *antiparticle*.

**precession** A slow change in the orientation of a rotating body's axis caused by the gravitational influence of neighboring bodies. Earth's axis precesses around in a conical pattern over a period of 25,800 years.

**prominence** A flamelike plume of gas that follows magnetic field lines in the solar atmosphere. An **active** or **eruptive prominence** undergoes rapid changes, whereas a **quiescent prominence** remains suspended in the solar atmosphere for a prolonged period.

**proper motion** The angular rate at which a star changes its observed position on the celestial sphere. **Annual proper motion** is the angle (seldom more than a small fraction of 1 second of angular measurement) through which a star appears to shift in the course of one year.

**protogalaxy** A progenitor of a normal galaxy. The building blocks from which galaxies were assembled through a process of collisions and mergers, protogalaxies are believed to have formed a few hundred million years after the Big Bang when clouds of gas collapsed under the action of gravity.

**proton** An elementary particle, composed of three quarks, that has a positive electrical charge and is a constituent of every atomic nucleus. See also *atom*.

**proton–proton chain (pp chain)** A sequence of reactions that fuse together hydrogen nuclei (protons) to create helium nuclei. The net result of the process is to convert four protons into one helium nucleus, which consists of two protons and two neutrons. The proton–proton reaction is the dominant hydrogen-burning process in stars similar to, or less massive than, the Sun. See also *fusion, hydrogen burning, neutron, proton*.

**protoplanetary disk** A flattened disk of dust and gas surrounding a newly formed star and within which matter may be aggregating together to form the precursors of planets. See also *planetesimal*.

**protostar** A star in the early stages of formation. It consists of the central part of a collapsing cloud that is heating up and is accreting matter from its surroundings, but within which hydrogen fusion reactions have not yet commenced.

**pulsar** A rapidly rotating neutron star from which we receive brief pulses of radiation, at short and precisely timed intervals, as it spins around its axis.

**pulsating variable** see *variable star*.

# Q

**quantum** see *photon*.

**quark** A fundamental particle, the main matter constituent of all atomic nuclei. Quarks join in bunches of three to make baryons (for example, protons and neutrons) or in quark–antiquark pairs to form particles called mesons. See also *antiparticle, baryon*.

**quasar** A very compact but extremely powerful source of radiation that is almost starlike in appearance but is believed to be the most luminous kind of active galactic nucleus. The name is an abbreviation for quasi-stellar radio source, but is also applied to **quasi-stellar objects (QSOs)**, which are not strong radio emitters.

**quiescent prominence** see *prominence*.

# R

**radial velocity** The component of a body's velocity that is along the line of sight directly toward, or away from, an observer. The radial velocity of a celestial body can be obtained by measuring the Doppler effect in its spectrum. See also *Doppler effect, red shift, spectrum*.

**radiant** The point in the sky from which the tracks of meteors that are members of a particular meteor shower appear to radiate. See also *meteor*.

**radio galaxy** A galaxy that is exceptionally luminous at radio wavelengths. A typical radio galaxy contains an active galactic nucleus from which jets of energetic charged particles are being propelled toward huge clouds of radio-emitting material that in many cases are much larger than the visible galaxy. See also *active galaxy*.

**radio telescope** An instrument that is designed to detect radio waves from astronomical sources. The most familiar type is a concave dish that collects radio waves and focuses them onto a detector.

**red dwarf star** A cool, red, low-luminosity star that, when plotted on a Hertzsprung–Russell diagram, is located toward the bottom end of the main sequence. See also *Hertzsprung–Russell diagram, main sequence*.

**red-giant star** A large, highly luminous star with a low surface temperature and a reddish color. A red giant has evolved away from the main sequence, is "burning" helium in its core rather than hydrogen, and is approaching the final stages of its life. See also *helium burning, Hertzsprung–Russell diagram, main sequence*.

**red shift** The displacement of spectral lines to longer wavelengths that is observed when a light source is receding from an observer. The shift in wavelength is proportional to the speed at which the source is receding. **Cosmological red shift** is a wavelength shift that is caused by the expansion of the universe. See also *blue shift, Doppler effect, spectral line*.

**red supergiant star** An extremely large star of very high luminosity and low surface temperature. Stars of this kind are located toward the top-right corner of the Hertzsprung–Russell diagram. See also *Hertzsprung–Russell diagram*.

**reflecting telescope (reflector)** A telescope that uses a concave mirror to collect light, reflect light rays to a focus, and form an image of a distant object.

**reflection nebula** see *nebula*.

**refracting telescope (refractor)** A telescope that uses a lens to refract (bend) light rays in order to bring them to a focus and form an image of a distant object.

**regolith** A layer of loose rock, rocky fragments, and dust that covers the surface of a planet or planetary satellite.

**regular cluster** see *galaxy cluster*.

**relativity** Theories developed in the early part of the 20th century by Albert Einstein to describe the nature of space and time and the motion of matter and light. The **special theory of relativity** describes how the relative motion of observers affects their measurements of mass, length, and time. One of its consequences is that mass and energy are equivalent. The **general theory of relativity** treats gravity as a distortion of space-time associated with the presence of matter or energy. One of its consequences is that massive bodies deflect rays of light. See also *gravitational lensing*, *space-time*.

**resonance** A gravitational interaction between two orbiting bodies that occurs when the orbital period of one is an exact, or nearly exact, simple fraction of the orbital period of the other. For example, Jupiter's moon Io is in a 1:2 resonance with another of Jupiter's moons, Europa (Io's period is half of Europa's period). When a small object is in resonance with a more massive one, it experiences a periodic gravitational tug each time one of the bodies overtakes the other, the cumulative effect of which gradually changes its orbit.

**retrograde motion** **(1)** The apparent backward motion of a planet, from east to west relative to the background stars. For most of the time, a planet such as Mars or Jupiter will move from west to east relative to the stars (**direct motion**), but it will appear to reverse direction each time it is being overtaken by Earth (around the time of opposition). See also *opposition*. **(2)** Orbital motion in the opposite direction of that of Earth and the other planets of the solar system. **(3)** The motion of a satellite along its orbit in the opposite direction to that in which its parent planet is rotating.

**retrograde rotation** The rotation of a body around its axis in the opposite direction to the rotational motion of Earth, the Sun, and the majority of the planets. Viewed from above its North Pole, Earth rotates around its axis and revolves around the Sun, in a counterclockwise direction (**direct rotation**), whereas a planet with retrograde rotation spins in the opposite (clockwise) direction. The planets Venus, Uranus, and Pluto exhibit retrograde rotation.

**rich cluster** see *galaxy cluster*.

**right ascension (RA)** The angular distance, measured eastward, between the **first point of Aries** (where the Sun's path around the sky crosses the celestial equator from south to north) and a celestial body. It is expressed in hours, minutes, and seconds of time, where 1 hour is equivalent to an angle of 15°. Together with declination, it specifies the position of a body on the celestial sphere. See also *celestial sphere*, *declination*, *ecliptic*, *equinox*.

**ring** A flat distribution of small particles and lumps of material that revolves around a planet, usually in the plane of its equator. A **ring system** consists of a number of concentric rings surrounding a planet. The planets Jupiter, Saturn, Uranus, and Neptune each have a ring system.

**rocky planet** A planet (also called a terrestrial planet) that is composed mainly of rocks and has basic characteristics similar to Earth. Within the solar system, there are four rocky planets: Mercury, Venus, Earth, and Mars. See also *gas planet*.

**rupes** Scarps or cliffs on the surface of a planet or a satellite. See also *moon*.

# S

**satellite** A body that revolves around a planet, otherwise known as a "moon." An **artificial satellite** is an object deliberately placed in orbit around Earth or around another solar system body.

**Schmidt–Cassegrain telescope** A type of catadioptric telescope. Light enters the telescope tube through a thin corrector lens and is reflected from a concave mirror at the bottom of the tube toward a small convex mirror fixed to the inner face of the correcting lens. It is then reflected back down the tube, through a hole in the concave mirror, to a focus. This is a popular, compact design for small and moderate-sized telescopes. See also *catadioptric telescope*.

**Schwarzschild radius** see *black hole*.

**semimajor axis** see *ellipse*.

**Seyfert galaxy** A spiral galaxy with an unusually bright, compact nucleus that in many cases exhibits brightness fluctuations. First identified by American astronomer Carl Seyfert in 1943, Seyfert galaxies comprise one of the several categories of active galaxies. See also *active galaxy*.

**shepherd moon** A small natural satellite that, through its gravitational influence, confines orbiting particles into a well-defined ring around a planet. A pair of shepherd moons, where one is slightly closer to the planet than the other, can squeeze particles into particularly narrow rings.

**sidereal orbital period** see *orbital period*.

**sidereal time** A time system based on the apparent rotation of the celestial sphere. **Local sidereal time** is defined to be 0 hours at the instant the first point of Aries crosses an observer's meridian. The **sidereal day** corresponds to Earth's axial rotation period measured relative to the background stars, and is equal to 23 hours 56 minutes 4 seconds of mean (civil) time. See also *equinox*, *right ascension*.

**singularity** A point of infinite density into which matter has been compressed by gravity, and a point at which the known laws of physics break down. Theory implies that a singularity exists at the center of a black hole. See also *black hole*.

**solar cycle** A cyclic variation in solar activity (for example, the production of sunspots and flares), which reaches a maximum at intervals of about 11 years. Because the polarity pattern of magnetic regions on the Sun reverses every 11 years or so, the overall duration of the cycle is 22 years. The sunspot cycle is the 11-year variation in the number (and overall area) of sunspots. See also *solar flare*, *sunspot*.

**solar eclipse** see *eclipse*.

**solar flare** A violent release of huge amounts of energy—in the form of electromagnetic radiation, subatomic particles, and shock waves—from a site located just above the surface of the Sun.

**solar mass** A unit of mass equal to the mass of the Sun, which provides a convenient standard for comparing the masses of stars. One solar mass is equivalent to $1.96 \times 10^{27}$ tons ($1.989 \times 10^{30}$ kg). Stellar masses range from about 0.08 solar masses up to about 100 solar masses.

**solar nebula** The cloud of gas and dust from which the Sun and planets formed. As the cloud collapsed, most of its mass accumulated at the center to form the Sun, whereas the rest flattened out into a disk within which planets were assembled by the process of accretion. See also *accretion*, *protoplanetary disk*.

**solar system** The Sun together with everything that revolves around it (the planets and their satellites, asteroids, comets, meteoroids, gas, and dust).

**solar wind** A stream of fast-moving, charged particles (predominantly electrons and protons) that escapes from the Sun and flows outward through the solar system like a wind.

**solstice** One of the two points on the ecliptic at which the Sun is at its maximum declination north or south of the celestial equator. On or around June 21 each year, the Sun reaches its greatest northerly declination. This is the Northern Hemisphere **summer solstice** (the **winter solstice** in the Southern Hemisphere). On or around December 22 each year, the Sun reaches its greatest southerly declination. This is the Northern Hemisphere winter solstice (the summer solstice in the Southern Hemisphere). See also *celestial equator*, *declination*, *ecliptic*.

**space-time** The four-dimensional combination of the three dimensions of space (length, breadth, and height) and the dimension of time. The concept that time and space are intimately linked, rather than (as Newton had believed) being separate entities, was proposed in 1908 by Hermann Minkowski and was incorporated into Albert Einstein's theories of relativity. See also *relativity*.

**special theory of relativity** see *relativity*.

**spectral line** A feature that appears at a particular wavelength in a spectrum. An **emission line** is a bright feature corresponding to the emission of light at that wavelength, whereas an **absorption line** is a dark feature corresponding to the absorption of light at that wavelength. See also *spectrum*.

**spectral type** A class into which a star is placed according to the lines that appear in its spectrum. The principal spectral types, arranged in decreasing order of temperature, are labeled O, B, A, F, G, K, M and are subdivided into numbers from 0 to 9. For example, the spectral type of the Sun is G2. See also *luminosity*, *spectral line*, *spectrum*.

**spectroscopic binary** see *binary star*.

**spectroscopy** The science of obtaining and studying the spectra of objects. Because the detailed appearance of a spectrum is influenced by factors such as chemical composition, density, temperature, rotation, velocity, turbulence, and magnetic fields, spectroscopy can reveal a wealth of information about the physical and chemical properties of, and processes occurring in, planets, stars, gas clouds, galaxies, and other kinds of celestial bodies. See also *spectrum*.

**spectrum** A beam of electromagnetic radiation spread out into its constituent wavelengths. A **continuous spectrum** is the unbroken spread of wavelengths emitted by a hot solid or liquid or a dense gas (the continuous spectrum of sunlight appears to human eyes as a rainbow band of colors). A hot, low-density gas emits light at particular wavelengths only; the resulting spectrum consists of bright **emission lines**, each of which corresponds to one of the wavelengths at which emission takes place. If a low-density gas is silhouetted against a source of a continuous spectrum, it absorbs light at certain wavelengths to produce a series of dark **absorption lines**. A typical star has an **absorption-line spectrum** (a continuous spectrum with dark lines superimposed by its atmosphere), whereas an emission nebula has an **emission-line spectrum**. See also *spectral line*.

**spiral arm** A spiral-shaped structure extending outward from the central bulge of a spiral or barred spiral galaxy. It consists of gas, dust, emission nebulae, and hot young stars,

**spiral galaxy** A galaxy that consists of a spheroidal central concentration of stars (the **nuclear bulge**) surrounded by a flattened disk composed of stars, gas, and dust, within which the major visible features are clumped together into a pattern of spiral arms. See also *galaxy, spiral arm.*

**star** A self-luminous body of hot plasma that generates energy by means of nuclear fusion reactions.

**starburst galaxy** A galaxy within which star formation is taking place at an exceptionally rapid rate.

**star cluster** A group of between a few tens and around 1 million stars held together by gravity. All the member stars of a particular cluster are thought to have formed from the same original massive cloud of gas and dust. There are two principal types of cluster: open clusters and globular clusters. See also *globular cluster, open cluster.*

**stellar-mass black hole** see *black hole.*

**stellar parallax** see *parallax.*

**stellar wind** An outflow of charged particles from the atmosphere of a star. See also *solar wind.*

**sun dog** One of a pair of colored patches of light that sometimes may be seen on either side of the Sun, separated from the Sun by an angle of about 22°. Otherwise known as a **parhelion** or **mock sun**, a sun dog is formed when ice crystals in Earth's atmosphere refract sunlight. A **moon dog**, or **parselene**, is a patch of light that sometimes forms by the same process on either side of the Moon. A **parhelic circle** is a large, faint ring of white light, produced by the reflection of sunlight from atmospheric ice crystals, which crosses the Sun, passes through a pair of sundogs, and extends around the sky. Although a complete circle may be seen occasionally, more usually it is only possible to see arcs of light extending outward from the sundogs.

**sunspot** A patch on the surface of the Sun that appears dark because it is cooler than its surroundings. Sunspots occur in regions where localized concentrated magnetic fields impede the outward flow of energy from the solar interior. See also *solar cycle.*

**supergiant** An exceptionally luminous star with a very large diameter. Supergiant stars appear at the top of the Hertzsprung–Russell diagram. See also *Hertzsprung–Russell diagram.*

**superior conjunction** see *conjunction.*

**superior planet** A planet that travels around the Sun on an orbit that is outside the orbit of Earth. The superior planets are Mars, Jupiter, Saturn, Uranus, Neptune, and Pluto. See also *inferior planet.*

**supermassive black hole** see *black hole.*

**supernova (plural: supernovae)** A catastrophic event that destroys a star and causes its brightness to increase, temporarily, by a factor of around 1 million. A **type II supernova** occurs when the core of a massive star collapses and the rest of the star's material is blasted away; the collapsed core usually becomes a neutron star. A **type Ia supernova** involves the complete destruction of a white dwarf. The expanding cloud of debris from a supernova is called a **supernova remnant**. See also *neutron star, white dwarf.*

**synchrotron radiation** Electromagnetic radiation that is emitted when electrically charged particles (usually electrons) gyrate at very high speed around lines of force in a magnetic field. Synchrotron radiation has a characteristic continuous spectrum that is different from that which is emitted by a star or a black body. Astronomical sources of synchrotron radiation include supernova remnants and radio galaxies. See also *black body, electromagnetic radiation, spectrum.*

**synchronous rotation** The rotation of a body around its axis in the same period of time that it takes to orbit another body. Synchronous rotation, which is also known as **captured rotation**, is caused by tidal forces acting between the two bodies. Because its rotational and orbital periods are the same, the orbiting body always keeps the same face turned toward the object around which it is revolving. Like most of the planetary satellites, Earth's moon displays synchronous rotation. See also *orbital period, satellite.*

## T

**tail (of a comet)** A stream, or streams, of ionized gas and dust that is swept out of the head of a comet (the coma) when it approaches, and begins to recede from, the Sun. A **type I tail** (or **gas tail**) consists of ionized gas driven out of the coma by the solar wind. A **type II tail** (or **dust tail**) is composed of dust particles that have been swept out of the coma by the pressure of sunlight. See also *comet.*

**tectonic plate** One of the large, rigid sections into which Earth's lithosphere (which comprises the crust and the rigid uppermost layer of Earth's mantle) is divided. Carried along by slow convection currents in the mantle, tectonic plates drift very slowly across the surface of the planet. Their relative motions give rise to phenomena such as earthquakes, volcanic activity, and mountain-building. The term "tectonic" is sometimes also used to refer to large-scale geological structures, and features resulting from their movement, on planets other than Earth. See also *convection, crust, mantle.*

**tektite** A small, rounded, glassy object formed when a large meteorite or asteroid strikes a rocky planet, melting the surface rocks and throwing molten drops of rock into the atmosphere. Typically a few inches (centimeters) across, tektites have been shaped by their flight through the atmosphere. On Earth's surface, they are found in a number of specific locations, called **strewn fields**. See also *asteroid, meteorite.*

**terrestrial planet** see *rocky planet.*

**transit (1)** The passage of a particular celestial body across an observer's meridian. **(2)** The passage of a body in front of a larger one (for example, the passage of the planet Venus across the face of the Sun, or a satellite across the face of a planet).

**T Tauri star** A young star, surrounded by gas and dust, that varies in brightness and usually shows evidence of a strong stellar wind (a stream of gas flowing away from the star). T Tauri stars are believed still to be contracting toward the main sequence. They are named after the first star of this kind to be identified. See also *main sequence, protostar.*

## UV

**ultraviolet radiation** Electromagnetic radiation with wavelengths shorter than visible light but longer than X-rays. The hottest stars radiate strongly at ultraviolet wavelengths.

**umbra (1)** The dark, central cone of the shadow cast by an opaque body. The illuminating source will be completely hidden from view at any point within the umbra. **(2)** The darker, cooler central region of a sunspot, where the temperature is about 2,700–3,600°F (about 1,500–2,000°C) cooler than the average for the solar surface. See also *eclipse, penumbra, sunspot.*

**vacuum energy** see *cosmological constant.*

**Van Allen belts** Two concentric doughnut-shaped zones that contain charged particles (electrons and protons) trapped in Earth's magnetic field. They were discovered in 1958 by American space scientist James Van Allen.

**variable star** A star that varies in brightness. A **pulsating variable** is a star that expands and contracts in a periodic way, varying in brightness as it does so. An **eruptive variable** is a star that brightens and fades abruptly. A **cataclysmic variable** is a star that suffers one or more major explosions (for example, a nova). See also *Cepheid variable, nova.*

**vernal equinox** see *equinox.*

**volcanic crater** see *crater.*

## W

**wavelength** The distance between two successive crests or between two successive troughs in a wave motion.

**WIMP** The acronym for **Weakly Interacting Massive Particle**, one of a range of postulated elementary particles that have high masses (tens or hundreds of times as great as that of a proton) but interact so exceedingly weakly with ordinary matter that they have not yet been directly detected. WIMPs are widely considered to comprise the major part of the dark-matter content of the universe. See also *dark matter.*

**white dwarf star** A star of low luminosity but relatively high surface temperature that has ceased to generate energy by nuclear-fusion reactions, that has been compressed by gravity to a diameter comparable to that of Earth, and that is slowly cooling and fading. See also *black dwarf, Hertzsprung–Russell diagram.*

**Wolf–Rayet star** A very hot star from which gas is escaping at an exceptionally rapid rate, which is surrounded by an expanding gaseous envelope, and which has emission lines in its spectrum. See also *emission line, spectrum.*

## XYZ

**X-ray burster** An object that emits strong bursts of X-rays, lasting from a few seconds to a few minutes. The bursts are believed to occur when gas drawn from an orbiting companion star accumulates on the surface of a neutron star and triggers a nuclear-fusion chain reaction. See also *fusion, neutron star.*

**X-ray radiation** Electromagnetic radiation with wavelengths shorter than ultraviolet radiation but longer than gamma rays. X-rays are emitted by extremely hot clouds of gas, such as the solar corona.

**zenith** The point on the sky directly above an observer (that is, 90° above the observer's horizon).

**zodiac** A band around the celestial sphere that extends for 9° on either side of the ecliptic, and through which the Sun, Moon, and naked-eye planets appear to travel. The zodiac contains part or all of 24 constellations. In the course of the year, the Sun passes through 13 of these constellations, 12 of which correspond to the astrological "signs of the zodiac." See also *ecliptic.*

**zodiacal light** A faint, cone-shaped glow that extends along the direction of the ecliptic from the western horizon after sunset or from the eastern horizon before sunrise. Most easily seen from tropical skies, it is caused by the scattering of sunlight by particles of interplanetary dust that lie close to the plane of the ecliptic.

# INDEX

Page numbers in **bold** indicate feature profiles or extended treatments of a topic. Page numbers in *italic* indicate pages on which the topic is illustrated.

# ACKNOWLEDGMENTS

**Dorling Kindersley** would like to thank the following people for their help in the preparation of this book: Anne Brumfitt and her colleagues at the European Space Agency for editorial advice; Stephen Hawking for permission to reproduce the quotation on p.21; Giles Sparrow for advice on the contents list; Gillian Tester and Andrew Pache for DTP support; Dave Ball, Sunita Gahir, and Marilou Prokopiou for additonal artwork; Malcolm Godwin of Moonrunner Design; Rajeev Doshi of Combustion Design and Advertising; Philip Eales and Kevin Tildsley of Planetary Visions; Tim Brown and Giles Sparrow of Pikaia Imaging; Tim Loughhead of Precision Illustration; John Plumer of JP Map Graphics; Richard Tibbitts of Antbits; and Greg Whyte of Fanatic Design.

For their help in preparing the revised edition, Dorling Kindersley would like to thank: Ian Ridpath for planning the updates and providing most of the new text; Robin Scagell, Giles Sparrow, and Robert Dinwiddie for additional text; Carole Stott for advice on picture selection, as well as additional text; Professor Derek Ward-Thompson for helping to plan the sections on galaxy evolution and galaxy superclusters and for his comments on the text; Professor Carlos Frenk and Rob Crain for their images of simulations of galaxy formation; Andy Lawrence for providing an original image from the UKIDSS project; Lili Bryant and Laura Wheadon for editorial assistance; Natasha Rees for design assistance; Mik Gates for new illustrations; and Anita Kakar, Rupa Rao, Priyaneet Singh, Alka Ranjan, Ivy Roy, Bimlesh Tiwary, Tanveer Zaidi, Tarun Sharma, Pushpak Tyagi, Aashirwad Jain, Tina Jindal, Deepak Negi, Farhan Patwary, Surya Sarangi, and Arani Sinha at DK Delhi; and Suhita Dharamjit (Senior Jacket Designer), Rakesh Kumar (DTP Designer), Priyanka Sharma (Jackets Editorial Coordinator), and Saloni Singh (Managing Jackets Editor).

**Smithsonian Enterprises**
Kealy Gordon, Product Development Manager
Jill Corcoran, Director, Licensed Publishing
Brigid Ferraro, Vice President, Consumer and Education Products
Carol LeBlanc, President

**PICTURE CREDITS**
Dorling Kindersley would like to thank the following for their help in supplying images: Till Credner; Robin Scagell at Galaxy Picture Library; Romaine Werblow in the DK Picture Library; Anna Bond at Science Photo Library.

**Key:**
t = top; b = bottom/below; c = center; f = far; l = left; r = right; a = above.

**Abbreviations:**
**AAO** = Anglo Australian Observatory; **ASU** = Arizona State University; **BAL** = Bridgeman Art Library (www.bridgeman.co.uk); **Caltech** = California Institute of Technology; **Chandra** = Chandra X-Ray Observatory; **Credner** = Till Credner www.allthesky.com; **DSS** = Digitized Sky Survey; **ESA** = European Space Agency; **ESO** = © European Southern Observatory, licensed through the Creative Commons Attribution 3.0 license—http://creativecommons.org/licenses/by/3.0/; **GPL** = Galaxy Picture Library; **GSFC** = Goddard Space Flight Center; **HHT** = The Hubble Heritage Team; **HST** = Hubble Space Telescope; **JHU** = John Hopkins University; **JPL** = Jet Propulsion Laboratory; **JSC** = Johnson Space Center; **KSC** = Kennedy Space Center; **DMI** = David Malin Images; **MSFC** = Marshall Space Flight Center; **NASA** = National Aeronautics and Space Administration; **NOAO** = National Optical Astronomy Observatory/Association of Universities for Research in Astronomy/National Science Foundation; **NRAO** = Image courtesy of National Radio Astronomy Observatory/AUI; **NSSDC** = National Space Science Data Center; **SPL** = Science Photo Library; **SOHO** = Courtesy of SOHO/EIT Consortium. SOHO is a project of international cooperation between ESA and NASA; **STScI** = Space Telescope Science Institute; **TRACE** = Image courtesy of the Lockheed Martin team of NASA's TRACE Mission; **USGS** = U.S. Geological Survey.

SIDEBAR IMAGES
**© CERN Geneva** (Introduction); **SOHO** (The Solar System); **NASA:** HST/ESA, HEIC and HHT (STScI/AURA) (Milky Way); HST/HHT (STScI/AURA) (Beyond our Galaxy); **SPL:** Kaj R. Svensson (The Night Sky).

**1 NASA:** JPL-Caltech/K. Su (University of Arizona).
**2–3 Processed image © Ted Stryk:** Raw data courtesy NASA/JPL.
**4 Corbis:** Roger Ressmeyer (tc). **5 NASA:** JPL (tr); JPL/STScI (cla). **4–5 NASA and The Hubble Heritage Team (AURA/STScl):** NASA, ESA, N. Smith (University of California, Berkeley), and The Hubble Heritage Team (STScI/AURA) (b).
**6–7 NOAO:** Adam Block (background).
**8 Corbis:** Digital Image © 1996 Corbis; Original image courtesy of NASA, **9 Corbis:** (ca); **Landsat 7 satellite image courtesy of NASA Landsat Project Science Office and USGS National Center for Earth Resources Observation Science:** (tc); **NASA:** JSC (bc).
**10 Corbis:** Roger Ressmeyer (cla); **SPL:** ESA (tl); Jisas/Lockheed (cl); **SOHO:** (clb). **11 SPL:** Scharmer et al./Royal Swedish Academy of Sciences.
**12 GPL:** JPL. **13 ESA:** DLR/FU Berlin (G. Neukum) (tr); **NASA:** JPL (cra), (crb); JPL/STScI (trb).
**14 Chandra:** NASA/CXC/MIT/F.K. Baganoff et al. (tl); **NOAO:** Eric Peng (JHU), Holland Ford (JHU/STScI), Ken Freeman (ANU), Rick White (STScI) (cla); T. A. Rector and Monica Ramirez (clb). **15 © 2005 Russell Cromon** (www.rcastro.com).
**16 2MASS:** T. H. Jarrett, J. Carpenter, & R. Hurt (cla); **Chandra:** X-Ray: NASA/CXC/ESO/P. Rosati et al.; Optical: ESO/VLT/P. Rosati et al. (clb); **SPL:** Carlos Frenk, Univ. of Durham (tl). **17 NASA:** ESA, A. van der Wel (Max Planck Institute for Astronomy, Heidelberg, Germany), H. Ferguson and A. Koekemoer (STScI), and the CANDELS team.
**18–19 Corbis:** Roger Ressmeyer.
**20–21 NASA:** HST/HHT (STScI/AURA).
**22 NASA:** ESA/STScI/B. Salmon (br). **23 NASA:** HST/ESA, Richard Ellis (Caltech) and Jean-Paul Kneib (Observatoire Midi-Pyrenees, France) (tc).
**24 NASA:** HST/ESA and J. Hester (ASU) (b); **NOAO:** Nathan Smith, Univ. of Minnesota (tr). **Tapio Lahtinen:** (cla). **25 Alamy Stock Photo:** Cristian Cestaro (ca); **Corbis:** (tcr); **ESO:** ALMA (NAOJ/NRAO)/E. O'Gorman/P. Kervella (cla); **GPL:** Nigel Sharp, NSF REU/AURA/NOAO (cr); STScI (tr); **NASA:** GSFC (bc); HST, HHT (STScI/AURA) (cl); JPL (cb/Europa), (cb/Ganymede), (cb/Io); JPL/DLR (German Aerospace Center) (cb/Callisto); **SPL:** Pekka Parviainen (br).
**26 Alamy Stock Photo:** Event Horizon Telescope Collaboration/UPI (br); **Chandra:** NASA/CXC/U. Amsterdam/S. Migliari et al. (bl); **NASA:** JPL—Caltech/ASU/Harvard–Smithsonian Center for Astrophysics/NOAO (cl); **SPL:** NOAO (c); STScI/NASA (tl). **26–27 NASA:** HST/H. Ford (JHU), G. Illingworth (UCSC/LO), M. Clampin (STScI), G. Hartig (STScI), the ACS Science Team and ESA (tc). **27 Gemini Observatory/Association of Universities for Research in Astronomy:** GMOS–South Commissioning Team (tl); **NASA:** HST/N. Benitez (JHU), T. Broadhurst (The Hebrew Univ.), H. Ford (JHU), M. Clampin (STScI), G. Hartig (STScI), G. Illingworth (UCO/Lick Obs.), the ACS Science Team and ESA (ca); **SPL:** Max-Planck-Institut für Astrophysik (crb); **XENON Collaboration:** (br).
**28 Alamy Stock Photo:** Andrew Dunn (cl). **NOAO:** Todd Boroson (ca). **29 DK Images:** Andy Crawford (cr); Clive Streeter/Courtesy of the Science Museum, London (crb); Colin Keates/Courtesy of the Natural History Museum, London (cb); Harry Taylor (ca); **SPL:** Lawrence Berkeley Laboratory (cra).
**30 Corbis:** Raymond Gehman (cla); **SPL:** Alfred Pasieka (tr); CERN (br). **31 © CERN:** Mc Cauley/Thomas (cl); **SOHO:** (bc).
**32–33 Courtesy of the National Science Foundation:** B. Gudbjartsson.
**34 DK Images:** (clb). **34– 35 NASA:** HST/HHT (STScI/AURA) (tc). **35 SPL:** (bl).
**36 ESA:** AOES Medialab (ca); **ESO:** Steven Beard (UKATC)/ESO (tr); C. Malin (cla); J. Emerson/VISTA./Cambridge Astronomical Survey Unit (crb); ALMA (ESO/NAOJ/NRAO) (clb); **thecmb.org | Damien P. George:** (cb). **37 Alamy Stock Photo:** NG Images (crb); **Chandra:** NASA/SAO/CXC/G. Fabbiano et al. (cbr); NGST (car); **GPL:** Robin Scagell (tl); **NASA:** General Dynamics (cra); CXC/ESO/F.Vogt et al.); Optical (ESO/VLT/MUSE & NASA/STScI) (tr, tc, tc/Neutron Star); JPL-Caltech (ca/Galaxy Evolution Explorer, cb); **NOAO:** (clb)
**38 Corbis:** (bl); **NASA:** JSC (br). **39 Alamy Images:** Kolvenbach (br); **NASA:** JPL (t).
**40 Corbis:** (bl); Lester Lefkowitz (cl). **41 Corbis:** (ca); **Dreamstime.com:** Andrey Armyagov (bl).
**42–43 SPL:** W. Couch and R. Ellis/NASA (bc).
**43 Laser Interferometer Gravitational Wave Observatory (LIGO):** T. Pyle (br).
**44 NOAO:** Todd Boroson (bc). **45 NASA:** HST/ESA, J. Blakeslee and H. Ford (JHU) (tc); **SPL:** Sanford Roth (cra).
**46–47 NASA:** HST/H. Ford (JHU), G. llingworth (UCSC/LO), M. Clampin (STScI), G. Hartig (STScI), the ACS Science Team, and ESA.
**49 Getty Images:** Fabrice Coffrini/Afp (ca).
**50 Corbis:** Bettmann (tr). **51 Corbis:** Bettmann (tc); **NASA:** HST/HHT (STScI/AURA) (tl).
**52–53 © CERN:** Maximilien Brice.
**54 ESA:** Planck Collaboration (ca); **Image courtesy of Andrey Kravstov:** Simulations were performed at the National Center for Supercomputing Applications (Urbana-Champaign, Illinois) by Andrey Kravtsov (The Univ. of Chicago) and Anatoly Klypin (New Mexico State Univ.). Visualizations by Andrey Kravtsov (b). **54–55 NASA:** HST/K.L. Luhman (Harvard–Smithsonian Center for Astrophysics, Cambridge, Mass.); and G. Schneider, E. Young, G. Rieke, A.Cotera, H. Chen, M. Rieke, and R. Thompson (Steward Obs., ASU, Tuscon, Ariz.) (c). **55 ESO:** Radio: NRAO/AUI/NSF/GBT/VLA/Dyer, Maddalena & Cornwell, X-ray: Chandra X-ray Observatory; NASA/CXC/Rutgers/G. Cassam-Chenaï, J. Hughes et al., Visible light: 0.9-meter Curtis Schmidt optical telescope; NOAO/AURA/NSF/CTIO/Middlebury College/F. Winkler and Digitized Sky Survey. (crb); **NASA:** ESA, and G. Bacon (STScI); science—ESA, P. Oesch (Yale University), G. Brammer (STScI), P. van Dokkum (Yale University), and G. Illingworth (University of California, Santa Cruz) (tr); **SPL:** NASA (c, crb).
**56 Corbis:** Roger Ressmeyer (br); **NASA:** Provided by the SeaWiFS Project, NASA/GSFC, and ORBIMAGE (tr); **SPL:** Dr Linda Stannard, UCT (c); John Reader (bl); MSFC/NASA (clb). **57 Courtesy of the NAIC–Arecibo Observatory, a facility of the NSF:** (bl); **NASA:** JPL/AUS (cl); **SETI League photo, used by permission:** (br).
**58 courtesy of Saul Perlmutter and The Supernova Cosmology Project:** (bl). **59 SPL:** Royal Obs., Edinburgh/AATB (bc).
**60–61 Corbis:** Roger Ressmeyer.
**63 BAL:** Bibliothèque des Arts Decoratifs, Paris, France/Archives Charmet (cr); **SPL:** David Nunuk (tl).
**64 British Library, London:** shelfmark: Or.5259, folio: f.29 (cr); **The Picture Desk:** The Art Archive/British Library, London (cl); **SPL:** Frank Zullo (tr).
**64–65 Corbis:** Paul A Souders (b).
**66 Alamy Images:** Robert Harding Picture Library (cl); **SPL:** John Sanford (b). **67 Corbis:** Jeff Vanuga (tr); Royalty–Free (cb); **DMI:** Akiri Fujii (cr); **The Picture Desk:** The Art Archive/Biblioteca d'Ajuda, Lisbon/Dagli Orti (cla).
**68 SPL:** Pekka Parviainen (tr); Sheila Terry (clb); **Tunc Tezel:** (br). **69 Corbis:** Carl and Ann Purcell (bc); **GPL:** Jon Harper (cr); **The Picture Desk:** The Art Archive/British Library, London, UK (br); **SPL:** Eckhard Slawik (tl, tr); John Sanford (bl).
**70 SPL:** ESA (cl). **71 AAO:** Photograph by David Malin (l); **SPL:** John Chumack (cr); Rev. Ronald Royer (c).
**72 courtesy of the Archives, California Institute of Technology:** (bl); **Corbis:** Stapleton Collection (tl).
**73 BAL:** Private Collection/Archives Charmet (bl); **NOAO:** (bcl, br); Jeff Hageman/Adam Block (cr); Joe Jordan/Adam Block (cbr); N. A. Sharp (cbl); Peter Kukol/Adam Block (tr); Yon Ough/Adam Block (bcr).
**74 Corbis:** Digital image © 1996 Corbis; original image courtesy of NASA (cla); **SPL:** Chris Madeley (r); Stephan J Krasemann bl. **75 Credner:** (bc, tcl); **NAOJ:** H. Fukushima, D. Kinoshita, and J. Watanabe (tr); **Nature Publishing Group (www.nature.com):** Victor Pasko (bcl); **Polar Image/Pekka Parviainen:** (cr); **SPL:** Magrath/Folsom br.
**76 DK Images:** (bl); **GPL:** Dave Tyler (c, ca); Robin Scagell (r); **NASA:** C. Mayhew and R. Simmon (NASA/GSFC), NOAA/NGDC, DMSP Digital Archive (cl); **SPL:** Frank Zullo (clb). **77 DK Images:** Andy Crawford (tr).
**78-79 Novapix:** S.Vetter.
**80 DK Images:** (cl); **courtesy of John W. Griese:** (bl); **SPL:** Frank Zullo (r). **81 Credner:** (cbl); **DK Images:** (tl, tr); **GPL:** Robin Scagell (cl, cr, bcl, bcr, br).
**82 Corbis:** Bettmann (cra); **DK Images:** (bl, bc, crb); courtesy of the Science Museum, London/Dave King (ca); **Science and Society Picture Library:** Science Museum, London (cl). **83 DK Images:** (t, bl); **Dreamstime.com:** Fotum (crb/Magnification); **GPL:** Robin Scagell (cr/Aperture).
**84 DK Images:** (clb, bl, bcr, br); **Dreamstime.com:** Vinicius Tupinamba (cbl/finderscope view, bcl/red dot view); **GPL:** Celestron International (tr). **85 Corbis:** Roger Ressmeyer (cla, cal); **DK Images:** (tc, trb, clb); **GPL:** Rudolf Reiser (cbl); Robin Scagell (car, cra); **Getty Images:** SSPL/Babek Tafreshi (b).
**86–87 DK Images**.
**88 Corbis:** Science Faction/Tony Hallas (cl); **DK Images:** (tr, bl, br); **Dreamstime.com:** Neutronman (cr); **Will Gater:** (bc). **89 DK Images:** (tl, c, cr); **GPL:** Philip Perkins (bl); Dave Tyler (br); **SPL:** J-P Metsavainio (cra).
**90 Corbis:** Roger Ressmeyer (tr, cl); **ESO:** (b); **Gemini Observatory:** Observatory/AURA/Manuel Paredes (cr).
**91 Corbis** Dusko Despotovic (cla); **ESO:** G Hüdepohl/www.atacamaphoto.com (tr); **Getty Images:** Photolibrary/Robert Finken (br); **W. M. Keck Observatory:** UCLA Galactic Center Group (crb); **Photo courtesy of the Large Binocular Telescope Observatory:** The LBT is an international collaboration among institutions in the United States, Italy and Germany (clb).
**92–93 ALMA Observatory:** Babak Tafreshi.
**94 ESA:** C. Carreau (bl); **ESA/Hubble:** NASA, ESA, G. Illingworth, D. Magee, and P. Oesch (University of California, Santa Cruz), R. Bouwens (Leiden University), and the HUDF09 Team (cb); **NASA:** HST (c); ESA (r).
**95 ESA:** C. Carreau (crb); LFI & HFI Consortia (cla); CNES-Arianespace/Optique Vidéo du CSG—L. Mira (tr); Khosroshani, Maughan, Ponman, Jones (bl).
**96–97 NASA:** JPL/STScI.
**98–99 TRACE**.
**100 akg-images:** (c); **NASA:** JPL (cb). **101 NASA:** JPL (c).
**102 SPL:** (tr). **103 Corbis:** Yann Arthus-Bertrand (ca); **NASA:** Erich Karkoschka (ASU Lunar and Planetary Lab) and NASA (tcl).
**105 NASA:** (br); GSFC (crb); **SOHO:** (l, cr).
**106 SPL:** John Chumack (cr); NOAO (tr); **SOHO:** (b, cl). **107 Alamy Images:** Steve Bloom Images (bl); **Science and Society Picture Library:** Science Museum, London (cla); **SPL:** Chris Butler (cra); Jerry Rodriguess (tr); **SOHO:** (clb); **TRACE:** (c); A. Title (Stanford Lockheed Institute) (cr).
**108–109 © Alan Friedman/avertedimagination.com**
**110 NASA:** John Hopkins University Applied Physics Laboratory/Carnegie Institution of Washington (r); University of Colorado/John Hopkins University Applied Physics Laboratory/Carnegie Institution of Washington (ca). **111 NASA:** University of Colorado/John Hopkins University Applied Physics Laboratory/Carnegie Institution of Washington (cb); **SPL:** A.E. Potter and T. H. Morgan (crb).
**112 NASA:** Johns Hopkins University Applied Physics Laboratory/Carnegie Institution of Washington (clb, crb); **GPL:** NASA/JPL/Northwestern Univ. (cl); **NSSDC/GSFC/NASA:** Mariner 10 (cr). **113 NASA:** John Hopkins University Applied Physics Laboratory/Carnegie Institution of Washington (cr, clb, crb).
**114 GPL:** NASA/JPL/Northwestern Univ. (tr); **NASA:** JPL/Northwestern Univ. (cl); Image reproduced courtesy of Science/AAAS (c); John Hopkins University Applied Physics Laboratory/Carnegie Institution of Washington (br, cra, bc); **SPL:** NASA (cr). **115 David Rothery:** NASA/JHU APL/Carnegie Institute of Washington (ca, cra); **GPL:** NASA/JPL/Northwestern Univ. (bl); **NASA:** Johns Hopkins University Applied Physics Laboratory/Carnegie Institution of Washington (tr, br).
**117 ESA:** VIRTIS-Venus Express/INAF-IAPS/LESIA-Obs. Paris/G. Piccioni (bc); **SPL:** NASA: (l).
**118 ESA:** (bl); **NASA:** Ames Research Center (cl); JPL (tr, c, cr); **NSSDC/GSFC/NASA:** Magellan (cra); Venera 13 (clb); Venera 4 (tl). **119 NASA:** JPL (tl, cla); Goddard Space Flight Center Scientific Visualization Studio. (cr); **NSSDC/GSFC/NASA:** Magellan (tcr); **SPL:** NASA (trb).
**120 NASA:** JPL (bl, cr, tr); Goddard Space Flight Center Scientific Visualization Studio. (cl); **NSSDC/GSFC/NASA:** Magellan (c, ca); **SPL:** David P. Anderson, SMU/NASA (br). **121 NASA:** JPL (bc, br, cla, cal, cb); **NSSDC/GSFC/NASA:** Magellan (car, tr).
**122 NASA:** JPL (tc, tr); **SPL:** David P. Anderson, SMU/NASA (b). **123 NASA:** JPL (tl, c, crb, bl, bc, br); **SPL:** David P. Anderson, SMU/NASA (cra).
**124 NASA:** JPL (tl, cl, cr, br); **NSSDC/GSFC/NASA:** Magellan (c, bc). **125 NASA:** JPL (tr, cl, ca, bl, br); **NSSDC/GSFC/NASA:** Magellan (cb).
**127 NASA:** GSFC. Image by Reto Stöckli, enhancements by Robert Simmon (l); **SPL:** Emilio Segre Visual Archive/American Institute of Physics (crb).
**129 Corbis:** Jamie Harron/Papillio (tc); **DK Images:**

(cb/fungi); Andrew Butler (cb/plants); Geoff Brightling (cb/animals); M.I.Walker (cb/protists); **FLPA—Images of Nature:** Frans Lanting (tl); **SPL:** Scimat (cb/monerans).
**130 Alamy Images:** FLPA (crb); **Corbis:** image by Digital image © 1996 Corbis; original image courtesy of NASA (cra); Lloyd Cluff (tl); Robert Gill/Papilio (cl); Sygma/PierreVauthey (bcr); **National Geographic Image Collection:** Image from Volcanoes of the Deep, a giant screen motion picture, produced for IMAX Theaters by the Stephen Low Company in association with Rutgers Univ. Major funding for the project is provided by the National Science Foundation (bl). **131 Corbis:** (br); Jon Sparks (bl); Kevin Schafer (t); Michael S Yamashita (crb); **NASA:** ASF/JPL (c);
**132 Corbis:** Craig Lovell (tc, cb); Macduff Everton (bl); **Landsat 7 satellite image courtesy of NASA Landsat Project Science Office and USGS National Center for Earth Resources Observation Science:** (cla). **133 NASA:** JSC—Earth Sciences and Image Analysis.
**134 Corbis:** Elio Ciol (cal); image by Digital image © 1996 Corbis; original image courtesy of NASA (bl); Layne Kennedy (br); Tom Bean (tl); **NASA:** GSFC/JPL, MISR Team (crb); JSC—Earth Sciences and Image Analysis (cra).
**135 Corbis:** (tl); Galen Rowell (cr, b); Marc Garanger (tr); **NASA:** JPL (cl).
**136–137 Michael Light (www.projectfullmoon.com):** (c).
**138 akg-images:** (cla); **Corbis:** Roger Ressmeyer (cra); **NASA:** JSC (c); MSFC (b); **SPL:** ESA, Eurimage (trb).
**139 Corbis:** Roger Ressmeyer (cr); **ESA:** Space-X, Space Exploration Institute (bc); **Galaxy Contact:** NASA (ca) **GPL:** Thierry Legault (tr); **NSSDC/GSFC/NASA:** Lunar 3 (crb); **Scala Art Resource:** Biblioteca Nazionale, Florence, Italy (clb); **USGS:** (cbr).
**140 NASA:** LRO/LOLA Science Team (tr).
**142–143 NASA**.
**144 NASA:** JSC (bl); JPL (tl); **NSSDC/GSFC/NASA:** Apollo 11 (br); Galileo (crb); Lunar Orbiter 5 (cl); **SPL:** John Sanford (cra). **145 GPL:** Damian Peach (bl); **NASA:** (bc); JPL (cla); **NSSDC/GSFC/NASA:** Apollo 17 (tc); Lunar Orbiter 5 (br); Ranger 9 (cra); USGS/Clementine (crb).
**146 NASA:** JSC (tc, cra). **146–147 Michael Light (www.projectfullmoon.com): (b)**. **147 NASA:** JSC (tl); **NSSDC/GSFC/NASA:** Apollo 17 (tr).
**148 ESA:** Space-X, Space Exploration Institute (cl); **GPL:** NASA (cr, bc); **NSSDC/GSFC/NASA:** Apollo 15 (tc); Lunar Orbiter 3 (c); **U.S. Department of Energy:** (br). **149 NASA:** GSFC (cbr); GSFC/ASU/Lunar Reconnaissance Orbiter (bl); JPL/USGS (tr); Lunar Prospector (bc); **NSSDC/GSFC/NASA:** Lunar Orbiter 4 (tc).
**151 NASA:** JPL (br); JPL-Caltech/University of Arizona (cr); **USGS:** (l).
**152 ESA:** DLR/FU Berlin (G. Neukum) (cr); **NASA:** Cornell University, JPL and M. Di Lorenzo et al. (ca); JPL (tr, cla, cl, bl); JPL/Cornell Univ./Mars Digital (clb). **153 NASA:** JPL-Caltech/MSSS (tr); JPL-Caltech/University of Arizona (tl, cla, cra).
**154–155 NASA:** JPL-Caltech/ASU.
**156 ESA:** DLR/FU Berlin (G. Neukum) (tr); **NASA:** JPL/ASU (cl); JPL/MSSS (cal, cra, crb, bl, br). **157 ESA:** DLR/FU Berlin (G. Neukum) (bl); **NASA:** JPL (trb); JPL-Caltech/University of Arizona (br); JPL/MSSS (cal, cla).
**158 NASA:** JPL/MSSS (tc, ca). **158–159 NASA: JPL/USGS b**. **159 ESA:** DLR/FU Berlin (G. Neukum) (tc, cla, cra); **NASA:** JPL/MSSS (tcb).
**160 ESA:** DLR/FU Berlin (G. Neukum) (tc, tr, ca, cr); **NASA:** JPL/ASU (tlb); JPL/MSSS (c, bl); JPL-Caltech/University of Arizona (br). **161 ESA:** DLR/FU Berlin (G. Neukum) (b); **NASA:** JPL/MSSS (cl); JPL/USGS (tr); JPL-Caltech/University of Arizona (cra).
**162 ESA:** DLR/FU Berlin (G. Neukum) (bl); **NASA:** JPL (cl); JPL/MSSS (tc); JPL-Caltech/University of Arizona (cr); JPL/UArizona (br). **163 ESA:** DLR/FU Berlin (G. Neukum) (bl); OMEGA (bc); **NASA:** JPL/Cornell (tc, ca); JPL-Caltech/University of Arizona (crb).
**164 ESA:** DLR/FU Berlin (G. Neukum) (tr, br); **NASA:** JPL/ASU (cbl); JPL/Cornell (cla); JPL/MSSS (cra); JPL-Caltech/University of Arizona (clb); Mars Orbiter Laser Altimeter (MOLA) Science Team (crb).
**165 ESA:** DLR/FU Berlin (G. Neukum) (cra, bc); **NASA:** JPL/MSSS (tr, cb, br); JPL/USGS (tl); Mars Global Surveyor/USGS (bl).
**166 NASA:** JPL/Cornell (tr, cla). **166–167 NASA:** JPL/Cornell (b). **167 NASA:** JPL (tl); JPL/Cornell (tc, tr).
**168–169 NASA:** JPL-Caltech/MSSS.
**170 NASA:** HST/R. Evans and K. Stapelfeldt (JPL) (cl); JPL (bc); JPL-CalTech/UCLA/MPS/DLR/IDA (bc/Occator Crater, bc/Vesta).
**171 DK Images:** (tc).
**172 ESA:** © 2008 MPS for OSIRIS Team MPS/UPD/LAM/IAA/RSSD/INTA/UPM/DASP/IDA (cbr, bc, bl); **NASA:** JPL-Caltech (c); JPL/JHU/APL (tr, br); JPL/USGS (cl); **NSSDC/GSFC/NASA:** Goldstone DSC antenna-radar (cr). **173 GPL:** NASA/JPL (b); **NASA:** JPL (tr).
**174 Corbis:** R Kempton (tr); **NASA:** JPL-Caltech/UCLA/MPS/DLR/IDA (ca, b). **175 Japan Aerospace Exploration Agency (JAXA):** (crb, bl, br); **courtesy of Osservatorio Astronomico di Palermo Giuseppe S.Vaiana:** (cra); **NASA:** JPL-Caltech/UCLA/MPS/DLR/IDA (cla, ca).
**176 NASA:** JPL/JHU/APL (cla, cr, crb); Goddard/University of Arizona (clb, bc); Goddard/University of Arizona/Lockheed Martin (bl); **SPL:** NASA (tc).
**177 GPL:** NASA/JPL/JHU/APL (t); **Rex by Shutterstock:** Uncredited/AP/Shutterstock (br).
**179 GPL:** NASA/JPL/ASU (l); **NASA:** HHT (STScI/AURA); NASA/ESA, John Clarke (Univ. of Michigan) (cr).
**180 NASA:** HST/ESA, and E.Karkoschka (ASU) (cb); JPL/STScI (tl). **181** JPL/Cornell (crb); **NASA:** JPL-Caltech/SwRI/MSSS/Gerald Eichstadt/Sean Doran © CC NC SA (tr); JPL-Caltech/SwRI/MSSS (clb).
**182 Laurie Hatch Photography/Lick Observatory:** (cbl); **NASA:** JPL/Cornell Univ. (cal, car, cl, bl); JPL/Lowell Obs. (cra); JPL/ASU (tl); **courtesy of Scott S. Sheppard, University of Hawaii:** (bcr).
**183 DK Images:** Andy Crawford (cra); **NASA:** JPL/DLR (German Aerospace Center) (tcl, tcr, cla); JPL/University of Arizona (b).
**184 GPL:** NASA/JPL (tc); **NASA:** JPL/PIRL/ASU (cl); JPL/ASU (bl); JPL/ASU/LPL (br). **185 GPL:** NASA/JPL/USGS.
**186 GPL:** NASA/JPL/DLR (German Aerospace Center) (t); **NASA:** JPL (bl); JPL/Brown Univ. (crb, br).
**187 BAL:** Private Collection (crb); **GPL:** NASA/JPL (b); **NASA:** JPL/ASU (tr); JPL/DLR (German Aerospace Center) (tc, cla).
**189 NASA:** HST/ESA, J. Clarke (Boston Univ.), and Z. Levay (STScI) (bl); JPL/STScI (tl).
**190 NASA:** JPL (tl); JPL-Caltech/STScI (tr); JPL/STScI (c, cb). **191 NASA:** JPL-Caltech/University of Virginia (cra); JPL-Caltech/R. Hurt (SSC) (crb); JPL/STScI (clb, cbl); JPL/Univ. of Colorado (tr).
**192 NASA:** JPL (tr); JPL/STScI (cra, crb, bl); JPL-Caltech/Space Science Institute (cl, br). **193 NASA:** JPL/STScI (cla, clb, r).
**194 NASA:** JPL (br); JPL/STScI (tr, cl, bl, bc); JPL/STScI/Universities Space Research Association/Lunar & Planetary Institute (c); JPL-Caltech/Space Science Institute (tc). **195 NASA:** JPL (tc, tr, cbr); JPL/STScI (cl, bl); JPL/STScI/Universities Space Research Association/Lunar & Planetary Institute (br).
**196 NASA:** (br); ESA/JPL/University of Arizona (bl); JPL/Cassini is a cooperative project of NASA, the ESA, and the Italian Space Agency. The JPL, a division of Caltech, manages the Cassini mission for NASA's Office of Space Science, Washington, D.C. (tr); JPL/STScI (cla, clb); JPL-Caltech/ASI (cra). **197 NASA:** JPL/STScI (tl, tr, cbl, cr, br); JPL/ASU (bl).
**198–199 NASA:** JPL-Caltech/Space Science Institute.
**201 Corbis:** Roger Ressmeyer (crb); **GPL:** JPL/STScI (l); **W. M. Keck Observatory:** Courtesy Lawrence Sromovsky, UW-Madison Space Science and Engineering Center (tr); **NASA:** JPL (cb).
**202 NASA:** HST/Erich Karkoschka (ASU) (tl); JPL (cl, c, bl, br); JPL/USGS (cb); **NSSDC/GSFC/NASA:** (cra). **203 Corbis:** Sygma (c); **Brett Gladman, Paul Nicholson, Joseph Burns, and JJ Kavelaars, using the 200 inch Hale Telescope:** (br); **NASA:** JPL (tl, tr, bl, clb, bc).
**205 GPL: NASA/JPL (l); NASA: JPL (crb); JPL/HST (cra).**
**206 Corbis:** Roger Ressmeyer (bc); **NASA:** JPL (tl, bl, br); **NSSDC/GSFC/NASA:** Voyager 2 (cl, c/left); **SPL:** NASA (c/right). **207 Liverpool Astronomical Society:** With thanks to Mike Oates (br); **NASA:** JPL/USGS (t, clb); **courtesy of A.Tayfun Oner:** (bc).
**208 NASA:** JHUAPL/SwRI (tr, clb). **208–209 NASA:** JHUAPL/SwRI. **209 ESO:** (crb); **Lowell Observatory Archives:** (br); **NASA:** JHUAPL/SwRI (cra); Johns Hopkins University Applied Physics Laboratory/Southwest Research Institute/Lunar and Planetary Institute/Paul Schenk (tl).
**210 Corbis:** Bettmann (br); **NASA:** ESA and P. Kalas (University of California, Berkeley) (bl); HST/M. Brown (Caltech) (clb); **NSSDC/GSFC/NASA:** Denis Bergeron, Canada (cl). **211 NASA:** (br); Johns Hopkins University Applied Physics Laboratory/Southwest Research Institute/Roman Tkachenko (bc).
**212 W. M. Keck Observatory:** Mike Brown (California Institute of Technology) (clb); **NASA:** ESA and M. Brown (California Institute of Technology) (bc); HST/Mike Brown (California Institute of Technology) (cla). **213 Corbis:** Jonathan Blair (cra); **GPL:** Michael Stecker (c); **NASA:** JPL-Caltech (cb, br).
**214 SPL:** Pekka Parviainen (tr). **215 Corbis:** Jonathan Blair (cra); **DK Images:** (b); **NASA:** JPL/Brown Univ. (cl); JPL/USGS (cl); **SOHO:** (clb).
**216 akg-images:** (c); **DK Images:** (cr); **NOAO:** Roger Lynds (bl); **SPL:** Pekka Parviainen (bc); Detlev van Ravenswaay (tr); Rev. Ronald Royer (crb). **217 Corbis:** Gianni Dagli Orti (br); **Damian Peach:** (bl); **DMI:** Akira Fujii (cl); **ESA:** MPAE, 1986, 1996 (cr); **SPL:** John Thomas (tl); Richard J. Wainscoat, Peter Arnold Inc. (tr).
**218 ESA:** Rosetta/NAVCAM, CC BY-SA IGO 3.0 (cra); **NASA:** ESA/H. Weaver and E. Smith (STScI) (br); JPL (ca); JPL-Caltech (clb); JPL-Caltech (bc/left); **courtesy of Lowell Observatory:** (cbr); **SPL:** Frank Zullo (cla). **219. NASA:** JPL/UMD (ca); JPL-Caltech/UMD (tr, cb); JPL-Caltech/LMSS (cla); Dan Burbank (ISS) (crb); Solar Dynamics Observatory (SDO) (br).
**220 © The Natural History Museum, London:** (crb, bc, br); **SPL:** (cr, bl); David McLean (ca); Sputnik (cl).
**221 Corbis:** Jonathan Blair (bl); **DK Images:** Harry Taylor (cb, bc); **Getty Images:** NASA/AFP (t); **NASA:** Carnegie Mellon Univ./Robotic Antarctic Explorer (LORAX) (br).
**222 Corbis:** Matthew McKee/Eye Ubiquitous (bl); **DK Images:** courtesy of the Natural History Museum, London/Colin Keates (c); **GPL:** UWO/Univ. of Calgary (cl); Muséum National d'Histoire Naturelle, Paris: Département Histoire de la Terre (bc); **© The Natural History Museum, London:** (br); **SPL:** D. van Ravenswaay (cbr); Michael Abbey (cr); Pascal Goetgheluck/Francois Robert (tr). **223 Alamy Images:** H. R. Bramaz (cla); **NASA:** JSC (br); KSC (crb); **© The Natural History Museum, London:** (tr, bl).
**224–225 John P. Gleason, Celestial Images**
**226 SPL:** Chris Butler (cra); Tony and Daphne Hallas (tr); **Planetary Visions:** (b). **227 Corbis:** Image by © National Gallery Collection; by kind permission of the Trustees of the National Gallery, London (cr); **NASA:** JPL-Caltech (tl); Goddard Space Flight Center (cl).
**228 NASA:** HST/Jeff Hester (ASU) (tr); **NOAO:** Adam Block (b). **229 NASA:** Goddard/Adler/U. Chicago/Wesleyan (cra); **NRAO:** (cr, cr/inset); **SPL:** (bl); **B.J. Mochejska (CfA), J. Kaluzny (CAMK), 1m Swope Telescope:** (bc).
**230–231 NASA:** X-Ray: CXC/UMass/D. Wang et al.; Optical: ESA/STScI/D. Wang et al.; IR: JPL-Caltech/SSC/S. Stolovy.
**233 Corbis:** Bettmann (br); **GPL:** Andrea Dupree, Ronald Gilliland (STScI)/NASA/ESA (cra); Robin Scagell (cr); **Shutterstock:** Sami Ghumman (bl); **SOHO:** (tl).
**234 NASA:** HST/Wolfgang Brandner (JPL/IPAC), Eva K. Grebel (Univ. Washington), You-Hua Chu (Univ. Illinois Urbana-Champaign) (tr). **235 NASA:** HST/C. A. Grady (NOAO, NASA, GSFC), B. Woodgate (NASA, GSFC), F. Bruhweiler and A. Boggess (Catholic Univ. of America), P. Plait and D. Lindler (ACC, Inc., GSFC), and M. Claupin (STScI) (br).
**236 Courtesy of Andy Steere:** (bl). **237 Chandra:** NASA/STScI/R. Gilliand et al. (tl).
**238 AAO:** Photograph by David Malin (car); **ESO:** APEX/DSS2/SuperCosmos/Deharveng (LAM)/Zavagno (LAM) (tr); **NASA:** HST/J. Hester and P. Scowen (ASU) (cr). **238–239 ESO:** VPHAS+ team/N. J. Wright (Keele University) (b). **239 courtesy of Armagh Observatory:** (bc/left); **ESA/Hubble:** NASA, D. Padgett (GSFC), T. Megeath (University of Toledo), and B. Reipurth (University of Hawaii) (crb); **NASA:** HST/ESA and HHT (STScI/AURA) (cl); HST/J. Hester (ASU) (tr); HST/Kirk Borne (STScI) (tc).
**240 ESA/Hubble:** NASA, ESA, and the Hubble Heritage Team (AURA/STScI) (cb); **ESO:** J. Alves (ESO), E. Tolstoy (Groningen), R. Fosbury (ST–ECF), and R. Hook (ST–ECF) (VLT) (cl); Leonardo Testi (Arcetri Astrophysical Obs., Florence, Italy (NTT + SOFI) (tl).
**241 ESO:** J. Emerson/VISTA/Cambridge Astronomical Survey Unit (l); Mark McCaughrean (Astrophysical Institute, Potsdam, Germany (VLT, ANTU, and ISAAC) (tc); **© Smithsonian Institution:** (b).
**242 NASA:** HST/H. Ford (JHU), G. Illingworth (UCSC/LO), M. Clampin (STScI), G. Hartig (STScI), the ACS Science Team and ESA (tc); **NOAO:** Michael Gariepy/Adam Block (br); T. A. Rector (NRAO/AUI/NSF and NOAO) and B. A. Wolpa (NOAO) (cl). **243 ESA/Hubble:** NASA/STScI (bc); **ESO:** (cb); **Geert Barentsen & Jorick Vink (Armagh Observatory) & the IPHAS Collaboration:** (tr); **Richard Crisp (www.narrowbandimaging.com):** (tc); **NASA:** JPL—Caltech/S. Carey (Caltech) (bl).
**244 NASA:** HST/ESA, STScI, J. Hester, and P. Scowen (ASU) (bl); **NOAO:** T. A. Rector (NRAO/AUI/NSF and NOAO) and B. A. Wolpa (NOAO) (tc). **245 ESO:** (VLT, ANTU + ISAAC).
**246 ESO:** (tr); **NASA:** HST/HHT (STScI/AURA) (cla); **NOAO:** Todd Boroson (bl); **SPL:** National Optical Astronomy Observatories (br). **247 2MASS:** E. Kopan (IPAC)/Univ. of Massachusetts (tc); **NASA:** ESA and M. Livio and the Hubble 20th Anniversary Team (STScI) (bl); JPL-Caltech/Spitzer Space Telescope (br); JPL-Caltech/Univ. of Wisconsin (tr).
**248–249 ESO:** T. Preibisch.
**250 GPL:** Gordon Garradd (cl); **SOHO:** (b); **TRACE:** (tr). **251 Corbis:** Bettmann (cl); **SOHO:** (tr).
**252 GPL:** Duncan Radbourne (bl); **DMI:** Akira Fujii (tr); **ESA/Hubble:** NASA (cl, c); **SPL:** Dr. Fred Espenak (cr); Eckhard Slawik (br). **253 GPL:** Deep Sky Survey (clb); **DMI:** Akira Fujii (tr, cla); **ESO:** ALMA (NAOJ/NRAO)/L. Matrà/M. A. MacGregor (tc); **courtesy of Joe Orman:** (cb); **SPL:** Eckhard Slawik (bc).
**254 ESO:** B. Bailleul (tr); **Matt BenDaniel (http://starmatt.com):** (bl); **Credner:** (cla). **255 NASA:** HST/Bruce Balick (Univ. of Washington), Jason Alexander (Univ. of Washington), Arsen Hajian (U.S. Naval Obs.), Yervant Terzian (Cornell Univ.), Mario Perinotto (Univ. of Florence, Italy), Patrizio Patriarchi (Arcetri Obs. Italy) (bc); HST/Bruce Balick (Univ. of Washington) Vincent Icke (Leiden Univ., The Netherlands), Garrett Mellema (Stockholm Univ.) (crb); HST/HHT (STScI/AURA) (l); HST/HHT (STScI/AURA) (c); HHT (STScI/AURA); D. Garnett (University of Arizona) (cra).
**256 ESO:** P. Kervella (cr). **Haubois et al., A&A, 508, 2, 923, 2009, reproduced with permission © ESO/Observatoire de Paris:** (cra); **NASA:** HST/Jon Morse (Univ. of Colorado) (tl); **H. Olofsson (Stockholm Observatory) et al:** (br); **SPL:** Eckhard Slawik (cla); Royal Obs., Edinburgh/AAO (bc). **257 ESA/Hubble:** NASA/Judy Schmidt (br); **NASA:** ESA (bc); HST/NOAO, ESA, the Hubble Helix Nebula Team, M. Meixner (STScI), and T. A. Rector (NRAO) (tr, cr).
**258 R. Corradi (Isaac Newton Group), D. Goncalves (Inst. Astrofisica de Canarias):** (cb); **NASA:** HST/ESA/Hans van Winckel (Catholic Univ. of Leuven, Belgium) and Martin Cohen (Univ. of California Berkely) (t); HST/ESA, HEIC, and HHT (STScI/AURA) (bc); HST/HHT (STScI/AURA); W. Sparks (STScI) and R. Sahai (JPL) (br). **259 W. M. Keck Observatory:** U.C. Berkeley Space Sciences Laboratory (clb); **NASA:** HST/Andrew Fruchter and ERO Team (Sylvia Baggett (STScI), Richard Hook (ST–ECF), and Zoltan Levay (STScI) (br); STScI (cla); **SPL:** NOAO (cra).
**260–261 NASA:** ESA and the Hubble SM4 ERO Team.
**262 ESA/Hubble:** NASA/N. Smith (University of Arizona, Tucson), and J. Morse (BoldlyGo Institute, New York) (br); **NASA:** ESA and Valentin Bujarrabal (Observatorio Astronomico Nacional, Spain) (cla); CXC/GSFC/K. Hamaguchi, et al. (c); ESA/STScI (tr). **263 NASA:** HST/Raghvendra Sahai and John Trauger (JPL), the WFPC2 Science Team.
**264 ESA/Hubble:** NASA (crb); **ESO:** B. Bailleul (tr); **NASA:** ESA/Hubble (clb); **© Observatoire de Paris:** bl); **SPL:** Celestial Image Co. (tc). **265 ESA/Hubble:** NASA, ESA, R. O'Connell (University of Virginia), F. Paresce (National Institute for Astrophysics, Bologna, Italy), E. Young (Universities Space Research Association/Ames Research Center), the WFC3 Science Oversight Committee, and the Hubble Heritage Team (STScI/AURA) (cb); **NASA:** HST (br); HST/HHT (AURA/STScI) (t).
**266 Chandra:** NASA/U. Mass/D. Wang et al. (c); **ESO:** L. Calçada (bl); **NASA:** ESA and L. Bedin (STScI) (tr).
**268 Chandra:** NASA/CXC/SAO (tr); NASA/SAO/CXC (cl); **ESO:** M. van Kerkwijk (Institute of Astronomy, Utrecht), S. Kulkarni (Caltech), VLT Kueyen (cr); **NASA:** Compton Gamma Ray Obs. (cbl); HST/Fred Walter (State Univ. of New York at Stony Brook) (cra); HST/HHT (AURA/STScI) (crb). **269 Alamy Stock Photo:** Stocktrek Images, Inc./Stocktrek Images, Inc. (cb); **NASA:** HST/Jeff Hester (ASU) (cra); William P. Blair and Ravi Sankrit (JHU) (t); CXC/Univ of Toronto/M. Durant et al. (br, bc/Vela Pulsar Jet, bc, bc/Vela Pulsar Jet 4).
**270–271 NASA:** ESA, G. Dubner (IAFE, CONICET-University of Buenos Aires) et al.; A. Loll et al.; T. Temim et al.; F. Seward et al.; VLA/NRAO/AUI/NSF; Chandra/CXC; Spitzer/JPL-Caltech; XMM-Newton/ESA; and Hubble/STScI (c). **271 NASA:** CXC/MSFC/M. Weisskopf et al. (bc); ESA, J. Hester (Arizona State University) (tr); ESA (cr).
**272 Corbis:** (bl); **GPL:** Michael Stecker (cra); **NASA:** HST/ESA, CXO, and P. Ruiz-Lapuente (Univ. of Barcelona) (bc); HST/H. Richer (Univ. of British Columbia) (cla); JPL-Caltech/STScI/CXC/SAO (bl); **SPL:** Dr S. Gull and Dr J. Fielden (crb); Royal Greenwich Obs. (ca). **273 NASA:** JPL-Caltech/STScI/CXC/SAO (bl); ESA, R. Sankrit, and W. Blair (JHU) (tr); HST/Dave Bennett (Univ. of Notre Dame, Indiana) (cbr); HST/ESA and HHT (STScI/AURA) (tc); HST/NOAO, Cerro Tololo Inter-American Obs. (br); **NOAO:** Doug Matthews and Charles Betts/Adam Block (c).
**274 Science Photo Library:** (cra). **275 ESO:** Mark McCaughrean (Astrophysical Institute Potsdam, Germany) (VLT ANTU + ISAAC).
**276 GPL:** Damian Peach (cra, crb); Robin Scagell (bc); **courtesy of Padric McGee, University of Adelaide:** (cl); **NASA:** HST/K.L. Luhman (Harvard–Smithsonian Center for Astrophysics, Cambridge, Mass.), G. Schneider, E. Young, G. Rieke, A. Cotera, H. Chen, M. Rieke, and R. Thompson (Steward Obs., ASU) (tl); **SPL:** Eckhard Slawik (cr, bl). **277 Benjamin Fulton:** (cl); **GPL:** Damian Peach (c); **NOAO:** (cra); **SPL:** Dr. Fred Espenak (b); John Sanford (tr).
**278 AAO:** Photograph by David Malin. **279 GPL:** Damian Peach (crb); **The Picture Desk:** The Art Archive/National Library, Cairo/Dagli Orti (bc); **SPL:** Tony and Daphne Hallas (cr).
**280 SPL:** Celestial Image Co. (tc, b). **281 ESO:** (bc); **GPL:** Duncan Radbourne (bl); **SPL:** George Fowler (bc/right); John Sanford (cl, br); **Matthew Spinelli:** (c); **courtesy of Thomas Williamson, New Mexico Museum of Natural History and Science:** (cr).
**282 Credner:** (tr); **NASA:** HST/F. Paresce, R. Jedrzejewski (STScI) and ESA (c). **282–283 NASA:** HST/ESA and HHT (STScI/AURA) (b). **283 NASA:**

ESA and H. Bond (STScI) (br/2006); HST (tr); JPL-Caltech (cl).

**284 Alamy Stock Photo:** Steven Milne (crb); **GPL:** DSS (N) (tl); **infoastro.com/Victor R. Ruiz:** (c); **SPL:** John Sanford (cl, cl/insert); **courtesy of Jerry Xiaojin Zhu, Carnegie Melon University:** (br).

**285 Credner:** (tr); **courtesy of Mark Crossley:** (cl/left); **GPL:** Damian Peach (cl/right); **NASA:** HST/Margarita Karovska (Harvard–Smithsonian Center for Astrophysics) (cra); JPL-Caltech (cr); **NOAO:** Tom Bash and John Fox/Adam Block (b); **SPL:** Eckhard Slawik (crb);

**286 NASA:** JPL-Caltech/Iowa State (cr); **GPL:** DSS (N) (c); DSS (S) (cl); Robin Scagell (crb); **SPL:** (cra). **287 Matt BenDaniel (http://starmatt.com):** (tr); **GPL:** Martin Mobberley (crb); **NASA:** HST/F. Paresce, R. Jedrjejewski (STScI), ESA (br); JPL-Caltech/Iowa State (cr); **SPL:** NOAO (bc).

**288 © 2005 Loke Tan (www.starryscapes.com):** (b); **NOAO:** Heidi Schweiker (tr). **289 Credner:** (bl); **ESO:** (ANTU UT1 + TC) (tl); **NASA:** (ca).

**290 NOAO:** (bl, bc); N. A. Sharp, REU Program (crb); **SPL:** Celestial Image Co. (cra); Eckhard Slawik (cl); Jerry Lodriguss (cr); **P. Seitzer (Univ. Michigan):** (tl). **291 ESO:** Das Universumisteine Scheibe (cr); **NASA:** HST/HHT (STScI/AURA) (tc); **SPL:** Eckhard Slawik (cl); Tony and Daphne Hallas (b).

**292–293 NASA:** ESA and HHT (STScI/AURA).

**294 AAO:** Photograph by David Malin (br); **ESO:** Y. Beletsky (cal); **NASA:** ESA (cra); HST/HHT (STScI/AURA) (bc); **NOAO:** (cl); **SPL:** Dr. Fred Espenak (tr); **© 2005 Loke Tan (www.starryscapes.com):** (cb).

**295 ESA/Hubble:** NASA (br); **ESO:** (c); **NASA:** HST/HHT (STScI/AURA) (crb); **NOAO:** (tr, bc); Bruce Hugo and Leslie Gaul/Adam Block (bl); Michael Gariepy/Adam Block (cl).

**296 ESA:** ESA/Hubble/STScI; CC BY 4.0 (tr); **ESA/Hubble:** C. Burrows (STScI & ESA)/J. Hester (Arizona State University), J. Morse/STScI (cra); **ESO:** ALMA (NAOJ/NRAO), S. Andrews et al./NRAO/AUI/NSF/S. Dagnello (crb); ALMA/NAOJ/NRAO (bl); B. Saxton (NRAO/AUI/NSF); ALMA (NAOJ/NRAO) (crb/Spiral.).

**297 Canadian Space Agency:** MOST (cra); **ESA:** CNES/D. Ducros (crb); **ESO:** (NACO + VLT) (bl); **NASA:** JPL-Caltech/R. Hurt (SSC) (cr).

**298 ESA:** Alfred Vidal-Madjar (Insitute d'Astrophysique de Paris, CNRS, France) (cla); **ESO:** M. Kornmesser/Nick Risinger (skysurvey.org) (cra); **NASA:** JPL-Caltech/H. Knutson (Harvard-Smithsonian CfA) (ca);
ESA and A. Feild (STScI) (bc); NOAJ/Subaru/J. Carson, College of Charleston (tr). **299 ESO:** M. Kornmesser (clb, fcla, bc); **NASA:** (ca); Ames/JPL-Caltech (cra); JPL-Caltech (cla/Earth); NASA Ames/JPL-Caltech/T. Pyle (cla); Ames Research Center (tc).

**300–301. ESO**

**302 ESO:** (tr, br); **NASA:** HST/HHT (STScI/AURA) (cr/Sb, bc); **NOAO:** (cra, cr/E0, cr/E6, cl; Adam Block (cr/E2, cr/Sa, cr/SBa, cr/SBb); Jeff Newton/Adam Block (cr/S0); Jon and Bryan Rolfe/Adam Block (cr/Sc); Nicole Bies and Esidro Hernandez/Adam Block (cr/SBc); P. Massey (Lowell), N. King (STScI), S. Holmes (Charleston), G. Jacoby (WIYN) (clb); **SPL:** Royal Obs., Edinburgh (cla). **303 NASA:** ESA, and the Hubble Heritage (STScI/AURA)-ESA/Hubble Collaboration (t).

**304 AAO:** Photograph by David Malin (cr); **ESO:** NASA, ESA, and The Hubble Heritage Team (STScI/AURA) (cl); **NASA:** HST/HHT (STScI/AURA) (bc, br); **NOAO:** (tc, c). **304–305 NASA:** ESA, A. Aloisi (STScI/ESA), and HHT (STScI/AURA)-ESA/Hubble Collaboration (tc). **305 Chandra:** NASA/SAO/G. Fabbiano et al. (br); **courtesy of D. A. Harper, University of Chicago:** (bl); **NASA:** HST/ESA and D. Maoz (Tel-Aviv Univ. and Columbia Univ.) (tr); HST/R. de Grijs (Institute of Astronomy, Cambridge, UK) (cr); HST/HHT (STScI/AURA) (cl).

**306 ESA:** AOES Medialab (br); SPIRE/Herschel-ATLAS/S.J. Maddox) (cl). **306–307 NASA:** ESA, and HHT (STScI/AURA) (tc). **307 Robert A. Crain, Ian G. McCarthy, Carlos S. Frenk, Tom Theuns & Joop Schaye:** (c/Row 3). **Image courtesy of Rob Crain (Leiden Observatory, the Netherlands), Carlos Frenk (Institute for Computational Cosmology, Durham University) and Volker Springel (Heidelberg Institute of Technology and Science, Germany),** partly based on simulations carried out by the Virgo Consortium for cosmological simulations: (c/Rows 1 and 2); **NASA:** AURA/STScI and WikiSky/SDSS (bl).

**308 ESO:** (crb); **R. Jay GaBany, Cosmotography.com:** Blackbird Observatory, D. Martínez-Delgado (IAC, MPIA), J. Peñarrubia (U.Victoria), I. Trujillo (IAC), S. Majewski (U.Virginia), M. Pohlen (Cardiff) (clb); **NASA:** J. English (University of Manitoba), S. Hunsberger, S. Zonak, J. Charlton, S. Gallagher (PSU) and L. Frattare (STScI) (cla). **308–309 NASA:** ESA, HHT (STScI/AURA)—ESA/Hubble Collaboration and K. Noll (STScI) (tc). **309 NASA:** ESA and HHT (STScI/AURA)-ESA/Hubble Collaboration/B. Whitmore (STScI) and James Long (ESA/HST) (bl, bc); H. Ford (JHU), G. Illingworth (UCSC/LO), M. Clampin (STScI), G. Hartig (STScI), the ACS Science Team and ESA (cr).

**310 ESO:** (br); **NASA:** ESA and B. Schaefer and A. Pagnotta (Louisiana State University, Baton Rouge)/CXC, SAO, HHT (STScI/AURA) and J. Hughes (Rutgers University); **NOAO:** (tl); **SPL:** Max-Planck-Institut für Radioastronomie (bl); **courtesy of www.seds.org:** (cra). **310-311 ESO:** VMC Survey (b). **311 ESA/Hubble:** Digitized Sky Survey 2 (tc); **ESO:** (cr); **NASA:** X-ray: NASA/CXC/CfA/R. Tuellmann et al. (bc); **Mary Evans Picture Library:** (cra).

**312 Chandra:** NASA/CXC/SAO (bc); **ESA/Hubble:** NASA/T. Lauer (National Optical Astronomy Observatory) (clb). **313 NASA:** ESA and HHT (STScI/AURA) (bc); **SPL:** Tony and Daphne Hallas (t).

**314 Chandra:** NASA/CXC/SAO/PSU/CMU (cb); **NASA:** CXC/Wisconsin/D. Pooley & CfA/A. Zezas (tc); ESA and HHT (STScI/AURA) (tr); HST/HHT (STScI/AURA) (br); JPL-Caltech/STScI/CXC/UofA/ESA/AURA/JHU (clb); **NOAO:** N. A. Sharp (bl); **SPL:** GSFC (cla). **315 NASA:** ESA, S. Beckwith (STScI), and HHT (STScI/AURA) (b); JPL-Caltech/R. Kennicutt (University of Arizona)/DSS (tc/left); **SPL:** (cr); George Bernard (cra); Los Alamos National Laboratory (tc/right).

**316 Corbis:** Bettmann (br); **NASA:** HST/HHT (STScI/AURA) (c); X-Ray: UMass/Q.D. Wang et al.; Optical: STScI/AURA/HHT; Infrared: JPL-Caltech/University of Arizona/R. Kennicutt/SINGS Team (cb); **NOAO:** George Jacoby, Bruce Bohamanm, and Mark Hanna (tr); **SPL:** Kapteyn Laboratorium (tc). **317 Adam Block:** (cla); **ESO:** ALMA (ESO/NAOJ/NRAO; Visible Light Image: NASA/ESA HST (bc); **NASA:** ESA and HHT (STScI/AURA)-ESA/Hubble Collaboration/B. Whitmore (STScI) and James Long (ESA/HST) (br); ESA (cra); **SPL:** Celestial Image Co. (bl).

**318 NASA:** HST/H. Ford (JHU), G. Illingworth (UCSC/LO), M. Clampin (STScI), G. Hartig (STScI), the ACS Science Team and ESA (br); HST/HHT (STScI/AURA) (ca, bl); **SPL:** Max-Planck-Institut für Astrophysik (crb); NOAO (bc). **319 NASA:** ESA/Hubble (cra); HST/HHT (STScI/AURA) (bc); **SPL:** STScI/NASA (cla); **Sloan Digital Sky Survey (SDSS):** Giuseppe Donatiello (br).

**320 NOAO:** Adrian Zsilavee and Michelle Qualls/Adam Block (br); **NRAO:** (cb); **SPL:** STScI (crb); Sloan Digital Sky Survey (SDSS): (bc). **320–321 Credner:** (c/background). **321 ESO:** S. Gillessen et al. (br); **NASA:** W. Purcell (NWU) et al., OSSE, Compton Obs. (crb); **NOAO:** Eric Peng, Herzberg Institute of Astrophysics/NRAO/AUI (tr); **NRAO:** (c).

**322 Chandra:** X-ray (NASA/CXC/M. Karovska et al.); radio 21-cm image (NRAO/VLA/J. Van Gorkom/Schminovich et al.); radio continuum image (NRAO/VLA/J.Condon et al.); optical (DSS U.K. Schmidt Image/STScI) (cr); **ESO:** Optical: WFI; Submillimeter: MPIfR/ESO/APEX/A. Weiss et al.; X-Ray: NASA/CFX/CfA/R. Kraft et al. (cl); X-ray (NASA/CXC/Columbia/F. Bauer et al.); Visible light (NASA/STScI/UMD/A. Wilson et al.) (cra); **NASA:** HST/E. J. Schreier (STScI) (br); **NRAO:** (tl). **323 ESA/Hubble:** NASA (br); **courtesy of Vanderbilt Dyer Observatory:** (cra); **NASA:** HST/L. Ferrarese (JHU) (bc/right); HST/HHT (STScI/AURA) (ca); HST/Walter Jeffe/Leiden Obs., Holland Ford/JHU/STScI (bl); CXC/A. Zezas et al. (bc); JPL-Caltech/IPAC/Event Horizon Telescope Collaboration (cl).

**324 NASA:** ESA and Andy Fabian (University of Cambridge, UK) (tl); HST/J. Holtzman (cra); X-ray: NASA/CXC/SAO; Optical: NASA/STScI; Radio: NSF/NRAO/AUI/VLA (crb). **325 GPL:** DSS (crb); STScI (cra); **NASA:** JPL-Caltech/Yale University (bl); HST/A, Martel (JHU), H. Ford (JHU), M. Clampin (STScI), G. Hartig (STScI), G. Illingworth (UCO/Lick Obs.), the ACS Science Team and ESA (cbl); **courtesy of Cormac Reynolds, Joint Institute for VLBI in Europe, The Netherlands:** (cla); **SPL:** © Estate of Francis Bello (br).

**326 ESA/Hubble:** NASA (br); **NASA:** HST/N. Benitez (JHU), T. Broadhurst (The Hebrew Univ.), H. Ford (JHU), M. Clampin (STScI), G. Hartig (STScI), G. Illingworth (UCO/Lick Obs.), the ACS Science Team, ESA (cr); **SPL:** Royal Obs., Edinburgh (bc). **326–327 ESA:** Hubble & NASA, RELICS, CC BY 4.0 (c). **327 AAO:** AURA/Royal Obs., Edinburgh/UK Schmidt Telescope, Skyview (bl); Royal Obs. Edinburgh. Photograph from UK Schmidt plates by David Malin (clb); **Chandra:** NASA/CXC/UCI/A. Lewis et al. (tlb); Pal. Obs. DSS (tl); **ESO:** (VLT UT1 + ISAAC) (cl); **NASA:** CXC/AIfA/D. Hudson & T. Reiprich et al. (crb). **NRAO:** F.N. Owen, C. P. O'Dea, M. Inoue, and J. Eilek (br).

**328 Matt BenDaniel (http://starmatt.com):** (b); **GPL:** Robin Scagell (cla); **NASA:** ESA, and The Hubble Heritage Team (STScI/AURA) (tl); **NOAO:** Local Group Galaxies Survey Team (cr); **SPL:** Celestial Image Co. (c). **329 Chandra:** NASA/CXC/Columbia U./C. Scharf et al. (bl); **ESO:** FORS Team, 8.2 meter VLT Antu (cbr); **NASA:** ESO (bc); HST/J. English (U. Manitoba), S. Hunsberger, S. Zonak, J. Charlton, S. Gallagher (PSU), and L. Frattare (STScI) (br); **NOAO:** Doug Matthews/Adam Block (cl); **SPL:** Jerry Lodriguss (cla); Tony and Daphne Hallas (cr).

**330–331 Rogelio Bernal Andreo (Deep Sky Colors)**.

**332 AAO:** Photograph by David Malin (cl); **NASA:** ESA and The Hubble SM4 ERO Team (tr, cb); ESA, the Hubble Heritage Team (STScI/AURA)-ESA/Hubble Collaboration, and W. Keel (University of Alabama) (clb); JPL-Caltech/GSFC/SDSS (br). **333 courtesy of the Archives, California Institute of Technology:** (cra); **Chandra:** NASA/CXC/U. Mass/Q.D. Wang et al. (bc/left); NASA/STScI and NOAO/Kitt Peak (bc/right); **ESO:** INAF-VST/OmegaCAM. (cla); **NASA:** HST/W. Keel (Univ. Alabama), F. Owen (NRAO), M. Ledlow (Gemini Obs.), and D. Wang (Univ. Mass.) (br); ESA, E. Jullo (JPL), P. Natarajan (Yale), & J. P. Kneib (LAM, CNRS) (ca); **NOAO:** Jack Burgess/Adam Block (bl).

**334 GPL:** NASA—MSFC/Chandra/M. Bonamente et al. (br). **334–335 ESA/Hubble:** NASA/Johan Richard (Caltech, USA). **335 Bell Labs, Lucent Technologies:** Greg Kockanski, Ian Dell'Antonio, and Tony Tyson (br).

**336 © Smithsonian Institution:** (ca); **NASA:** ESA/Hubble (crb); **Science Photo Library:** Mark Garlick (bc). **337 Courtesy of NASA/WMAP Science Team:** (c); **NRAO:** Rudnick et al./NASA (crb); **SPL:** Max Planck Institute for Astrophysics/Volker Springel (tl).

**338–339 Sloan Digital Sky Survey:** (c). **339 2dF Galaxy Redshift Survey Team (www2.aao.gov.au/2dFGRS):** (c); **Alamy Images:** Richard Wainscoat (cr).

**340–341 NASA:** 2MASS/T. Jarrett (IPAC/Caltech).

**342–343 NASA:** JPL.

**344–345 Credner**.

**346 Alamy Stock Photo:** Prisma by Dukas Presseagentur GmbH (br); **British Library, London:** shelfmark: Harley 647, folio: f.13 (c); **Science & Society Picture Library:** Science Museum (clb). **347 akg-images:** Musée du Louvre, Paris (br); **BAL:** Private Collection, The Stapleton Collection (la, lb); **British Library, London:** shelfmark: Maps.C.10.c.10, folio: 5 (tr).

**354 BAL:** National Gallery of Art, Washington D.C., USA/Lauros/Giraudon (br); **Courtesy of Peter Wienerroither:** (cb); **GPL:** Robin Scagell (t). **355 BAL:** Palazzo Vecchio (Palazzo della Signoria) Florence, Italy (br); **Credner:** (tr); **NOAO:** Adam Block (cl).

**356 Credner:** (tr); **GPL:** Michael Stecker (tc); **SPL:** Harvard College Obs. (br). **357 Credner:** (br); **NOAO:** (tcb); Hillary Matthis, N. A. Sharp (tcl); **courtesy of Ian Ridpath:** (cra).

**358 Credner:** (bl); **Digital Library of Dutch Literature (www.dbnl.org):** (br); **GPL:** Robin Scagell (tr); **NOAO:** Fred Calvert/Adam Block (cb).

**359 Credner:** (c, br); **NOAO:** Adam Block (cr).

**360 NOAO:** Gary White and Verlenne Monroe/Adam Block (tr); Jeff Cremer/Adam Block (br); Joe Jordan/Adam Block (cr). **361 British Library, London:** shelfmark: Or. 8210/S. 3326 (br); **Credner:** (c); **GPL:** Damian Peach (tc).

**362 Credner:** (br); **NOAO:** Elliot Gellam and Duke Creighton/Adam Block (bl); Jon and Bryan Rolfe/Adam Block (tc); N. A. Sharp (cb). **363 Corbis:** The Stapleton Collection (tc); **Credner:** (br); **GPL:** Damian Peach (cr).

**364 Credner:** (tr); **GPL:** Eddie Guscott (crb); **NOAO:** Burt May/Adam Block (br). **365 akg-images:** Hessisches Landesmuseum (bc); **Credner:** (br); **NOAO:** Adam
Block (cl).

**366 Credner:** (br); **GPL:** Damian Peach (tc). **367 Corbis:** Allinari Archives/Mauro Magliani (tr); **GPL:** Philip Perkins (cbr); **NOAO:** Adam Block, Jeff and Mick Stuffings, Brad Ehrhorn, Burt May, and Jennifer and Louis Goldring (br); Heidi Schweiker (ca).

**368 Corbis:** Archivo Iconografico, S. A. (br); **Credner:** (bl); **NOAO:** Adam Block (c); **SPL:** Tony Hallas (cl).

**369 Credner:** (ca, bc); **NOAO:** T. A. Rector (NRAO/AUI/NSF and NOAO) and M. Hanna (br).

**370 Corbis:** Massimo Listri (tr); **Credner:** (cr); **SPL:** Jerry Lodriguss (br). **371 akg-images:** (bl); **Credner:** (r); **GPL:** Robin Scagell (tc).

**372 NOAO:** Adam Block (bl); **SPL:** John Sanford (tc).

**373 akg-images:** © Sotheby's (br); **Credner:** (t).

**374 Credner:** (b); **NOAO:** N. A. Sharp (trb); Sharon Kempton and Karen Brister/Adam Block (tc). **375 BAL:** Private Collection, The Stapleton Collection (cra); **Credner:** (br); **GPL:** Robin Scagell (car); NOAO: Nigel Sharp, Mark Hanna (c).

**376 Credner:** (tc, br); **GPL:** Nik Szymanek/Ian King (car); **NOAO:** (cr). **377 Corbis:** Arte and Immagini sr (bc); **Credner:** (bl); **GPL:** Damian Peach (cbr); **NOAO:** REU Program (cr).

**378 Corbis:** Archivo Iconografico, S. A. (tc); **Credner:** (br); **NOAO:** Adam Block (bl); Morris Wade/Adam Block (clb). **379 akg-images:** Museum of Fine Arts Boston/Erich Lessing (br); **Credner:** (ca, bc).

**380 Credner:** (br); **NOAO:** Bill Schoening (bl); Hillary Matthis, REU Program (cl). **381 Corbis:** Gianni Dagli Orti (br); **Credner:** (bl); **GPL:** Michael Stecker (cb); **NOAO:** N. A. Sharp, Vanessa Harley/REU Program (bc).

**382 Credner:** (tr, br); **NOAO:** N. A. Sharp, REU Program (ca); **SPL:** John Sanford (cb). **383 akg-images:** Erich Lessing (tr); **Credner:** (br); **GPL:** Robin Scagell (ca, cl).

**384 Corbis:** Bettmann (br); **Credner:** (bl); **GPL:** Nik Szymanek (cl); Robin Scagell (c). **385 Credner:** (tr, br); **GPL:** Damian Peach (tc); **courtesy of Osservatorio Astronomico di Palermo Giuseppe S. Vaiana:** (cl).

**386 Corbis:** Richard T. Nowitz (br); **Credner:** (tr); **NOAO:** (bc). **387 Credner:** (b); **courtesy of William McLaughlin:** (clb, crb).

**388 BAL:** Palais du Luxembourg, Paris, France/Giraudon (crb); **Credner:** (c); **GPL:** Robin Scagell (tr); **NOAO:** Todd Boroson (br). **399 Corbis:** The Stapleton Collection (br); **Credner:** (bl); **NOAO:** Francois and Shelley Pelletier (tc).

**390 Credner:** (r); **GPL:** Duncan Radbourne (cl); **The Picture Desk:** The Art Archive/Bodleian Library, Oxford (bl). **391 GPL:** Michael Stecker (tr, br); **NOAO:** Jim Rada/Adam Block (tc).

**392 Credner:** (cr, br); **GPL:** Pedro Rè (tc); **The Picture Desk:** The Art Archive/Private Collection/Marc Charmet (c). **393 Credner:** (bl); **GPL:** Michael Stecker (cr); **NOAO:** Michael Gariepy/Adam Block (tr).

**394 Corbis:** Todd Gipstein (tc); **NOAO:** Adam Block (cb); Allan Cook/Adam Block (bl). **395 Credner:** (t).

**396 Daniel Verschatse (www.astrosurf.com):** (cla); **Credner:** (cb); **GPL:** Gordon Garradd (bcr); Yoji Hirose (cra). **397 Credner:** (ca, br); **Mary Evans Picture Library:** (cra); **NOAO:** Bob and Bill Twardy/Adam Block (c).

**398 Corbis:** Araldo de Luca (cr); **GPL:** (crb, bl). **399 BAL:** Bibliothèque Nationale, Paris, France/Archives Charmet (bl); **Credner:** (br); **GPL:** Gordon Garradd (ca).

**400 GPL:** Michael Stecker (tr, clb); **NOAO:** Todd Boroson (bl). **401 Credner:** (t); **NOAO:** (bl).

**402 Corbis:** Archivo Iconografico, S. A. (br); **Credner:** (bl); **SPL:** Rev. Ronald Royer (tr). **403 Corbis:** Andrew Cowin (cr); **Credner:** (ca, bc); **GPL:** Pedro Rè (tcl).

**404 Credner:** (cra, bc); **NOAO:** T.A Rector (br). **405 AAO:** Royal Obs. Edinburgh. Photograph from UK Schmidt plates by David Malin (tc); **Credner:** (cl, br); **ESO:** (VLT UT1 + FORS1) (trb).

**406 Credner:** (bl); **GPL:** Gordon Garradd (tr); **NOAO:** Nicole Bies and Esidro Hernandez/Adam Block (br). **407 BAL:** Musée Conde, Chantilly, France/Giraudon (br); **Credner:** (bl); **GPL:** DSS (cr); **NOAO:** Adam Block (c).

**408 akg-images:** Museo Capitular de la Catedral, Gerona/Erich Lessing (c); **Credner:** (tr, bc). **409 Credner:** (bl); **GPL:** Michael Stecker (bcr); Pedro Rè (cl); **NOAO:** (crb).

**410 Credner:** (tr); **GPL:** Chris Pickering (br); Gordon Garradd (tc); **The Picture Desk:** The Art Archive/Museo Civico Padua/Dagli Orti (bl). **411 Credner:** (r); **GPL:** Robin Scagell (cb); **NOAO:** (bl).

**412 Corbis:** Bettmann (bcr); **Credner:** (cr); **GPL:** Yoji Hirose (cbl); **NOAO:** (bl). **413 Credner:** (tr, bc); **GPL:** Gordon Garradd (tc).

**414 Credner:** (tr, br); **GPL:** Gordon Garradd (tc). **415 akg-images:** Pergamon Museum, Berlin/Erich Lessing (cl); **Credner:** (tr, br); **GPL:** Gordon Garradd (c).

**416 BAL:** Cheltenham Art Gallery and Museums, Gloucestershire, UK (cr); **Credner:** (tr, bl). **417 akg-images:** Coll.Archiv f. Kunst and Geschichte (crb); **Credner:** (tr, bc).

**418 Credner:** (br); **GPL:** Chris Livingstone (tcr); Michael Stecker (ca). **419 Credner:** (tr, cb); **GPL:** Gordon Garradd (bcl).

**420 Credner:** (tc, br); **ESO:** Jean-Luc Beuzit, Anne-Marie Lagrange (Observatoire de Grenoble, France), and David Mouillet (Observatoire de Paris-Meudon, France) (c). **421 BAL:** © The Trustees of the Chester Beatty Library, Dublin (bl); **Credner:** (br); **GPL:** Chris Livingstone (tr); **NOAO:** Marcelo Bass/CTIO (tc).

**422 Alamy Images:** Chris Cameron (bcr); **Credner:** (tr, cb); **SPL:** (br). **423 Credner:** (tr, br).

**424 Credner:** (bl); **GPL:** Gordon Garradd (c); **Volker Wendel and Bernd Flach-Wilken (www.spiegelteam.de):** (cra). **425 Alamy Images:** Adam van Bunnens (tr); **Till Credner/AlltheSky.com:** (c); **DK Images:** Courtesy of the Science Museum, London/Dave King (crb).

**426–427 Alamy Stock Photo:** NG Images.

**429 DK Images:** (tr).

**430 GPL:** Robin Scagell (cr). **431 GPL:** Robin Scagell (tr); **NOAO:** Ryan Steinberg and family (tc).

**436 DMI:** Akira Fujii (cra). **437 Credner:** (tr).

**442 Corbis:** Roger Ressmeyer (cra). **443 GPL:** Gordon Garradd (tl).

**448 Credner:** (cra). **449 GPL:** Yoji Hirose (cra); **NOAO:** (tl).

**454 Credner:** (cra). **455 DMI:** Akira Fujii (cra).

**460 Alamy Images:** Pixonnet.com (cra). **461 DMI:** Akira Fujii (tr).

**466 DMI:** Akira Fujii (cr). **467 DMI:** Akira Fujii (tl).

**472 Corbis:** Reuters/Ali Jarekji (bl). **473 Credner:** (tr); **NOAO:** Svend and Carl Freytag/Adam Block (ca).

**478 Alamy Images:** Gondwana Photo Art (bc). **479 GPL:** Chris Livingstone (tl).

**484 GPL:** Yoji Hirose (br). **485 GPL:** Robin Scagell (cla).

**490 Credner:** (tr). **491 GPL:** Robin Scagell (tr).

**496 Credner:** (br). **497 GPL:** Yoji Hirose (tr).

ENDPAPERS **NOAO:** Nathan Smith, Univ. of Minnesota.

All other images © Dorling Kindersley
For further information see: www.dkimages.com

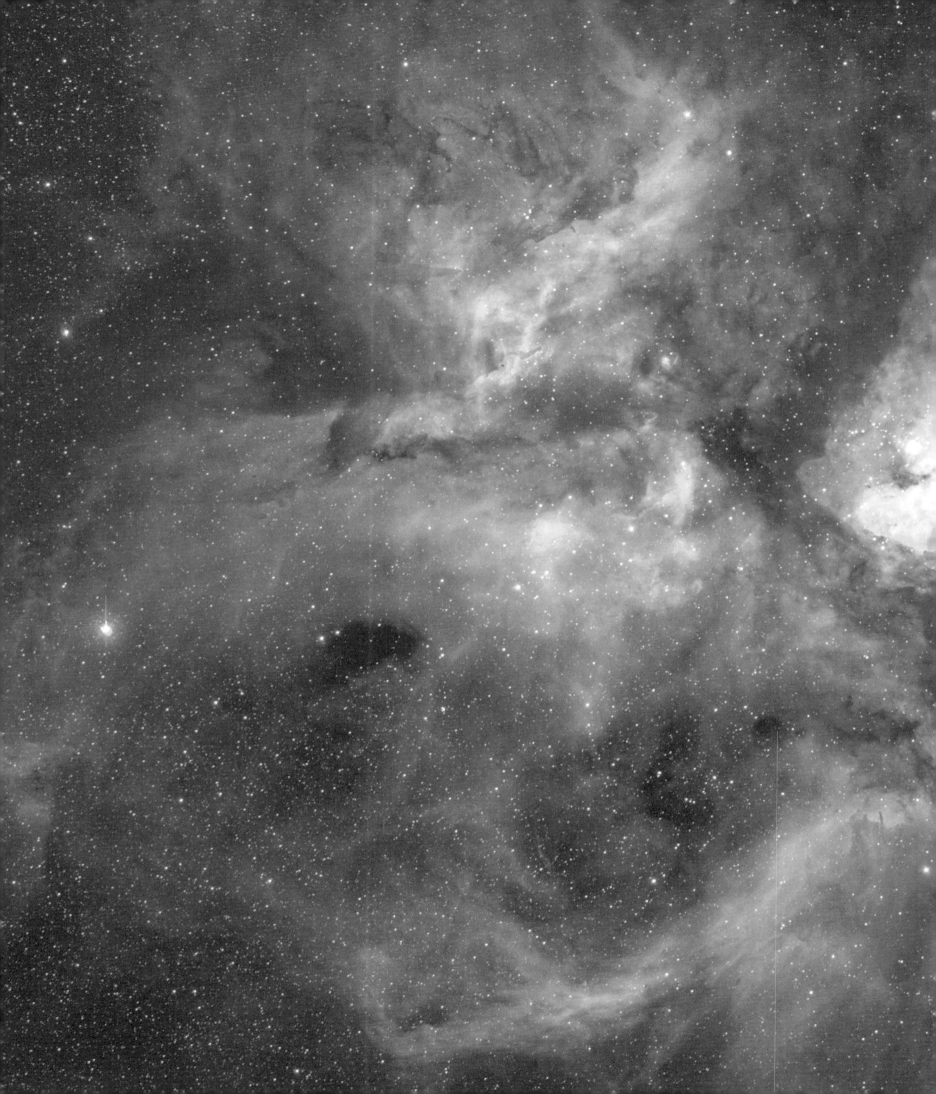